INTEGRATED OPTOELECTRONICS

QUANTUM ELECTRONICS—
PRINCIPLES AND APPLICATIONS

EDITED BY

PAUL F. LIAO

Bell Communications Research, Inc.

Red Bank, New Jersey

PAUL L. KELLEY

Lincoln Laboratory

Massachusetts Institute of Technology

Lexington, Massachusetts

A complete list of titles in this series appears at the end of this volume.

INTEGRATED OPTOELECTRONICS

Edited by

Mario Dagenais

Department of Electrical Engineering
University of Maryland
College Park, Maryland

Robert F. Leheny

Bellcore
Red Bank, New Jersey

John Crow

IBM, T. J. Watson Research Center
Yorktown Heights, New Jersey

ACADEMIC PRESS

San Diego Boston New York
London Sydney Tokyo Toronto

Front cover: This is a photo of a 1 Gigabit/sec per channel, 4-channel module for both the transmission and receiving of data. It has a transmitter OEIC consisting of a laser, control photodiode, and impedance matching array; and a receiver OEIC consisting of a photodiode + amplifier array. The chips are flip bonded to the carrier, and self-aligned to the lightguides on the carrier. Photo courtesy of IBM.

This book is printed on acid-free paper. ♾

Academic Press, Inc.
A Division of Harcourt Brace & Company
525 B Street, Suite 1900, San Diego, California 92101-4495

United Kingdom Edition published by
Academic Press Limited
24-28 Oval Road, London NW1 7DX

Library of Congress Cataloging-in-Publication Data

Integrated optoelectronics / {edited by} Mario Dagenais, Robert F. Leheny, John Crow.
p. cm. -- (Quantum electronics--principles and applications)
Includes bibliographical references and index.
ISBN 0-12-200420-5 (alk. paper)
1. Optoelectronic devices. 2. Integrated optics. 3. Integrated circuits. I. Dagenais, Mario, date. II. Leheny, Robert F. III. Crow, John. IV. Series.
TA1750.I5885 1994
621.381'045--dc20 93-29840
CIP

PRINTED IN THE UNITED STATES OF AMERICA
94 95 96 97 98 99 EB 9 8 7 6 5 4 3 2 1

Contents

PART II
MATERIALS GROWTH

PART III
DEVICE PROCESSING

Contributors

Numbers in parentheses indicate the pages on which the authors' contributions begin.

E. Arthurs (3), *Bellcore, Morristown, New Jersey 07960*

R. Baets (529), *University of Gent-IMEC, Department of Information Technology, Gent, Belgium*

R. Bates (489), *IBM, T. J. Watson Research Center, Yorktown Heights, New York 10598*

R. Bhat (121), *Bellcore, Red Bank, New Jersey 07701*

D. Botteldooren (529), *University of Gent-IMEC, Department of Information Technology, Gent, Belgium*

P. Buchmann (257), *IBM Research Division, Zurich Research Laboratory, Switzerland*

Joe C. Campbell (419), *Microelectronics Research Center, University of Texas, Austin, Texas 78712*

Winston K. Chan (297), *Bellcore, Red Bank, New Jersey 07701*

Larry A. Coldren (339), *University of California at Santa Barbara, Santa Barbara, California 93106*

John Crow (39, 447, 489, 627), *IBM, T. J. Watson Research Center, Yorktown Heights, New York 10598*

J. Ewen (489), *IBM, T. J. Watson Research Center, Yorktown Heights, New York 10598*

S. F. Fang (145), *Department of Electrical and Computer Engineering, Coordinated Science Laboratory, Materials Research Laboratory, Univeristy of Illinois, Urbana, Illinois 61801*

M. S. Goodman (3), *Bellcore, Morristown, New Jersey 07960*

L. R. Harriott (213), *AT&T Bell Laboratories, Murray Hill, New Jersey 07974*

Izuo Hayashi (645), *Optoelectronics Technology Research Corporation, Ibaraki 300-26, Japan*

John R. Hayes (317), *Bellcore, Red Bank, New Jersey 07701*

J. M. Jaffe (39), *IBM, T. J. Watson Research Center, Yorktown Heights, New York 10598*

T. L. Koch (557), *AT&T Bell Laboratories, Holmdel, New Jersey 07733*

U. Koren (557), *AT&T Bell Laboratories, Holmdel, New Jersey 07733*

P. Lagasse (529), *University of Gent-IMEC, Department of Information Technology, Belgium*

F. Libbrecht (529), *University of Gent-IMEC, Department of Information Technology, Gent, Belgium*

Hadis Morkoç (145), *Department of Electrical and Computer Engineering, Coordinated Science Laboratory, Materials Research Laboratory, University of Illinois, Urbana, Illinois 61801*

G. Morthier (529), *University of Gent-IMEC, Department of Information Technology, Gent, Belgium*

D. Rogers (489), *IBM, T. J. Watson Research Center, Yorktown Heights, New York 10598*

M. W. Sachs (39), *IBM, T. J. Watson Research Center, Yorktown Heights, New York 10598*

H. Temkin (213), *AT&T Bell Laboratories, Murray Hill, New Jersey 07974*

Charles W. Tu (83), *Department of Electrical and Computer Engineering, University of California at San Diego, La Jolla, California 92093*

P. Vettiger (257), *IBM Research Division, Zurich Research Laboratory, Switzerland*

O. Voegeli (257), *IBM Research Division, Zurich Research Laboratory, Switzerland*

Osamu Wata (447), *Fujitsu Laboratories, Ltd., Atsugi 243-01, Japan*

D. J. Webb (257), *IBM Research Division, Zurich Research Laboratory, Switzerland*

Eli Yablonovitch[1] (297), *Bellcore, Red Bank, New Jersey 07701*

[1]Present address: Department of Electrical Engineering, University of California at Los Angeles, California 90024.

Preface

Optoelectronic technology is associated with the generation, transmission, routing, and detection of optical signals. Today optoelectronics is finding widespread applications ranging from fiber optic telephone lines to notebook computer displays, CD audio systems and memories, and laser printers. These applications are building blocks for the emerging information age and optoelectronics is rightly considered a critical enabling technology. Recent studies by both private and government agencies have identified the importance of optoelectronics and have encouraged expanded investment in this technology as a means for insuring the continued development of information-age applications. As the economic vitality and variety of personal opportunities in our society become more and more dependent on effortless access to information the role that continued advancement of optoelectronics will have in pacing progress becomes even more critical.

This book is about one key technological challenge in expanding the versatility and functionality of optoelectronics: research aimed at the integration of multiple-function devices onto a single substrate or chip. The goal of this research is to enhance the performance and reliability of optoelectronic modules while lowering their manufacturing cost through the development of OptoElectronic Integrated Circuits (OEICs). The vision is to achieve the advantages that integration has demonstrated for electronic components. Over a period of 30 years the electronics industry has moved from discrete transistors to Very Large Scale Integrated Circuits (VLSI) and the cost and reliability advantages achieved through these developments have largely been responsible for enabling all the advancements in electronic information transmission and processing that have been so instrumental in transforming our society today. Electronic IC technology exploits the similarity of function of the basic building blocks (MOSFETs, CMOS, or bipolar transistors) and gains functionality and performance from the development of complex on-wafer interconnection of these devices.

Optoelectronic devices are largely based on similar materials, typically III-V semiconductor materials, and are processed or fabricated using many of the same technologies that form the basis for silicon-integrated circuit fabrication. However, optical components rely on both the optical and electronic properties of the materials they are fabricated from and they often require heteroepitaxial material with accurately controlled composition to achieve optimum performance. As a result OEICs are typically more complex in structure than silicon ICs. A further complication is that in their most versatile form OEICs are built up from components that are very different in functionality—light emitter, waveguide, intensity or phase modulator, and/or detector. Each of these different components requires different material structures to achieve optimized performance. Given this complexity OEICs require flexible processing technologies and OEIC fabrication is typically very different from the much simpler (in concept) doping operation used to realize silicon devices.

Today monolithically integrated optical receivers are already in use in compact disk and CD-ROM products (Si OEICs) and are beginning to appear in data communication products (GaAs and Si OEICs). Also, the integration of photonic components (Photonic Integrated Circuits, PICs), such as waveguide modulators or lasers with waveguides and the use of flip-chip bonding to hybrid integrate two chips together, are emerging from the development phase. The application of the technology covered by the book is therefore just beginning. The hybrid techniques are expected to migrate in the future to monolithic integration, motivated by the desire for lower cost, smaller size, and higher reliability for the same function, with the expectation that ultimately our experiences with integration will produce inventions with totally new functions not yet envisioned.

Over the past decade a number of research groups have focused on developing and demonstrating the integration of multiple optical functions with electronics and their efforts have identified and begun to address the development of an integrated optoelectronics technology. At the same time the complexity of design for discrete components to meet advanced optical systems requirements has increased. Examples include wavelength controlled laser PICs incorporating passive waveguides and gratings with quantum well active region as well as vertical cavity surface emitting lasers (VCSELs) which incorporate very high reflectivity mirrors (99%) within the laser diode structure. Progress in material deposition control, processing, and device design required to meet these device needs has led to the development of many manufacturing processes that can readily be adapted to OEICs. Today we are poised to exploit these developments.

This book brings together a group of acknowledged university and industry research experts from around the world to share their insights on the common theme of optoelectronic integration. These experts report not only on the state-of-the-art, but also on the physics and design experience that go into implementing integrated chips and modules. The editors have tried to create a cohesive set of articles that includes a discussion of the tradeoffs between materials growth, device processing, state-of-the-art discrete component design, multifunctional chip design, packaging, and systems requirements. Levels of integration discussed encompass electrical, optoelectronic, and all optical devices in both monolithic and hybrid form. The editors have targeted the book for researchers, practicing engineers, and graduate students in electronic or optical science who are working with optoelectronics technology and its applications. The editors hope that newcomers to the field can use this text to find the information required for understanding the major issues involved in optoelectronic integration.

This book then is a summary of where optoelectroncis technology is at this point in time from the perspective of emerging OEIC technology. The text is organized into five parts, each covering a broad technology area important to OEICs. Each part is then subdivided into Chapters authored by experts on specific topics related to the technology area being covered. In this way the editors have endeavored to provide within each part a self-contained overview of these critical enabling technologies. The part titles provide a guide to the topics covered: system requirements for OEICs, materials growth, device processing, state-of-the-art discrete components, and finally a six-Chapter part dealing specifically with the main theme of optoelectronic integrated circuits. It is the hope of the editors that an interested reader will be able to become familar with all the enabling technologies important to optoelectronics and thereby gain insight into the field beyond what is required to understand OEICs.

Mario Dagenais
Robert F. Leheny
John Crow

PART I

System Requirements for OEICs

Chapter 1

TELECOMMUNICATIONS SYSTEM APPLICATIONS FOR OPTOELECTRONIC INTEGRATED CIRCUITS

M. S. Goodman and E. Arthurs

Bellcore
Morristown, New Jersey 07960

1. WHY OPTOELECTRONIC INTEGRATION?

The rapid advances in optoelectronics have caused this technology to be a crucial element driving the transition from the Industrial Age to the Information Age [1]. For example, the development of semiconductor laser diodes and photodiodes was critical for the realization of cost-effective optical fiber communications networks. These optical fiber systems are in widespread use today for "long-haul networks" spanning oceans and continents. They are widely used in the United States and some other countries for "interoffice networks," namely the networks that interconnect major telecommunications

ISBN 0-12-200420-5

switching centers (telephone company central offices). Today we are at the threshold of another revolutionary development in the transition to the Information Age, namely the integration of optical and electronic components into optoelectronic integrated circuits (OEICs) combining features from each of these significant technologies.

The primary objectives for optoelectronic integration are similar to those for electronic integration: increase functionality and decrease cost per component. The development of electronic ICs over the past 20 years has exhibited growth from a few transistors per chip to more than 10 million transistors per chip today. Electronic integration was strongly motivated by the "connection bottleneck"—the difficulty of connecting efficiently discrete electronic components. However, integration is perhaps more important for optical technologies than it was for electronic technologies, because one of the key attributes of optics is its *inherent parallelism* (simultaneously acting on many beams or wavelengths of light, as a waveguide or lens would), as opposed to the serial nature of transistors. Unlike the development of electronic integrated circuits, which involved the replication of huge numbers of identical components, optoelectronic integration promises the aggregation of rather varied functions from fundamentally different types of components, including some based on different materials. In addition, OEICs can provide multifunction single components. Although a distinction is sometimes made between OEICs that incorporate "primarily photonic" components (such OEICs being called photonic integrated circuits) and those integrating both optical and electronic functionality, that distinction is not emphasized in this chapter.

Optoelectronic integration uses materials with very high electron mobility, ranging from about 3800 cm^2/V s in InP and 8500 cm^2/V s in undoped GaAs to more than 11,000 cm^2/V s in artificially lattice-matched quaternaries ($Ga_xIn_{1-x}As_yP_{1-y}$). Thus, the OEIC electronic components will be able to operate at the requisite speeds (Gbits/s) for broadband communications. Mass production of such circuits will result in much lower costs and potentially smaller sizes (per unit component, per user, per functional component, and per system) than are possible with the use of discrete components and may offer other significant advantages as well [2]. Optoelectronic integration may allow increasing complexity and functionality along with the high reliability achievable with integration. For example, the precise alignment of a single optical fiber to an OEIC that combines many laser transmitters is much simpler (and thus more reliable) than the use of multiple discrete component transmitters to many fibers. OEICs will be necessary to take advantage of the inherent parallelism of optical systems, particularly multiwavelength systems, where OEICs may have their most significant advantages over discrete components. OEICs offer the ability to utilize the noise immunity characteristics of on-chip optical waveguides and interconnections. Finally, integration of functions will yield, through increasing complexity, new functionality unavailable and perhaps unattainable without integration.

The purpose of this chapter is to discuss possible OEIC applications that could enhance the performance and functionality of lightwave telecommunications systems and networks. The chapter is divided into three parts. In the first part, a vision of the future broadband networks is presented. The second part discusses various approaches to accessing broadband networks. The third part briefly indicates additional lightwave network applications. Specific optical network architectures are described from a functional perspective, as examples of where potential OEICs may be appropriate. In addition, a catalog of "generic OEIC chip concepts" is developed from a functional perspective, with the objective of motivating device research with a focus on potential applications (see Appendix, Figures Cat-1 to Cat-10).

2. A VISION OF THE FUTURE BROADBAND NETWORK

We begin the discussion with a brief review of some network terminology. By *network* we mean the organization of transmission facilities serving a group of nodes. These nodes represent the users; they may be individuals, computers, broadcast video sources, or entire "subnetworks." Because the number of communications links required for full connection of all the nodes increases as N^2, *switching* is required to provide connectivity among the nodes economically. Historically, transmission bandwidth has been a scarce resource. Since the advent of optical fibers, this has been changing. The transmission bandwidth of silica glass fiber is abundant (more than 10 Tbit/s for the low-loss wavelength regions from 1100 to 1600 nm in single-mode fibers), and departures from the traditional telecommunications approaches used with copper cable–based systems may be required to take advantage of this huge bandwidth.

Networks may be classified into several topological groups, such as buses, stars, meshes, rings, and trees. Switching among the nodes in these networks can be divided into two main types, *circuit switching* and *packet switching*. In circuit-switched networks, fixed bandwidth is allocated between an origin and a destination at the call setup time. Networks with circuit switching generally have a few common transmission rates and do not have network overloads—if all available bandwidth is in use, new calls are simply blocked from entering the network. On the contrary, networks employing packet switching can accommodate a diverse set of transmission rates. The data to be transmitted are divided into packets (of fixed length, or with some fixed maximum length) and then transmitted across the network. Whereas circuit switching allocates bandwidth end to end, packet techniques allocate bandwidth by link as the packet traverses the network. Temporary network overloads are allowed and are handled by each node's memory buffers. Whereas circuit switching techniques are appropriate for continuous bit-rate calls (such as voice or broadcast video traffic), packet techniques are generally more appropriate for bursty bit-rate calls (for traffic such as data or image communications) so that repetitive call setup time can be avoided.

As optical fiber technology advanced during the 1980s and optical fibers became the transmission medium of choice for the public telecommunications network, and as powerful computers based on highly integrated microelectronics were also developing, a new vision of telecommunications emerged. In this vision, a future "intelligent broadband network" will enable people, computers, and other data devices to communicate with each other, in any medium (voice, data, image, video, or multimedia), and at reasonable cost. Fulfilling this vision will require flexible topology broadband backbone optical networks (Fig. 1) linking computers and other data devices, along with voice networks and video networks, encompassing a hierarchy of networks of networks (Fig. 2).

The global importance of a broadband telecommunications infrastructure for commerce, education, and entertainment should not be underestimated. Once widely deployed, such networks will enable nearly everyone to access a broad range of information services—for example, to have access to the educational materials of the great text, image, and video libraries and databases of information. Surely the realization of this vision through broadband networks will rank as one of the major achievements of our society.

2.1. Evolution to the Broadband Network

The public circuit-switched voice network today is based on time-multiplexed 64 kbit/s channels (fixed time slots) in T1 (1.5 Mbit/s) and T3 (45 Mbit/s) backbone network trunks. These trunk groups are often multiplexed to form higher-rate point-to-point optical fiber links between telephone company switching centers. The trunk groups are then demultiplexed from the fiber links and the individual 64 kbit/s channels are switched for the end users.

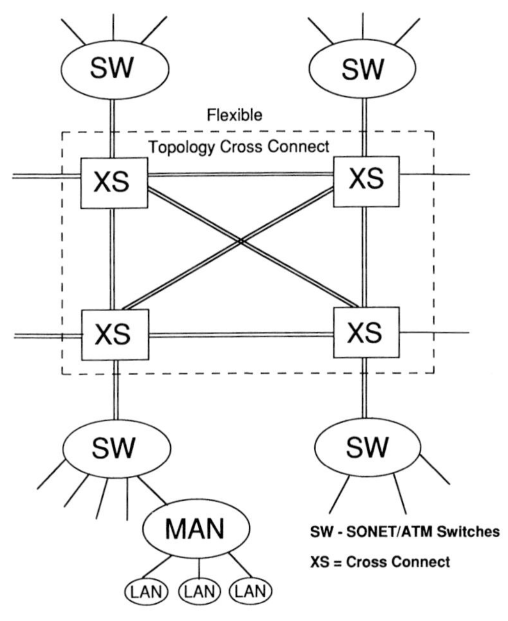

Fig. 1. Flexible topology broadband network. The figure shows central elements of a broadband network, including SONET/ATM switches (SW) and digital cross-connect systems (XS).

As this network evolves toward the broadband future, optical fiber links will be connected to digital cross-connect systems capable of reconfiguring the entire network or portions of it to provide routing diversity and fault tolerance (Fig. 1). This reconfiguration may involve switching from one set of optical fibers to another or regrouping of trunk groups. A North American standard has been adopted for an optical transmission signal hierarchy, known as synchronous optical network (SONET). SONET is also the basis for an international standard known as SDH. The SONET optical transmission rates are multiples of a basic building block of 51 Mbit/s (Table 1), rather than the 64 kbit/s rate of the current public circuit-switched network. The 64 kbit/s "atomic unit" for the current network was motivated by the characteristics of a "dominant service," namely voice traffic. Similarly, the 51 Mbit/s and higher data rates are motivated by the expectation that high-quality, full-motion video and multimedia services will be the dominant traffic for the future networks. This high-rate atomic unit required a new approach to the fundamental structure—even at only 155 Mbit/s, there are more than two thousand 64 kbit/s time slots. Thus, asynchronous transfer mode (ATM) was developed [3–6]. In this approach, specific periodic time slots are not assigned to a channel. Bandwidth is segmented into fixed-size cells, which

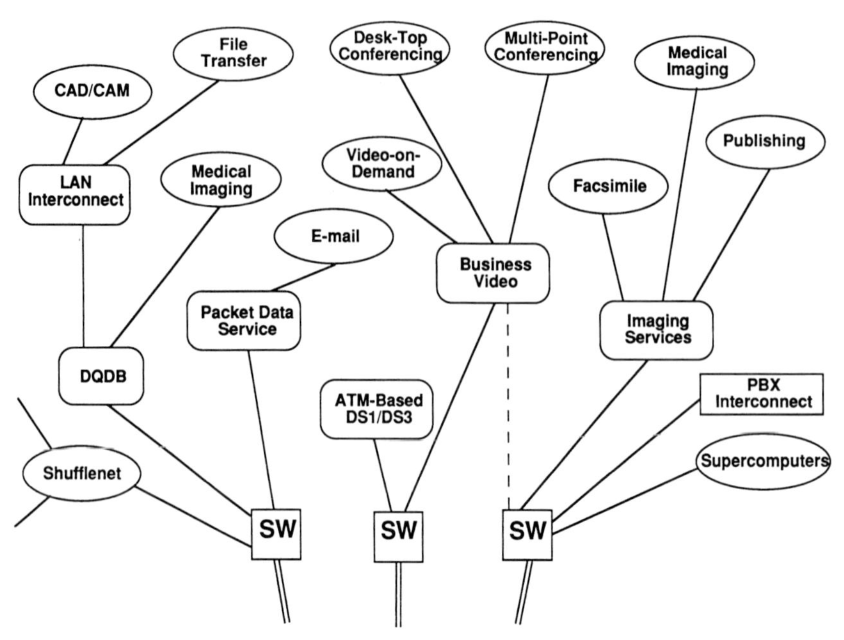

Fig. 2. Broadband network hierarchy and supported services. The broadband network will incorporate a hierarchy of networks supporting a diverse array of services.

Table 1.

SONET Optical Transmission Rates

Level	Transmission Rate (Mbit/s)
OC-1	51.84
OC-3	155.52
OC-9	466.56
OC-12	622.08
OC-18	933.12
OC-24	1244.16
OC-36	1866.24
OC-48	2488.32
OC-96	4976.64
OC-192	9953.28

can be allocated to services on demand. ATM is not restricted to a single data rate but is intended to be flexible for high-speed network growth, with higher data rate frame structures formed from the basic SONET rate. The SONET standard currently relies on optical direct detection at a single wavelength. OEICs could play meaningful roles in the implementation of integrated functions for the SONET transmitters and receivers if they could provide (1) better noise immunity than discrete components have, due to their on-chip interconnections (lack of stray capacitance, etc.); (2) more cost-effective functionality than discrete components, due to the integration; or (3) better reliability (e.g., due to alignment of a single fiber for several transmitters) (Fig. Cat-1). The SONET standard will provide network access at high transmission rates (including 155 Mbit/s, 622 Mbit/s, and beyond).

SONET transmission rates planned for the broadband network infrastructure initially will be 155 Mbit/s, 622 Mbit/s, 1.2 Gbit/s, and 2.4 Gbit/s. However, as the use of truly broadband services emerges, a growing demand will be placed on the broadband highways. This need may be met by increased utilization of the optical fiber capacity through the use of higher data transmission rates or through the use of multiple wavelengths. Multi-wavelength systems are particularly attractive for expansion of capacity [7, 8], because they may utilize the same fiber links and electronic time division multiplexing and demultiplexing equipment used for single-wavelength systems. Furthermore, a multiwavelength approach does not suffer from increasingly stringent transmission tolerances, which would be associated with use of higher data rates (e.g., effects due to optical dispersion, impedance matching in the multiplexing electronics, power budget, optical reflections). It is for this future multiwavelength broadband network that OEICs may have their greatest impact.

3. ACCESS TO THE BROADBAND NETWORK

In this section we describe methods for accessing the broadband network, including examples of centralized switched access, using time-multiplexed channels, add/drop multiplexers, and very high capacity switches. We then discuss distributed access to the broadband highway, using local and metropolitan area networks (MANs and LANs). Finally we introduce Local Subscriber Loop Access. This discussion, which is principally focused on telecommunications networks and systems, is followed by a discussion of broadcast video distribution, from a head-end station to remote sites, typical of distribution in the cable TV industry.

3.1. Switched Access to the BISDN

Access to the broadband highway can be provided through very high capacity switches that can process the ATM cells being presented to the network by the user. An international agreement has been adopted, specifying that ATM cells should be used for transmitting the information in the broadband integrated services digital network (BISDN). This method allows asynchronous presentation of information to the network at a variety of different incident data rates, so that the backbone network is not sensitive to the type of service supported. The purpose of a packet switch is to route and transport a data packet from an input port on the switch to one or more output ports. ATM packet switches today are based on electronic switch architectures; however, the large capacity of optical fibers means that these packet switches could become information bottlenecks in the future, as large numbers of users access the network at high data rates. Although there are designs for very high speed electronic ATM switches, a possible alternative is to use an optoelectronic packet switch to avoid this switching capacity bottleneck. This alternative could become especially attractive if switched multicast services become widespread.

A number of possible optoelectronic switch architectures have been proposed [9–19] for both unicast and multicast designs. We illustrate this problem with a discussion of a proposal for an optical multicast packet switch fabric known as StarTrack [6, 19].

The optical ATM switch fabric is shown schematically in Fig. 3. This fabric is formed from two internal networks, a multiwavelength optical star transmission network and an electronic control track. Each input port has associated with it a unique wavelength and each output port has a tunable receiver. At the beginning (and end) of the control track is a token generator that generates control packets (tokens) and synchronizes the operation of the switch fabric. The essential operation of the switch may be understood as follows: Packets arriving on optical fiber input trunks are stored in memory at an input port. A unique transmission wavelength is associated with each input port. The switch is controlled in a cycle having two phases, a write phase and a read phase. During the write phase, input ports sequentially write information into the control tokens, indicating the output port to which a given packet is to be sent (thereby reserving that output port). The token is then sent to the output ports of the fabric, which read the tokens to determine whether they have been reserved to receive packets, and they tune their receivers to listen at the appropriate wavelengths. Finally, data at the input ports are transmitted in parallel to the desired output ports. Multicast

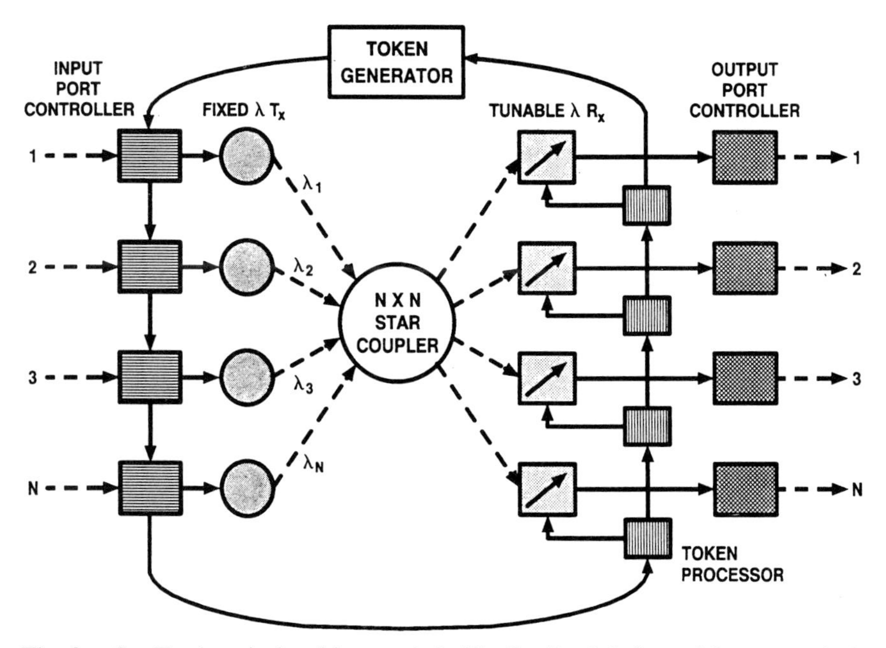

Fig. 3. StarTrack optical multicast switch. The StarTrack is formed from an optical star and an electronic control track. At the start of a cycle, input ports reserve their desired output ports writing into the token. The output ports read the token and "tune" their receivers to the appropriate wavelengths. This is followed by parallel transmission to the receivers.

transmission is easily accomplished because several output ports can simultaneously tune their receivers to the same input port wavelength. This switch fabric has a number of useful features, including easily implemented multilevel packet priorities and call splitting (partial completion of calls). Thus a 64-port StarTrack switch with a transmission rate at each port of about 2 Gbit/s could provide peak throughputs near 100 Gbit/s. Larger-capacity switches may be constructed from such StarTrack units.

In the StarTrack example, OEICs could play a decisive role in making an optical switch fabric competitive in cost and performance with more traditional electronic approaches. The optical switch fabric has the important potential advantage over its electronic counterparts that performing multicast operations is natural and packet duplication is not necessary. Thus, an OEIC (Fig. Cat-2) could implement the tunable StarTrack receivers with electronic selection from multiple fixed-wavelength receivers and incorporate an integrated wavelength demultiplexer (e.g., a miniaturized version of the design in refs. 20 and 21). It could further incorporate electronic memory buffers and "input/output port operations" such as packet accounting and control. This type of functionality could enable switch architectures like the StarTrack to be viable contenders with the next generation of electronic switch fabrics.

The performance of a StarTrack packet switch can be dramatically enhanced using multiple parallel token tracks (Fig. 4). In the multiple track StarTrack switch the input ports and output ports are divided so that each receiver needs to be tuned over only a fraction of the total wavelength range of the input ports. This significantly enhances the throughput performance of the switch (Fig. 5), while relaxing the constraints on the tunable receivers. However, this approach requires multiple tunable receivers at each output port and multiported buffer memories. The tuning time must be of the order of a packet transmission time (200–600 ns). Again, OEICs could be critical to the implementation of such a switch (Fig. Cat-3).

3.2. Distributed Access to the Broadband Network: LANs AND MANs

Data networking technology is in a rapid state of development. Local area networks (LANs) based on the Ethernet multiaccess standard at 10 Mbit/s

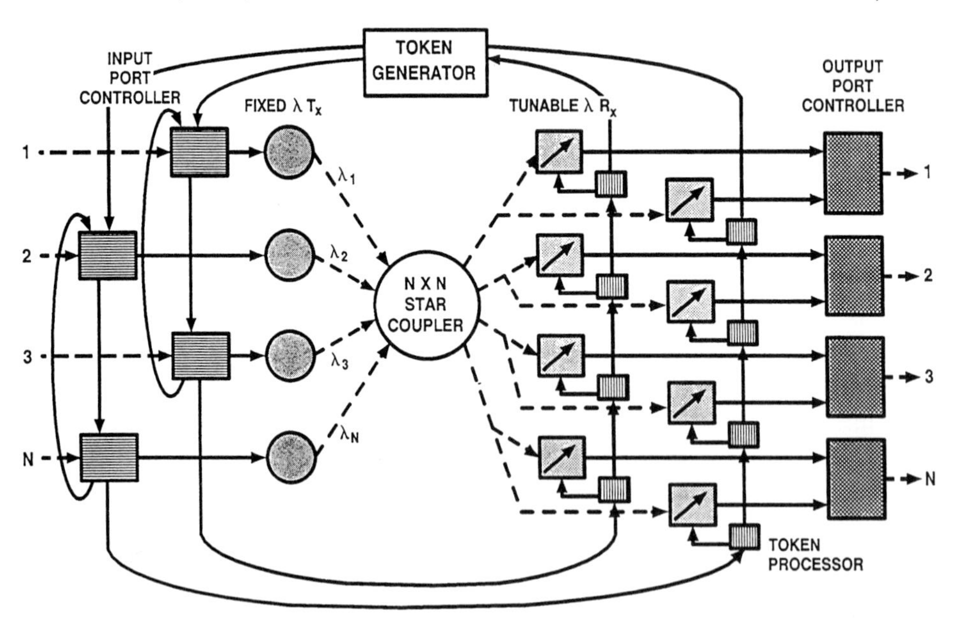

Fig. 4. Multitrack StarTrack architecture. An enhanced StarTrack with multiple control tracks. The receivers' required tuning range can be smaller than with a single track, and the switch performance increased because of the parallelism.

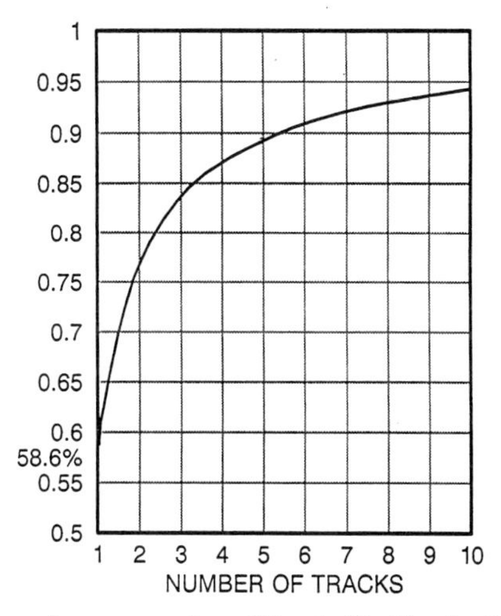

Fig. 5. Throughput performance of multitrack StarTrack. The theoretical performance of a multitrack StarTrack: performance increases rapidly with increasing numbers of control tracks.

are widespread. However, the Ethernet standard is not directly extendible to high data rates because its multiaccess protocol requires that the minimum packet duration be approximately twice the round-trip transmission delay, which requires very large (and inefficient) packets at high transmission rates. Data networks to support higher-speed LANs have been widely investigated and a standard has been adopted, the fiber distributed data interface (FDDI). The FDDI standard uses dual counterrotating rings to implement a 125 Mbit/s standard, based primarily on the use of multimode fiber and light-emitting diodes (LEDs), for compatibility with maximum internode distances of only 2 km. Again, the redundancy possible with OEICs could play a useful role, by providing replicated transceiver facilities at low cost for backup purposes. (An enhanced version of FDDI has been proposed, called FDDI-II, extending the data rate to about 600 Mbit/s, compatible with SONET rates.)

LANs can be interconnected to form metropolitan area networks (MANs), requiring more capacity and larger node distances than LANs can accommodate. A standard, distributed queue dual bus (DQDB) [22], has evolved for MANs that is useful in the 200 Mbit/s range (Fig. 6) for interconnecting LANs such as Ethernets or FDDI networks. The first offered broadband telecommunications service, SMDS (initially meaning switched multimegabit data service but today used with broader meaning), will be implemented on this platform. DQDB is based on two unidirectional buses that may be either open or loop structured. DQDB may be implemented with

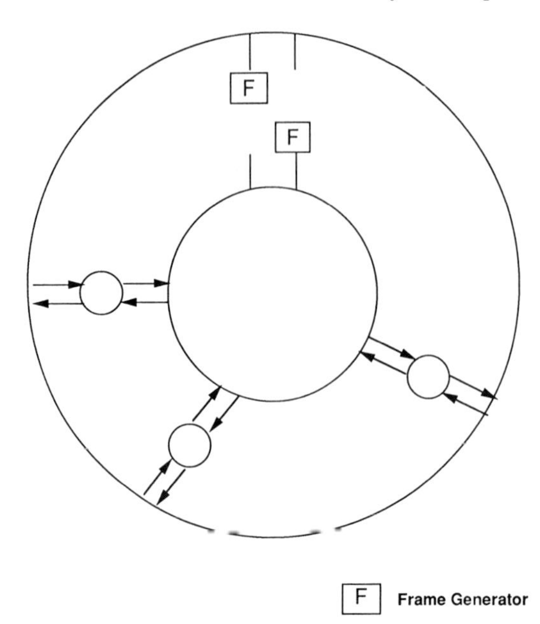

Fig. 6. DQDB architecture. The dual-bus structure with nodes access to each bus and frame generator is shown. The potential application for OEIC might be to enhance performance and reliability with redundant transmitters and more reliable connection to bus.

dual single-wavelength optical buses, where each port requires a transmitter and receiver pair for each bus. For LANs and MANs the network interface costs are critical. Thus, the opportunity for OEICs in this arena would arise principally from their potential for cost reduction compared with discrete components and from the potential reliability advantages due to integration.

FDDI and DQDB are examples of single-wavelength optical networks that do not tap the large bandwidth potential of optical fibers. In multiwavelength applications, the potential of OEICs really becomes clear. Not only do they provide redundancy or a reduced packaging cost and reliability advantage, they may even be required for any of these applications to be practical. An example multiwavelength MAN architecture is the ShuffleNet architecture, which was initially proposed by a group at AT&T Bell Laboratories [23, 24]. In the ShuffleNet architecture, users interface to a unidirectional multiwavelength optical bus through a set of network interface units (NIUs). In a simple case, each NIU has two fixed-wavelength optical transmitters and two fixed-wavelength optical receivers. The input and output ports of the NIUs are connected to the multiwavelength optical bus as indicated schematically in Fig. 7, which is assumed to have a total length of a few hundred

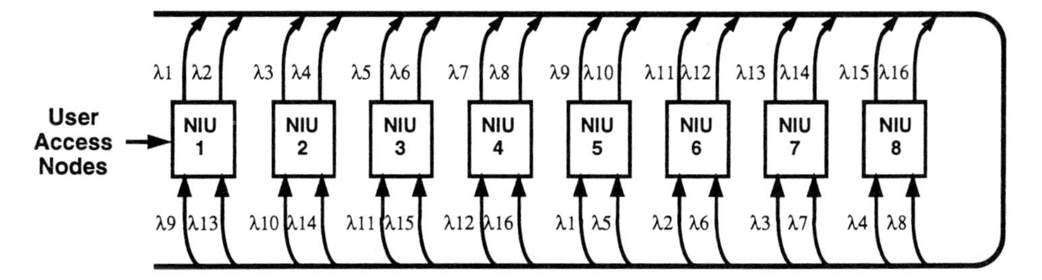

Fig. 7. ShuffleNet multihop network. The AT&T ShuffleNet LAN architecture. Users access the network interface units (NIUs), which access the multiwavelength bus at two wavelengths and receive at two different wavelengths. Each NIU incorporates a 2 × 2 packet switch. This idea can be explaned (with larger packet switches) to several wavelengths per NIU (©IEEE, 1990).

meters to a few kilometers. The data packet information loops around the NIUs, which filter the data by listening at their specific wavelength and passing data packets on, if necessary, regenerating them at their output wavelengths. The fixed-wavelength interconnects in this architecture may be arranged in a particular pattern to enhance the network performance. For example, in the ShuffleNet, the wavelengths are arranged in a "perfect shuffle" as shown in Fig. 8 for an eight-node network (hence the name ShuffleNet). At each node (NIU) this network has a 2 × 2 electronic packet switch to route the packet from the input to the output of the NIU if the received packet is not destined for this node. In this example, it is possible to get a packet from any input node to any output node in a path requiring three "hops" or less. Thus a packet from the input port at NIU-1 that is destined for the user interfaced to node NIU-3 may be transmitted on wavelength 2 to NIU-6, where it is received, and regenerated on wavelength 11 to NIU-3. A nice feature of this type of network is that alternative routes are available—for example, if a particular node is congested. In this example, a data packet may get from NIU-1 to NIU-3 by an alternative route: first use wavelength 1 to get to NIU-5, then wavelength 10 to NIU-2, then wavelength 4 to NIU-8, and finally wavelength 15 to NIU-3. The ShuffleNet is an example of a large class of multihop networks which have been extensively studied (see, for example, ref. 24).

The ShuffleNet combines the features of an optical star network (see the discussion of LAMBDANET later) and the features of a multihop ring network. Obvious extensions of this network involve larger electronic switches at each NIU and the possibility of multiple optical buses; for example, one

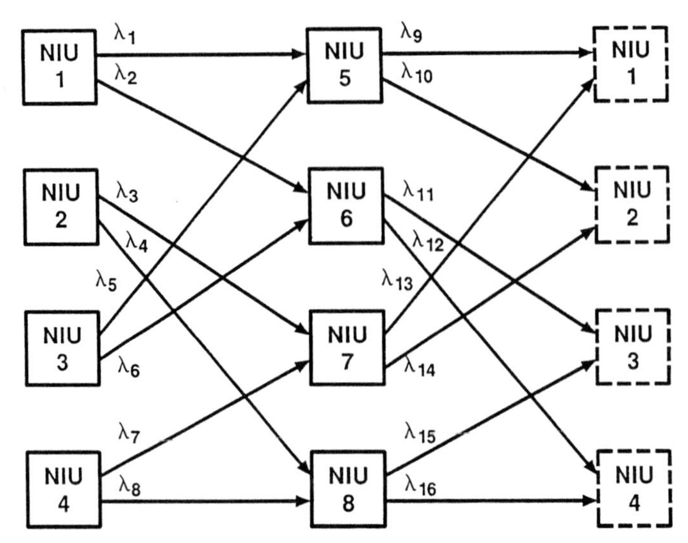

Fig. 8. Perfect shuffle routing. For routing in a ShuffleNet a perfect shuffle is chosen because of its theoretical performance characteristics. An advantage of this approach is that there are different paths to each node in case of congestion.

could consider 8 × 8 packet switches in a larger multihop network. Although such a network may have internal queueing delays, it has very attractive features from the viewpoint of OEIC application. The designs described involve optical transmitters at two (or several) fixed wavelengths and optical receivers that require wavelength demultiplexers or filters for receiving two (or several) wavelengths. Not only is the transmitter/receiver electronics a likely candidate for integration, but also the electronic packet switch at the NIU could be a candidate for inclusion in the OEIC. The maximum link distances in such a network are typically less than a kilometer and the transmission rates are only of the order of 1 Gbit/s, so reasonably low-power lasers could be used, and receiver sensitivity is not a serious constraint. Indeed, all the NIU functions could be envisioned as a single OEIC, which we will call the multihop OEIC (Fig. Cat-4). Note that for the multihop OEIC, the wavelength tuning could be performed using temperature tuning or some other relatively slow tuning mechanism. The multiwavelength optical bus needs to support a relatively small number of discrete wavelengths (less than eight) and direct detection can be used.

There are many potential variations and enhancements for the multihop architecture. For example, dynamic wavelength allocation allows the architecture to be varied so that wavelengths could be reused and real-time alternative routing strategies may be imposed. One example is the proposal for a scalable, modular, multihop wavelength routing network by Brackett [25]. This architecture (Fig. 9) shows network access nodes interconnected by a transparent optical network, with wavelength translation and wavelength routing. It directly addresses two important features, namely scalability (the ability to add more nodes) and modularity (the ability to add one node at a time). Networks with simple fixed-wavelength schemes may not always be easily expanded, particularly in small incremental units. A salient feature of this architecture is that the number of wavelengths is independent of the number of nodes. As shown in Fig. 9, the proposed network consists of access nodes (the square elements), wavelength routing cross-connection elements (the diamond-shaped elements), and a common network control to permit the dynamic rearrangeability. One way of realizing such an all-optical approach is to utilize acousto-optic switches. The rearrangeability allows the dynamic reuse of wavelength and bandwidth throughout the network to meet changing traffic, service, or performance requirements and to provide a robust, fault-tolerant network. Each network access node can transmit to and receive from several other nodes by selecting the appropriate wavelength.

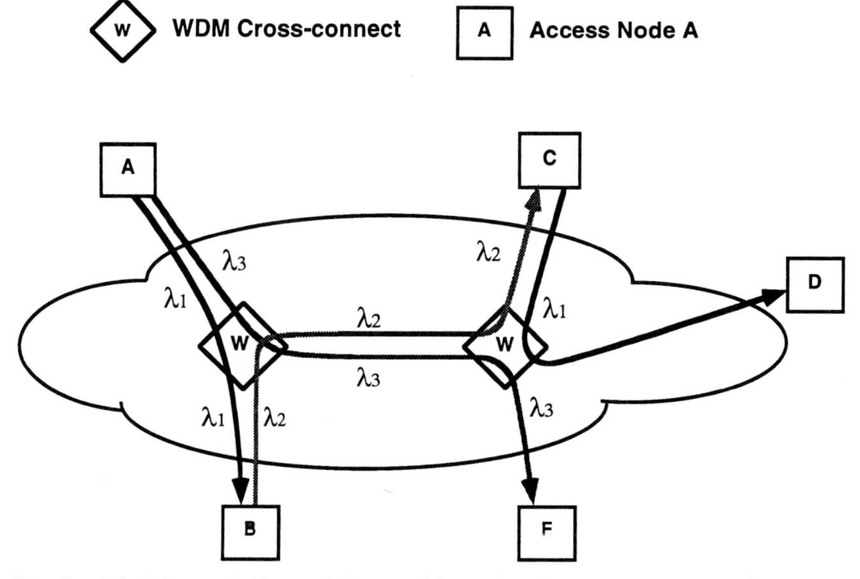

Fig. 9. Modular, scalable, multihop multiwavelength network. A MAN/WAN architecture that is both scalable and modular. The diamonds indicate the WDM cross-connects and the squares indicate the network access nodes. The internal portion of the network is all optical. This network decouples the number of wavelengths used from the number of nodes and features wavelength reuse, reducing the number of required wavelengths (©IEEE, 1992).

With a specific configuration of the wavelength routing cross-connects, this transparent optical transmission may extend over large distances. In the figure, wavelength 3 carries a one-hop signal (no intermediate detections or translations) from node A to node E, while a signal from node A to node C is carried in two hops, using wavelength 1 from A to B and wavelength 2 from B to C. Wavelength 1 is reused to carry a signal from C to D.

An expanded view of the network (Fig. 10) shows details of the transmitter and receiver elements, as well as the wavelength interchanging cross-connect portion of the network. This type of network provides several types of opportunity for OEIC technology (Fig. Cat-5), including transmitter arrays incorporating optical couplers and optical amplifiers and receiver arrays with on-chip wavelength multiplexing. Such a network can become practical only with optoelectronic integration, which is required to simplify the packaging, and alignment necessary for the laser transmitter arrays, the optical amplifiers, and the wavelength multiplexing and demultiplexing components. It is important that this network architecture is "transparent" in the

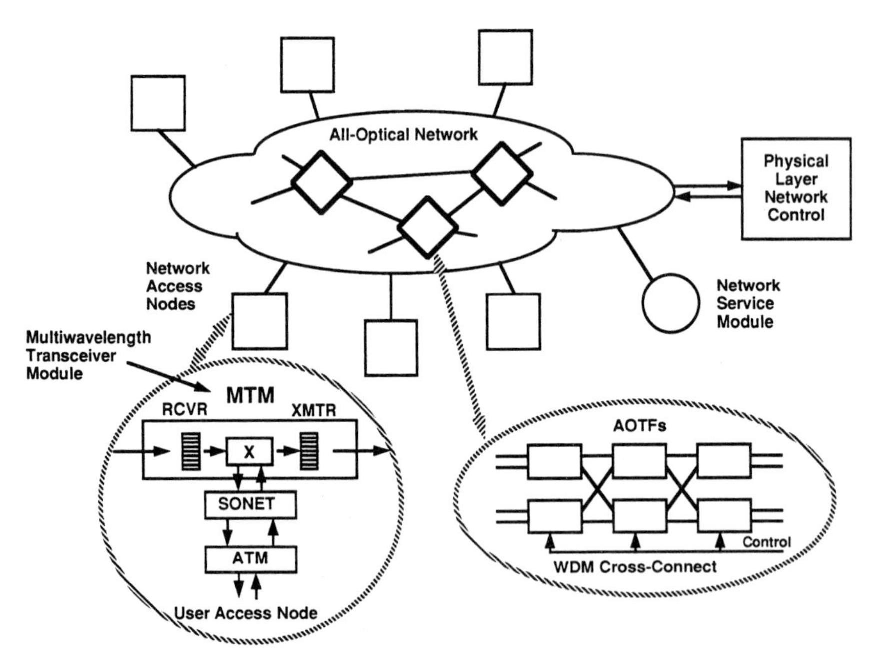

Fig. 10. Expanded view of modular, scalable, multihop multiwavelength network. An expanded schematic showing functions within the WDM cross-connects and within the network access nodes. The WDM cross-connects use acousto-optic tunable filters. The network access nodes perform various functions, including wavelength translation, using multiwavelength transmitter and receiver arrays (©IEEE, 1992).

sense that the wavelength routing that occurs does not depend on the detailed format of the data being transported. While the terminating equipment will have some dependence on the detailed format of the data, the switching of wavelengths will not.

The potential market for OEIC technology for LANs and MANs may be significant. As an indication of this market size, electronic interfaces to LANs currently number in the millions worldwide.

3.3. Fiber in the Subscriber Loop: OEICs for the Masses

In order to provide universal service, these broadband networks must extend to the individual subscribers' residences and businesses. As the network infrastructure evolves toward BISDN, optical fiber will enter the local business and residential markets or "subscriber loops." The optical fiber–based subscriber loops (fiber in the loop or FITL) will probably start as fiber to the curb (FTTC), with metallic drops from each optical network unit (ONU) serving several homes, initially to provide voice-grade service. This approach has already reached approximate cost parity (for installed first costs) with copper distribution networks for some applications, particularly in newly constructed neighborhoods. This cost parity is considered significant for the subscriber loop economics. Even though the optical fiber has more capability than the copper-based subscriber loop pairs, initial deployment will probably be based primarily on the economics of voice-grade services. Then, as *broadband services* start to penetrate into the subscriber loops, these initial systems may be simply upgraded for broadband access. Since high-performance optical systems are not required for voice-grade FTTC, these systems are more strongly driven by cost than by performance. In order to make these FITL systems cost-competitive with conventional copper loops for telephone services, very few fibers are installed (typically one or two fibers serving a small group of customers, approximately 4 to 64 homes). Initially such fiber systems are likely to require only discrete low-data-rate components. There have already been more than 100 field trials of FITL systems worldwide with more planned, and mass deployment is expected to begin in 1993 [26]. Eventually, systems with fiber extending all the way to the home (FTTH) will be developed, providing the full benefit of the fiber bandwidth and reliability.

An important problem will be how to upgrade the initial FITL systems for broadband services without costly installation of additional fibers. Many approaches are being investigated. Exactly how this is done will depend on the transmission rates required at the subscribers' premises, whether transmission along the fiber is unidirectional or bidirectional (whether one or two fibers are used), whether only point-to-point services or both point-to-point and point-to-multipoint services are to be supported, and a host of other issues such as network monitoring and control, network robustness and reliability, and powering. Advances in digital video compression technology, for example, have dramatically reduced the bandwidth required to provide digital video services to the home.

Upgraded FITL systems have been studied which enable individual subscribers access from 1.5 to 155 Mbit/s and even to 622 Mbit/s in architectures employing fiber to the curb and fiber to the home [27, 28]. A promising way to upgrade initial FITL systems deployed for voice-grade communications is to utilize multiwavelength networks. These systems may involve only two wavelengths initially—for example, to separate services: voice grade services at 1.3 μm and broadband services at 1.5 μm. As more and more broadband services become available, although time division multiplexing is likely to be employed, systems having several more closely spaced wavelengths may well be advantageous.

Figure 11 shows several versions of FITL systems for point-to-point and point-to-multipoint optical network architectures. An example of a two-fiber FITL system upgrade with wideband/broadband services to the ONU bidirectional transmission is shown in Fig. 12 [28, 29]. (A similar system employing individual array transmitter/receivers for use with a two-fiber system is also possible [28].) Finally, an upgrade to full broadband services with a broadband ONU at the subscribers' premises is shown in Fig. 13. Other variations and similar architectures are possible—the salient feature is that OEICs could play a critical role in the subscriber loops, in the cost-effectiveness of the systems, and in their enhanced performance. Typically, about 50% of the cost of the ONU is associated with the optoelectronic components and their packaging. Examples of potential generic OEICs for these example FITL architectures are shown in Figs. Cat-6 and Cat-7.

The subscriber loop presents a greater potential opportunity for optoelectronics and OEICs than almost any other area of telecommunications networks. The device performance criteria (small required number of devices per chip, transmitter power, receiver sensitivity, etc.) are less stringent than for most other telecommunications applications, and the potential number of OEIC units is very large because of the number of residential and business subscriber locations (the number of households in the United States alone is nearly 100 million). It has been estimated that by the year 2015 there should be extensive penetration of fiber in the loop; thus the mass market

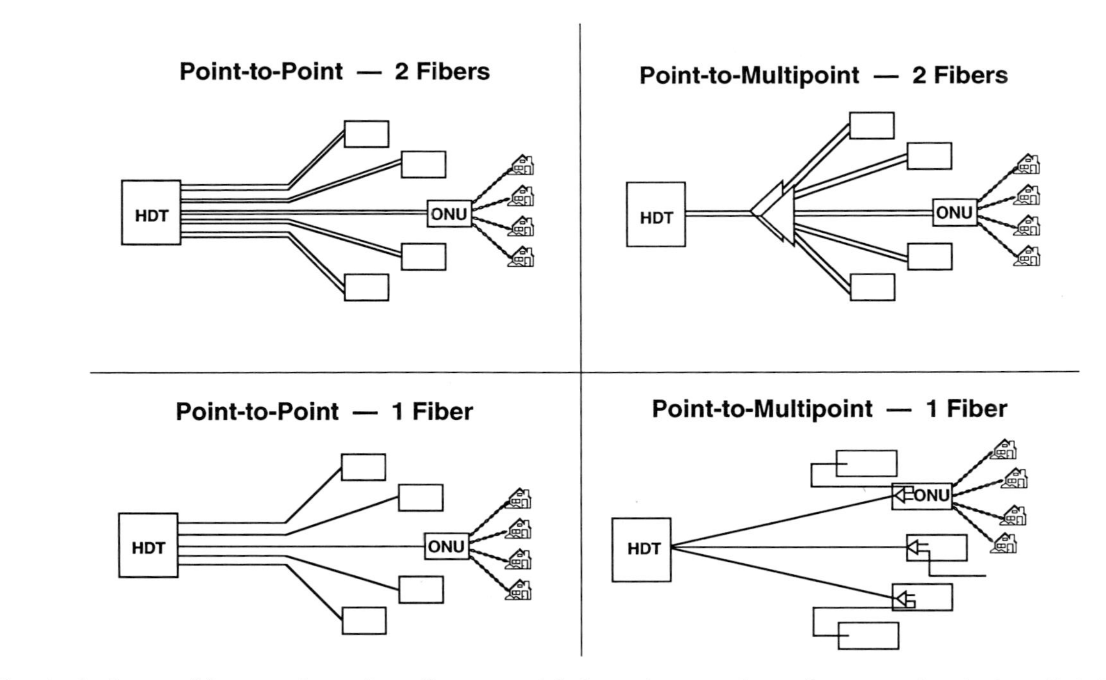

Fig. 11. Fiber-in-the-loop architecture alternatives. Four potential alternatives are shown for connecting the host digital terminal (HDT) to the optical network units (ONUs). The HDT may be located in a central office or in a "remote node" in the loop plant. ONUs may serve several residences (or business locations) in fiber-to-the-curb systems or may be located at the home for fiber-to-the-home.

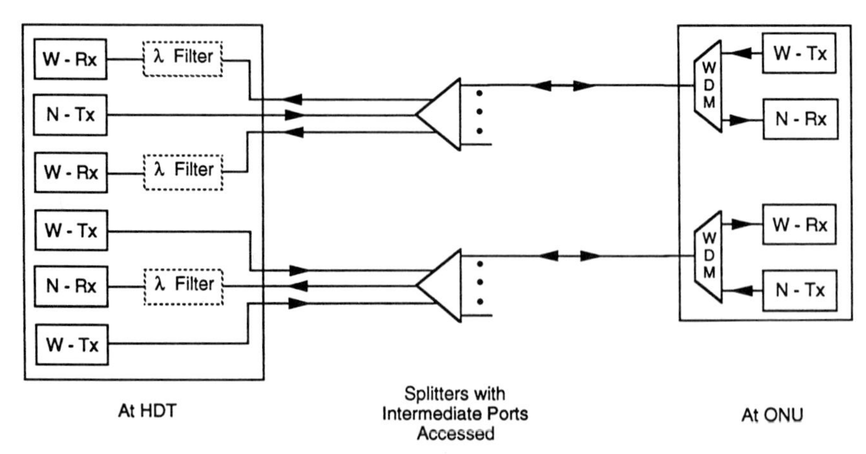

Fig. 12. Potential two-fiber WDM upgrade for fiber-in-the-loop. In this approach separate wavelengths are used for narrowband and wideband services and intermediate splitter ports are used to reduce the split ratio for the wideband signals. The large number of ONUs and the multiple transmitters and receivers at each ONU make such systems candidates for integration.

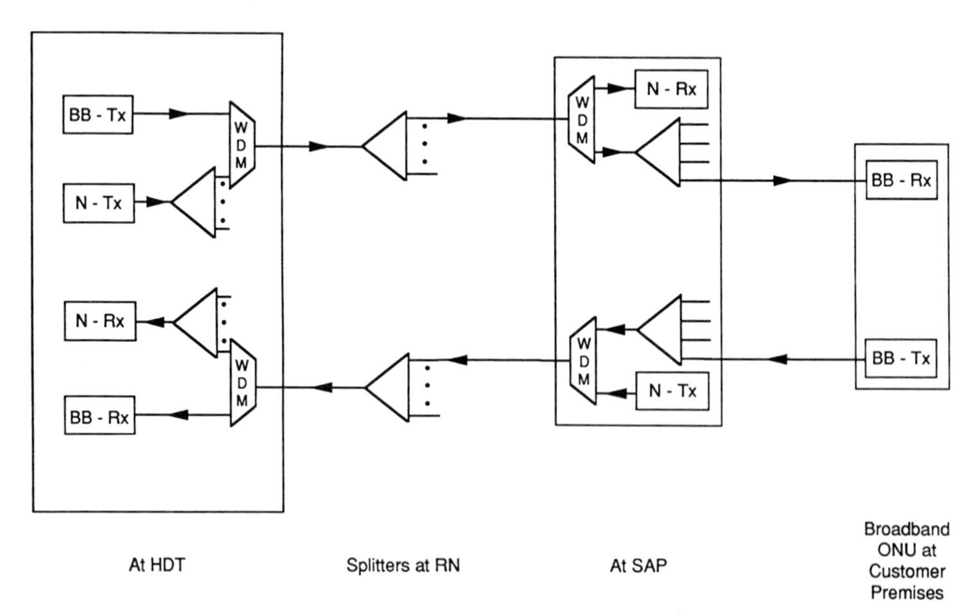

Fig. 13. Upgrade of a fiber in the loop system for broadband services. A potential FTTC upgrade to provide broadband services over additional fiber links to the home.

potential is considerable. As local loop standards develop, this market should present a significant opportunity for OEIC technology.

3.4. Local Video Distribution with Multiwavelength Optical Networks

Video information is becoming an essential ingredient of the information age. In addition to entertainment video, there is a growing number of other one-way video services—for example, video travelogues, video shopping, and subscriber-ordered pay-per-view video. An example of an optical network supporting a broadcast distribution application is based on the LAMBDANET architecture [30]. The LAMBDANET architecture is composed of a cluster of N communication nodes (Fig. 14), with each node transmitting its information on a unique wavelength using single-frequency lasers and each node receiving all the information from the other nodes. The nodes are optically coupled together using a transmissive optical star coupler at a hub location, which broadcasts information on all wavelengths to all the nodes. In addition, each node can have a optical transceiver tuned to a common "control wavelength." The LAMBDANET, which is similar to a number of optical star networks, can be applied to different system applications, including wide-area information distribution and video distribution. A schematic example of a broadcast video backbone LAMBDANET is shown in Fig. 15. In this configuration, all the transmitters are colocated in the central node and each of the remote nodes has access continuously to all the broadcast data. The figure shows a video center (head-end station) that receives

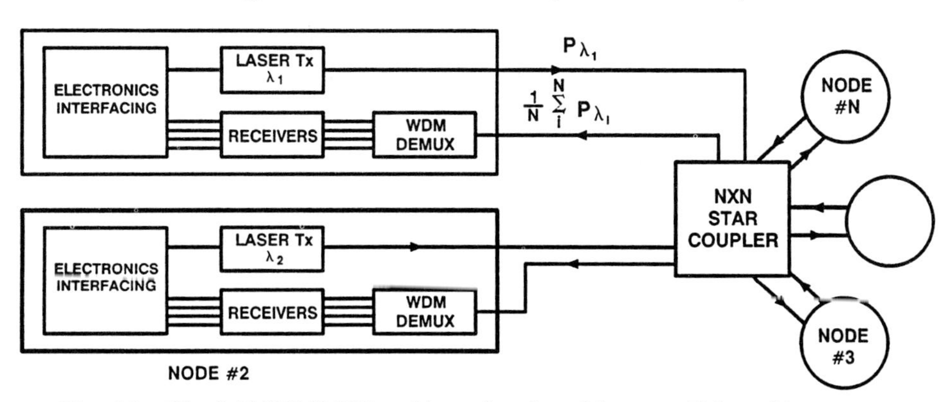

Fig. 14. The LAMBDANET multiwavelength architecture. This architecture associates a unique wavelength with each node. Information is broadcast to all nodes. Each node receives information from all the other nodes simultaneously and in parallel (©IEEE, 1990).

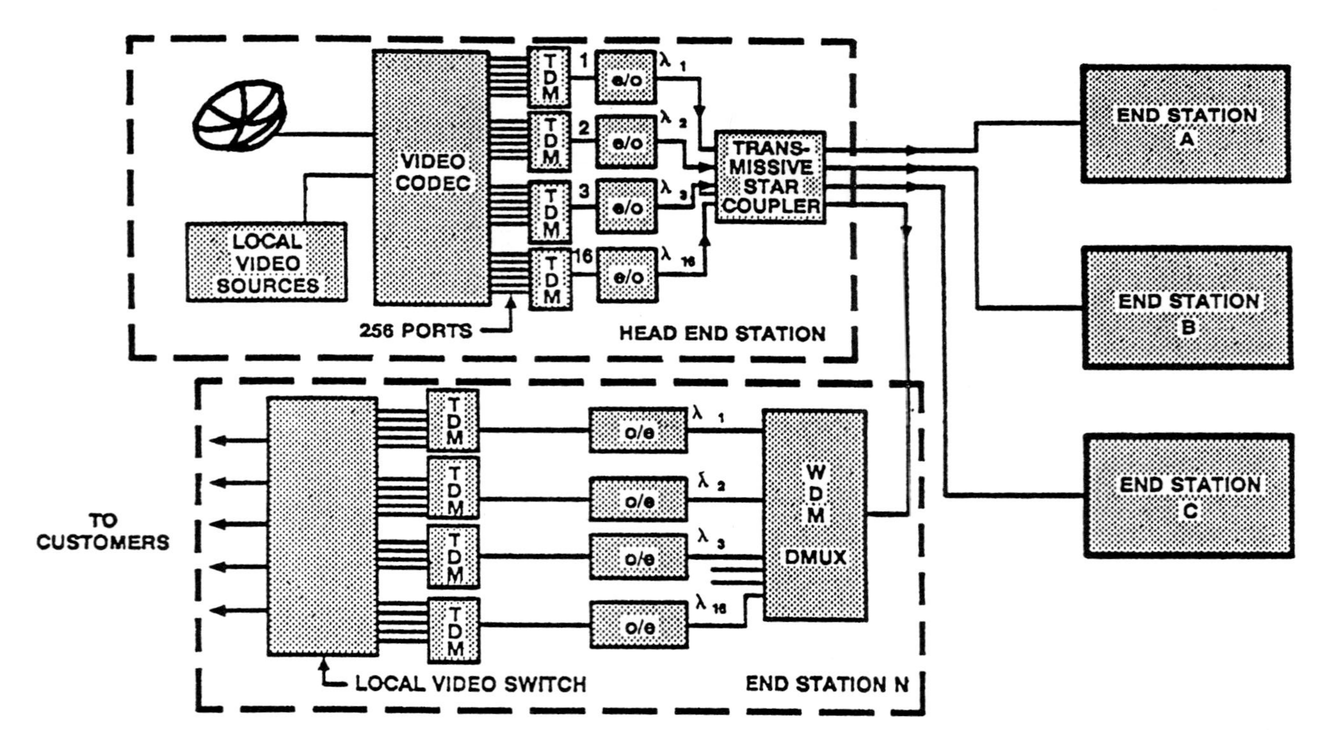

Fig. 15. Video distribution using a LAMBDANET. Diagram showing a video broadcast architecture in which the transmissive star coupler is located in one node (the head end). Very large capacities for digital broadcast information are possible in such architectures (©IEEE, 1990).

video source information from satellite and alternative video sources (for example, a videotape library or local video studios). In the LAMBDANET video broadcast example, at least 256 video channels (each at a data rate of approximately 155 Mbit/s for highest quality or incorporating time-multiplexed channels within the 155 Mbit/s stream for even more video channels) are optically multiplexed together to form 16 high-speed data channels at 2.5 Gbit/s, which are transmitted to the remote stations. Wavelength and time demultiplexing separates the individual channels at the remote stations. These remote stations may have further local switching or direct distribution using either fiber or coaxial cable to the end users. Although this example is a digital time-multiplexed system, a number of variations are possible. For example, analog broadcast systems (with or without subcarrier multiplexing) could be implemented instead or in combination with the TDM approach just described.

Multiwavelength optical star networks for video distribution could readily utilize OEICs. In particular, electronic time multiplexing and transmitters for the video centers could be useful and multiwavelength receivers incorporating wavelength demultiplexing could be very useful. Because this is a broadcast multiwavelength network, the number of receivers is proportional to $N(N - 1)$. Thus OEICs incorporating wavelength and time demultiplexing and optical receivers would be welcome for video distribution applications. A functional diagram of an OEIC for video distribution networks is shown in Fig. Cat-8.

A further application might involve self-healing SONET rings, useful in the interoffice telephone plant for network robustness [31, 32]. One such approach would be to use a LAMBDANET-like node to form a multiwavelength survivable ring network with switch consolidation [33] as shown in Fig. 16. It is clear that the use of $N - 1$ receivers along with wavelength multiplexers/demultiplexers and electronic logic for channel selection may be a candidate for OEIC implementation.

4. LINEAR LIGHTWAVE NETWORKS

An active area of research has centered on linear lightwave networks, in which the nodes perform only linear operations on the lightwave signals [34]. The operations, such as power splitting and combining and amplification, are all based on linear combinations of the optical power. In these networks, physical connectivity between end points is established by lightwave channels, implemented using multiwavelength systems. This flexible topology can be

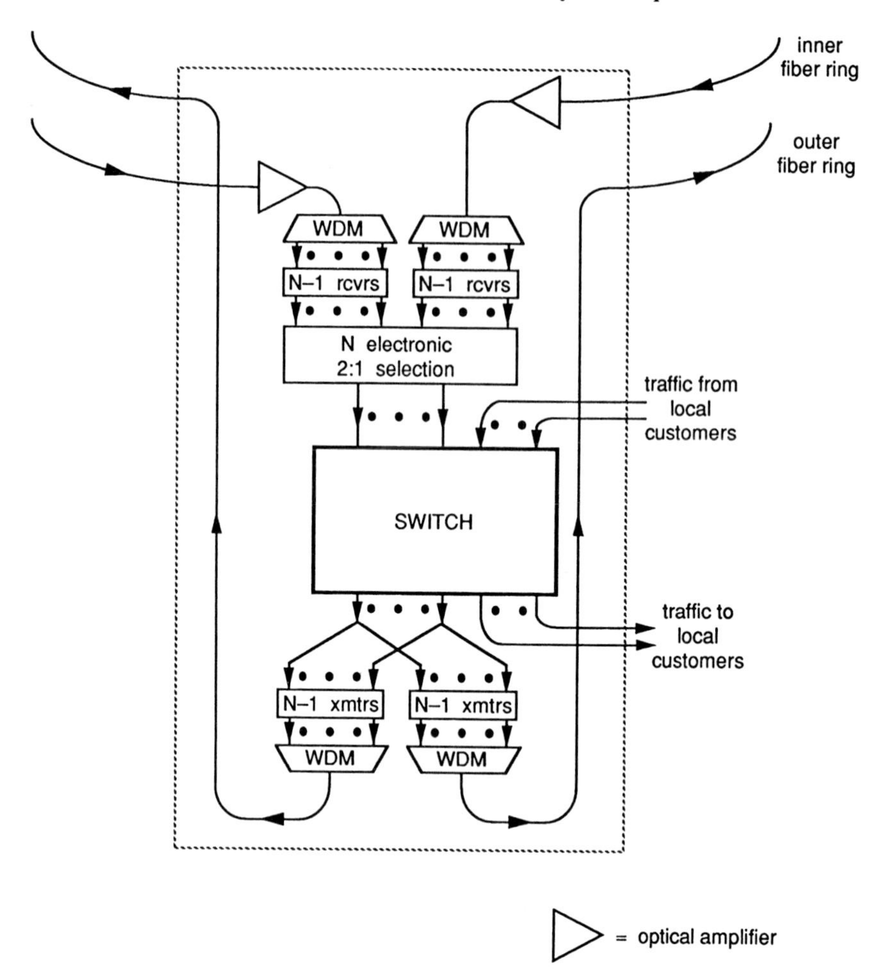

Fig. 16. Switch node for multiwavelength ring switch-consolidation architecture. The switch node is connected with a survivable multiwavelength ring architecture to $N - 1$ local central offices (COs), which share the switching resources. One wavelength is used to communicate between the switch and each local CO (©IEEE, 1990).

changed either by tuning transmitters and receivers as in other multiwavelength networks or by appropriate control of splitting (and combining) ratios of the incoherent signals with broadband coupler devices. Therefore there are no optical-to-electronic conversions internal to the network—all this processing is performed in the optical domain. Thus this type of network allows not only topological rearrangement as performed with digital cross-connect systems but also the exploitation of additional capacity attainable in multiwavelength systems. In general, these networks may permit reuse of wavelengths, similar to frequency reuse in cellular radio systems.

An example of a linear lightwave network is shown in Fig. 17, where the rectangular boxes represent the transmitters and receivers and the circles represent the internal network routing nodes at which optical splitting and combining occur. Figure Cat-9 depicts the functional block diagram of a 2 × 2 internal routing node that could be implemented using electro-optic (or acousto-optic) waveguide switches on an OEIC. The values X_1 to X_4 (which can be varied from 0 to 1) represent the power splitting and combining ratios, which control the proportion of the input power that can be sent to the output links. That is,

$$P^{o}(1) = (1 - X_1)(1 - X_3)\,P^{i}(1) + (1 - X_2)X_3P^{i}(2)$$

and

$$P^{o}(2) = X_1(1 - X_4)P^{i}(1) + X_2X_4P^{i}(2).$$

5. CONCLUSIONS

Although the goal of most research on OEICs is to produce the best performance possible, it is our view that the strength of OEIC technology lies more in the important functionality attained through their unique ability to provide enhanced capability at low cost (since most lightwave networks are

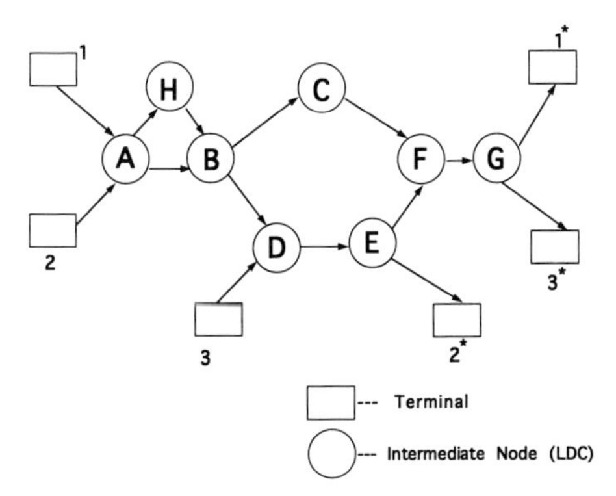

Fig. 17. A linear lightwave network. An all-optical internal network is used to route and distribute multiwavelength signals across the network. The intermediate nodes may use waveguide switches, for example, to route the signals (©IEEE, 1991).

driven by cost) than in individual component performance. Historically, this was the experience in electronics as well: IC technology lagged far behind discrete component technology for record performance, during the entire period of its meteoric rise to electronic market dominance! However, electronic IC technology did superbly address the connection bottle-neck.

We have presented a number of potential application areas in telecommunications in which OEICs could be usefully employed. For some of these example architectures OEICs present a way to extend or upgrade performance, and in others OEICs would make the crucial difference between a viable architecture or none at all. In all cases, a potential advantage of OEICs will be their cost/component effectiveness.

The potential telecommunications market in which OEICs can have a major impact is growing. Figure 18 shows a projection of the revenue growth due to optoelectronic transmitter–receiver pairs for telecommunications assuming discrete components [35]. This projection, dominated by the new market for fiber in the loop, indicates revenues due directly to optoelectronic components of more than $1 billion within about a decade and rising rapidly thereafter. This time frame is an appropriate time target for the transfer of OEIC technology from research laboratories to production lines, so that OEICs may be used in many of the applications described as the anticipated growth occurs. There have already been impressive laboratory demonstrations of possible OEIC technology [36–46]. The functional catalog of possible OEICs provided here is intended as a generic guide to the thinking of the systems architects and is not intended to be exhaustive. It is our hope that this type of catalog will continue to grow in functionality and will soon be realized in practice. Figure Cat-10 presents a "wish list" of potential OEIC functional units for building systems (the detailed specifications will, of course, depend on the application).

OEICs offer possible advantages in price and reliability for single-wavelength optical networks. They may also enable cost-effective access to parallel transmission on many fibers. However, it seems that sophisticated multiwavelength networks will not be possible without them. The history of communications is filled with examples of an "enabling technology" leading to a revolutionary advance in the effectiveness of the communications. Optoelectronic integrated circuit technology is, in our view, the enabling technology that will be a driver for broadband communications, especially multiwavelength optical network applications, and thus will play a key role in making universal broadband communications possible and ushering in the Information Age.

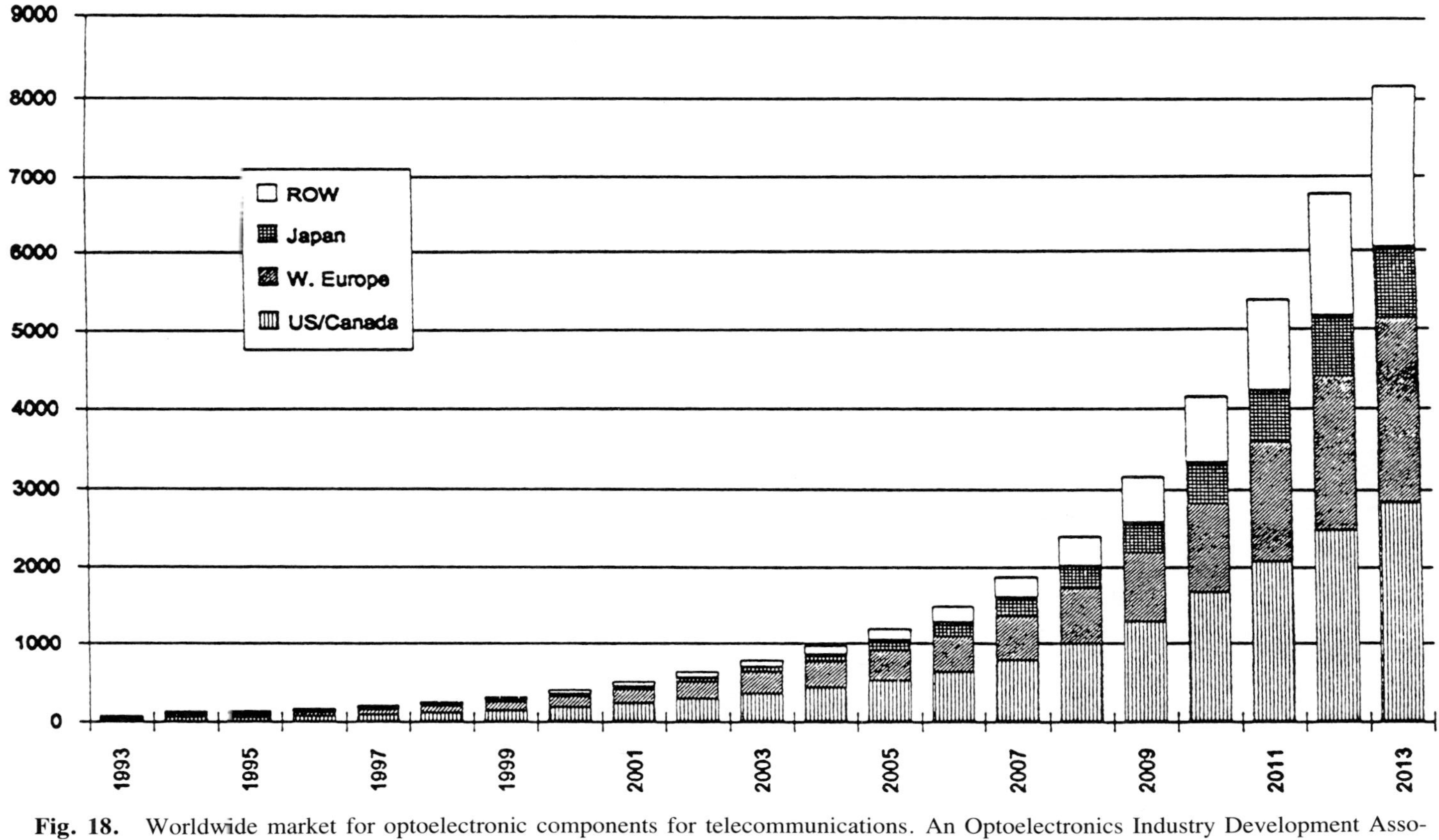

Fig. 18. Worldwide market for optoelectronic components for telecommunications. An Optoelectronics Industry Development Association (OIDA) projection of the worldwide market for optoelectronics components for telecommunications. This is projected to become a multi-billion-dollar market as OEIC technology matures for volume production.

APPENDIX: FUNCTIONAL OEIC CATALOG

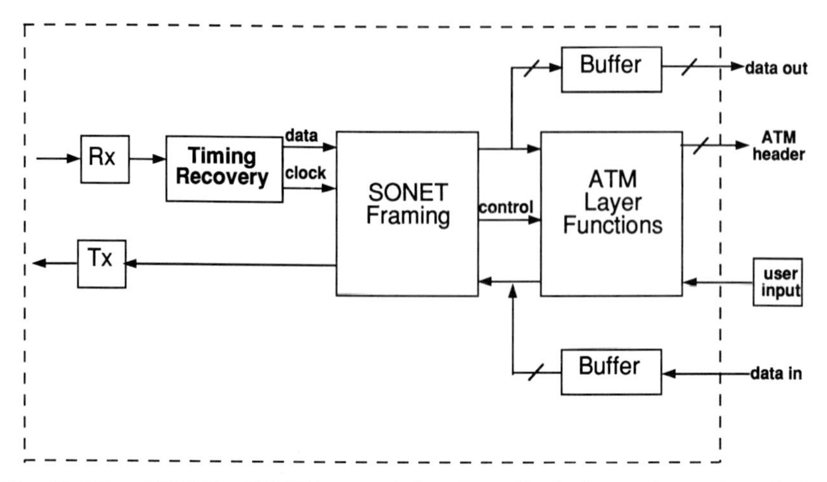

Fig. Cat-1. OEIC for SONET network interface. Typical optoelectronic and electronic functions needed for SONET interfaces are shown. Integration could provide cost reduction, reliability, and performance enhancements.

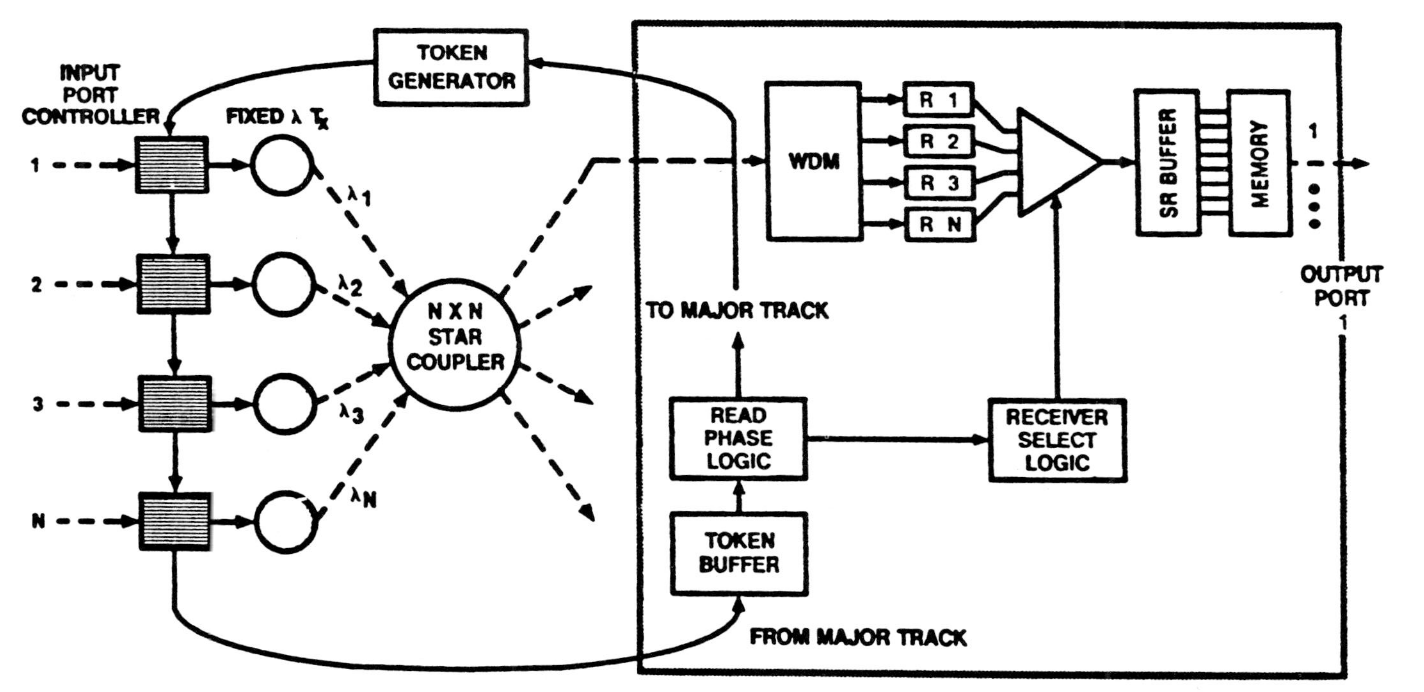

Fig. Cat-2. Multiple receiver OEIC for StarTrack switch. Possible OEIC for a single output of the switch is shown in the box. "Wavelength tuning" is accomplished by electronic selection of a fixed-wavelength receiver array. OEIC could include simple control logic and the fast shift register (SR) and memory module.

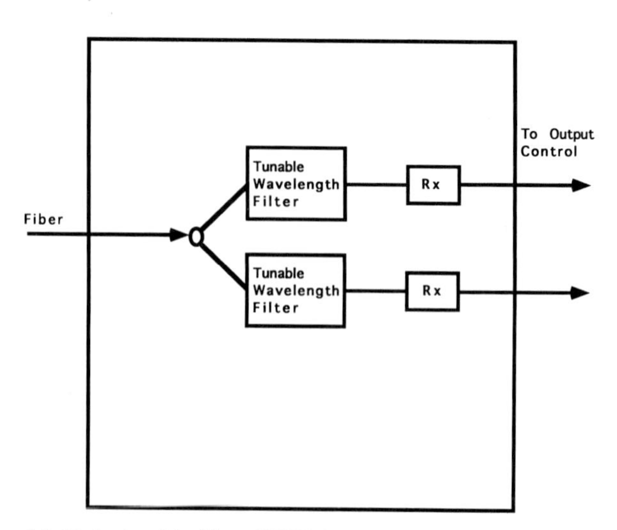

Fig. Cat-3. Multiple tunable filter OEIC for multiple track StarTrack switch. Possible tunable receiver with rapidly tunable (few ns) elements. Tuning range for multitrack StarTrack is reduced relative to single-track StarTrack.

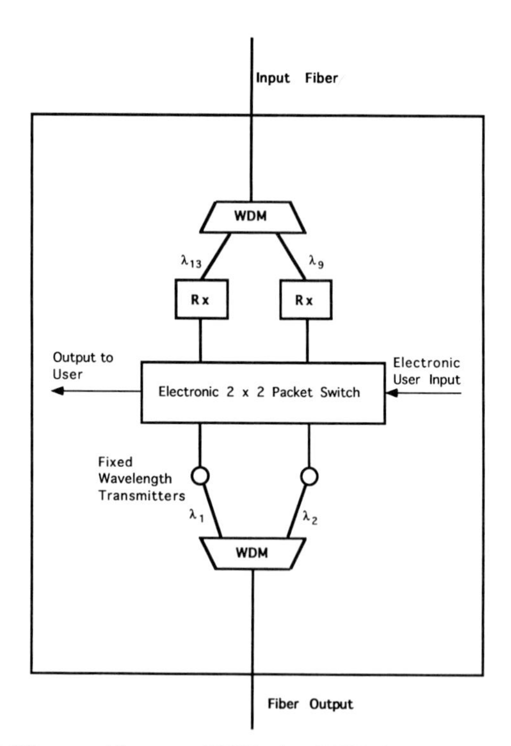

Fig. Cat-4. Multihop architecture OEIC. An OEIC for ShuffleNet-type architectures. Possible OEIC (for NIU-1 of Fig. 8) could include the transmitter and receiver arrays and associated electronics and potentially could incorporate integrated electronics for small (2 × 2) electronic packet switches (including memory buffers).

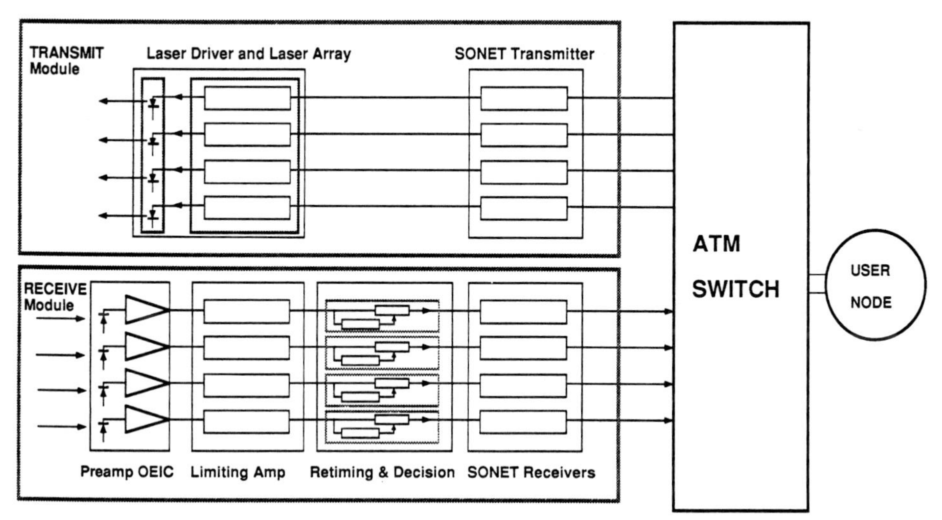

Fig. Cat-5. Modular, scalable, multiwavelength, multihop OEICs. Potential OEICs for a modular scalable multiwavelength multihop network could include portions of the transmit modules and receive modules. For many wavelengths, OEICs will probably be critical for this network.

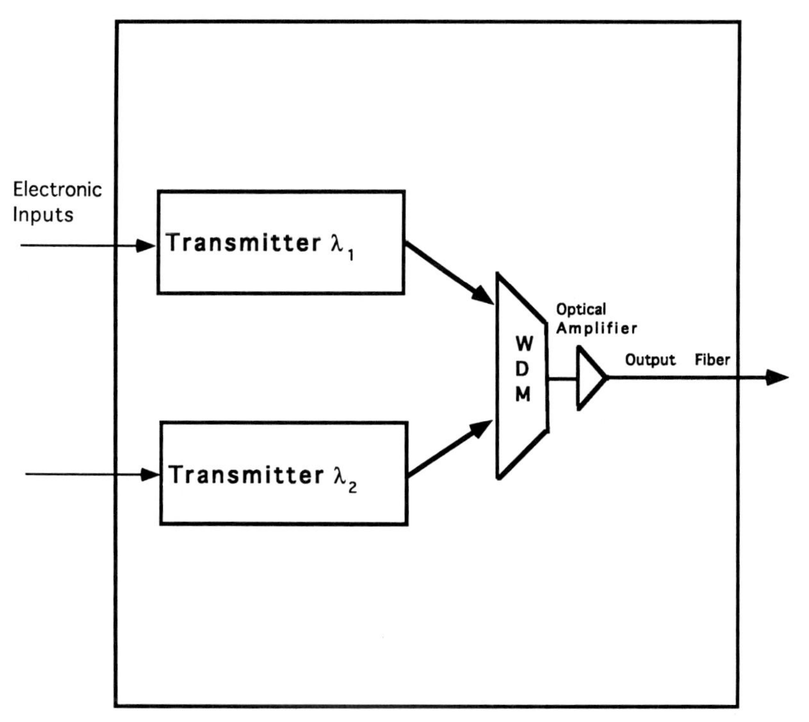

Fig. Cat-6. Fiber-in-the-loop OEIC I. Multiwavelength OEIC for fiber-in-the-loop upgrade. Integration may provide cost advantages as well as improved reliability compared with discrete components. An integrated optical amplifier might alleviate power budget constraints.

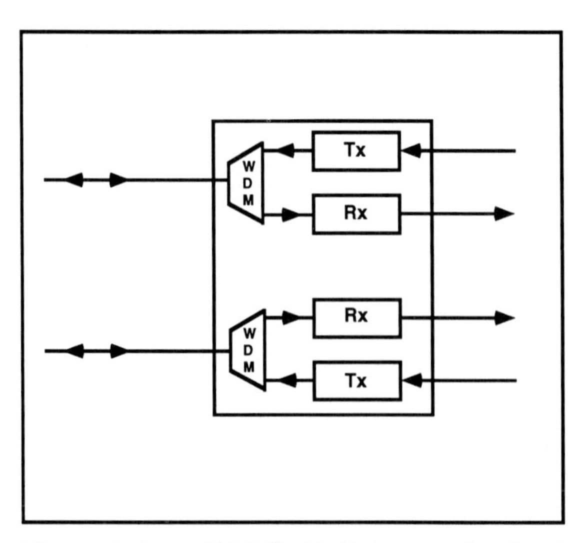

Fig. Cat-7. Fiber-in-the-loop OEIC II. Multiple transmitter/receiver array with integrated wavelength multiplexers and demultiplexers may provide cost advantages necessary for broadband fiber-to-the-home architectures.

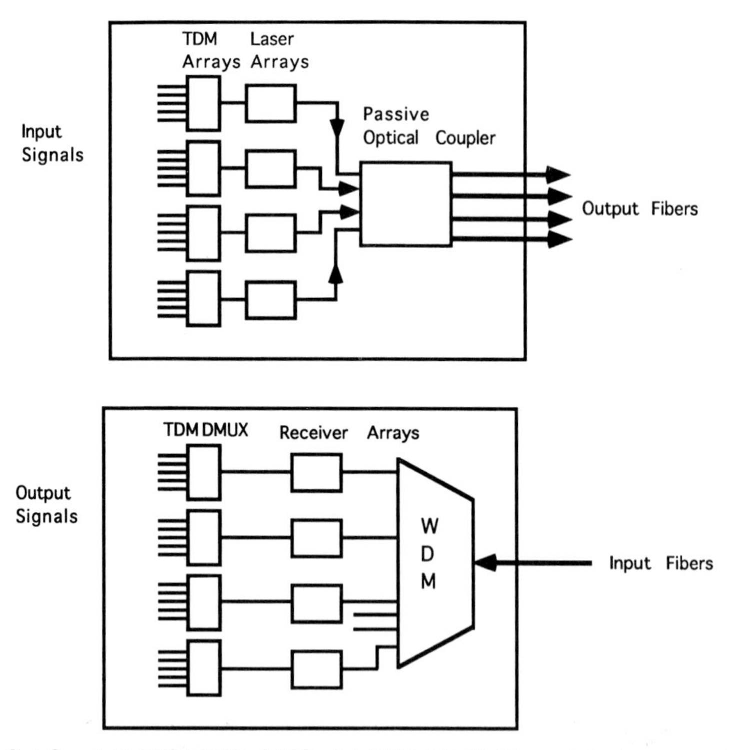

Fig. Cat-8. LAMBDANET OEIC. LAMBDANET-like architectures require on the order of N fixed-wavelength receivers for each node. Integration is important for this type of architecture. An OEIC implementation might also provide for TDM time division multiplexing/demultiplexing of input/output signals.

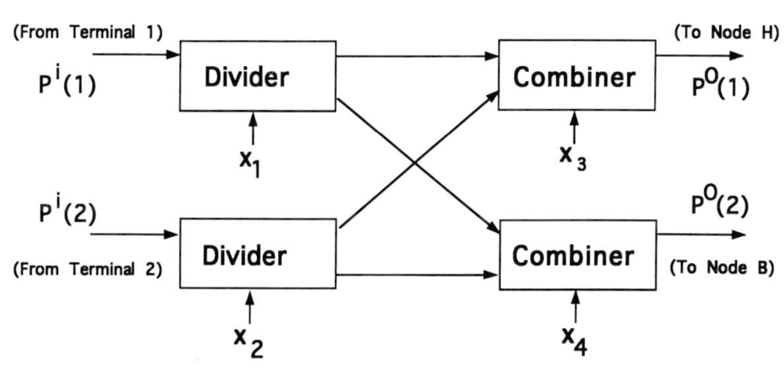

Fig. Cat-9. Linear lightwave network OEIC. An OEIC that incorporates, for example, waveguide switches which varies the splitting and combining ratios to provide routing. Signals x_1 to x_4 vary these ratios from 0 to 1.

Device Function (Optical)	Device Function (Electronic)
Laser Arrays (Fixed)	Laser Drivers
Laser Arrays (Wavelength Tunable)	Receiver preamps and electronics.
Detector Arrays	Packet Buffers and RAM Memories
Tunable Filters	Electronic Mux/Demux
Passive Waveguides (Couplers, etc.)	Electronic Switches
Active Waveguides (Switches, etc.)	Random Logic Arrays
Optical Amplifiers	Read Only Memories
Spatial Light Modulators	Microprocessors
Acousto-optic Transducers	

Fig. Cat-10. Catalog of modular functions for OEICs. A list of functional building blocks for OEICs. The system architect should be able to choose parameters associated for each function from a catalog to construct future telecommunications systems.

REFERENCES

1. A. Bergh, The Current Status and Future Prospects of Optoelectronics Technology in the USA, *IEEE Circuits Devices* **6**(3), 32–39 (1991) and private communication (1991).
2. M. Dagenais, R. F. Leheny, H. Temkin, and P. Bhattacharya, Applications and Challenges of OEIC Technology, *J. Lightwave Technol.* **8**, 846–862 (1990).
3. F. A. Tobaggi, Fast Packet Switch Architectures for Broadband Integrated Services Network, *Proc. IEEE* **77**, 133–167 (1990).
4. R. Ballert and Y. C. Ching, SONET: Now It's the Standard Optical Network, *IEEE Commun. Mag.* **28**, 8–15 (1990).
5. J. Y. Hui, "Switching and Traffic Theory for Integrated Broadband Networks." Kluwer Academic Publishers, Boston, 1990.
6. I. Chlamtac and W. R. Franta, Rationale, Directions, and Issues Surrounding High Speed Networks, *Proc. IEEE* **77**, 94–120 (1990).
7. C. A. Brackett, Dense Wavelength Division Multiplexing Networks: Principles and Applications, *IEEE J. Selected Areas Commun.* **8**(6), 948–964 (1990).
8. G. R. Hill, Wavelength Domain Optical Network Techniques. *Proc. IEEE* **77**, 121–132 (1990).
9. H. Toba, K. Oda, K. Nosu, and N. Takato, Factors Affecting the Design of Optical FDM Information Distribution Systems, *IEEE J. Selected Areas Commun.* **8**(6), 965–972 (1990).
10. T. T. Lee, M. S. Goodman, and E. Arthurs, A Broadband Optical Multicast Switch, *ISS'90*, Stockholm (1990).
11. M. S. Chen, N. R. Dono, and R. Ramaswami, A Media Access Protocol for Packet-Switched Wavelength Division Multiaccess Metropolitan Area Networks, *IEEE J. Selected Areas Commun.* **8**(6), 1048–1057 (1990).
12. K. Y. Eng, M. A. Santoro, T. L. Koch, J. Stone, and W. W. Snell, Star-Coupler-Based Optical Cross-Connect Switch Experiments with Tunable Receivers, *IEEE J. Selected Areas Commun.* **8**(6), 1026–1031 (1990).
13. H. Kobrinski, M. P. Vecchi, M. S. Goodman, E. L. Goldstein, T. E. Chapuran, J. M. Cooper, M. Tur, C. E. Zah, and S. G. Menocal, Jr., Fast Wavelength Switching of Laser Transmitters and Amplifiers, *IEEE J. Selected Areas Commun.* **8**(6), 1190–1202 (1990).
14. H. Kobrinski, Crossconnection of Wavelength-Divison Multiplexed High Speed Channels, *IEE Electron. Lett.* **23**, 974–976 (1987).
15. K. W. Cheung, S. C. Liew, and C. N. Lo, Experimental Demonstration of Multiwavelength Optical Network with Microwave Subcarriers, *IEE Electron. Lett.* **25**, (1990).
16. E. Arthurs, M. S. Goodman, H. Kobrinski, and M. P. Vecchi, Hypass: An Optoelectronic Hybrid Packet Switching System, *IEEE J. Selected Areas Commun*, **6**, 1500–1510 (1988).
17. H. S. Hinton, F. B. McCormick, T. J. Cloonan, F. A. P. Tooley, A. L.

LENTINE, AND S. J. HINTERLONG, Photonic Switching Fabrics based on S-SEED Arrays, in "*Photonic Switching II*" (K. Tada and H. S. Hinton, eds.). Springer-Verlag, Heidelberg, 1990.

18. A. CISNEROS AND C. A. BRACKETT, A Large ATM Switch based on Memory Switches and Optical Star Couplers, *Proc. ICC '91*, 721–727, Denver (1991).
19. M. S. GOODMAN, Multiwavelength Networks and New Approaches to Packet Switching, *IEEE Commun. Mag.* **27**, 27–36 (1989).
20. J. B. D. SOULE, A. SCHERER, H. P. LEBLANC, N. C. ANDREADAKIS, R. BJAT, AND M. A. KOZA, Monolithic InP/InGaAsP/InP Grating Spectrometer for the 1.48–1.56 Micron Wavelength Range, *Appl. Phys. Lett.* **58**, 1949–1952 (1991).
21. J. B. D. SOOLE, A. SCHERER, Y. SILVERBERG, H. P. LEBLANC, N. C. ANDREADAKIS, AND C. CANEAU, WDM Detection Using Integrated Grating Demultiplexer and High Density *p-i-n* Array, *Proc. LEOS Summer Topical Meeting on Integrated Optoelectronics*, Paper WB-1, Santa Barbara (1992).
22. IEEE Metropolitan Area Network Standard 802.6 (1989).
23. A. S. A*CAMPORA*, M. J. K*AROL*, *AND* M. G. H*LUCHYJ*, Multihop Lightwave Networks: A New Approach to Achieve Terabit Capabilities, *Proc. ICC '88*, Philadelphia (1988).
24. I. BAR-DAVID AND M. J. KAROL, Nested Multi-connected rings for Large High Capacity LANs and MANs, *Infocom '90 Conf. Proc.* 1172–1189 (1990).
25. C. A. BRACKETT, Scalability and Modularity in Multiwavelength Optical Networks, LEOS Summer Topical Meeting on Optical Multiple Access Networks, Paper TuB1, Santa Barbara (1992).
26. P. W. SHUMATE AND R. K. SNELLING, *IEEE Commun. Mag.* 68–74 (March 1991).
27. D. L. WARING, D. S. WILSON, AND T. R. HSING, Fiber Upgrade Strategies Using High-bit-rate Copper Technologies for Video Delivery, *J. Lightwave Technol.* **10**, 1743–1750 (1992).
28. T. E. CHAPURAN, S. S. WAGNER, AND R. C. MENENDEZ, Fiber-in-the-Loop Upgrades to Support Wideband and Broadband Services, *Proc. 4th Workshop on Optical Local Networks*, pp. 203–208, Versailles (1992).
29. T. E. CHAPURAN, S. S. WAGNER, AND R. C. MENENDEZ, Wideband and Broadband Upgrades for Near-Term Fiber-in-the-Loop Systems, *NFOEC '92*, paper 24.1 (April 1992).
30. M. S. GOODMAN, H. KOBRINSKI, M. P. VECCHI, R. M. BULLEY, AND J. L. GIMLETT, The LAMBDANET Multiwavelength Network: Architecture, Applications, and Demonstrations, *IEEE J. Selected Areas Commun.* **8**(6), 995–1004 (1990).
31. T-H. WU AND R. C. LAU, A Class of Self-Healing Ring Architectures for SONET Network Applications, *Proc. Globecom '90*, San Diego (1990).
32. E. L. GOLDSTEIN, Optical Ring Networks with Distributed Amplification, *IEEE Photon. Technol. Lett.* **3**, 390–393 (1991); E. L. GOLDSTEIN, L. ESKILDSEN, T. H. WU, M. ANDREJCO, A. E. WILNER, V. SHJAH, L. CURTIS, D. MAHONEY, W.

C. Young, A. Yiyan, H. Izadpanah, C. E. Zeh, N. Andreadakis, F. Favire, B. Pathak, T. P. Lee, and C. Lin, Use of Quasi-Distributed Optical Amplification in SONET Self-Healing Inter-Exchange Networks, *Proc. OFC '92*, p. 254, San Jose (1992); A. F. Elrefaie, Self Healing WDM Ring Networks with All Optical Protection Path, *Proc. OFC '92*, p. 255, San Jose (1992).

33. S. S. Wagner and T. E. Chapuran, Multiwavelength-Ring Networks for Switch Consolidation and Interconnection, *ICC '92*, Chicago (June 1992).
34. K. Bala, T. E. Stern, and K. Bala, Algorithms for Routing in a Linear Lightwave Network, *IEEE Infocom '91*, Bal Harbor, Miami (April 1991).
35. A. A. Bergh and R. F. Leheny, Optoelectronics Industry Association (OIDA), private communication.
36. T. Ikegami, Optoelectronic IC's for Telecommunications, *Proc. LEOS Summer Topical Meeting on Integrated Optoelectronics*, Santa Barbara (1992).
37. N. Takato, T. Kominato, A. Sugita, K. Jinguji, H. Toba, and M. Kawachi, Silica-Based Integrated Optic Mach–Zehnder Multi/Demultiplexer Family with Channel Spacing of 0.01–250 nm, *J. Selected Areas Commun.* **8**(6), 1120–1127 (1990).
38. D. A. Smith, J. E. Baran, J. J. Johnson, and K. W. Cheung, Integrated-Optic Acoustically-Tunable Filters for WDM Networks, *J. Selected Areas Commun.* **8**(6), 1151–1160 (1990).
39. G. Winzer, W. Doldissen, C. Cremer, F. Fiedler, G. Heise, R. Kaiser, R. Marz, H. F. Mahlein, L. Morl, H. P. J. Nolting, W. Rehbein, M. Schienle, G. Schulte-Roth, G. Unterborsch, H. Unzeitig, and U. Wolff, Monolithically Integrated Detector Chip for a Two-Channel Unidirectional WDM Link at 1.5 Microns, *J. Selected Areas Commun.* **8**(6), 1183–1189 (1990).
40. U. Koren, T. L. Koch, B. I. Miller, G. Eisenstein, G. Raybon, An Integrated Tunable Light Source with Extended Tunability Range, *IOOC '89*, Kobe (1989).
41. H. Matsueda, T. P. Tanaka, and H. Nakano, An Optoelectronic Integrated Device Including a Laser and Its Driver Circuit, *IEE Proc.* **131**, 299–303 (1984).
42. H. Matsueda, "*The Physics of Optoelectronic Integrated Circuits.*" Shokabo, Tokyo, 1990.
43. U. Koren, Optoelectronic Integrated Circuits, in "*Optoelectronic Technology and Lightwave Communication Systems*" (C. Lin, ed.). Van Nostrand Reinhold, New York, 1989.
44. See also papers in *Proc. LEOS Summer Topical Workshop on Integrated Optoelectronics*, Santa Barbara (August 1992).
45. C. E. Zah, P. S. D. Lin, F. Faivre, B. Pathak, R. Bhat, C. Caneau, A. S. Gozdz, N. C. Andreadakis, M. A. Koza, and T. P. Lee, 1.5 Micron Compressive-Strained Multiple Quantum Well 20 Wavelength Distributed Feedback Laser Array, *Proc. OFC92*, pp. 40–41, San Jose (1992).
46. C. Cremer, N. Emeis, M. Schier, G. Heise, G. Ebbinghaus, and L. Stoll, Grating Spectrograph Integrated with Photodiode Array in InGa AsP/InGaAs/InP, *Photon. Technol. Lett.* **4**, 108–110 (1992).

Chapter 2

COMPUTING SYSTEM APPLICATIONS

John Crow, J. M. Jaffe, and M. W. Sachs

IBM, T. J. Watson Research Center
Yorktown Heights, New York 10598

1. INTRODUCTION

This chapter describes the application of optoelectronics (OE) to computing. Compared to the telecommunication industry's use of fiber optics and optoelectronics, the computer industry is about a decade behind in the commercial utilization of OE technology. However, applications are emerging which have the potential of using large numbers of OE components ($>10^6$/yr). An optoelectronic market forecast by the Opto-Electronic Industry Association predicts that the computer industry applications for interconnect, storage, and display will be even larger (in equipment revenue) than those

ISBN 0-12-200420-5

of telecommunications and will represent a $50 billion business by the turn of the century. For an OE module technology to be used pervasively in these applications it must be cost effective, reliable, and compact, similar to the integrated circuit (IC) electronic technology already used in computer equipment. These are all the promised features for an integrated OE technology and specifically the OEIC chip. Thus, the computer industry is expected to be more than just a user of today's discrete device OE modules; it should be the motivating force for this new generation of highly integrated optoelectronic technology.

The broad set of present-day computer applications which can take advantage of optoelectronics includes communications, input/output (I/O), storage, display, and printing. (Figure 12 illustrates these elements in a computing system and highlights where optoelectronics is or might be used.) Computer systems use the global telecommunications networks, and these requirements were covered in Chapter 1. The local area networks (LANs) use both fiber optics and copper wiring in products today, but there is a trend toward optics, exemplified by emerging U.S. standards such as the Fiber Distributed Data Interface (FDDI) that specify optoelectronic technology. The I/O networks, which primarily support interprocessor and processor-to-storage data transfers, also use fiber-optic wiring in products today. Again, there are emerging standards, such as the proposed American National Standards Institute Fibre Channel standard (FC), which specify optical interconnections. Although intraprocessor interconnections and parallel processor networks use electrical wiring harnesses today, there is considerable R&D activity in exploring the advantages of optics. In data storage products, optical storage media are playing a bigger role for both archival and "almost on-line" storage systems. Multimedia and image processing applications are expected to drive this need. Optoelectronic read/write heads are a key technology. In printing, the laser-driven electrophotographic printer has become the pervasive office printer. In displays, the liquid crystal flat panel display with integrated thin-film transistor drivers is a fast-growing market because of the rapid growth of portable computers. Thus, it is apparent that optoelectronic technology is used pervasively in computing systems and is playing an increasingly important role. This chapter reviews each of these areas and delineates the particular optoelectronic requirements of each area.

However, advances in optoelectronics are having a more fundamental impact on computing systems. The radical changes in computer design and computing, in particular the growth of distributed processing, are facilitated by the explosion in communications bandwidth made possible by optical communications. These changes are enabling entirely new classes of applications, such as image processing, and will enable as yet unforeseen types of application, which will in turn foster further revolutionary development of technology. Surely there are other factors that drive these trends, but optical communications is an enabler without which these new applications would not be possible.

The key ingredients in computer design (processor performance and memory capacity and speed) have exponential growth curves. Figure 1 illustrates that in mainframes, processor speed is growing by approximately a factor of 10 per decade [1]. Similar curves apply to memory capacity and speed, as well as to other classes of processor. To continue to support increased demands for existing applications, I/O bandwidth must grow on this scale, or an unbalanced system will result with insufficient I/O performance to support its computational power.

In fact, the total I/O bandwidth available in a system should grow nonlinearly with the growth of the other basic commodities. For example, if the speed and memory of a uniprocessor are doubled, a straightforward extrapolation of a balanced system would require a doubling of I/O bandwidth. However, the growth of parallelism adds a further component of required growth of performance of the interconnection system.

Another new element is that I/O and communications no longer involve just the traditional data traffic. Use of voice, image, and video in computer applications is growing. Large organizations wish to integrate these types of information in a common information transport system. There is, in par-

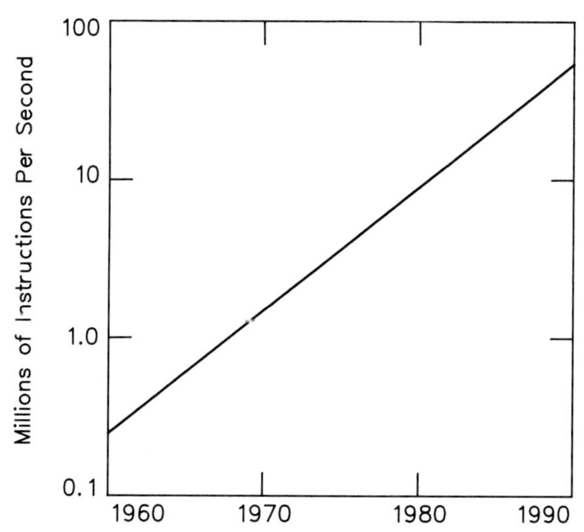

Fig. 1. Mainframe performance growth with time.

ticular, growing use of image data, in additional to traditional computer data formats, in transaction processing applications [2]. This vastly increases the amount of information which must be moved and stored.

2. TRENDS IN THE COMPUTER INDUSTRY

A number of trends in the computer industry are driving optoelectronics technologies. These include

- Continuing rapid increases in processing power at all levels of cost
- Demand for central system processing power that continues to outrun the increases in the power available in a single processor, resulting in the need to couple processors to obtain higher central system processing power
- Growth in the use of distributed processing, fueled by the availability of inexpensive microprocessors
- New applications such as image and multimedia which require massive amounts of on-line data storage

In this section, we discuss each of these trends in more detail.

2.1. Increasing Processing Power

The continuing rapid growth of computer processing power requires similar increases in the rate at which data can be moved between a computer and its input/output and storage devices as well as between computers. In the past, the data rates of I/O and storage devices have not kept up with processing power and designs have focused on providing increasing numbers of devices and data transfer paths (channels) to make up the difference. A high-performance processor may require more than 100 I/O channels to provide adequate aggregate bandwidth using traditional electrical transmission. Configurations with such large numbers of channels and I/O devices are difficult to plan and manage. Increased bandwidth will enable significant reductions to be made in the number of I/O channels. Furthermore, as we shall see, new forms of data and new applications of the increased processing power have led to the need for tremendous increases in the bandwidth of at least some of the interconnection paths. The computer industry is looking to fiber optics to provide the increased bandwidth.

This growth in processing power, along with increases in storage device capacity and performance, makes possible new applications that place further demands on the performance of computer interconnections. In the commercial transaction processing environment, a user at a terminal has expected subsecond response time for an inquiry involving a few hundred bytes of information to be displayed. The vast storage capacity of optical storage devices, compared with magnetic storage devices, permits using images of documents in place of, or in addition to, traditional forms of computer data.

Integration of document images into commercial transaction processing allows an enterprise to make drastic reductions in the cost of handling and storing paper documents while enabling much more rapid processing of transactions that cannot easily be encoded in traditional computer data formats. A document image stored on the computer system is accessible in seconds, whereas days may be required to retrieve a paper document from its file.

As a result of integrating images into transaction processing, a commercial transaction may involve the movement of tens of thousands of bytes of compressed image data with the same desired subsecond response time as for transactions involving only a few hundred bytes of data in traditional format.

Medical image processing (for example, digitized x-ray pictures) involves images with resolutions on the order of 1000 × 1000 picture elements (PELs) and 8 to 16 bits of gray-scale data per PEL, i.e., one to two million bytes (Mbytes) per picture. Scientific visualization (display of the results of scientific computations) involving color and motion requires display of images which may have as many as 2000 × 2000 PELs with 24 bits of color and intensity information per PEL, repeated 60 times per second. This requires a data transfer rate of 720 Mbyte/s and is to be compared with data rates for "traditional" I/O, which are typically in the range of 1 to 10 Mbyte/s. (It should be noted that practical deployment of data rates in excess of 100 Mbyte/s is still in the future but is desired for certain applications.) See ref. 3. for examples of applications of scientific visualization.

The computer industry is responding to these bandwidth challenges. Arrays of magnetic disks provide both capacity and data transfer bandwidth which are considerably higher than in the past. The FDDI standard [4] is a high-performance, token-ring, local area network backbone which uses optical transmission at 125 Mbit/s. Major mainframe manufacturers in Japan, the United States, and Europe have introduced high-speed I/O channels using fiber optics. See, for example, ref. 5, which describes IBM's serial fiber-optic channel interface, with a transmission bandwidth of 200 Mbit/s for attachment of I/O and storage devices to its processors and for processor-to-processor communication.

A contemporary goal is a data transfer rate of 100–200 Mbyte/s for large sets of data. To provide data transfer at this rate, the ANSI X3T9.3 Committee has introduced the High Performance Parallel Interface (HIPPI) standard [6] and numerous vendors are providing it on computers, I/O devices, and workstations. HIPPI provides up to 200 Mbyte/s of bandwidth, at distances up to 25 m, on a parallel point-to-point interconnect, and switching systems are available which allow multiple HIPPI-capable computers and devices to be interconnected. The X3T9.3 Committee is now in the process of designing the Fibre Channel [7], a proposed standard for fiber-optic interconnections which will provide up to 100 Mbyte/s of useful bandwidth in a geography of dimensions on the order of 10 km, using serial fiber-optic links with bandwidth of approximately 1 billion bits per second (1 Gbit/s).

2.2. Coupled Processors

When demand for central processor power outstrips the capability of a single processor, further power is obtained by coupling processors into a single system. There are many coupling schemes, each with different requirements of interconnection topology and performance. The coupling scheme must provide some combination of processor-to-processor communication and sharing of data. Three common coupling schemes are known as loose coupling, tight coupling, and massive parallelism.

2.2.1. Loose Coupling

With loose coupling, individual systems are coupled using, primarily, I/O interconnections to share data. The interconnection system provides both access to shared data and rapid exchange of messages among the processors. A variant of loose coupling, sometimes known as close coupling, includes a high-speed shared random-access storage unit which provides rapid access to the portion of the shared data which is actively being used [8]. An example is the Fujitsu Corp. System Storage [9]. Figure 2 illustrates a loosely coupled complex with a shared storage unit.

Each processor's I/O channels are interconnected through an interconnection network to the shared devices. The devices are shared through some combination of multiple interfaces on the devices and a multipoint interconnection network such as a parallel bus or a crosspoint switch as used with the IBM fiber-optic channels [10]. The processors can exchange messages, for such purposes as work scheduling and control of access to shared data, through either the I/O network or the shared store. The shared store is typically connected to the processors by means of separate high-speed

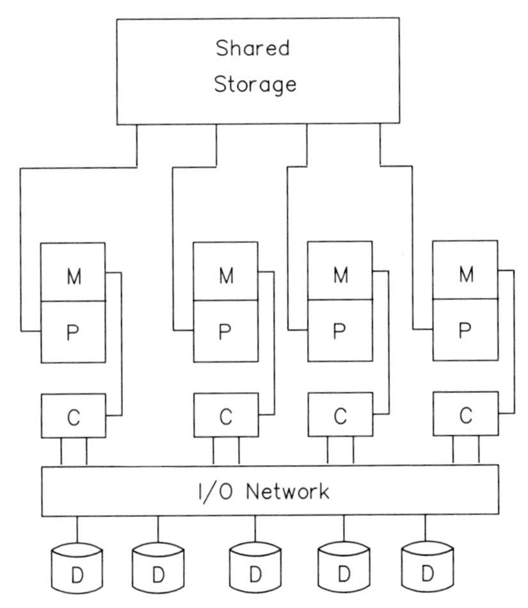

Fig. 2. Loosely coupled system with shared storage. M, Memory; P, processor; C, I/O channels; D, disk drive or other I/O device.

interconnections in order to obtain latency appropriate to storage of the data in a random-access memory. As processor speeds continue to increase, the loosely coupled system will benefit from the use of fiber optics to provide higher bandwidth and increased distance among the elements of the complex. Furthermore, fiber optics can enable data centers which are a few kilometers apart to be combined into a single complex as long as the latency in the long-distance links does not produce unacceptably low performance. These distances are unattainable, at typical I/O bandwidth, with traditional copper bus technology, which is limited to distances of the order of 100–200 m.

2.2.2. *Tight Coupling*

Figure 3 illustrates a tightly coupled system. With tight coupling, some number of processors (typically two to eight) share a common main memory which contains both instructions and data. The complex is under control of a single operating system and set of application programs. A common set of I/O channels provides access to data. Dispatching software schedules work for concurrent execution on all processors. Either each processor may be working on a separate task or a group of them may be working on a common task. Since the processors may be sharing common data in main

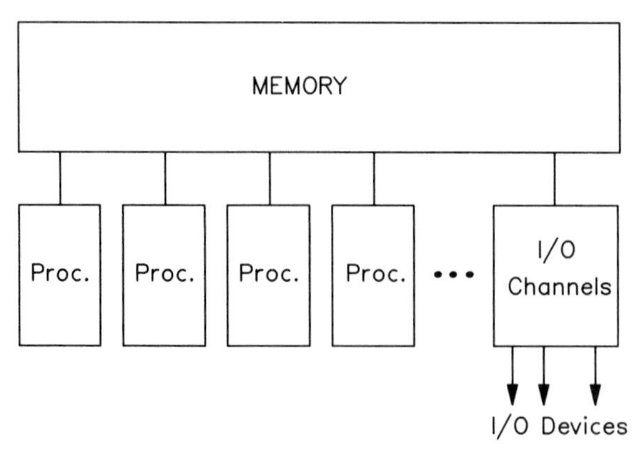

Fig. 3. Tightly coupled system.

memory, they must communicate to ensure orderly sharing of the data without interfering with each other. The communication is through a combination of special instructions and shared words in memory.

Today, the interconnections among the individual processors in the tightly coupled complex are electrical, using traditional parallel bus, switch, or point-to-point approaches. The individual processors are in close proximity, with transmission distances on the order of a few meters or less. As the processing speed of the individual processors and of the aggregate continues to increase, electrical transmission, even at these short distances, will be stressed. It will become more and more difficult to increase the number of bits which can be transferred in parallel, both because of pin limitations in chips, modules, and boards and because of signal skew in parallel transmission. Serial fiber optics offers a potential solution to the speed limitations of parallel electrical interconnection in this environment. The challenge is to develop a system organization which will overcome the latency inherent in serial transmission.

2.2.3. Massive Parallelism

The advent of the inexpensive microprocessor has led to an additional coupling configuration called *massive parallelism*. Massive parallelism refers to the use of a very large number (hundreds or thousands) of interconnected microprocessors to provide very high computing power at less cost than a single high-performance processor or tightly coupled processor complex.

Today, massively parallel systems are generally used for specific problems which are amenable to parallel execution. Specific massively parallel

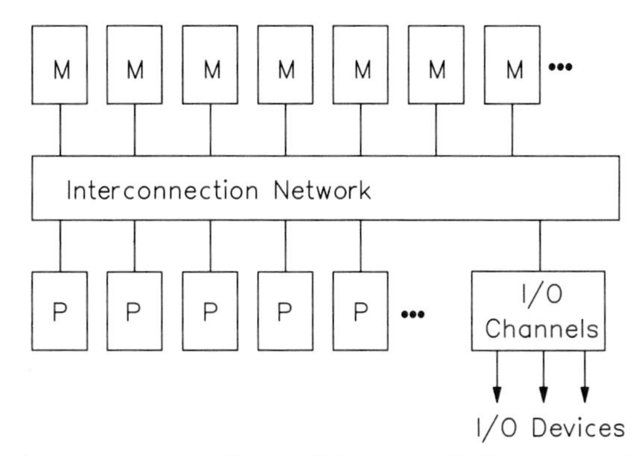

Fig. 4. Shared-memory massively parallel system. P, Processor; M, memory module.

systems are optimized for specific classes of problem. However, it is anticipated that as new programming techniques are developed and as more experience is gained in using these systems, their applicability will broaden. Examples of massively parallel machines which are commercially available are the Connection Machine [11] and hypercube-based systems [12].

There are two basic coupling schemes in massively parallel systems: shared memory and message passing. The shared-memory system is shown in Fig. 4. In this system, as in the tightly coupled system, all processors share a common memory which contains both programs and data. The processors and memory are connected to a high-speed interconnection network. All data sharing and communications between the processors take place through the shared memory. To allow parallel access to the memory, the memory is divided into a large number of modules which are individually connected to the high-speed network. A set of I/O channels connected to the high-speed network moves the data between the I/O devices and the shared memory.

The message-passing system is illustrated in Fig. 5. In a message-passing system, each processor has its own memory and the processors exchange

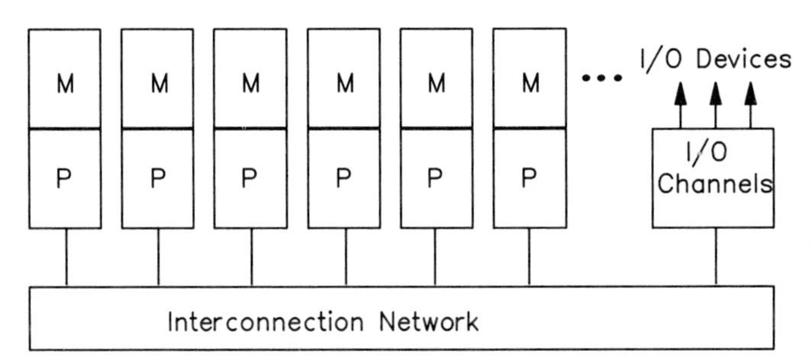

Fig. 5. Message-passing massively parallel system. P, Processor; M, memory.

data and control information by sending messages through the interconnection network. I/O is performed by channels connected to the same interconnection network. Although this is superficially like loose coupling, in this scheme a large number of processors work together on the same problem and the interconnection network must have speed and latency comparable to memory speed rather than comparable to I/O speed.

A particular case of massive parallelism is its use in the processing of a complex query against a database. An example of such a parallel database machine is that of Teradata Corp. [13], illustrated in Fig. 6. This is an example of a message-passing system. It can have up to 1024 processors. As shown in the figure, a few of the processors, called interface processors, communicate with the host computer to which the database machine is attached. Each of the rest, called access module processors, is attached to one or more disk drives which contain a portion of the database. The interface processor receives a database query from the host computer and broadcasts it to the access module processors. The query is processed by all access module processors at the same time, each processing against its own portion of the database. The separate results are then combined to give the final result of the query. Typically, these database machines must support applications that process a large amount of data. This requires high-performance connectivity between processors and high-performance I/O to ensure that the I/O subsystem does not exacerbate concurrency control problems for data-contention intensive applications.

Many different interconnection network schemes are used in massive par-

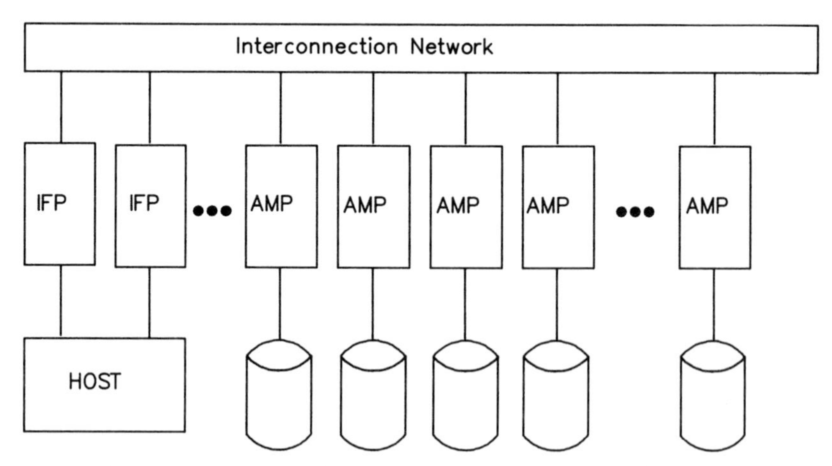

Fig. 6. Teradata database machine. IFP, interface processor; AMP, access module processor.

allelism, depending on the characteristics of the problems for which the system is designed. An overview of various interconnection schemes may be found in ref. 14. All such schemes require very high throughput and low latency in order to obtain aggregate system performance which approaches the total processing performance of the individual microprocessors in the system. Today, these interconnection networks are electrical, using parallel data transmission at distances of a few meters or less. As discussed earlier, in conjunction with tightly coupled processors, increased performance will eventually be limited by the number of bits which can, in practice, be transferred in parallel at high speed. Optical transmission in a serial format, or with a relatively small number of parallel paths, offers a potential solution if interconnection networks can be developed in which the latency for serial transmission is not a problem. Optical crossbar networks are discussed in ref. 15.

An additional factor in interconnection networks for massive parallelism is that these networks generally consist of multiple levels of active switching elements. For optics to play a practical role in massive parallelism, it will be essential to develop either all-optical high-speed switching elements or very low cost, high-speed optoelectronic conversion devices to enable the switching to be accomplished electrically at both low cost and low connection latency.

2.3. Microprocessors and Distributed Processing

One of the most significant factors in computing in recent years is the ubiquity of the low-cost, powerful microprocessor and its emergence as the heart of intelligent workstations. As a result, the individual computing paradigm is shifting from shared central systems to personal workstations. The distribution of intelligence throughout an enterprise is fueling a huge requirement for high-bandwidth communications networks. This is supported by several key configurations. One configuration consists of workstations and workstation servers connected by a LAN. Figure 7 depicts a typical configuration, several workstations and a file server interconnected by a coaxial cable bus.

Advanced applications, such as image processing and multimedia, demand very high LAN performance and require fiber optics to supply the required bandwidth. This is one of the applications for the FDDI LAN mentioned earlier. In larger enterprises, LANs that traverse longer distances or backbone LANs that connect individual LANs are needed to allow the same applications to be used as in the more restricted geography of a single LAN.

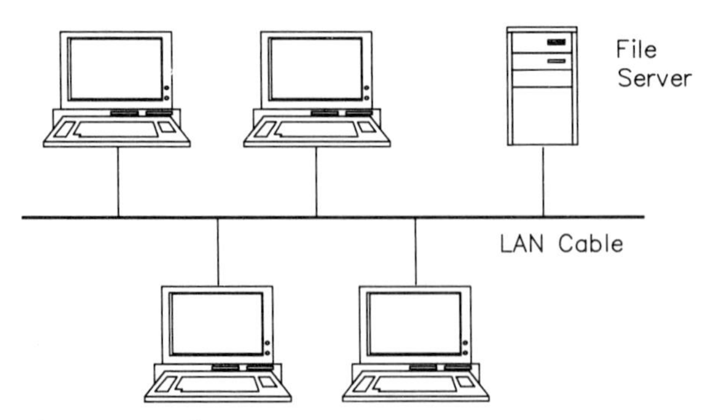

Fig. 7. Local area network

Such a system is illustrated in Fig. 8. Several communications servers and a file server are interconnected by a high-speed fiber-optic backbone LAN. Each communications server is also connected to a local LAN, which provides service to a number of workstations.

A related configuration of particular importance involves supercomputers. Results of complex supercomputer calculations generate huge traffic requirements to individual visualization workstations which far exceed the requirements of typical client–server configurations. Today, these interconnections are supplied by the HIPPI interfaces. In future years, optical channels, defined by the proposed FC standard, will provide these local, very high bandwidth, interconnections, utilizing switches for interconnecting many nodes.

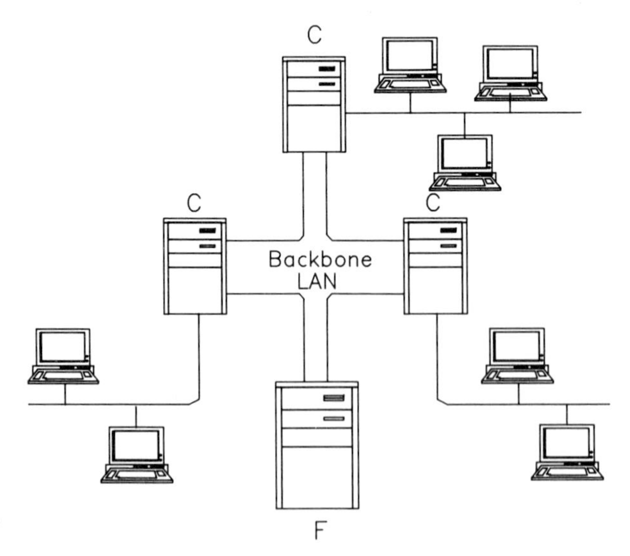

Fig. 8. Backbone LAN. C, communications server; F, file server.

An equally important trend is the geographic distribution of computing. This is driven by the pervasiveness of computing and increasing needs to share data, image, facsimile, and so forth between multiple locations of an enterprise and between enterprises. Metropolitan area networks (MANs) (<50 km) of very high bandwidth are resulting from several aspects of this trend. First, many enterprises have multiple sites in a metropolitan area. Second, most enterprises have significant interenterprise relations with nearby enterprises. In the coming "wired world," these enterprises could jointly build networks, or public and private network service providers could provide a common MAN resource.

A particularly important MAN application is disaster recovery. Valuable enterprise data which are critical to the business' operation must be stored in a location with physical failure characteristics different from those of the main site, to protect against power failures and other unforeseeable situations. There is a tension between locating the backup too far away (more costly communications and more latency) or too close (not separately failing). A good compromise is the largest MAN distance. This is also helped by the fact that a repeaterless point-to-point optical link could be installed over such distances.

Finally, the same trends are causing growth in the need for wide-area network communications. Here, too, is a demand for increased bandwidth. First, for existing applications, network performance must scale to stay in balance with growth in other commodities. Second, the growth in the number of systems, on-line transaction processing as a way of doing business, interenterprise communications, and other new applications add a nonlinear component to this requirement. Third, if network performance is high enough, an enterprise can place backup data and systems at wide area distances to protect against risks, such as earthquakes, for which MAN distances do not provide adequate separation between the main site and the backup site.

2.4. Data Storage

New applications of computers that involve image, full-motion video, voice, and high-resolution graphics, as well as combinations of these, called multimedia, are making unprecedented demands for information storage capacity. Although magnetic disk storage capacities continue to increase, they are being outrun by the storage demands of these applications.

Both in the central system and in the workstation, optical storage technology is providing the increased storage capacity. In addition, removable optical storage media, such as the compact-disk read-only memory (CD-

ROM), an adaptation of the audio compact disk, provide means of distributing both programs and reference information with storage density and reliability unmatched by magnetic diskettes and with random-access capability unavailable with magnetic tape. A single 12-cm-diameter CD-ROM disk holds approximately 500 megabytes of information. Emerging technology can provide mixed data in traditional format, motion video, and voice on a single disk. Write-once removable media provide archival storage as well.

Present optical storage technology does not, however, provide either the random-access speed or the data transfer rate of contemporary magnetic disk storage. As a result, applications, such as on-line transaction processing, which require rapid access to data on optical storage must provide temporary storage on magnetic disk for data which are actively in use, thus increasing software complexity and data storage costs. Furthermore, although some optical storage devices provide the capability of erasing and rewriting information, these capabilities are cumbersome and do not provide an unlimited number of erase and rewrite cycles. Therefore optical storage is not the medium of choice for temporary data which are actively being revised. Obviously, an optical storage technology which combines very high storage capacity with the random-access performance, data transfer rates, and alterability of today's magnetic disk storage would be a valuable addition to computer system technologies.

Because of these limitations of optical storage, data-intensive applications continue to drive improvements in both capacity and performance of magnetic storage devices. Certain supercomputer applications, including processing of seismic data and remote-sensing data, involve the storage and analysis of massive amounts of data. The growth and reduced cost of supercomputers are making a greater number of such applications cost-effective; however, the bandwidth and latency required to serve the data to supercomputers are introducing a data transfer rate requirement not previously so dominant for such systems.

An emerging answer to the need for high data transfer rate and very high capacity in magnetic storage is the disk array. A disk array consists of some number of disk drives which are interconnected so as to have the appearance of a single drive with the total capacity of the individual drives. An example is shown in Fig. 9.

With information transfer to and from all drives concurrently, the data transfer rate of the array is a multiple of the rate of an individual drive. In a configuration most useful for the large volumes of data used in supercomputer applications, the individual drives are byte-interleaved. In other words, successive data bytes are placed on the different drives. For best perfor-

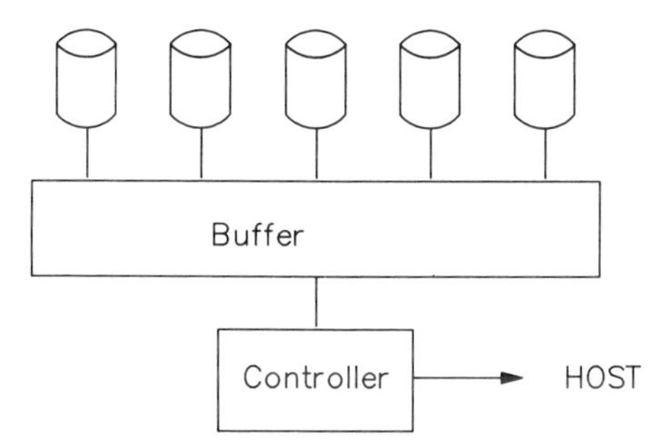

Fig. 9. Disk array

mance, the rotational positions of the drives are synchronized so that the successive bytes are transferred to and from the drives at approximately the same time. The performance advantage of this configuration is realized when the amount of data to be transferred is large enough that the data transfer time dominates the time to access the first byte of data. Typical arrays of this type have data transfer rates on the order of 100 Mbyte/s and require fiber-optic connections to the processor if the disk system is more than 25 m or so from the processor.

Work at the University of California at Berkeley [16] has demonstrated that disk arrays have much more general applicability than just to the supercomputer applications. The Berkeley work has introduced the general concept of redundant arrays of independent disks (RAID), which is the data storage analog of the use of multiple microprocessors to obtain high computing power. In a RAID system, large numbers of inexpensive disk drives replace high-performance, expensive disks. Although an individual drive in a RAID system is not necessarily as reliable as the traditional disk drives used with large processors, the reliability of the array can be made as good as, or better than, the reliability of the traditional large disk drives by using additional disks to store redundant information, which allows the data on a failed drive to be recreated. Reference 16 analyzes various drive configurations and data organizations and shows that certain configurations provide higher reliability and higher performance than traditional disk drives. As a result, it is expected that high-performance, high-reliability disk arrays will become very widely used, further increasing the demand for high-performance optical interconnections to attach the arrays to computer systems.

3. SYSTEM-LEVEL ISSUES RELATED TO TECHNOLOGY REQUIREMENTS

All of the basic characteristics of fiber-optic data transmission make this technology very attractive to computer system designers, as described in the previous section. Its high bandwidth facilitates new applications which require high processor performance and involve large amounts of data. High bandwidth also enables reduction of the number of I/O channels, even for lower-bandwidth applications, which potentially reduces cost and simplifies system setup and management. Its long-distance transmission capability allows placement of high-performance I/O devices near their users throughout a campus, instead of within the confines of a computer room, and facilitates interconnecting multiple sites with high-bandwidth connections. The low bit error rate of fiber optics, compared with other serial transmission technologies, facilitates application of the high-bandwidth and long-distance capabilities of the technology by reducing both the complexity of error recovery and the performance consequences of recovery actions.

However, the high-bandwidth and long-distance transmission made possible by fiber optics creates a need for changes in other aspects of system design in order to take advantage of these characteristics. Some of these changes in turn put further requirements on the basic technologies. In this section, we review some of the system-level issues in the use of high-bandwidth optical transmission.

3.1. Interconnection Topology

Today, the dominant interconnection topology in the optical domain is the point-to-point link, although extensive research activity is taking place on multipoint topologies such as the optical passive star coupler and the distributed bus. More complex topologies are formed by interconnecting point-to-point link segments electronically. For example, in the ring (Fig. 10), the nodes of the network are interconnected in the form of a ring using point-to-point links. Each node relays the messages to the next node in the ring. In high performance I/O, the preference is for centralized switches, as shown in Fig. 11. In this topology, a fiber pair connects each node to the central switch. One fiber is used for transmitting, the other for receiving.

One of the advantages of the switched topology is that a crossbar switch may be used to enable all pairs of nodes to communicate at the same time. If n nodes are connected to the switch, there may be $n/2$ simultaneous interchanges of information. An example of this is the switched topology of the IBM fiber-optic channels [5]. In contrast, the ring topology generally permits only one message (or at most a few messages) to be transmitted at the same time.

Switches can be categorized as circuit switches and packet switches. In a circuit switch, a dedicated connection is created by an explicit protocol,

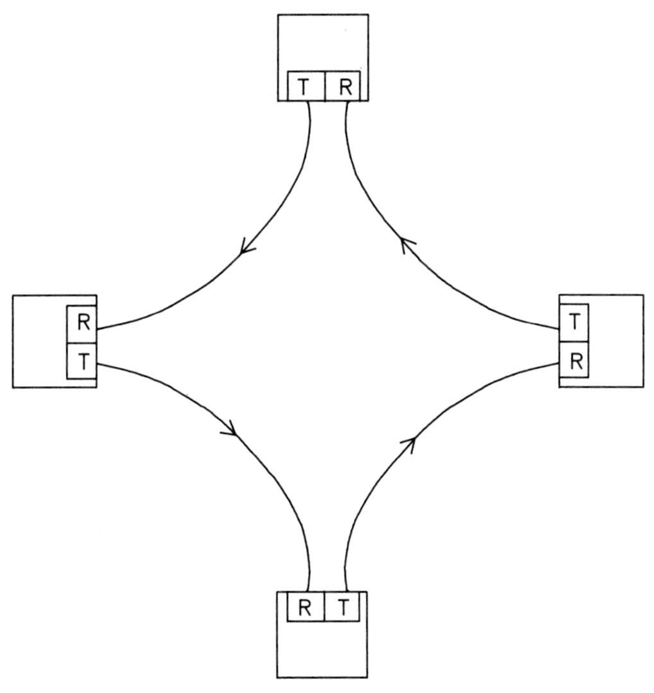

Fig. 10. Fiber-optic ring. T, transmitter; R, receiver.

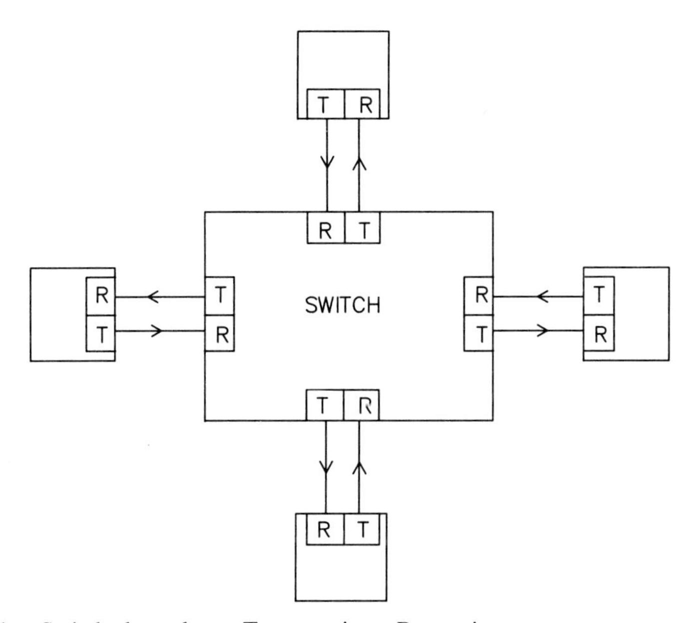

Fig. 11. Switched topology. T, transmitter; R, receiver.

after which information is exchanged between the two connected nodes. The connection is then explicitly broken after the information exchange is completed. Today's telephone network is an example of circuit switching. In a packet switch, the information flow is divided into short messages, called packets, each of which contains the address of its destination. Each packet is individually routed through the switch to its destination without first creating an explicit connection.

Circuit switching provides relatively simple, high-performance interconnection, provided that the times to make and break the connection are short compared with the duration of the information transfer. Packet switching permits multiplexing of short messages and does not have the overhead of the protocol to create an explicit connection before each message can be sent. However, the function to route each message separately through the switch has its own complexity and performance concerns.

In circuit switching, the overhead to create a circuit generally involves one control message in each direction to create the circuit before the data can be transferred. Thus, at a minimum, this overhead involves one round-trip delay between the end points. For distances on the order of a few kilometers or more, this round-trip delay dominates the connection set-up time of the switch itself. In the switch used with the IBM fiber-optic channels [10], the connection set-up time within the switch is approximately 2 μs in the absence of contention.

Assume that the sender and receiver are 10 km apart. For fiber-optic transmission, the signal propagation velocity is approximately 2×10^8 m/s, the delay for 10 km is 0.5×10^{-4} s, and the round-trip delay is 10^{-4} s. Thus the connection set-up time is approximately 10^{-4} s. At a bit rate of 1 Gbit/s, 10^5 bits can be transmitted from the sender to the receiver during this period. For messages of this length or longer, circuit switching is simple and effective. However, for many applications, data transfer lengths may not be long enough for good performance with circuit switching at Gbit/s bandwidths; in this case, high-speed packet switching is an attractive alternative. Needed are packet switches that are capable of packet throughput commensurate with the bandwidth of the attached optical links.

Another technology that is currently being explored is wavelength division multiplexing (WDM), often in conjunction with optical passive star couplers. Use of multiple wavelengths potentially multiplies the useful bandwidth of the fiber by the number of wavelengths in use. WDM, used with an optical passive star coupler [17], can also provide an optical switching function equivalent to that of a crossbar switch.

For the switching applications of WDM, the technology challenge is to develop inexpensive tuning and demultiplexing components to provide some combination of selecting the emitter wavelength, selecting the receiver wavelength, and demultiplexing multiple wavelengths at the receiver, as well as to develop wavelength-insensitive optical amplifiers to restore the signal strength lost in going through a passive network. Fast reconfiguration for this type of network, including the address decoding and wavelength tuning, is another technical challenge. As network architects have yet to show that multiple independent optical channels really result in any significant new system-level functions or any enhancement in network latency, addressing ease, or robustness (as evidenced by the earlier investigations and products using CATV technology), there is not a serious concern that WDM technology development is behind system need.

Yet another topology consideration is the use of multiple fibers to obtain higher bandwidth than for a single fiber, a technique sometimes referred to as striping. Earlier, we mentioned a scientific visualization application which could use a data transfer rate of 720 Mbyte/s if it were available. This data rate can be achieved with 1 Gbit/s transmission links by using eight links and transferring a portion of the data on each link at the same time. Of course, striping potentially multiplies the cost and complexity of a single link by somewhat more than the number of fibers being used. Inexpensive integrated arrays of emitters and receivers are needed to reduce the cost of this means of achieving very high bandwidth.

3.2. Link Errors and Packet Losses

One of the attractive features of fiber optics for data transmission is its low bit error rate compared with other serial transmission media. Bit error rates in the range 10^{-15} to 10^{-12} are realizable in practice. Although these error rates are far from low enough to be ignored, they do mean that most messages are transmitted without errors. As a result, it is possible to use simplified error handling in which errors are detected in the link hardware but recovered through higher-level protocols. This allows reduction of link hardware complexity and potentially increases useful bandwidth. The trade-off is that error recovery may take longer (e.g., more data retransmitted), but it occurs seldom enough that there is a net gain in system performance.

At very long distances, the time to perform error recovery by retransmission may be unacceptable because of the amount of data to be retransmitted and the round-trip delay incurred by the receiver's requesting retransmission from the sender. For such situations, forward error correction may be used [18]. With forward error correction, enough redundant information is transmitted with the data to enable the receiver to calculate the correct information when an error is detected. Although the redundant information costs bandwidth, it may be worth spending the bandwidth to avoid the retransmissions. For any given system, the bandwidth costs of the forward error correction must be balanced against the throughput costs of retransmission protocols. Another possible application of forward error correction is to enable the use of less expensive transmitting and receiving components, which result in links with increased bit error rate (before the forward error correction is applied). Here, the cost savings in the transmitting and receiving components must be balanced against the increased complexity of the forward error correction hardware. As before, the performance benefits of avoiding retransmission must be balanced against the bandwidth costs of the forward error correction information transmitted with the data.

With high-speed packet switching, it is possible for packets to be lost when congestion occurs, e.g., when the aggregate rate of packet arrivals at a switch exceeds its throughput. System-level error recovery procedures have to provide for recovery from these losses. In order to get the full benefits of the low bit error rate of fiber optics, the high-speed packet switches must be designed to have packet loss rates no higher than the loss rate due to the bit error rate of the fiber optics. For example, in a system with links whose bit error rate is 10^{-12} and packet lengths are 10,000 bits, the packet loss rate due to the bit error rate is 1 in 10^8. For the system designer to take advantage of this low bit error rate, the packet switches must also lose no more than approximately 1 packet in 10^8 per link attached to the switch. Alternatively, if a higher packet loss rate is acceptable, then a higher bit error rate may be tolerated, which allows some combination of less expensive link components and longer transmission distances than would otherwise be allowable.

4. OPTOELECTRONICS TECHNOLOGY REQUIREMENTS AND STATUS

Diverse computer applications involving optoelectronics have been described in Sections 2 and 3 and illustrated together in Figure 12. In this section, we will describe the implications of the trends in these applications for optoelectronic component implementations.

Just as the applications require diverse architectures for cost-effective performance, more than one version of OE technology needs development if the systems are to be optimally (and even competitively) implemented. Some

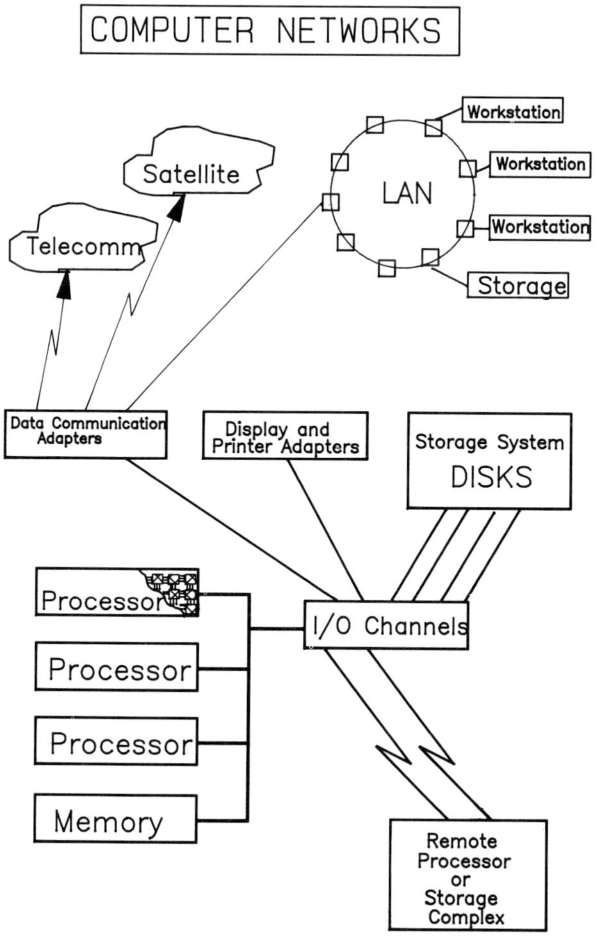

Fig. 12. General types of computer networks. The examples at the top are data communications networks; the example at the bottom is a data processing or interconnect network.

of these applications can benefit greatly from an OEIC technology, whereas others may not justify the development investment over a more primitive, yet more flexible, discrete device technology.

The established technology for the longer-distance data communications applications is already optical, having been developed by the telecommunications industry. Most of the features of and requirements for optical links, and specifically OEICs, in these applications will be the same as those discussed in Chapter 1. The differences in the requirements for these links between telecommunications and data communications applications are primarily architectural, with data links needing faster access protocols, lower error rates, different frame structures, etc. These discussions [19] are outside the scope of this book.

As the networks become more localized (in the LAN environment, for example, the network is in a building or on a single enterprise campus), the ratio of cost for network attachment components to the cost for cabling increases dramatically. In addition, the frequency of reconfiguring of the network is higher than in telecommunications. Therefore, hardware optimizations which are different from those in telecommunications become attractive. In the LAN environment, the established wiring technology today uses electrical wires (twisted pairs or coax), with growing use of fiber optics for backbones or in special noisy environments.

In the I/O channel networks—that is, the connections between the processor and its peripheral storage and input/out devices—the environment is moving from a computer room to a specialized version of a LAN. Implementation technology is moving from a parallel wire electrical bus to a serial optical fiber circuit-switched network and eventually to a packet-switched network.

At even shorter distances, the networks move from a communications orientation to an interconnection orientation, connecting a processor to another processor, a processor to remote memory. The interconnections are more integral to the computing process itself. Here, dense electrical wiring, in bus formats of multibyte-wide parallel lines, for latency and data bandwidth, is firmly entrenched today as the most cost-effective technology. Lower processor cycle times may reveal electrical limitations in the data rate possible on a line under acceptable density, crosstalk, and cost constraints, leading to opportunities for optical implementations of these buses. Currently, there is considerable research activity on optical buses.

The ultimate short-distance connection is at the intrachip level, where there is a merging of the interconnection with the processing (or logic) element itself. As optical-to-electrical conversions would probably add too much latency to the transmission of information (with 100-ps delays on chip and 2–3 ns between chips being required for electrical connections), researchers are exploring optical devices to replace the electronic transistor so that all operations can be performed with optics. Unfortunately, such optical devices do not come close to meeting the performance requirements for the data processing element [20]. Without some realistic candidates for discrete optical logic elements, it is premature to discuss integration issues or OEIC requirements, so "optical logic" will not be discussed further in this chapter, although progress in device research is covered in later chapters of this book.

Finally, in the peripheral area—data storage, printing, and display—there are possible roles for OEICs. The primary storage medium today is the magnetic disk, which is used in systems from mainframes to personal computers (PCs). The potential for larger storage density with optical media and the capability to remove the storage medium from the computer have resulted in optical storage devices being introduced into products today. They are used primarily for access to large data bases or for archival storage. As mentioned in Section 2.4, for increased usage as on-line, direct access devices, it will be necessary to develop faster access to the data and the ability to stack disks on top on each other to gain improved volume storage density over magnetics. Of course, areal density must also continue to increase, implying ever smaller data bits on the disk. Silicon-based OEICs have been introduced into read/write heads for optical products in the consumer audio industry. This was done to increase the compactness of the head, to improve the reliability over hybrid packages, and to lower the cost of head assembly. It is expected that these highly integrated heads will also find use in optical data storage products, for example, the CD-ROM.

In printing, the laser printer has become a pervasive I/O device in the PC and workstation environment. The evolution of these printers is toward faster printing, higher resolution, and color. The array capability of OEICs could play a role here.

In display devices, the cathode ray tube is the established medium and there seem to be no opportunities for OEICs in this implementation. The flat-panel liquid crystal display is growing in importance, especially as the laptop PC market grows. Displays approaching a million addressable picture elements are being developed, placing demands for a high density of elements over large areas. The electrical drivers for the picture elements are fabricated in arrays and attached to the liquid crystal array. As the display becomes more compact, the integration of the drivers onto the display will become attractive. This application area will not be covered in this book.

4.1. Local Area Networks

The primary local area networks used today are the Institute of Electrical and Electronic Engineers (IEEE) standards, token ring [21] and CSMA/CD (carrier sense multiple access with collision detection) [22]. The physical layers use electrical links based on coaxial cable or twisted-wire cable, with data rates between 4 and 16 Mbit/s. The link drivers and receivers for these networks are packaged in printed circuit card–mountable modules. The number of installed nodes is expected to grow to 28 million (CSMA/CD) and 19 million (token ring) by the mid-1990s [23]. The costs per node of these networks have decreased to the point (about $250 per card) where they are used out to the PC on the desk. Fiber-optic links are used in electrically adverse environmental conditions. There will probably not be a convergence on only one LAN, as users have different cost, performance, and network support objectives, and so adapters are available to get from one network to another.

The FDDI standard has been the expected optical technology entry to LAN applications. The data rate on the link is 125 Mbit/s, with maximum internode spacing of 2 km specified. Usage is just beginning and is expected to grow to only 800,000 nodes by the mid-1990s. Applications are likely to be confined to building backbones, connections between buildings, and some high-performance applications. The reasons are the high cost of the specified InGaAsP/InP light-emitting diode (LED) and photodiode (PD) discrete devices and the cost of the electronics to process the frames of data arriving from the network. When one considers the number of cable backbones in a building compared with the number of connections between the backbone and the offices and laboratories of the building, it becomes obvious that optical technologies have not captured the volume markets of the token ring or CSMA/CD. The introduction and acceptance of twisted-wire options into the FDDI standard for these short-distance, backbone-to-office connections illustrates the challenge for optical technology to get on a volume-driven learning curve, even if link bandwidth demand increases.

An FDDI optical physical-layer standard based on existing, volume-manufactured, OE components from the industrial and commercial sector (e.g., the compact disk laser or the GaAs red LED) would be more cost-competitive against electrical wiring today. Examples of these modules based on this type of OE technology and very low-cost assembly techniques are emerging from the PC intraconnection and automotive application environment [24]. In 1994, OE links using this inherently low-cost technology were introduced as products [24b] at a lower installed cost than data-grade twisted wire electrical links. With this GaAs-based discrete technology as a low-cost entry, cost reductions and density and performance enhancements might be obtained using OEIC link front-end components from the GaAs IC industry. As the OEIC technology matures on this learning curve, further cost–performance advantages would arise from the ability to integrate the full link-adapter functions on the same chip, to speed up link access and frame handling, and to provide link servicing and safety functions.

The OEIC could also play a role in reducing the cost for internetwork adapters, either between short-distance and long-distance segments of an optical network or for transitions between optical and electrical segments of a network. An example would be an OEIC with many OE receivers from local optical networks, multiplexing and framing electronics, and an electrical driver for a special discrete OE chip (e.g., a narrow-linewidth laser) which drives a long-distance telecommunications link.

A typical FDDI transmitter–receiver module is shown in Fig. 13. It is a PC card-mountable module operating from standard card power supplies. The module is mounted on the card edge, and network cables are attached directly to the module. Therefore, the module must strain-relieve this load onto the card. The size of the module is typically 30 by 50 mm, limited primarily by the size and robustness of the cable connector. Thus, it consumes significant "real estate" on the card. With this bulky connector, the size-reduction potential of OEICs cannot be attained.

Robustness of the components and ease of use by relatively untrained personnel are primary requirements for network technologies in the LAN environment. This implies modules and cabling with large alignment tolerances and susceptibility to the environment. Multimode fibers are generally used for these reasons. The OE components should have low noise and distortion properties even with slightly misaligned components, or many connectors in the link, or very short links. Again, the ability to integrate special front-end safety and fault recovery functions at small extra cost would be an attractive OEIC feature. An example would be multiple lasers and photodetectors on the same chip providing alternative paths around failed OE devices, or the ability to self-test the link adapter.

The need for low cost and robustness has led to a renewed interest in very large core plastic fibers for short links of a LAN (<100 m). These PMMA-based fibers have low loss in the visible region of the spectrum, allowing the use of red LED emitters (variations of the LEDs manufactured in large volume for the indicator lamp industry) as line drivers. The visible wavelengths are also well matched to silicon-based photodiodes and OEICs, which would be adapted from the optical storage–consumer audio industries.

4.2. I/O Channels

The trend of connections between the processor and the input/output devices (primarily magnetic storage devices) was described in Section 2.1 as an ever-increasing number of links connecting ever-faster (and thus more densely packaged) processors and DASD. Thus, there are many port adapters on a card and a growing number of I/O cards needed in a system. The migration from a parallel coaxial wire bus to a serial optical link and from multidrop

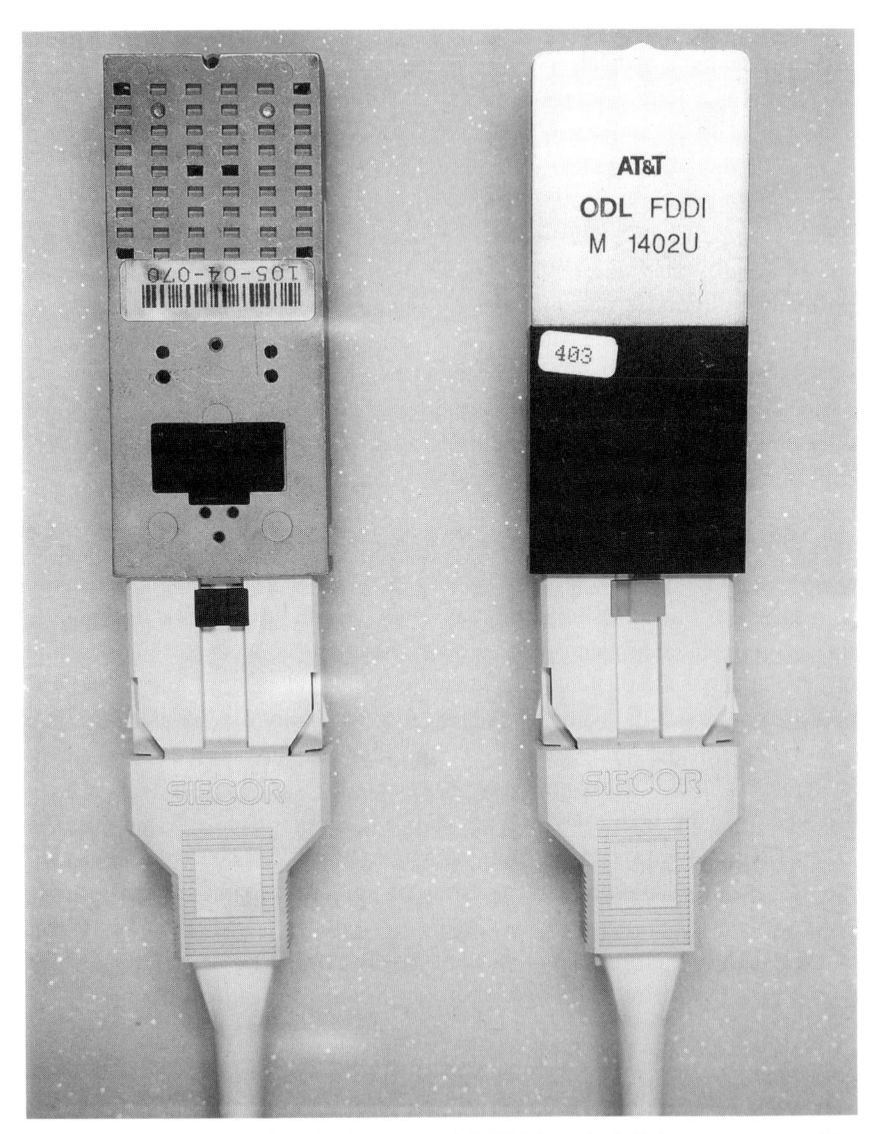

Fig. 13. Photograph of typical commercial FDDI optical link components. The modules contain OE devices and their drivers and amplifiers, but not timing, coding, or protocol-handling ICs for the FDDI interface. (From Hewlett-Packard, ATT Technologies, and Siecor Corp.)

to switched networks (for increased connectivity and higher availability) relieves much of this congestion at the nodes, but compounds it in the switch. With bus performance goals in the 100–200 MByte/s range, the serial link rates could easily grow beyond 1 Gbit/s. The overwhelming majority of links (more than 90%) are installed in the "extended" computer room, i.e., over distances of the dimension of a building (<1 km). Nevertheless, as discussed in Section 2, longer runs are also required for remote data storage and a distance of 10 km is included in the ANSI FC. As new computer applications utilizing long-distance channels evolve (examples mentioned earlier) and architectures change to accommodate them, the long-distance runs are expected to take on increasing importance.

Most mainframe and minicomputer makers today, worldwide, have introduced fiber-optic link technology for the channel connections. InGaAsP LEDs or GaAs injection lasers driving multimode fibers are used for the short-distance connections, while InGaAsP lasers and single-mode fibers are required for long distance. The typical channel link adapter module looks similar to the FDDI module of Fig. 13 and is assembled on a printed circuit card with Si bipolar IC or CMOS IC single-chip modules (SCMs) containing the port adapter functions.

As the number of channel connections per system is growing into the hundreds and the number of systems sold per year is in the tens of thousands, module volume is expected to grow into the millions per year. The smaller systems have fewer connections per system, of course, but proportionately more systems are generally sold. Another trend, already implemented in mainframes and increasingly in minicomputers and workstations, is to integrate the port adapter functions into fewer, higher-density CMOS chips, with the chips mounted in multichip modules (MCMs). Higher chip and package integration for channel port adapters will make the expensive card "real estate" occupied by discrete device, optical link adapters (or, in the case of switches, the number of cards consumed by link adapters) increasingly undesirable. The potential component volumes in this application, coupled with the high speed of the links and the need for higher-density packaging in the box, make an OEIC implementation both attractive and potentially economically justifiable to develop.

Two versions of optical technology are utilized for I/O channels, one using the InP long-wavelength (1.3 μm) technology with its long-distance potential and telecommmunications industry base and one using GaAs short-wavelength (0.78 and 0.85 μm) technology with its low device cost and consumer industry base. An example of a link adapter card at 266 Mbit/s, using lasers selected from the compact video disk player industry manufacturers, is illustrated in Fig. 14. This card is capable of 2-km distances using standard multimode fiber cables, yet it costs two to five times less than InP LED or laser-based modules. Note that the card of Fig. 14 contains clocking and serialization functions as well as the transceiver functions contained in the modules in Fig. 13. However, it is still fabricated with many discrete components and thus occupies about 50% more space than the FDDI modules. Clearly, the GaAs IC industry, which is growing to meet a demand from the consumer radio and video industry for fast analog chips (as well as a digital IC demand for high-speed, low-power ICs), could provide an OEIC migration path and reduce card size.

From an architectural viewpoint, two link technologies are not necessarily incompatible and both can be cost effective in their respective environments. For long-haul links, a more expensive link adapter technology is made affordable by the dominance in the link overall cost of the long cables, the tariffs, and the extra network services. For short-distance links, with their simpler protocols and large numbers of nodes clustered together, link adapter

Fig. 14. Link adapter module. This card contains the link adapter functions for a 265-Mbit/s T/R full duplex link, using consumer-based OE technology. (From International Business Machines Corp.)

cost and robustness are more important. There will probably always be active interfaces between these environments anyway, as the long-distance runs (out of the building and off the site) will go through gateways to adapt to common carriers or provide special security or fault recovery features. Having a gateway makes optical link hardware conversions easy to implement.

The Committee that developed the physical layer serial link standards for the Fibre Channel Standard accomodated the cost-performance tradeoffs between the various versions of optical technology, as well as the use of copper technology, in its physical-layer options. Table 1 illustrates the options currently under consideration for standardization. As in the LAN environment, electrical link solutions are also considered where optical links are not cost-effective.

Both InP- and GaAs-based OEICs are being explored for the channel application and are discussed in this book.

Table 1.

Technology Options in the ANSI Fiber Channel Standard

1.062 Gbit/s	532 Mbit/s	266 Mbit/s	133 Mbit/s
Single-mode fiber 1300 nm Laser 2 m–10 Km	Single-mode fiber 1300 nm Laser 2 m–10 Km	Single-mode fiber 1300-nm laser 2 m–10 km	Multimode fiber 1300 nm LED 0–500 m
Single-mode fiber 1300 nm Laser 2 m–2 km	Multimode fiber 780-nm Laser 2 m–1 km	Single-mode fiber 1300-nm laser 2 m–2 km	75-ohm CATV Coaxial cable ECL, single ended Drv/Rec 0–50 m
75-ohm CATV Coaxial cable ECL, single ended Drv/Rec 0–50 m	75-ohm CATV Coaxial cable ECL, single ended Drv/Rec 0–50 m	Multimode fiber 1300-nm LED 0–1 km	75-ohm miniature Coaxial cable 0–10 m
		Multimode fiber 780-nm laser 2 m–2 km	
		75-ohm CATV Coaxial cable ECL, single ended Drv/Rcv 0–50 m	

From ANSI X3T9.3 Fiber Channel Physical Layer Specification FC-0, Rev. 2.1, May, 25, 1991 [7].

4.3. Processor Interconnections

Processor data buses today are 4–32 bytes wide (1 byte is nominally eight to nine lines). For example, the HIPPI interface is 4 or 8 bytes wide. The coupling concepts discussed in Chapter 2 imply that low latency is a major requirement in transferring data between nodes. This means that the time taken to establish a connection over the bus should be small with respect to the processor's cycle time (e.g., 1–100 cycles), and the time taken to transfer a useful amount of data should likewise be short. For example, a processor transferring a 4-Kbyte page in 4 μs requires a data rate of 1 Gbyte/s. The link distance is limited to 100 m if a round-trip request and reply is only to add another 2 μs. For a 10-ns cycle time processor, this transfer still takes over 600 machine cycles, which could be unacceptable in some closely coupled processor configurations.

The interprocessor and processor–memory interconnections are implemented today with electrical buses, and there are growing chip and package I/O pin congestion, unacceptable signal distortion, and noise at the higher bus speeds. To solve performance problems, electrical buses utilize more ground and shield lines between signal lines, better termination for the lines, and differential modes of signaling. All of this adds cost and lowers I/O density. Table 2 outlines the expected trends in electrical buses at the chip, card, and board levels [25].

Table 2.

Capabilities of Electrical Interconnection Technologies

Interconnect type	Rise time (ps)	Delay (ps/cm)	Density (lines/mm)	Drive power (mW)	Maximum interconnect length (cm)
One-chip	50–250	100	250–500	2–5	0.7–1.4
Thin-film carriers	200–500	65	13–40	25	20–45
Ceramic MCM	400–1000	80	2–3	25	20–50
Printed circuit board	1000–2000	70	1.6–4.5	25	40–70
Flex ribbon cable	500–1000	62	0.8–2.5	25	30–100

This table shows the capabilities, in the near future, of electrical interconnections for various levels of computer packaging. The estimates are for manufacturable components, qualified (e.g., with respect to signal distortion, noise immunity, power dissipation) for the computer environment.

For tightly coupled processors, the buses are inside a single equipment frame, so that distances are only a few meters at the most. Generally, it is felt that existing electrical solutions are adequate if multichip modules and multilevel card technology are affordable [26]. However, the trend toward looser coupling between multiprocessor clusters leads to a desire for buses extending between equipment frames, over a few tens of meters. The earliest application of optical links to the processor bus is expected in this area.

The functions required to attach a bus to a multiprocessor network are shown in Fig. 15. The buffers speed-match and align the bus data with the network, the coder function is for transmission reliability and data verification, and the serialization function is used to reduce the number of lines over the network. The complexity of these functions is determined by the network architecture requirements and the implementation cost. These functions are generally implemented with digital ICs operating on parallel lines of data at lower speed. The timing and OE line driver and receiver functions are analog in nature and operate at network line speed. An attractive OEIC would capture all these

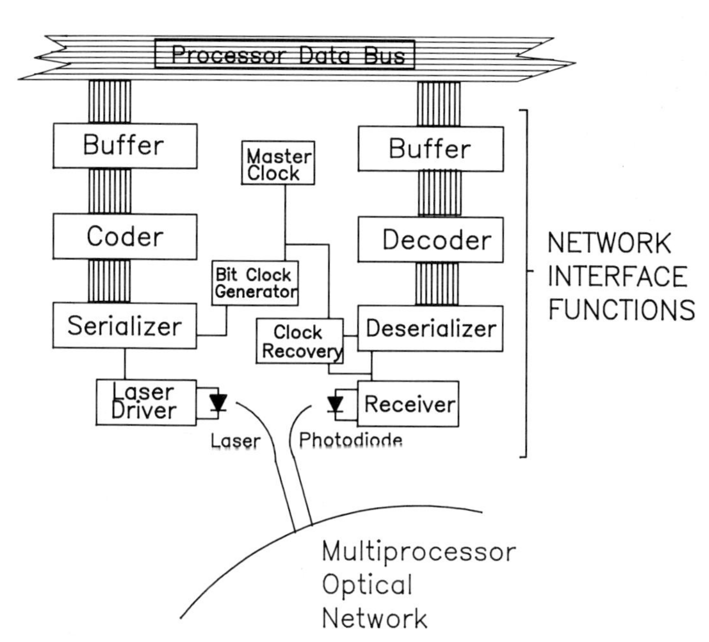

Fig. 15. Typical link adapter functions for a data bus.

functions on a single chip. However, this high level of integration would customize the chip to one specific application, and may not be cost-effective if that application's volume demand were small or the chip yield were adversely affected by the level of integration. Another function partitioning, in this case, would be to integrate the OE and front-end analog functions onto one OEIC, with generic characteristics to many applications and integrate the protocol specific digital functions onto another IC chip.

4.4. Optical Storage

The challenge in optical storage technology is to improve the access times of the read/write head, while continuing to improve the storage density of the optical media and cost per bit of data stored at a rate competitive with magnetic disk storage (improvements of about 20%/yr).

A read/write (R/W) optical head is shown schematically in the upper part of Fig. 16. The beam from the laser traverses a series of optical elements which shape, polarize, and focus it onto the optical disk, where it either reads or writes the data. In the case of magneto-optic media, for example, writing is performed by inducing a phase change in the medium by high optical energy density; reading is performed by a polarization rotation in the reflected optical beam caused by the phase-altered disk medium. The reflected beam from the disk traverses a second optical path through lenses and beamsplitters, where a "sensing" photodiode reads the data signal and a quadrant "servo" photodiode provides tracking and focusing signals for the head focusing optics servo system. Today's head consists of an assembly of these discrete parts and is bulky and heavy compared with a magnetic recording head. Use of this type of head in the large-volume consumer audio industry helps keep the cost down even though many discrete parts are assembled.

Note also in Fig. 16 the electronics that data-modulate (and in many cases also high-frequency dither) the laser and amplify the photodiode signals. These IC chips are often placed on the head to keep noise-susceptible, analog signal leads short. Marchant [27] describes in detail the optical storage read/write head and quantifies these requirements.

Storage density is expected to increase through the use of shorter-wavelength, visible lasers, implying a laser development opposite to that of the communications industry. However, one approach being investigated is to achieve shorter wavelength by frequency doubling a laser in a nonlinear medium. This has been demonstrated using small discrete infrared lasers and nonlinear crystal assemblages [28], but it could evolve into a compact single chip by the integration of a nonlinear lightguide onto a laser chip OEIC.

The Evolution of Optical Storage Read/Write Heads

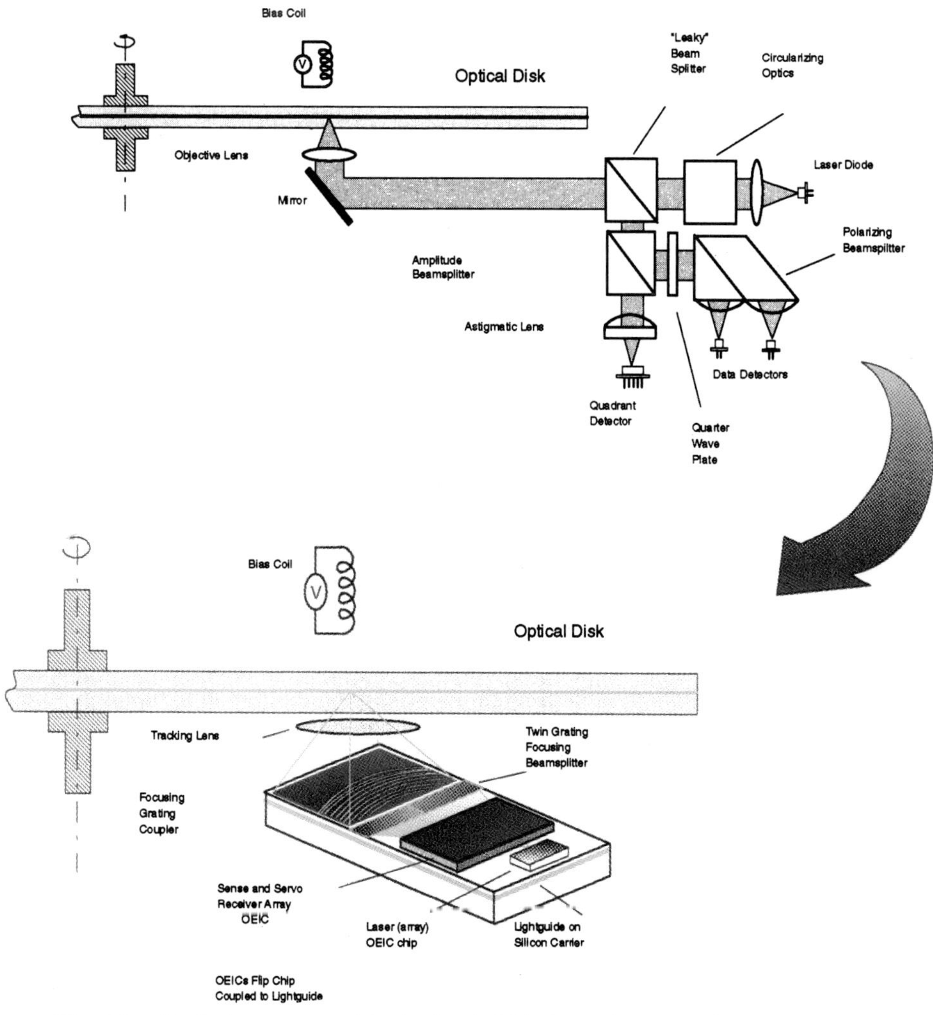

Fig. 16. Optical storage read/write head. This is a schematic of the elements in a read-and-write head for an optical storage system.

Another way to increase storage density is to pack tracks on a tighter pitch. Today, this result in unacceptable crosstalk from adjacent tracks. However, multiple beams reading parallel tracks can be used in schemes to process the signals together and suppress the noise from the adjacent tracks. As parallel track reading configurations are developed, the need for integrated heads becomes even more severe. Low-profile R/W heads will allow disks to be stacked, improving volumetric storage density. Key to this low profile is the alteration (or better yet removal) of the optical elements in the head. Lightguide technology with embedded diffraction gratings has been investigated with these objectives [29], and an OEIC or an OIC would also play a critical role in implementing a low-profile head.

Access speed can be improved by using more powerful lasers (migrating from 30–50 mW to more than 100 mW) and laser arrays. Thus, the laser-array OEIC (with integrated laser monitor photodiodes to control each laser in the array) is an attractive development. Work on monolithic, processed-mirror, integrated-photodiode laser-array OEICs is discussed in more detail in Parts III and V. Also described is the enhancement in laser power obtainable from low-threshold quantum-well lasers together with facet passivation coatings.

Speed can also be improved by use of more sensitive photoreceivers and photoreceiver OEIC arrays, as this can allow the beamsplitters in the retro-reflective optical path to direct more laser power to the disk and thus speed up writing and reading (given that the requirements for optical energy density do not change with disk rotation speed). Part V of this book describes some early work on the design of an OEIC containing both a sense amplifier with its photodetector and a servo amplifier with its quadrant photodetector all integrated on one 1×2 mm chip. Increased noise immunity and much lower mass are expected from this monolithic approach.

The lower part of Fig. 16 illustrates schematically how an integrated opto-electronic head might reduce the number of parts, the number of assembly and testing steps, and the size and weight of the package and promote the introduction of multitrack read/write operations. It shows a laser chip (could be an array) and a receiver (photodiode and amplifier) flip-chip mounted to a carrier which has the optical elements (lenses, lightguides, coupling gratings, etc.) processed onto it along with the electrical connection lines. In Part V, research experiments on this type of assembly are discussed in more detail.

4.5. Optical Printing

The requirements in printing for laser writing are similar, although not the same, to those in storage. Laser wavelength is matched to the sensitivity peak of the photoconductor on the drum, and drum write power is generally less than in storage. Spot size on the drum is in the range 50–100 μm, so diffraction-limited optics is not required, but the beam scan is over areas comparable to that of a sheet of paper. It is expected that laser arrays can speed up the printer, as in the storage case. This book will not cover the specific issues of this application.

5. OPPORTUNITIES AND REQUIREMENTS FOR AN OEIC TECHNOLOGY

As discussed earlier, optoelectronic components are used in computer system products today, and the trend is to use this technology increasingly in the future. The OE components in the technology today are predominately discrete devices, with precision-assembled packages and extensive testing of the modules at many stages of assembly and before use in the product. Compared with electrical line driver/receiver modules, they are very expensive components and thus have not found widespread use in computer applications. The obvious applications to long-distance lines have been made, and the computer industry is finding itself a major user of this technology, as shown in the market projections shown in Fig. 17. Thus, business conditions are emerging for the computer industry to take a leadership role in developing the new generations of optoelectronic technology.

The potential applications discussed in the previous section could generate the million-parts-per-year volume of chips needed to justify an integrated chip technology like an OEIC or an OIC. The performance demands of the chips are for hundreds to thousands of Mbit/s, at which the distortion and noise susceptibility of hybrid packaging can become a problem that is expensive to solve.

5.1. General Motivations for an OEIC Technology

Although there is no widespread product use of OEICs in computer industry applications, there are two notcable product implementations which can serve to illustrate some of the motivations for using OEICs. In 1993, Sony introduced the Minidisc portable audio disk player which used a silicon chip-based receiver OEIC in the disk reading head. By integrating a number of electrical and optical functions into one chip and subassembly, Sony realized both a size and a cost reduction. In 1994, IBM introduced a GaAs receiver OEIC into their 531 Mbit/sec Fibre Channel Data link. More uniform test-

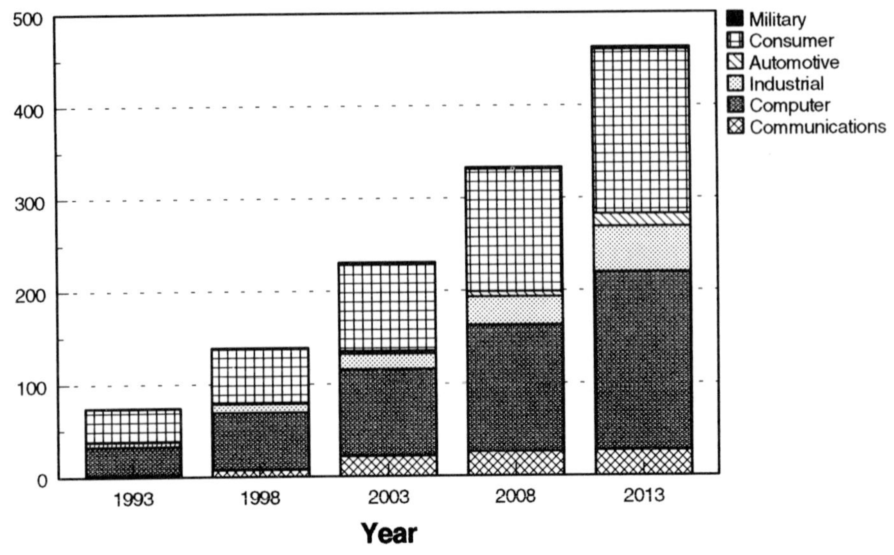

Source: OIDA, 1993

Fig. 17. Relative market share projections for OE components. This shows the relative market share projections from various industry sectors. Telecom: SONET, ISDN, ATM; Datacom: FDDI, HIPPI; Other: CATV, auto, etc. (From Electronicast, 8/90.)

ing and assembly leading to a lower cost Optical Subassembly was one benefit realized by using the OEIC.

Some general speculations can be made about what will be important in developing OEICs. A general requirement is that OEICs must offer technology features or application features justifying their development expense. Furthermore, they must be aimed at the applications that can generate the volumes of components needed to offset the high initial development cost associated with any complex-process, complex-function chip. Although these statements seem obvious, they have largely been ignored by developers of OEICs.

With respect to hardware technology features, performance and/or cost enhancements must be demonstrated over competing discrete-device implementations. The performance enhancements might come from the higher speed possible due to lower and more controllable parasitics, higher reliability due to fewer parts to assemble, better noise immunity, or some novel new functions arising from interactions between optical and electrical devices on the same chip. The cost enhancements might come from smaller modules (thus, more modules per card), the volume-manufacturing ease of a planar-processed chip (over the assembly of many chips), or the ease of design and quick turnaround time associated with a controlled fabrication process. For lower cost to be realized, the yield of the OEIC chip must be high. This implies that optical and electrical devices must be developed with simple and reproducible structures and that compatibility between the optical and electrical device processes is important.

The application-driven reasons for the OEIC are that the optical interfaces should be easier to design into the system (because of the chip models and design tools available) and easier to use (because of the ability to offer more functions and more robustness at comparable size, power, and cost). Again, providing these enhancements at costs comparable to or lower than those of discrete-device implementations is essential. Note that here the feature of the OEIC is the superior function offered by the chip; this may even be realized by compromising the performance of a particular device.

In many cases, the major cost factor in an OE module is in its packaging. For example, a $1 to $2 CD laser can easily become a $50 optical link driver when assembled with optics and connector housing and qualified for link performance and reliability requirements. Packaging consists of attachment of the chips and passive components to the chip carrier, testing the performance of the module during assembly, and final qualification of the module for the application. For optical packages, many of the assembly steps require precision mechanical alignments and ultrastable attachment techniques. Although not usually recognized as an important feature, easier packaging would be a good reason to develop an OEIC technology. An OEIC would reduce the number of components in a package. It could allow one chip alignment to yield many simultaneous optical component alignments. The ability to add electronic function at relatively low cost would allow the addition of self-testing functions to module assembly or qualification.

The density advantage of an OEIC over discrete devices is likewise dependent on advanced packaging. For example, it is feasible to integrate four channels of link adapter function onto a single chip, allowing four channels to occupy less card space than one link adapter module does today [30]. But to realize this size reduction, the packaging technology must be able to bring arrays of optical lines up to the module and fan them into and out of the chips. The chip carrier needs enough electrical I/O pins (e.g., about 20–25 per line to serialize a high-speed byte-wide data bus) to provide data and control functions to all the channels and permit module testing. This is well within the capabilities of electrical IC technology but beyond the capabilities of today's optical module technology. These packaging issues are evident not only for link applications but also for storage and printer/display applications.

5.2. General Requirements for an OEIC for Computer System Use

OEICs should offer enough functions to make systems easier to design. Front-end functions can be realized on chips with a few hundred devices. For a full link adapter, about 10K devices will be needed. For applications which can benefit from arrays of the function on a single chip, 50K-device OEICs will be needed. Chips of this complexity require a robust and mature technology for both electrical and optical devices. This book discusses the relative complexity of chip processes and device complexity as they pertain to yielding complex OEIC chips. This book also discusses the process control needed to obtain functional high-speed circuits.

There will be a performance, reliability, and cost trade-off between adding large amounts of electronic function to the OEIC or introducing an additional VLSI IC chip to the package. The trade-off crossover will move in the direction of the OEIC with the maturity of the OEIC process.

OEICs should be targeted to applications in the range 100 to 1000+ Mbit/s, where the cost of hybrid packaging becomes significant and the major growth in module volume is expected in the coming decade.

The robustness of the OEIC must be designed in at the beginning. Robustness can come from simple device structures, devices which are tolerant to process variations or environmental variations, or devices and circuits which are immune to electrical or optical noise sources from within or without the chip. Chapters of this book which discuss the practicality of OEICs consider robustness to be comparable in importance to performance.

Module robustness is related to low bit error rate (e.g., $<10^{-15}$), fault tolerance to component failure, safety of use (e.g., Class 1 OSHA optical radiation standards), and immunity to electrical and optical noise sources in the environment. This reflects back into the design of the OEIC itself.

As the requirement for complexity on chip increases, the ability to design the chip and assure that it will have high yield in an integrated process is increasingly dependent on good device and circuit models and a good chip design, simulation, and verification system. These models and design tools are more complex than today's IC design systems in the sense that they must integrate both high-speed analog and digital functions and both optical and electrical functions. The noise sources are more than a digital IC design system accounts for, and distortion and crosstalk between a large set of devices and circuits need models. The models must include the effects of optical as well as electrical wiring on the chip and of feedback and noise from the package. These design systems do not exist today and are a challenging development in themselves. This book discusses the status of modeling and design tools for OEICs. It is an absolutely essential development for OEICs to be applied to computer systems. It gives the potential for the OEIC to be simultaneously complex, high-yielding, and designed and built on a timely schedule (less than 1 year).

OEICs must be designed with packaging in mind. Light sources which must be placed on the edge of the OEIC may be restrictive to proper chip layout. Optical components which are surface emitters or detectors require different packaging than edge emitters/detectors. This may result in cost and reliability advantages or disadvantages. If OEICs are flip-mounted onto carriers, as is done in many IC multichip modules today, the optical wiring escapes must be taken into account in the OEIC design. Inclusion of optical or mechanical alignment marks in the OEIC may have to be accommodated to permit automated packaging.

The OEIC designer may thus develop different lasers or photodiodes than the discrete device designer, to accommodate a more complex package. Optical wiring (light guides) may be included on the chip give flexibility in device placement. Different coupling techniques may be developed to permit easy packaging of the OEIC on the chip carrier.

Other packaging aspects influencing OEIC design are compatibility with the power supplies of ICs and the thermal characteristics of the package and environment. OE devices should operate from 5-volt supplies today, migrating to lower voltages (e.g., 3.3 volts) with time or as the speed increases. Most of the computer applications permit only air-cooled modules, meaning that the functions integrated on the chip cannot generate more than a few watts of heat. Thus the device's heat sensitivity and chip heat dissipation capability may become major influences to the OEIC device designer. To date, many of these practical issues in OEIC development have not been addressed in the technical community.

The foregoing list of requirements describes a technology which is truly advanced over the optoelectronic technology of today, and the challenges in achieving it are discussed in this book. The challenges, after all, are similar to the challenges that electronics had to overcome to go from discrete transistors, capacitors, and resistors to ICs. The rewards of IC technology development are still being felt in this information age. One hopes the OEIC will make a similar impact.

6. SUMMARY

We have described a number of trends in the computer industry which can benefit from further developments in optoelectronics. Increases in computer processing power require at least proportionate increases in I/O bandwidth; fiber optics is already playing a role in these increases. Coupled systems and massively parallel systems require increasing performance of their interconnection networks. Distributed processing requires both increased transmission distance and increased bandwidth, both of which are best provided by fiber optics. Various applications require movement and storage of large amounts of data; optical storage and optical data transmission are the best ways to satisfy these requirements.

Although Gbit/s transmission at distances of tens of kilometers is in routine use in the telecommunications industry, the technologies used in the telecommunications industry are not necessarily suitable for use in computer systems. To maintain the proper cost balance in the computer system, low-cost link hardware, packaged in a way which allows it to be integrated into a computer system, is required. This is essential to the highest-performance central systems and even more essential in the coming generations of low-cost, but very high performance, intelligent workstations.

Applying optoelectronics technology to the interconnections in a high-performance multiprocessor or massively parallel system will undoubtedly require parallel optical transmission systems; low-cost arrays of emitters and receivers will be essential. These arrays will also be essential to the development of very high speed I/O links using parallel transmission. Some scientific visualization problems already require a data transmission bandwidth which is a multiple of the bandwidth of the highest-speed optical transmission systems available today.

In the switching area, high-speed packet switches are required for many I/O applications that involve data transfer quantities which are too small for good performance of circuit switching systems. These switches must have packet throughput commensurate with the bandwidths of the attached optical links and packet loss rates comparable to or lower than the packet losses due to the bit error rates of the attached links. Massively parallel systems involve multistage switching networks with large numbers of interconnections. As these networks migrate to optical interconnections to obtain further performance increases, either very low cost optoelectronic components or inexpensive all-optical switching nodes will be required.

One of the key evolutions of OE technology is to go from discrete-device chips and precision-assembled packaging into an integrated-function chip and lower-cost, denser package. The optoelectronics industry today is experiencing difficulty in competing with established electrical interconnect technology in all but the long-distance applications and the low-speed industrial applications. More advanced modules, ones which are cost-effective as well as high-performance, and which offer sufficient function and robustness to be "user friendly" to the computer system designer, can be realistically expected by using a combination of integrated optoelectronic chips and loose alignment tolerance optical components. These modules would have significant useage in the multibillion-dollar data communications and data storage markets. This book reviews the status of these technologies—not only the chips but also the underlying materials and processes, design and modeling tools, and packaging. For the computer industry, it appears that the major applications for optoelectronics are still in the future, which is fortunate because the major system leverages will come from a new generation of optoelectronics, optimally based on the OEIC.

Today, high-performance optoelectronics is enabling entirely new classes of computer applications. As optoelectronics technology evolves to meet the challenges described in this chapter, computer systems will also evolve and as yet unforeseen applications will further challenge optoelectronics, fostering further revolutionary developments.

REFERENCES

1. A. Peled, *Sci. Am.* **257**(4), 56 (1987).
2. R. F. Dinan, L. D. Painter, and R. R. Rodite, *IBM Syst. J.* **29**, 421 (1990); see also accompanying articles in same issue.
3. *Computer* **22**(8) (1989), special issue on scientific visualization.
4. Fiber Distributed Data Interface (FDDI) Token Ring Media Access Control (MAC), ANSI X3.139–1987, American National Standards Institute, New York (1987).
5. J. C. Elliott and M. W. Sachs, *IBM J. Res. Dev.*, **36**, 577 (1992).
6. High Performance Parallel Interface Mechanical, Electrical, and Signalling Protocol Specification, American National Standard X3.183–1991, American National Standards Institute, New York (1992). Computer and Business Equipment Manufacturers Association, Washington, DC.
7. Fibre Channel—Physical and Signalling Interface, rev. 3.0, proposed American National Standard X3.230–199x, Computer and Business Equipment Manufacturers Association, ANSI, New York (June 1992).
8. D. M. Dias, B. R. Iyer, J. T. Robinson, and P. S. Yu, *IEEE Trans. Software Eng.* **15**, 437 (1989).
9. K. Shimizu, H. Tsunoda, and T. Chiba, *Fujitsu* (Japan) **42**(2), 113 (1991).
10. C. J. Georgiou, T. A. Larsen, P. W. Oakhill, and B. Salimi, *IBM J. Res. Dev.*, **36**, 593 (1992).
11. L. W. Tucker and G. G. Robertson, *Computer* **21**(8), 26 (1988).

12. W. C. Athas and C. L. Seitz, *Computer* **21**(8), 9 (1988).
13. P. M. Neches, *Proc. Compcon Spring 86*, p. 374, IEEE Computer Society Press, Washington, DC, 1986.
14. *Computer* **20**(6), 1987, special issue on interconnection networks.
15. A. A. Sawchuk, B. K. Jenkins, C. S. Raghavendra, and A. Varma, *Computer* **20**(6), 50 (1987).
16. D. A. Patterson, G. Gibson, and R. H. Katz, *Proc. Sigmod International Conference on Management of Data*, Chicago, June 1–3, 1988, p. 109; Association for Computing Machinery, New York, 1988.
17. N. P. Dono, P. E. Green, Jr., K. Liu, R. Ramaswami, and F. F.-K. Tong, *IEEE J. Selected Areas Commun.* **8**, 983 (1990).
18. M. S. Roden, "Digital Communication Systems Design," Prentice Hall, Englewood Cliffs, NJ, 1988. See especially chapter 4.
19. The interested reader is referred, for example, to the critique by I. Cidon, J. Derby, I. Gopal, and B. Kadaba, *Proc. 10th International Conference on Computer Communications*, Narosa Publishing, New Delhi, 1990, p. 315; also IBM Research Report RC15572, Yorktown Heights, New York, April 1990.
20. R. W. Keyes, "The Physics of VLSI Systems," Addison Wesley, Reading, MA, 1987; R. W. Keyes, *Science* **230**, 138 (1985).
21. 802.5–1989 LANs: Token-Ring Access. IEEE Computer Society Press, Washington, DC, 1989.
22. 8802.3–1989 CSMA/CD Access Method and Physical Layer Specifications. IEEE Computer Society Press, Washington, DC, 1989.
23. L. Doyle and S. Frankle, "Local Area Networks,"International Data Corp. Communications Market Planning Service Publ. 4969, Vol. 1, July 1990; Publ. 4986, Vol. 1 August 1990; Publ. 5015, Vol. 1, August, 1990.
24. D. Hanson, "High Speed, Low Cost Plastic Peripheral Links," *Proc. IEEE LEOS'91 Annual Meeting* Nov. 1991, *San Jose*, CA. D. Hanson, "High Speed Plastic Fiber Optic Links," *IEEE Lightwave Commun. Syst. Mag.* **3**, 51 (February 1992).
24b. S. Joiner, "Cost Effective Packaging of LEDs for Fiber Optics," *Proceedings of IEEE/LEOS Annual Meeting*, p. 31, 11/93, San Jose, CA.
25. A. Deutsch, *IBM J. Res. Dev.* **34**, 601 (1990).
26. G. Arjavalingam and B. Rubin, *Fiber Integrated Opt.* **8**, 235 (1989).
27. A. B. Marchant, "Optical Recording—A Technical Overview," Ch. 8. Addison-Wesley, New York, 1990.
28. W. Kozlovsky, W. Lenth, and W. Risk, *Conf. Digest—IEEE/LEOS Topical Meeting on Optical Data Storage*, p. 111 (1990).
29. S. Ura, T. Suhara, H. Nishihara, and J. Koyama, *Trans. IECE Jpn.* **J39-C** 609 (1986).
30. J. D. Crow, *IEEE Circuits Devices Mag.* **7** 20 (March 1991).

PART II
Materials Growth

Chapter 3

MOLECULAR BEAM EPITAXY WITH GASEOUS SOURCES

Charles W. Tu

Department of Electrical and Computer Engineering
University of California at San Diego
La Jolla, California 92093

1. INTRODUCTION

The success of any technology, including optoelectronic integrated circuits (OEICs), depends crucially on the quality of the materials in which devices and circuits are fabricated. Integrated optoelectronic devices involve heterostructure layers, quite often complex, and they have to be grown by advanced thin-film techniques, such as vapor-phase epitaxy (VPE) or molecular beam epitaxy (MBE). The vapor-phase technique includes hydride, chloride, and metalorganic VPE (MOVPE), the last being covered in the next chapter.

ISBN 0-12-200420-5

Molecular beam epitaxy is a very versatile thin-film growth technique, which was pioneered by Cho and Arthur in the late 1960s [1]. It has been proved capable of controlling layer thickness and composition precisely and of growing uniform, ultrathin layers with abrupt interfaces. Many review articles, conference proceedings [2–5], and books [6–11] have been devoted to MBE. This chapter therefore emphasizes recent results, especially concerning MBE with gaseous sources, i.e., gas-source or hydride-source MBE (GSMBE or HSMBE), metalorganic MBE (MOMBE), and chemical beam epitaxy (CBE). As in other fields, acronyms are commonly used in the MBE field, but not all of them are adopted by everyone. For the purpose of discussion in this section, Fig. 1 summarizes these variants of MBE according to the sources used and shows their relations to each other and to VPE according to the operating pressure [12]. Figure 2 shows a schematic diagram of an MBE system with either solid or gaseous sources [13]. Cryopump, turbomolecular pump, and diffusion pump all can be used. A combination of the first two is more common.

Conventional solid-source MBE has been studied extensively and used for growing various arsenide heterostructures, e.g., AlGaAs/GaAs, GaAs/InGaAs, and InAlAs/InGaAs. The growth of mixed group V compounds, such as InGaAsP, is important for long-wavelength optical communication and OEIC, but it cannot be achieved easily and controllably by conventional solid-source MBE because of the high vapor pressures of solid arsenic and

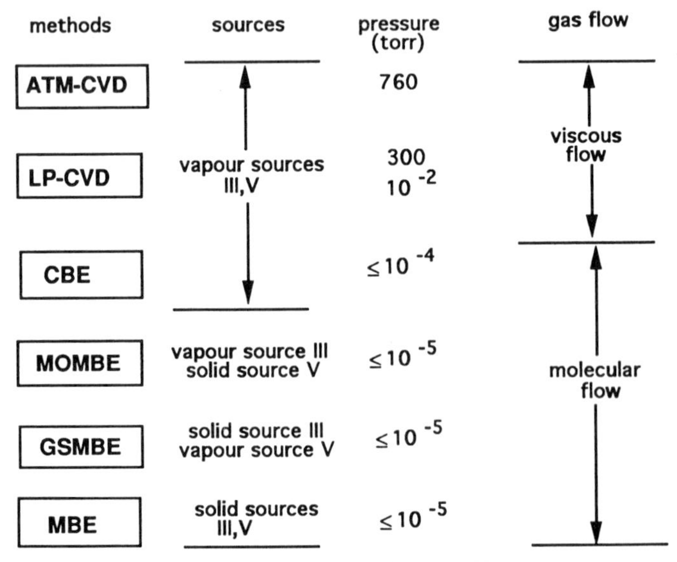

Fig. 1. Comparison of various epitaxy techniques [12].

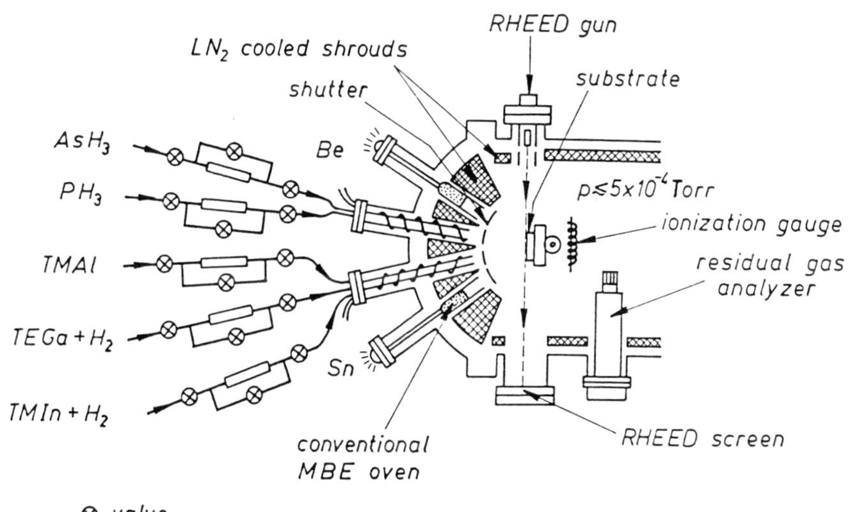

Fig. 2. Schematic diagram of an MBE system with solid and gaseous sources. (From Ref. 13.)

phosphorus. Therefore, gaseous sources, such as phosphine and arsine, through a high-temperature cracker, were first used by Panish [14] and Calawa [15], respectively, to obtain a well-controlled flux of phosphorus and arsenic species, mainly P_2 and As_2. This technique is called GSMBE or HSMBE [16]. Since the group III elements are derived from solid sources, as in MBE, the growth mechanism is the same as in conventional solid-source MBE. It should be noted that valved crackers [17, 18] for solid As_4 and P_4 have been used successfully to control the As_2 and P_2 fluxes.

Veuhoff *et al.* [19] first reported the replacement of the group III elemental sources by gaseous metalorganic compounds, such as trimethylgallium (TMGa), triethylgallium (TEGa), and trimethylindium (TMIn). In their experiment they used cracked arsine and phosphine as well, and they called their technique MOMBE. However, Tsang [20] called the all-gaseous approach CBE and demonstrated that high-quality InP and related compounds could be grown and high-performance devices could be achieved. The term MOMBE is limited to the situation where MO group III sources and a solid arsenic source are used. Fraas *et al.* [21] have used such sources for high-throughput production of GaAs solar cells in a hot-wall configuration, called vacuum chemical epitaxy (VCE) [21].

The growth of the arsenides is very well understood and controlled, but the growth and composition of compounds containing P and As have been determined mainly by trial and error. In this chapter we shall present a tutorial, for the first time, on *in situ* control of the growth and composition of phosphides and mixed group V compounds, i.e., InP, GaAsP, InAsP, and InGaAsP. Then we shall discuss the MOMBE and CBE growth kinetics, selective-area growth, and the use of alternative gaseous precursers.

2. GAS-SOURCE MBE

Because the MBE process occurs in a vacuum environment, a variety of *in situ* characterization technique can be used, e.g., reflection high-energy-electron diffraction (RHEED) and quadruple mass spectroscopy. The main advantage of MBE, compared with other growth techniques, is the easy use of *in situ*, real-time monitoring of RHEED intensity oscillations for growth rate and composition calibration. Figure 3 shows a sequence of the formation of the first two complete monolayers of GaAs in relation to RHEED intensity oscillations [22]. Because the sticking coefficient of As (or P) is very small unless there is a free Ga atom and the group III metal atoms have unity sticking coefficient at the normal MBE growth temperature, growth is controlled by the arrival rate of the group III atoms when an overpressure of arsenic is provided. Therefore, only group III–induced oscillations are observed for growth rate calibrations. The arsenic flux is not usually controlled as precisely. However, materials properties would depend on the V/III ratio, as it would influence the point-defect concentrations. A few groups have reported arsenic-induced oscillations on a Ga-rich surface, where the growth rate is limited by the arsenic flux [23, 24]. Because of the high background pressure of hydrogen from cracking arsine and phosphine in GSMBE and CBE, Chin *et al.* [25] show that it is particularly important to use the group V–induced oscillations to calibrate the group V incorporation rates and V/III incorporation ratios.

It should be noted that group V–induced oscillations are used only for calibration. The growth of actual heterostructures is performed with a V/III incorporation ratio greater than unity to maintain a smooth surface.

2.1. Control of Incorporation Rates of Group V Species

Figure 4a illustrates a typical RHEED oscillation pattern during the growth of GaP at 610°C. It starts with a phosphorus-stabilized surface. Then Ga-induced oscillations are observed after the Ga shutter is opened. The RHEED intensity decreases after the phosphine shutter is closed because Ga atoms

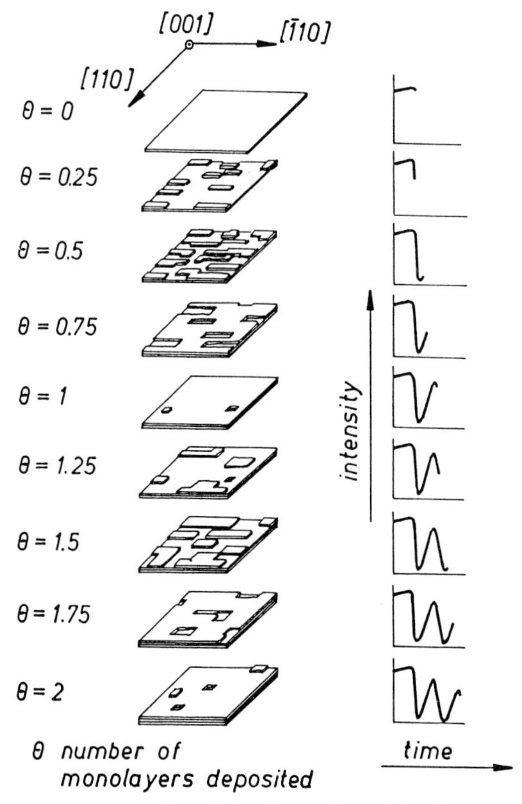

Fig. 3. Real-space representation of the formation of the first two complete monolayers of GaAs (001) in relation to RHEED intensity oscillations. θ is the fractional layer coverage. (From Ref. 22.)

accumulate on the substrate. The final step involves closing the Ga shutter and opening the phosphine shutter again. Here the oscillation resumes with incoming phosphorus species reacting with accumulated Ga atoms on the surface. This represents phosphorus-controlled growth because there is an excess amount of Ga atoms on the surface and the available amount of phosphorus limits the growth rate. To avoid the inevitable decrease in the RHEED intensity after growth commences, one can deposit excess Ga atoms on the surface without phosphine right after the surface is stabilized with phosphorus. Then the Ga shutter and the phosphine shutter are closed and opened, respectively, at the same time. The resultant P-induced oscillation with a larger amplitude is shown in Fig. 4b. Figure 5 shows linear behavior of phosphorus-limited growth rate, or incorporation rate, versus the phosphine flow rate in GSMBE of GaP and AlP at a particular growth temperature and

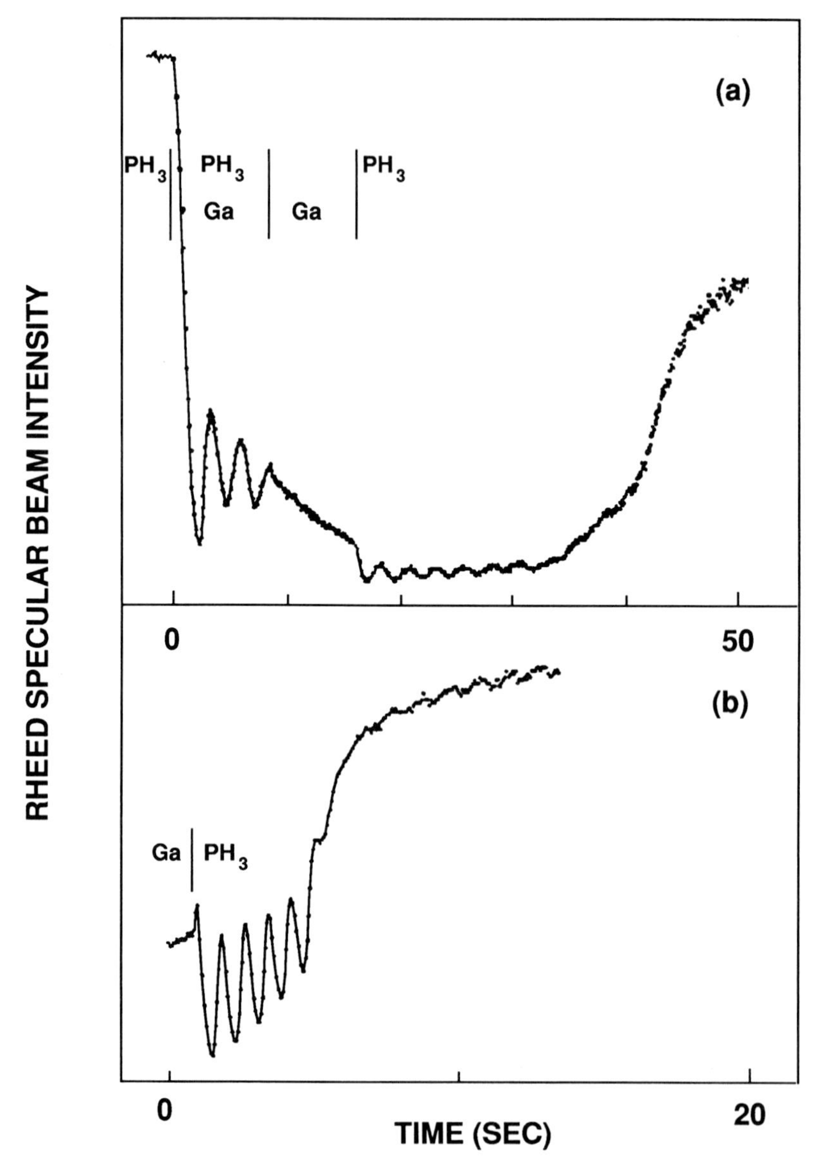

Fig. 4. (a) RHEED oscillations of Ga- and phosphorus-controlled growth on GaP(100). The intensity decreases when the Ga shutter is opened. (b) Phosphorus-controlled RHEED intensity oscillations. The intensity recovers after excess Ga atoms are consumed. The vertical scale and the sample azimuthal angle are different from those in (a). (From Ref. 25.)

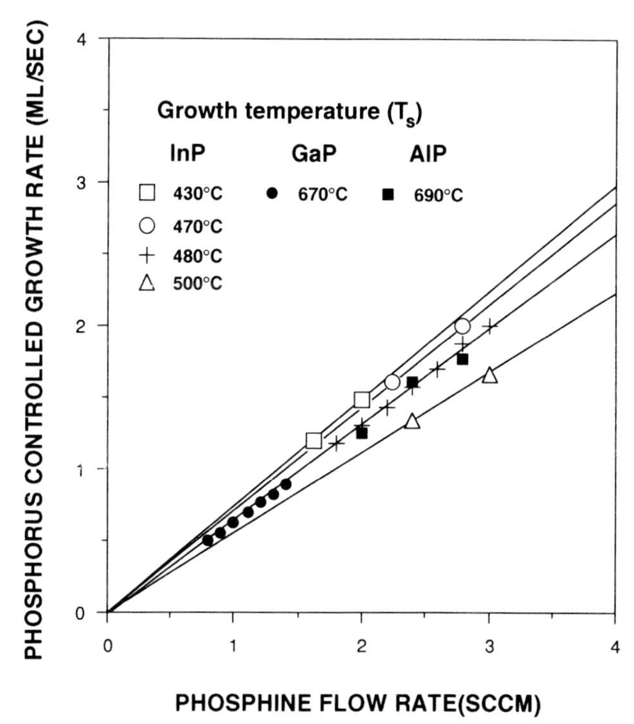

Fig. 5. Phosphorus-controlled growth rate vs. phosphine flow rate. (From Ref. 25.)

InP at different growth temperatures. It is important to note that the incorporation rate depends on the growth temperature, even if the flux is kept constant. Therefore, the incorporation rate gives a more meaningful calibration than the ion gauge reading of the flux. The incorporation rate is independent of variations due to ion gauges and MBE systems.

Figure 6 shows how one can establish the V/III incorporation ratio, an important growth parameter that determines the material quality. In Fig. 6 the growth rates of (a) GaP, (b) AlP, and (c) InP are plotted as a function of the reciprocal group III cell temperatures (T_{III}). The phosphorus-controlled growth rates, as measured by the procedure just mentioned, are shown with thick lines. They are independent of group III fluxes. With the phosphine flow rate fixed, the group III–controlled growth rate is measured under various group III fluxes. As the open circles show in Fig. 6c, the growth rate increases exponentially with increasing In cell temperature; i.e., it increases linearly as $1/T_{III}$ decreases on the semilog plot. This linear dependence stops at a certain point, and then the growth rate becomes constant even as the In flux increases further because there are more In atoms than available P

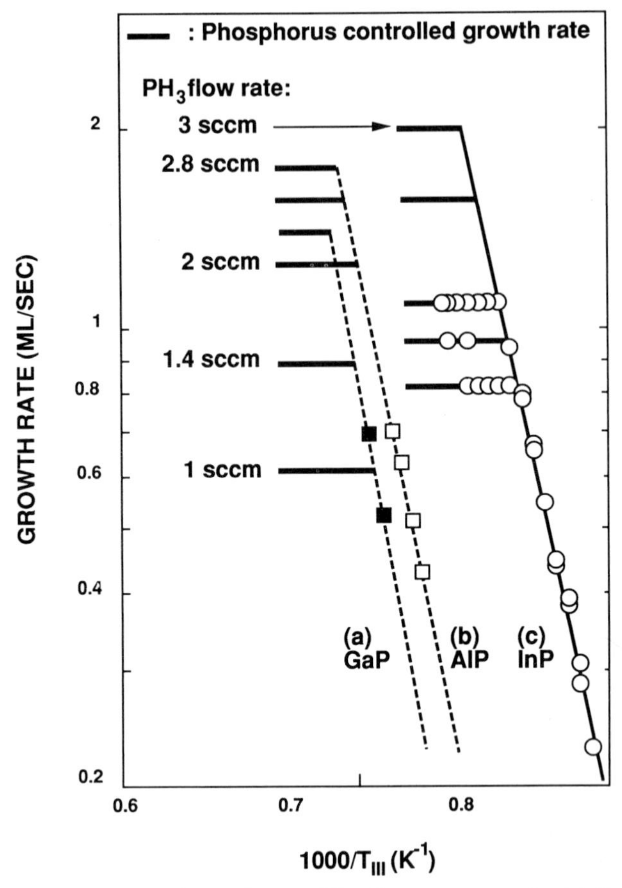

Fig. 6. Growth rate vs. reciprocal group III cell temperatures for (a) GaP, (b) AlP, and (c) InP. (From Ref. 25.)

atoms in the P beam flux that can be incorporated at the growth temperature. The growth rate in the plateau region is therefore limited by the phosphine flow rate, independent of the In flux. The point at which the growth rate saturates indicates a unity V/III incorporation ratio. It should be emphasized that this approach of studying group V–controlled growth rate is for calibrating the group V flux and establishing the V/III ratio. Of course, during GSMBE of a heterostructure, a V/III flux incorporation ratio >1 should be used to prevent a hazy surface from In or Ga droplets. Samples with the highest purity should be grown as close to stoichiometry as possible [26]. The best InP layer with 77 K mobility of 112,000 cm^2/V-s and background carrier concentration N_d–N_a of 2×10^{14} cm^{-3} have been achieved [27].

2.2. Composition Control and a Growth Kinetic Model for Mixed Group V Compounds

Normally the group V composition in mixed group V compounds is determined by trial and error. That is, several samples are grown with different group V flow rates or fluxes at different substrate temperatures, and then their compositions are determined *ex situ* by x-ray rocking curves or photoluminescence (PL). Hou *et al.* [28] have found that under certain conditions group V–controlled RHEED oscillations can also be used *in situ* to determine the group V composition in GaAsP and InAsP.

2.2.1. $GaAs_{1-x}P_x$

A Ga-rich surface is again intentionally formed by depositing a several-layer amount of Ga, as discussed before. Figures 7a and 7b show group V–induced RHEED intensity oscillations for GaAs and $GaAs_{1-x}P_x$, respectively [28]. The (As + P)-limited growth rate is found to be higher than the As-limited growth rate. In principle, because both As_2 and P_2 have high vapor pressure and can desorb easily from the surface, the As incorporation rate

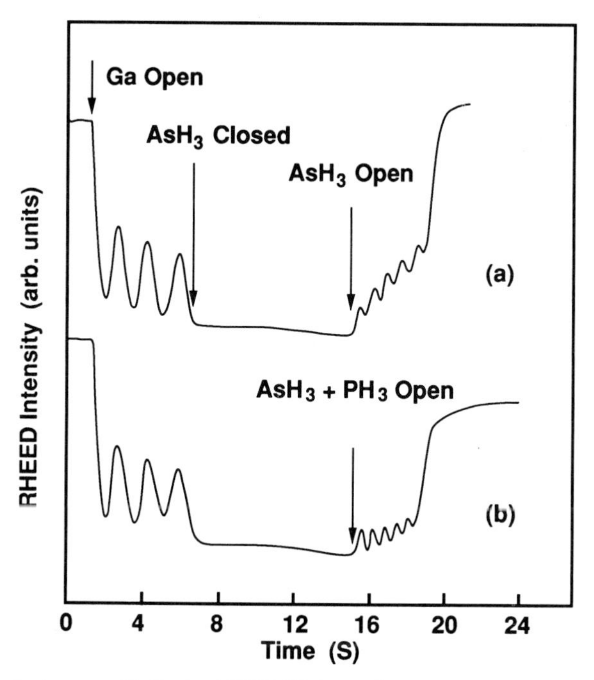

Fig. 7. Typical group V–limited RHEED intensity oscillations on a Ga-rich GaAs surface. The oscillation is induced by (a) As and (b) As and P. (From Ref. 28.)

into GaAsP would be modified by the injection of P due to the displacement of As by P. However, since excess Ga atoms are deposited on the surface before the group V–limited RHEED oscillation measurements, the displacement of As by P may be negligible, at least at relatively low substrate temperatures when group V desorption is not significant. It is reasonable then to assume that the increase in the growth rate shown in Fig. 7b is uniquely due to the addition of P. Therefore, from the difference in the group V–limited growth rates the phosphorus composition can be deduced.

As mentioned before, however, in normal growth the V/III ratio should be greater than unity. Therefore, we should check whether the *in situ* P composition determination by RHEED (when V/III < 1) is applicable to the normal growth condition (when V/III > 1) if the same flow rates and substrate temperature are used. A series of GaAs/$GaAs_{1-x}P_x$ strained layer superlattices (SLS), typically consisting of 15 periods of GaAs(95 Å)/ GaAsP(95 Å) layers, were grown with the same fixed arsine flow rate of 1.6 sccm as in the RHEED measurements and a phosphine flow rate ranging from 0.8 to 2.8 sccm. The thickness of the $GaAs_{1-x}P_x$ layer is less than the critical layer thickness, so pseudomorphic $GaAs_{1-x}P_x$ growth is to be expected. In such a case, the structural parameters can be accurately determined from the x-ray diffraction simulation based on the dynamical theory.

Figure 8 shows *in situ* RHEED-determined P compositions in $GaAs_{1-x}P_x$ (indicated by open circles) and P compositions determined *ex situ* by x-ray rocking curves and simulations (filled circles) versus the phosphine flow-rate fraction (phosphine flow rate over the total hydride flow rate). Typical errors in determining x from the RHEED oscillation are indicated for $0.3 < x < 0.6$. This uncertainty is significant when the phosphorus composition is greater than 0.3 because the relatively large surface strain reduces the number of oscillations and causes an error in reading the time scale. Two points are to be noted concerning Fig. 8. First, the composition of phosphorus incorporated in the solid is different from the PH_3 flow-rate fraction in the gas phase. This is attributed to different sticking coefficients and arrival rates of arsenic and phosphorus (a factor of the square root of the As_2-to-P_2 molecular mass ratio). Second, it appears that the *in situ* determination of P compositions by RHEED oscillations and *ex situ* determination by x-ray diffraction are in very good agreement for $x < 0.35$. The deviation for samples with the highest phosphorus composition may be due to partial strain relaxation in SLSs and a higher phosphorus sticking coefficient on a GaP-like surface compared with a GaAs-like surface.

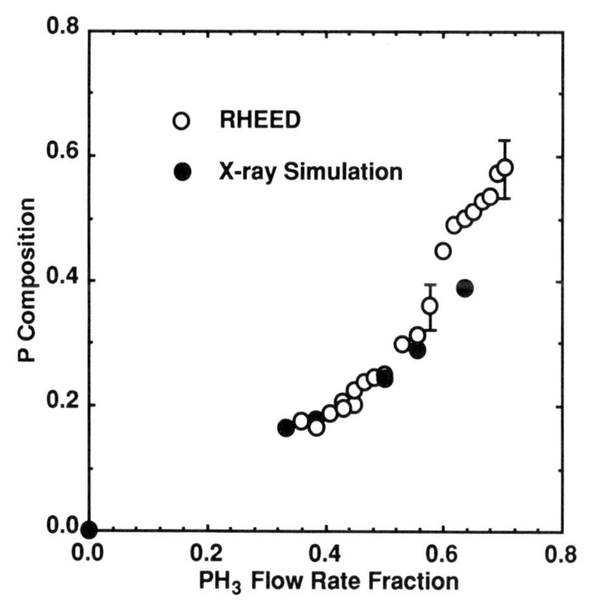

Fig. 8. Phosphorus composition in $GaAs_{1-x}P_x$ as a function of the PH flow-rate fraction (over the total hydride flow rate). (○) RHEED oscillation data, and (●) x-ray simulation data for $GaAs_{1-x}P_x$/GaAs SLSs. (From Ref. 28.)

How can group V–induced RHEED oscillations, performed on a Ga-rich surface, be used to determine the composition of structures grown under normal conditions, using excess group V fluxes? For these two cases Liang and Tu [29] propose growth kinetic models that consider both physical and chemical processes, such as desorption, adsorption, and decomposition, on the surface during growth. Figure 9 shows two potential energy diagrams as a function of reaction coordinates for the two cases. Under normal growth conditions with excess As_2 and P_2 (V/III > 1) the As_2 or P_2 species in the gas phase, V_2(g), from cracked AsH_3 or PH_3, are first physisorbed on the surface [surface physisorption, V_2(s.p.)] and then become chemisorbed [surface chemisorption, V_2(s.c.)] before incorporation into the lattice. See Fig. 9a. Because of high vapor pressures of As_2 and P_2, their desorption processes from physisorbed and chemisorbed states, with activation energies E_D and E_e, respectively, have to be considered. For the case of a Ga-rich surface for group V–induced RHEED experiments, because of the infinite amount of Ga atoms on the surface, the As_2 and P_2 species can become chemisorbed first before incorporation into the lattice. See Fig. 9b. For the two models,

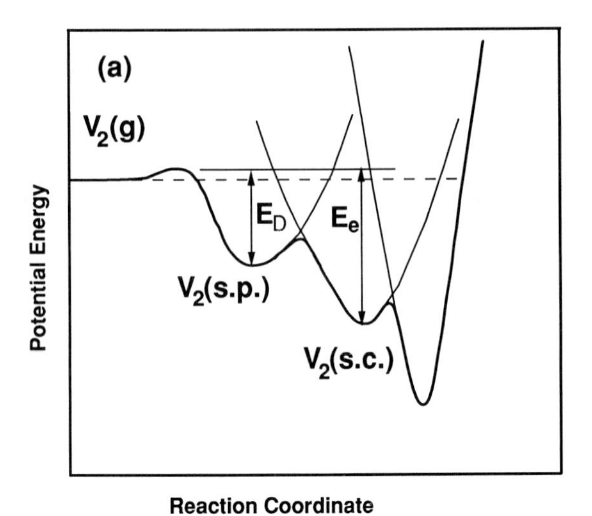

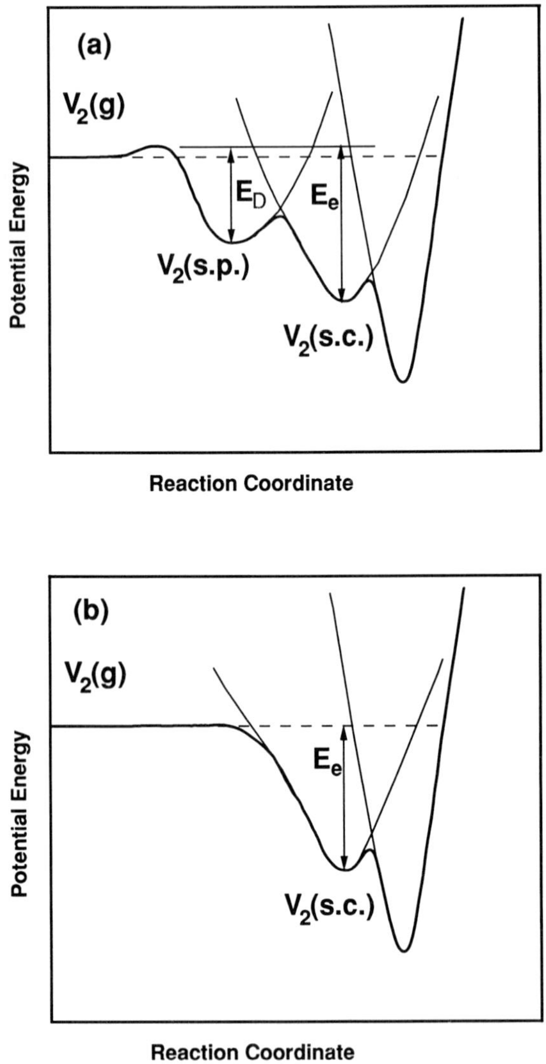

Fig. 9. Schematic diagrams of group V adsorption and desorption reaction coordinates: (a) Ga-rich surface (for group V–induced RHEED oscillations) and (b) group V–rich (normal growth) condition. (From Ref. 30.)

rate equations for As_2 or P_2 species on the surface can be written down. Then, in a steady-state condition, the growth rate and the P composition x in the two cases can be derived. When all the desorption phenomena can be ignored, as at relative low substrate temperatures, the two composition equations are identical [29]:

$$x = \frac{1}{1 + \dfrac{S_{As}}{S_P}\sqrt{\dfrac{M_{P_2}}{M_{As_2}}\dfrac{f_{AsH_3}}{f_{PH_3}}}},$$

where S is the sticking coefficient, M mass, and f flow rate.

To determine the temperature range of validity for *in situ* determination of P compositions, we compare the P compositions determined from group V–induced (group III–rich) RHEED oscillations and from *ex situ* measurements of layers grown under normal (group V–rich) conditions, as a function of substrate temperature. Figure 10 shows the temperature dependence of P compositions for GaAsP/GaAs SLSs on GaAs substrates [31]. We see that at temperatures higher than about 520°C the results of *in situ* and *ex situ* measurements start to deviate from each other because of desorption of As_2 or P_2 [32]. From the low-temperature data we can determine the ratio

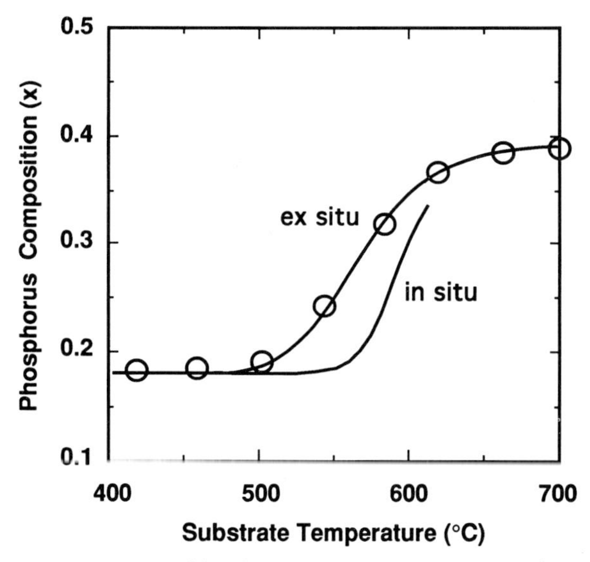

Fig. 10. Phosphorus composition in the GaAs layer of GaAsP/GaAs SLSs. The *ex situ* compositions were determined from x-ray rocking curves and dynamic simulations. The line marked *in situ* was obtained from group V–induced RHEED oscillations. (From Ref. 30.)

of the sticking coefficients of As_2 to P_2, which is 3.8 for the case of GaAsP. Therefore, we can determine the P composition simply from the flow-rate ratio of arsine to phosphine. It is interesting to note that the *in situ* and *ex situ* methods seem to converge again at higher substrate temperatures. However, it is difficult to obtain RHEED oscillations at those temperatures.

Using the models as depicted in Fig. 9, Liang and Tu [29] have studied the P composition for a variety of normal growth conditions for GaAsP. Activation energies for the desorption processes are obtained independently by measuring the As and P temperature-dependent incorporation rates and studying the desorption behavior of As from GaAs and P from GaP by monitoring the RHEED intensity [33]. The desorption activation energies are comparable to the heats of vaporization of As_2 and P_2 [33]. Using only four preexponential factors as adjustable parameters, they fit the *ex situ*–determined P composition curve in Fig. 10. Then, fixing these parameters, they calculated the P compositions as a function of other growth parameters, such as flow rates and flow-rate fractions, for different substrate temperatures. Figure 11 shows that the calculations (lines) agree well with experimental results (data points). From such calculations, one can see that because of desorption of As_2 and P_2 the solid-phase P composition at a high substrate temperature can be simply found from the phosphine-to-arsine flow-rate ra-

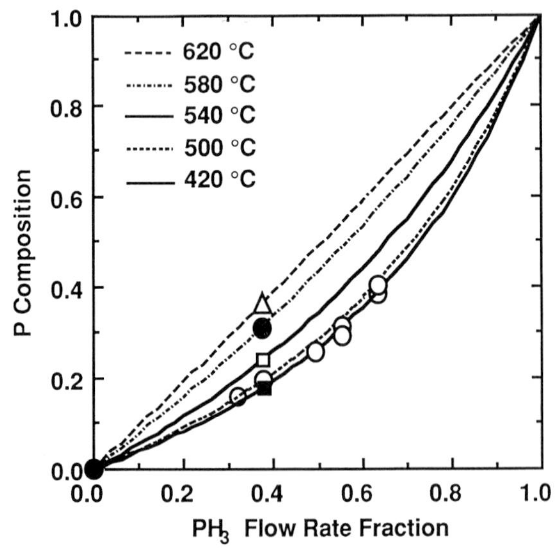

Fig. 11. The P composition in GaAsP as a function of the PH_3 flow rate fraction at different growth temperatures. The symbols represent experimental data and the lines were calculated from the kinetic model. (From Ref. 29.)

tio. It is also interesting to note that similar curves have been obtained for MOVPE-grown GaAsP except that the MOVPE growth temperatures are much higher [34] because AsH_3 and PH_3 are not precracked.

2.2.2. $InAs_xP_{1-x}$

The situation with $InAs_xP_{1-x}$ is somewhat different because As_2 incorporates much more efficiently than P_2 in $InAs_xP_{1-x}$. Once the As and In incorporation rates, R_{As} and R_{In}, respectively, at a desired growth temperautre are calibrated by As- and In-induced RHEED oscillations, respectively, Hou and Tu [35] show that when the ratio R_{As}/R_{In} is less than unity, all of the As_2 dimers react with In atoms on the surface and the remaining In atoms react with P_2 dimers. The excess P_2 molecules desorb from the surface. Thus, the As composition x in $InAs_xP_{1-x}$ is just the ratio R_{As}/R_{In}. Based on this idea of *in situ* controlling the As composition in $InAs_xP_{1-x}$, a series of $InAs_xP_{1-x}$/InP SLSs were grown at 2 sccm PH_3 flow rates and different R_{As}/R_{In} ratios. Shown in Fig. 12 is the As composition determined *ex situ* from x-ray rocking curves of SLSs, versus the R_{As}/R_{In} ratio. Excellent agreement can be seen when $x < 0.5$. The discrepancy for $x > 0.5$ is similar to that in the case of $GaAs_{1-x}P_x$ discussed earlier. A possible reason

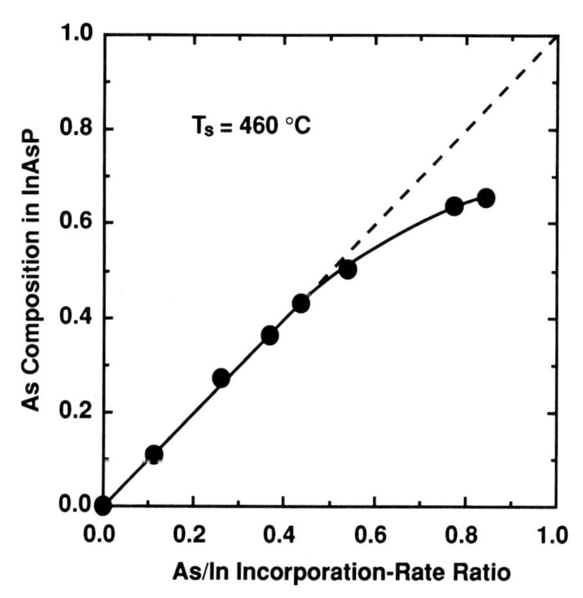

Fig. 12. The As composition in $InAs_xP_{1-x}$ determined from x-ray rocking curve measurements of $InAs_xP_{1-x}$/InP SLSs vs. the incorporation ratio of As to In obtained from RHEED oscillations. (From Ref. 35.)

is that the large surface strain may affect the incorporation behavior of As and P. Another possibility is that partial strain relaxation in InAsP/InP SLS samples may lead to underestimation of the As composition.

Of course, there is an upper limit to the P_2 flux for this simple *in situ* method of composition determination. When the PH_3 flow rate is much higher than the AsH_3 flow rate (by about a factor of 5), more P_2 will be incorporated. As a result, the As composition in InAsP would start to decrease as the P_2 flux increases [35]. Therefore, to keep this simple *in situ* method valid, it is important to use as small a V/III incorporation ratio as possible. Again, note that at higher substrate temperatures desorption of As_2 and P_2 is significant and the incorporation of one species can be influenced easily by the presence of the other. Therefore, the *in situ* method is applicable only for substrate temperatures lower than about 500°C [35].

Thus, to obtain a layer of $InAs_xP_{1-x}$ with small x, it is necessary to use a small flow rate of AsH_3. However, the mass-flow-controller output usually becomes unstable when a small setting is chosen. Therefore, a relatively large PH_3 flow rate, compared with AsH_3, has to be used in order to dilute the arsenic fraction on the growth front, but then the V/III incorporation ratio would become large. An alternative is to use comparable AsH_3 and PH_3 flow rates and pulse the AsH_3 flux, effectively growing an $InAs_yP_{1-y}$/InP short-period superlattice (SPSL) [36]. In this case, because the As incorporation is much greater than that of P, this superlattice is basically an InAs/InP SPSL. Thus, the As composition is determined by the time intervals of growing $InAs_yP_{1-y}$ (essentially only InAs) with respect to InP [36].

Highly strained multiple quantum wells (SMQWs) can be grown by either of these two methods. Figure 13 shows that excitonic absorption at 1.06, 1.3, and 1.55 μm can be obtained [37]. The advantage of using InAsP, as opposed to InGaAsP, is the simpler and independent control of the growth rate and layer thickness (by the In flux only) and composition (by the AsH_3 and PH_3 flow-rate ratio). Woodward *et al.* [38] have fabricated a 1.06-μm modulator from an InAsP/InP SMQW structure grown by CBE. Hou *et al.* [39] have also observed a quantum-confined Stark effect in InAsP/InP SMQWs at 1.3 μm.

In principle, we can apply the same growth kinetics model to InAsP. However, in this case, because of the very narrow temperature range in which the P composition changes, it is difficult to apply the same procedure for GaAsP to InAsP. We have to consider that the sticking coefficient ratio of As_2 and P_2 will depend on temperature. For GaAsP we consider it to be a constant in the temperature range studied, whereas in the case of InAsP, for temperatures above 520°C, As_2 preferentially desorbs, so the sticking

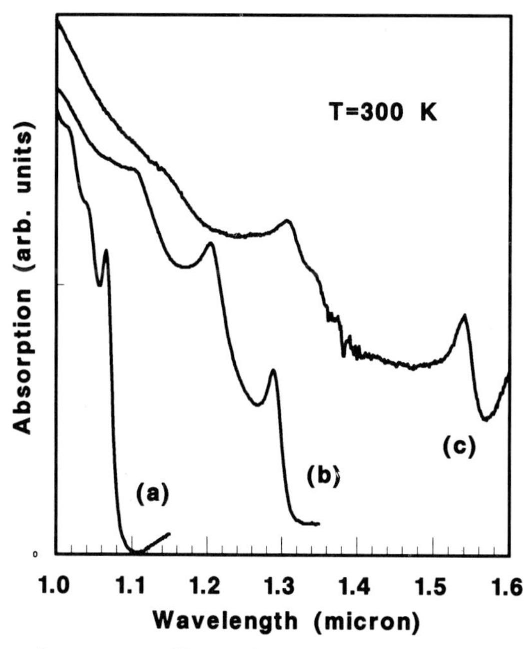

Fig. 13. Absorption spectra of InAsP/InP strained MQW structures at room temperature. First-subband heavy-hole excitonic absorptions are obtained at 1.06, 1.3 and 1.55 mm, respectively. (Reprinted with permission from *Journal of Electronic Materials* Vol. 21, p. 137, 1992, a publication of The Minerals, Metals and Materials Society, Warrendale, Pennsylvania 15068.)

coefficients of As_2 and P_2 are different [30]. Similar behavior in P composition of InGaAsP has been observed by Tappura *et al.* [40].

2.3. Growth of $In_yGa_{1-y}As_xP_{1-x}$ and InGaAsP/InP MQWs

Although the As_2 and P_2 incorporation behavior is different for InAsP and GaAsP growth, we can control the composition in InGaAsP by combining the knowledge of the *in situ* determination for both [41]. The growth rates of Ga and In are calibrated first by measuring group III–induced RHEED oscillations on GaAs and InP substrates, respectively. The Ga composition in $In_{1-y}Ga_yAs_xP_{1-x}$ is then obtained as normally done for $III_{1-y}III_yV$ compounds. Because the As composition in $In_{1-y}Ga_yAs_xP_{1-x}$ derives from As bonding with both In and Ga, it can be estimated individually for InAsP and GaAsP as discussed earlier. It is reasonable to assume that the respective distribution of arsenic flux (F_{As}) to In and Ga is $(1 - y)F_{As}$ and yF_{As}. The arsenic bonds with indium with almost 100% efficiency, independent of the presence of phosphorus at low phosphorus flux, but arsenic bonds with Ga with an efficiency *p*, that depends on the presence of phosphorus. Once *p* is determined for a given substrate temperature and PH_3 flow rate, the arsenic composition in $In_{1-y}Ga_yAs_xP_{1-x}$ can be found from $(1 - y)R_{As}/R_{In} + ypR_{As}/R_{Ga}$. With this *in situ* method, proper composition control of InGaAsP lattice-matched to InP can be easily achieved.

Figure 14 shows a high-resolution x-ray rocking curve of a 20-period $In_{0.7}Ga_{0.3}As_{0.65}P_{0.35}$ (100 Å)/InP (150 Å) undoped MQW, which is only the second growth run with the described method [41]. Very sharp and distinct satellite peaks are observed, suggesting good periodicity of the multilayered structure. The zero-order peak from this MQW is only 38 arc seconds away from that of the InP substrate, correponding to a lattice mismatch of 0.1%. The compositions and layer thicknesses agree well with a simulation based on the dynamical theory. It is notable that some satellite peaks are missing in the rocking curve. This can be attributed to interdiffusion at the InGaAsP/InP interface [42]. Low-temperature PL (~20 K) of this InGaAsP/InP MQW

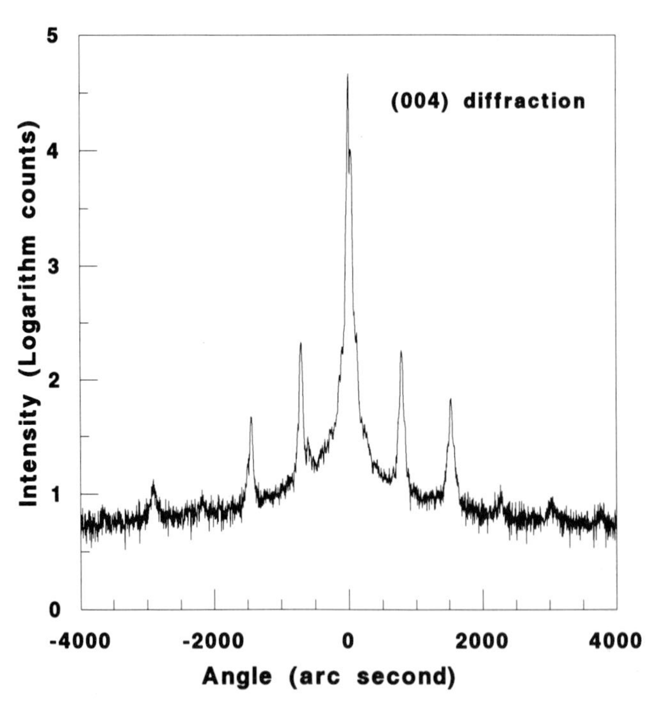

Fig. 14. An x-ray rocking curve taken with the (004) diffraction from a 20-period $In_{0.7}Ga_{0.3}As_{0.65}P_{0.35}$ (101.5 Å)/InP (152 Å) MQW structure. Some satellite peaks are missing because of intermixing at the InGaAsP/InP interface. (From Ref. 41.)

shows very luminescent and sharp exciton peaks, with a full width at half-maximum (FWHM) ranging from 6 to 9 meV [41].

3. MOMBE AND CBE

The next level of complexity in MBE using gaseous sources is MOMBE/CBE. MOMBE and CBE have attracted increasing interest not only for fundamental studies of growth kinetics but also for optoelectronic device applications owing to their potential for low surface defect density, excellent uniformity, high throughput, and selective-area growth. One of the main origins of oval defects in MBE is Ga spitting from Ga droplets condensed near the tip of the Ga crucible or from convection in the Ga melt. By using a gaseous source near room temperature, such as trimethylgallium (TMGa) or triethylgallium (TEGa), the oval defect density can be drastically reduced [43]. Furthermore, by appropriate design of the diffuser plate at the mouth of the alkyl injector, a large-area, very uniform beam flux can be generated. The uniformity in MOMBE/CBE can be comparable to or better than that in MOVPE or MBE with solid group III sources [44]. This uniformity can be achieved even without substrate rotation [45], thus simplying the design of a multiple-wafer reactor. Although carbon incorporation could be an obstacle to achieving high-purity materials, especially in Al-containing compounds, the ability to grow high-quality, highly carbon-doped *p*-type GaAs and InGaAs makes MOMBE/CBE a suitable growth technique for AlGaAs/GaAs and InP/InGaAs heterojunction bipolar transitors (HBTs). Furthermore, with appropriate growth temperatures and flow rates, the gas molecules can decompose on a semiconductor surface but desorb on an insulator surface, such as SiO_2, MOMBE/CBE is ideal for selective-area growth, which will be discussed later. However, because the growth mechanism in MOMBE/CBE would be more complicated than in MBE and GSMBE, more stringent control of substrate temperature and even group V flux is required, as discussed in the following.

3.1. Growth of Binary Compounds

Because the mean free paths of source molecules are very long compared to the source-to-substrate distance in MOMBE/CBE, one can expect that gas-phase reactions are unimportant, unlike MOVPE, but surface reactions would play an important role in determining the growth process. Because the operating pressure in the growth chamber is between 10^{-5} to 10^{-4} torr, one can also use RHEED and other *in situ* analytical techniques, in particular modulated-beam mass spectroscopy (MBMS), to monitor the surface and growth. Tsang *et al.* [46] first studied the behavior of CBE growth of GaAs by RHEED oscillations. Subsequently, a number of groups reported RHEED investigations of growth kinetics. In addition, MBMS and temperature-programmed desorption spectroscopy (TDS) can provide directly additional information on the evolution of chemical species during CBE as a function of various growth parameters [47].

Growth behaviors and kinetic mechanisms of growth of GaAs, InAs, and GaSb by MOMBE or CBE have been studied by several groups [48–51]. The first kinetic model proposed by Robertson *et al.* [48] considers only TEGa decomposition paths. Subsequently, group V species, unlike the situation with MBE or GSMBE, were found to have a strong effect on the growth rate [52–54]. The Robertson model has been extended to explain the effects of arsenic species [49–51]. Figure 15a shows a three-dimensional plot for GaAs, summarizing the dependence of the CBE growth rate on the substrate temperature and surface arsenic concentration [55]. Using MBMS, Martin and Whitehouse [56] showed that the TEGa precursor decomposes readily to form diethyl gallium (DEGa) radicals between room temperature and 350°C, where growth of GaAs is still insignificant. Over the temperature range 350–500°C, GaAs growth commences, as indicated by RHEED, and the rate of desorbing DEGa radicals decreases, as indicated by MBMS. In fact, the desorbing DEGa flux has a minimum value around 500°C, which corresponds to the maximum in the growth rate. Thus, the growth of GaAs corresponds to the decomposition of the DEGa radicals, which confirms the

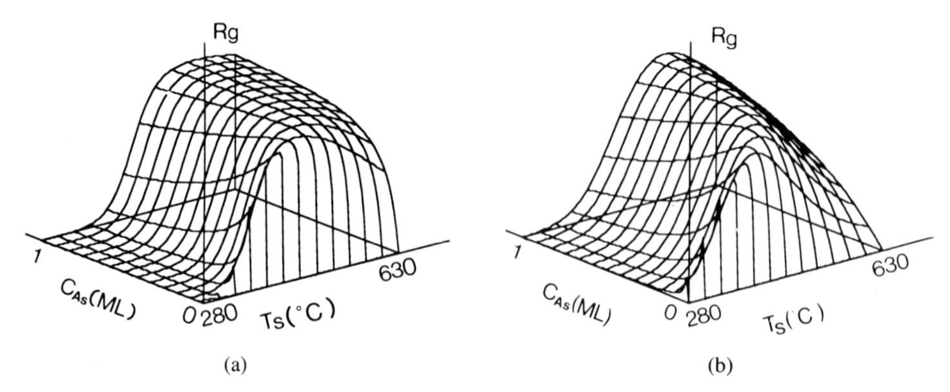

Fig. 15. Growth rates (R_g) of (a) GaAs and (b) InAs grown by MOMBE or CBE as a function of substrate temperature (T_s) and arsenic surface concentration (C_{A_s}) in monolayer (ML) units. (From Ref. 55.)

main assumption of the Robertson model. At higher substrate temperatures, competition between decomposition and desorption of DEGa takes place, resulting in lowering of the growth rate. Above a growth temperature of 660°C, further reduction of growth rate is a result of elemental Ga desorption.

Besides the substrate temperature, the arsenic species, or V/III ratio, also has a significant effect on the growth kinetics. The growth rate of GaAs rises rapidly as the As flux increases from a very small value, changing the growth mode from being limited by As species (V/III incorporation ratio <1) to being limited by the Ga species (V/III incorporation ratio ≥ 1). At a low substrate temperature the growth rate decreases with increasing arsenic flux, unlike the case of MBE or MOVPE, because of blocking of TEGa adsorption sites or enhanced desorption of DEGa, as seen also in the rise of the DEGa desorption signal in MBMS [56]. However, at high substrate temperature and large arsenic flux, the growth rate is independent of arsenic surface concentration, making the CBE process similar to solid-source MBE. Very fast decomposition of TEGa molecules at high substrate temperatures generates many free Ga atoms on the surface, as in solid-source MBE. In this temperature range, however, the surface diffusion length of Ga adatoms is markedly influenced by the presence of the derivatives of TEGa or TMGa molecules, as seen by *in situ* scanning microprobe RHEED (μ-RHEED) [57].

For MOMBE/CBE growth of InAs, control of substrate temperature is even more important because the dependence on temperature of decomposition and desorption of In species is even steeper than that of Ga species, as shown in Fig. 15b [55]. This will make the growth of ternary InGaAs more complicated, as discussed in the following. Note also that the InAs growth rate decreases with arsenic flux.

3.2. Ternary Compounds

A complication in growing ternary compounds is that, due to the nature of chemical reactions on the surface, the growth rate of the binary components in a ternary compound is affected by the presence of other molecular species on the surface, resulting in composition changes. Figure 16 shows the GaAs growth rate and the GaAs component of the growth rate in AlGaAs and InGaAs as a function of substrate temperature [58]. Curve (a) is the result of Chiu *et al.* [52] and curve (b) a result of Kobayashi *et al.* [59], for GaAs growth by CBE and MOMBE, respectively. Curve (c) is the GaAs growth rate in AlGaAs grown with TEGa, Al, and As_4. The decrease in temperature sensitivity in AlGaAs is a result of stronger bonding in AlGaAs than in GaAs. Curve (d) is the GaAs growth rate in InGaAs grown with TMIn,

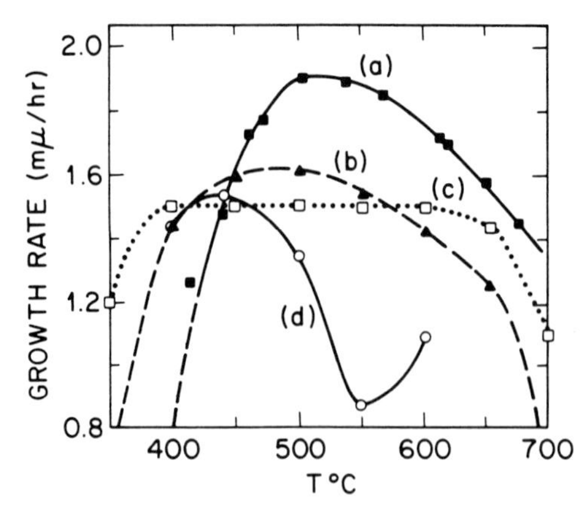

Fig. 16. Growth rate vs. substrate temperature [16]. Curve (a) is for GaAs grown with TEGa. (From Ref. 43.) Curves (b)–(d) are scaled from the plotted data of Kobayashi *et al.* [59]. Curve (b) is for GaAs grown with TEGa and As_4. Curves (c) and (d) are for the GaAs component in AlGaAs grown with Al, TEGa, and As_4 and in InGaAs grown with TEGa, either In or TMIn, and As_4, respectively.

TEGa, and AsH_3. The minimum in the equivalent GaAs growth rate of InGaAs at about 550°C is due to the evaporation of In above that temperature. The strong inhibition of the GaAs growth rate arises from the influence of indium atoms in the interfacial region. Using TMIn, TEIn, or elemental In to grow InGaAs on InP also produces the same growth behavior, confirming that it is indium itself in the growing film that influences the growth rate rather than any alkyl indium or gallium species [60]. Surface spectroscopic studies and MBMS now provide insight into this process. The nature of the inhibition has been identified by comparing the surface scattering of TEGa from GaAs and GaAs dosed with In [61]. On the former surface most of the impinging TEGa is converted to Ga with the production of a limited amount of volatile DEG ($<$10%), whereas on the latter surface the reflected TEGa flux was found to increase to 40% of the incident TEGa. Data from MBMS also shows that the scattered DEGa signal at high temperatures is significantly more intense when indium is present in the lattice [61]. These results demonstrate that the indium on the surface alters the nature of the potential energy surface governing surface scattering in such a way that desorption of DEG becomes much more favored than decomposition.

Furthermore, the transient change in the surface indium segregation is also significant. Iimura *et al.* [62] observed RHEED intensity oscillations during growth of GaAs and InGaAs. A number is assigned for each period of the oscillations. In spite of addition of TMIn to the TEGa flow, the growth rate of InGaAs is actually lower than that of GaAs at the eightth period. This behavior is different from that in MBE or GSMBE, where the group III incorporation rates are additive. Increasing the growth temperature makes the decrease of the InGaAs growth rate more pronounced, as shown in Fig. 17. Therefore, the transient behavior of the InGaAs growth rate results from increasing indium segregation at the surface.

Because of these complications due to surface reactions, it is not easy to achieve *in situ* composition determination as in GSMBE. Nevertheless, for growing ternary and quaternary compounds, trial-and-error calibration for the proper conditions is used. Figure 18 shows the variation of $In_xGa_{1-x}As$ lattice mismatch, determined by x-ray diffraction, as a function of substrate temperature for compositions close to lattice match with InP (about x = 0.53). The positive mismatch here means In-rich (or Ga-deficient) InGaAs. The small plateau region in substrate temperature between 490 and 515°C indicates that the growth of lattice-matched InGaAs on InP can be achieved only in a narrow temperature window. The composition varies rapidly with temperature outside this temperature regime, underlining once more the importance of controlling the substrate temperature.

One should note here that high-quality materials can be achieved by CBE. High-purity InP has been reported, with background impurities as low as 2

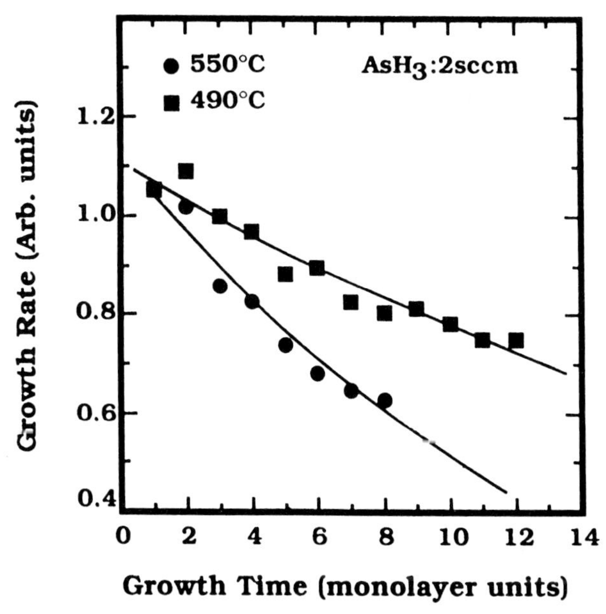

Fig. 17. Transient behavior of the InGaAs growth rate, which decreases faster at a higher growth temperature. (From Ref. 62.)

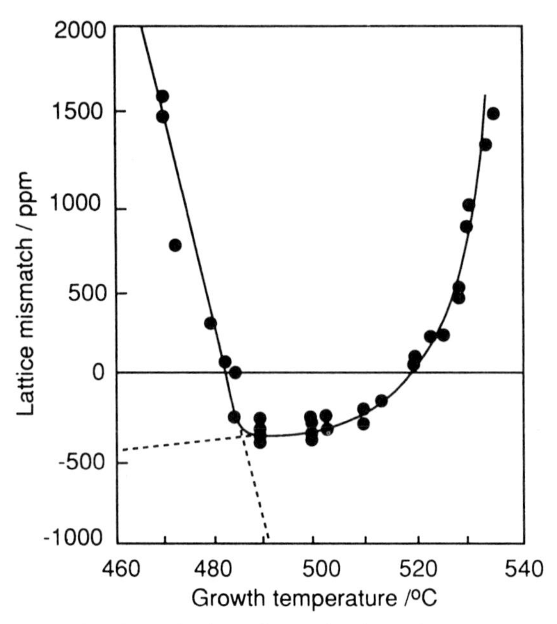

Fig. 18. Variation of InGaAs lattice mismatch with substrate temperature for compositions close to lattice match with InP. A 100-ppm change in mismatch is equivalent to a 0.14% change in In content. (From Ref. 60.)

$\times\ 10^{14}$ cm^{-3} and 77 K Hall mobilities as high as 132,000 cm^2/V-s [63]. The combined 4 K PL linewidth for donor-bound exciton recombinations can be as narrow as <0.1 meV [63]. Very high quality $In_{0.53}Ga_{0.47}As$ has also been grown by CBE, with a 2 K PL linewidth of 1.2 meV [64]. Intense PL peaks were also observed for InGaAsP, lattice-matched to InP, with a PL linewidth of about 5 meV [41]. Quantum wells of InGaAs/InP grown by CBE also exhibit linewidths comparable to or narrower than those of quantum wells grown by MOVPE or GSMBE, e.g., 5 meV for a 4-nm-wide single quantum well [65].

3.3. Atomic Layer Epitaxy

It is important here to point out some unique advantages of MOMBE/CBE, despite the problem of controlling the substrate temperature. They are atomic layer epitaxy (ALE), carbon-doped *p*-type GaAs, and selective-area growth. The last topic will be discussed in Section 4.

Atomic layer epitaxy, in which monolayers of group III and group V atoms are alternately deposited, has the potential for the ultimate control of thickness and composition uniformity. Ideally, a limiting mechanism is required so that the adsorption of Ga or As_2 will saturate at exactly one monolayer coverage. For As_2, the sticking coefficient at typical growth temperatures decreases from unity to being negligible when the surface changes from a Ga- to an As-saturated state. For Ga, this saturation behavior can be achieved only with molecular gallium species, such as TMGa [66,67] and diethyl-GaCl (DEGaCl) [68]. Slow growth rate is inevitable in this process, however. Two mechanisms have been proposed for ALE. One is that when a monolayer of TMGa has been adsorbed on a GaAs surface, no more TMGa can be adsorbed. The other is that a monolayer of Ga metal is formed on the surface by the decomposition of TMGa, and additional TMGa subsequently supplied cannot be adsorbed on the Ga-covered surface. Although there is still controversy about the two mechanisms, ALE has been actively pursued in MOVPE and VPE [69]. The condition for ALE in MOMBE/CBE can be studied easily by monitoring the RHEED intensity under different growth conditions [70, 71].

In Fig. 19a the deposition time of TMGa in the absence of an As_2 flux is sufficient for the growth of 6 and 12 monolayers of GaAs, but only one monolayer of GaAs is obtained as seen from the As-induced oscillation. When TEGa is used for a similar ALE experiment, shown in Fig. 19b, growth of several monolayers of GaAs is observed after the As_2 supply is resumed. These results confirm that ALE can be achieved more readily with TMGa.

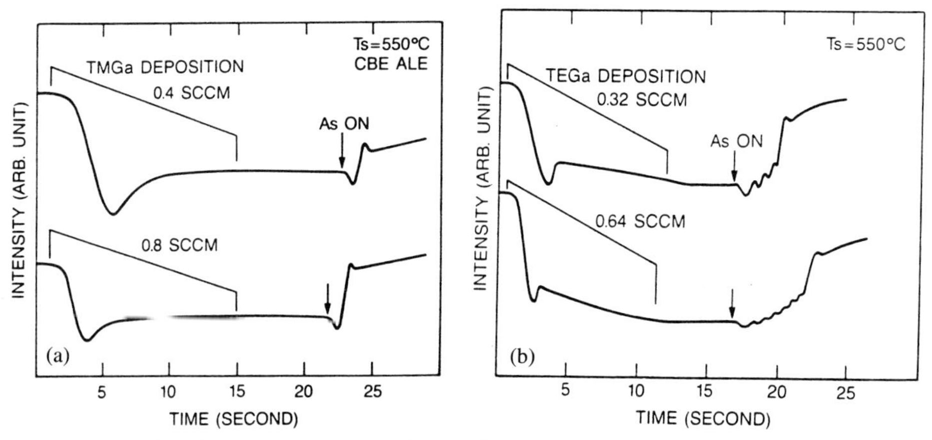

Fig. 19. RHEED intensity scans for (a) TMG and (b) TEGa on GaAs. The time interval in which TMG and TEGa deposition occurs is marked. (From Ref. 70.)

Subsequent experiments by Chiu *et al.* [72] indicate that the limiting mechanism exists only for a short period of time at 550°C.

3.4. Carbon-Doped *p*-type GaAs and InGaAs

Despite the problem of carbon incorporation in Al-containing compounds grown by MOMBE/CBE, carbon doping using TMGa for GaAs epilayers is of great importance. Electrically active doping concentrations as high as 10^{21} cm^{-3} have been obtained [73]. The interest in highly carbon-doped *p*-type doping is due to the fact that the diffusivity of carbon is at least one order of magnitude lower than that of the commonly used beryllium [74], which is important for GaAs/AlGaAs HBTs. Using trimethylamine alane, TEGa and AsH_3, along with TMGa for C doping and TESn for *n*-type doping, Abernathy and co-workers [75] reported very-high-performance *Pnp* and *Npn* HBT devices [75] and first CBE integrated circuits (decision circuits and laser drivers).

For InGaAs, carbon can become an *n*-type dopant owing to the amphoteric nature of C in III–V compounds. $In_{0.53}Ga_{0.47}As$ layers (lattice-matched to InP) grown by CBE with organometallic sources and AsH_3 all show *n*-type conductivity, whereas those grown by MOMBE with TMGa, In, and As_4 show *p*-type conductivity up to 5.5×10^{18} cm^{-3} [76]. de Lyon *et al.* [77] reported higher carbon doping efficiency in GaAs with CBr_4, $CHBr_3$, or CCl_4 than with TMGa. For InGaAs, using GSMBE and MOMBE (TEGa) with CCl_4, Chin *et al.* [78] reported a hole concentration as high as 5×10^{19} cm^{-3}. For the same doping level by CBE, using TMGa, much lower substrate temperature (~380°C) is required, which results in low growth rate (~0.2 μm/hr) [79].

4. SELECTIVE-AREA GROWTH

Most of the work on OEIC carried out thus far has relied on the growth of a complete epitaxial structure containing, for example, both the transistor and laser layers, followed by processing. An alternative is selective-area growth (SAG) and regrowth combined with either *in situ* or *ex situ* processing. For example, Berger *et al.* [80] have demonstrated an MBE-grown GaAs HBT integrated with a quantum well laser; Gailhanou *et al.* [81] used SAG by GSMBE to fabricate planar buried heterostructure laser diodes; and Hamm *et al.* reported HBTs made from SAG by CBE [82]. Four cases are possible for SAG by MBE/GSMBE and MOMBE/CBE: (1) growth through a mechanical mask, (2) growth on etched wells in the substrate through a self-aligned semiconductor mask, (3) growth on insulator-patterned substrates, and (4) growth enhanced locally by an external energy source. The beam nature of MBE allows the first two methods easily. Refractory metal (W, Ta, or Mo) [83] and Si masks [84] can be interposed in front of the substrate. For self-aligned masking, selective etches and photolithographic techniques are used to fabricate undercut mesa structures on a substrate. When a layer is subsequently grown on such a substrate, the regions under the overhanging parts of the mesa are masked from the beams, and no deposit is formed there. Growth on etched wells has been investigated extensively for MBE [85, 86], GSMBE [87], and MOMBE/CBE and MOVPE [88]. A challenging task is the control of the cross-sectional shape of such structures. For this reason the growth profiles at the edges of the deposits have received considerable attention. The growth rates are different for different crystallographic orientations, and the selectively grown areas are bounded by well-defined crystallographic planes [86]. This behavior offers the possibility to fabricate buried heterostructures in a single epitaxial step [84]. Under appropriate conditions, etched wells can be planarized in MOVPE and even in GSMBE of InP [87]. In this section we concentrate on the last two techniques, which are unique to epitaxial techniques using gaseous sources, i.e., MOMBE/CBE and MOVPE.

4.1. SAG on an Insulator-Patterned Substrate

Selective-area growth, in which the epitaxial layers grow only on an unmasked area in a substrate partially covered with an insulator, has been investigated for MOMBE/CBE and MOVPE [88]. These insulators include SiO_2 and SiN_x, patterned *ex situ* by conventional photolithographic techniques, and SiO_2 [89], and GaAs or InP oxide [90], patterned in the same vacuum system by a combination of focused ion beam lithography and gaseous Cl_2 etching. The latter method of *in situ* patterning and overgrowth is discussed in Chapter 6 of this book. In this section we compare the conventional approach for MOMBE/CBE and MOVPE.

In order to achieve good selectivity, impinging group III atoms or molecules, with a sufficiently low flux to avoid formation of a stable adsorbed phase on the insulator surface, should adsorb on the opening but desorb completely from the insulator surface. Once the metal atoms are adsorbed, they do not readily desorb, unless at a much higher temperature. Thus, high substrate temperatures promote selectivity because desorption of the adsorbed group III species prevails over their decomposition. In conventional

MBE, where elemental Ga and As_4 are used, SAG without polycrystalline GaAs deposition on SiO_2 can be obtained above 700°C [91]. In MOVPE the SAG temperature is in the range 600–670°C and that for MOMBE/CBE is in the range 490–540°C [92, 93].

Selective-area growth is not determined solely by the group III species, however. Kayser [92] compared selective growth of InP/GaInAs in low-pressure MOVPE and CBE. The striking difference is that SAG for MOVPE is independent of V/III ratios, whereas SAG for MOMBE/CBE decreases with increasing V/III ratios. The reason is as follows. In CBE, P_2 is produced by cracking PH_3 in a high-temperature cell, whereas in MOVPE the production of the elemental groupV species is promoted *selectively* on the semiconductor surface, not on the insulator surface. Because the presence of phosphorus species enhances the deposition of In and Ga, in CBE the increasing presence of the elemental group V species enhances the reaction with the group III species on the insulator surface. Thus, the selectivity in CBE decreases with the V/III ratio.

Selective-area growth of ternary InGaAs is possible, but it is more difficult because SAG of both InAs and GaAs is required simultaneously [92]. For InGaAs, nucleation on SiO_2 is more sensitive to an increase of the TMIn pressure than to that of the more stable TMGa.

For the conditions under which growth on a dielectric mask is inhibited, it is important to know the interfaces between the selectively grown and insulator areas and the extent to which material rejected by the mask contributes to the growth on the adjacent unmasked semiconductor. On a GaAs or InP(100) surface, the side-wall growth is smooth when the stripe is aligned in the $[\bar{1}10]$ direction and rough when aligned along the [110] direction [94]. In the latter case, the interfaces are smooth with a clearly defined crystallographic plane, believed to be [111]B (As or P terminated) [94]. For MOVPE and MOMBE/CBE, materials can have a pile-up at the interfaces. This pile-up depends on the V/III ratios and can be eliminated in MOMBE/CBE, but not so easily in MOVPE. A schematic diagram for SAG by MOVPE and MOMBE/CBE is shown in Fig. 20 [92]. Kayser shows that at low V/III ratios the reactants leave the mask by reevaporation in MOMBE/CBE, but surface migration of the group III species over the mask in MOVPE results in accelerated growth in openings near the mask. Cotta *et al.* [89] also reported that in CBE of InGaAs with a thin (~5 nm) oxidized Si mask on InP, the precursor material is transferred from the slow-growing {111} planes to the (100) plane and migration of species from the Si mask to the growing areas is negligible. MOMBE/CBE, therefore, exhibits the best selectivity [92].

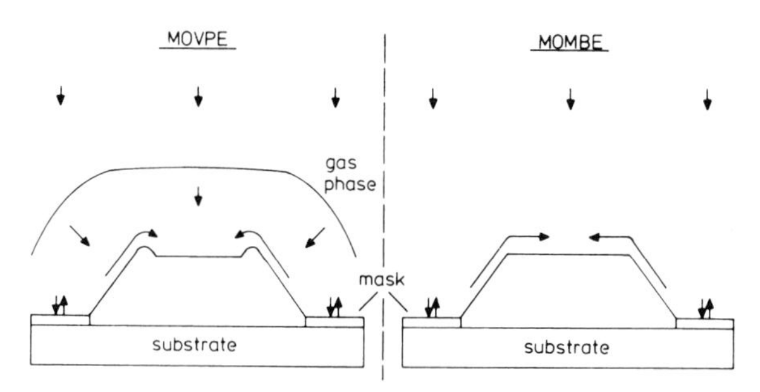

Fig. 20. Schematic diagram of SAG in (a) MOVPE and (b) MOMBE/CBE. (From Ref. 92.)

For lateral integration of devices, it is desirable to eliminate the often observed {111} planes at the side walls of the growth structures. Heinecke [95] found that the sticking coefficient of phosphorus on exactly oriented (100) wafers is substantially lower than on misoriented substrates. By using slightly misoriented substrates [e.g., 2° off toward (110)] they achieved SAG of planar InGaAsP/InP structures having nearly perfectly vertical side walls even for 2-μm-thick layers or narrow stripes.

4.2. Photon-Assisted SAG

Laser-assisted MOVPE is an active area of research because it can extend the growth window for ALE [96] and can be applied to quantum-well lasers [97]. Laser-assisted MOMBE/CBE can also be used to achieve SAG and localized compositiion change in ternary or quaternary compounds for device applications. Donnelly *et al.* [98] reported laser-assisted MOMBE using a 193-nm ArF excimer laser. GaAs is grown from TEGa and As_4 at low temperature, where growth due to pyrolytic decomposition of TEGa on the substrate surface is very slow. The growth rate enhancement in the center of the film, in the region of highest laser intensity, is shown to be near unity down to a substrate temperature of 200°C. Laser-enhanced MOMBE, with an argon ion laser, occurs only for substrate temperatures above 350–400°C [99, 100]. For both cases, above about 450°C, the growth of GaAs resulting from pyrolysis of TEGa occurs with maximum efficiency, hence laser enhancement does not occur. The range of temperature indicates the difficulty in measuring the substrate temperature in MBE, especially at the low range.

Laser-induced decomposition of adsorbed TEGa could occur by a photolytic process (photodissociate TEGa), an electronic process (creat electron–hole pairs in GaAs, which aid in TEGa decomposition), or a pyrolytic process (thermally decompose TEGa with transient heating of GaAs by the pulsed laser). In the case of excimer laser irradiation, the strong dependence of the deposition rate on fluence precludes both the photolytic and electronic mechanisms. Therefore, the enhanced pyrolysis of adsorbed TEGa molecules is due to transient heating of the GaAs surface by pulsed excimer laser irradiation. Based on a numerical solution of the one-dimensional heat flow equation, an estimated peak temperature rise is about 700°C at a fluence of 0.1 J cm^{-2} or about 10^{16} photons cm^{-2}. Subsequently, Donnelly and coworkers [101] confirmed the pyrolytic nature of the excimer laser–enhanced growth at large laser fluence (but photolytic nature at low laser fluence) by studying laser-assisted decomposition of TEGa molecules on GaAs surfaces for different wavelengths (193 nm versus 351 nm of XeCl).

With an Ar ion laser, the temperature rise is small, about 25°C. In addition, enhanced growth is observed for both GaAs and GaP [102], the latter being transparent to argon ion laser irradiation. Therefore, the enhanced growth for an argon ion laser is due mainly to a photolytic effect. A small portion of the TEGa molecules may be fully decomposed and most of the molecules may be partly decomposed. These partially decomposed TEGa molecules may absorb the light of the Ar ion laser. Dong *et al.* [100] also find that desorption of arsenic under laser irradiation plays a role in the growth process. Nagata *et al.* [103], monitoring the RHEED intensity behavior on a Ga-terminated and an As-terminated surface, conclude that the laser irradiation has greater effect on an As-terminated surface, compared with a Ga-terminated surface. In addition, laser irradiation enhances surface migration of TEGa molecules; the RHEED oscillations, after TEGa is turned on, are more pronounced with laser irradiation [100, 103]. Laser irradiation also results in higher photoluminescence [99] or cathodoluminescence intensity [104]. In addition, Iga *et al.* [104] found that Ar^+-laser irradiation resulted in reduced carbon incorporation.

Perhaps the most promising application of laser-assisted CBE is in localized composition change for devices. For InGaAs grown by CBE at a normal substrate temperature, Iga *et al.* [105] reported a decrease in the growth rate in an area irradiated by an Ar ion laser due to local heating that reduces the Ga content in InGaAs. Reduction of Ga content predicts *in situ* growth of a smaller-bandgap material surrounded by larger-bandgap material. The same group has succeeded in fabricating a double-wavelength laser array with the InGaAsP/InGaAsP multiple-quantum-well active region grown by Ar^+ laser–assisted CBE [106]. The rest of the structure was grown by conventional CBE. Figure 21 shows a schematic diagram of the fabrication method for the separate confinement heterostructure (SCH) laser array. The laser diodes with irradiated MQW active layers operate at 1.40 μm, and

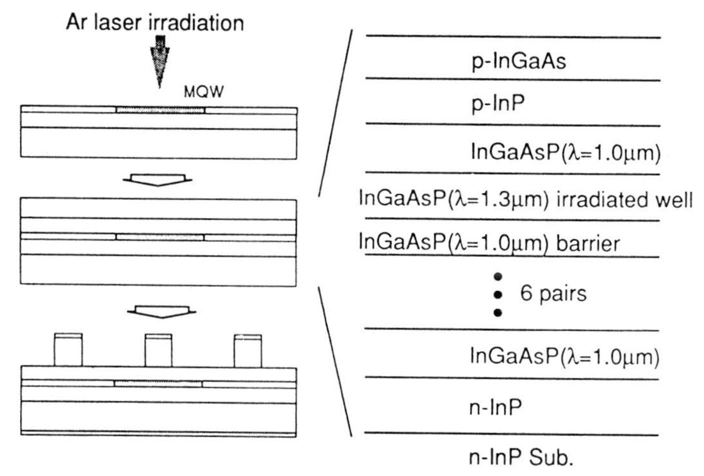

Fig. 21. Schematic diagram of the fabrication process in laser-assisted CBE with the dual-wavelength laser array. Only the MQW active region is irradiated with an Ar ion laser, as indicated by the shaded area. (From Ref. 106.)

those with nonirradiated MQW active layers operate at 1.28 μm. The threshold current of the former is actually lower than those of the latter, probably because the irradiated samples are strained.

5. ALTERNATIVE SOURCES

Until recently, the sources used in GSMBE and MOMBE/CBE have been the same as those used in MOVPE, such as TMGa, TEGa, TMIn, TMAl, AsH_3, and PH_3. However, because the operation and growth mechanisms for MOMBE/CBE and MOVPE are not the same, different gaseous sources may be optimal for each of these techniques [107, 108]. For example, in MOVPE the precursor molecules undergo many collisions before reaching the substrate, so the ambient may be important in considering pyrolysis pathways of the precursors [107]. On the other hand, in CBE the precursor molecules travel from the inlet to the substrate without collision. If the vapor pressure of a source is very low, that source may not be attractive for MOVPE but may be suitable for CBE. In addition, the group V precursors are not cracked in MOVPE but are cracked in CBE.

To reduce the danger of the highly toxic and high-pressure AsH_3 and PH_3, alternative group V sources, especially in liquid form, have been actively pursued. The hazard associated with the use of a liquid is much lower because of the slower dispersion in the atmosphere in case of an accident.

Tertiary butylarsine (TBAs) and tertiary butylphosphine (TBP) are the most promising for MOVPE, and they are also suitable for CBE [109]. Other alternatives for CBE are phenylarsine (PhAs) [110], monoethylarsine (EAs) [111], and dimethylaminoarsenic (DMAAs) [112, 113]. TBAs, PhAs, and EAs are similar in that only a single H on AsH_3 is replaced by an organic radical. The radicals are stable, so they should contribute little to carbon contamination. Dimethylaminoarsenic, like alane (discussed in the following), has no As—C bond but has As—N bonds, so it should result in even lower carbon incorporation.

For MOMBE/CBE carbon incorporation may be a serious problem for Al-containing compounds but not for Al-free compounds. To minimize carbon incorporation, compounds with less stable bonding of the hydrocarbon ligands or compounds without carbon bonds are desirable. The first alternative to TMAl or TEAl is triisobutylaluminum (TIBAl) [114]. The butyl alkyl, as a high-molecular-weight hydrocarbon alkyl, reduces the Al—C bonding strength compared with TEAl and TMAl. For TIBAl a reduction of the hole concentration in GaAlAs layers from 5×10^{17} to $1\text{–}2 \times 10^{17}$ cm^{-3} was achieved [114]. However, the low bonding strength, which is preferred with respect to the carbon uptake, may cause problems with the decomposition of TIBAl already in the storage container. The ideal compounds should be free of carbon bonds. Group III hydrides, however, are unstable, but forming an adduct can eliminate this problem. Trimethylamine alane, TMAAl ($AlH_3N(CH_3)_3$) has been applied successfully for CBE of AlGaAs, whose carbon level decreases from 3×10^{18} cm^{-3} (obtained when using TEAl and TEGa) to a value below the detection limit of the SIMS measurement ($\sim 2\text{–}3 \times 10^{17}$ cm^{-3}) [115]. The Ga analog, however, is too unstable to be useful at present. Kamp *et al.* [116] reported that TIBAl, rather than TMAAl, produces a higher PL response in AlGaAs with a carrier concentration below $p = 10^{16}$ cm^{-3}.

Although the question of which source materials are best suited for the growth of AlGaAs is still open, TEGa and TMIn are the standard organometallic precursors used for depositing high-quality InP and InGaAsP. Carbon incorporation is not a serious problem in In-containing and Al-free compounds.

Using elemental Si and Be as dopants in MOMBE/CBE can result in formation of thermally stable carbides, which reduce the doping efficiency. Therefore, gaseous dopant sources are desirable. Disilane [117], diethylzinc, triethylzinc, diethyltellurium, and others have been investigated [118]. An all-gaseous doping CBE system is particularly attractive for system manufacture and device production applications because, in principle, injection of dopants with the organometallic group III precursors automatically leads to uniform dopant distribution across the entire wafer (provided the substrate temperature is uniform, of course). With gaseous doping sources, laser-induced selective-area doping becomes possible [117]. The doping efficiency is found to increase by a factor of 2.5 with ArF excimer laser irradiation at 50 mJ/pulse.

6. CONCLUDING REMARKS

Although the techniques of MBE with gaseous sources are relatively recent developments with less accumulated effort than for solid-source MBE and MOVPE, they have produced excellent device-quality materials. Compared with MBE, gas-source MBE, with elemental group III and gaseous group V sources, has the advantage of phosphorus-handling capability, which is important for short-wavelength InGaAlP/InGaP and long-wavelength InGaAsP/InP optoelectronic applications. Compared with MOVPE, it has the advantages of a simple growth mechanism and *in situ* monitoring and control of the group V compositions by RHEED intensity oscillations. On the other hand, CBE with all-gaseous sources, although more complicated and costly to set up, offers the potential of low surface-defect density, excellent uniformity without substrate rotation, high throughput, and selective-area growth for novel device applications. With precursors that are more suitable for MOMBE/CBE, the problem of carbon incorporation has been steadily minimized, especially for Al-containing compounds. Because of low carbon incorporation in Al-free compounds, the major accomplishment of CBE is in the InGaAsP/InP system for long-wavelength optoelectronic devices and integrated circuits. As the demand on layer structures becomes more complex and stringent for advanced applications, GSMBE and MOMBE/CBE will become more important for OEICs.

ACKNOWLEDGMENTS

I would like to acknowledge my students T.P. Chin, H.K. Dong, H.Q. Hou, and B.W. Liang for many of the results presented here; various publishers and authors for permission to use their figures; and the Office of Naval Research, the Air Force Wright Laboratories, and the NSF I/UCRC on Integrated Circuits and Systems (ICAS) for support of the work at UCSD mentioned in this chapter.

REFERENCES

1. For a review of the early development of MBE, see A. Y. Cho and J. R. Arthur, *Prog. Solid-State Chem.* **10**, 157 (1975).
2. The proceedings of the US MBE Workshops are published in the March–April issues of *J. Vac. Sci. Technol. B* (1983–present, except 1985 and 1991).
3. The proceedings of the International Conference on MBE are published in *J. Cryst. Growth* **81** (1987); **95** (1989); **111** (1991); **127** (1993).
4. The proceedings of the International Conference on CBE and Related Growth Techniques are published in *J. Cryst. Growth* **105** (1990); **120** (1992); **136** (1994).
5. The proceedings of the International Conference on MOVPE, which recently includes a MOMBE/CBE Workshop, are published in *J. Cryst. Growth* **107** (1991).
6. B. R. Pamplin, ed., "Molecular Beam Epitaxy." Pergamon Press, Oxford, 1980.
7. K. Ploog, ed., "Molecular Beam Epitaxy of III–V Compounds: A Comprehensive Bibliography, 1958–1983." Springer-Verlag, Berlin, 1984.
8. L. L. Chang and K. Ploog, eds. "Molecular Beam Epitaxy and Heterostructures." Nijhoff, Dordrecht, 1985.
9. E. H. C. Parker, ed., "The Technology and Physics of Molecular Beam Epitaxy." Plenum Publishing, New York, 1985.
10. M. A. Herman and H. Sitter, "Molecular Beam Epitaxy: Fundamentals and Current Status." Springer-Verlag, Berlin, 1989.
11. M. B. Panish and H. Temkin, "Gas-Source Molecular Beam Epitaxy." Springer-Verlag, Berlin, 1993.
12. G. J. Davies, P. J. Skevinton, E. G. Scott, C. L. French, and J. S. Foord, *J. Cryst. Growth* **107**, 999 (1991).
13. W. T. Tsang, *J. Cryst. Growth* **81**, 261 (1987).
14. M. B. Panish, *J. Electrochem. Soc.* **127**, 2729 (1980).
15. A. R. Calawa, *Appl. Phys. Lett.* **38**, 701 (1981).
16. M. B. Panish and H. Temkin, *Annu. Rev. Mater. Sci.* **19**, 209 (1989).
17. D. L. Miller, S. S. Bose, and G. J. Sullivan, *J. Vac. Sci. Technol. B* **8**, 311 (1990)
18. G. W. Wicks, M. W. Koch, J. A. Varriano, F. G. Johnson, C. R. Wie, H. M. Kim, and P. Colombo, *Appl. Phys. Lett.* **59**, 342 (1991).
19. E. Veuhoff, W. Pletschen, P. Balk, and H. Luth, *J. Cryst. Growth* **55**, 30 (1981).
20. W. T. Tsang, *Appl. Phys. Lett.* **45**, 1234 (1984).
21. L. M. Fraas, E. Malocsay, V. Sundaram, R. W. Baird, B. Y. Mao, and G. Y. Lee, *J. Cryst. Growth* **105**, 35 (1990).
22. B. A. Joyce, P. J. Dobson, J. H. Neave, K. Woodbridge, J. Zhang, P. K. Larsen, and B. Boelger, *Surf. Sci.* **168**, 423 (1986).
23. B. F. Lewis and R. Fernandez, *J. Vac. Sci. Technol. B* **4**, 560 (1986).

24. R. Chow and R. Fernandez, *Mater. Res. Soc. Symp. Proc.* **145**, 13 (1989).
25. T. P. Chin, B. W. Liang, H. Q. Hou, M. C. Ho, C. E. Chang, and C. W. Tu, *Appl. Phys. Lett.* **58**, 254 (1991).
26. H. Rothfritz, G. Tränkle, R. Müller, F. Herrmann, and G. Weiman, *J. Cryst. Growth* **120**, 130 (1992).
27. M. Lambert, A. Perales, R. Vergnaud, and C. Stark, *J. Cryst. Growth* **105**, 97 (1990).
28. H. Q. Hou, B. W. Liang, T. P. Chin, and C. W. Tu, *Appl. Phys. Lett.* **59**, 292 (1991).
29. B. W. Liang and C. W. Tu, *J. Appl. Phys.* **74**, 255 (1993).
30. C. W. Tu, B. W. Liang, and H. Q. Hou, *J. Cryst. Growth* **127**, 251 (1993).
31. H. Q. Hou, B. W. Liang, M. C. Ho, T. P. Chin, and C. W. Tu, *J. Vac. Sci. Technol. B* **10**, 953 (1992).
32. B. W. Liang and C. W. Tu, *J. Cryst. Growth* **128**, 538 (1993).
33. B. W. Liang and C. W. Tu, *J. Appl. Phys.* **72**, 2806 (1993).
34. L. Samuelson, P. Omling, and G. Grimmeiss, *J. Cryst. Growth* **61**, 425 (1983).
35. H. Q. Hou and C. W. Tu, *Appl. Phys. Lett.* **60**, 1872 (1992).
36. H. Q. Hou, C. W. Tu and S. N. G. Chu, *Appl. Phys. Lett.* **58**, 2954 (1991).
37. H. Q. Hou, T. P. Chin, B. W. Liang, and C.W. Tu, *J. Electronic Mater.* **21**, 137 (1992).
38. T. K. Woodward, T. Sizer, and T. H. Chiu, *Appl. Phys. Lett.* **58**, 1366 (1991).
39. H. Q. Hou, A. N. Cheng, H. H. Wieder, W. S. C. Chang, and C. W. Tu, *Appl. Phys. Lett.* **63**, 1833 (1993).
40. K. Tappura, T. Hakkarainen, K. Rakennus, M. Hovinen, and M. Pessa, *J. Cryst. Growth* **112**, 27 (1991).
41. H. Q. Hou and C. W. Tu, *J. Cryst. Growth* **120**, 167 (1992)
42. J. M. Vandenberg, M. B. Panish, R. A. Hamm, and H. Temkin, *Appl. Phys. Lett.* **56**, 910 (1990).
43. W. T. Tsang, *Appl. Phys. Lett.* **46**, 1086 (1985).
44. H. Heinecke, B. Baur, N. Emeis, and M. Schier, *J. Cryst. Growth* **120**, 144 (1992).
45. W. T. Tsang, *J. Appl. Phys.* **58**, 1415 (1985).
46. W. T. Tsang, T. H. Chin, J. E. Cunningham, and A. Robertson, *Appl. Phys. Lett.* **50**, 1376 (1987).
47. T. Martin, C. R. Whitehouse, and P. A. Lane, *J. Cryst. Growth* **107**, 969 (1991).
48. A. Robertson, T. H. Chiu, W. T. Tsang, and J. E. Cunningham, *J. Appl. Phys.* **64**, 877 (1988).
49. B. W. Liang and C. W. Tu, *Appl. Phys. Lett.* **57**, 689 (1990).
50. T. Kaneko, H. Asahi, and S. Gonda, *J. Cryst. Growth* **120**, 39 (1992).
51. C. L. French and J. S. Foord, *J. Cryst Growth* **120**, 63 (1992).

52. T. H. Chiu, J. E. Cunningham, and A. Robertson, *J. Cryst. Growth* **95**, 136 (1988).
53. C. W. Tu, B. W. Liang, and T. P. Chin, *J. Cryst. Growth* **105**, 195 (1990).
54. T. Kaneko, H. Asahi, Y. Okuno, T. W. Kang, and S. Gonda, *J. Cryst. Growth* **105**, 69 (1990).
55. B. W. Liang and C. W. Tu, *J. Cryst. Growth* **111**, 550 (1991).
56. T. Martin and C. R. Whitehouse, *J. Cryst. Growth* **105**, 57 (1990).
57. T. Isu, M. Hata, Y. Morishita, Y. Nomura, S. Goto, and Y. Katayama, *J. Cryst. Growth* **120**, 45 (1992).
58. M. B. Pannish and H. Temkin, *Annu. Rev. Mater. Sci.* **19**, 209 (1989).
59. N. Kobayashi, L. J. Benchimol, F. Alexandre, and Y. Gao, *Appl. Phys. Lett.* **51**, 1907 (1987).
60. N. K. Singh, J. S. Foord, P. J. Skevington, and G. J. Davies, *J. Cryst. Growth* **120**, 33 (1992).
61. E. T. Fitzgerald and J. S. Foord, *Surf. Sci.* **278**, 121 (1992).
62. Y. Iimura, K. Nagata, Y. Aoyagi, and S. Namba, *J. Cryst. Growth* **105**, 230 (1990).
63. H. Heinecke, B. Baur, R. Hoger, and A. Miklis, *J. Cryst. Growth* **105**, 143 (1990).
64. E. F. Schibert and W. T. Tsang, *Phys. Rev. B* **34**, 2991 (1986).
65. W. T. Tsang and E. F. Schubert, *Appl. Phys. Lett.* **49**, 220 (1986).
66. S. M. Bedair, M. A. Tischler, and T. Katsuyama, *Appl. Phys. Lett.* **47**, 51 (1985).
67. M. Yu, U. Memmert, and T. F. Kuech, *Appl. Phys. Lett.* **55**, 1011 (1989).
68. K. Mori, M. Yoshida, A. Usui, and H. Terao, *Appl. Phys. Lett.* **52**, 27 (1988).
69. A. Usui, *Proc. IEEE* **80**, 1641 (1992).
70. T. H. Chiu, W. T. Tsang, J. E. Cunningham, and A. Robertson, *J. Appl. Phys.* **62**, 2302 (1987).
71. B. W. Liang, T. P. Chin, and C. W. Tu, *J. Appl. Phys.* **67**, 4393 (1990).
72. T.H. Chiu, J. E. Cunningham, A. Robertson, and D. L. Malm, *J. Cryst. Growth* **105**, 155 (1990).
73. M. Konagai, T. Yamada, T. Akatsuka, S. Nozaki, R. Miyake, K. Saito, T. Fukamachi, E. Tokumitsu, and K. Takahashi, *J. Cryst. Growth* **105**, 359 (1990).
74. B. T. Cunningham, L. J. Guido, J. E. Baker, J. S. Major, Jr., N. Holonyak, Jr., and G. E. Stillman, *Appl. Phys. Lett.* **55**, 687 (1989).
75. C. R. Abernathy, F. Ren, S. J. Pearton, T. R. Fullowan, R. K. Montgomery, P. W. Wisk, J. R. Lothian, P. R. Smith, and R. N. Nottenburg, *J. Cryst. Growth* **120**, 234 (1992).
76. E. Tokumitsu, J. Shirakashi, M. Qi, T. Yamada, S. Nozaki, M. Konagai, and K. Takahashi, *J. Cryst. Growth* **120**, 301 (1992).
77. T. J. de Lyon, N. I. Buchan, P. D. Kirchner, J. M. Woodall, G. J. Scilla, and F. Cardone, *Appl. Phys. Lett.* **58**, 517 (1991).

78. T. P. Chin, P. D. Kirchner, J. M. Woodall, and C. W. Tu, *Appl. Phys. Lett.* **59**, 2865 (1991).
79. J. I. Shirakashi, A. Miyano, R. T. Yoshioka, M. Konagai, and K. Takahashi, *J. Cryst. Growth* (1993).
80. P. R. Berger, N. K. Dutta, D. L. Sivco, and A. Y. Cho, *Appl. Phys. Lett.* **59**, 2826 (1991).
81. M. Gailhanou, C. Labourie, J. L. Lievin, A. Perales, M. Lambert, F. Poingt, and D. Sigogne, *Appl. Phys. Lett.* **58**, 796 (1991).
82. R. A. Hamm, A. Feygenson, D. Ritter, Y. L. Wang, H. Temkin, R. D. Yadrish, and M. B. Panish, *Appl. Phys. Lett.* **61**, 592 (1992).
83. A. Y. Cho and F. K. Reinhert, *Appl. Phys. Lett.* **21**, 355 (1972).
84. W. T. Tsang and A. Y. Cho, *Appl. Phys. Lett.* **32**, 491 (1978).
85. E. Kapon, S. Simhony, J. P. Harbison, and L. T. Florez, *Appl. Phys. Lett.* **56**, 1825 (1990).
86. S. Guha, A. Madhukar, K. Kariani, L. Chen, R. Kuchibhotla, R. Kapre, M. Hyngaji, and Z. Xie, *Mater. Res. Soc. Symp. Proc.* **145**, 27 (1989).
87. J.-L. Lievin, D. Bonnevie, F. Poingt, C. Starck, D. Sigogne, O. Le Gouezigou, and L. Goldstein, *Appl. Phys. Lett.* **59**, 1407 (1991).
88. R. Bhat, E. Kapon, S. Simhony, E. Colas, and M. Koza, *J. Cryst. Growth* **107**, 716 (1991).
89. M. A. Cotta, L. R. Harriott, Y. L. Wang, R. A. Hamm, H. H. Wade, J. S. Weiner, D. Ritter, and H. Temkin, *Appl. Phys. Lett.* **61**, 1936 (1992).
90. Y. Hiratani, Y. Ohki, Y. Sugimoto, and K. Akita, *J. Cryst. Growth* **111**, 570 (1991).
91. A. Okamoto and K. Ohata, *Appl. Phys. Lett.* **51**, 1512 (1987).
92. O. Kayser, *J. Cryst. Growth* **107**, 989 (1991).
93. N. Furuhata and A. Okamoto, *J. Cryst. Growth* **112**, 1 (1991).
94. D. A. Andrews, A. Z. Rejman-Greene, B. Wakefield, and G. J. Davies, *J. Cryst. Growth* **95**, 167 (1989).
95. H. Heinecke, *J. Cryst. Growth* **120**, 376 (1991).
96. A. Doi, Y. Aoyagi, and S. Namba, *Appl. Phys. Lett.* **49**, 785 (1986).
97. Q. Chen, J. S. Osinski, and P. D. Dapkus, *Appl. Phys. Lett.* **57**, 1437 (1990).
98. V. M. Donnelly, C. W. Tu, J. C. Beggy, V. R. McCrary, M. G. Lamont, T. D. Harris, F. A. Baiocchi, and R. C. Farrow, *Appl. Phys. Lett.* **52**, 1065 (1988).
99. H. Sugiura, R. Iga, T. Yamada, and M. Yamaguchi, *Appl. Phys. Lett.* **54**, 335 (1989).
100. H. K. Dong, B. W. Liang, M. C. Ho, S. Hung, and C. W. Tu, *J. Cryst. Growth* **124**, 181 (1992).
101. J. A. McCaulley, V. R. McCrary, and V. M. Donnelly, *J. Phys. Chem.* **93**, 1148 (1988).
102. H. Sugiura, R. Iga, and T. Yamada, *J. Cryst. Growth* **120**, 389 (1992).
103. K. Nagata, Y. Iimura, Y. Aoyagi, and S. Namba, *J. Cryst. Growth* **105**, 52 (1990).

104. R. Iga, H. Sugiura, T. Yamada, and K. Wada, *Appl. Phys. Lett.* **55**, 451 (1989).
105. R. Iga, H. Sugiura, and T. Yamada, *Appl. Phys. Lett.* **61**, 1423 (1992).
106. T. Yamada, R. Iga, and H. Sugiura, *Appl. Phys. Lett.* **61**, 2449 (1992).
107. G. B. Stringfellow, *J. Cryst. Growth* **105**, 260 (1990).
108. M. Weyer, *J. Cryst. Growth* **107**, 1021 (1991).
109. E. A. Beam, T. S. Henderson, A. C. Seabaugh, and J. Y. Yang, *J. Cryst. Growth* **116**, 436 (1992).
110. P. Kaul, A. Schutze, D. Hohl, A. Brauers, and M. Weyer, *J. Cryst. Growth* **123**, 411 (1992).
111. D. M. Speckman and J. P. Wendt, *Appl. Phys. Lett.* **56**, 1134 (1990).
112. S. Salim, J. P. Lu, K. P. Jensen, and D. A. Bohling, *J. Cryst. Growth* **124**, 16 (1992).
113. C. R. Abernathy, P. W. Wisk, D. A. Bohling, and G. T. Muhr, *Appl. Phys. Lett.* **60**, 436 (1992).
114. B. J. Lee, Y. M. Houng, J. N. Miller, and J. E. Turner, *J. Cryst. Growth* **105**, 168 (1990).
115. C. A. Abernathy, A. S. Jordan, S. J. Pearton, W. S. Hobson, D. A. Bohling, and G. T. Muhr, *Appl. Phys. Lett.* **56**, 2654 (1990).
116. M. Kamp, F. Konig, G. Morsch, and H. Luth, *J. Cryst. Growth* **120**, 124 (1992).
117. K. Kimura, S. Horiguchi, K. Kamon, M. Shimazu, M. Mashita, M. Mihara, and M. Ishii, *J. Cryst. Growth* **81**, 276 (1987).
118. J. Musolf, D. Marx, A. Kohl, M. Weyers, and P. Balk, *J. Cryst. Growth* **107**, 1043 (1991).

Chapter 4

ORGANOMETALLIC CHEMICAL VAPOR DEPOSITION FOR OPTOELECTRONIC INTEGRATED CIRCUITS

R. Bhat

Bellcore
Red Bank, New Jersey 07701

1. INTRODUCTION

The use of organometallic chemical vapor deposition (OMCVD) for the growth of epitaxial layers of III–V compounds became popular with the work of Manasevit [1–3] and Dupuis [4]. The technique had, however, been proposed earlier by Ruehrwein [5] and Didchenko *et al.* [6]. In spite of the wide usage today of this technique, there is no agreement on a unique name; it is also called metal organic chemical vapor deposition (MOCVD), organometallic vapor phase epitaxy (OMVPE), and so on. III–V compound epitaxial layers can also be deposited by liquid-phase epitaxy (LPE), chloride or hydride vapor-phase epitaxy (VPE), and various forms of molecular beam epitaxy (MBE). MBE traditionally uses elemental sources. However, variants of the MBE technique have been introduced, such as gas-source MBE

ISBN 0-12-200420-5

(GSMBE) in which the group V elements are produced from gaseous sources such as arsine and phosphine, organometallic MBE (OMMBE or MOMBE) in which organometallic compounds are used as group III sources, and chemical beam epitaxy (CBE) in which organometallic compounds are used as group III sources and either organometallic compounds or hydrides are used as group V sources (the definitions of these variants are not universally agreed upon). The most popular technique, however, is OMCVD because of its relatively simple growth apparatus (at least for small reactors), versatility, and the expectation that the knowledge gained in the scale-up of silicon chemical vapor deposition reactors would be applicable to III–V OMCVD reactors. Considering the papers presented at the Institute of Electrical and Electronic Engineers (IEEE) Semiconductor Laser Conference, a breakdown of the techniques used for fabricating lasers yields: OMCVD 48%, LPE 36%, and MBE and related techniques 15% for the 1988 conference, and OMCVD 63%, LPE 15.5%, and MBE and related techniques 21% for the 1990 conference. If such a trend in laser fabrication is taken as an indication of the trend in optoelectronic integrated circuit (OEIC) growth techniques as well, it can be seen that OMCVD is clearly the technique of choice.

A schematic drawing of a typical OMCVD reactor is shown in Fig. 1. In the most common form of OMCVD, vapors of organometallic compounds of the group III elements are reacted with the hydrides of the group V elements on a monocrystalline substrate placed on a hot silicon carbide–coated graphite susceptor. Because of the high vapor pressures and toxicities of the group V hydrides and the increasing concerns about our environment, there has been significant work on replacing these hydrides with less toxic and less volatile organometallic compounds of the group V elements [7–11]. The carrier gas used is ultrahigh-purity hydrogen obtained by diffusing hydrogen through a palladium–silver alloy membrane. The flows of gases are regulated and monitored by electronic mass flow controllers. The susceptor is heated using a radio-frequency generator or infrared radiation from tungsten–halogen lamps. The susceptor (substrate) temperature is monitored with a thermocouple or an infrared pyrometer. The reactor can be operated at atmospheric pressure or low-pressure (usually 10–150 torr). For low-pressure operation a mechanical vacuum pump is used, along with a throttle valve and feedback control. The design of the gas manifold that inputs reagents into the reactor is critical if abrupt interfaces are to be achieved. Two types of manifolds are in use today—a linear and a cylindrical manifold. In both cases, reagents are either swept into the main carrier gas stream going to the reactor or into the vent, the switching being accomplished by valve

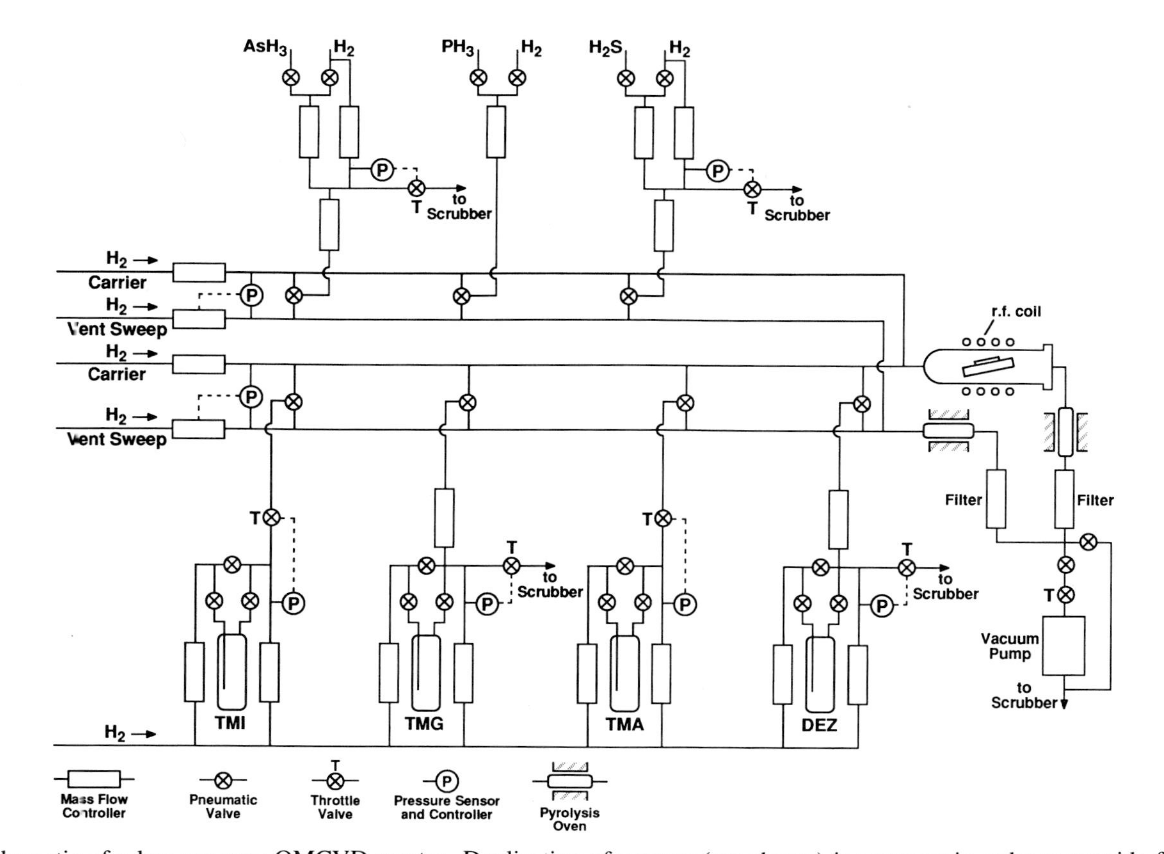

Fig. 1. Schematic of a low-pressure OMCVD reactor. Duplication of sources (not shown) is common in order to provide flexibilty in the growth of complex structures.

arrangements that result in zero or nearly zero dead volumes in order to obtain abrupt interfaces. The difference between the manifolds is that in the cylindrical manifold the substrate is equidistant from the point at which each reagent is switched into the carrier gas stream, but it is not in the case of a linear manifold. Additional practices that have been adopted are to keep the total gas flow into the reactor constant and to have a zero differential pressure between reactor and vent lines in the switching manifold. These precautions are taken to suppress oscillations in the gas input to the reactor upon valve switching. Such oscillations give rise to undesirable composition and doping fluctuations at interfaces. The reactor cell also must be designed to avoid dead volumes and recirculating flows. This is in addition to the other design criteria (see, for example, [12–16]) that address uniformity of deposit.

In this chapter we address only the growth issues that are relevant to the fabrication of OEICs. The reader is referred to the excellent text by Stringfellow [17] for a thorough understanding of the theory and practice of OMCVD. We will concentrate on the GaAs/AlGaAs and InP/InGaAsP materials systems. To the crystal grower, the latter system presents an additional challenge. Whereas the lattice mismatch between GaAs and AlGaAs cannot exceed 1600.9 ppm and is usually neglected, the composition of InGaAsP must be carefully controlled to ensure lattice matching (or a chosen amount of strain in the case of strained layers) between this quaternary alloy and its InP substrate.

2. MATERIALS CAPABILITIES FOR OEIC FABRICATION

For OEIC fabrication, it is first of all necessary to establish a basic materials growth capability that includes (1) the capability to grow the range of III–V compounds of interest; (2) capability to grow heterostructures, both strained and unstrained; (3) control of layer purity; (4) control of layer composition; (5) doping control over a wide range; (6) interface control; and (7) control of uniformity of layer thickness, composition, and doping. In achieving this capability, some of the factors that must be considered are design of the reactor cell, choice of reagents, growth parameters such as the substrate temperature and V/III ratio, and reagent switching sequence. The design of the reactor cell is of the utmost importance in ensuring uniformity. In addition, dead volumes and recirculating flows in improperly designed reactor cells result in nonabrupt interfaces.

The nature of the reagents is also important. For example, the choice of triethylgallium (TEG) for the growth of GaAs at atmospheric pressure will result in poor thickness uniformity [18]. If carbon incorporation in GaAs and AlGaAs is a concern, then the ethyl or isobutyl compounds of Ga and Al are to be preferred over the methyl compounds. Furthermore, in selective-area epitaxy the choice of compounds such as diethylgallium chloride (DEGCl) will result in more flexibility than if trimethylgallium (TMG) or TEG is used [19, 20]. This is because DEGCl decomposes into volatile GaCl, whereas TMG or TEG decomposes into nonvolatile Ga, which can lead to the formation of polycrystalline deposits on the mask.

The choice of the doping species has to be based on many factors: some dopants, such as zinc and sulfur, have a high vapor pressure and their incorporation efficiency will be poor at high growth temperatures, whereas the decomposition of other dopant species (silane, for example) has a high activation energy and their doping efficiency will be low for low growth temperatures. Other factors, such as the diffusivity of the dopants and possible memory effects (adsorption on the walls of the reactor or the tubing and slow desorption), also need to be considered.

Growth temperature, growth rate, and the V/III ratio influence material purity, morphology, other properties such as the photoluminescence efficiency, and growth behavior on patterned substrates.

The reagent switching sequence is a key parameter in achieving abrupt interfaces. In the growth of an As compound on top of a P compound and vice versa, particular attention has to be paid to the switching sequence in order to avoid significant incorporation of As in the P compound, because As is incorporated in preference to P.

Having established a basic materials growth capability, several approaches may be taken for OEIC fabrication. An OEIC may comprise many noncompatible devices such as lasers, waveguides, modulators, detectors, and electronic components consisting of high-electron-mobility transistors (HEMTs) or heterojunction bipolar transistors (HBTs) on the same chip. Ideally, it would be convenient if all the devices needed for the OEIC could be grown where wanted simultaneously. However, this ideal approach is far from what can be achieved with today's technology. Some of the other possible approaches toward integration are:

1. Growth on nonplanar substrates
2. Regrowth: selective epitaxy
3. Sequential deposition of devices by selective-area epitaxy using laser-enhanced deposition

We will consider each of these approaches and address some of the materials growth issues in each case.

2.1. Growth on Nonplanar Substrates

This approach can be divided into two categories. In the first category, one uses the nonplanarity of the substrate to shift the levels of the devices so that, after selective etching, nearly planar processing is possible. This approach is best understood by referring to Fig. 2, which shows the sequence of fabrication steps leading to the MSM-HEMT OEIC. All layers required for processing the OEIC are deposited in one growth. Selective etching is then used to remove layers in places where they are not required. As another

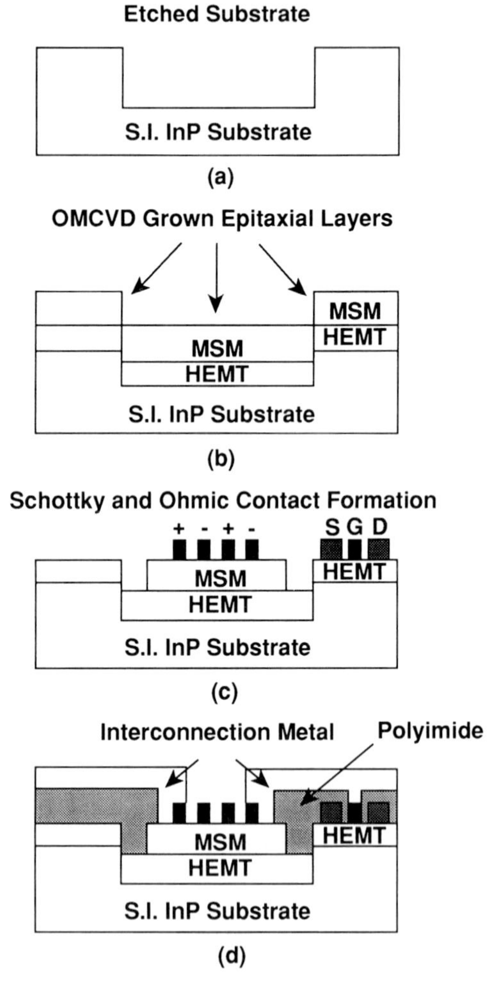

Fig. 2. Schematic showing the steps in the fabrication of an MSM-HEMT OEIC. (a) Etched well formation, (b) OMCVD growth, (c) Schottky and ohmic contact formation, and (d) polyimide and interconnect metal deposition.

example of this approach [21] we show schematically in Fig. 3 a method for butt coupling a laser with a waveguide. This approach to OEIC fabrication is clearly the most direct route, as one can take discrete device structures that have been optimized on planar substrates and combine them together on the same wafer by growth on a nonplanar substrate. The disadvantages of this approach are a possible increase in parasitics due to the presence of unwanted layers below a device, possible degradation of some devices due to additional diffusion in the extra time at the growth temperature, and the deviation in doping concentration as well as the deviation in composition of the ternary and quaternary layers near and on the side walls of the etched wells. The latter two problems may impose limits on the density of the devices, because the wells must be substantially larger than the devices for the active area of the devices to be far from the side walls. Furthermore, if regions of material with sufficiently high composition deviation and lattice mismatch are present after device fabrication, dislocations generated in these regions may propagate, with time, to the active regions of the devices and cause premature failure.

In the other category, the growth characteristics on nonplanar substrates are exploited to integrate different devices. In order to do this it is necessary to understand the growth rate, composition, and doping variation on different crystallographic planes. The growth rate variation that occurs and the various crystallographic facets which appear when growth takes place on a nonplanar substrate are dependent on, among other parameters, the crystal growth technique, the growth temperature and pressure, the material being

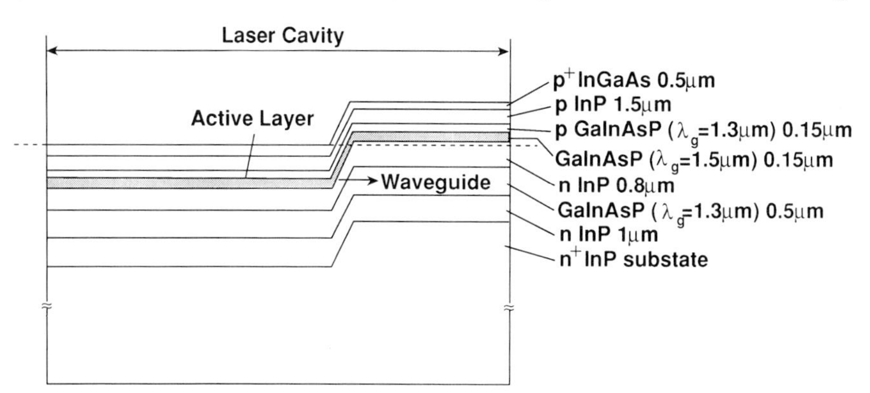

Fig. 3. Schematic of a laser butt-coupled with a waveguide formed by growth on a nonplanar substrate (CNET patent no. 05033, March 30, 1984, awarded to L. Menigaux, A. Carenco, and P. Sansonetti). The dashed line indicates the final level of the structure after planarization. (Reproduced by permission of the authors [21] and the *Journal of Crystal Growth*.)

grown, and the depth and width of the patterned features on the substrate. As a result, it is hard to predict the nature of the overgrowth ahead of time, although some modeling has been done for amorphous material deposition in chemical vapor deposition [22], and this makes this technique difficult to use for OEIC fabrication. Figure 4a and b show the schematic cross section of a GaAs/AlGaAs single quantum well graded index separate confinement heterostructure (GRINSCH) laser grown in a $[01\bar{1}]$ oriented groove in a (100) GaAs substrate by atmospheric-pressure OMCVD [23] and by MBE [24], respectively. In OMCVD growth the AlGaAs layer forms a sharp corner at the bottom of the groove and defines {111} planes on the side walls, while

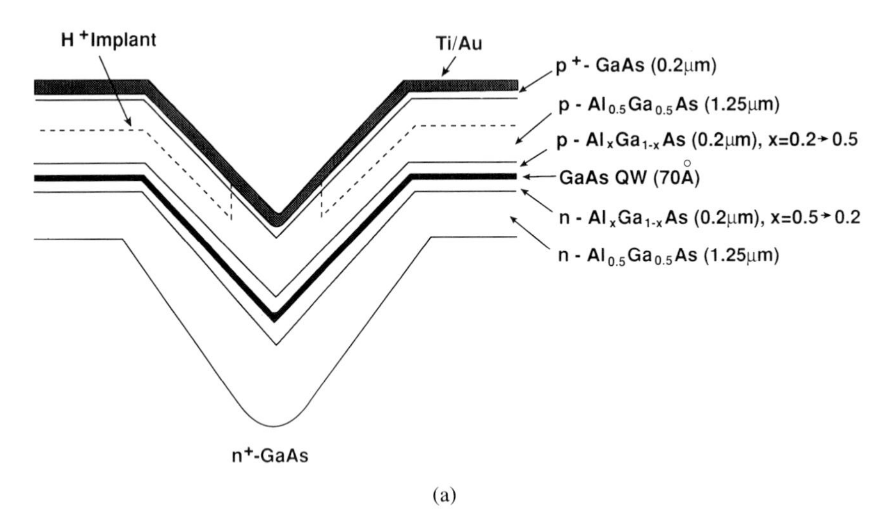

(a)

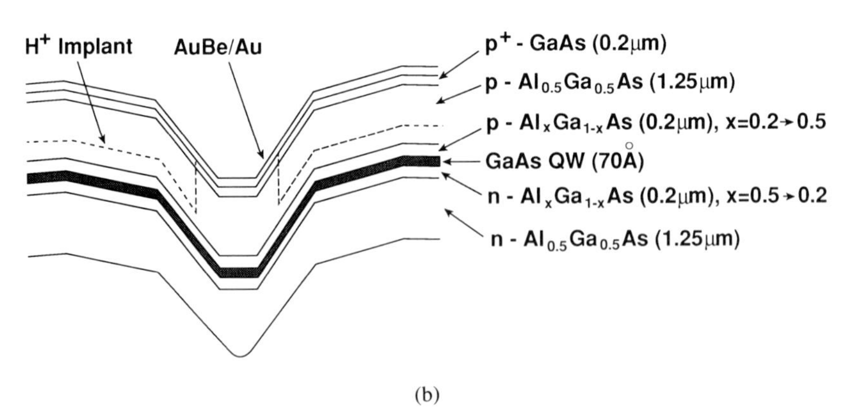

(b)

Fig. 4. Schematic cross section of a GaAs/AlGaAs graded index separate confinement heterostructure (GRINSCH) laser grown in a V-groove by (a) OMCVD at atmospheric pressure and (b) MBE.

the GaAs layer grows faster at the bottom of the groove than on the side walls, developing a crescent shape at the bottom of the groove. In contrast, when the same structure is grown by MBE, a (100) plane always develops at the bottom of the groove for both GaAs and AlGaAs, with growth being faster there than on the side walls. In addition, {311} facets appear at the top corners between the {111} side walls and the (100) top surface. Similar features have been observed when such structures are grown by OMMBE [25]. The major reason for the differences in the growth features obtained by OMCVD and MBE appears to be the fact that in MBE the growth species are in the form of a beam and have only one chance to stick to the surface, with those that don't stick being lost to vacuum. In OMCVD, however, species that arrive by diffusion through the concentration boundary layer have many chances to stick to the growing surface, because those that desorb can collide with molecules in the gas phase and be redirected toward the surface. Furthermore, a nonplanar surface makes the concentration boundary layer also nonplanar, with the result that the diffusion is no longer one-dimensional but is two- or three-dimensional [26].

As an example of an OEIC that can be fabricated by taking advantage of the growth rate variation and hence the growth morphology, we show schematically in Fig. 5 the structure of a GaAs/AlGaAs integrated waveguide–detector. Because of the growth morphology, if growth takes place for a sufficiently long time, the material growing in the region above the groove

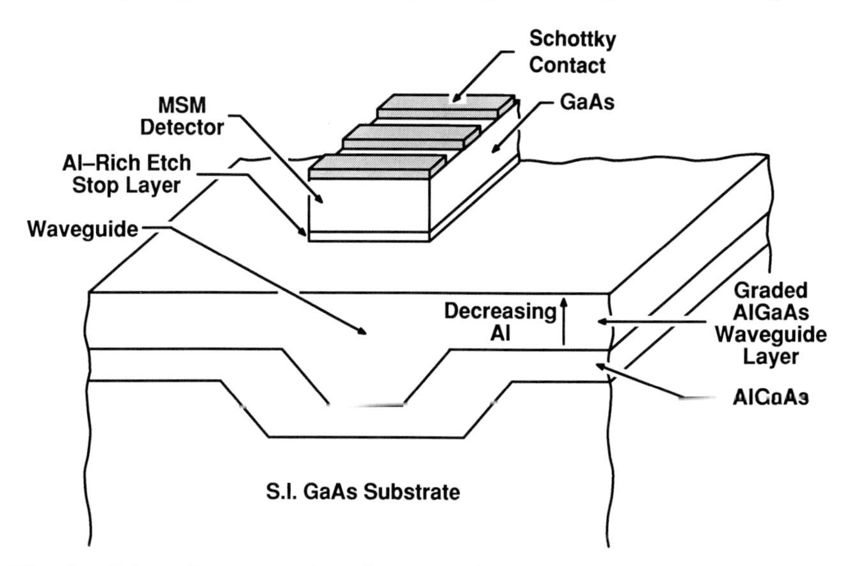

Fig. 5. Schematic cross section of a waveguide-detector OEIC formed by growth of GaAs/AlGaAs on a nonplanar substrate.

is bounded on both sides by material with a higher aluminum content grown earlier. The light is, therefore, guided in the higher-refractive-index region above the groove. The top GaAs layer can be used to form a metal–semiconductor–metal (MSM) detector. A high aluminum content etch-stop layer is used to define the detector. If the same concept could be made to work in the InP-based materials system, the device would be even more useful because a wider choice of energy bandgaps is available in this system. However, some materials growth problems need to be solved before this can be realized, as indicated in the following.

When an alloy semiconductor layer is grown on a nonplanar substrate, the composition of the layer growing on surfaces other than (100) can be substantially different from that on (100). This is probably due to the different sticking probabilities of the various species on different surfaces. In addition, the presence of neighboring regions that incorporate the various growth species differently can give rise to concentration gradients of growth species, both in the gas phase and on the surface, which would not be present on a planar substrate. The "communication" via the gas phase and the surface between neighboring planes of a nonplanar substrate can, therefore, give additional variations in composition compared with those expected for planar substrates having surface orientations corresponding to those present on the nonplanar substrate. Such compositional deviations can be a severe problem in alloy systems in which lattice matching is achieved only for specific compositions, because exceeding the critical layer thickness can give rise to dislocations. As an example, we show in Fig. 6 a cross-section transmission electron micrograph (TEM) of a multiple quantum well InGaAsP/InP laser structure grown in a $[01\bar{1}]$-oriented V-groove [27]. The InGaAsP was lattice-matched on a (100) substrate, but the large composition deviation on the side walls of the groove results in a dense network of dislocations. However, such compositional variations on a nonplanar substrate can probably be made use of to create novel devices and device integration.

In addition to compositional variation, there can be significant lateral variation of doping on a nonplanar substrate. It was shown [28] that the incorporation of some dopants in InP and GaAs can vary by more than two orders of magnitude in going from an $\{h11\}$A to an $\{h11\}$B plane (h = 1, 2, or 3) as seen in Figs. 7 and 8. Thus n- and p-type dopants which show large differences in incorporation between the A and B faces are useful for obtaining lateral patterning of doping by growth on nonplanar substrates. Figure 9 shows a cross section of InP layers, alternately doped with S and Zn, grown on a (100)-oriented InP substrate with a V-groove etched along $[01\bar{1}]$. The doping levels in the layers were $n = 1 \times 10^{18}$ cm^{-3} and $p = 5$

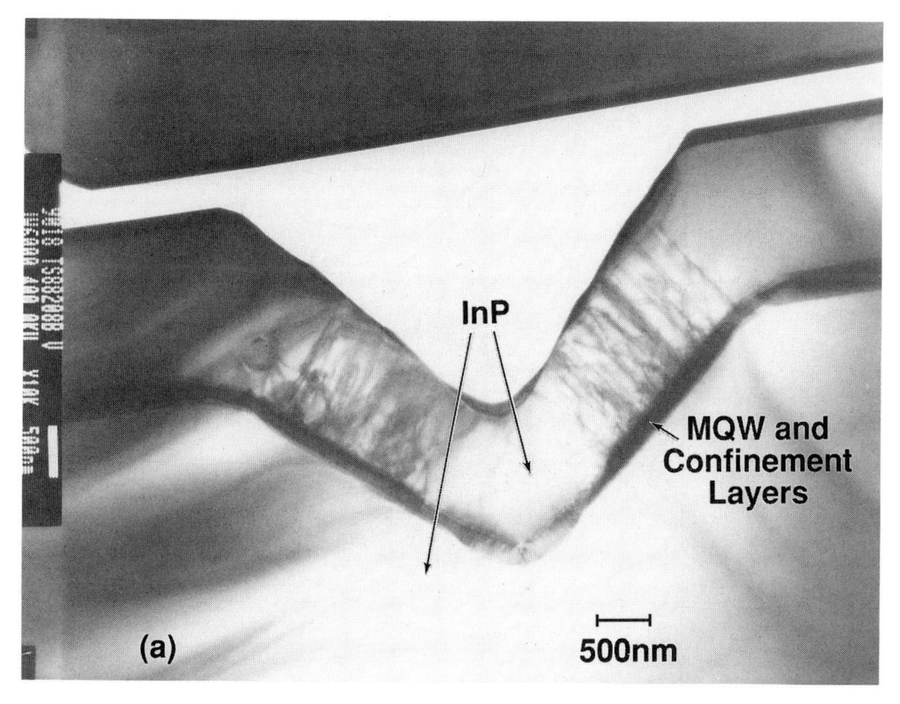

Fig. 6. (002) cross-section TEM of a multiple quantum well InP/InGaAsP laser grown in a V-groove showing the dense network of dislocations on the side walls.

$\times\ 10^{17}\ cm^{-3}$ on the (100) surface. It can be seen that the p/n layers are preserved on the (100) surface, but the layers in the groove are all p-type. This effect is due to the fact that, on the A-type side walls of the groove, the p doping increases and the n doping decreases with respect to the levels on the (100) surface, as seen in Fig. 7. Because of the n- and p-doping levels chosen, the p-doping level greatly exceeds the n-doping level on the V-groove side walls. Diffusion of Zn during the growth converts the n-type layers in the groove to p-type, resulting in only p-type layers in this region. This technique of selectively growing alternating p/n layers was used to fabricate lasers with current blocking layers in a single growth step [29]. The orientation dependence of doping can also be used to fabricate lateral p–n junction arrays by growth on nonplanar substrates and doping simultaneously with n- and p-type dopants. An example [29] of such a lateral p–n junction array in InP is shown in Fig. 10. Like the composition variation, the doping variation could be exploited by the device designer to create novel integrated devices. In any case, the composition and doping variations that occur when the substrate is nonplanar must be taken into consideration in other OEIC fabrication approaches as well.

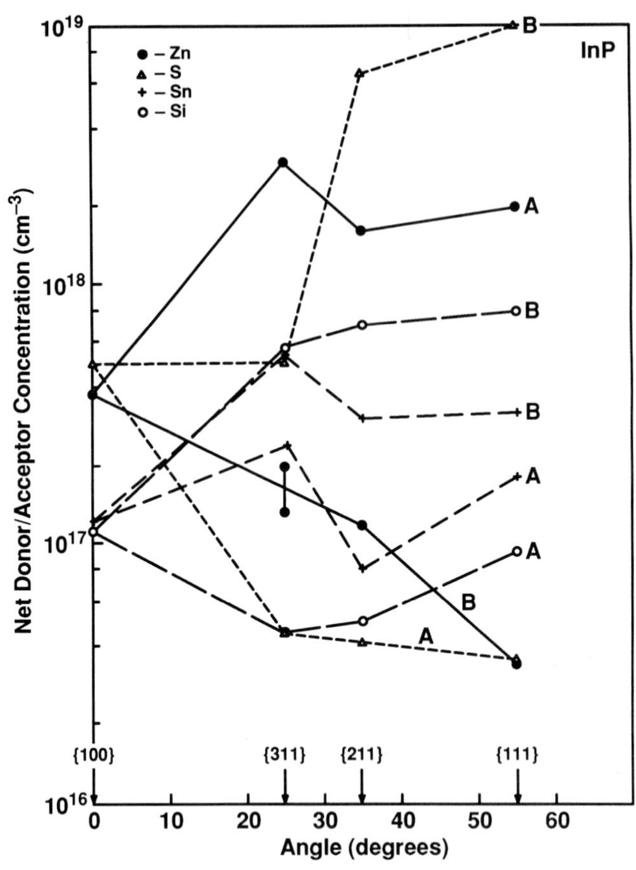

Fig. 7. Variation of net acceptor/donor concentration with orientation for S, Zn, Sn, and Si in InP. The growth temperature was 625°C except in the case of Si doping, which was done at 580°C.

2.2. Selective Epitaxy and Regrowth

Another useful approach for OEIC fabrication is selective epitaxy and regrowth. In essence, device structures which have been optimized on planar substrates are deposited sequentially in selected areas by using silicon dioxide or silicon nitride masks. Most studies of selective epitaxy [30–33] have concentrated on the conditions under which selective epitaxy is obtained, the morphology of the selective deposits, and the thickness variation of the deposit across the openings in the mask. In summary, better selectivity, better morphology, and lower thickness variation are obtained at lower pressures, because of the higher mean free path, and higher temperatures, because of the larger surface diffusivity and volatility of the growth species.

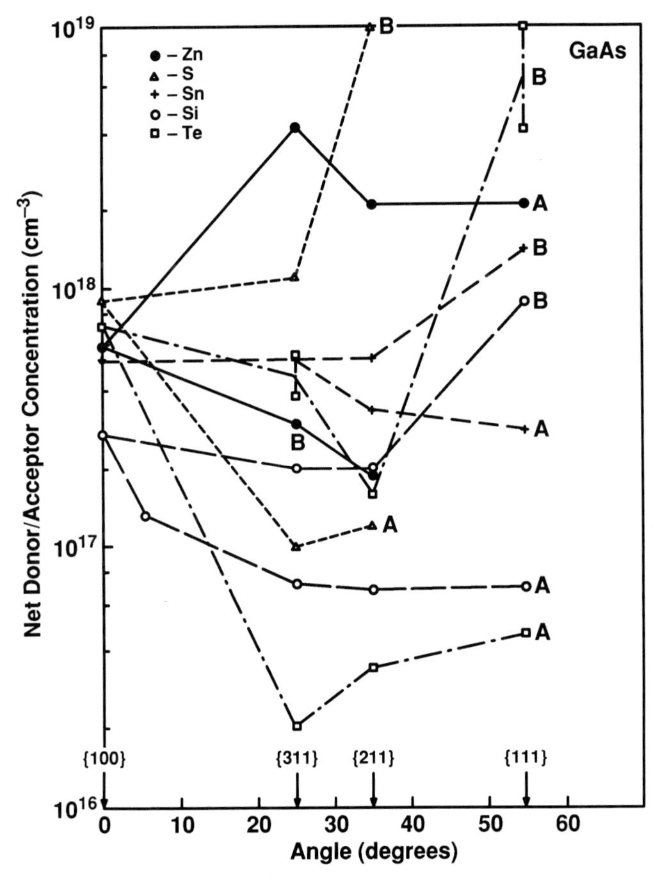

Fig. 8. Variation of net acceptor/donor concentration with orientation for S, Zn, Sn, and Te in GaAs. The growth temperature was 625°C except in the case of Si doping, which was done at 675°C.

The thickness variations that can be obtained in selective epitaxy can be used to make novel device structures such as tapered waveguides [34] as shown in Fig. 11. The extent of the growth rate enhancement in the open areas depends on the width of the masked region from which additional growth species diffuse (mostly via the gas phase, as indicated by the large distance—more than 300 μm at a growth pressure of 30 torr—over which tapering takes place).

Composition variation in the case of selective alloy semiconductor deposition has received far less attention [35, 36]. Most earlier studies were on GaAs/AlGaAs, and composition deviation in the selective epitaxy of AlGaAs probably went unnoticed because lattice matching is not a problem

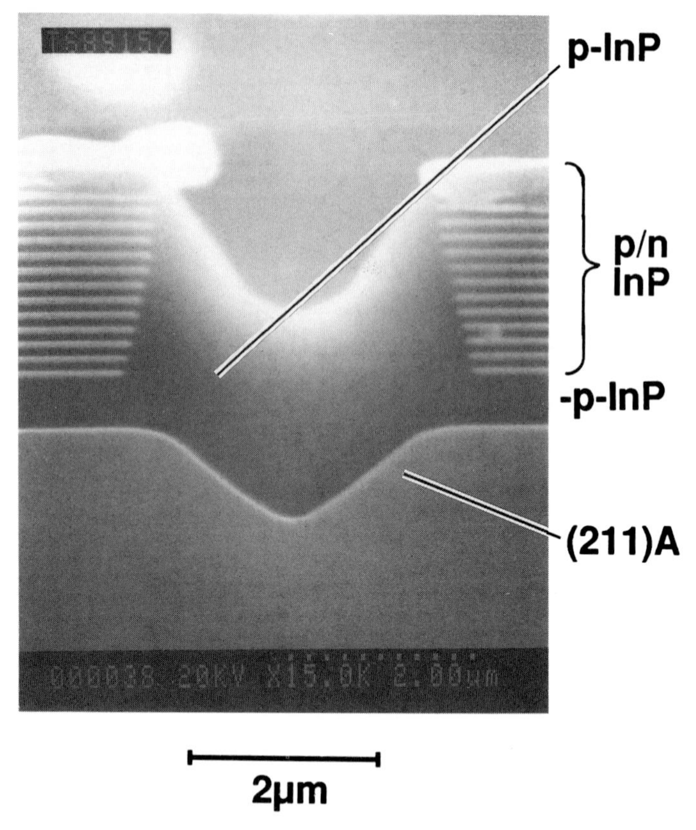

Fig. 9. Scanning electron micrograph of a stained cleaved edge showing the growth of p/n alternating layers on a (100)-oriented InP substrate with a $[01\bar{1}]$-oriented groove. A p-type epitaxial layer was grown on the substrate prior to etching of the groove.

in this system. A substantial deviation in composition can pose severe problems in selective epitaxy of materials such as InGaAsP. Figure 12 shows a selectively deposited InGaAs layer (under conditions that would give lattice-matched InGaAs layers on a planar InP substrate) that is severely mismatched because of a large composition deviation caused by the presence of masked areas. Figure 13 shows the mole fractions of InAs and GaAs as a function of distance away from the mask edge for InGaAs selectively deposited inside 50×500 μm rectangular openings in a silicon nitride mask [35]. It is seen that significant composition deviation can occur up to 80 μm away from the mask edge for the growth conditions used. Although such a composition deviation poses a challenge to the crystal grower, it could be utilized in device structures. The composition deviation is believed to be

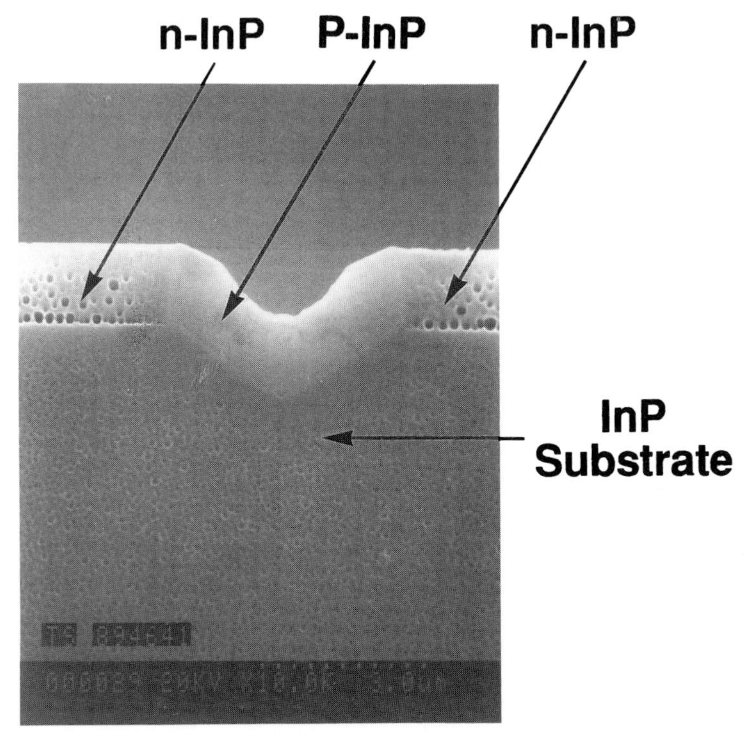

Fig. 10. Scanning electron micrograph of a stained cleaved edge showing the formation of lateral *p–n* junctions by growth of an InP layer in a $[01\bar{1}]$-oriented groove while codoping with Zn and S.

due to different contributions from the growth species diffusing from the masked areas to the open regions either on the surface [35, 36] or via the gas phase [34, 37] or both. There is also a contribution from species diffusing from non-(100) facets near the edge of the mask to the (100) surface [37]. The major contribution, however, is believed to be due to species diffusing via the gas phase [34, 37]. Since such a phenomenon is absent in OMMBE and CBE, these techniques may be more useful for selective-area growth than OMCVD.

For selective regrowth in the InP/InGaAsP materials system, it is often necessary to preserve several materials such as InP, InGaAs, and InGaAsP during the initial heat-up to growth temperature. It has been found [38] that InP is unstable in AsH_3, and InGaAs in PH_3. InGaAsP is unstable in AsH_3 + PH_3 mixtures which deviate substantially from the composition used in growing it. However, it has been shown that InP and InGaAsP can be preserved in a phosphine plus As_4 mixture, with the As_4 emanating from a GaAs

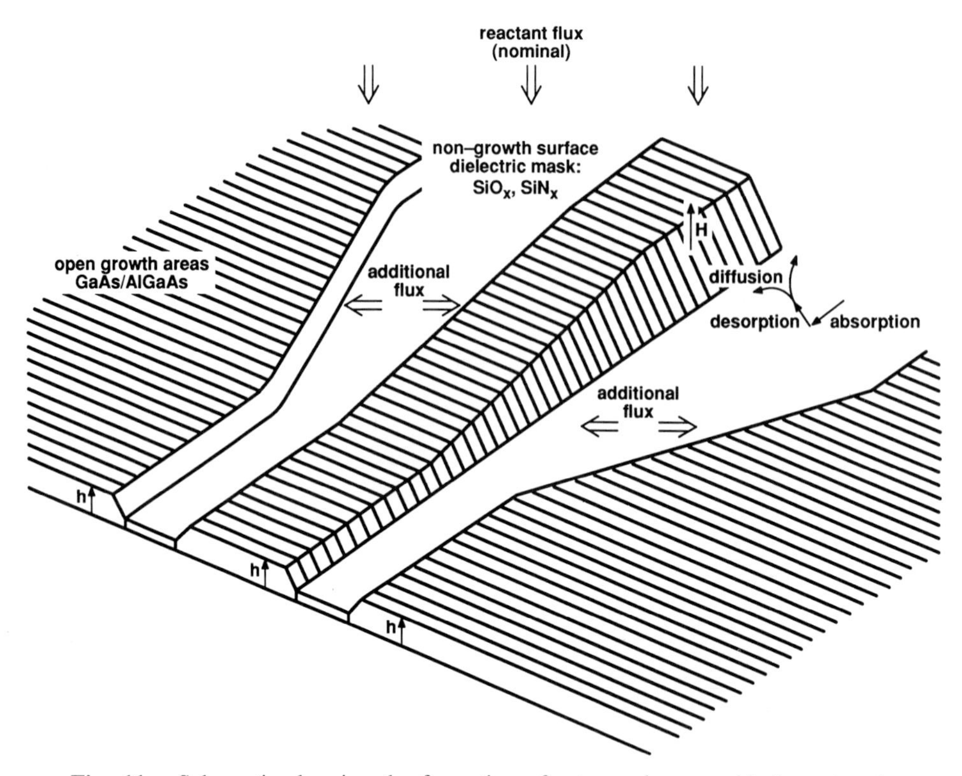

Fig. 11. Schematic showing the formation of a tapered waveguide by using the diffusion of additional growth species from neighboring masked areas.

or InAs wafer placed ahead of the wafers to be preserved in a horizontal reactor [38]. InGaAs reacts slightly with this mixture, forming a thin layer of InGaAsP. To preserve all three materials at the same time it is necessary to use $As_4 + P_4$ mixtures [38], P_4 being generated from an InP wafer placed upstream of the wafers to be preserved. To obtain more controlled generation of As_4 and P_4 species and to preserve the large areas that are likely to be needed for OEICs, it is necessary to modify the OMCVD reactor. Such a modification might consist of having a separate susceptor (placed upstream of the susceptor on which the wafers to be preserved and grown are kept) that can be heated independently and on which the GaAs (or InAs) and InP wafers used for generating the elemental vapors are placed. This susceptor can be kept at a temperature higher than that of the substrates to be preserved during the heat-up prior to growth, and its temperature can be lowered on commencement of growth. Alternatively, a plasma may be used to crack the arsine and/or phosphine to generate elemental vapors, upstream of the wafers to be preserved, during the initial heat-up prior to growth.

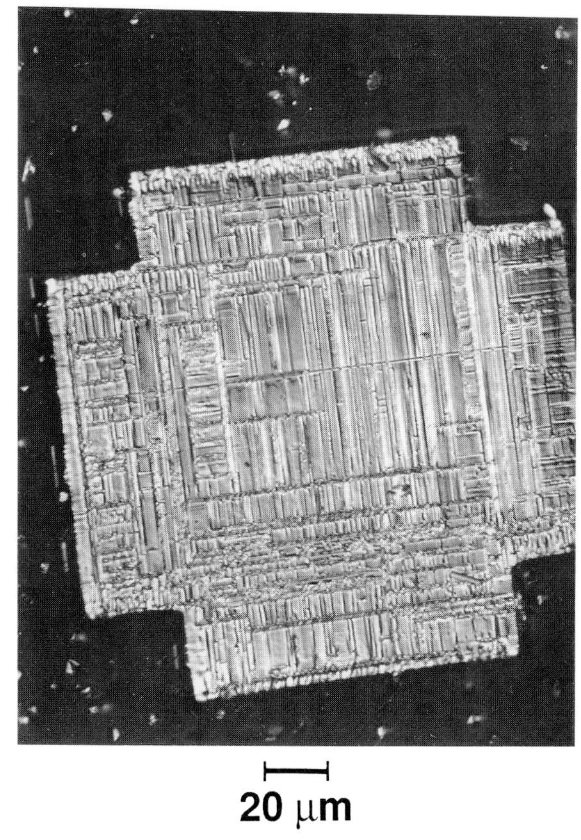

Fig. 12. Nomarski interference contrast photomicrograph of an GaInAs layer grown in an opening in a dielectric mask. The vapor-phase composition was such that a lattice-matched layer would have grown on an unmasked substrate.

The ability to etch *in situ* before regrowth is probably important for obtaining clean interfaces. Although *in situ* etching of GaAs [39, 40] and InP [41–44] has been studied, more work has to be done in this area to develop processes that are compatible with OEIC structure growth. In particular, the capability to etch other materials such as AlGaAs, AlInAs, InGaAs, and InGaAsP must be developed. Furthermore, the effects of *in situ* etching on device performance must be established.

In addition to cleaning surfaces before growth/regrowth, *in situ* etching can be used for etching wells or grooves in a masked substrate prior to selective-area growth. This not only reduces processing steps but also enables one to take advantage of characteristics of gas-phase etching [44]. Finally, the process would become even more attractive if the mask material could be etched inside the reactor.

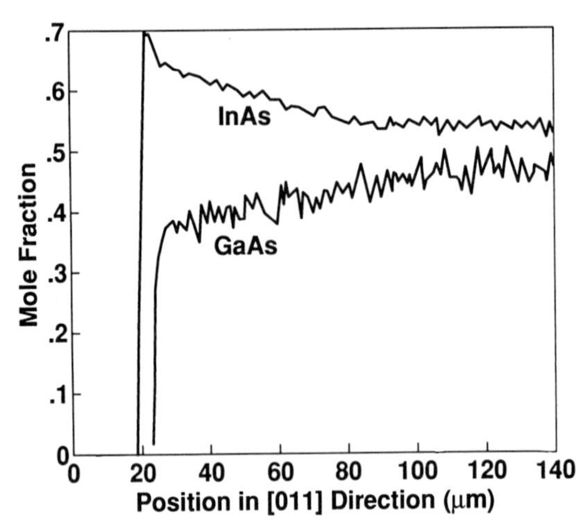

Fig. 13. GaAs and InAs mole fractions versus distance from the mask edge in the [011] direction, for a GaInAs epilayer grown on a partially masked substrate at 600°C and a reactor pressure of 76 torr. The mole fractions were measured by scanning Auger electron spectroscopy. The measurements were done along the length ([011] direction) of a 50 × 500 μm GaInAs island, which had a large field of silicon nitride butting up against its shorter side. (Reprinted with permission from the authors [35] and the *Journal of Electronic Materials*, Vol. 19, p. 345, 1990, a publication of The Minerals, Metals and Materials Society, Warrendale, Pennsylvania 15086.)

2.3. Sequential Deposition of Device Structures by Laser-Assisted Selective Epitaxy

Laser-assisted selective epitaxy (LASE) promises to be an attractive technique for OEIC structure growth. Some work has been done in this area using either a thermal [45] or a photochemical effect [46]. In the thermal process, a laser is used to create local regions of high temperature, where deposition occurs. This process is likely to lead to defect generation because of the large thermal gradients and hence stresses created. If the temperature gradient is kept sufficiently small, by heating the substrate uniformly and using a laser only to produce a slight local increase in temperature, the process can be used to enhance growth in the illuminated regions without generating defects. However, this mode of operation does not give complete growth selectivity. The thermal process also does not give sharply defined regions of deposition because the temperature decreases gradually away from the illuminated area as a result of thermal conduction. Furthermore, the gaussian beam profile of the laser will cause the temperature to vary across the illuminated area. The thickness profile will depend on the temperatures achieved, with the thickness being nearly constant over the area where the temperature is in the diffusion-controlled regime of growth.

In the photochemical process the deposition reaction is activated by a laser beam having a specific wavelength, with activation being either in the gas phase or on the wafer surface. Activating the reaction in the gas phase is less desirable because it will result in diffuse deposition resulting from diffusion of the activated species in the gas phase. Even in the case of surface activation, surface diffusion and desorption with subsequent readsorption of the activated species can lead to blurring of the edges of the deposit. The beam profile will also lead to a thickness variation across the illuminated area and, unlike the case of the thermal process, a flat-topped region will not be obtained.

Laser-assisted atomic layer epitaxy (LALE) has been used [47–49] to obtain locally enhanced growth. In atomic layer epitaxy (ALE) the group III and V species flow into the reactor alternately, with these flows being separated by hydrogen flushes. For a narrow range of temperatures (usually substantially lower than the temperatures used in conventional OMCVD), one cycle of alternate exposure to the group III and V species leads to the growth of one monolayer, if the group III and V fluxes are high enough and the exposure time sufficiently large [50]. The range of temperatures over which a monolayer per cycle is obtained can be widened by laser illumination [51]. The thickness profile in selective-area LALE resembles that of the thermal process. The process is useful only where thin layers are needed, because the growth rates obtained with ALE are low. In addition, the process does not lead to complete selectivity.

Whether a thermal or photochemical process is used, control of composition (in the case of alloy semiconductors), doping, and thickness uniformity must be developed. In addition, such issues as the minimum size of the deposited area and the minimum separation between deposited regions need to be addressed. However, Bedair *et al.* [52] have demonstrated that a GaAs field effect transistor (FET), a GaAs *p-i-n* detector, and GaAsP/GaAs quantum wells can be deposited on the same wafer.

Laser-assisted etching and ablation may also be useful in OEIC fabrication. Selectively etched grooves [53] and selective ablation of quantum wells [54] to make laser arrays with selected wavelengths, as well as *in situ* lateral patterning of current confinement in lasers by selective laser ablation [55], have been demonstrated.

3. UNIFORMITY AND SCALE-UP

For cost-effective production of OEICs it is necessary to have an OMCVD reactor that can deposit layers with uniform thickness, doping, and composition over large areas. Thickness and doping uniformity better than 1% and composition uniformity better than 1×10^{-4} over one or several 2- to 3-inch-diameter wafers will be needed as integration increases from the present level of less than 100 devices per chip (for InP-based OEICs). A reactor with a horizontal, rectangular reaction cell can be used for up to two 2- to 3-inch-diameter wafers. For higher throughput a vertical geometry high-speed rotating-disk reactor [56], a barrel reactor [57], and a planetary rotating-disk reactor [58] have been considered. Depending on the materials system and the device structure, uniformity of thickness, doping, doping–thickness product, and composition of less than 1% [57, 58] to a few percent [56, 59, 60] have been achieved to date.

4. SAFETY

The OMCVD process, like most semiconductor processes, involves the use of gases and chemicals that are toxic, pyrophoric, or both. It is necessary to handle them properly. Furthermore, the reactor exhaust gases are toxic and need to be scrubbed before they are allowed to escape into the atmosphere. In order to ease the hazard in handling highly toxic arsine and phosphine gases, some organoarsines and organophosphines have been developed [7–11]. However, these sources do not yet perform as well as arsine and phosphine for the deposition of the entire range of compounds that are of interest for OEICs, although some of the sources have been shown to be adequate in a few cases [61–65]. The large variation with growth temperature of the composition of GaInAs grown using tertiary butylarsine [66] and the tendency toward parasitic reactions [67] are of concern. However, in the long term it is expected that viable alternatives to the group V hydrides will be found. It is also possible that the hydrides will continue to be used but will be stored and transported in a safer manner in a molecular sieve [68] or generated on demand [69]. Those starting out in OMCVD should consult with established practitioners regarding safety issues. The reader is also referred to Stringfellow's book on OMCVD [17] for further details of toxicity and safety issues and to the American Vacuum Society publication on proper procedures to follow when pumping hazardous gases [70].

5. CONCLUSIONS

In this chapter we have outlined several approaches to the growth of structures for OEIC fabrication and addressed the growth issues concerning each of these approaches. Although rudimentary OEICs are presently fabricated by OMCVD, it is clear that many growth issues need to be addressed before the full potential of the technique can be utilized in fabricating complex OEIC structures. Finally, for those who wish to obtain additional information on OMCVD, we include references to several earlier reviews [71–79]. In addition, the reader is referred to the special issues of the *Journal of Crystal Growth* (every other year) that contain the proceedings of the International Conference on Metalorganic Vapor Phase Epitaxy.

6. ACKNOWLEDGMENTS

The author wishes to thank his collegues at Bellcore for having provided the motivation for the work on which much of this chapter is based. He also particularly wishes to thank C. Caneau for critical reading of the manuscript.

REFERENCES

1. H. M. Manasevit, *Appl. Phys. Lett.* **12**, 156 (1968).
2. H. M. Manasevit and W. I. Simpson, *J. Electrochem. Soc.* **116**, 1725 (1969).
3. H. M. Manasevit, *J. Cryst. Growth* **13/14**, 306 (1972).
4. R. D. Dupuis and P. D. Dapkus, *Appl. Phys. Lett.* **31**, 466 (1977).
5. R. A. Ruehrwein, U.S. Patent 3,312,570 (1967).
6. R. Didchenko, J. E. Alix, and R. H. Toeniskoetter, *J. Inorg. Nucl. Chem* **14**, 35 (1960).
7. R. Bhat, M. A. Koza, and B. J. Skromme, *Appl. Phys. Lett.* **50**, 1194 (1987).
8. C. A. Larsen, C. H. Chen, M. Kitamura, G. B. Stringfellow, D. W. Brown, and A. J. Robertson, *Appl. Phys. Lett.* **48**, 1531 (1986).
9. C. H. Chen, C. A. Larsen, and G. B. Stringfellow, *Appl. Phys. Lett.* **50**, 218 (1987).
10. A. Brauers, O. Kayser, R. Kall, H. Heinecke, P. Balk, and H. Hofmann, *J. Cryst. Growth* **94**, 663 (1989).
11. D. M. Speckman and J. P. Wendt, *J. Cryst. Growth* **105**, 275 (1990).
12. H. Moffat and K. F. Jensen, *J. Cryst. Growth* **77**, 108 (1986).
13. C. A. Wang, S. H. Groves, and S. C. Palmateer, *J. Cryst. Growth* **77**, 136 (1986).

14. E. J. Thrush, C. G. Cureton, and A. T. R. Briggs, *J. Cryst. Growth* **93**, 870 (1988).
15. J. van de Ven, G. M. J. Rutten, M. J. Raaijmakers, and L. J. Giling, *J. Cryst. Growth* **76**, 352 (1986).
16. D. I. Fotiadas, S. Kieda, and K. F. Jensen, *J. Cryst. Growth* **102**, 441 (1990).
17. G. B. Stringfellow, "Organometallic Vapor-Phase Epitaxy: Theory and Practice." Academic Press, New York, 1989.
18. R. Bhat and V. G. Keramidas, *Proc. SPIE* **323**, 323 (1982).
19. Y. Nakayama, S. Ohkawa, and H. Ishikawa, *Fujitsu Sci. Tech. J.* **53** (1977).
20. T. F. Kuech, M. A. Tischler, N. I. Buchan, and R. Potemski, *J. Cryst. Growth* **99**, 324 (1990).
21. B. Rose, D. Remiens, H. Hornung, and D. Robein, *Fifth Int. Conf. on MOVPE*, Aachen, Germany (1990).
22. C. H. J. Van den Brekel and A. K. Jansen, *J. Cryst. Growth* **43**, 488 (1978).
23. R. Bhat, E. Kapon, D. M. Hwang, M. A. Koza, and C. P. Yun, *J. Cryst. Growth* **93**, 850 (1988).
24. E. Kapon, J. P. Harbison, C. P. Yun, and N. G. Stoffel, *Appl. Phys. Lett.* **52**, 607 (1988).
25. W. Quinn, C. Caneau, S. Simhony, E. Kapon, and R. Bhat, *Second Int. Conf. on Chemical Beam Epitaxy and Related Techniques*, Houston, Texas (1989).
26. P. Demeester, P. Van Daele, A. Ackaert, and R. Baets, *J. Appl. Phys.* **63**, 2284 (1988).
27. R. Bhat, E. Kapon, J. Werner, D. M. Hwang, N. G. Stoffel, and M. A. Koza, *Appl. Phys. Lett.* **56**, 863 (1990).
28. R. Bhat, C. Caneau, C. E. Zah, M. A. Koza, W. A. Bonner, D. M. Hwang, S. A. Schwarz, S. G. Menocal, and F. G. Favire, *Fifth Int. Conf. on MOVPE*, Aachen, Germany (1990).
29. R. Bhat, C. E. Zah, C. Caneau, M. A. Koza, S. G. Menocal, S. A. Schwarz, and F. J. Favire, *Appl. Phys. Lett.* **56**, 1691 (1990).
30. K. Kamon, M. Shimazu, K. Kimura, M. Mihara, and M. Ishii, *J. Cryst. Growth* **77**, 297 (1986).
31. H. Heinecke, A. Brauers, F. Gafahrend, C. Plass, N. Putz, K. Werner, M. Weyers, H. Luth, and P. Balk, *J. Cryst. Growth* **77**, 303 (1986).
32. A. R. Clawson, C. M. Hanson, and T. T. Yu, *J. Cryst. Growth* **77**, 334 (1986).
33. K. Hiruma, T. Haga, and M. Miyazaki, *J. Cryst. Growth* **102**, 717 (1990).
34. E. Colas, A. Shahar, J. Soole, W. J. Tomlinson, J. R. Hayes, C. Caneau, and R. Bhat, *Fifth Int. Conf. on MOVPE*, Aachen, Germany (1990).
35. J. S. C. Chang, K. W. Carey, J. E. Turner, and L. A. Hodge, *J. Electron. Mater.* **19**, 345 (1990).
36. Y. D. Galeuchet, P. Roentgen, and V. Graf, *J. Appl. Phys.* **68**, 560 (1990).
37. O. Kayser, *Fifth Int. Conf. on MOVPE*, Aachen, Germany (1990).
38. R. Bhat, M. A. Koza, C. E. Zah, C. Caneau, C. C. Chang, S. A. Schwarz,

A. S. Gozdz, P. S. D. Lin, and A. Yi-Yan, *Fifth Int. Conf. on MOVPE*, Aachen, Germany (1990).

39. R. Bhat, B. J. Baliga, and S. K. Ghandhi, *J. Electrochem. Soc.* **122**, 1378 (1975).
40. R. Bhat and S. K. Ghandhi, *J. Electrochem. Soc.* **124**, 1447 (1977).
41. A. R. Clawson, *J. Cryst. Growth* **69**, 346 (1984).
42. P. D. Agnello and S. K. Ghandhi, *J. Cryst. Growth* **73**, 453 (1985).
43. K. Pak, Y. Koide, K. Imai, A. Yoshida, T. Nakamura, Y. Yasuda, and T. Nishinaga, *J. Electrochem. Soc.* **133**, 2204 (1986).
44. C. Caneau, R. Bhat, M. A. Koza, J. R. Hayes, and R. Esagui, *Fifth Int. Conf. on MOVPE*, Aachen, Germany (1990).
45. S. M. Bedair, J. K. Whisnant, N. H. Karam, D. Griffis, N. A. El-Masry, and H. H. Stadelmaier, *J. Cryst. Growth* **77**, 229 (1986).
46. P. Balk, H. Heinecke, and N. Putz, *J. Vac. Sci. Technol. A* **4**, 711 (1986).
47. S. Iwai, T. Meguro, and Y. Aoyagi, *Fifth Int. Conf. on MOVPE*, Aachen, Germany, (1990).
48. N. H. Karam, H. Liu, I. Yoshida, and S. M. Bedair, *Appl. Phys. Lett.* **52**, 1144 (1988).
49. Q. Chen, J. S. Osinski, and P. D. Dapkus, *Appl. Phys. Lett.* **57**, 1437 (1990).
50. W. G. Jeong, E. P. Menu, and P. D. Dapkus, *Appl. Phys. Lett.* **55**, 245 (1989).
51. T. Meguro, T. Suzuki, K. Ozaki, Y. Okano, A. Hirata, Y. Yamamoto, S. Iwai, Y. Aoyagi, and S. Namba, *J. Cryst. Growth* **93**, 190 (1990).
52. S. M. Bedair, H. Liu, J. C. Roberts, and J. Ramdani, *Fifth Int. Conf. on MOVPE*, Aachen, Germany (1990).
53. P. D. Brewer, D. McClure, and R. M. Osgood, Jr., *Appl. Phys. Lett.* **47**, 310 (1985).
54. J. E. Epler, D. W. Treat, H. F. Chung, T. Tjoe, and T. L. Paoli, *Appl. Phys. Lett.* **54**, 881 (1989).
55. J. E. Epler, D. W. Treat, and T. L Paoli, *Appl. Phys. Lett.* **56**, 1828 (1990).
56. G. S. Tompa, M. A. McKee, C. Beckham, P. A. Zawadzki, J. M. Colabella, P. D. Reinert, K. Capuder, R. A. Stall, and P. E. Norris, *J. Cryst. Growth* **93**, 220 (1988).
57. J. Komeno, H. Tanaka, N. Tomesakai, H. Itoh, T. Ohori, M. Takikawa, M. Suzuki, and K. Kasai, *J. Cryst. Growth* **105**, 30 (1990).
58. P. M. Frijlink, *J. Cryst. Growth* **93**, 207 (1988).
59. A. Mircea, A. Ougazzaden, P. Daste, Y. Gao, C. Kazmierski, J.-C. Bouley, and A. Carenco, *J. Cryst. Growth* **93**, 235 (1988).
60. C. G. Cureton, E. J. Thrush, and A. T. R. Briggs, *Fifth Int. Conf. on MOVPE*, Aachen, Germany (1990).
61. H. Tanaka, T. Kikkawa, K. Kasai, and J. Komeno, *Jpn. J. Appl. Phys.* **28**, L901 (1989).
62. S. G. Hummel, C. A. Baylor, Y. Zou, P. Grodzinski, and P. D. Dapkus, *Appl. Phys. Lett.* **57**, 695 (1990).

63. B. I. Miller, M. G. Young, U. Koren, M. Oron, and D. Kisker, *Appl. Phys. Lett.* **56**, 1439 (1990).
64. W. J. Duncan, D. M. Baker, M. Harlow, A. English, A. C. Burness, and J. Haigh, *Electon. Lett.* **25**, 1603 (1989).
65. D. M. Baker, W. J. Duncan, M. D. Learmouth, and T. G. Lynch, *Electron. Lett.* **25**, 1598 (1989).
66. M. Abdalla, D. G. Kenneson, W. Powazinik, and E. S. Koteles, *Appl. Phys. Lett.* **57**, 494 (1990).
67. T. R. Omstead, P. M. Van Sickle, P. W. Lee, and K. F. Jensen, *J. Cryst. Growth* **93**, 20 (1988).
68. R. S. Sillmon and J. A. Freitas, Jr., *Appl. Phys. Lett.* **56**, 174 (1990).
69. D. N. Buckley, C. W. Seabury, J. L. Valdez, G. Cadet, J. W. Mitchell, M. A. Digiuseppe, R. C. Smith, J. R. C. Filipe, R. B. Bylsma, U. K. Chakrabarti, and K. W. Wang, *Appl. Phys. Lett.* **57**, 1684 (1990).
70. J. F. O'Hanlan and D. B. Fraser, *J. Vac. Sci. Technol A* **6**, 1226 (1988).
71. S. D. Hersee and J. P. Duchemin, *Annu. Rev. Mater. Sci.* **12**, 65 (1982).
72. P. D. Dapkus, *Annu. Rev. Mater. Sci.* **12**, 243 (1982).
73. G. B. Stringfellow, "Semiconductors and Semimetals," vol. 22, "Lightwave Communications Technology" (W. T. Tsang, ed.), p. 209. Academic Press, New York, 1985.
74. M. Razheghi, "Semiconductors and Semimetals," vol. 22, "Lightwave Communications Technology" (W. T. Tsang, ed.), p. 299. Academic Press, New York, 1985.
75. J. J. Coleman and P. D. Dapkus, in "Gallium Arsenide Technology" (D. K. Ferry, ed.), p. 79. Howard W. Sams, Indianapolis, 1985.
76. M. J. Ludowise, *J. Appl. Phys.* **58**, R31 (1985).
77. L. M. Miller and J. J. Coleman, *CRC Crit. Rev. Solid State Mater. Sci.* **15**, 1 (1988).
78. P. D. Dapkus and J. J. Coleman, in "III–V Semiconductor Materials and Devices" (R. J. Malik, ed.), p. 147. North Holland, Amsterdam, 1989.
79. M. Razheghi, "The MOCVD Challenge," vol. 1, "A Survey of GaInAsP- InP for Photoic and Electronic Applications." Adam Hilger, Philadelphia, 1989.

Chapter 5

LATTICE-MISMATCHED HETEROEPITAXY

S.F. Fang and Hadis Morkoç

Department of Electrical and Computer Engineering
Coordinated Science Laboratory
Materials Research Laboratory
University of Illinois
Urbana, Illinois 61801

1. INTRODUCTION

The interest in the deposition of various compound semiconductors on Si substrates stems from the well-established Si-based electronics technology,

ISBN 0-12-200420-5

the abundance of silicon and its low cost, and the excellent properties of compound semiconductors compared with Si. Early efforts were focused on the exploration of high-efficiency compound semiconductor solar cells on Si. This is to take advantage of large absorption coefficients, large open circuit voltages, and high junction operating temperatures afforded by compound semiconductors. The fact that the native substrates for these compounds are expensive, heavy, fragile, and unavailable in large areas and quantities makes the heteroepitaxy all the more attractive. Therefore, depositing GaAs on Si for solar cells was readily justified. In addition, the thermal conductivity of Si is better than that of GaAs by almost threefold. In fact, when space applications are contemplated, the light weight of Si, almost one-third that of GaAs, also becomes a strong motivator. If high-quality tunnel junctions could be obtained, high-efficiency tandem solar cells utilizing both compound semiconductors and Si solar cells and taking advantage of a larger portion of the Sun's spectrum can also be expected.

Against this background, researchers set out to find ways to grow GaAs on Si. Early efforts were limited to schemes where Ge was grown on Si first, which is relatively easy to accomplish, followed by the GaAs growth. The argument was that it would be easier to grow GaAs on Ge because of the excellent lattice and thermal match between them and the large body of experimental data available in the literature. Despite the progress made, the presence of Ge turned out to be more of a limitation than a promoter because of the chemical instability between GaAs and Ge and ease with which they diffuse into each other, changing the composition as well as acting as dopants.

With the advent of molecular beam epitaxy (MBE) and other refined crystal growth techniques, unprecedented control over growth process and parameters became feasible. As a result, researchers began experimenting with growth process parameters, particularly those having to do with the initiation of growth. When this was coupled with the *in situ* growth-monitoring capabilities afforded by MBE, it became possible to unravel the less understood features of heteroepitaxy. In particular, the use of miscut substrates along with the presurface treatments made it possible to grow material with a single domain, lacking, for example, the two phases of GaAs rotated with respect to another by 90° (otherwise called antiphase defects).

The developments in the crystal growth and detailed surface studies paved the way for a better understanding of the Si surface and its interaction with As and Ga. In the process, researchers were able to grow single-domain GaAs and other compound semiconductors readily. The results obtained in one laboratory were soon repeated in other laboratories, indicating the level of understanding. With the antiphase domain issue behind, the effort was soon turned to the ever important issue of dislocations, inevitable in the epitaxial films because of the 4% lattice mismatch between GaAs and Si. Much progress was made in this area, and it was discovered that proper miscut from the major (001) plane leads to the formation of perfect edge dislocations which are confined to the interface, not degrading the quality of the epitaxial layers; also, fewer of them are required as they are capable of relieving more strain than the troublesome threading dislocations. Additional methods, such as use of strained layer superlattices to bend out the dislocations and thermal cycling during and after growth, have led to drastic reduction in the number of dislocations to about 10^5 cm^{-2} [1].

With the advances in the quality of the epitaxial films, efforts were made to fabricate discrete electronic devices and optical devices. Although much progress has been made in electronic devices, optical devices have been slow in coming. Low-threshold lasers have been obtained, but their lifetimes have not been encouraging. These shortcomings have been attributed to the remaining dislocations and the residual strain caused by the thermal mismatch between GaAs and Si. Although preliminary in nature, optical devices driven by Si circuitry on Si substrates have been successfully demonstrated.

This chapter is arranged to address the issues involved in and the principles of the crystal growth. This is followed by a detailed discussion of optical devices—passive devices, such as modulators, and active devices, such as lasers.

2. STRAINED LAYERS AND RELATED GROWTH ISSUES

Developments in epitaxial growth techniques such as MBE and organometallic vapor phase epitaxy (OMVPE) have made it possible to grow multicomponent epitaxial heterostructures with precise composition and thickness control. These new methods have paved the way for growing metastable lattice-mismatched strained structures different from those possible at equilibrium. Consequently, even artificially layered structures that are lattice mismatched can be realized and investigated.

The mechanisms describing epitaxial growth on a substrate with a different lattice constant are very much dependent on the materials involved and details of the overall structure. The lattice-mismatched layer is assumed to be strained coherently without the generation of misfit defects if the layer thickness is kept below a critical thickness. The strain in a layer with a lattice deformation of Δa is

$$\varepsilon = \frac{\Delta a}{a_0} \tag{1}$$

where a_0 is an undeformed lattice constant.

In a semiconductor with a zincblende or diamond structure, as an isotropic continuum, the strain energy density is [2]

$$\mu = 2\mu\varepsilon^2 \frac{1 + \nu}{1 - \nu} \tag{2}$$

where μ is the shear modulus and ν is Poisson's ratio. The misfit strain energy increases with the thickness of a strained layer. When the thickness of a strained layer reaches a critical value, it becomes energetically favorable for misfit dislocations to be generated. The strain will be gradually relaxed by misfit dislocations as the layer thickness increases. Dislocations in an active layer degrade device properties; therefore, it is important to grow strained layers within the critical thickness, otherwise they would be partially or completely relaxed.

The critical thickness has been calculated by several workers on the basis of energy minimization and force balance, each with different results [2–6]. Here we will attempt to summarize the principles behind those calculations. Consider a heterostructure in which an epitaxial layer of thickness h is grown on a thick substrate and with a thick capping layer. The substrate and the capping layer are of the same material. The lattice mismatch rate between the epitaxial layer and substrate is f. From the minimum-energy concept, the critical thickness h_c is derived as [3]

$$h_c = \frac{b(1 - \nu \cos^2 \alpha)}{4\pi f(1 + \nu) \cos \gamma} \ln\left(\frac{B}{b}\right) \tag{3}$$

where b is the Burgers vector, α is the angle between b and the direction of dislocation (here, α is 0 for a screw dislocation, 90° for an edge dislocation, and 60° for a 60° dislocation), γ is the angle between b and $\bar{h}$, with $\bar{h}$ a unit vector that lies in the interface plane and is orthogonal to the direction of dislocation. B represents the extent of the stress field radius due to a dislocation whose core is within a cylinder of radius b, the Burgers vector. There are two important forces on dislocation lines: F_e, the force exerted by the misfit strain, and F_1, the tension in the dislocation line. From the imbalance of these forces, the critical thickness is derived as [4]

$$h_c = \frac{b}{2\pi f} \frac{(1 - \nu \cos^2 \alpha)}{(1 + \nu) \cos \gamma} \left(\frac{\ln h_c}{b + 1}\right) \tag{4}$$

The critical thickness resulting from the energy minimization method has been found to agree well with the experimental result in the case of Si_xGe_{1-x} grown on Si substrates [5]. For this heterosystem it is predicted that the critical thickness will range from about 100 Å for 2% mismatch to approximately 1000 Å for a 1% mismatch.

In many cases, metastable strained layers exceeding the critical thickness can be grown. Strained as-deposited layers in this metastable state may relax if the thermal treatment associated with device fabrication is too high. It is thus important to know whether a given strained-layer structure is stable or metastable. Tsao *et al.* [6] proposed that strained-layer breakdown is most directly determined by an excess stress (the difference between that due to misfit strain and that due to dislocation line tension) and temperature. Figure 1 shows strain relief for different layer thicknesses and misfits associated with a series of Si_xGe_{1-x} films grown at about 500°C [6]. Using the concept of excess stress, a stability criterion for strained heterostructures is discussed in the context of two kinds of processes [7]. The process in which a misfit section is formed at a single interface is called a "single-kink" mechanism,

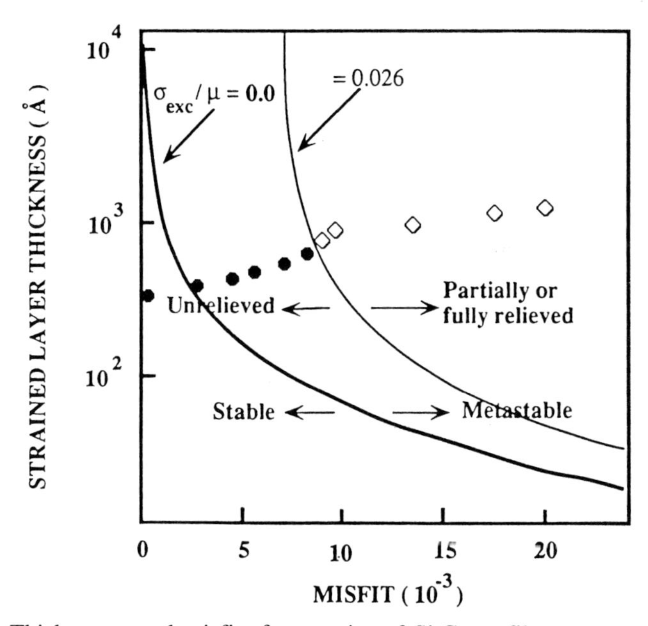

Fig. 1. Thicknesses and misfits for a series of Si_xGe_{1-x} films grown at approximately 500°C. Fully strained (filled circles) and partially strain relieved (open circles) films are separated by the d_{exc} = 0.026 isobar (d_{exc} is the difference between the stress due to misfit strain and that due to dislocation line tension). The d_{exc} = 0 isobar separates absolutely stable from metastable films. (After Ref. 6.)

and that in which misfit sections are formed at two interfaces is called a "double-kink" process. If the excess stress in either process exceeds zero anywhere within the structure, the structure is unstable or metastable to strain relief.

3. HETEROEPITAXY

Heteroepitaxy has a general problem of inherence: lattice mismatch, thermal expansion mismatch, and polar with nonpolar for heteroepitaxy of compound materials on element materials. Defects are induced into the epitaxial layer because of lattice mismatch. We define the lattice mismatch of the growth layer (lattice constant a_1) and the substrate (lattice constant a_0) as

$$f = \frac{a_1 - a_0}{a_0} \tag{5}$$

If the heteroepitaxially grown layer is completely relaxed by dislocation generation, and the Burgers vector of the dislocations is assumed to be parallel to the GaAs/Si interface, then an estimate of the dislocation density in the epilayer is

$$D = \left(\frac{a_1}{f}\right)^{-2} \tag{6}$$

For GaAs on Si, a_0, a_1, and f are 5.65 Å, 5.43 Å, and 4%, respectively. Therefore, the dislocation density in unstrained GaAs grown on Si would be as high as 5×10^{11} cm^{-2} before mutual annihilation. Epitaxial layers are usually grown at high temperatures. During cooling of the grown heterostructure down to room temperature, dislocations can also be induced because of differences in thermal expansion coefficients. For compound semiconductors grown on elemental semiconductors or vice versa—that is, polar structure on nonpolar structure—additional defects called antiphase domains (APDs) may also form. APDs are discussed in Section 3.1.

A heteroepitaxial system of GaAs on Si is representative of lattice-mismatched and polar-on-nonpolar semiconductor heterostructures. The heteroepitaxy of GaAs on Si is of great interest and has been extensively studied [8]. The advantages of GaAs on Si are the large size, high quality, low cost, high thermal conductivity, and high mechanical strength offered by Si substrates. Furthermore, one of the most important features of GaAs on Si is the possibility of monolithic integration of optical devices made of GaAs and GaAs buffered compound semiconductors with electrical devices made of Si. In this section, we discuss mainly the heteroepitaxy of GaAs on Si. The knowledge gained from this system will also be useful for other heteroepitaxial systems.

For epitaxial growth of high-quality GaAs on Si, the inherent problems—antiphase disorder due to polar-on-nonpolar epitaxy, 4% lattice mismatch between GaAs and Si, and 60% mismatch in thermal expansion coefficients—must be dealt with and/or alleviated.

3.1. Antiphase Domains (APDs)

Antiphase disorder used to be the biggest problem in the earlier studies. The surface of Si (100) exhibits an orthogonal 2×1 unit cell due to the dimerization of Si atoms at the surface [9]. An Si (100) surface misoriented toward [011], which is usually used for GaAs-on-Si growth, has basically two different single steps on it, one with the step edge parallel and the other with the step edge perpendicular to the dimerization direction. If a terrace of the Si surface exhibits 2×1 reconstruction, the terrace one step higher or lower than it exhibits 1×2 reconstruction, so, through superposition, the surface shows double-domain 2×1 reconstruction which appears as symmetric 2×2 reconstruction when observed from two orthogonal directions (Fig. 2). However, the edge atoms at the steps are similar to the surface atoms and have dangling bonds, and some of these dangling bonds may

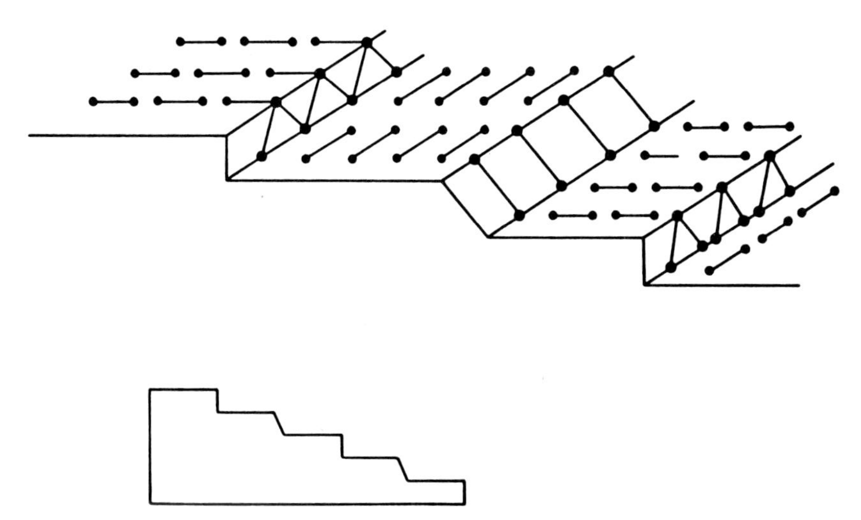

Fig. 2. Top and side views of a Si surface with monolayer-high steps. A terrace of the surface shows 2×1 reconstruction and the terrace one step higher or lower than that shows 1×2 reconstruction.

rebond at the step edge. The rebonding of dangling edge bonds influences the formation energy and the stability of steps and can be altered by high temperature or chemical treatment. Chadi [10] used a tight-binding-based total energy calculation approach to evaluate the step formation energy and found that a double step with a step edge perpendicular to the dimerization direction is energetically more favored than two types of single steps combined.

Si and GaAs crystalline structures consist of two interpenetrating face-centered cubic Bravais lattice space groups. In the case of Si, the two fcc lattices are the same. It is invariant to a rotation of $\pi/2$ and the [011] and $[01\bar{1}]$ directions are equivalent. In the case of GaAs, one fcc lattice is occupied by Ga and the other by As, and the [011] and $[01\bar{1}]$ directions are not equivalent. This distinction is visible in the (100) planes of GaAs (zinc-blende structure), which consist of alternating layers of gallium and arsenic.

When GaAs is grown on a GaAs substrate, the Ga and As atoms experience no ambiguity in choosing lattice sites; the epilayer just mimics the crystalline structure of the substrate. However, when epilayers of GaAs are grown on Si (100), Ga and As atoms can exhibit ambiguity in choosing lattice sites. The lattice sites on the (100) planes are indistinguishable, so there are no preferential nucleation sites for Ga and As. If the growth is started with simultaneous exposure to gallium and arsenic molecular beams, gallium may form the initial layer on some areas of the substrate and arsenic on rest of the surface; i.e., the first monolayer is part gallium and part arsenic on the Si surface. This results in arsenic–arsenic or gallium–gallium bonds as shown in Fig. 3a. Such arsenic–arsenic or gallium–gallium bond boundaries are called an-

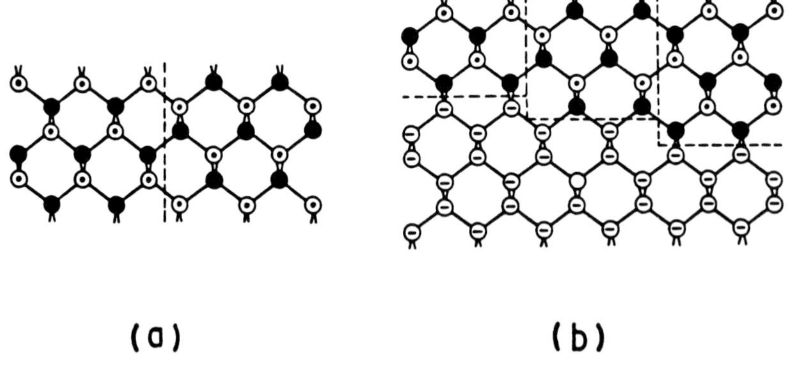

Fig. 3. (a) Antiphase boundary formation on silicon surface with no steps due to a nonuniform initial monolayer. (b) The presence of monolayer-high steps can lead to the formation of APBs even when the initial monolayer is uniform. (For clarity, the lattice mismatch is not shown.)

tiphase boundaries (APBs). The formation of antiphase disorders in the GaAs epilayers is aided by monolayer-high steps (or steps that are an odd number of atomic layers high). Since the GaAs (100) planes alternate between cation and anion planes, the single steps (monolayer or odd number of atomic layers high) cause a perturbation of the order of the (100) planes. This is illustrated in Fig. 3b. However, formation of APDs requires large energies, and the inherent growth kinetics may not favor them. So it is possible to grow APD-free epilayers in spite of the presence of monolayer-high steps and/or a nonuniform initial monolayer on the surface.

Antiphase boundaries are charged structural defects. To aid in understanding their ill-effects, we can make simple calculations based on the number of bonding orbitals and the number of valence electrons available to fill them. Ga—Ga bonds have electron deficiency and so act as acceptors. As—As bonds have excess electrons and so act as donors. In general, there are equal numbers of As—As and Ga—Ga bonds. Thus, the epilayer behaves like a highly compensated semiconductor, degrading the performance of devices fabricated on it. The APDs on Si (100) or the lack of them has been investigated [11–17], and two distinct solutions have been found to suppress the formation of APDs. They are described in the following.

3.1.1. Si (100) Misoriented Surface

We saw from the simple model just discussed that the formation of APDs is due to nonuniform surface coverage of the initial monolayer and monoatomic-layer-high steps on the surface. Preexposure of either Ga or As for an adequate time to ensure uniform coverage of the surface and the use of a tilted substrate to ensure steps that are two (or an even number of) atomic layers high should eliminate the formation of APDs.

Gallium and arsenic have different optimum temperatures for bonding to silicon. As_2 has a higher sticking coefficient to silicon than As_4. So in MBE growth, it is preferable to use a cracker for the arsenic source, which otherwise produces As_4 tetramers. The initial substrate temperature depends on the choice of prelayer. After much experimentation, the optimal initial temperature for an As_2 prelayer was found to be around 460°C in MBE growth [11, 18]. Corresponding to this initial temperature, the optimal growth rate is about 0.1–0.2 μm/h. The arsenic prelayer on a Si surface forms As—As dimers [19] and is very unreactive; it is 10^{-11} times as reactive as a clean Si surface with oxygen [20]. Arsenic does not stick well to the almost inert As layer over silicon, so the preexposure duration is not critical for arsenic. On the other hand, Ga sticks to a layer of Ga on top of silicon and "pile-up" can occur. When Ga prelayers are used, precise control of the duration of preexposure is required to make sure that on the average just a monolayer of Ga is deposited on the Si surface and the substrate temperature is consistent with such a goal. In addition, the surface migration of Ga on a Si surface is high, hence a low initial substrate temperature (around 500°C) and low initial growth rates are also used for the Ga prelayer.

It is important to make sure that the substrate heating is uniform. Otherwise, two effects can lead to the formation of APDs. One effect is that different initial growth temperatures may lead to two distinct kinds of Si (100)/As prelayer structures, which lead to two different crystal orientations of GaAs layers. This is discussed in Section 3.2. The other effect is that an As initial layer may form on cooler areas of the surface and a Ga initial layer on hotter areas, leading to antiphase boundaries. A heating gradient, as shown in Fig. 4, was used on a Si sample to force the formation of APDs in the GaAs epilayer. The sticking coefficient of As is apparently high in cool regions and almost null in the hotter regions. If the growth is started with an As prelayer, the hot part will actually start with the Ga prelayer. This is depicted in Fig. 5, where on the left-hand side of Fig. 5a, the growth is started with a Ga prelayer, whereas the right-hand side indicates the As prelayer region (where the two meet, APBs form). Magnified views of these regions are shown in Fig. 5b–d. Near the As prelayer side of the transition region, the As prelayer domains are larger than the Ga prelayer islands. Near the Ga prelayer transition region, the situation is reversed. In the middle of the mixed-domain region the island sizes are expected to be comparable.

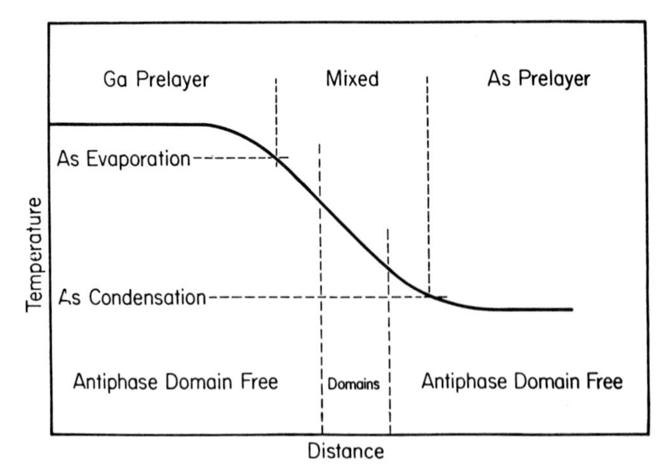

Fig. 4. Phenomenological description of As and Ga prelayer determined domains on either side of a temperature gradient introduced during the initiation of the growth. In the regions where the domains merge an antiphase structure is formed.

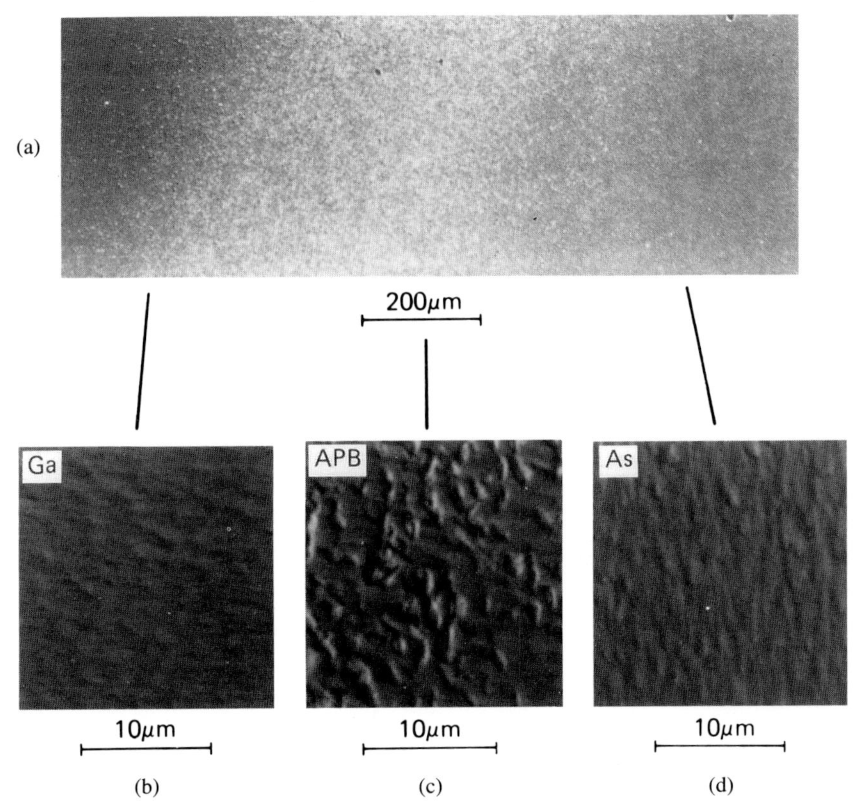

Fig. 5. Photomicrograph of a surface of GaAs film (2 μm thick) on Si near the temperature gradient described in Fig. 4. (a) All three regions are shown: Ga prelayer on the left, As prelayer on the right, and mixed domains in the center. (b–d) All three regions at a much larger magnification.

The important conclusion to be drawn from Fig. 5 is the narrowness of the transition region (~150 μm). This should be contrasted with the fact that the temperature gradient was induced over a much larger width (several mm); the actual value is not known accurately. This simply means that intrinsic growth kinetics are not favored. This point is further explored in the following discussion of the growth of GaAs on an exact (100) Si surface.

3.1.2. Si (211) Surface

Si (211)–oriented substrates have been used to prevent the formation of antiphase domains [13, 14, 21]. On the Si (211) surface a bonding site in one sublattice has two dangling bonds, while the site in the other sublattice has only one dangling bond. This distinction between the sublattices enables anions and cations to have preferential bonding sites and thus avoid antiphase boundaries. Because GaAs epilayers with superior device quality have been achieved in the more commonly used (100) plane, it may not be necessary to switch to the (211) orientation. The surface atomic density is lower on the (211) planes than the (100) planes. This alters the growth rate, dopant incorporation, and so forth on the two different surfaces.

APD-free GaAs epilayers can also be obtained on exact (100) Si substrates [22]. Heat treatment of the substrate at 1000°C for 30 min before initiation of growth was speculated to be crucial for APD-free epilayers on exact (100) surfaces. It has been reported that high-temperature treatment leads to step doubling [23, 24]. The formation of APDs requires large energies and the inherent growth kinetics may not favor their formation. When smaller domains are formed, massive atomic transfer may take place to eliminate the small domains, e.g., As prelayer domains in or surrounded by a Ga prelayer and vice versa. Naturally, other defects form during this process. Sometimes two APBs inclined toward each other may annihilate one another as illustrated in Fig. 6. This may also contribute to the suppression of APDs.

It has been established experimentally that exactly oriented (100) Si substrates can be used to get APD-free GaAs layers. However, this results in a much higher dislocation density in the epilayer than on misoriented substrates [25].

APD-free GaAs is now routinely grown on Si (100) and Si (100) mis-

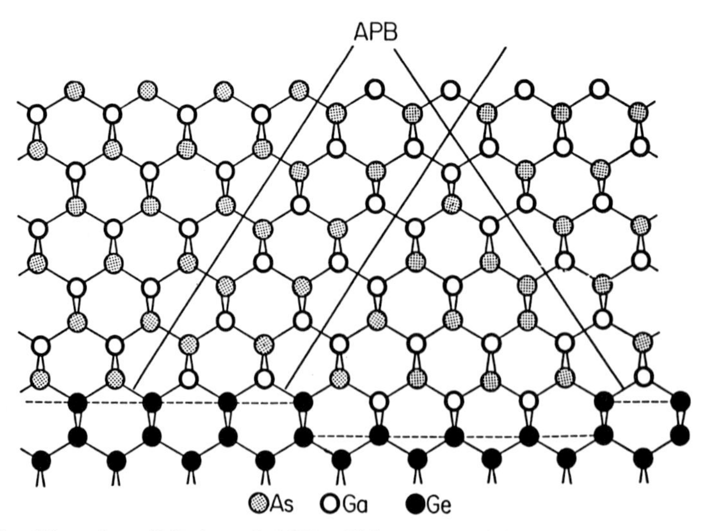

Fig. 6. Mutual annihilation of APBs. This may explain the successful APB-free growth on exact (100) silicon surfaces which have down steps and up steps.

oriented 4° toward [011] [26]. This was initially attributed to the ability to generate a surface consisting entirely of double steps by proper pregrowth annealing of the Si substrates [26]. However, contradictory work by several researchers shows that APD-free GaAs can be grown on Si (100) in the presence of steps of odd height [27–32]. Mass transfer [29] and APD self-annihilation [30, 32] have been suggested to explain the elimination of APDs; however, the growth mechanism of APD-free GaAs on Si is still not fully understood.

3.2. Growth Initiation

Growth initiation from the viewpoint of initial sublattice allocation and antiphase domain suppression was discussed in the last section. Exactly what happens on the surface of the Si substrate during growth initiation is discussed in detail in this section. The kinematics of initial nucleation, size of nucleated islands, morphology of initial epilayers, nature and strength of Si–Ga and Si–As bonds, and so forth are also discussed in depth in this section.

3.2.1. Arsenic Prelayer

Recall from the last section that the As-terminated Si surface is highly unreactive. The nature of the As-terminated Si surface is explored further in this section. Depending on the growth (substrate) temperature, an arsenic-terminated silicon surface exhibits two different surface reconstructions [33, 34]. At substrate temperatures above 450°C, the Si (100)/As surface shows dimerized 1×2 reconstruction. It is important to bear in mind that the Si (100) surface exhibits a 2×1 reconstruction. For substrate temperatures below 450°C, the Si (100)/As surface exhibits the same reconstruction as the Si substrate, i.e., the 2×1 reconstruction. The bonds formed at lower temperatures, when heated to temperatures above 450°C, change their reconstruction. This change is permanent; when the surface is cooled, it does not go back to the original 2×1 reconstruction. It has also been established experimentally that the Si (100)/As surface formed at higher temperatures is far less reactive than the Si (100)/As surface formed at lower temperatures [33, 34]. All these results have led to the conclusion that the Si—As bond formed at lower temperatures (<450°C) is weak and the one formed at higher temperatures is quite strong.

The two distinct kinds of Si (100)/As prelayer structure lead to two different crystal orientations of the GaAs epilayer. The strongly bonded (high-temperature) Si (100)/As structure leads to SiAs/GaAs stacking. The weakly bonded (low-temperature) Si (100)/As structure can lead to SiGaAs or SiAsAsGa stacking. Reflection high-energy electron diffraction (RHEED) studies have shown that SiGaAs stacking is more probable than SiAsAsGa stacking [33]. The initial stacking of atoms is also strongly dependent on surface orientation, strain relaxation, etc., and this area requires further investigation.

The As prelayer used, regardless of the growth temperature, affects the surface mobility of Ga and initial island formation [35, 36]. After the initial As monolayer has formed on Si, the additional As sticks exclusively to Ga on the surface. The Ga mobility is proportional to surface temperature and inversely proportional to excess As present. A relatively high surface temperature and/or less As flux leads to higher mobility of Ga on the surface. High mobility of Ga atoms due to high surface temperature interferes with nucleation processes; i.e., the atoms are too mobile to form islands in one place. However, the high mobility of Ga atoms due to low arsenic pressure (less As flux) at a relatively low surface temperature has been successfully exploited to grow good-quality epilayers of GaAs on Si [37–39]. This process is called migration-enhanced MBE.

3.2.2. Gallium Prelayer

Although most groups favor the use of an As prelayer because of the ease with which an As-stabilized surface can be obtained, a Ga prelayer also works. Naturally, the Ga-terminated Si surface has not been as extensively investigated as the As-terminated Si surface. Unlike As, Ga does not form a stable single-monolayer surface on Si [36, 40, 41]. At low monolayer (ML) coverage (less than 0.1 ML), virtually all of the Ga atoms are in rows perpendicular to the dimer rows (parallel to the Si dimerization direction). Most of these rows seem to start or stop at steps, growing perpendicular to A-type steps and parallel to B-type steps. At higher monolayer coverage (typically above 0.1 ML), large areas are covered by arrays of Ga atoms with 3×2 periodicity. The models for SiGa bonds propose that the Ga bonds to the surface as dimers, lying between the Si dimer rows. It is interesting to note that the adsorbed Ga forms rows perpendicular to the Si dimers, analogous to Ga dimer growth on GaAs (100). The submonolayer-thick Ga on Si (100) produces surface structures of 3×2, 5×2, 2×2, and 8×1 sequentially in the temperature range 350–680°C [40].

The orientation of the surface also has an effect on the type and strength of As/Ga bonding to Si. In the absence of Ga, a single monolayer of arsenic is very strongly bound to both Si (100) and Si (111). For the Si (111) orientation, the bonding appears to take place predominantly between Si and

As atoms. In this case, even when a Ga prelayer is used, the As atoms displace the predeposited Ga atoms and form As—Si (111) bonds [38].

3.3. Initial Nucleation

The shape of the initial nucleation after a few (four) seconds of growth is shown in Fig. 7. The area between the islands consists of As-terminated Si for GaAs on Si (111) and a thin layer consisting of both Ga and As for GaAs on Si (100) [36]. An increase in the distance between islands is observed with an increase in the growth temperature. This has been attributed to a decreasing nucleation density or increasing coalescence of growing supercritical nuclei or both [42]. It has been demonstrated that growth of the GaAs epilayer can be initiated at room temperature [43]. Room-temperature growth initiation leads to an amorphous, nonstoichiometric layer. This process results in an essentially two-dimensional nucleation. This is a new dem-

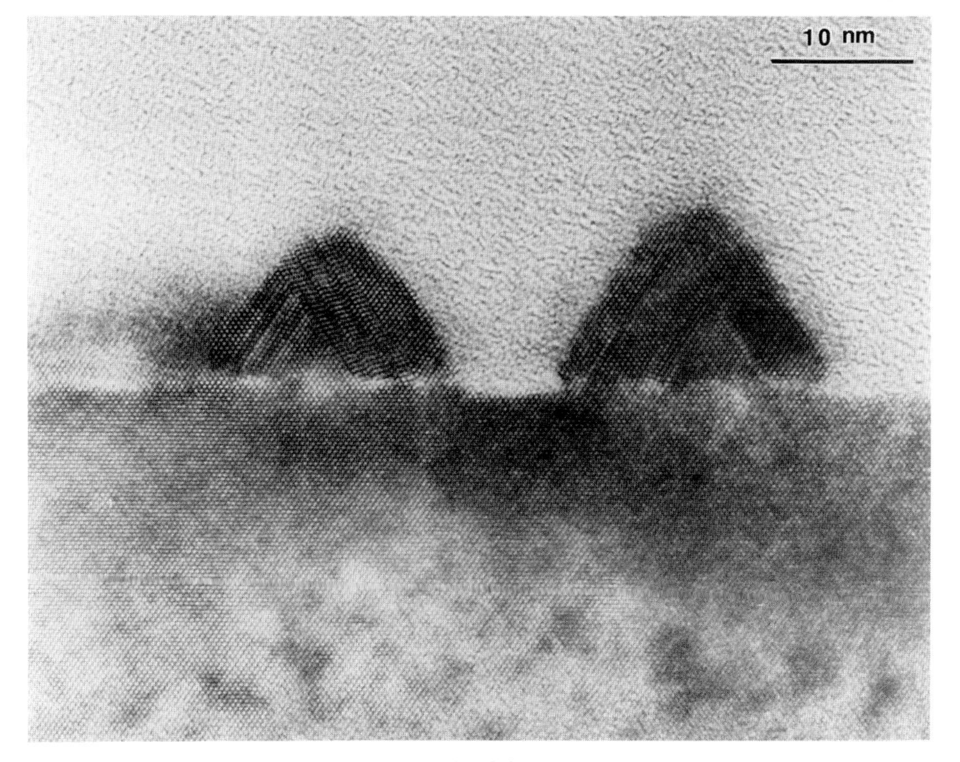

Fig. 7. TEM cross-section micrograph of GaAs on Si 4 seconds after growth initiation. (After Ref. 36.)

onstration, and further studies are required to evaluate its applicability and merits.

The GaAs cluster formation kinetics have also been numerically simulated and studied [44]. It has been shown that pyramidal as well as inverted pyramidal GaAs clusters can form with almost equal probability on As-terminated Si surfaces. However, the inverted pyramidal clusters grow at lower rates than pyramidal clusters. The shape of the initial cluster can affect the number of stacking faults (planar defects), and this is discussed in the section on dislocations.

3.4. Two-Step Growth Process

The growth of GaAs on a silicon substrate covered with a uniform layer of GaAs is similar to the growth of GaAs on a GaAs substrate. Therefore, the growth rate as well as the surface temperature can be increased after sufficient time has elapsed to ensure nucleation to cover the entire Si surface with GaAs epilayer. This growth initiation at low temperature and subsequent increase to a higher (normal) growth temperature are often referred to as the two-step growth process. We saw earlier how that initial temperature is chosen based on the choice of prelayer. A number of studies have been carried out to relate the choice of initial temperature to the quality of the initial (buffer) epilayer [18, 42, 45]. A low growth temperature of around 250°C leads to good surface morphology, but the quality of the initial buffer layer is not good at this temperature. As the initial temperature is increased, the surface morphology is degraded, tending toward white cloudiness of the surface [42]. However, the quality of the initial epilayer improves with increased temperature. When the second step of the two-step growth process is initiated, the higher temperature has an annealing effect on the initial buffer layer; i.e., the quality of the initial buffer layer is improved by subsequent high-temperature growth. This is due to the annealing of microtwins, but threading dislocations are generated in the process [46].

Some research groups favor stopping the growth a few minutes after the initiation, usually after the first 200 Å of growth, and raising the substrate temperature under arsenic overpressure [47, 48]. This technique allows the nucleation islands to coalesce and form a uniform layer of GaAs on silicon. RHEED patterns can be used to determine whether the coalescence has occurred.

3.5. Selective-Area Epitaxy

Selective area epitaxy (or patterned epitaxial growth) is required for cointegrations of silicon- and GaAs-based devices [49]. Processing temperatures for silicon devices are usually above 1000°C, which is much higher than that for GaAs devices (~600°C for MBE-GaAs, ~800°C for MOCVD-GaAs). Therefore, for cointegration of silicon- and GaAs-based devices, Si devices have to be fabricated and then epitaxial GaAs layers are grown on Si substrates in which device structures are already fabricated. In selective-area epitaxial growth on Si, the Si surface is protected by a dielectric coating (SiO_2 or Si_3N_4). The coating layers are lithographically patterned. Holes are usually etched onto the silicon substrate using buffered HF solutions. Epitaxial layers are deposited on the Si substrate through openings in the dielectric mask. In ideal selective growth, only epitaxial layers will grow in the unmasked areas of the pattern and on the coating mask. For MBE growth of GaAs on GaAs, selective growth requires high growth temperatures (e.g., 700°C) [50] that are well above the optimum temperatures for high-quality GaAs epilayers. Polycrystalline GaAs is normally deposited on the coating mask as a by-product of selective-area epitaxial growth. By selective-area epitaxy, strains and dislocations in the heterostructure were reduced by growth on small and separated areas [51–58]. Matthews *et al.* [59] pointed out that if threading dislocations were the source of misfit dislocations, one should be able to reduce the density of interface dislocations by limiting the lateral dimension of the sample before growth because too few threading dislocations would be present to nucleate a large number of misfit dislocations. Fitzgerald *et al.* [51] reported reduction of misfit dislocations from a mismatched InGaAs/GaAs interface by controlling the size of the growth area. They grew the layers on 2-μm-high pillars and showed by cathodoluminescence (CL) that misfit dislocations could be reduced to almost zero on very small pillars (lateral dimensions < 67 μm). Similar results were obtained by Hodson *et al.* [52], who demonstrated a factor of 20 reduction in the dislocation density of GaAs grown on silicon pillars as the distance to lateral free space was decreased.

Mismatches in thermal expansion of heteroepitaxial structures may cause wafer bowing [60] and cracks in thick epilayers [61]. These may present nontrivial problems for practical applications in industry—for example, in wafer handling and processing as in photolithography and in device reliability, especially when the device is subject to frequent thermal cycling. In selective-area epitaxy, these problems could be eliminated. In conventional whole area growth, GaAs layers grown on Si are in tensile biaxial stress due to the mismatch in thermal expansion, and photoluminescence (PL) peaks shift toward lower energy [62]. The thermal stresses in selective-area epitaxy for GaAs on Si have been investigated by Lingunis *et al.* [53, 54] and Karam *et al.* [55], using PL mapping and three-dimensional finite-element analysis. PL for selective epitaxial growth on opening windows of different sizes was investigated [54]. Monotonic reduction in the stress with decreased island size was observed for selective area epitaxy (SAE) by conventional MOCVD after oxide mask removal, as shown in Fig. 8. Yacobi *et al.* [57] observed the stress with SEM CL in GaAs grown on both InP and Si by selective-area epitaxy through patterned windows 10 to 200 μm wide. They found that the stress was uniaxial near the edge of a patterned region and changed to biaxial away from the edge. Ackaert *et al.* [58] demonstrated control of the location of microcrack formation through the design of a wedge mask. As shown in Fig. 9, the microcrack location and direction were controlled by the wedge position and wedge shape. There is a transition region between monocrystal grown on the open window and polycrystal grown on the coating mask. Dislocations may be generated in this region and penetrate into the monocrystal layer. However, more work is needed in order to better understand the phenomena and mechanism in selective-area epitaxy.

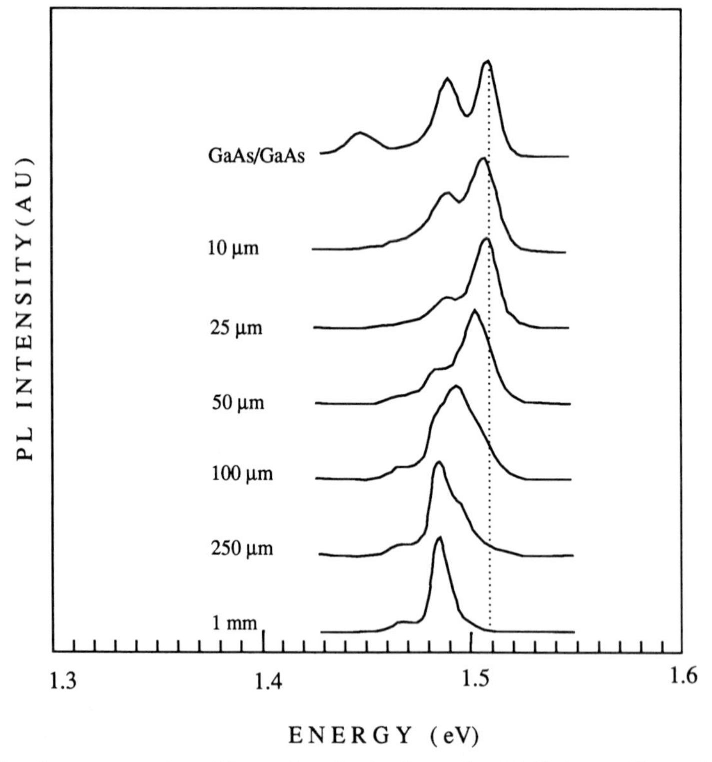

Fig. 8. Low-temperature PL spectra of selective epitaxial GaAs on Si as a function of island size, after oxide mask removal.

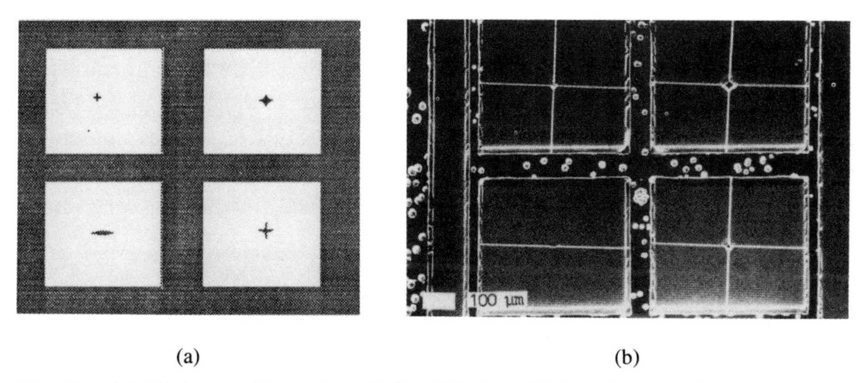

(a) (b)

Fig. 9. (a) Wedge configurations defined in the middle of a growth zone. (b) Microcrack formation in a 3-μm GaAs-on-Si layer structure. (After Ref. 58.)

3.6. Low-Temperature Epitaxial Growth

It is important to develop a low-temperature epitaxial growth scheme, especially in patterned epitaxy to avoid the degradation of devices that have already been formed. In OEICs, the patterned epitaxy is considered imperative, as we discussed in the last section.

Epitaxial growth of GaAs on Si commonly involves a high-temperature (>850°C) oxide desorption step. This is the highest temperature employed in conventional epitaxial growth of GaAs on Si. Various methods for low-temperature Si cleaning have been reported. Low-temperature Si cleaning was achieved by using an incident of Ga (at 800°C) [63], Si (at 800°C) [64], or Ge (at 625°C) [65]; by electron cyclotron resonance (ECR) hydrogen plasma (at 300°C) [66]; by a combination of incident Ga or In with hydrogen plasma (at <500°C) [67]; and more recently by employing hydrogen-passivated Si substrates (at ~500°C) [68, 69]. Epitaxial GaAs layers have been successfully grown on Si with low Si cleaning temperatures of 400°C [66], 500°C [67], and 600°C [69]. Hydrogen plasma has shown potential for low temperature epitaxial growth. However, residual impurities and damage induced by plasma bombardment need to be eliminated in order to obtain high-quality epitaxial layers.

The pregrowth Si cleaning temperature (T_{pre}) has been reduced to 600°C in MBE [69], almost the same as the epitaxial growth temperature. As part of the chemical preparation of the substrate, this method utilizes a final HF treatment whereby the Si surface dangling bonds are terminated by hydrogen with a resultant (1 × 1) bulklike surface structure. Upon medium-temperature heat treatment (~500°C), hydrogen leaves the surface, leading to the common orthogonal 2 × 1 surface reconstruction. Figure 10a shows a

(1 × 1) H RHEED pattern of Si as treated by HF, and Fig. 10b shows the two-domain (2 × 2) RHEED pattern of Si after the substrate temperature is increased above 530°C. Figure 11 shows the surface morphology of the GaAs layers grown under the same conditions but for various T_{pre} values, including that on the hydrogen-passivated surface. The GaAs layer, grown on Si still passivated by hydrogen [(1 × 1) H bulklike surface], showed a very rough

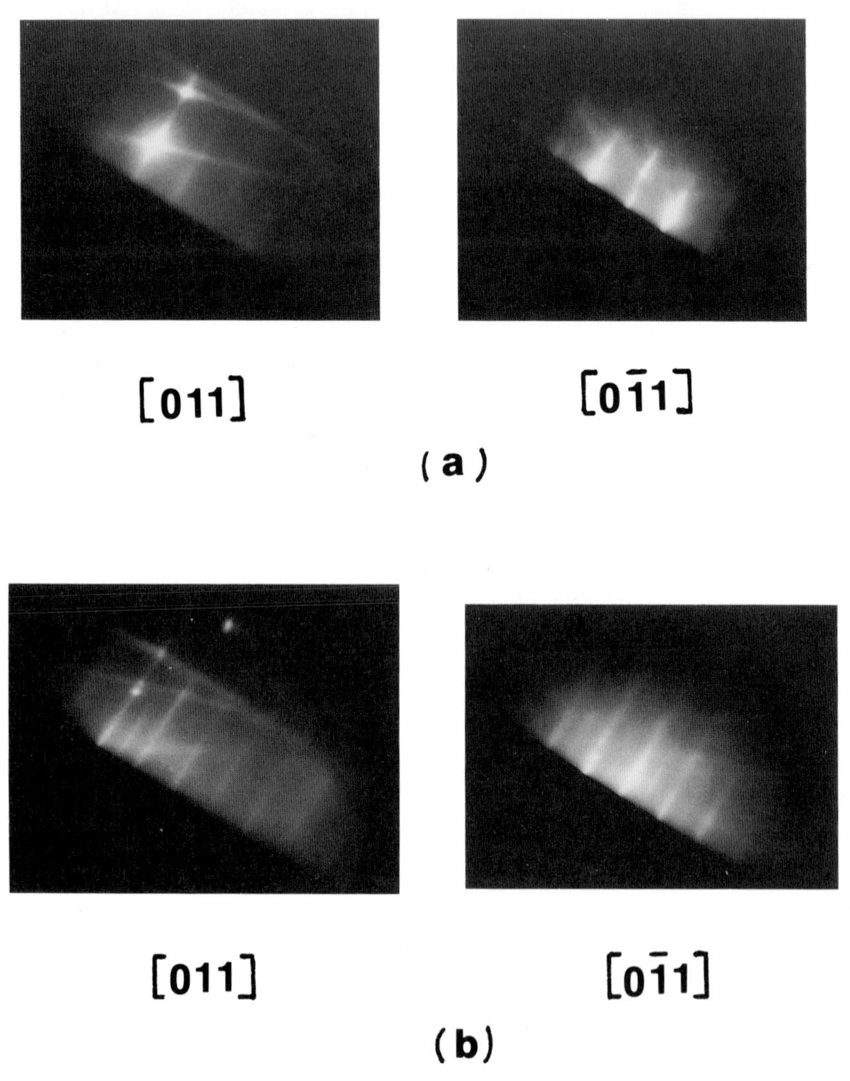

Fig. 10. RHEED patterns of (a) the 1 × 1 bulklike Si surface as HF etched and (b) the Si surface of a 2 × 2 reconstruction at temperatures above 530°C. (After Ref. 69.)

surface (Fig. 11a). For a T_{pre} of 530°C, although the surface hydrogen was mostly desorbed as determined by the observation of a two-domain 2×1 reconstruction, the GaAs layer was still rough (Fig. 11b). For T_{pre} higher than 600°C, the GaAs layers were of good morphology (Fig. 11c). Figure 12 shows the low-temperature (4 K) PL spectrum of two GaAs layers. The PL spectrum of GaAs on Si is shifted toward lower energies because of the tensile stress caused by the difference in thermal expansion coefficients during cooling from the growth temperature to room temperature. Peaks A and B at 826 nm (1.50 eV) and 837 nm (1.48 eV) associated with sample A are interpreted as the heavy-hole ($m_j = \pm 3/2$) and light-hole ($m_j = \pm 1/2$) bound exciton peaks. The $\pm 3/2$ and $\pm 1/2$ heavy- and light-hole bands are degenerate when the layers are stress free. The splitting observed here is caused by the residual biaxial tensile strain mentioned earlier [71, 72]. Full width at half-maximum (FWHM) for peak B is 5 meV for the GaAs layers grown with T_{pre} above 600°C, and the PL intensity is stronger for the sample grown with higher T_{pre}. This FWHM value compares well with the value for conventional GaAs-on-Si layers [72, 73], showing that good-quality GaAs epitaxial layers can be obtained on the 2×1 reconstructed Si surfaces with pregrowth substrate preparation temperatures as low as 600°C.

4. DISLOCATIONS AND OTHER DEFECTS

The lattice constant of gallium arsenide is larger than that of silicon. This mismatch causes strain in the epilayer. The energy associated with the strain is proportional to the thickness of the epilayer [5]. If the thickness of the epilayer is small (usually less than a critical thickness), the mismatch is accommodated by elastic deformation of the lattice; i.e., the GaAs lattice is compressed in the plane of growth and expanded in the plane perpendicular to the plane of growth. When the epilayers are thick, the associated strain energy is larger than the misfit dislocation energy. Then dislocations are generated to relieve strain energy as shown in Fig. 13.

Dislocations enable the growth and utilization of lattice-mismatched material systems and are not inherently harmful. Perfect edge dislocations are effective for release of strain energy, whereas inclined dislocations (60° in the case of GaAs on Si) thread to the surface and degrade the quality of the epilayer. Dislocations affect the quality of the epilayer in several different ways. Dislocations and other structural defects form nonradiative recombination centers. The number of these centers can be increased under intense photon fluxes, such as those in lasers. Another example is the increase in

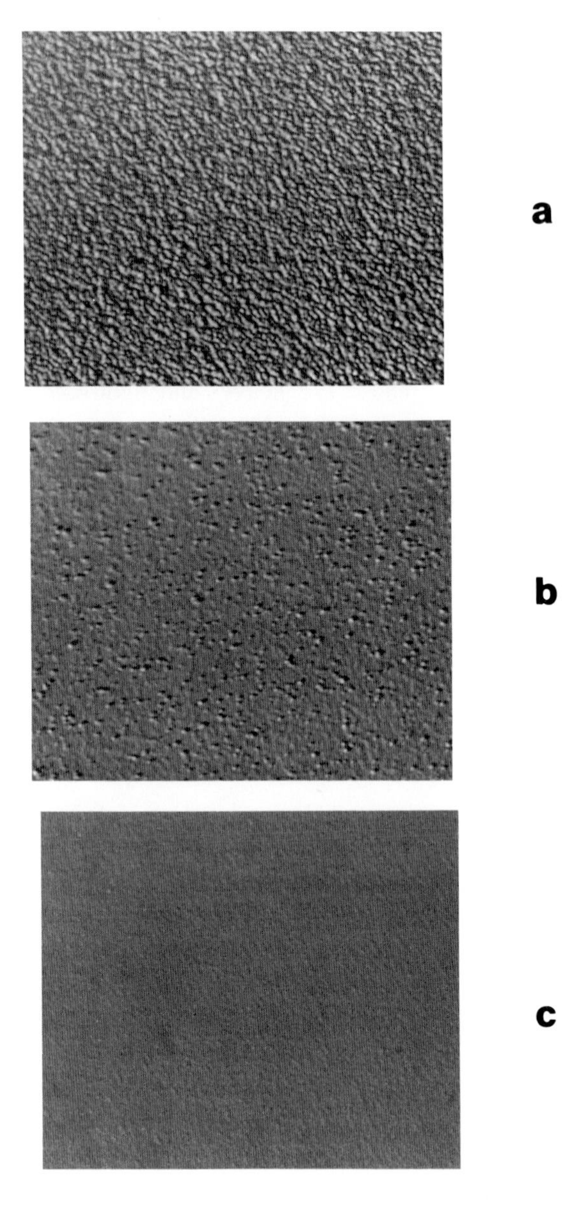

Fig. 11. Surface morphology from Nomarski optical microscopy of GaAs layers grown with T_{pre} values of (a) 430°C on 1 × 1 hydrogen terminated Si, (b) 530°C on 1 × 2 Si, and (c) >600°C.

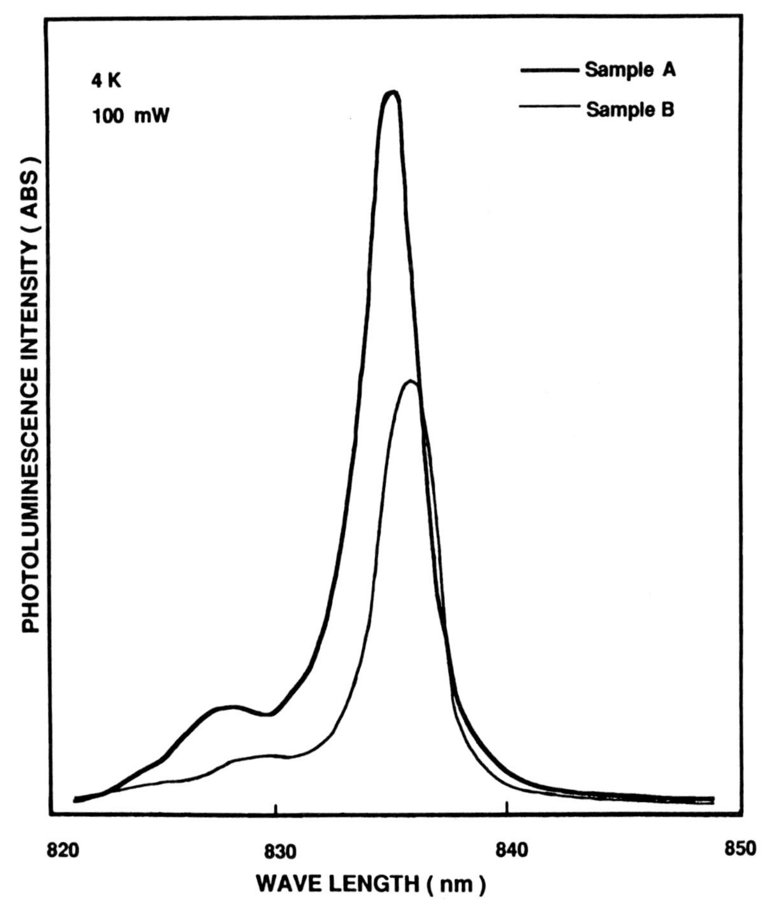

Fig. 12. Low-temperature (4 K) PL spectra of the GaAs layers grown with T_{pre} values of 675°C (sample A) and 600°C (sample B), respectively.

the ease of impurity diffusion along the line of dislocation thread; thus Si can diffuse from the substrate and contribute to autodoping in the epilayer. In the next few paragraphs, we review the classification scheme to distinguish among dislocations and the characteristics of different types of dislocations. Much emphasis will be placed on identifying parameters that aid the formation of benign dislocations. The different techniques used to reduce the threading of dislocations in the epilayer, namely the use of substrate tilt, pseudomorphic superlattices, and thermal treatment, are discussed in Sections 5.1, 5.2 and 5.3, respectively.

Dislocations are often specified in term of Burgers vectors. Knowledge about the characteristics of Burgers vectors aids better understanding of dis-

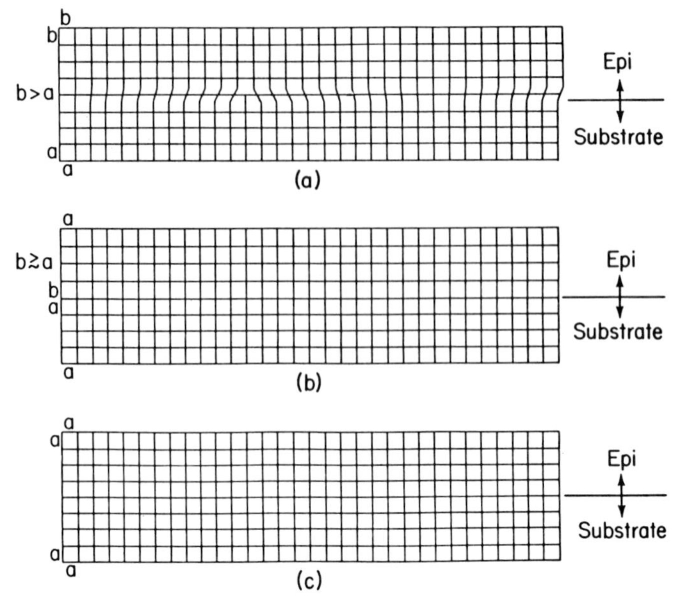

Fig. 13. Lattice arrangement of an epilayer on a substrate. (a) In the case of large lattice mismatch, e.g., 4% in the case of GaAs on Si, strain must be taken by the generation of misfit dislocations as shown. The dislocation shown is a perfect edge dislocation running perpendicular to the paper. (b) If there is a small mismatch, the mismatch is taken by coherent strain in the film. The example shows the film to be under compressive strain because of its larger lattice constant when relaxed. (c) The case of perfect mismatch.

locations. In a perfect crystal, there is no difficulty in choosing a closed path, where the path is defined by a series of displacements of the Bravais lattice. However, in a crystal with a defect, the same path fails to return to its starting point. This is illustrated in Fig. 14. The Bravais lattice vector that joins the starting point, B, and the end point, C, is the Burgers vector of the dislocation [74]. The Burgers vector is not dependent on the path chosen. The misfit accommodated by a dislocation is the projection of its Burgers vector onto the substrate plane. So any dislocation that has its Burgers vector parallel to the substrate plane is preferred over a dislocation whose Burgers vector is at an inclination to the substrate.

The commonly occurring dislocations in zincblende and diamond semiconductors can be classified in two different ways (1) those whose Burgers vectors are parallel to the growth (Si/GaAs interface) plane and (2) 60° dislocations (Fig. 15). The former is called a type I dislocation and the latter a type II [75, 76]. The type I dislocation has a Burgers vector of type 1/2 <011>, and

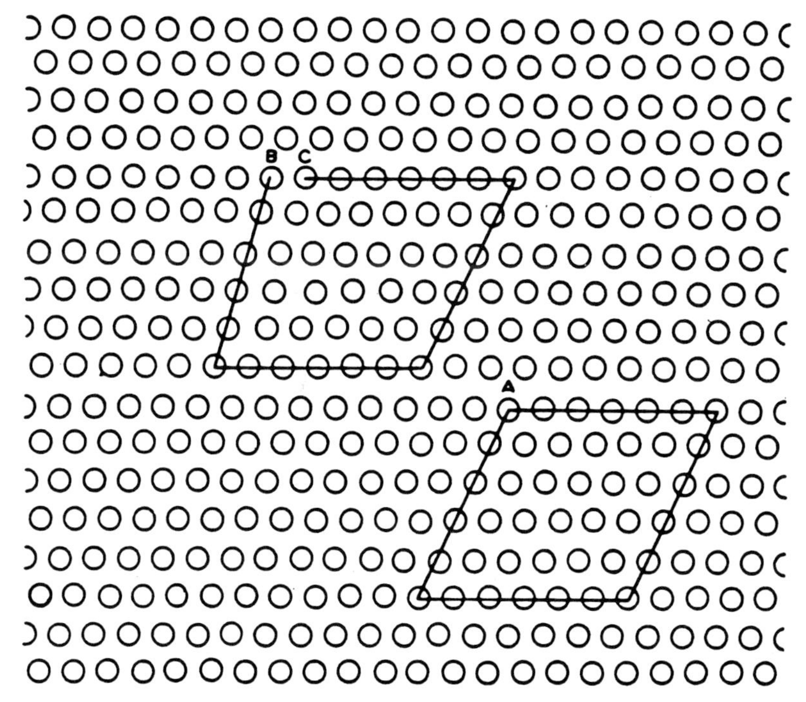

Fig. 14. An imperfect crystal illustrating the concept of a Burgers vector. The Burgers vector joins starting point B and end point C.

type II has a Burgers vector of type 1/2 <110>. For a [100] growth axis the symbol <011> refers to the four equivalent directions ±[011] and ±[0$\bar{1}$1]. These four directions are perpendicular to the growth axis. Similarly, the symbol <110> refers to eight equivalent directions that are not perpendicular to the growth axis, i.e., ±[110], ±[1$\bar{1}$0], ±[101], ±[10$\bar{1}$].

Type I dislocations are pure edge type. Otsuka has postulated that type I dislocations are generated at the terrace/riser edge of steps at the Si surface. TEM studies that have shown that the initial nucleation occurs on the terrace of Si steps [77] add confidence to Otsuka's postulate. Type I dislocations have their Burgers vector normal to the dislocation line. Both the dislocation line and the Burgers vector lie in the (100) plane. They have extra lattice fringes along both (111) and (1$\bar{1}\bar{1}$) planes and are efficient in accommodating mismatch because their Burgers vectors are parallel to the growth plane.

Type II dislocations have their Burgers vector at an angle of 45° to the plane of the substrate and 60° to the dislocation line. These are not usually generated at steps. They have an extra lattice fringe only along the (111) direction. Because their Burgers vectors are at an angle to the surface plane,

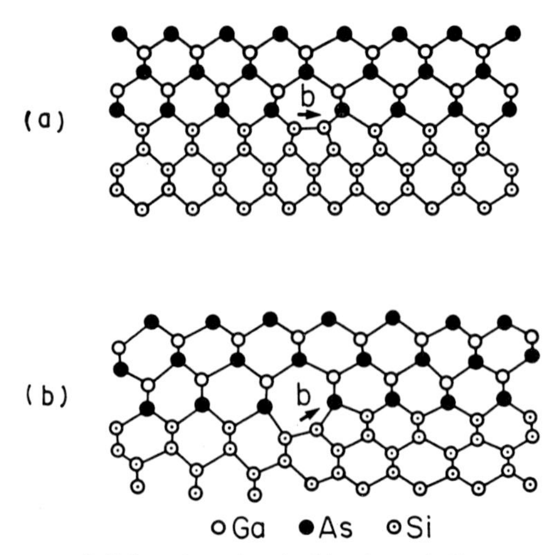

Fig. 15. Two types of dislocations in zincblende and diamond crystalline semiconductors: (a) type I with Burgers vector parallel to the interface and (b) type II with Burgers vector inclined at an angle to the interface. (After Ref. 75 or 76.)

more of these dislocations are required to accommodate a given mismatch than for type I.

Dislocations generally move by slip along crystallographic planes which contain both their Burgers vector and dislocation line, i.e., the (100) plane for type I dislocations and (111) plane for type II. The (111) planes are easy slip planes in GaAs. So type II dislocations can easily move through the GaAs epilayer and reach the surface. Such propagation of dislocations from the interface through the epilayer is called threading. Besides causing enhanced impurity diffusion, threading lines also cause partial short circuiting of *p–n* junctions and degradation of optical and electrical properties of epilayers.

Otsuka *et al.* [75, 76] undertook a study of dislocations at the GaAs/Si interface using high-resolution electron microscopy (HREM). The GaAs film used in his study was grown on the (100) Si substrate whose surface is tilted toward the $(0\bar{1}\bar{1})$ azimuth by 1°. Two types of cross-sectional specimens, one parallel and the other perpendicular to the tilt axis, were prepared by mechanical grinding and ion milling. The 1-MV ultra-high-vacuum, high-voltage electron microscope at the Tokyo Institute of Technology with a point resolution of 1.4 Å was used in this study.

Figure 16 is an HREM image of the GaAs/Si interface, which appears as a bright band in the micrograph. This image was taken from the cross-sectional specimen perpendicular to the tilt axis. Fine parallel fringes crossing the interface along two directions are (111) and $(1\bar{1}\bar{1})$ lattice planes in

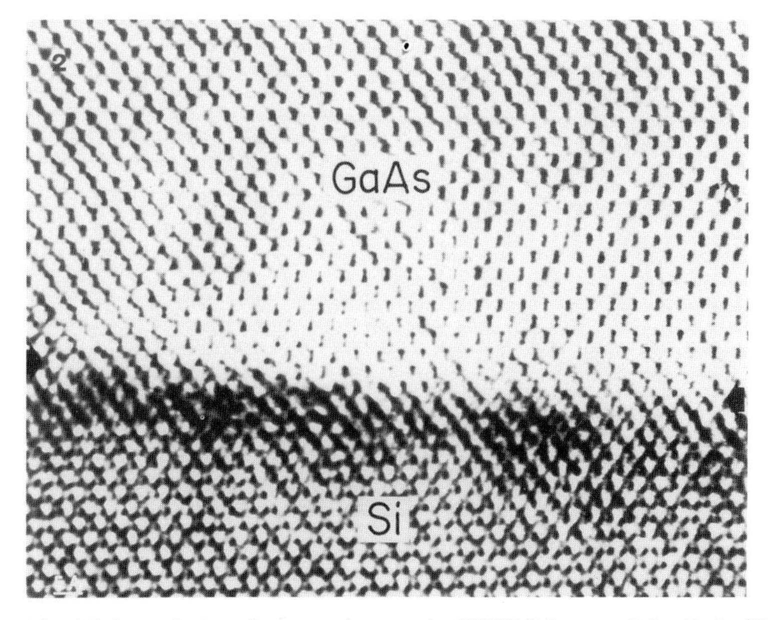

Fig. 16. High-resolution electron microscopic (HREM) image of the GaAs/Si interface. The interface is tilted by 4° about the [011] axis, which is perpendicular to the image plane.

GaAs and Si crystals. Several extra fringes can be seen in the Si side of the image. These extra fringes are terminated at the interface, causing bending of neighboring fringes. They are direct images of misfit dislocations which have formed as a result of the lattice mismatch between Si and GaAs. Figure 17 shows magnified images of these misfit dislocations. Bright spots in the images correspond to Si–Si atom pairs and Ga–As atom pairs in the Si and GaAs crystals, respectively. In both cases, the dislocation lines are perpendicular to the imaging plane. These images, therefore, show projections of core structures of misfit dislocations. As seen in the micrographs, two misfit dislocations apparently have different core structures. The dislocation shown in Fig. 17a (type I) has extra lattice planes along both (111) and $(1\bar{1}\bar{1})$ orientations, as indicated by arrows. The dislocation shown in Fig. 17b (type II), on the other hand, has an extra lattice plane only along the (111) orientation. A perfect one-to-one correspondence of lattice planes exists between Si and GaAs crystals along the $(1\bar{1}\bar{1})$ orientation.

The two types of misfit dislocations in HREM images of two cross-sectional specimens, one parallel and the other perpendicular to the axis of the substrate tilt, were counted. The analysis shows a remarkable difference between two specimens. In the former cross-sectional specimens, one-third of

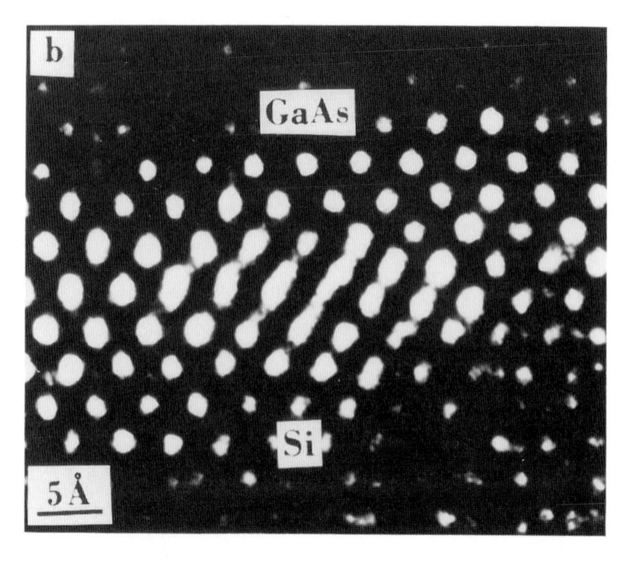

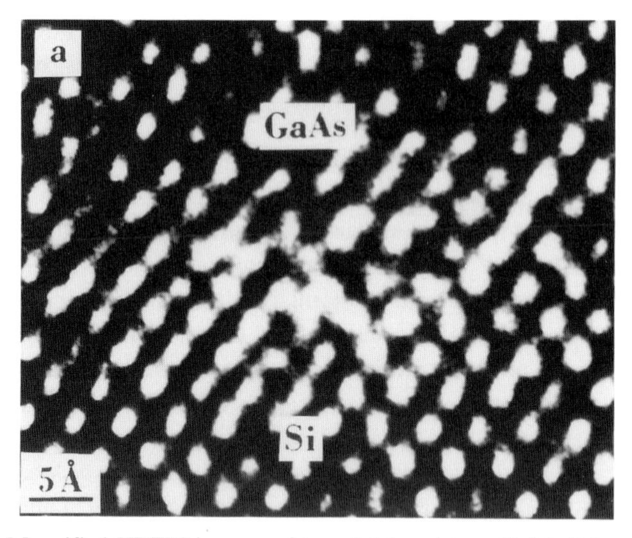

Fig. 17. Magnified HREM images of type I (a) and type II (b) dislocations.

the misfit dislocations are type II and the others are type I. In the latter specimens, almost all misfit dislocations are type I. It is, therefore, clear that the substrate surface tilt directly influences the formation of two types of misfit dislocations. Fischer *et al.* [25, 78] have also determined the density of dislocations at the Si/GaAs interface using TEM. They studied the (011) and $(01\bar{1})$ cross sections of a Si substrate tilted toward (011). For the

(011) cross-sectional specimen the average spacing (between dislocations) parallel to the risers was 78 Å, while the average spacing for the (01$\bar{1}$) cross-sectional sample was 96 Å. The breakdown of dislocation populations is shown in Table 1. Counting a particular type of dislocation at an interface is a difficult task. Even though the numbers in Table 1 may not be very accurate, the ratio of type I to type II dislocations gives us an indication of the advantages of using tilted substrates.

In summary, dislocations are required to accommodate the mismatch. Type I is preferable to type II for two reasons: (1) type I dislocations are more efficient in accommodating the mismatch, and (2) they do not propagate to the surface. The following section focuses on techniques used to induce preferentially type I dislocations.

5. DISLOCATION REDUCTION

Dislocations in lattice-mismatched heteroepitaxy of GaAs on Si have been reduced mainly through (1) using misoriented substrates, (2) inserting strained layer superlattices, and (3) annealing. They are discussed in this section.

5.1. Through Substrate Misorientation

In the last section we saw that type I dislocations occur at steps and type II do not. Recall from the section on the Si (100) surface that tilted substrates have double-layer-high steps. The steps reduce the number of active sources for the generation of type II dislocations. The 4.1% lattice mismatch causes a dislocation approximately every 25 atomic planes. A surface that has a step every 25 atomic planes will serve to induce the required number of (type I) dislocations. The minimum tilt angle required to produce a step every 25 planes is $q = 1.6°$. However, steps on the Si surface do not occur at regular intervals, so a slightly greater angle is required. Surfaces tilted toward some other directions, <001> for example, cause a double-staircase

Table 1.

Breakdown of Type I and Type II Dislocations

	Type I	Type II
(011) cross section	67%	33%
(01$\bar{1}$) cross section	99%	3%

shape on the surface. On such surfaces, type I dislocations are induced in both directions. The minimum angle of tilt required in this case is 2.3°. The effect of tilt direction on the surface step shape is illustrated in Fig. 18.

Fischer *et al.* [79, 25] have studied the effectiveness of substrate tilt in reducing the dislocation density. In their study, three different substrates were utilized: exact (no tilt) (100), 4° off toward [011], and 4° off toward [001]. The cross-sectional bright-field images of the three samples are shown in Fig. 19. The epilayer on the exact (100) surface contains massive dislocation networks which thread from the GaAs/Si interface to the surface. The epilayer thickness was about 2 mm. The dislocation density near the surface is as high as 10^{10} cm^{-2}. In addition to threading dislocations, a large number of stacking faults are present along {111}.

The epilayer on the substrate misoriented 4° toward [011] has fewer dislocations than the exact (100) sample. The dislocations are also confined to a region approximately 1 μm thick from the GaAs/Si interface. The dislocation density near the film surface is approximately two to three orders lower than that of the exact (100) sample. The tilt toward [001] reduces the surface dislocation density even further. The further reduction in this case is due to steps running in both [011] and [$01\bar{1}$] directions, which provide a greater number of sources (steps) for the generation of type I misfits to accommodate the lattice mismatch.

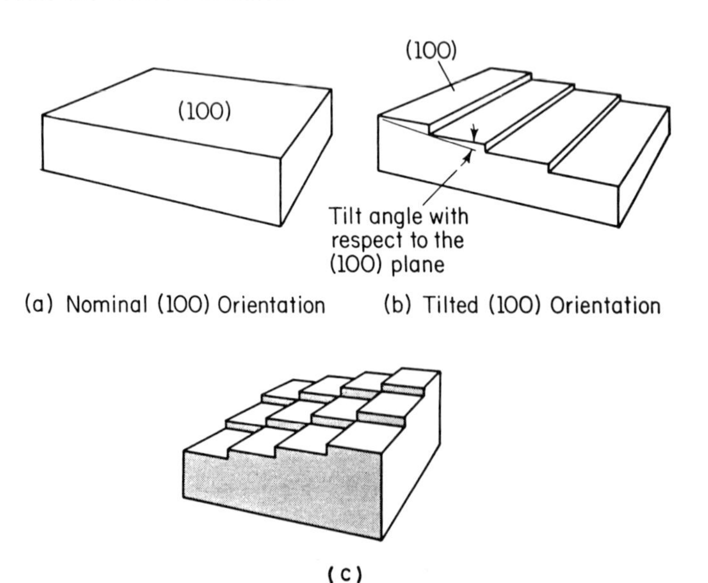

Fig. 18. Effect of tilt on the substrate surface: (a) nominal (100) surface, (b) (100) tilted toward (110), and (c) (100) tilted toward (001).

5.2. Through Strained Layer Superlattices

The use of misoriented substrates is a first step toward defending against type II dislocations. Defects such as threading dislocations and stacking faults can form in many different ways during the nucleation and growth of epitaxial films. For example, misoriented initial nucleation islands are also a source of threading dislocations. So, a significant density of threading dislocations is still found in the epilayer. A study of how dislocations are avoided in a pseudomorphic thin film enables one to come up with schemes to tackle threading dislocations that are still present in the epilayer. The threading dislocation line has associated with it a stress field. If additional stress fields are introduced in the epilayer, the stress fields of the epilayer and the thread line interact. The stress–strain field in the epilayer can be used to bend the dislocation line [80, 81]. One approach is to bend the thread line toward another thread so that they annihilate one another. Another approach is to bend the thread line away from the surface toward the edge and parallel to the growth plane; i.e., the thread line does not reach the surface at all.

A material with a larger lattice constant than the substrate material introduces compressive strain in the epilayer in directions parallel to the interface. Compressive strain repulses the dislocation away from the source of the strain. This force, called the Peach–Koehler force, depends on the strain (e) and the layer thickness (h). If the thickness of the larger lattice constant layer (h) is less than the critical thickness, no additional dislocations are introduced. $In_xGa_{1-x}As$/GaAs and $GaAs_{1-y}P_y$/GaAs, AlP, AlGaP, and GaP [82] strained layer superlattices (SLSs) have been used to reduce dislocation threading to the surface. The usual practice is to grow 0.2–0.5 μm of GaAs on top of a Si substrate, then grow the SLS, and then continue with the GaAs buffer layer. It is important to bear in mind that the SLSs discussed so far are not lattice matched to GaAs. Therefore extreme care must be taken to keep the thickness of the SLS less than the critical thickness to avoid generating further misfits [72]. The force required to bend the dislocations away from the surface introduces constraints on the minimum value of the SLS thickness. A very thin SLS may not produce enough strain to bend the dislocations effectively, and this is illustrated in Fig. 20a. Usually the mole fraction of the ternary alloy and/or the thickness is adjusted for an optimum design. The interaction between the SLSs and the threading dislocations, especially the mutual annihilation, bending of dislocations away from the surface and parallel to the interface, etc., are illustrated in Fig. 20 [11, 73, 84]. Figure 21 is a cross-sectional TEM micrograph showing these inter-

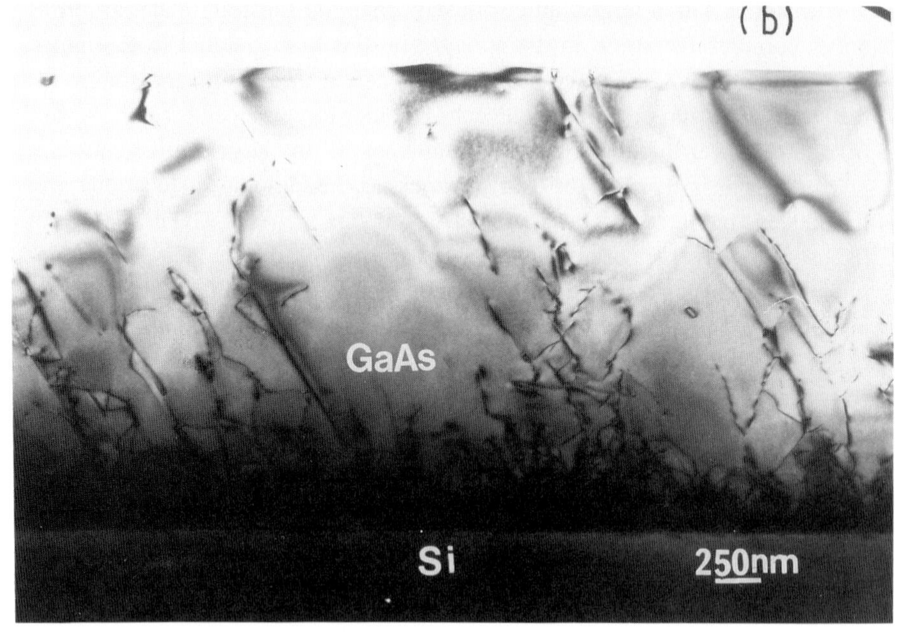

Fig. 19. Bright-field images of a GaAs layer grown on (a) an exact (100) silicon surface, (b) (100) Si tilted 4° toward <011>, and (c) (100) Si tilted 4° toward <001> and incorporating two five-period 100 Å GaAs/100 Å $In_{0.15}Ga_{0.85}As$ strained-layer superlattices.

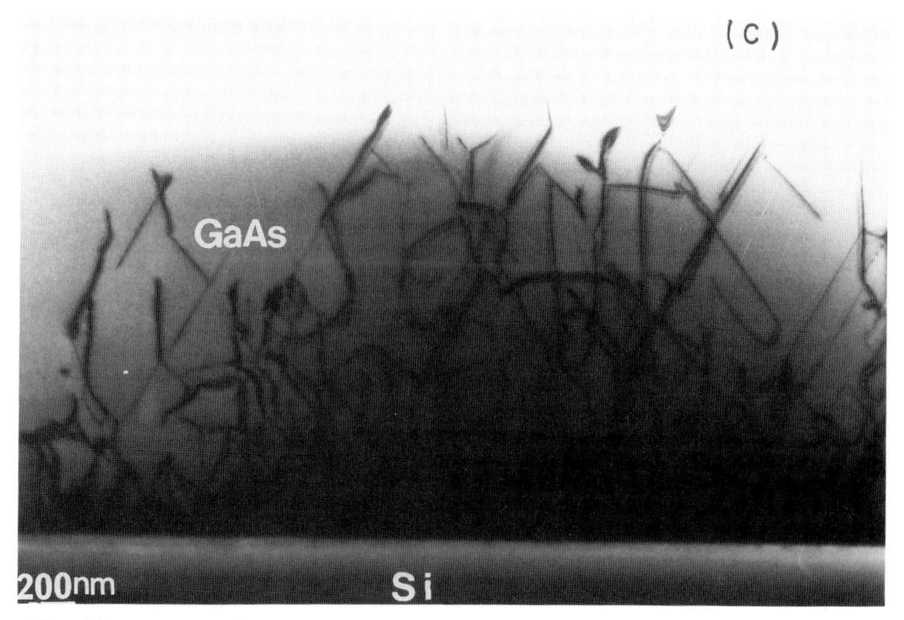

Fig. 19. *(continued)*

actions between the SLS and the threading dislocations in the GaAs epilayer on Si [84].

The lattice mismatch problem associated with the ternary–binary SLS discussed in the last paragraph can be avoided by using a different configuration. In this configuration, the SLS is composed of two materials having equal but opposite lattice mismatches, such that the average lattice constant is matched to that of GaAs. $In_xGa_{1-x}As/GaAs_{1-y}P_y$-based SLSs ($y = zx$) can be grown, on an average lattice matched to GaAs [84, 85]. Because of the lattice match, more periods as well as thicker layer SLSs can be grown. These systems are more efficient than the binary–ternary SLS in combating threading dislocations.

Modulation-doped AlAs/GaAs superlattices have also been used to reduce threading dislocations [86]. Recall that AlAs is nearly lattice matched to GaAs. In these lattice-matched superlattices the interaction between the built-in electric field and the threading dislocation is used to bend the threading dislocation. Dislocation bending occurs during growth. At growth temperature all the dopants (Si) in the AlAs are ionized. Furthermore, the AlAs acts as a barrier to electrons and all the electrons are confined to GaAs. Under these conditions it is reasonable to assume that electric charges give rise to an electric field that aids the bending of the threading line just like a strain field.

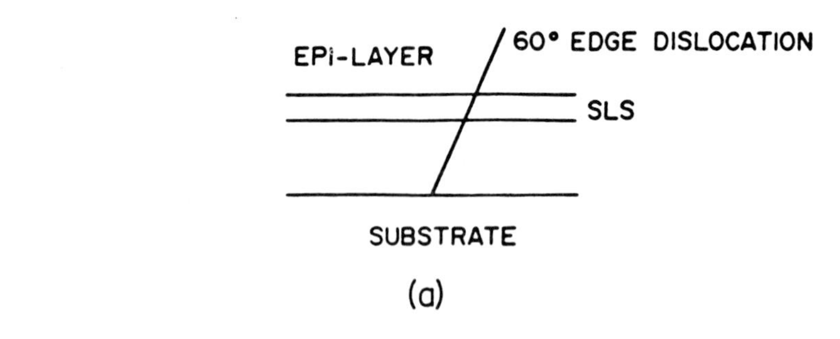

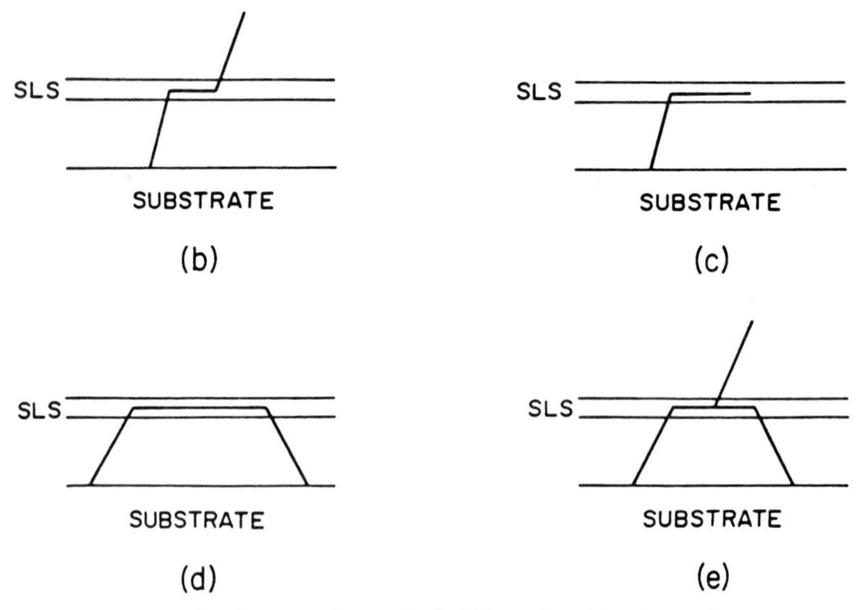

Fig. 20. Interaction between the strain field introduced by the strained layer superlattice (SLS) and the threading dislocation: (a) no interaction at all because the strain field is very weak, (b) strain field not strong enough to bend the threading dislocation away from the surface, (c) strong strain field bending the threading dislocation away from the surface and parallel to the interface, (d) mutual annihilation of threading dislocations, and (e) mutual annihilation with propagation. (After Ref. 84.)

5.3. Through Annealing

Recall that the strain due to lattice mismatch is accommodated by dislocations. Type I dislocations are energetically more favorable than type II dislocations. High-temperature annealing could allow sufficient strain relaxation by dislocation movement and annihilation to lower the energy of the

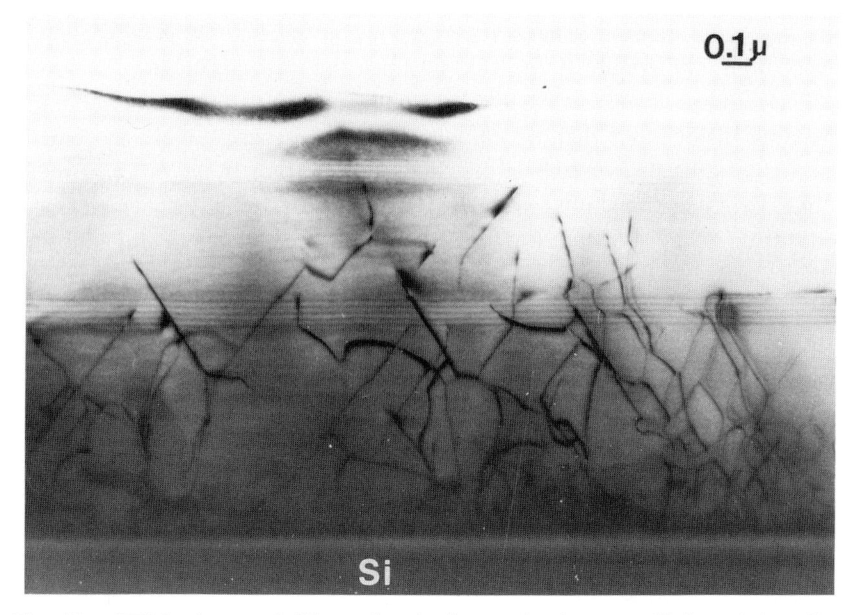

Fig. 21. TEM micrograph illustrating the interaction between SLSs and threading dislocations in GaAs on Si. (After Ref. 84.)

system. Thus annealing aids the conversion of type II dislocations to harmless type I misfit dislocations and greatly improves the quality of the epilayer [87–90].

The annealing process can be carried out during growth inside the growth chamber or outside in a separate furnace. The former is called *in situ* annealing and latter *ex situ* annealing. *In situ* annealing can be carried out in three distinct ways:

1. *Static annealing:* The growth is interrupted and the substrate temperature is raised under As overpressure.
2. *Dynamic annealing:* Instead of stopping growth, an AlGaAs layer is grown at a higher substrate temperature (around 700°C).
3. *Cyclic annealing* [91]: The growth and annealing steps are mixed. Cycles of growth at the normal standard temperature for a while, cooling down to room temperature, and annealing at a high temperature for 5 minutes are repeated several (usually around 10) times. The main idea behind this technique is that cooling as well as heating plays a part in dislocation reduction.

Ex situ annealing can be carried out in two different ways: prolonged annealing (typically for about 30 minutes) in a furnace under As overpressure and rapid thermal annealing (RTA), where the sample is subjected to a very high temperature (900–1000°C) for a few seconds.

TEM studies have shown that annealing is very effective in reducing threading dislocations that are caused by the free energy of the system. It also reduces the dislocations at the Si/GaAs interface by an order of magnitude. Cross-sectional images of an as-grown and externally annealed GaAs epilayer on Si are shown in Fig. 22. Similar pictures for the *in situ* annealed samples are shown in Fig. 23.

A typical plan-view bright-field image of a GaAs-on-Si sample near the film surface is shown in Fig. 24 [92]. The wavy dark lines in the image are bending contours, which indicate the presence of high residual strain in the epilayer. The plan-view weak-beam dark-field (WBDF) image of the GaAs/Si interface is shown in Fig. 25. The parallel dislocations (approximately to either [011] or [01$\bar{1}$]) in these images are misfit dislocations (type I) with an average spacing of about 100 Å. The irregularly arranged contours in the images are threading dislocations. Figure 25b clearly demonstrates the effect of thermal annealing in reducing threading dislocations.

Secondary-ion mass spectroscopic (SIMS) studies have been carried out on the annealed samples to optimize the annealing temperature and duration[93]. This study has shown that up to 850°C and 3 hours there is no measurable displacement in the interface region. However, higher temperatures (around 950°C) for 30 minutes resulted in complete interdiffusion of Ga and As atoms into the Si substrate. High-temperature rapid thermal annealing (around 900°C for 10 seconds) on the epilayer was carried out by Chand *et al.* [94, 95]. The interfacial atomic diffusion in this case is negligible.

6. OPTOELECTRONIC DEVICES ON SILICON SUBSTRATES

The most attractive application of GaAs and other compound semiconductors that have direct bandgaps on Si is in the integration of optoelectronic devices and Si VLSI on a single chip. It has been argued that availability of the on-chip compound semiconductor lasers, modulators, and detectors might highly optimize the partitioning of components for faster signal processing.

6.1. LEDs and Lasers

GaAs light-emitting diodes (LEDs) grown on Si and emitting at 870 nm have been demonstrated by MOCVD [96, 97] and MBE [78]. Other LEDs fab-

Fig. 22. TEM bright-field crosssection of a GaAs on Si sample (a) as grown and (b) after *ex situ* annealing at 850°C for 0.5 hour.

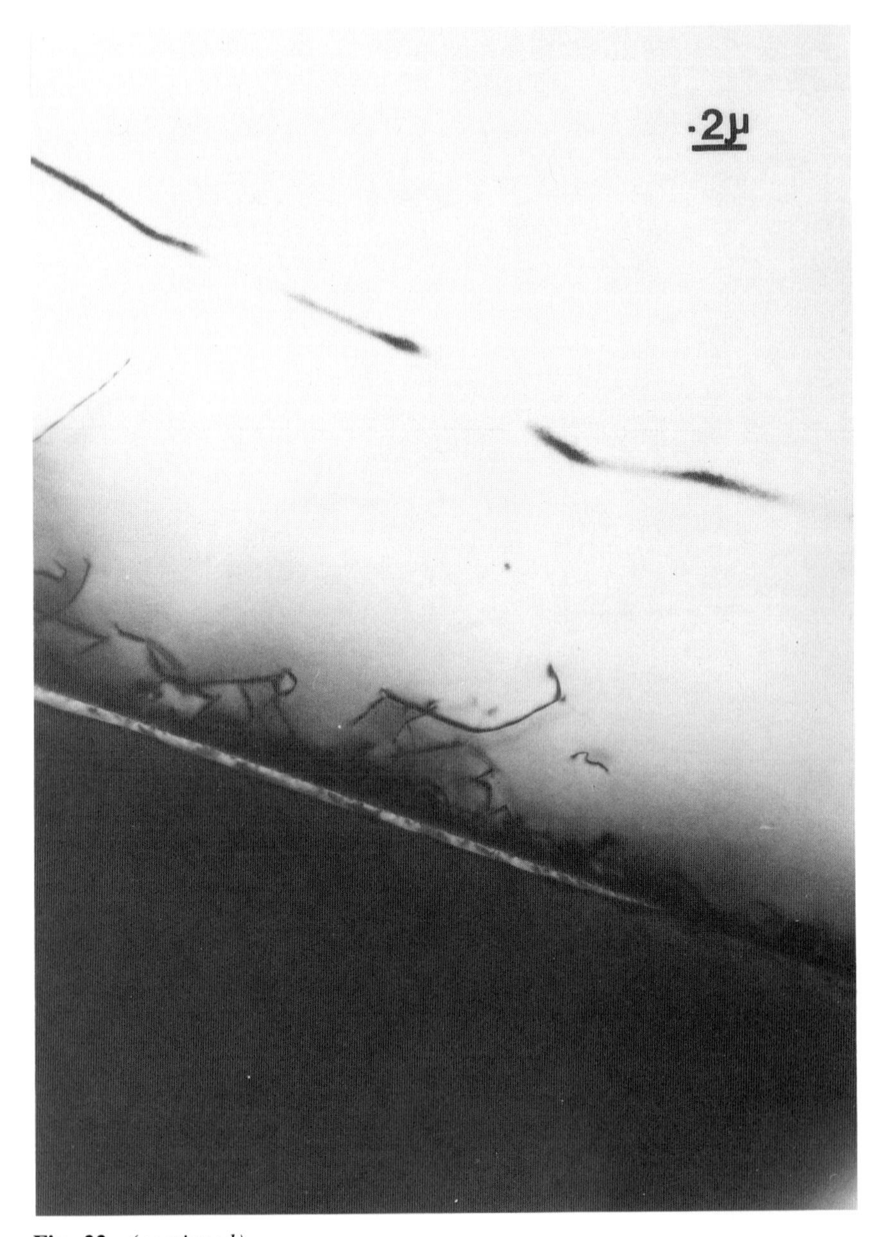

Fig. 22 *(continued)*

Fig. 23. TEM bright-field image of a GaAs-on-Si sample (a) as grown and (b) after *in situ* annealing at 700°C for 0.5 hour.

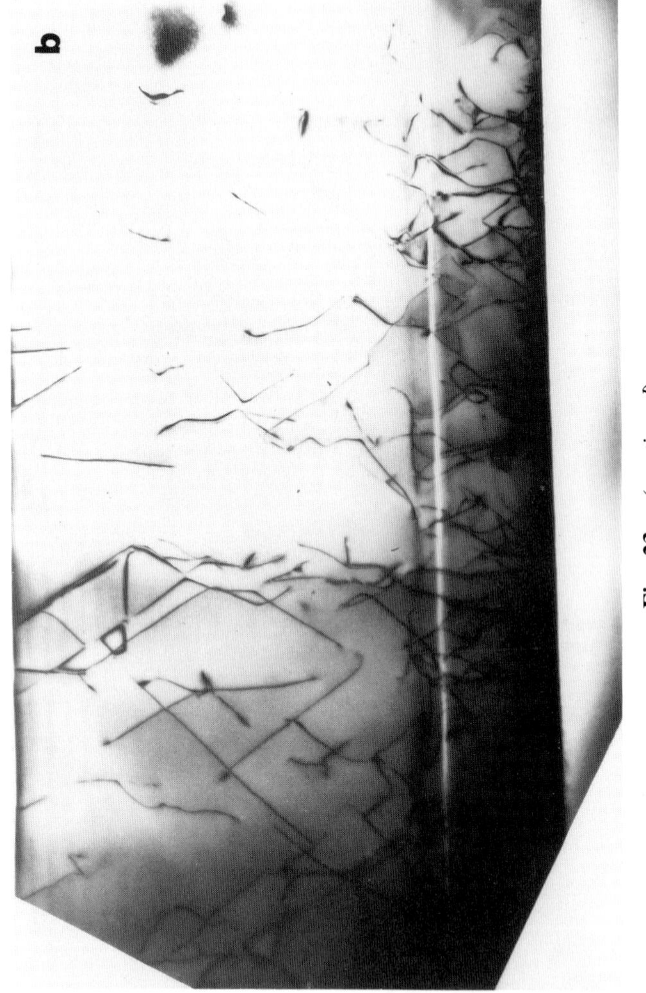

Fig. 23. *(continued)*

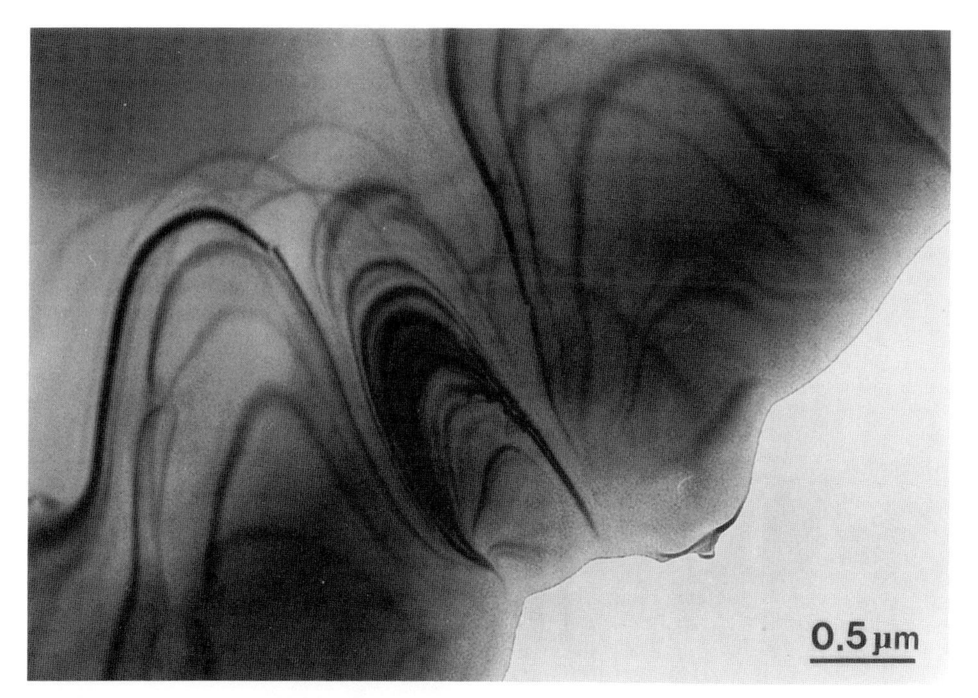

Fig. 24. Plan-view bright-field images of a GaAs-on-Si sample. The wavy dark lines in the image are bending contours indicating the presence of high residual strains. (After Ref. 92).

ricated from GaInAsP [98], AlGaAs [99], InGaP [100], and GaP [101], which emit in the infrared and visible parts of the spectrum at 1.15 μm, 700 nm, 600 nm, and 565 nm, respectively, have also been demonstrated on Si substrates. Figures 26 and 27 show an emission spectrum and output power–current characteristics of an InGaP LED grown on Si [ref. 100]. Despite the high dislocation density (10^7–10^8 cm^{-2}), no degradation was observed in the LED characteristics after 2000 hours of operation at an injection current of 500 A/cm^2.

Successful growth of GaAs/AlGaAs diode lasers on Si was first achieved on Ge-coated Si [102] and then by direct growth on Si by MBE [103]. This was followed by reports of MOCVD-grown lasers on Si [104, 105]. It has been long established that the development of room-temperature continuous-wave (CW) operation of a GaAs/AlGaAs-on-Si current injection laser is an important step toward potential realization of OEIC, as with components utilizing GaAs on Si. In the early stage of this development, high threshold current density was the main limiting factor preventing room-temperature CW lasers from being achieved. Efforts to reduce the threshold current den-

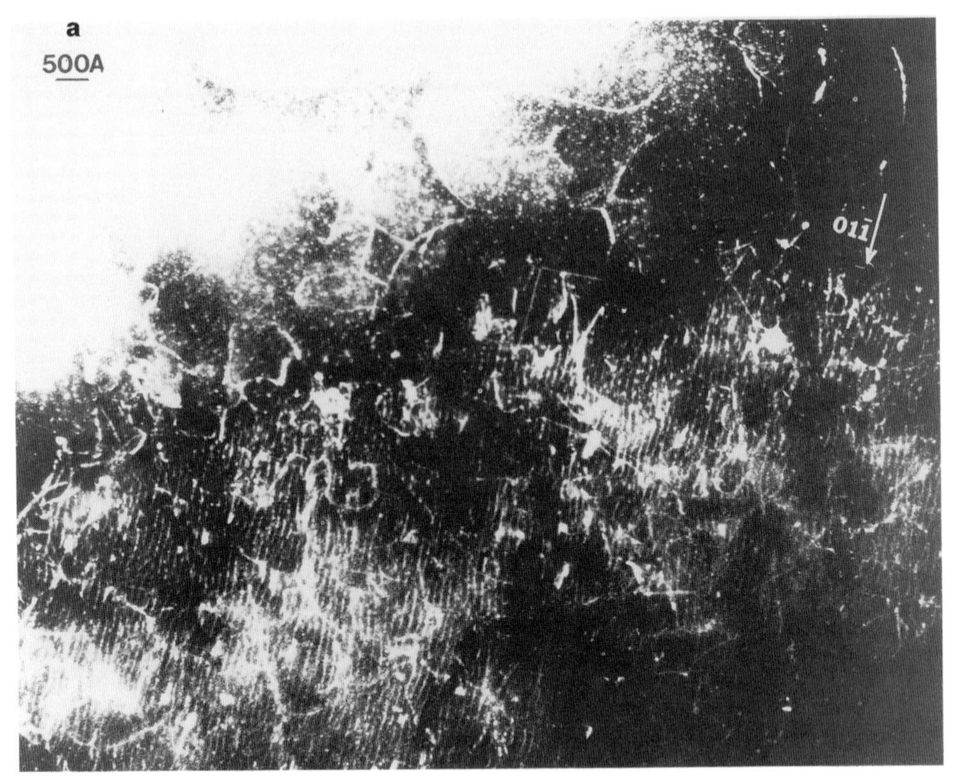

Fig. 25. Plan-view weak-beam dark-field (WBDF) images of GaAs on Si (a) before annealing and (b) after annealing. The parallel lines in the images are misfit dislocations and the irregularly arranged lines are the threading dislocations. (After Ref. 92.)

sity mostly centered around looking for ways to improve the quality of epitaxial growth.

This flurry of activity led to stripe geometry MBE-grown-on-Si lasers with room-temperature threshold current densities of 1000 A/cm^2 for a 10 × 380 μm^2 stripe [106]. Broad-area, graded-refractive-index MBE-grown GaAs/AlGaAs-on-Si lasers, having a width of 110–120 μm and a cavity length of 520 μm, exhibited a threshold current density of 600 A/cm^2 [107]. MOCVD-grown-on-Si GaAs/AlGaAs lasers were reported to have a room-temperature threshold current density of 3.3 × 10^3 A/cm^2 for 4 × 380 μm^2 stripes and a threshold current density of 300 A/cm^2 in 200 × 380 μm^2 broad-area lasers [108]. This particular result is comparable to threshold current densities as low as 214 A/cm^2 obtained by MBE in broad-area (120 × 1900 μm^2) GaAs/AlGaAs lasers on Si by Chen *et al.* [109]. These low

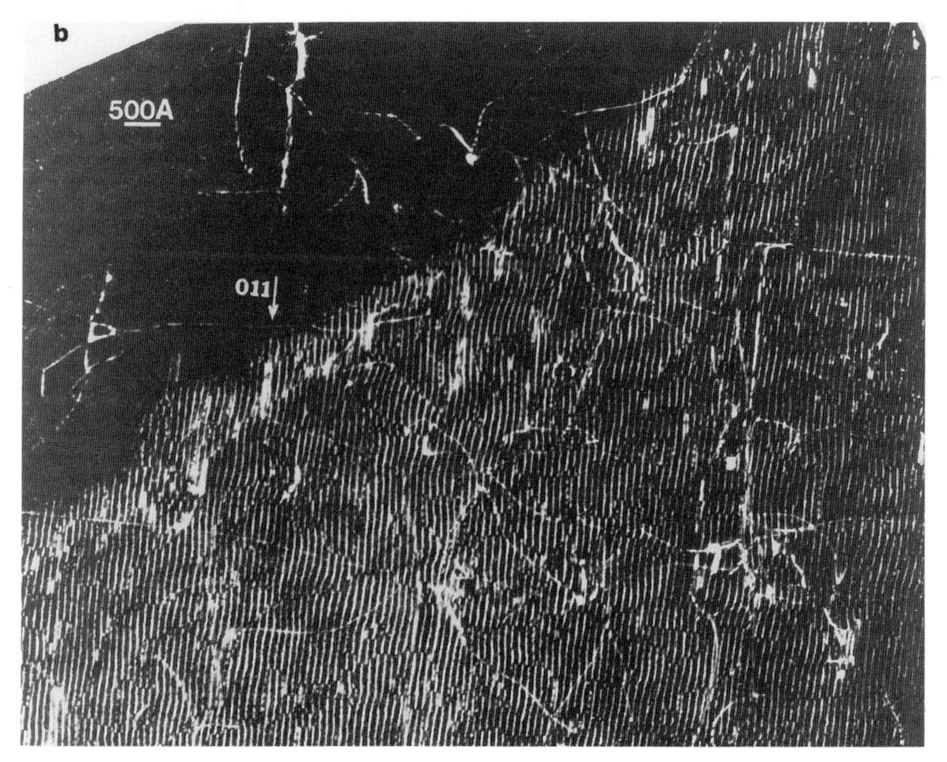

Fig. 25. *(continued)*

thereshold densities led to the first report of room-temperature CW operation of an AlGaAs/GaAs broad-area laser on Si [110]. Figure 28 shows the light-current (*I-L*) characteristics of this room-temperature CW GaAs-on-Si laser. Figure 29 shows a comparison of the threshold current densities of GaAs-on-Si and GaAs-on-GaAs lasers [109].

The first room-temperature CW operation of a GaAs/AlGaAs-on-Si laser grown by MOCVD in its entirety was reported by Egawa *et al.* [111] with a threshold current density of 1.41 kA/cm^2. Figure 30 shows the injection current versus light output power (*I-L*) characteristics of the GaAs-on-Si laser. Thermal cycle annealing (TCA) from 30 to 850°C was employed to improve the quality of GaAs-on-Si epilayers. The growth temperature was 750°C, and SiO_2 back-coated Si substrates were employed to decrease the Si incorporation, which was caused by high growth temperatures. As shown in Fig. 30, both TCA and back-coating Si substrates could reduce the threshold current density. However, the influence of the SiO_2 back-coating on the strain of the grown epilayer is not clear.

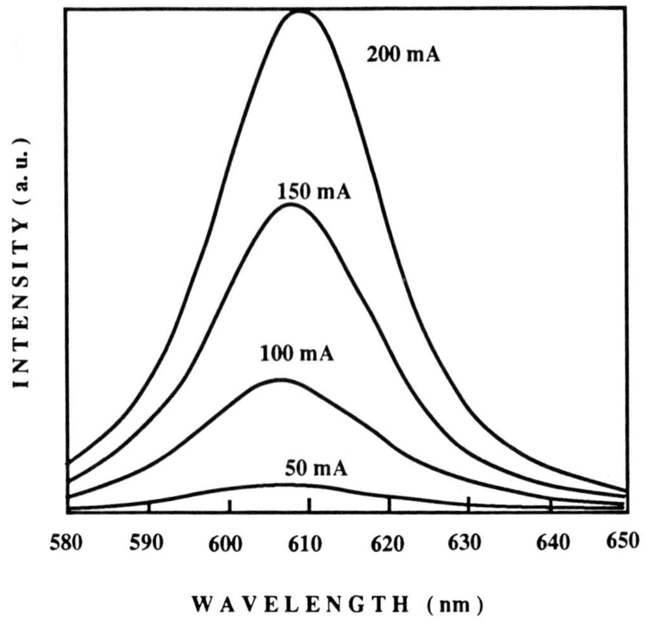

Fig. 26. Emission spectrum of the InGaP LED on a Si substrate at room temperature. (After Ref. 100.)

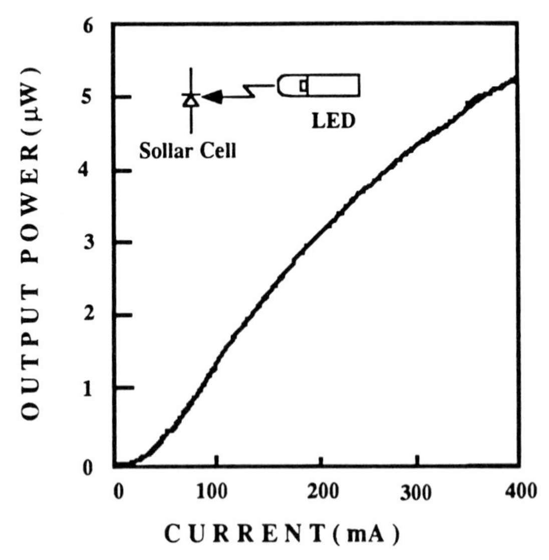

Fig. 27. Output power–current characteristics of an InGaP LED on a Si substrate at room temperature. (After Ref. 100.)

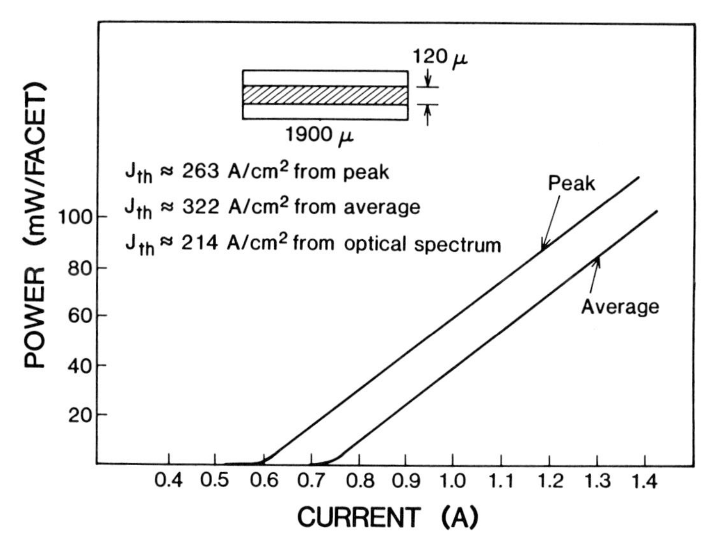

Fig. 28. Light–current characteristics of a long-cavity broad-area GaAs-on-Si laser.

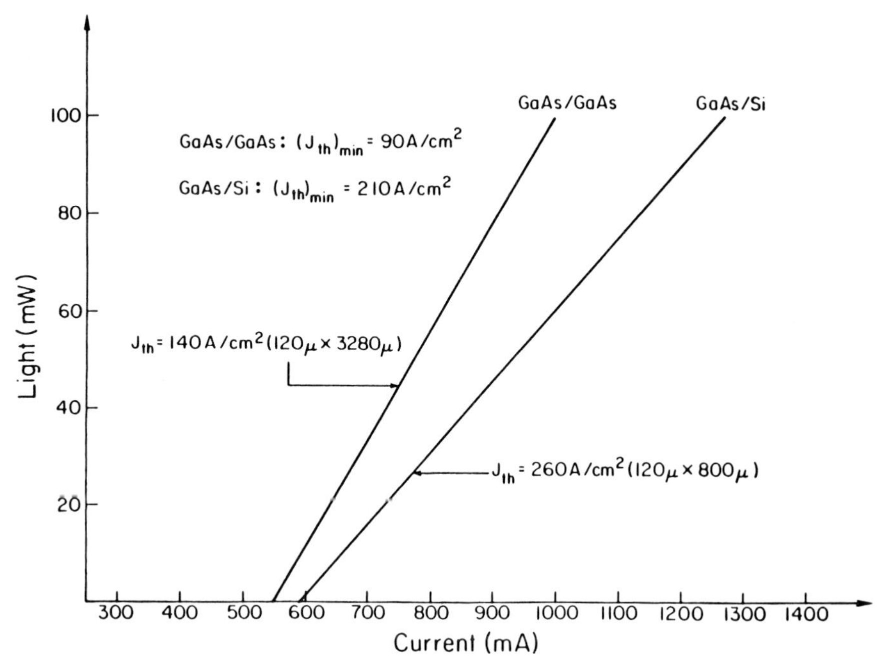

Fig. 29. Comparison of the threshold current densities of GaAs-on-Si and GaAs-on-GaAs lasers.

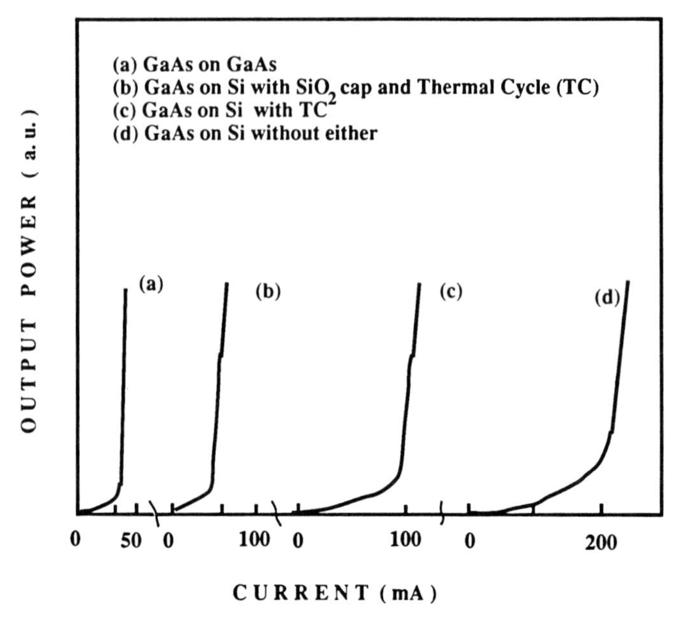

Fig. 30. Room-temperature *I-L* characteristics of $Al_{0.3}Ga_{0.7}As/GaAs$ SQW lasers on (a) GaAs substrate, (b) SiO_2 back-coated Si substrate with thermal cycle annealing, (c) Si substrate without SiO_2 back coating and with thermal cycle annealing, and (d) SiO_2 back-coated Si substrate without thermal cycle annealing. (After Ref. 111.)

Output power and quantum efficiency are also important parameters of a laser. GaAs/AlGaAs lasers grown on Si by MOCVD have reported output powers of 130 mW/facet [112]. Under pulsed conditions, output powers of 400 mW/facet have been achieved for GaAs/AlGaAs lasers grown on Si by a hybrid growth technique involving migration-enhanced MBE followed by MOCVD [113]. GaAs/AlGaAs lasers on Si have achieved total external quantum efficiencies of 70% for $125 \times 250\ \mu m^2$ broad-area pulse-operated MOCVD-grown lasers [114] and 87% for $10 \times 380\ \mu m^2$ ridge waveguide CW MBE-grown lasers [27].

One of the most important goals in GaAs-on-Si research is to modulate lasers at high speed with signals generated by a Si chip or a GaAs chip on the same Si wafer. Figure 31 shows the ridge waveguide laser used in modulation experiments, which was grown on Si in $10 \times 380\ \mu m^2$. The frequency response property is shown in Fig. 32 [27]. It had a corner frequency of 2.5 GHz when the laser was operated about 20% above the threshold. This result compares favorably with those obtained with similar GaAs-on-GaAs lasers which show a corner frequency of about 2 GHz [115].

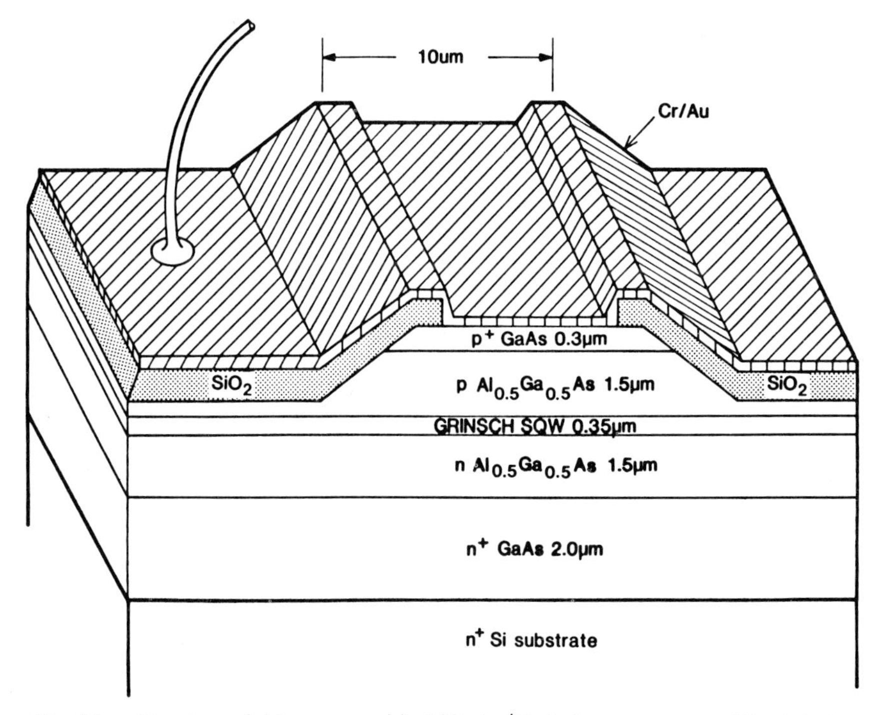

Fig. 31. Structure of ridge waveguide AlGaAs/GaAs lasers grown on Si substrates by MBE.

As for the longevity of lasers on Si substrates, a photo-pumped CW 300 K GaAs/AlGaAs laser on Si, operated near threshold, demonstrated a lifetime of more than 4 hours, but degraded much faster at higher light outputs [116]. Later, the same group demonstrated a lifetime of 17 hours in CW GaAs/AlGaAs lasers on Si at room temperature [117]. Using QW GaAs/AlGaAs lasers on Si with a strained InGaAs QW buffer layer, Choi *et al.* [118] demonstrated improved performance and lifetime. Lifetimes up to 56.5 hours at CW-RT operation have been achieved. Room-temperature CW GaInAsP/InP double-heterostructure lasers emitting at 1.27 μm have also been demonstrated on Si substrates [119] with threshold currents of 45 mA and lifetimes exceeding 5 hours. The laser structure and *L-I* characteristic are shown in Figs. 33 and 34, respectively. A much longer lifetime has been demonstrated by Sugo *et al.* [120] in InGaAs/InGaAsP lasers on Si operating at a longer wavelength. Following a 2-μm-thick GaAs buffer layer, a 10-μm-thick InP buffer layer was grown to reduce the dislocation density in the active region as well as to reduce the strain. A lifetime of over 2000 hours of RT-CW operation has been achieved, which is attributed by the

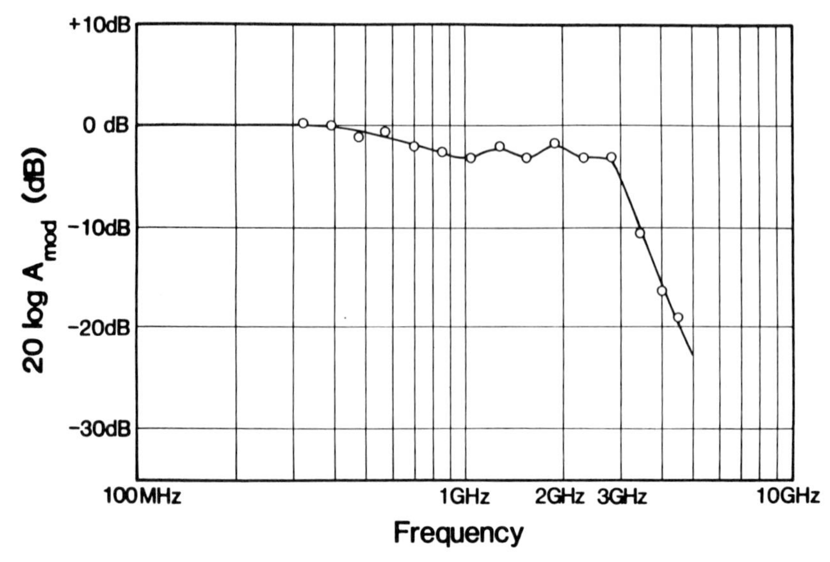

Fig. 32. Frequency response (microwave modulation) of a 10 × 380 μm ridge waveguide GaAs laser on Si.

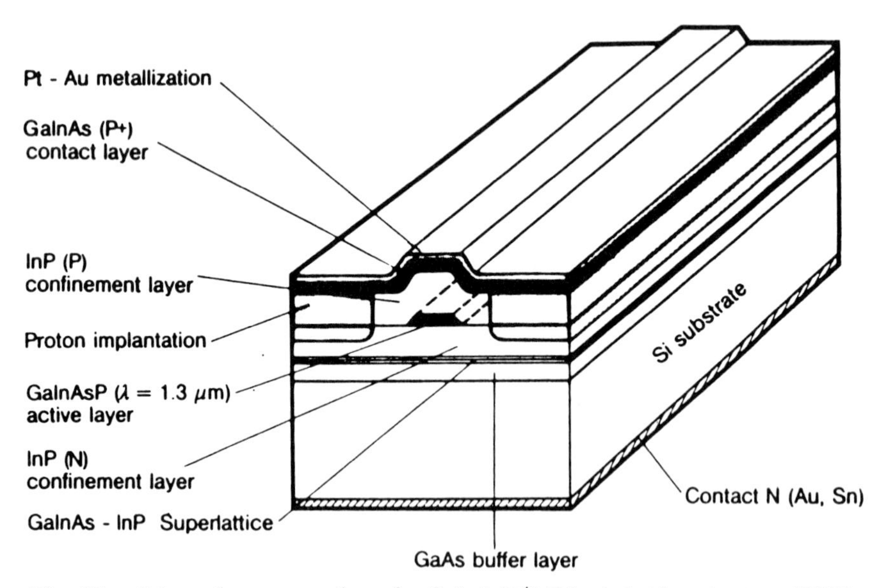

Fig. 33. Schematic cross section of a GaInAsP/InP buried ridge structure (BRS) laser. (After Ref. 119.)

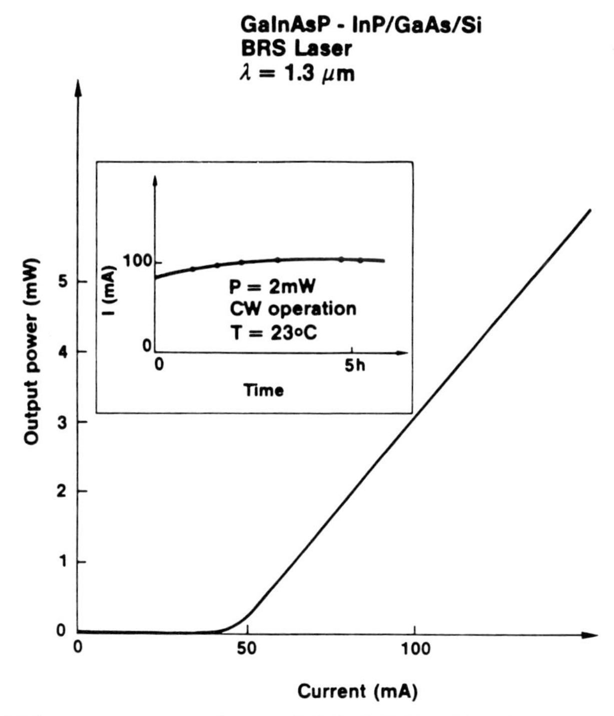

Fig. 34. Light output–current input of GaInAsP/InP BRS lasers on Si substrate. Inset: typical aging characteristics of the GaInAsP/InP lasers on Si substrates. (After Ref. 119.)

authors to the particulars of the buffer layer construction and the inherent properties of the InGaAsP system.

Meanwhile, graded-index separate confinement heterostructure (GRIN-SCH) GaAs/AlGaAs lasers on patterned Si substrates have been grown by selective-area MBE [121, 122]. Figure 35 shows the structure of selective-area grown stripe GaAs-on-Si lasers [121]. Reduced threshold current densities were achieved compared with similar lasers grown on planar Si substrates, as shown in Fig. 36 [121, 122]. The reduction of laser threshold current densities has been chiefly attributed to the current confinement effect by the high-resistivity polycrystalline GaAs/AlGaAs films surrounding the active devices.

Thermal mismatch between GaAs and Si is of much concern in relation to device reliability. The disparity is apparent from the drastically different linear thermal expansion coefficients of 6.0×10^{-6} and 2.3×10^{-6} K^{-1} for

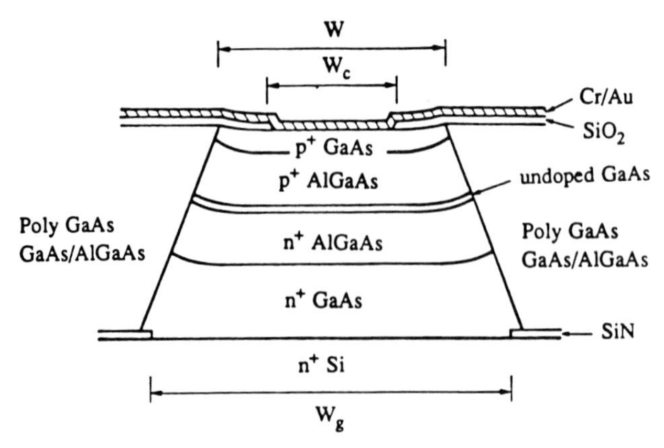

Fig. 35. Schematic cross section of a 10-μm-wide selective-area grown GaAs/Si laser. (After Ref. 121.)

GaAs and Si, respectively. If no additional dislocations were introduced and no substrate bowing occurred upon cooling from the growth temperature, the thermal mismatch induced strain would be

$$\varepsilon = \Delta T \left(\frac{\Delta L}{\mathrm{L}\ \Delta T} \bigg/ \mathrm{GaAs} - \frac{\Delta L}{L\ \Delta T} \bigg/ \mathrm{Si} \right) \tag{7}$$

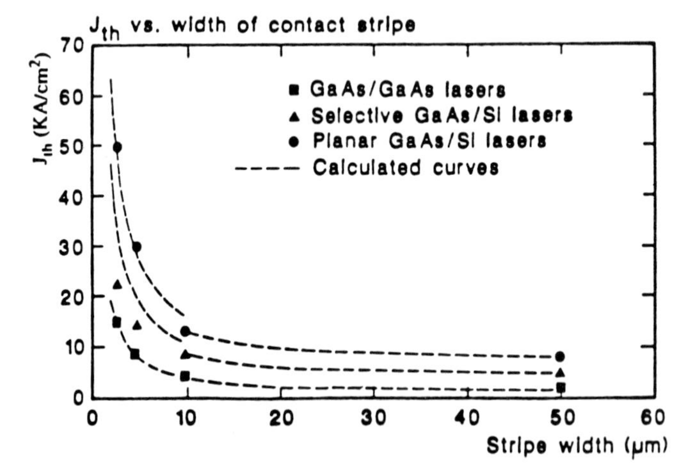

Fig. 36. Plot of J_{th} vs. stripe width (W_c) for DH lasers grown on planar, patterned Si substrates and on a GaAs substrate, respectively. The dashed lines are J_{th} vs. W_c calculated using a model described in Ref. 185 that takes into account effects of current spreading and carrier outdiffusion. (After Ref. 121.)

where ΔT is the temperature drop. For a 500°C temperature change, this strain amounts to about 0.2%. The stress can be expressed by

$$s = \varepsilon \frac{E}{1 - \nu} \tag{8}$$

where E is elasticity modulus (or Young's modulus) and ν is the Poisson's ratio, $C_{12}/(C_{11}+C_{12})$. For $\varepsilon = 2 \times 10^{-3}$, $\sigma = 5.3 \times 10^{8}$ dyn cm^{-2}, this is about twice that in AlAs/GaAs heterostructures. In 2-inch-diameter wafers, one can also expect concave bowing of about 5 μm, which increases the residual strain. For thicker layers, the bowing component may exceed the elastic limit of GaAs and cracking may occur during processing, if not before.

Strain has been shown to be a factor in the rapid degradation of lasers. Strain may be induced by bonding [123] and by the SiO_2 used for delineating stripes in GaAs lasers [124]. Other studies [125–128] have shown that device failure is also caused by dislocations present and/or generated under strong pumping such as that in lasers. The energy given by band-to-band recombination is thought to be released to the lattice through phonon generation as heat. In this process, to conserve momentum, vacancies and dislocations are generated.

In the case of GaAs on Si, there is not only a sizable residual strain that aids in the generation of defects but also dislocations propagating from the interface. Large photon flux in the pumped region can cause additional defects to form as well as decorating [127] the existing and newly created dislocations, causing degradation of lasers. Luckily, in quantum well lasers, the volume of the high-photon-density region is extremely small, so one would not expect large numbers of dislocations to be present. However, large photon flux can lead to generation of additional defects and dislocations seeded by a few initially present. To circumvent the contribution by the residual strain somewhat, selective epitaxy has been carried out [61], in which GaAs epitaxy is performed only where it is needed. This selective-epitaxy process reduces the residual strain from about one-half of that in the uniform epitaxy cases to possibly about 0.1%. The tendency for bowing is also reduced.

Other semiconductors, such as InP and its lattice-matched compounds with InAs and GaAs, present smaller thermal mismatches with Si. Therefore, longer lifetimes for InGaAsP/InP lasers on Si may be expected.

6.2. Optical Modulators and Waveguides

Modulators constitute another important component of potential OEICs. Modulators with local "onboard" lasers might be more advantageous than modulators with external lasers. GaAs/(Al,Ga)As optical reflector modulators [129, 130], optical waveguides, and phase modulators [131] have been prepared on Si substrates by MBE. The structure of a reflector modulator with no antireflection coating on the surface is shown in Fig. 37 [130]. GaAs multiple-quantum-well reflector modulators were grown on Si with an AlAs/AlGaAs dielectric mirror inserted into the device structure. Modulation ratios of 51.4% at an external bias of 8.5 V, and more recently of 80%, have been achieved at an external bias of 9 V (Fig. 38). These modulation ratios are comparable to those of similar devices grown on GaAs substrates. Figure 39 shows the structure of a GaAs/AlGaAs-on-Si ridge waveguide [131]. The measured near-field intensity profiles show that the guides are single mode and are well confined in both vertical (depth) and lateral (width) directions, as shown in Fig. 40. The phase shift efficiency of 3.5°/V mm was obtained in a phase modulator grown on Si [131]. An average TE mode propagation loss of 1.24 dB/cm was obtained for a single-mode GaAs/AlGaAs single-heterostructure ridge (ridge width of 6 μm) waveguide [131]. However, the losses of waveguides on Si are still about an order of magnitude higher than those on GaAs [132], which is most likely due to the high dislocation density (> 10^6 cm^{-2}) in GaAs-on-Si epilayers.

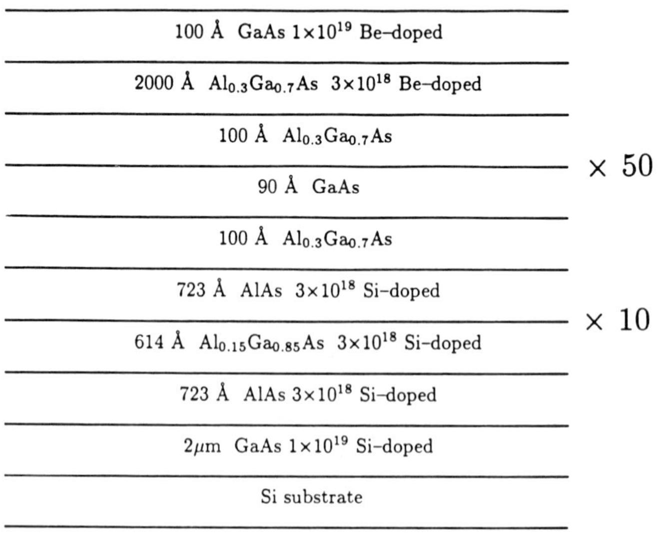

Layer	Repeat
100 Å GaAs 1×10^{19} Be-doped	
2000 Å $Al_{0.3}Ga_{0.7}As$ 3×10^{18} Be-doped	
100 Å $Al_{0.3}Ga_{0.7}As$	× 50
90 Å GaAs	
100 Å $Al_{0.3}Ga_{0.7}As$	
723 Å AlAs 3×10^{18} Si-doped	× 10
614 Å $Al_{0.15}Ga_{0.85}As$ 3×10^{18} Si-doped	
723 Å AlAs 3×10^{18} Si-doped	
2μm GaAs 1×10^{19} Si-doped	
Si substrate	

Fig. 37. Schematic of a GaAs/AlGaAs-on-Si reflector modulator structure.

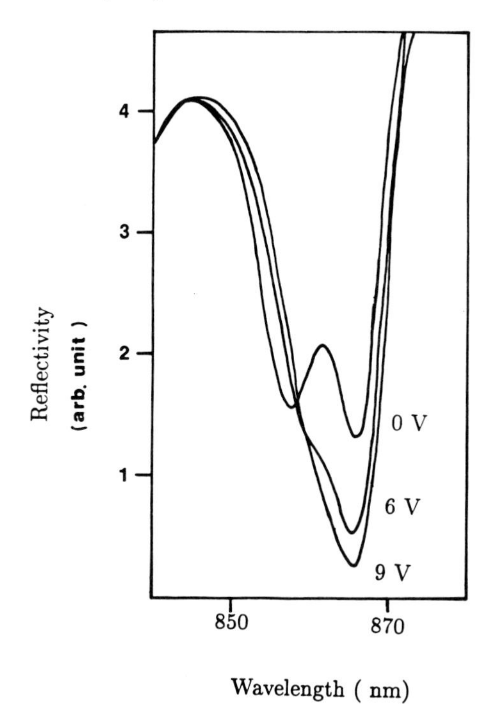

Fig. 38. Reflectivity of the modulator shown in Fig. 37 at 0 V, 6 V, and 9 V reverse biases.

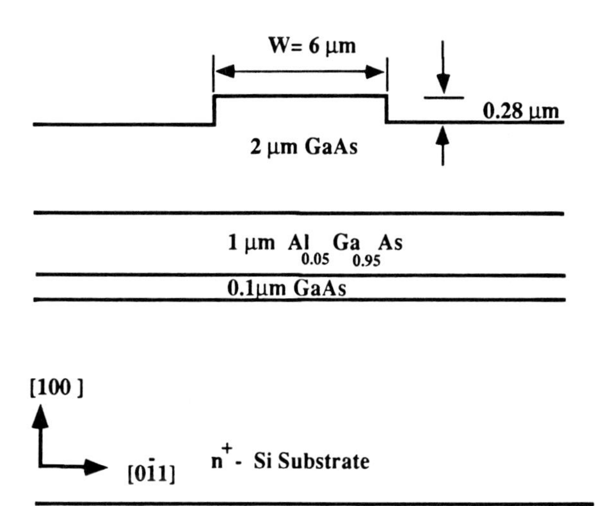

Fig. 39. Schematic cross section of the GaAs/AlGaAs ridge waveguide on a Si substrate. Widths W are 6–10 μm with the interval of 0.5 μm. (After Ref. 131.)

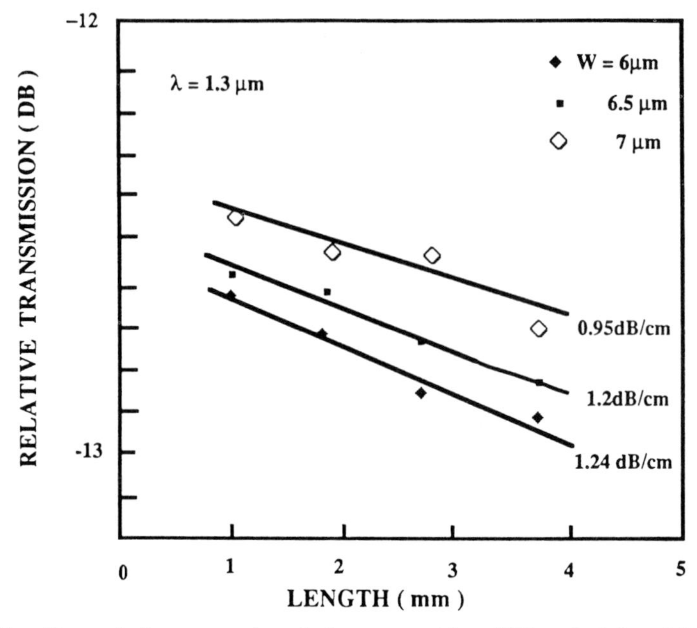

Fig. 40. Transmission versus length for waveguides of W = 6, 6.5 and 7 μm for the fundamental TE modes. Average propagation loss coefficients at 1.3 μm are shown by slopes of straight lines. (After Ref. 131.)

6.3. Detectors

For optoelectronic devices on Si, Si can be used directly as a detector medium at wavelengths shorter than 1.1 μm (the band gap of Si). However, Si has a low absorption coefficient (i.e. a large absorption depth) and thus a long sweep-out time proportional to this length will limit the device speed. The advantages of GaAs, such as its high mobility and high absorption coefficient, make it a superior material for high-speed optical detectors compared with Si. Furthermore, low band gap compound semiconductors grown on Si for photodetectors can be used for wavelengths longer than 1.1 μm.

Figure 41 shows the structure of a p-i-n GaAs-on-Si photodetector grown by MBE [133]. The frequency response of the same device is shown in Fig. 42. The detector with an i region of 2 μm shows a 3-dB corner frequency near 6 GHz. Compared with GaAs on GaAs, the photodetectors on Si exhibited softer reverse and, to some extent, slightly softer forward turn-on characteristics, attributed to the presence of defects in the high-field i region.

HgCdTe infrared focal plane arrays on CdTe/GaAs/Si grown by MOCVD have achieved resistance–area products of 10^7 W cm^2 for 320 × 5 arrays with a cutoff wavelength of 4.6 μm at 77K [134]. This is comparable to or

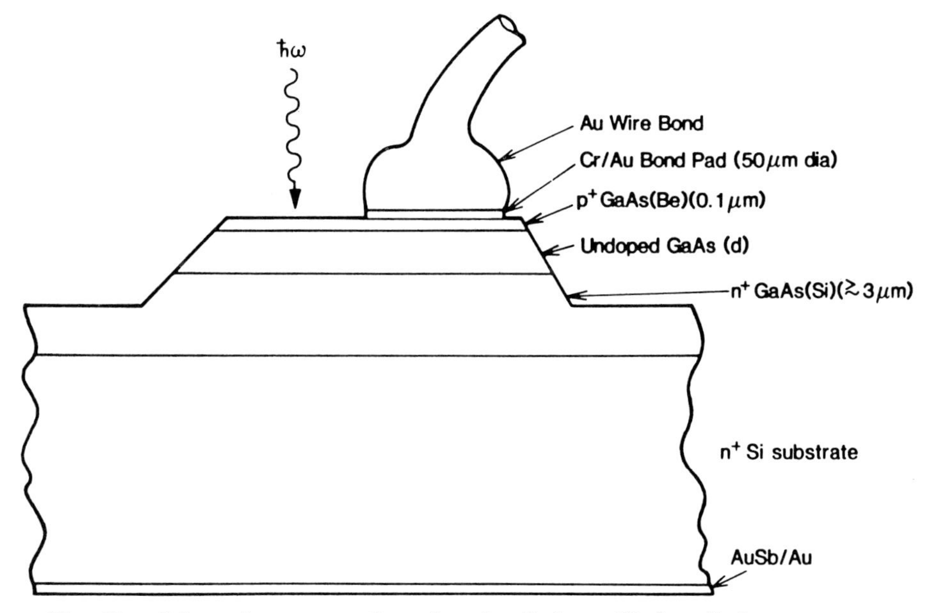

Fig. 41. Schematic representation of a p-i-n GaAs-on-Si photodiode.

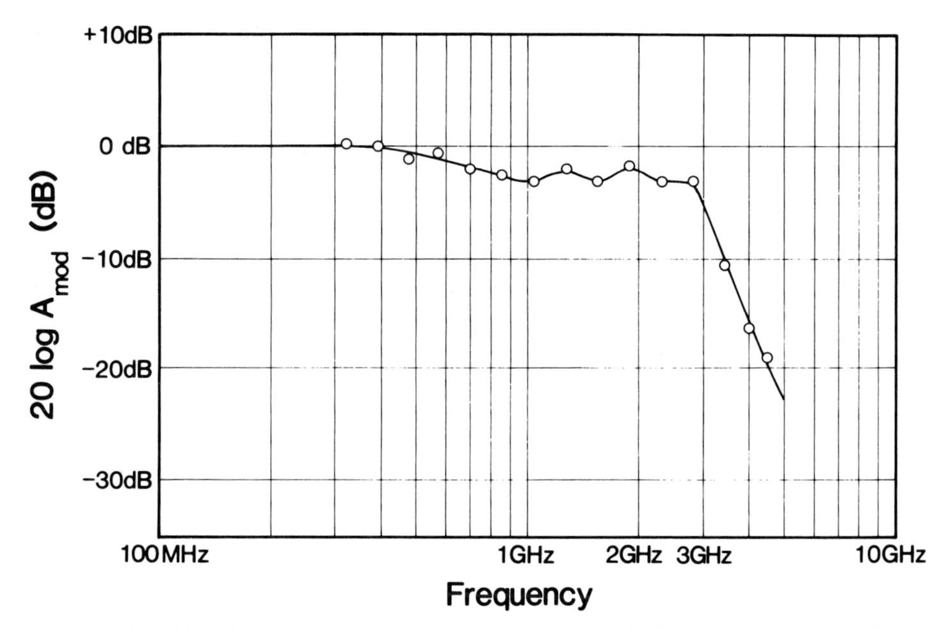

Fig. 42. Modulation response versus frequency of a GaAs-on-Si p-i-n photodetector showing a 3-dB bandwidth of 4 GHz.

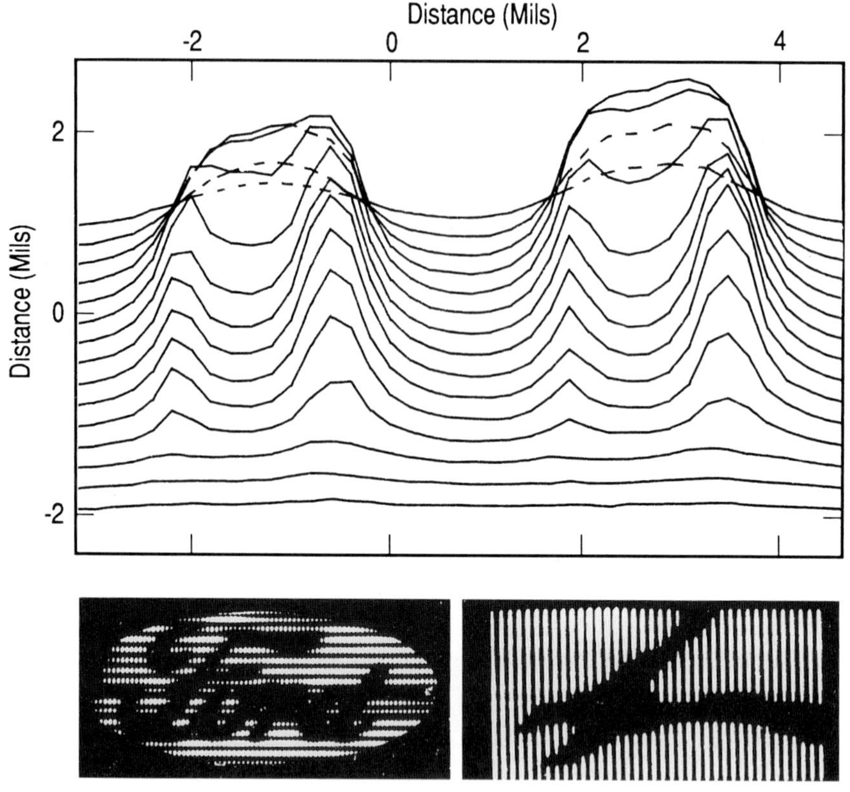

Fig. 43. Spot scan (top) for two adjacent pixels from a 1 × 32 HgCdTe/Si array. Thermal imagery is shown (bottom) for this array wire bonded to a CCD NMOS readout chip. (After Ref. 134.)

better than published results for MOCVD HgCdTe grown on any substrate. Several 1 × 32 arrays were also bonded to a 2 × 32 charge-coupled device (CCD) readout chip for thermal imaging. Figure 43 shows such a thermal image obtained by this array [134]. Liquid-phase epitaxy (LPE)-grown HgCdTe infrared focal plane arrays on MOCVD-grown CdZnTe/GaAs/Si have demonstrated resistance–area products of 6.0×10^4 W cm^2 for 128 × 128 arrays with a cutoff wavelength of 6.0 μm at 80 K [135].

6.4. Solar Cells

Thin-film III–V compound solar cells deposited on Si substrates show potential for developing high-efficiency, low-cost, and lightweight solar cells [91, 136–140]. Single-crystal cascades with Si as the bottom cell and III–V compounds on the top appear promising for high-performance systems, especially those destined for space-based operations. Solar cells, with each material absorbing in a different portion of the solar spectrum, could achieve conversion efficiencies as high as 36% [141].

Recombination losses at dislocations, which reduce the short-circuit current and increase the excess leakage current, are thought to be the limiting factor in determining the performance of GaAs/Si solar cells. The thermal cycle growth technique is commonly used in GaAs-on-Si solar cells to reduce dislocation density and thus to improve conversion efficiency. Ohmachi *et al.* [138] demonstrated high-efficiency GaAs-on-Si solar cells with efficiencies of 18.3% (AM0, 1 sun) and 20.0% (AM1.5, 1 sun) [139]. Vernon *et al.* [139] demonstrated efficiencies of 19.5% (AM0, 189 sun) and 21.7% (AM1.5, 237 sun) in MOCVD-GaAs/Si solar cells which have a defect density of $\sim 2 \times 10^7$ cm^{-2}.

7. MONOLITHIC DEVICES

There is great interest in the fabrication of GaAs and other compound semiconductor devices along with Si devices on the same Si chip, with the ultimate goal of fabricating OEICs. The combination of Si and GaAs processes must take into account the difference in processing temperatures for the two materials: Si processing temperatures are usually up to 1000°C, but GaAs processing temperatures normally have to be well below 850°C to prevent arsenic out-diffusion. Therefore, for monolithic device fabrication, high-temperature Si processing should be completed first, similar to the case in selective-area epitaxy, which has been discussed in Section 3.5.

Integration of a GaAs/AlGaAs LED and a single Si driver MOSFET on one chip was demonstrated by Choi *et al.* [142]. LED modulation at rates up to 27 MHz was accomplished by applying voltage pulses to the gate of the MOSFET. The same group also demonstrated monolithic integration of a GaAs/AlGaAs LED and a Si driver circuit composed of 10 MOSFETs [143]. The circuit diagram for the LED/driver circuit is shown in Fig. 44. A photograph of the completed circuit is shown in Fig. 45. LED modulation rates of more than 100 MHz have been demonstrated.

The high dislocation densities in GaAs-on-Si layers, although limiting the performance of most optoelectronic devices, reduce the carrier recombination lifetime while retaining reasonably high carrier drift mobility, which translates to a higher quantum efficiency [144]. Picosecond GaAs photoconductors on Si for local integration with processed Si circuits have also been demonstrated [145]. The GaAs-on-Si photoconductors with 15-μm gap

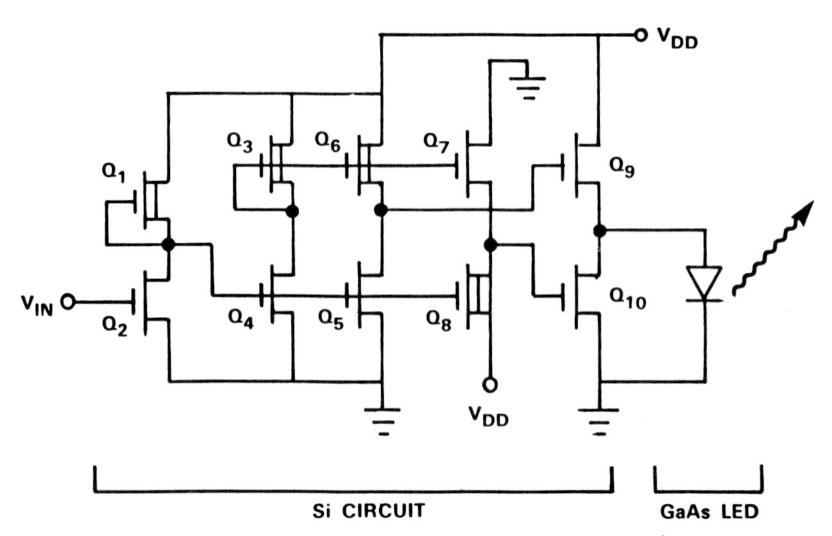

Fig. 44. Circuit diagram of the monolithically integrated LED/driver circuit.

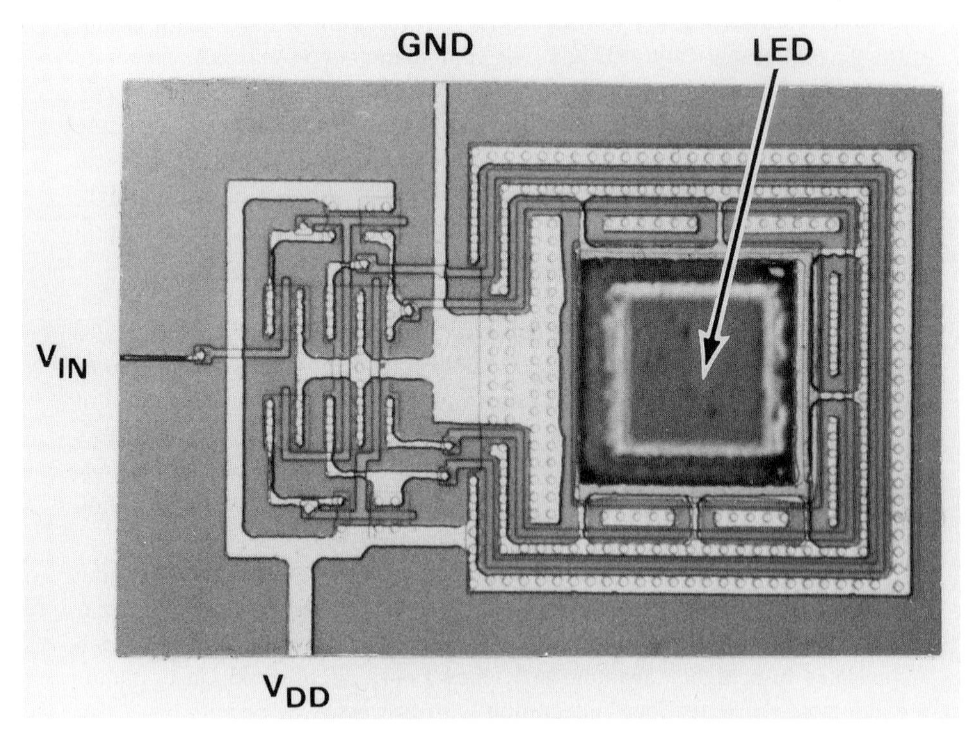

Fig. 45. Photomicrograph of the LED/driver circuit.

lengths have exhibited sampling oscilloscope limited responses larger than 16 mA, with electrical pulse widths less than 20 ps.

8. CONCLUSIONS

Promising results have been achieved in the heteroepitaxial growth of GaAs and other compound semiconductors on Si substrates. This has been demonstrated even by minority carrier devices such as lasers, which demand high crystalline quality. However, the dislocation densities and strain in heteroepitaxially grown layers must be reduced further to achieve high-performance optical devices on Si. The use of strained-layer-superlattice buffer layers and thermal cycling have been somewhat effective in reducing dislocation densities. Further investigations are necessary to understand these mechanisms and to optimize the buffer layer structure and the thermal cycling conditions. Unfortunately, the bimolecular nature of these defects implies that the lower their density the harder it is to reduce them further. New approaches in buffer structures of heteroepitaxy and bandgap material systems might be the way to success in heteroepitaxial growth of optical devices.

ACKNOWLEDGMENTS

The authors acknowledge the assistance of their colleagues J. Chen, Dr. J. Chyi, A. Salvador, S. Strite, D. Mui, J. Reed, and N. Lucas. Special thanks go to Mrs. S. White for her assistance in the preparation of the manuscript. The authors also express their gratitude to their collaborators, among whom are Professors H. Zabel, N. Otsuka, A. Yariv, R. Alfano, K. Shum, J. Mazur, and J. Washburn; Drs. C. Choi, H. Chen, D. Neumann, K. Zanio, C. Sparks, and E. Specht; and Mr. M. Longerbone and Mr. L.P. Erickson. This work is funded by the U. S. Air Force Office of Scientific Research under grant AFOSR-89–0239, and H.M. acknowledges support by the U.S. Department of Energy under contract DE-AC-02–76ER-01198.

REFERENCES

1. T. Soga, T. Jimbo, and M. Umeno, *Appl. Phys. Lett.* **56**, 1433 (1990).
2. C. Mailhiot and D. L. Smith, *Solid State Mater. Sci.*, **16**, 131 (1990).

3. J. H. VAN DER MERWE, *J. Appl. Phys.*, **34**, 123 (1962).
4. J. W. MATTHEWS AND A. E. BLAKESEE, *J. Cryst. Growth* **27**, 118 (1974).
5. R. PEOPLE AND J. C. BEAN, *Appl. Phys. Lett.* **47**, 322 (1985); **49**, 229 (1986).
6. J. Y. TSAO, B. W. PICRAUX, AND D. M. CORNELISON, *Phys. Rev. Lett.* **59**, 2455 (1987).
7. J. Y. TSAO AND B. W. DODSON, *Appl. Phys. Lett.* **53**, 848 (1988).
8. S. F. FANG, K. ADOMI, S. IYER, H. MORKOÇ, H. ZABEL, C. CHOI, AND N. OTSUKA, *J. Appl. Phys.* **68**, R31 (1990).
9. P. E. WIERENGA, J.A. KUBBY, AND J. E. GRIFFITH, *Phys. Rev. Lett.* **59,** 2169 (1987).
10. D. J. CHADI, *Phys. Rev. Lett.* **59**, 1691 (1987).
11. G. MUNNS, *M.S. Dissertation,* University of Illinois at Urbana-Champaign (1987).
12. W. I. WANG, *Appl. Phys. Lett.* **44**, 1149 (1984).
13. S. L. WRIGHT, M. INADA, AND H. KROEMER, *J. Vac. Sci. Technol.* **21**, 534 (1982).
14. S. L. WRIGHT, H. KROEMER, AND M. INADA, *J. Appl. Phys.* **55**, 2916 (1984).
15. K. MORIZANE, *J. Cryst. Growth* **30**, 249 (1977).
16. W.T. MASSELINK, T. HENDERSON, J. KLEM, R. FISCHER, P. PEARCH, H. MORKOÇ, M. HAFICH, P. D. WANG, AND G. Y. ROBINSON, *Appl. Phys. Lett.* **45**, 1309 (1984).
17. R. FISCHER, H. MORKOÇ, D. A. NEUMANN, H. ZABEL, C. CHOI, N. OTSUKA, M. LONGERBONE, AND L. P. ERICKSON, *J. Appl. Phys.* **60**, 1640 (1986).
18. S. J. ROSNER, S. M. KOCH, S. LADERMAN, AND J. S. HARRIS, *Proc. MRS 1986 Spring Meeting* **67**, 77 (1986).
19. R. I. G. UHRBERG, R. L. G. UHUBERG, R. D. BRIBGANS, R. Z. BACHRACH, AND J. E. NORTHRUP *Phys. Rev. Lett.* **56**, 520 (1986).
20. M. RAZEGHI, F. OMNES, M. DEFOUR, AND P. MAUREL, *Appl. Phys. Lett.* **52**, 209 (1988).
21. H. KROEMER, K.H. POLASKO, AND S.C. WRIGHT, *Appl. Phys. Lett.*, **36**, 763 (1980).
22. H. NOGE, H. KANO, T. KATO, M. HASHIMOTO, AND I. IGARASHI, *J. Cryst. Growth* **83,** 431 (1987).
23. R. KAPLAN, *Surf. Sci.* **93**, 145 (1980).
24. T. SAKAMOTO AND G. HASHIGUCHI, *Jpn. J. Appl. Phys.* **25**, L78 (1986).
25. R. FISCHER, D. NEUMAN, H. ZABEL, H. MORKOÇ, C. CHOI, AND N. OTSUKA, *Appl. Phys. Lett.* **48**, 1223 (1986).
26. M. RAZEGHI, R. BLONDEAU, M. DEFOUR, F. OMNES, AND P. MAUREL, *Appl. Phys. Lett.* **53**, 854 (1988).
27. H. Z. CHEN, J. PASLASKI, A. GHAFFARI, H. WANG, H. MORKOÇ, AND A. YARIV, *Tech. Dig. IEDM*, 238 (1987).
28. H. K. CHOI, J. P. MATTIA, G. W. TURNER, AND B. Y. TSAUR, *IEEE Electron Devices Lett.* **9**, 512 (1988).

29. F. K. Hopkins and J. T. Boyd, *Opt. Eng.* **27**, 701 (1988).
30. R. I. G. Uhrberg, R. D. Bringans, M. A. Olmstead, R. Z. Bachrach, and J. E. Northrup, *Phys. Rev. B* **35**, 3945 (1987).
31. C. C. Chang, *Surf. Sci.* **23**, 283 (1970).
32. J. Varrio, H. Asonen, J. Lammasniemi, K. Rakennus, and M. Pessa, *Appl. Phys. Lett.* **55**, 1987 (1989).
33. M. Kawabe and T. Shiraishi, *Fifth Int. Conf. on Molecular Beam Epitaxy*, Sapporo, Japan, 145 (1988).
34. M. Kawabe, T. Ueda, and H. Takasugi, *Jpn. J. Appl. Phys.* **26**, L114 (1987).
35. M. Zinke-Allmang, L. C. Feldman, and S. Nakahara, *Appl. Phys. Lett.* **52**, 144 (1988).
36. R. D. Bringans, M. A. Olmstead, R. I. G. Uhrberg, and R. Z. Bachrach, *Phys. Rev. B* **36**, 9569 (1987).
37. Y. Horikoshi, M. Kawashima, and H. Yamaguchi, *Jpn. J. Appl. Phys.* **25**, L868 (1986).
38. W. Stolz, Y. Horikoshi, M. Naganuma, and K. Nozawa, *Fifth Int. Conf. on Molecular Beam Epitaxy*, Sapporo, Japan, 129 (1988).
39. J. E. Palmer, G. Burns, C. G. Fonstad, and C. V. Thompson, *Appl. Phys. Lett.* **55**, 990 (1989).
40. T. Sakamoto and H. Kawanami, *Surf. Sci.* **111**, 177 (1981).
41. J. Nogami, S. Park, and C.F. Quate, *Appl. Phys. Lett.* **53**, 2086 (1988).
42. D. K. Biegelsen, F. A. Ponce, A. J. Smith, and J. C. Tramontana, *Proc. 1986 Spring MRS Meeting* **67**, 45 (1986).
43. J. Castagne, C. Fontaine, E. Bedel, and A. Munoz-Yague, *J. Appl. Phys.* **64**, 2372 (1988).
44. D. K. Choi, S. M. Koch, T. Takai, T. Halicioglu, and W. A. Tiller, *J. Vac. Sci. Technol.* **B6**, 1140 (1988).
45. S. J. Rosner, S. M. Koch, and J. S. Harris, Jr., *Proc. 1987 Spring MRS Meeting* **91**, 155 (1987).
46. C. W. Nieh, H. Z. Chen, L. Eng, A. Yariv, H. Morkoç, and N. Otsuka, unpublished results.
47. T. C. Chong and G. Fonstad, *J. Vac. Sci. Technol.* **B5**, 815 (1987).
48. S. M. Koch, S. J. Rosner, D. Schlom, and J. S. Harris, Jr., *MRS Symp. Proc.* **67**, 37 (1986).
49. R. J. Matyi and H. Shichijo, *Thin Solid Films* **181**, 213 (1989).
50. A. Okamoto and K. Ohata, *Appl. Phys. Lett.* **51**, 1512 (1987).
51. E. A. Fitzgerald, P. D. Kirchner, R. Proano, G. D. Pettot, J. M. Woodall, and D. G. Ast, *Appl. Phys. Lett.* **52,** 1496 (1988).
52. P. D. Hodson, P. Kightley, R. C. Goodfellow, T. B. Joyce, J. R. Riffat, P. R. Bradley, and R. J. M. Griffiths, *Semicond. Sci. Technol.* **3**, 715 (1988).
53. E. H. Lingunis, N. M. Haegel, and N. H. Karam, *Appl. Phys. Lett.* **59**, 3428 (1991).

54. E. H. Lingunis, N. M. Haegel, and N. H. Karam, *Solid State Commun.* **76**, 303 (1990).
55. N. H. Karam, V. Haven, S. M. Vernon, N. El-Masry, E. H. Lingunis, and N. Haegal, *J. Cryst. Growth* **107**, 129 (1991).
56. R. J. Matyi, H. Shichijo, T. M. Moore, and H.-L. Tsai, *Appl. Phys. Lett.* **51**, 18 (1987).
57. B. Yacobi, C. Jagannath, S. Zemon, and P. Sheldon, *Appl. Phys. Lett.* **52**, 555 (1988).
58. A. Ackaert, L. Buydens, D. Lootens P. Van Daele, and P. Demeester, *Appl. Phys. Lett.* **55**, 2187 (1989).
59. J. W. Matthews, S. Mader, and T. B. Light, *J. Appl. Phys.* **41**, 3800 (1970).
60. J. Vilms and D. Kerps, *J. Appl. Phys.* **53**, 1536 (1982).
61. B. Yacobi, S. Zemon, P. Norris, C. Jagannath, and P. Sheldon, *Appl. Phys. Lett.* **51**, 2236 (1987).
62. W. Stolz, F.E.G. Guimaraes, and K. Ploog, *J. Appl. Phys.* **63**, 492 (1988).
63. S. L. Wright and H. Kroemer, *Appl. Phys. Lett.* **36**, 210 (1980).
64. G. S. Higashi, Y. J. Chabal, G. W. Trucks, and K. Raghavachari, *Jpn. J. Appl. Phys.* **24**, L227 (1985).
65. J. F. Morar, B. S. Meymerson, U. O. Karlsson, F. J. Himpsel, F. R. McFeeley D. Rieger, A. Taleb-Ibrahimi, and J.A. Yarmoff, *Appl. Phys. Lett.* **50**, 463 (1987).
66. I. Suemune, Y. Kunitugu, Y. Tanaka, Y. Kan, and M. Yamanishi, *Appl. Phys. Lett.* **53**, 2173 (1988).
67. Q. Z. Gao, T. Hariu, and S. Ono, *Jpn. J. Appl. Phys.* **26**, L1576 (1987).
68. S. S. Iyer, M. Arienzo, and E. De Fresart, *Appl. Phys. Lett.* **57**, 893 (1990).
69. S. F. Fang, A. Salvador, and H. Morkoç, *Appl. Phys. Lett.* **58**, 1887 (1991).
70. S. H. Wolf, S. Wagner, J. C. Bean, R. Hull, and J. M. Gibson, *Appl. Phys. Lett.* **55**, 2017 (1989).
71. S. Zemon, S.K. Shastry, P. Norris, C. Jagannath, and G. Lambert, *Solid State Commun.* **58**, 457 (1986).
72. R. J. Nemanich, D. K. Biegelsen, R. A. Street, B. Downs, B. S. Krusor, and R. D. Yingling, *Mater. Res. Soc. Symp. Proc.* **116**, 245 (1988).
73. M. Bugajski, K. Nauka, S. J. Rosner, and D. Mars, *Mater. Res. Soc. Symp. Proc.* **116**, 233 (1988).
74. N. W. Ashcroft and N. D. Mermin, "*Solid State Physics.*" Holt, Rinehart, & Winston, New York, 634 (1976).
75. N. Otsuka, C. Choi, L. A. Kolodziejski, R. L. Gunshor, R. J. Fischer, C. K. Peng, H. Morkoç, Y. Nakamura, and S. Nagakura, *J. Vac. Sci. Technol.* **B4**, 896 (1986).
76. N. Otsuka, C. Choi, Y. Nakamura, S. Nagakura, R. Fischer, C. K. Peng, and H. Morkoç, *Proc. 1986 Spring MRS Meeting*, **67**, 85 (1986).
77. R. Hull and A. Fischer-Colbrie, *Appl. Phys. Lett.* **50**, 851 (1987).
78. R. Fischer, H. Morkoç, D. A. Neumann, H. Zabel, C. Choi, N. Otsuka, M. Longerbone, and L. P. Erickson, *J. Appl. Phys.* **60**, 1640 (1986).

79. D. A. Neuman, H. Zabel, R. Fischer, and H. Morkoç, *J. Appl. Phys.* **61**, 1023 (1987).
80. J. W. Matthews and A.E. Blakeslee, *J. Cryst. Growth* **29**, 273 (1975).
81. J. W. Matthews and A. E. Blakeslee, *J. Cryst. Growth* **32**, 265 (1976).
82. T. Soga, S. Hattori, S. Sakai, M. Takeyasu, and M. Umeno, *J. Appl. Phys.* **57**, 4578 (1985).
83. J. W. Matthews, A. E. Blakeslee, and S. Mader, *Thin Solid Films* **33**, 253 (1976).
84. N. A. El-Masry, J. C. Tarn, and N. H. Karam, *J. Appl. Phys.* **64**, 3672 (1988).
85. N. El-Masry, J. C. L. Tarn, T. P. Humphreys, N. Hamaguchi, N. K. Karam, and S. M. Bedair, *Appl. Phys. Lett.* **51**, 1608 (1987).
86. M. Shinohara, *Appl. Phys. Lett.* **52,** 543 (1988).
87. N. Chand, R. People, F. A. Baiocchi, K. W. Wecht, and A. Y. Cho, *Appl. Phys. Lett.* **49**, 815 (1986).
88. J. W. Lee, H. Shichijo, H. L. Tsai, and R. J. Matyi, *Appl. Phys. Lett.* **50**, 31 (1987).
89. C. Choi, N. Otsuka, G. Munns, R. Houdré, H. Morkoç, S. L. Zhang, D. Levi, and M. V. Klein, *Appl. Phys. Lett.* **50**, 992 (1987).
90. R. Houdré and H. Morkoç, *Crit. Rev. Solid State Mater. Sci.* **16**, 91 (1990).
91. Y. Itoh, T. Nishioka, A. Yamamoto, and M. Yamaguchi, *Appl. Phys. Lett.* **52**, 1617 (1988).
92. N. Choi, Ph.D. Dissertation, Purdue University (1988).
93. G. Radhakrishnan, O. McCullough, J. Cser, and J. Katz, *Appl. Phys. Lett.* **52**, 731 (1988).
94. N. Chand, R. Fischer, A. M. Sergent, D. V. Lang, S. J. Pearton, and A. Y. Cho, *Appl. Phys. Lett.* **51**, 1013 (1987).
95. N. Chand, J. Allam, J. M. Gibson, F. Capasso, F. Beltram, A. T. Macrander, A. L. Hutchinson, L. C. Hopkins, C. G. Bethea, B. F. Levine, and Y. Cho, J. *Vac. Sci. Technol. B* **5**, 822 (1987).
96. Y. Shinoda, T. Nishioka, and Y. Ohmachi, *Jpn. J. Appl. Phys.* **22**, L450 (1983).
97. R. M. Fletcher, D. K. Wanger, and J. M. Ballantyne, *Appl. Phys. Lett.* **44**, 967 (1984).
98. M. Razeghi, R. Blondeau, M. Defour, F. Omnes, P. Maurel, and F. Brillouet, *Appl. Phys. Lett.* **53**, 854 (1988).
99. A. Hashimoto, Y. Kawarada, T. Kamijo, M. Akiyama, N. Watanabe, and M. Sakuta, *Appl. Phys. Lett.* **48**, 1617 (1986).
100. S. Kondo, H. Nagai, Y. Itoh, and M. Yamaguchi, *Appl. Phys. Lett.* **55**, 1981 (1989).
101. H. Mori, M. Ogasawara, M. Yamamoto, and M. Tachikawa, *Appl. Phys. Lett.* **51**, 1245 (1987).
102. T. H. Windhorn, G. M. Metze, B. Y. Tsaur, and J. C. C. Fan, *Appl. Phys. Lett.* **45**, 309 (1984).

103. T. H. Windhorn and G. M. Metze, *Appl. Phys. Lett.* **47**, 1031 (1985).
104. S. Sakai, T. Sago, M. Takeyama, and M. Umeno, *Jpn. J. Appl. Phys.* **24**, L666 (1985).
105. S. Sakai, T. Sago, M. Takeyama, and M. Umeno, *Appl. Phys. Lett.* **48**, 413 (1986).
106. H. Z. Chen, J. Paslaski, A. Yariv, and H. Morkoç, *Appl. Phys. Lett.* **52**, 605 (1988).
107. H. Z. Chen, A. Ghaffari, T. Wang, H. Morkoç, and A. Yariv, *Appl. Phys. Lett.* **51**, 1320 (1987).
108. H. K. Choi, J. W. Lee, J. P. Salerno, M. K. Connors, B. Y. Tsaur, and J. C. C. Fan, *Appl. Phys. Lett.* **52**, 1114 (1988).
109. H. Z. Chen, A. Ghaffari, H. Morkoç, and A. Yariv, *Opt. Lett.* **12**, 812 (1987).
110. H. Z. Chen, A. Ghaffari, T. Wang, H. Morkoç, and A. Yariv, *Workshop on Future Opportunities through GaAs on Si*, Marina Del Rey, CA, June 18–19, 1987.
111. T. Egawa, H. Tada, Y. Kobayashi, T. Soga, T. Jimbo, and M. Umeda, *Appl. Phys. Lett.* **57**, 1179 (1990).
112. J. Connolly, N. Dinkel, R. Menna, D. Gilbert, and M. Harvey, *Appl. Phys. Lett.* **53,** 2552 (1988).
113. J. H. Kim, R. J. Lang, G. Radhakrishnan, J. Katz, A. A. Narayanan, and R. R. Craig, *Appl. Phys. Lett.* **55**, 1492 (1989).
114. R. D. Dupuis, J. P. Van der Ziel, R. A. Logan, J. M. Brown, and C. J. Pinzone, *Appl. Phys. Lett.* **50**, 407 (1987).
115. K. Lau, and Y. Yariv, in *"Semiconductor and Semimetals"* (W. T. Tsang, ed.) Academic Press, Orlando, FL, Vol. **22,** p. 69, 1985.
116. D. G. Deppe, D. W. Nam, N. Holonyak, Jr., K. C. Hsieh, R. J. Matyi, H. Shichijo, J.E. Epler, and H. F. Chung, *Appl. Phys. Lett.* **51**, 1271 (1987).
117. D. G. Deppe, D. C. Hall, N. Holonyak, Jr., R. J. Matyi, H. Shichijo, and J. E. Epler, *Appl. Phys. Lett.* **53**, 874 (1988).
118. H. K. Choi, C. A. Wang, and N. H. Karam, *Appl. Phys. Lett.* **59**, 2634 (1991).
119. M. Razeghi, M. Defour, F. Omnes, Ph. Maurel, J. Chazelas, and F. Brillouet, *Appl. Phys. Lett.* **53**, 725 (1988).
120. M. Sugo, H. Mori, Y. Sakai, and Y. Itoh, *Appl. Phys. Lett.*, **60**, 472 (1992).
121. H. P. Lee, X. Liu, and S. Wang, *Appl. Phys. Lett.* **56**, 1014 (1990).
122. G. F. Burns, H. Blanck, and C. G. Fonstad, *Appl. Phys. Lett.* **56**, 2499 (1990).
123. R. L. Hartman and A. R. Hartman, *Appl. Phys. Lett.* **23**, 147 (1973).
124. A. R. Googwin, P. A. Kirkby, I. G. A. Davies, and R. S. Baulcomb, *Appl. Phys. Lett.* **34**, 647 (1979).
125. B. C. DeLoach, Jr., B. W. Hakki, R. L. Hartman, and L. A. D'Asaro, *Proc. IEEE* **61**, 1042 (1973).

126. P. Petroff and R. L. Hartman, *J. Appl. Phys.* **45**, 3899 (1974).
127. W. D. Johnston, jr. and B. I. Miller, *Appl. Phys. Lett.* **23**, 192 (1973).
128. P. Petroff and R. L. Hartman, *Appl. Phys. Lett.* **23**, 469 (1973).
129. W. Dobbelaere, D. Huang, M.S. Ünlü, and H. Morkoç, *Appl. Phys. Lett.* **53**, 94 (1988).
130. A. Salvador, K. Adomi, K. Kishino, M.S. Ünlü, and H. Morkoç, *J. Appl. Phys.* **69**, 534 (1991).
131. Y. S. Kim, S. S. Lee, R. V. Ramaswamy, S. Sakai, Y. C. Kao, and H. Shichijo, *Appl. Phys. Lett.* **56**, 802 (1990).
132. E. Kapon and R. Bhat, *Appl. Phys. Lett.* **50**, 1628 (1987).
133. J. Paslaski, H.Z. Chen, H. Morkoç, and A. Yariv, *Appl. Phys. Lett.* **52**, 1410 (1988).
134. K. Zanio, R. Bean, R. Mattson, P. Vu, S. Taylor, D. McIntyre, C. Ito, and M. Chu, *Appl. Phys. Lett.* **56**, 1207 (1990).
135. S. M. Johnson, M. H. Kalisher, W. L. Ahlgren, J. B. James, and C. A. Cockrum, *Appl. Phys. Lett.* **56**, 946 (1990).
136. M. Yamaguchi and C. Amano, *J. Appl. Phys.* **58**, 3601 (1985).
137. J. C. C. Fan, B. Y. Tsaur, and B. J. Palm, *Proc. 16th IEEE Photovolt. Spec. Conf.*, 692 (1982).
138. Y. Ohmachi, Y. Kadota, Y. Watanabe, and H. Okamoto, *Mater. Res. Soc. Symp. Proc.* **144**, 297 (1989).
139. S. M. Vernon, S. P. Tobin, V. E. Haven, L. M. Geoffroy, and M. M. Sanfacon, *Proc.* **22nd** *IEEE PV Spec. Conf.*, 353 (1991).
140. M. Yamaguchi, C. Amano, and Y. Itoh, *J. Appl. Phys.* **66**, 915 (1989).
141. B. Y. Tsaur, J. C. C. Fan, G. W. Turner, G. M. Davis, and R. P. Gale, *Proc. IEEE Photovolt. Spec. Conf.*, 1143 (1982).
142. H. K. Choi, C. W. Turner, T. H. Windhorn, and B. Y. Tsaur, *IEEE Electron Devices Lett.* **EDL-7**, 500 (1986).
143. H. K. Choi, J. P. Mattia, G. W. Turner, and B. Y. Tsaur, *IEEE Electron Devices Lett.* **EDL-9**, 512 (1988).
144. A. M. Johnson, R. M. Lum, W. M. Simpson, and J. Klingert, *IEEE J. Quantum Electron.* **23**, 1180 (1987).
145. J. D. Morse, R. Mariella, G. D. Anderson, and R. W. Dutton, *IEEE Electron Devices Lett.* **10**, 7 (1989).

PART III
Device Processing

Chapter 6

FOCUSED ION BEAM FABRICATION TECHNIQUES FOR OPTOELECTRONICS

L. R. Harriott and H. Temkin

AT&T Bell Laboratories
Murray Hill, New Jersey 07974

1. INTRODUCTION

In this chapter, we review applications of finely focused ion beams to fabrication of optoelectronic structures and devices. The examples given cover a wide range of processing techniques. Micromachining by direct physical sputtering has been used to form laser facets, lenses, beam splitters, reflectors, and other optical components. Focused ion beams have also been used as primary pattern generators in a number of ways. Maskless implantation doping and lithography are of interest. In particular, lithographic and implantation processes which occur entirely within a high vacuum have been combined with molecular beam epitaxy for three-dimensional fabrication of

ISBN 0-12-200420-5

the III–V materials used for optoelectronics. We begin the chapter by reviewing the basics of finely focused ion beam systems and then describe applications to optoelectronics including micromachining, vacuum lithography, and maskless implantation.

2. FOCUSED ION BEAM SYSTEMS

Current focused ion beam technology is based on the development of liquid metal ion sources (LMISs). LMISs offer high brightness and small virtual source size, making them suitable for probe-forming systems. The limiting factor in the performance of finely focused ion beam systems is generally chromatic aberration due to the moderate (10–40 eV) energy spread in the ions from the source. The performance of a probe-forming optical system is most generally characterized by the current density in the focused spot. In most applications the processing throughput is a direct function of the total ion current on the target. Since the beam size is usually dictated by the application requirements, current density achievable is a good measure of the overall performance of the ion optical column.

Figure 1a shows an example of a simple focusing system used mainly for applications employing sputtering, such as mask repair [1–6], circuit restructuring [7, 8, 9], and micromachining [10]. Ions from the LMIS are extracted at an energy of 3–6 keV and pass through a beam-limiting aperture to a three-element asymmetric electrostatic lens designed for low chromatic aberration [11]. The ion column also employs an electrostatic beam deflector. A system of this type operates at 20–50 keV with 100- to 200-nm spot size and a beam current of 100–300 pA. The beam can be scanned over fields of up to 1 mm usually with computer control for imaging with secondary electrons (or ions) and patterned exposures. The doses required for sputtering applications are quite high, 10^{17} to $10^{18}/\text{cm}^2$, limiting practical applications to those in which the area to be exposed to the ion beam is small, such as defect repair of lithographic masks.

For typical electrostatic ion focusing lenses chromatic aberration is dominant and is the limiting factor in the performance of ion microbeam systems [1]. The current density (J) in the focused spot is given by

$$J = \frac{dI}{d\Omega}\left(\frac{E/\Delta E}{C}\right)^2$$

Where $dI/d\Omega$ is the angular current density of the ion source, C the chromatic aberration of the lens, ΔE the energy spread of the ions, and E the

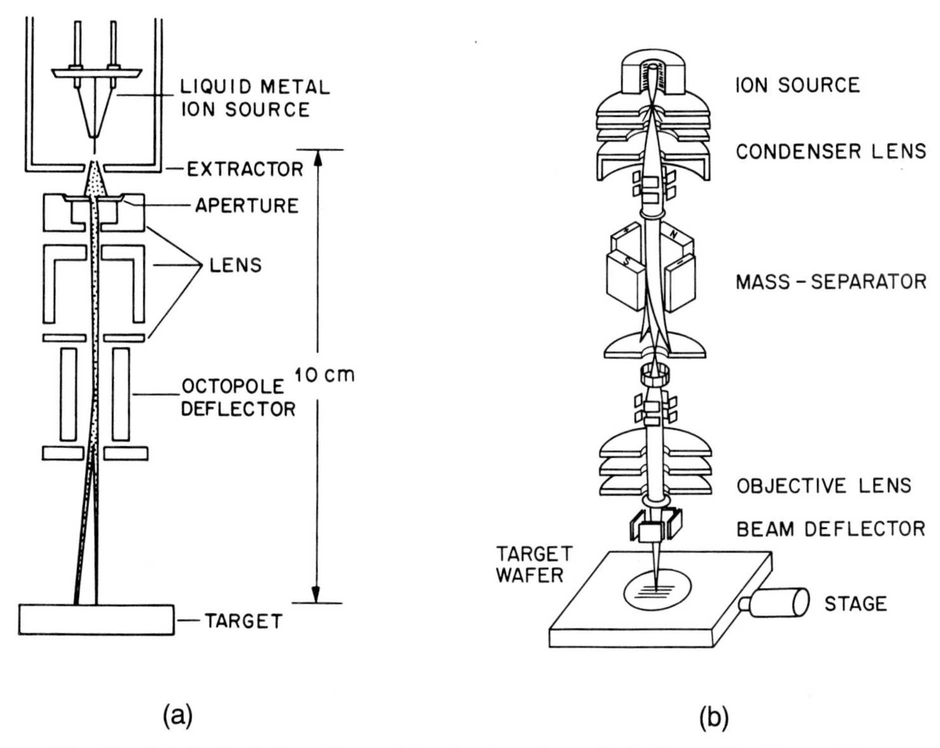

Fig. 1. (a) A single-lens focused ion beam column including a liquid metal ion source, lens, and beam deflector (Ref. 1). (b) A two-lens focusing column including an EXB mass filter for use with alloy dopant sources. (From Ref. 12.)

beam energy. Note that the current density is independent of the lens acceptance angle. In other words, it is constant for a given focusing system and does not depend on the focal spot size. The focal spot size increases with the solid angle of acceptance of the lens in such a way as to keep the current density in the spot the same. As can be seen from the preceding equation, the current density increases with increasing beam energy. This improvement is limited, however, by the maximum voltage of the particular lens. As the maximum lens voltage is increased by using larger gaps, the quantity (E/C) tends to remain constant due to increases in C [1]. Furthermore, increasing the source emission current also increases the energy spread. These factors combine to limit the maximum current densities in practical LMIS round lens focusing systems to about 1 A/cm^2. Beam spot sizes in state-of-the-art ion focusing columns are in the range 50 to 200 nm, with the smaller spot sizes possible at the price of reduced beam current. Current densities of 1 A/cm^2 or higher are usually achieved at beam energies of 20 keV or higher.

Applications for finely focused beams cover a wide range of processing techniques in industry and research [2–5]. These applications generally exploit one or more of three basic aspects of the ion–solid interactions. The desired effect of the ion beam may be produced by (1) pseudoelastic collisions between the energetic ions and target atoms where momentum transfer results in displacements (damage) and sputtering, (2) inelastic scattering of the ions with electrons in the solid producing excitations which may cause chemical changes such as in polymer resist exposure, or (3) the presence of the ion in the solid after it has come to rest such as in semiconductor doping. The design of ion focusing columns and choice of LMIS species are usually dictated by the application. The two major classes of focused ion beam (FIB) systems are the low-energy (10–15 keV) type with heavy ion species (usually Ga) in which momentum transfer and sputtering are important and the higher energy (50–200 keV) type in which electronic excitation and doping are the desired effects.

Figure 1b illustrates the other major type of ion focusing column [12]. It is designed to operate with an alloy LMIS so that many ion species can be produced in the beam. Ions from the source are focused to a crossover at an intermediate aperture. Crossed electric and magnetic ($E \times B$) fields act as a velocity filter, effectively separating the ion species by mass. The $E \times B$ filter is adjusted to allow only the desired ion species to pass through the intermediate aperture and the rest of the ion column. The objective lens then focuses the mass-selected beam onto the target.

This type of mass-separating column is usually designed to run at energies from 50 to 200 keV [13, 14]. These systems are used for maskless implantation [15] and lithography, producing minimum feature sizes of 100 nm [16, 17]. Again, there is a trade-off between beam diameter and current, with the minimum practical spot size of such a system about 50 nm. Eutectic alloy sources such as AuSi or AuSiBe are commonly used in such systems. The AuSiBe LMIS can provide beams of both *n*- and *p*-type dopants for III–V semiconductors.

3. MICROMACHINING

A very simple and straightforward focused ion beam patterning method is micromachining by localized physical sputtering. A typical system for this application uses a Ga liquid metal ion source and a single electrostatic focusing column and produces a 100- to 200-nm beam at 20 to 30 keV. Material removal is accomplished by scanning the ion beam by computer control in the desired pattern. Arbitrary shapes in the (X–Y) scanning directions can be produced with the beam control software. By varying the beam dwell time and thus the ion dose at each point in the pattern, the shape can be controlled in the height (Z) direction as well. Fairly complex patterns can thus be formed in virtually any material by software control of the beam scanning, making focused ion beam micromachining an attractive technique for prototyping of optical structures for optoelectronic integrated circuits (OEICs) [18].

The first application of FIB micromachining to optoelectronics was in creating facets in InP lasers [19]. The initial demonstration of the feasibility of this technique involved measuring light output characteristics of cleaved-facet lasers and then using the focused ion beam to form a new facet on the same device and remeasure its characteristics. Figure 2a and 2b illustrate the process used to form the laser facet. The dose required for the cut is delivered in several passes of the beam such that each pass removes material to a depth lower than the beam diameter. This is done to reduce redeposition effects due to the local geometry of the sputtering. The scan direction is also reversed after each pass to remove any redeposition asymmetry. Side-wall redeposition is partially compensated by the angular dependence of the sputtering yield. Furthermore, nonuniform sputtering of the material due to grain boundaries or other inhomogeneities is magnified by the increase of the sputtering yield with angle of incidence and can produce a rough bottom surface. All of these effects are material dependent, with redeposition effects being the limiting factor in high-aspect-ratio cuts.

Beam scanning during micromachining is accomplished under computer control with an address grid less than or equal to one beam radius to ensure uniform dose distribution. Figure 2b illustrates the dose delivery method for generating inclined surfaces or prism structures. These cuts are made with a series of successively smaller overlapping rectangular cuts. Smooth inclined surfaces result when each rectangle is decreased by one beam radius or less. The incline angle is arbitrary and is determined by the depth of the cut (h) of each rectangle and the decrement size (x) as $\tan(\theta) = h/x$. Each rectangle is made by several passes of the beam with alternating scan directions, as already described.

Figure 3 shows the laser facet formed by Ga ion beam micromachining. The facet is nearly vertical and very smooth. In this experiment a new facet was formed in a working InP laser. In this way the laser performance was measured with as-cleaved facets and again after the micromachined facet was formed. The resulting light–current curves are shown in Fig. 4. These data indicate an increase in the laser threshold current of at most 5% and

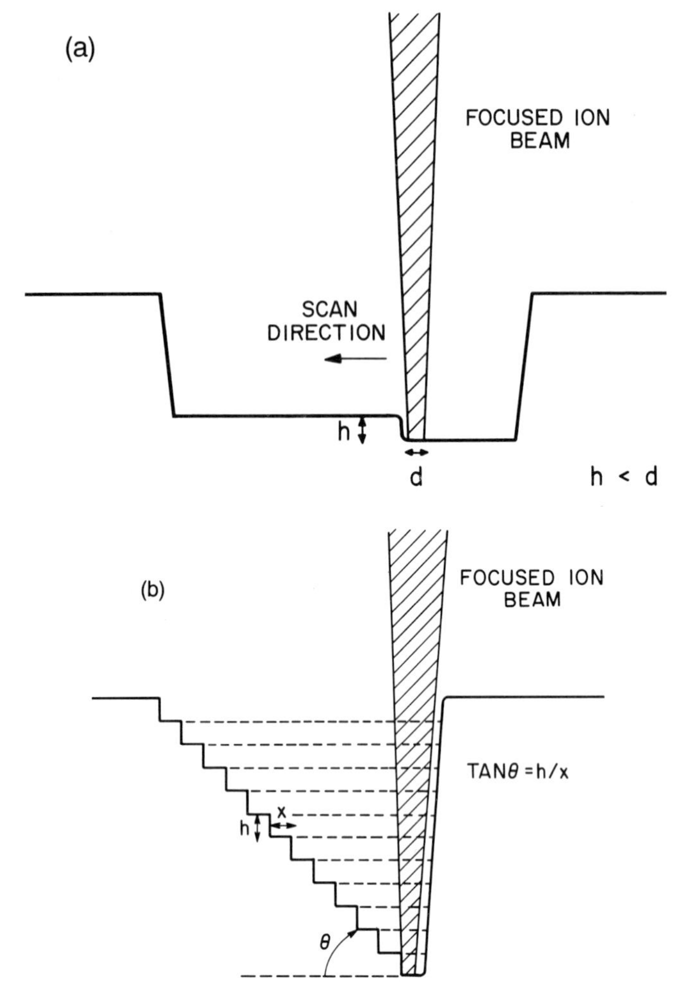

Fig. 2. The focused ion beam micromachining process. (a) Uniform dose delivery for rectangular cross section cuts. The dose is delivered in several beam passes to minimize redeposition effects. (b) Dose delivery method for producing angled features. A series of successively smaller rectangles is made to form the angled surface. If the decrement in rectangle size is less than or equal to the beam radius, the slope will be smooth. (From Ref. 20.)

only a slight decrease in the quantum efficiency, indicating the high quality of the facet.

The three-dimensional capability of FIB micromachining was demonstrated by fabricating a prism-shaped structure within the active stripe of an InP laser [20] as illustrated in Fig. 5. The vertical wall of the structure acted

Fig. 3. A facet formed in an InP laser by FIB micromachining. The laser originally had cleaved facets. (From Ref. 19.)

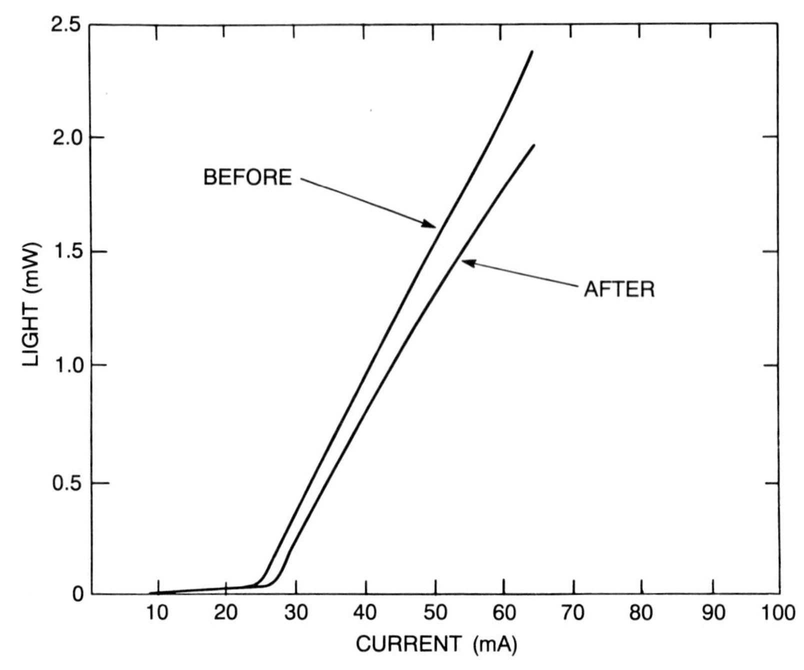

Fig. 4. Light–current output curves for the laser shown in Fig. 3 before and after the formation of the FIB machined facet. Only a slight increase in threshold current and slight decrease in slope efficiency is evident. (From Ref. 19.)

as the facet to define the optical cavity, while the 45° mirror acted as a reflector for the laser output, resulting in vertical light emission. Again, the threshold current and quantum efficiency were virtually unchanged by the addition of the FIB milled facet and reflector.

A subtle modification of the laser facet reflectivity by an asymmetric reflector written directly on the mirror coating was described by Bobeck *et al.* [21]. The idea is illustrated in Fig. 6. Antireflective facet coatings are used in distributed feedback (DFB) lasers to ensure single longitudinal mode operation. In the process the selectivity between the TE and TM transverse modes is reduced. The suppression of the TM modes may be incomplete, and occasionally the laser will switch to operation in a TM mode or in both a TM and TE mode. The random occurrence of a TM mode is a statistical process and represents a form of noise. Each occurrence of a TM mode constitutes an error in systems incorporating a laser. The number of such errors is detected in digital systems by the number of pulses which are incorrectly detected as logical "one" when logical "zero" was transmitted, or conversely. The number of such erroneously detected pulses is known as the bit error rate (BER).

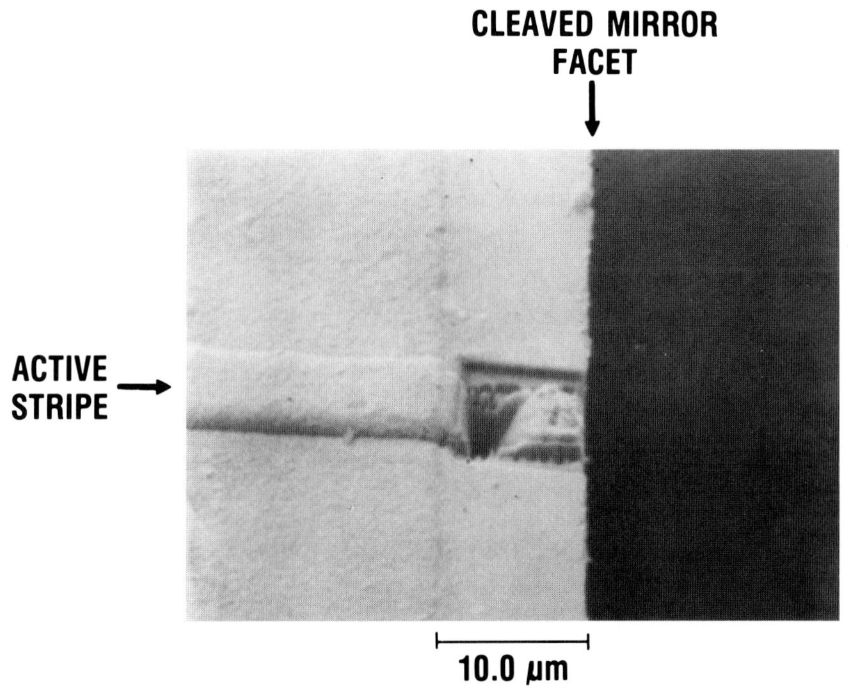

Fig. 5. A prism-shaped facet–reflector structure made in the active strip of an InP laser by FIB micromachining. The vertical face acts as the laser mirror, while the sloped surface reflects the laser light normal to the original direction. (From Ref. 20.)

A focused Ga ion beam system was used to micromachine a reflector in the facet coating of a DFB laser. The reflector consisted of a series of parallel lines with a period of 0.4 μm. The grooves were located in the region 25 μm wide and 8 μm high which overlapped the end of the active layer, as shown in Fig. 6. The grooves were aligned parallel to the plane of the active layer in order to enhance the reflectivity of the TE polarized mode and suppress the TM polarization. A small change in the facet reflectivity is expected to result in a large effect on the laser. This is shown in Fig. 7, which presents the laser spectra before and after the fabrication of a polarization-dependent reflector. Figure 7a shows the longitudinal mode spectrum of a DFB laser after the deposition of an antireflection coating. The spectrum is dominated by a single TE polarized mode at 1.554 μm. In addition, a weak TM mode is observed at 1.547 μm. This mode is approximately 30 dB weaker than the main peak. Nevertheless, the presence of the TM polarized mode is sufficient to cause an unacceptably high BER. Figure 7b shows the spectrum of the same laser after the fabrication of the polarizing

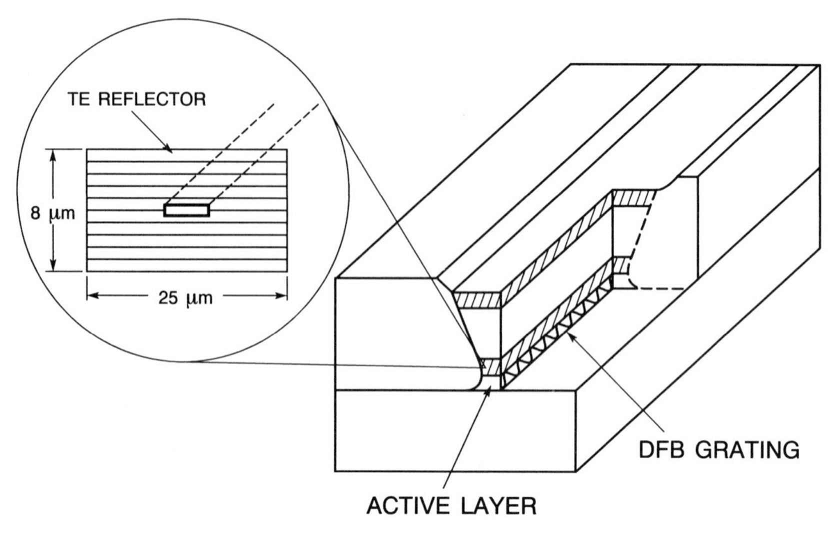

Fig. 6. Drawing of a polarization-sensitive reflector formed in the facet of an InP DFB laser by FIB micromachining. The reflector consisted of a 0.4-μm-pitch grating directly over the active output area. (From Ref. 21.)

reflector. The TM mode was completely suppressed and the main peak became considerably narrower due to the suppression of partition noise. The corrected laser was modulated at 1.7 Gb/s with a BER of less than 10^{-9}.

Although this example illustrates the ability of the FIB to perform surgical repairs of the laser facet, the ability to produce integrated polarization-sensitive reflectors on any surface is of wider interest in optoelectronics.

The use of FIB techniques for the fabrication of total internal reflection mirrors in grating surface emitting (GSE) laser arrays was described by Bossert *et al.* [22]. The FIB mirrors were used to couple two GSE linear arrays into a ring configuration, as shown in Fig. 8. In the GSE laser the second-order diffraction grating provides feedback to the laser and, in the first order, surface emission in the direction normal to the wafer plane. The surface emission has a very narrow far-field pattern. The laser consists of the alternating gain and feedback sections. In the gain section, a number of stripe lasers can be coupled together in the lateral direction through the evanescent field or y-branch coupling. Such arrays can be then coupled longitudinally to obtain mutual injection locking. The linear arrays are capable of high power output in a single longitudinal mode.

Coupling between the adjacent linear arrays can be achieved through the FIB-machined turning mirrors. Two pairs of such mirrors were used to form a two-column, eight-section, ring laser containing 80 individual stripe lasers,

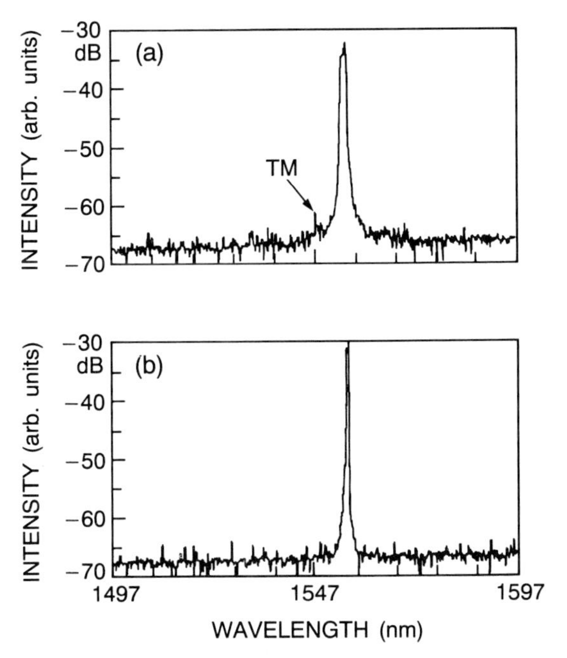

Fig. 7. (a) The longitudinal mode spectrum of a DBF laser showing a dominant TE mode with a TM mode present at roughly −30 dB. (b) Longitudinal mode spectrum of the same laser after the fabrication of the polarization-sensitive reflector, showing only a TE mode with a narrower linewidth. (From Ref. 21.)

as shown in Fig. 8. The machining was carried out by scanning the Ga beam focused to a 0.25-μm spot along a line inclined at 45° to the axis of each laser column. The scanning electron microscope top image of the turning mirror is shown in Fig. 9. The lower part shows the individual laser stripes (each 5 μm wide and covered with the contact metallization) of a gain section. A high degree of spatial coherence between the two columns was obtained, as judged by the narrow far-field pattern and single-mode output, confirming the quality of the FIB mirrors.

Another approach to high output power, narrow far field, devices is the unstable resonator illustrated in Fig. 10. High power output can be obtained from wide-stripe, essentially broad-area lasers, but with the far field greatly exceeding the diffraction limit. This is because, in the absence of any mode-stabilizing structure, the laser tends to operate in spatially incoherent filaments. The unstable resonator, produced by fabricating a divergent laser mirror, introduces high losses for incoherent modes and thus prevents formation of the filaments. The difficulty is in producing a monolithic mirror with the required curvature and smoothness. The FIB methods which rely

Fig. 8. A grating surface emitting (GSE) laser array with four gain sections, DBR gratings, and turning mirrors fabricated with FIB micromachining. (From Ref. 22.)

on software control for beam steering, and are capable of preparing very smooth mirror surfaces, are suitable for this purpose. The application of FIB micromachining to the preparation of such curved mirrors was described by Tilton *et al.* [23]. Mirrors with a surface smoothness better than 0.1 μm and a vertical slope of 2 to 3° were prepared.

The introduction of an unstable resonator mirror resulted in an expected increase in the threshold current and decrease in the slope efficiency. Nevertheless, power output greater than 600 mW/facet (pulsed) was reached and the far-field patterns exhibited near-diffraction-limited widths. The success is attributed to the smoothness of the curved mirror. The FIB technique is maskless and the quality of the mirror is not degraded by any imperfections in the masks used by more conventional methods.

An example of a passive component is the beamsplitter fabricated in a GaAs/GaAlAs waveguide shown in Fig. 11 [24]. Two perpendicular intersecting ridge waveguides were formed by conventional wet chemical etching. A precisely aligned groove was then micromachined at the intersection at an angle of 45°. The groove must penetrate only partially into the top waveguide layer and must be made accurately in terms of the angular and translational alignment, width, depth, and smoothness for best device performance. In addition, the slot must be narrow (less than 1 μm) to reduce

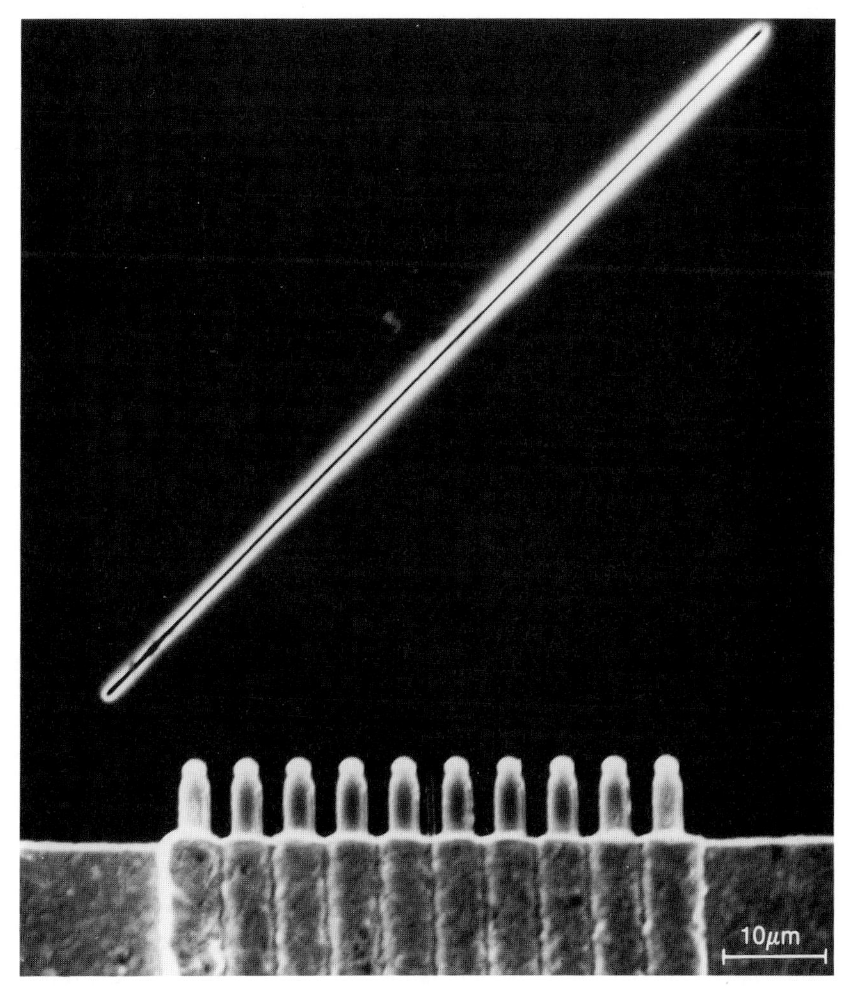

Fig. 9. Scanning electron micrograph of one of the turning mirrors used to form the GSE into a ring laser. (From Ref. 22.)

insertion loss. At the groove, the propagating mode is partially reflected from the micromachined wall and partially transmitted. The relative amounts of reflection and transmission were controlled accurately by the depth of the groove.

An example of a more complex shape fabricated by FIB micromachining is shown in Fig. 12. The structure consists of rings milled into the substrate around a previously fabricated conventional dome-shaped lens. The Fresnel-like structure was designed to reflect and refract light emitted by an InGaAsP light-emitting diode [25]. Such a complex structure is designed to couple

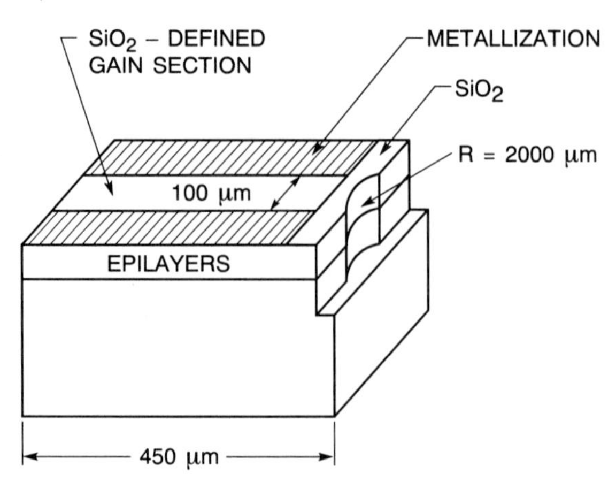

Fig. 10. Drawing of an unstable resonator structure with wide active area and curved mirror formed by FIB micromachining. (From Ref. 23.)

out internally reflected light. A larger conventional refracting lens would not work in this application because of the high index of refraction of InP. The micromachined structure consisted of concentric rings with varying radii and slope angles. The structure was milled directly onto the top surface of an LED.

Although most of the applications of FIB to optoelectronics deal with the active devices themselves, the micromachining techniques are also useful in device packaging. An interesting example was discussed by Yeh and Harriott [26]. One of the most difficult problems encountered in laser packaging is the thermal stability of the alignment between the laser and the optical fiber. Laser packages are made of several materials which are soldered together. This may include Si submount, BeCu heat sinks, and the semiconductor laser material itself, as well as the fiber attachment. Submicrometer lateral stability and angular displacement less than 10^{-3} radian are required for optimum alignment to the fiber. In many applications high-temperature operation is desired as well as immunity to repeated temperature cycling. The temperature extremes may be as great as −40°C to +85°C. It is important to know which part of this complex package arrangement is most susceptible to motion during these temperature excursions.

A FIB machine was used to write a reference grid structure across the soldered joints between the Si submount and BeCu heat sink, as shown in Fig. 13. The grids covered an area of 100 × 100 μm^2 with a pitch of 2 μm. The width of each line was 0.2 μm. The depth of the grid varied across the interface because of the different sputtering yields for the materials used. The soldered interfaces were examined with a scanning electron microscope

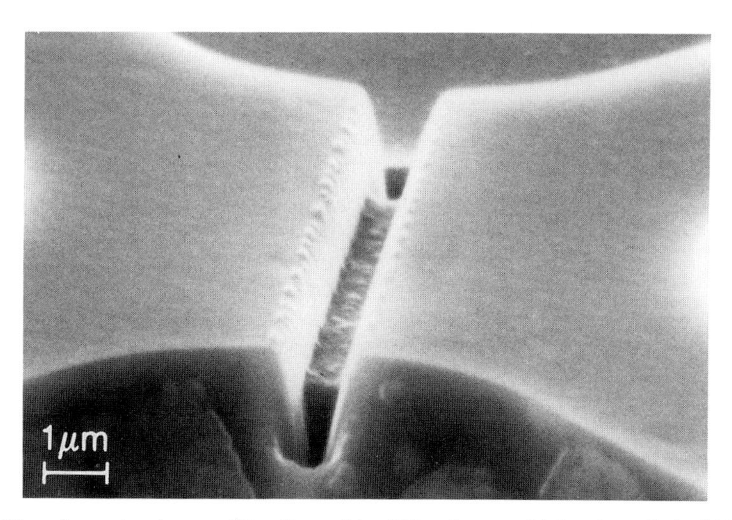

Fig. 11. A passive beamsplitter formed by FIB micromachining in a GaAs/AlGaAs ridge waveguide structure. (From Ref. 24.)

before and during thermal cycling. Relative motion of the parts was measured by observing the distortion in the grid. Figure 14 shows SEM images of a grid structure before (a), during (b), and after (c) a thermal cycle. In this experiment the temperature was raised to 150°C. A large displacement was observed at the peak temperature. Apparently the shear stress exceeded the strength of the solder at this temperature. When the temperature was

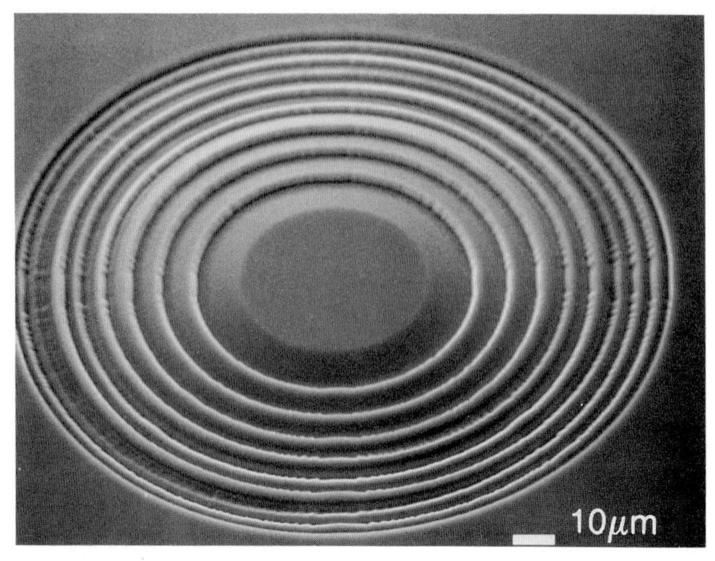

Fig. 12. Scanning electron micrograph of a lenslike structure formed in the output face of a light-emitting diode by FIB micromachining. The concentric rings each have a different inclination angle. (From Ref. 25.)

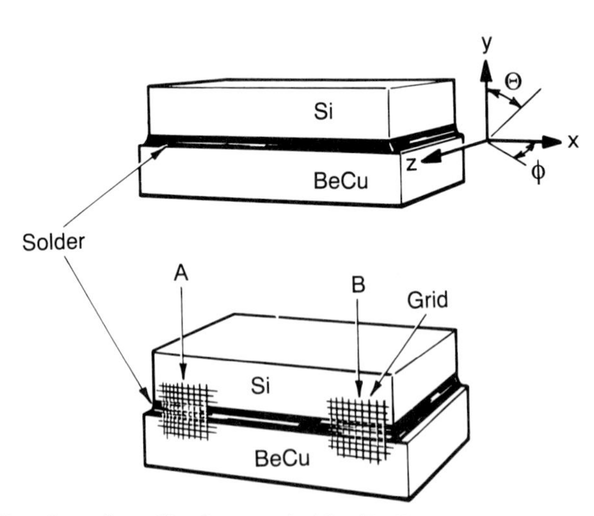

Fig. 13. Drawing of an Si-submount/solder/BeCu structure used in laser packaging indicating areas where reference grids were written with a focused ion beam. (From Ref. 26.)

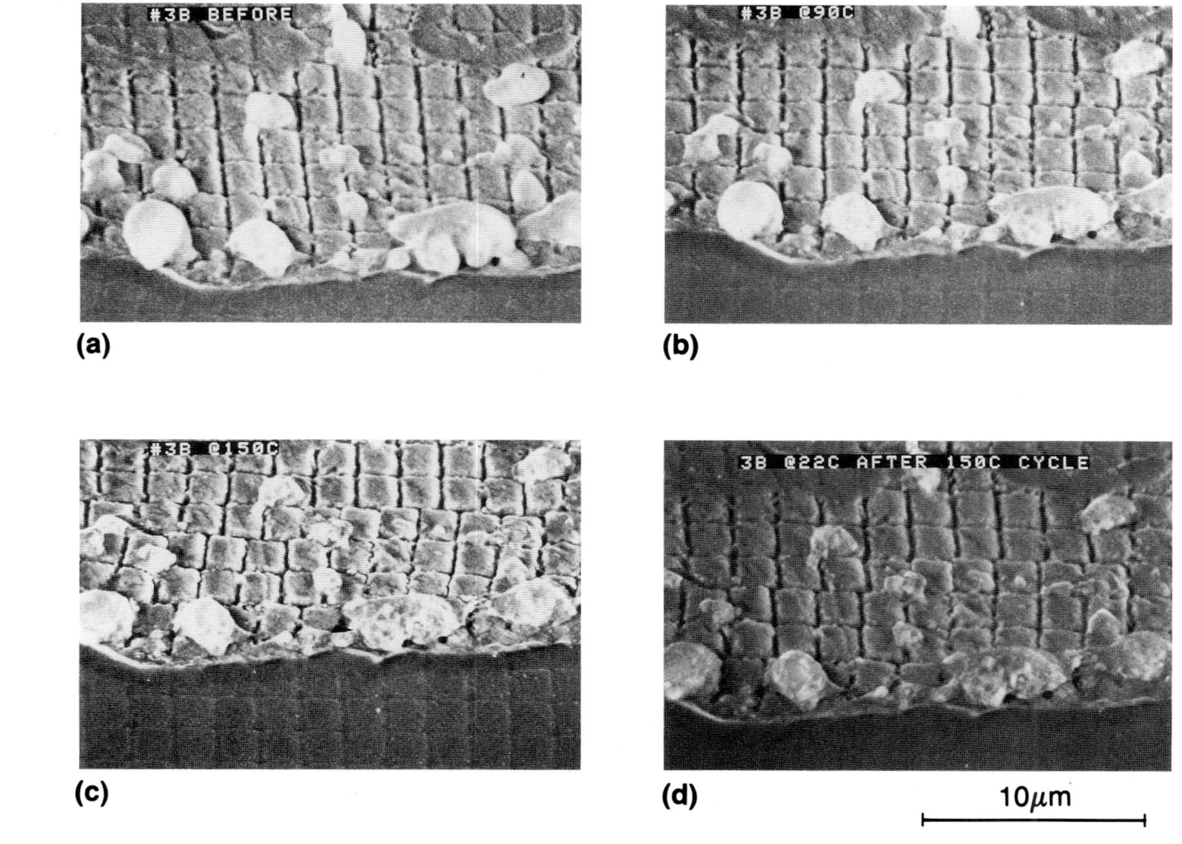

Fig. 14. Scanning electron micrographs of the Si/solder/BeCu interface with reference grids (a) before temperature cycling, (b) at 150°C, (c) at 90°C, and (d) after temperature cycling showing a permanent displacement of the solder. (From Ref. 26.)

reduced to room temperature again, the grids did not recover completely to their original shape; i.e., a permanent displacement of the solder occurred. This magnitude of displacement (about 0.5 μm) is difficult to measure by any other technique but is, nonetheless, significant in laser packaging.

The structures described in this section were used to evaluate designs of prototype structures, most of which can then be fabricated by conventional batch-processing methods. Some of the structures described here cannot, at present, be fabricated by other techniques. This illustrates the usefulness of the micromachining approach as a flexible prototyping tool.

3.1. Chemically Enhanced Etching

The throughput limitation for straight physical sputtering can be improved somewhat by introducing a chemically active etchant gas such as Cl_2 into the ion beam chamber in the vicinity of the target [27] as shown in Fig. 15. The gas-jet inlet nozzle in close proximity to the substrate is necessary so that the molecular flux of the etchant can be supplied to the target at a rate per unit area similar to that of the ions in the beam. A typical focused ion beam has a current density of about 1 A/cm^2, so gas pressures approaching 1 torr are necessary to achieve good etch rate enhancements over physical sputtering. Enhancements over physical sputtering of as much as a factor of 100 have been obtained with elevated substrate temperatures during etching. The chemical etching component also eliminates the problem of Ga droplets

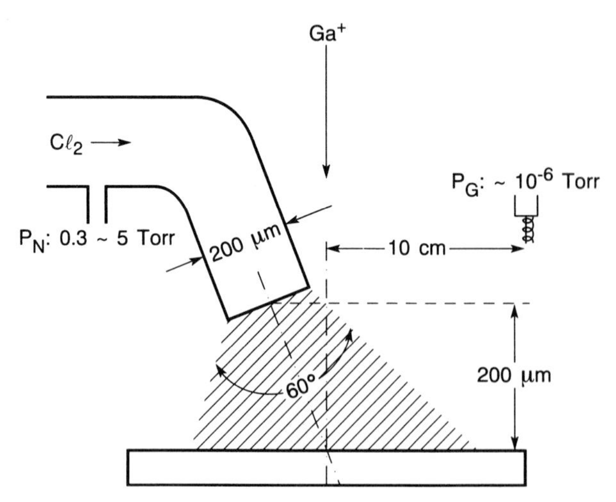

Fig. 15. A gas jet injection scheme used for chemically enhanced ion and electron beam etching. The gas jet allows high gas pressures in the vicinity of the beam while minimizing the effect on the rest of the vacuum system. (From Ref. 27.)

formed on the surface of GaAs seen when only physical sputtering is used. Although this method offers a large speed improvement over just physical sputtering, the ion doses required for typical features are still in the 10^{16} ions/cm^2 range, probably too high for large-area patterning.

Perhaps a more significant benefit of gas-assisted etching is in reducing redeposition of sputtered material. For defect repair in high-aspect-ratio structures such as x-ray lithography masks or in micromachining of three-dimensional optical structures in III–V compounds, reduction or elimination of redeposited material is crucial to producing the desired feature profile and generally outweighs throughput improvement considerations. An example of reduced redeposition is shown in Fig. 16a and b [28]. In this case the beam was programmed to produce a sawtooth structure in an InP substrate. The structure shown in Fig. 16a was made by straightforward micromachining. Redeposition of the sputtered material is evident, particularly at the bottom of the teeth, which appear rounded. The structure shown in Fig. 16b was made in the same way, except that XeF_2 was introduced into the chamber during the machining. While the overall etch rate enhancement in this case was very slight, less then 10%, the reduction in redeposition is evident. This is attributed to a lower sticking coefficient of the fluorides formed at the surface and sputtered by the Ga beam.

Surface damage is a major concern in nanostructure device fabrication, as discussed subsequently, and is severe for patterning by sputtering or chemically enhanced sputtering [29]. Because fairly high-energy (10–50 keV) heavy (usually Ga) ions are used in this process, the region below the etched surface will contain a high defect density. The effects of this damage can extend nearly 1 μm below the etched surface. Such damage is not recoverable by annealing in compound semiconductors and precludes epitaxial overgrowth.

Lower-energy (<50 eV) finely focused ion beam systems have been developed to address this problem [30]. Landing energies of only a few electron volts can be achieved by decelerating the beam just before striking the target. Patterning and overgrowth have recently been demonstrated in the GaAs/AlGaAs system using low-energy focused ion beam stimulated Cl_2 etching [31]. The *in situ* processing apparatus for this experiment is shown in Fig. 17. It was found that if the Ga ion beam energy was reduced to 1 keV, high-quality overgrowth of an AlGaAs/GaAs heterostructure could be achieved as illustrated by morphology and photoluminescence intensity of the overgrown areas. The overgrowth of structures with good luminescence intensity by this technique was demonstrated only for large areas (tens of micrometers) and it is not clear that ion beams with sufficiently low energy

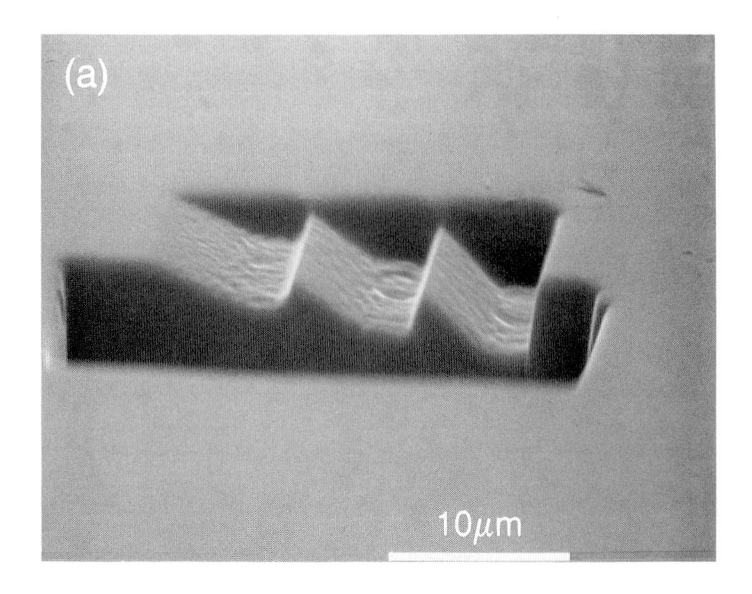

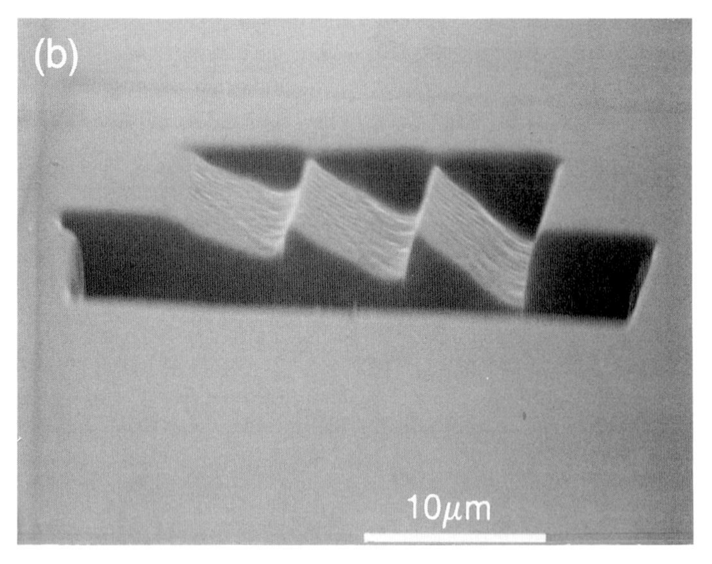

Fig. 16. Scanning electron micrographs of a sawtooth structure formed in an InP substrate (a) by direct micromachining and (b) by micromachining in the presence of 1 torr of XEF_2. The redeposition of material is significantly reduced by the gas-assisted etching. (L. R. Harriott, unpublished.)

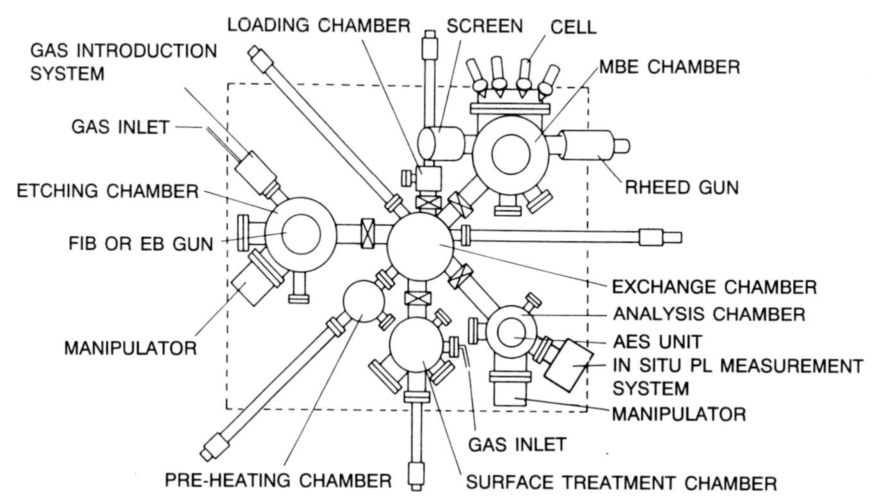

Fig. 17. The *in situ* processing apparatus used for chemically ion and electron beam etching and MBE overgrowth. (From Ref. 31.)

to avoid significant damage can be focused to very fine deep submicrometer dimensions required for quantum-effect devices.

Work has also been done on electron beam–stimulated etching of GaAs [32] with Cl_2 in a manner similar to that depicted in Fig. 17. This process, however, is not very efficient and requires very large electron doses in the range of 10^{17} to 10^{18} e^-/cm^2 (10^{-2} to 10^{-1} C/cm^2), 10,000 times the dose for e-beam resist exposure. The mechanism for this e-beam etching is not entirely clear but may be related to the one described in the following section on oxide masking.

3.2. Material Deposition

A gas inlet system such as the one shown in Fig. 15 can also be used to form deposits by using an appropriate precursor gas such as some of those used for chemical vapor deposition (CVD) of metals [33]. Metals such as Al, W, and Au have been deposited with such a scheme but generally are not very pure, with 20–30 atomic percent carbon included from the precursor. In addition, the deposition yield is typically only a few (<10) deposited metal atoms per incident ion. This means that the doses required for forming metal contacts or interconnects would be quite high (10^{17} to 10^{18} ions/cm^2), making coverage of large areas time consuming (for a dose of 10^{17} ions/cm^2 a typical submicrometer focused ion beam would require 10^8 seconds to cover a square centimeter).

4. MASKLESS IMPLANTATION

Maskless implantation for the preparation of diffraction gratings was described by Wu *et al.* [34]. This process offers high spatial resolution, and it is possible to vary the dose, species, and implantation energy simply by software control. The GaAs/AlGaAs distributed Bragg reflector (DBR) laser structure used is shown in Fig. 18. The second-order grating with a period of 2300 Å was produced by maskless implantation with Si^{2+} directly into the exposed passive waveguide structure. An implantation energy of 100 keV was used. The dose was varied from 4 to 8×10^{14} ions/cm^2 by changing the number of scans over the stripe. The dose needed for the formation of the grating was low compared to the ordinary dopant implants or direct micromachining. This results in much better throughput. Si was chosen because it is a stable dopant with low diffusivity, a feature important in structures with fine periodicity when high-temperature anneals are needed. After implantation the damage was removed by rapid thermal annealing at 950°C for 10 seconds.

The implanted diffraction grating is formed by periodic modulation of the free-carrier density. The local variation in the carrier density shifts the absorption edge and the free-carrier plasma frequency. The induced change in the index of refraction (Δn) can be as large as 5%. This is smaller than the change obtained in conventional gratings which depend on compositional modulation, and the design of the laser must take this into consideration. Laser operation in the Bragg mode was observed in a range of currents and

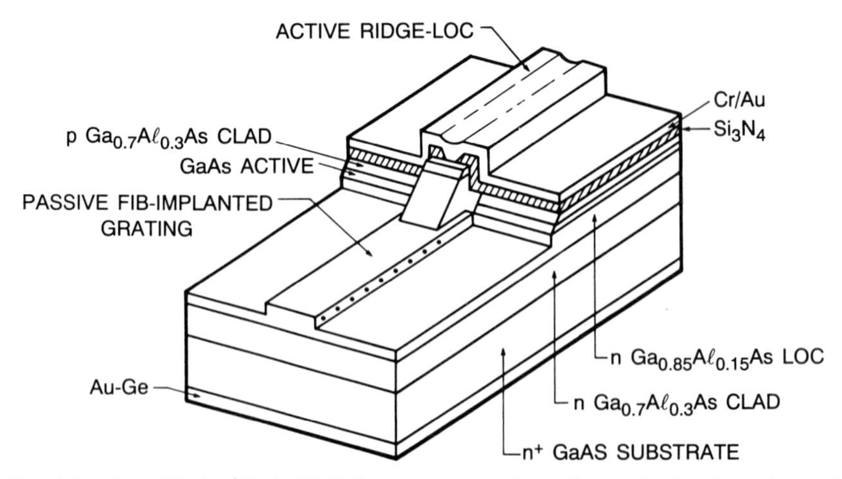

Fig. 18. An AlGaAs/GaAs DBR laser structure where the grating has been formed by maskless FIB implantation. (From Ref. 34.)

temperatures shown in Fig. 19. At low temperatures the Fabry–Perot modes of the active cavity were dominant. At higher temperatures, as the gain peak shifted to longer wavelength closer matched to the periodicity of the implanted grating, the DBR mode was observed. The laser operated in a single mode from 20 to 40°C. Further improvements could be obtained by the use of a first-order grating, requiring a smaller ion beam, and the optimization of the passive waveguide structure.

The process is compatible with molecular beam epitaxy and can, in principle, be done *in situ* in a common vacuum environment. This may allow regrowth on the grating structure.

One of the earliest ideas for combining focused ion beams with MBE

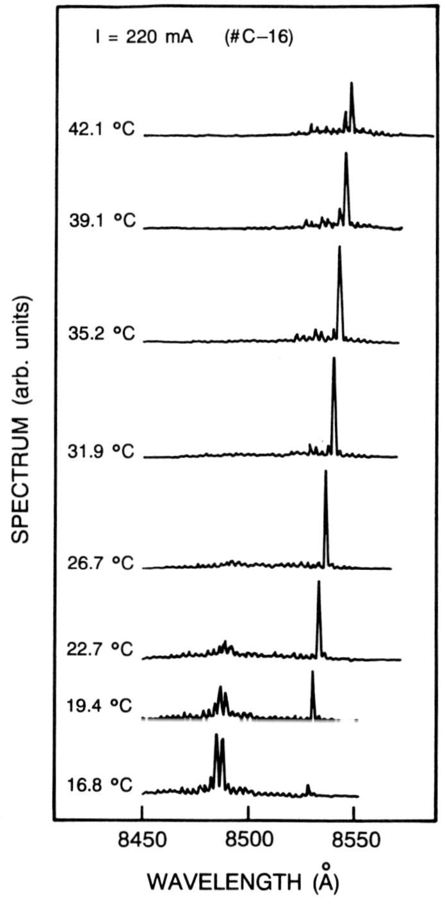

Fig. 19. Output spectra at various temperatures of the DBR laser with FIB implanted grating. (From Ref. 34.)

crystal growth came from the Optoelectronics Joint Research Laboratory (OJRL). A schematic diagram of their apparatus is shown in Fig. 20 [34, 35]. The system consisted of a 100-keV focused ion beam implanter and an MBE growth chamber connected by an ultrahigh-vacuum (UHV) sample transfer tube. The ion focusing column included an $E \times B$ momentum filter to separate the species produced in the alloy liquid metal ion source. For implantation into III–V compounds, a eutectic source of Au-Si-Be can be used to produced beams of both n and p dopants. The beam diameter was typically 100 nm at 100 kV (Si^{2+} and Be^{2+} ions were also produced from the source so that the beam energy was 200 keV).

The purpose of the type of system illustrated in Fig. 20 is to produce GaAs-based OEICs by implanting both electrical (FET) and optical (LED and laser) structures in multilayer structures with atomically abrupt interfaces. The importance of the UHV vacuum environment is demonstrated in Fig. 21. In this demonstration, epitaxial layers on two samples were implanted with Be ions and then overgrown. One sample was kept in the UHV environment while the other was exposed to air before the overgrowth. Figure 21a shows the photoluminescence (PL) intensity as a function of depth for the air-exposed sample. A large (10×) decrease in the PL intensity is seen at the interface for both the implanted and nonimplanted areas, while for the sample kept in UHV (Fig. 21b) there is no such dip at the interface.

A planar AlGaAs/GaAs double-heterostructure laser was fabricated using

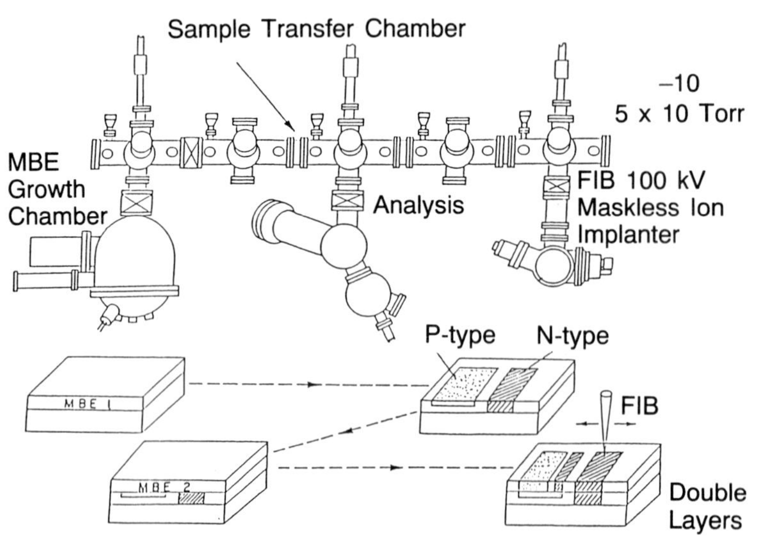

Fig. 20. The *in situ* processing apparatus used for the combination of maskless FIB implantation and MBE overgrowth. (From Ref. 35.)

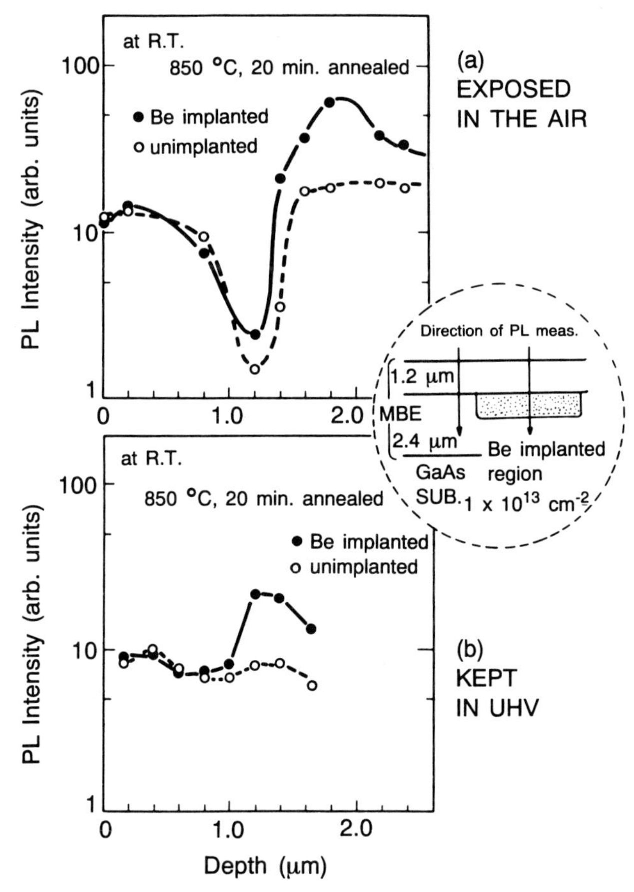

Fig. 21. Photoluminescence spectra of overgrown samples measured as a function of depth using sequential wet chemical etching. (a) The sample exposed to atmosphere prior to overgrowth shows a dip in PL at the interface; (b) the sample which remained in UHV shows no loss in PL at the interface. (From Ref. 35.)

the focused ion beam–MBE *in situ* processing system shown in Fig. 20 [36]. The buried conductive region for current confinement was formed in the undoped AlGaAs cladding layer by maskless Si ion implantation followed by overgrowth.

The *in situ* processing apparatus shown in Fig. 20 has been used to fabricate 2DEG patterns by maskless Si implantation and overgrowth [37]. The structure is shown in cross section in Fig. 22. A GaAs channel layer was grown on an AlGaAs layer that had been implanted selectively with an Si focused ion beam. A multilayer structure was overgrown in order to block Si outdiffusion during overgrowth and annealing. FIB implants were per-

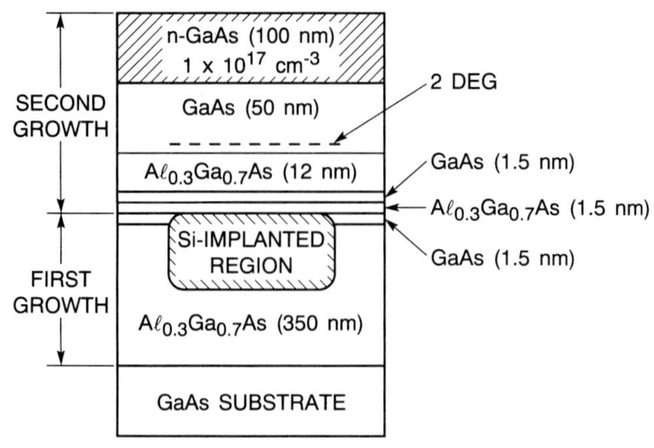

Fig. 22. Cross-sectional diagram of the multilayer sample used to fabricate 2DEG by maskless Si implantation and overgrowth. (From Ref. 37.)

formed in 400-μm-square areas at 40 and 80 keV with doses varying from 2×10^{12} to 1×10^{13} ions/cm^2. The samples were annealed at 1000°C for 6 seconds after overgrowth for electrical activation. The 2DEG characteristics were observed using Van der Pauw Hall measurements at 77 K and Shubnikov–de Haas (SdH) measurements at 4.2 K.

The best results were obtained for 40-keV implants with a dose of 2×10^{12} ions/cm^2. In this case, the mobility at 77 K was 32,000 cm^2/V with a sheet carrier concentration of 1×10^{11} cm^2. In general, an increase in electron mobility with decreasing Si implantation dose was observed. The idea is that lower doses produce less damage and result in higher-quality overgrowth and that not all of the damage can be annealed out. Lower-energy focused ion beams would result in a shorter implant ion range and larger dopant concentration in the surface layer while at the same time reducing damage effects (the limiting case is similar to delta doping). It is estimated that focused ion beam Si implants at a few keV could produce 2DEG with carrier concentrations of 3×10^{11}/cm^2 for an implant dose of 8×10^{11}/cm^2 and result in mobilities in excess of 50,000 cm^2/V at 77 K.

Although this work was done on fairly large areas, it is expected that maskless implantation and overgrowth could produce patterned 2DEG structures with much higher spatial resolution. The limit on the spatial resolution is determined by the focused ion beam size and lateral scattering of the ions as they are implanted. The results of calculations of the effective size of implant areas as a function of beam diameter and energy are shown in Fig. 23. The calculation shows that for a given beam size the effective implant area is reduced by reducing the beam energy and hence the lateral scattering

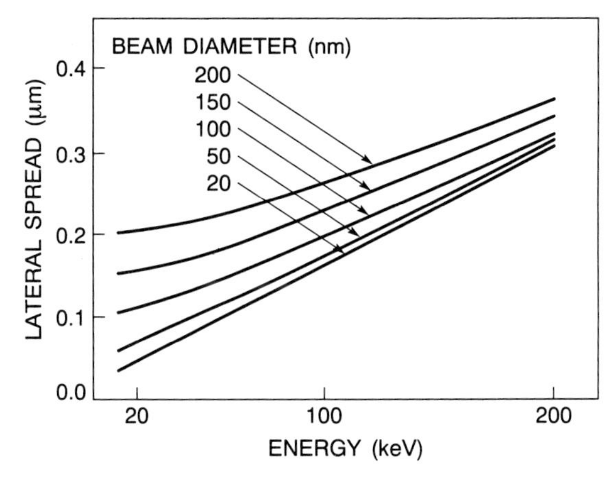

Fig. 23. Calculations of the effective size of focused ion beam implanted areas as a function of energy and beam diameter consideringa lateral scattering of the ions. (From Ref. 37.)

of the ions and projected range. For maskless doping of structures such as 2DEG it should be possible to fabricate quantum confinement size (less than 50 nm) structures by using beam energies less than 10 keV if the ion focusing column can be designed to produce a 20–50-nm spot.

Low-energy focused ion beam implants seem to be particularly attractive in forming 2DEG structures for several reasons. The lower energy results in less radiation damage, higher-quality overgrowth, and thus higher mobility. The shorter range of the ions allows a given sheet carrier concentration to be obtained at lower ion doses and therefore improved throughput. Finally, less lateral scattering of the lower-energy ions may allow scaling of maskless implantation to quantum-size dimensions to allow fabrication of 1DEG and zero DEG structures.

In addition to implantation doping, device isolation is an important application for focused ion beams. High-resistivity regions can be obtained with implantation either through defect-induced compensation or through carrier trapping by deep-level impurities. Point defects created by ion implantation are subject to annealing and therefore defect-induced compensation is not robust in the subsequent high-temperature processing. Carrier trapping caused by deep-level impurities, on the other hand, is stable in high-temperature processing.

Bulk InP can be doped with Fe acting as a deep acceptor level with an energy located about 0.65 eV below the conduction band edge to produce semi-insulating substrate material. Focused ion beam implantation of Fe into InP for the formation of semi-insulating regions has been demonstrated [38].

Doubly charged Fe ions were produced using an FeGe liquid metal ion source. A beam of 200-keV Fe was used to implant into a 1000-Å layer of *n*-type InP grown on a semi-insulating InP substrate. Four point resistance measurements were made on implant areas of $5 \times 30\ \mu m^2$ before implantation, as implanted, and after a 400°C rapid thermal anneal for 10 seconds. The resistance of the implanted areas changed from roughly 2.5 kΩ to approximately 10 MΩ as implanted and further increased to 20 mΩ after annealing. This result demonstrates the doping effect of the implanted Fe in producing high-resistivity regions and its stability to high (for InP) temperature processing.

4.1. Implanted Quantum Wire Structures

Figure 24 shows a very simple method for producing one-dimensional structures by focused ion beam implantation [39]. A *p*-type conductive area was made by Be ion implantation into a semi-insulating (SI) GaAs substrate. Then the focused ion beam was used to implant Si at 100 keV and 100-nm beam diameter, followed by a rapid thermal anneal forming a line *p–n* junction. The depletion region was increased by reverse biasing the junction and therefore decreasing the electrical size of the wire. Effective sizes of 100 nm can be obtained despite the lateral spreading of the ion-implanted junction due to scattering and diffusion.

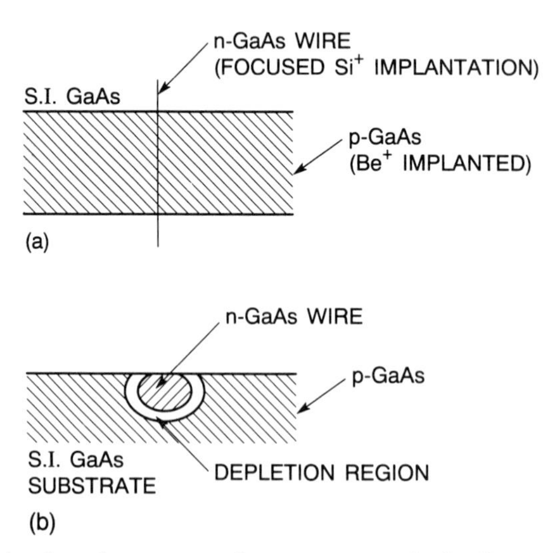

Fig. 24. An implanted quantum wire structure made by focused ion beam implantation of Si into a *p*-type area. (From Ref. 39.)

4.2. Damage Patterning

In the previous example, the dopant implanted by the ion beam was activated by annealing to form a junction device. The damage caused by the ion beam can also be used for patterning by introducing defects which destroy the conductivity of a previously doped semiconductor layer. A damage-dependent method for producing one-dimensional conductors is shown in Fig. 25 [39]. In this case a conductive current bar is first created with ion implantation and annealing as before. Nonconducting regions are then created by a second implant where the damage is not annealed out. The conducting channel between the damaged areas is further narrowed by the depletion region and can be made as small as 20 nm (even though the ion beam was 100 nm in diameter) by adjusting the distance between the damage areas and the dose. A variation of this damage patterning method has been used to produce a lateral gate transistor device [40].

4.3. Quantum Well Disordering

Impurity-induced compositional disordering of multi-quantum-well (MQW) structures of GaAs/AlGaAs has been studied with impurities such as Zn [41] as well as implantation of the constituent atom Ga [42]. Fine-scale lateral patterning using a focused Ga ion beam has also been demonstrated [43]. When the MQW structure is implanted with high-energy (100-keV) Ga ions, the defects produced cause enhanced interdiffusion of the Al and Ga within the implanted area. This occurs at moderate ion doses in the range 10^{13} to 10^{15} ions/cm^2. After annealing, the implanted areas are structureless,

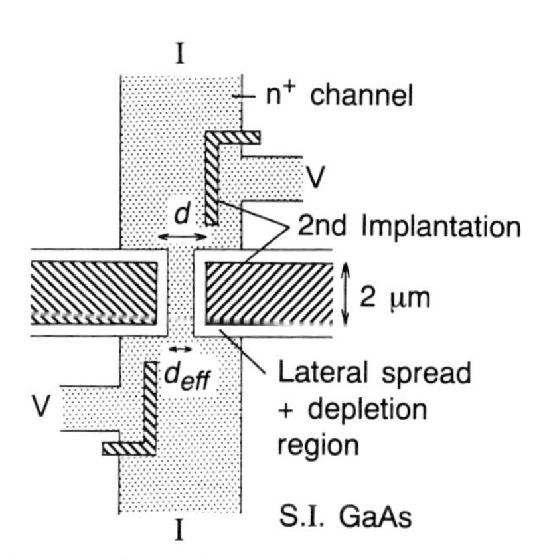

Fig. 25. Example of restriction of conductive regions by focused ion beam implantation damage. (From Ref. 39.)

leaving intact the unimplanted MQW structures. The limitations on the spatial resolution are due to ion scattering and diffusion. MQW structures with lateral dimensions of about 100 nm have been produced so far and were probably limited by the ion beam size. It is expected that features 50 nm or smaller can be produced in this way.

At the energy of 150 keV used in the focused Ga-beam experiments, the ion range calculated for amorphous material is about 58 nm. In most device applications this shallow range requires a cycle of implantation and overgrowth, as discussed for the formation of a 2DEG structure. The implant depth profile is mainly Gaussian, where the density peaks at the projected range. For crystalline substrates a longer range tail is also present, attributed to channeling. This is generally a nuisance effect and samples are implanted off-axis to avoid it. Laruelle *et al.* [44] investigated the depth and lateral extent of the quantum well disordering by focused Ga ion beam implantation and used the channeling effect to greatly extend the effective range.

The multi-quantum-well structure used in the channeling experiments is shown in Fig. 26. The sample consisted of five GaAs wells with thicknesses varying from 10 to 130 Å and separated from each other by 100 nm of GaAlAs. The thinnest well is buried 500 nm deep in the sample. The different well widths permit luminescence monitoring of each individual well. The implantation was performed at room temperature with a beam 80 nm in diameter and a low dose of 4×10^{13} ions/cm^2. The samples were then annealed at 900°C for 4 minutes. Spectrally resolved cathodoluminescence (CL) was used to detect any changes in the well width or composition. The spectra of nonimplanted material showed five lines (dashed lines of Fig. 26), each corresponding to the electron–heavy hole transition of the respective well, at the expected energy. Five lines were also observed in the implanted sample, but each blue shifted from its original position. The well–barrier interdiffusion length can be deduced from the magnitude of the shift. The surprising finding was that even the deepest well was affected.

The extent of lateral scattering was investigated by implanting a line pattern consisting of 11 lines with a 2-μm period. The width of the implanted lines was varied from 400 to 80 nm. After the annealing the sample was examined by CL line scans. The effect of Ga implantation was evidenced by the spatial modulation of the CL signal. The CL scan at the energy corresponding to the deepest well showed intensity modulation and 11 peaks were observed. It was estimated that at the largest depth of interest the lateral ion straggling is only about 35 nm or the same as the beam broadening at the projected range. The channeling process acts to extend the useful range and preserve the ion beam size well below the surface.

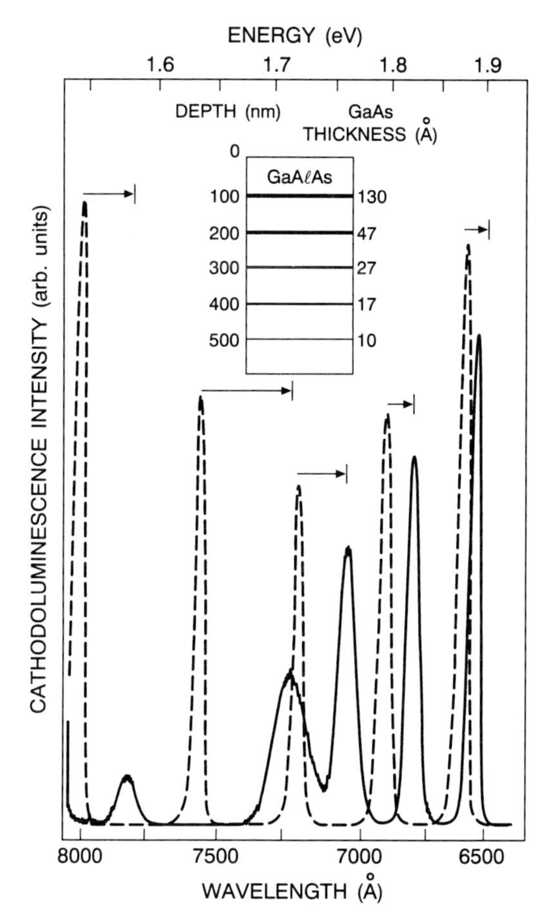

Fig. 26. A multi-quantum-well structure in GaAs/AlGaAs with five wells of different widths at different depths. The spectra show the luminescence (dashed line) of each of the wells at different wavelengths before and after (solid line) FIB implantation. (From Ref. 44.)

The channeling effect can be used to prepare 2DEG structures by implantation into an As-grown material [45]. This is a possible alternative to the implantation and regrowth process discussed earlier. A cross section of the epitaxial structure used is shown in Fig. 27a. The implantation was done with 100-keV Ga ions. In addition to the uniformly implanted sample, a multiple wire pattern with a 1-μm pitch, shown in Fig. 27b, was written. In this sample, two uniformly implanted areas were added at the ends of the wires to act as the electron reservoirs. The top GaAlAs layer was designed to be thicker than the projected ion range. This acts to keep the ion

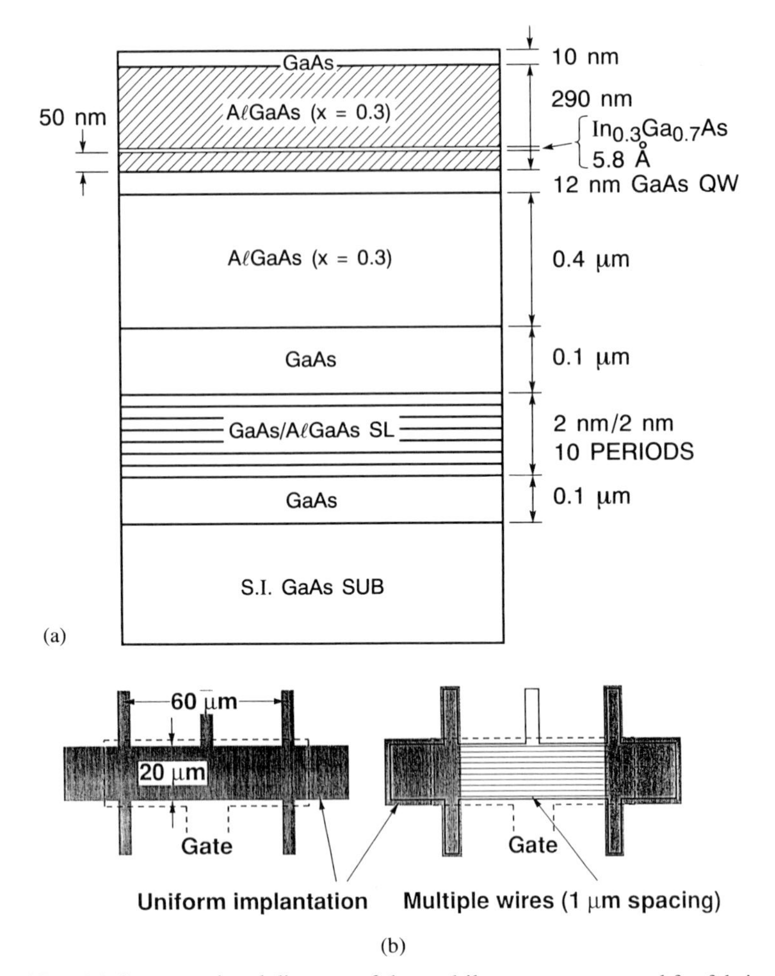

Fig. 27. (a) Cross-sectional diagram of the multilayer structure used for fabricating a 2DEG by FIB implantation into the as-grown structure. (b) Schematic diagrams of uniformly implanted and 1-μm-pitch 2DEG samples. (From Ref. 45.)

damage well away from the quantum well. Only the channeled ions can penetrate deep enough to modulation dope the active layer of GaAs. A very thin strained layer of $In_{0.3}Ga_{0.7}As$ was intended for dechanneling the implanted ions and controlling the depth profile.

Low-temperature magnetotransport measurements confirmed the presence of a good-quality 2DEG. The 2DEG mobility ranged from 0.5 to 10^4 $cm^2/V\ s$. The wire-implanted sample was examined by CL imaging to show formation of 2DEG wires deep below the surface. Careful examination of the SdH measurements, in which the oscillations start to occur at 1 T, showed that the wires had an effective width of the order of 0.2 μm. This value is larger than the diameter of the beam used for implantation by approximately 0.08 μm, consistent with the channeling model discussed earlier.

5. LITHOGRAPHY FOR *IN SITU* PROCESSING

The motivation for all vacuum processing of compound semiconductors stems from results of crystal growth methods such as molecular beam epitaxy (MBE). With MBE, layered structures of III–V compounds can be grown epitaxially with essentially atomically abrupt interfaces. The material composition can be changed from layer to layer to vary the bandgap of the material, resulting in spatial variations in electrical and optical properties of the structure. Heterostructures of InP/GaInAs/InP or AlGaAs/GaAs/AlGaAs can be fabricated with very thin layers to produce quantum mechanical confinement in two dimensions, leading to new device structures and improved performance of existing devices such as semiconductor lasers. Devices made from these quantum well materials are generally fabricated by growing a planar structure with MBE, or other crystal growth technique, followed by lateral patterning using conventional lithography and etching. The patterning is sometimes followed by overgrowth for surface passivation and/or optical and current confinement, as in a buried heterostructure laser. Conventional patterning processes introduce damage and contamination to the surfaces of the device. The result is an inactive layer at the regrown surfaces of the structure.

The role of the surface and any damage layers becomes more important as device dimensions approach that for lateral quantum confinement effects (~10 nm). Typical thicknesses of the damage layers for patterned III–V materials are in the range 10–50 nm. Therefore, structures patterned at dimensions on the order of twice the damage layer thickness or less cannot be functional. This problem of damaged layers and their inherent uncontrollability has made realization of lateral quantum confinement structures very difficult, especially at noncryogenic temperatures. In addition to the problems caused by patterning damage, the depletion layers formed at the air–semiconductor interface act to shrink the volume of active device material in small-scale structures. The condition of the surface is also obviously important for epitaxial regrowth of sharp vertical interfaces comparable to the conventional planar interfaces on unpatterned surfaces. Overgrowth is an important surface passivation method, particularly in small-scale structures, as well as being needed in assuring optical and current confinement in larger-scale devices such as lasers.

The main goal for *in situ* processing of compound semiconductor heterostructures is then to provide damage-free patterning which can be performed on a very fine lateral scale and which is compatible with high-quality epitaxial overgrowth [46].

Finely focused ion beams offer many possibilities for high-resolution patterning which is compatible with the high-vacuum environment of MBE crystal growth. When an energetic ion strikes a target, it loses its kinetic energy by both nuclear and electronic collisions and thus transfers momentum to the target atoms, causing displacements (damage) as well as electronic excitement. In addition, when the ion comes to rest in the substrate, its presence changes the chemistry of the target, as in semiconductor doping.

5.1. Photochemical Etching

Ion beam damage followed by photochemical etching is a two-step patterning method which exploits the semiconductor nature of the material to be patterned [47]. Finely focused ion beams were first used in this method with photoelectrochemical wet etching [48]. During the etching process, the photons incident on the substrate generate electron–hole pairs. The electrical bias drives the holes toward the surface, where they drive the etching chemistry. In areas irradiated by the ion beam, the defects created by the beam act as recombination sites for the holes so that the etching is inhibited in those areas. The ion doses required to inhibit the etching completely are on the order of $10^{11}/cm^2$ for 20-keV Ga on InP.

The wet chemical process is obviously not appropriate for vacuum processing, but it has been extended for use with photochemical dry etching of GaAs [49]. The mechanism responsible is also thought to involve recombination of photogenerated holes to inhibit the etching. One difficulty with this process for use in vacuum processing involving overgrowth on the patterned substrates is the fact that the ion-damaged areas remain after the etching. An additional annealing step is required before overgrowth, but it is not clear that the damage could be repaired to the extent that high-quality epitaxy would be possible.

5.2. Oxide Masking

The idea for a two-step ion beam patterning process using a thin vacuum-compatible imaging layer was first presented using an MBE-grown ternary

(InGaAs) layer (30 Å) on a binary (InP) substrate [50]. The ion beam damaged the ternary layer and partially removed it by physical sputtering at doses in the range of 10^{13} to 10^{15} ions/cm^2. Then material selective etching was used to transfer the pattern by etching the exposed binary substrate, leaving intact the areas of the ternary masking layer which had not been exposed to the ion beam. Wet chemical etchants are available with an extremely high degree of selectivity between the binary and ternary III–V materials, but so far the analogous dry etching schemes have not been found.

This idea has been extended, however, for use with dry etching, where ultrathin (20–30 Å) oxide layers are used for the imaging layer [51–53]. The apparatus for this scheme is shown in Fig. 28. The system is used for InP-based compounds with Cl_2 etching at 200°C substrate temperature with or without a low-energy Ar^+ flood beam during etching. It was found that if the Ar^+ energy used for etching was kept below 50 eV the damage introduced by etching was minor and high-quality MBE overgrowth was possible.

The model [54] proposed for the patterning mechanism is the following: The ion beam at doses of 10^{13} to 10^{15} ions/cm^2 modifies the sample surface by physical processes, resulting in some sputter removal of the oxide layer

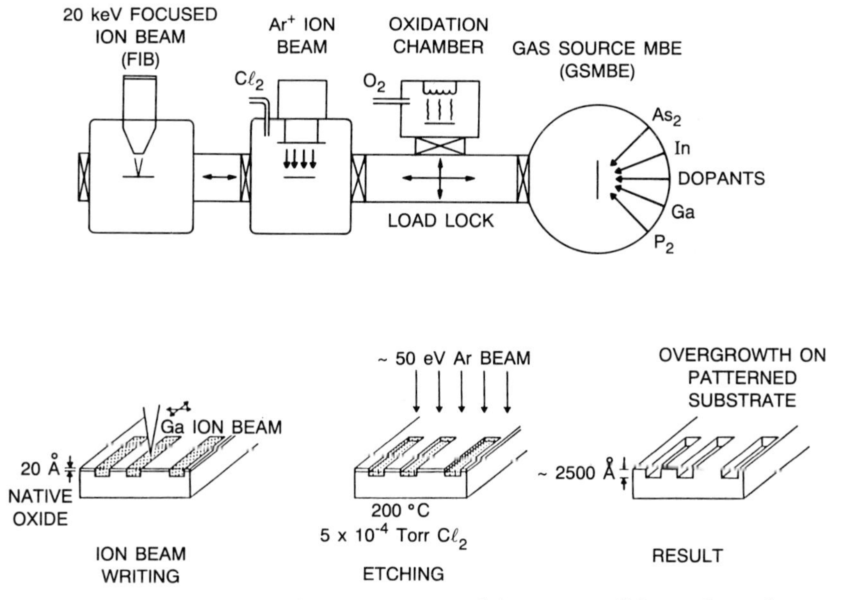

Fig. 28. The *in situ* processing apparatus used for vacuum lithography and overgrowth. The process used FIB patterning of a thin surface oxide imaging layer followed by dry etching and MBE overgrowth. (From Ref. 51.)

and damage to the underlying substrate to a depth that scales (1.5 to 2×) with the ion range because the ion range is usually 10 times longer than the oxide thickness. The Cl_2 etching is inhibited in the areas where the oxide remains intact and proceeds only in the ion-irradiated areas. The ion-damaged material is removed by the etchant, exposing clean undamaged substrate. The etching rate for the damaged material is probably slightly higher than that for the undamaged material. The surface step features develop in depth with time as long as the oxide layer remains. If the Ar^+ beam is used during the etching, the anisotropy is improved, but it also acts to sputter erode the oxide mask and limit the ultimate feature depth. If low-energy Ar^+ or no Ar^+ at all is used in the etching, the patterned surface is clean and undamaged, suitable for epitaxial overgrowth.

This process has demonstrated that features can be produced at the spatial resolution of the focused ion beam (200 nm in this case) and be overgrown with material which shows excellent morphology and high luminescence efficiency. For smaller ion beam diameters, the spatial resolution of this process could be improved to 50 nm or less since a thin imaging layer is used. Furthermore, by performing the oxidation *in situ,* multiple applications of this vacuum lithography process are possible. The result is a vacuum process analogous to conventional resist lithography without the associated damage or contamination. Here, the substrate surface is oxidized (spin-on resist), patterned by the focused ion beam (optical or e-beam resist exposure), the pattern is transferred by dry etching, the oxide layer is thermally desorbed in the growth chamber (resist stripping), and a new layer(s) is overgrown. This process can be repeated as many times as necessary without removing the sample from the vacuum.

It is also possible to pattern the oxide layer with something other than a focused ion beam. Electron-stimulated desorption of the oxide locally would accomplish the same end without the underlying substrate damage [31, 32]. This is likely to be a slow process, however, due to the large electron doses required. Photon beam desorption is also possible but probably would not result in high-resolution patterns on the nanometer scale.

A serious problem of a practical nature, for multichamber systems such as those shown in Figs. 17, 20, and 28, is the basic incompatibility of high-resolution patterning and other large vacuum process equipment such as MBE. Frequently, the limit of the spatial resolution of these systems is dictated by mechanical vibrations and electrical noise coupled to the beam writer from the other processing equipment rather than beam diameters or some fundamental limit of the processing. For the smallest features produced with vacuum lithography and overgrowth [54], separate apparatus was used for the patterning, etching, and growth. The samples were transferred between apparatus in vacuum with a specially designed actively pumped chamber. This type of coupling eliminated blurring of the patterns by noise and vibration sources in the other processes, allowing patterns at the resolution limit (50 nm in this case) of the focused ion beam to be formed. Figure 29 shows a transmission electron micrograph (TEM) of an InP/InGaAs multi-quantum-well sample that was patterned and overgrown using the oxide masking method. The original sample consisted of three 50-Å-wide InGaAs quantum wells with 50-Å InP spacer layers and a 200-Å InP cap layer. The samples were oxidized at 350°C in an O_2 atmosphere forming a layer roughly 30 Å thick. The oxide layer was patterned with a 100-keV Ga focused ion beam at a dose of 5×10^{14} ions/cm^2. The pattern was transferred into the substrate by Cl_2 etching at 200°C. An InP layer was then overgrown on the patterned sample using GSMBE. All of the sample transfers were accomplished via a portable vacuum transfer chamber.

A further advantage of separate processing chambers coupled via a vacuum transfer vessel is flexibility. New processes can be incorporated or

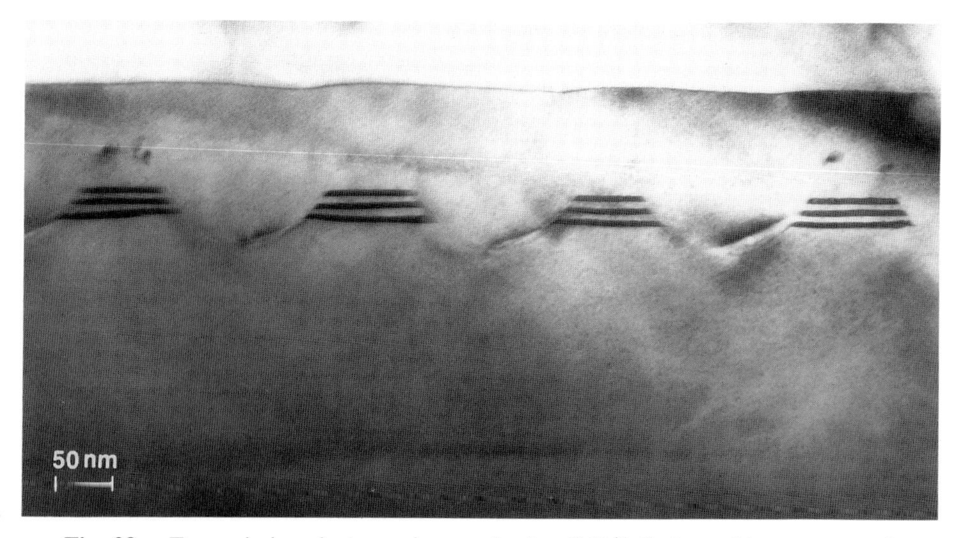

Fig. 29. Transmission electron micrograph of an InP/InGaAs multi-quantum-well structure made by vacuum lithography and overgrowth. The patterning and growth chambers were separate to avoid vibration problems. The sample was transferred with a portable vacuum chamber. (From Ref. 54.)

changed with a minimum effort, a nontrivial consideration in processing research.

5.3. Selective-Area Epitaxy

An alternative approach for creating buried heterostructures is patterned growth rather than growth on patterned substrates. In selective-area epitaxy, crystal growth is restricted to the desired areas and no material is deposited on the remainder of wafer. This restriction is obtained by covering the substrate with a mask that does not catalyze the decomposition of the metalorganic compounds used in MOMBE. Selective growth is not possible with elemental MBE or gas source MBE methods other than MOMBE, since these techniques do not rely on the catalyzed surface decomposition of metalorganics and invariably produce amorphous coverage on the mask. The selective growth technique has the potential for preparing complex, self-aligned structures that cannot be achieved otherwise and greatly simplifies post-growth processing of semiconductor devices. Although selective-area growth is used for the fabrication of active silicon devices such as bipolar transistors, its application to III–V compounds is still in its infancy. Selective growth by LPE and MOCVD has been used for the growth of blocking layers around the active mesa structure of heterostructure lasers, but aside from the work described next, no selective growth by any method has succeeded for active regions of III–V devices.

Oxides of InP and GaAs have been used as etch masks for vacuum lithography, as described in the previous section, and were considered for selective-area growth. Native oxides of GaAs [55, 56] have been used successfully as a mask in selective-area MOMBE growth. In this case, the oxide layer was formed in an O_2 atmosphere by light irradiation from a halogen lamp. The layer was estimated to be about 30 Å thick. Patterning was performed with a 10-keV electron beam in a Cl_2 atmosphere using the apparatus shown in Fig. 17. The results of these experiments showed that the trimethylgallium (TMG) was reflected from the GaAs surface and did not decompose for temperatures below 550°C, while TMG was seen to decompose at temperatures above 350°C on the exposed GaAs surface. Although only large (tens of micrometers) features were demonstrated, a high degree of selectivity was observed and it is expected that this process can be scaled to finer dimensions limited by the electron beam used for patterning.

For InP compounds, native oxide masks, including high-temperature oxides, yield poor growth selectivity, although selective growth of large-scale features using SiO_2 masks and conventional lithography has been demon-

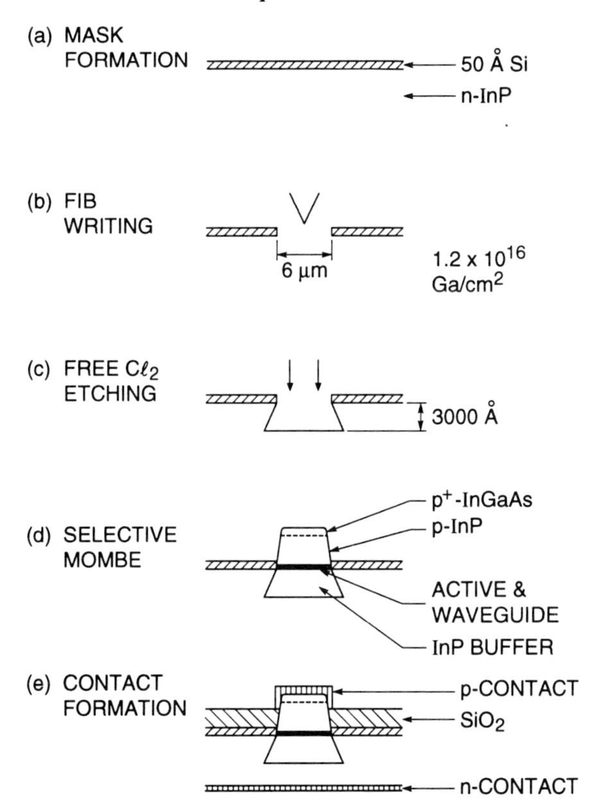

Fig. 30. The five-step process for the fabrication of GaInAsP/InP lasers by selective-area epitaxy. (From Ref. 58.)

strated [57]. Si has been used as the mask material for *in situ* selective growth of InP compounds because it does not react with the InP substrate, can be deposited in vacuum, is resistant to Cl_2 etching, and provides excellent selectivity in MOMBE growth [58]. The layer thickness (3–5 nm) was chosen as a trade-off between the thickness required to ensure complete surface coverage and minimization of the (Ga) ion dose required for patterning.

Openings in the Si mask were fabricated using a 20-keV Ga focused ion beam described in the previous section on oxide masking. The ion beam locally sputter-removed the Si and underlying native oxide layers which grew spontaneously on InP exposed to air. A dose of 1.2×10^{16} ions/cm^2 was used for effective patterning of the Si layer. The sample was then transferred, in vacuum, to the etching chamber, heated to 200°C and exposed to 5×10^{-4} torr Cl_2 gas. The etching was required to remove approximately

2000 Å of substrate material damaged in the ion writing process. The thin Si mask layer was not attacked in the etching.

The device potential of selective-area epitaxy with MOMBE was demonstrated by fabricating lasers. The five-step process for the fabrication of GaInAsP/InP heterostructure lasers is illustrated schematically in Fig. 30 (it should be noted that the layer thickness and lateral dimensions are not drawn to the same scale and the actual structure is much flatter). Stripe openings were 6 μm wide and 1.5 mm long. The stripes were aligned along the [011] direction to permit facet cleaving. The relatively large stripe width, compared to the width of 1.5 μm used for buried heterostructure and −4 μm for ridge waveguide lasers, was chosen mainly to minimize the relative importance of the edge growth and to avoid changes in composition encountered in narrower stripes [59]. The laser structure was grown at 530°C. The active layer of GaInAsP with the composition corresponding to a 1.3-μm bandgap was 500 Å thick. Details are shown in Fig. 31a. The cleaved and strained cross section of the entire laser structure, 1.4 μm thick, is shown in Fig. 31b. It is interesting to note that the edges of the active layer stripe are completely buried because the GaInAsP does not grow on the {111} planes of InP formed as a result of lower growth rate in this direction. The laser formed in a single step by selective-area epitaxy is a self-buried heterostructure device. The light–current characteristics of these lasers were measured and showed room-temperature laser threshold currents between 40 and 45 mA, which is lower than would be expected for a conventional buried heterostructure laser of the same width. The improved performance is attributed to the self-buried active area, which has no semiconductor–air interface or process-related damage.

6. SUMMARY

Finely focused ion beams have proved to be useful for two purposes in fabrication of optoelectronics. The simplest is micromachining. The direct nature of focused ion beam machining by sputtering and its large depth of focus allow fabrication of three-dimensional structures directly on wafers or completed devices. The nonselective nature of the sputtering means that virtually any material can be patterned without developing any special process. Micromachining is a slow process because of its serial nature and beam current density limitations. It is probably most useful in prototyping new devices and structures. Once a new device structure has been demonstrated using focused ion beam machining, a batch processing method can be de-

(a)

1000 Å	GaInAs	p^+
8000 Å	InP	p(Be) ~ 10^{18}
500 Å	InP	n ~ 10^{16}
750 Å	GaInAsP	(1.1 μm)
500 Å	GaInAsP	(1.3 μm)
750 Å	GaInAsP	(1.1 μm)
2500 Å	InP	n ~ 10^{16}

n–InP SUBSTRATE

(b)

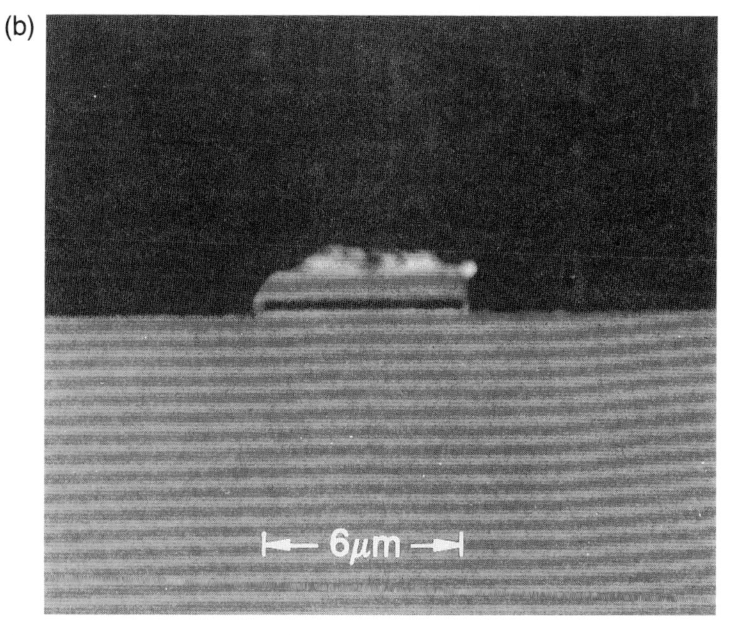

Fig. 31. (a) Detailed structure of the laser structure grown by selective-area epitaxy and (b) a strained cross-sectional image of the laser structure showing the self-buried active region. (From Ref. 58.)

veloped for larger-scale production. The flexibility of the focused ion beam allows many design ideas to be tried before commitment to production.

On the other end of the spectrum, focused ion beams have been used extensively in research into *in situ* processing of optoelectronic materials and devices. The large momentum transfer afforded by ion beams and the fine dimensions of the beam have made it an attractive candidate for patterning and overgrowth in the UHV environment. Ion beam damage is a concern with this type of processing, but developments in low-energy finely focused ion beams as well as damage-free etching address this problem. Using this approach, features at the beam diameter resolution limit of the ion beam have been fabricated and overgrown. The importance of these techniques will be more evident with decreasing feature size.

REFERENCES

1. A. Wagner, *Nucl. Instrum. Methods* **218**, (1983).
2. W. L. Brown, *Microelectron. Eng.* **9**, 269 (1989).
3. J. Melngailis, *J. Vac. Sci. Technol. B* **5**, 469 (1987).
4. L. R. Harriott, *Appl. Surf. Sci.* **36**, 432 (1989).
5. L. R. Harriott, in "VLSI Electronics: Microstructure Science," Vol. 21. Academic Press, New York, 1989.
6. A. Wagner and J. P. Levin, *Nucl. Instrum. Methods* **B37/38**, 224 (1989).
7. L. R. Harriott, A. Wagner, and F. Fritz, *J. Vac. Sci. Technol.* **B4**, 181 (1986).
8. J. Melngailis, C. R. Musil, E. H. Stevens, M. Utlaut, E. M. Kellogg, R. T. Post, M. W. Geis, and R. W. Mountain, *J. Vac. Sci. Technol.* **B4**, 176 (1986).
9. D. C. Shaver and B. W. Ward, *J. Vac. Sci. Technol.* **B4**, 185 (1986).
10. L. R. Harriott, R. E. Scotti, K. D. Cummings, and A. F. Ambrose, *J. Vac. Sci. Technol.* **B5**, 207 (1987).
11. J. Orloff and L. W. Swanson, *J. Appl. Phys.* **50**, 2494 (1979).
12. E. Miyauchi, H. Arimoto, H. Hashimoto, T. Furuya, and T. Utsumi, *Jpn. J. Appl. Phys.* **22**, L287 (1983).
13. R. Aihara, H. Sawaragi, B. Thompson, and M. H. Shearer, *Nucl. Instrum. Methods* **B37/38**, 212 (1989).
14. N. W. Parker, W. P. Robinson, and J. M. Snyder, *Proc. SPIE* **632**, 76 (1986).
15. W. M. Clark and M. W. Utlaut, *J. Vac. Sci. Technol.* **B6**, 1014 (1988).
16. T. Kato, A. Yasuoka, and K. Fujikawa, *Nucl. Instrum. Methods* **B37/38**, 218 (1989).
17. S. Matsui, Y. Kojima, and Y. Ochiai, *Appl. Phys. Lett.* **53**, 868 (1988).

18. L. R. Harriott, *Proc. SPIE* **773**, 190 (1987).
19. L. R. Harriott, R. E. Scotti, K. D. Cummings, and A. F. Ambrose, *Appl. Phys. Lett.* **48**, 1704 (1986).
20. L. R. Harriott, R. E. Scotti, K. D. Cummings, and A. F. Ambrose, *J. Vac. Sci. Technol.* **B5**, 207 (1986).
21. A. H. Bobeck, L. R. Harriott, R. L. Hartman, D. R. Kaplan, G. S. Przybylek, and W. J. Tabor, U.S. Patent No. 4,870,649 (1990).
22. D. J. Bossert, R. K. Defreez, H. Ximen, R. A. Elliott, J. M. Hunt, G. A. Wilson, G. A. Evans, N. W. Carlson, M. Lurie, J. M. Hammer, D. P. Bour, S. L. Palfrey, and R. A. Mantea, *Appl. Phys. Lett.* **56**, 2068 (1990).
23. M. L. Tilton, G. C. Dente, A. H. Paxton, J. Cser, R. K. DeFreez, C. Moeller, and D. Depatie, *IEEE J. Quantum Electron.*, **27**, 2098 (1991).
24. K. D. Cummings, L. R. Harriott, J. S. Osinski, C. E. Zah, R. Bhat, R. J. Contolini, **E. D. Beege, and T. P. Lee,** *Electron. Lett.* **23**, 1156 (1987).
25. W. M. MacDonald and L. R. Harriott, unpublished.
26. Y. L. Yeh and L. R. Harriorr, *Precision Engineering* (1989).
27. N. Takado, K. Asakawa, T. Yuasa, S. Sugata, E. Miyauchi, H. Hashimoto, and M. Ishii, *Appl. Phys. Lett.* **50**, 1891 (1987).
28. L. R. Harriott, unpublished.
29. M. Taneya, Y. Sugimoto, and K. Akita, *J. Appl. Phys.* **66**, 1375 (1989).
30. T. Kosugi, R. Mimura, R. Aihara, K. Gamo, and S. Namba, *Jpn. J. Appl. Phys.* **29**, 2295 (1990).
31. Y. Sugimoto, M. Taneya, and K. Akita, *J. Vac. Sci. Technol.* **B9**, 2703 (1991).
32. M. Taneya, Y. Sugimoto, H. Hidaka, and K. Akita *Jpn. J. Appl. Phys.* **28**, 515 (1989).
33. K. Gamo and S. Namba, *Proc. Mater. Res. Soc.,* **131**, 531 (1989).
34. M. W. Wu, M. M. Boenke, S. Wang, W. M. Clark, E. H. Stevens, and M. W. Utlaut, *Appl. Phys. Lett.* **53**, 265 (1988).
35. E. Miyauchi, T. Morita, A. Takamori, H. Hashimoto, Y. Bamba, and H. Hashimoto, *J. Vac. Sci. Technol.* **B4**, 189 (1986).
36. A. Takamori, E. Miyauchi, H. Arimoto, T. Morita, Y. Bamba, and H. Hashimoto, *Proc. 12th Int. Symp. on GaAs and Related Compounds,* p. 247. Hilger, Karuizawa, 1989.
37. H. Arimoto, A. Kawano, H. Kitada, A. Endoh, and T. Fujii, *J. Vac. Sci. Technol.* **B9**, 2675 (1991)
38. C. H. Chu, D. L. Barr, L. R. Harriott, and H. H. Wade, *J. Vac. Sci. Technol.* **B10**, 1273 (1992).
39. T. Hiramoto, K. Hirakawa, and T. Ikoma, *J. Vac. Sci. Technol.* **B6**, 1014 (1988).
40. A. D. Wieck and K. Ploog, *Appl. Phys. Lett.* **56**, 928 (1990).
41. W. D. Ladig, N. Holonyak, Jr., M. D. Camras, K. Hess, J. J. Coleman, P. D. Dapkus, and J. Bardeen, *Appl. Phys. Lett.* **38**, 776 (1981).

42. H. Sumida, H. Ahahi, S. Jae, Y. Kumiko, S. Gonda, and H. Tanoue, *Appl. Phys. Lett.* **54**, 520 (1989).
43. Y. Hirayama, Y. Suzuki, and H. Okamoto. *Surf. Sci.* **174**, 98 (1986); Hirayama and H. Okamoto, *J. Vac. Sci. Technol.* **B6**, 1018 (1986).
44. F. Laruelle, A. Bagchi, M. Tsuchiya, J. Merz, and P. M. Petroff, *Appl. Phys. Lett.* **56**, 1561 (1990).
45. S. Sasa, Y. J. Li, M. Miller, Z. Xu, K. Ensslin, and P. M. Petroff, *J. Vac. Sci. Technol. B,* to be published.
46. I. Hayashi, in "Emerging Technologies for In-Situ Processing" (D. Ehrlich and V. T. Nguyen, eds.). Nijoff, Dordercht, 13–22, 1988.
47. G. C. Chi, F. W. Ostermayer, K. D. Cummings, and L. R. Harriott, *J. Appl. Phys.* **60**, 4012 (1986).
48. K. D. Cummings, L. R. Harriott, G. C. Chi, and F. W. Ostermayer, *Appl. Phys. Lett.* **48**, 659 (1986).
49. H. Arimoto, M. Kosugi, H. Kitada, and E. Miyauchi, "Microcircuit Engineering," Vol. 88, p. 321. North Holland, Vienna, 1988.
50. H. Temkin, L. R. Harriott, and M. B. Panish, *Appl. Phys. Lett.* **52**, 1478 (1988).
51. H. Temkin, L. R. Harriott, R. A. Hamm, J. Weiner, and M. B. Panish, *Appl. Phys. Lett.* **54**, 1463 (1989).
52. Y. L. Wang, L. R. Harriott, R. A. Hamm, and H. Temkin, Appl. Phys. Lett. **56**, 749 (1990).
53. L. R. Harriott, H. Temkin, Y. L. Wang, R. A. Hamm, and J. S. Weiner, *J. Vac. Sci. Technol.* **B8**, 1380 (1990).
54. L. R. Harriott, H. Temkin, C. H. Chu, Y. L. Wang, Y. F. Hsieh, R. A. Hamm, M. B. Panish, and H. H. Wade, *Proc. SPIE* **1465**, 57 (1991).
55. K. Akita, Y. Sugimoto, M. Taneya, Y. Hiratani, Y. Ohki, H. Kawanishi, and Y. Katayama, *Proc. SPIE* **1392**, 576 (1990).
56. Y. Hiratani, Y. Ohki, Y. Sugimoto, and K. Akita, *J. Crystal Growth* **111**, 570 (1991).
57. H. Sugiura, R. Iga, T. Yamada, and T. Toriyama, *Jpn. J. Appl. Phys.* **6B**, 1089 (1991).
58. Y. L. Wang, H. Temkin, R. A. Hamm, R. Yadvish, D. Ritter, L. R. Harriott, and M. B. Panish, *Electron. Lett.* **27**, 1324 (1991).
59. Y. L. Wang, A. Feygenson, R. A. Hamm, D. Ritter, J. S. Weiner, H. Temkin, and M. B. Panish, *Appl. Phys. Lett.* **59**, 443 (1991).

Chapter 7

FULL-WAFER TECHNOLOGY FOR LARGE-SCALE LASER FABRICATION AND INTEGRATION

P. Vettiger, P. Buchmann, O. Voegeli, and D.J. Webb

IBM, Research Division
Zurich Research Laboratory
Switzerland

1. INTRODUCTION

Today's fabrication techniques for semiconductor lasers are based primarily on scribing followed by cleaving of the substrate along the crystallographic planes of the material to form the laser mirrors. Although this technique

ISBN 0-12-200420-5

provides smooth (atomically flat) mirrors, it has a number of drawbacks. First, even when discrete lasers are to be fabricated, important steps in the process, such as mirror coating and laser testing, must be performed at the bar/chip level. The individual handling of these very delicate parts is a considerable obstacle to the realization of high-yield processes with high throughput. Although lasers are now mass produced, especially for compact disc player applications, considerable improvements in yield and throughput would be achieved if the entire process were performed before the wafer is diced into chips. Another serious drawback is related to the fabrication of optoelectronic integrated circuits (OEICs). In this case one would like to integrate optical and electronic devices, together with the relevant interconnects (i.e., electrical wiring and optical waveguides) onto a single substrate, so that a complete circuit or system can be built. The evolution of microelectronics has followed a similar path, from the mounting of discrete transistors on circuit boards to the integration of many transistors and their interconnections monolithically on one substrate. This has led to today's very large scale integrated (VLSI) circuit fabrication technology. Since the fabrication of cleaved mirrors requires that those parts of the wafer at either end of the lasers be cleaved off, it is practically impossible to collect light from the laser and direct it to other devices on the chip for further processing. Furthermore, the size of the chip is strictly limited in one dimension to the length of the laser, which is usually a fraction of a millimeter. This compares with chip sizes of more than 1 cm on a side for typical electronic integrated circuits. The major motivation of this research is to develop VLSI-type fabrication and testing techniques which allow full-wafer processing and testing (FWP, FWT) and which provide opportunities for integration of the lasers with other optical and electrical devices [1–3]. The key step to FWP is the replacement of cleaving by a dry-etching technique so that the mirrors can be formed without it being necessary to break up the wafer. The mirrors are now the side walls of grooves, several micrometers deep, etched into the wafer. Mirror coating can be applied to all the lasers on the wafer at once, which results in a substantial reduction of bar and chip handling. The lasers can of course be operated on wafer because they now have mirrors. They can therefore be tested and screened on the full wafer using conventional automatic semiconductor testers to which some optical equipment has been added. Lasing characteristics of several thousand devices and data from test sites for process characterization and diagnostics can be collected much more quickly than for cleaved lasers. In addition, etched mirrors are not limited by crystallographic orientation, and hence nonplanar facets such as shaped mirrors are possible. Short-cavity lasers, groove-coupled-cavity lasers, beam deflectors, and surface emitters are other applications of etched mirrors that show great potential.

This chapter describes the successful development of a full-wafer technology [4][1] with potential for large-scale laser fabrication and optoelectronic integration.

2. CLEAVED MIRROR TECHNOLOGY

As an introduction to etched mirror technology, this section describes the design and fabrication of lasers with cleaved mirrors. Among the many known semiconductor laser structures [5, 6] only one type of laser is described, which served as the basic laser structure for the work discussed in this chapter. However, lasers with etched mirrors can be made with almost any of the structures employed for lasers with cleaved mirrors.

These lasers are single quantum well, graded index, separate confinement heterostructure (SQW-GRINSCH) lasers [7]. The basic structure is shown in Fig. 1. The quantum well confines the minority carriers in a direction perpendicular to the junction and modifies the density of states so that very efficient lasing action is realized. The core of the optical waveguide is formed by the GRIN region in which the quantum well is embedded. In this study, index guiding for optical confinement parallel to the junction plane is provided by a ridge waveguide [7–10]. Here the top contact layer and most of the upper cladding of the epitaxial structure are etched away except where

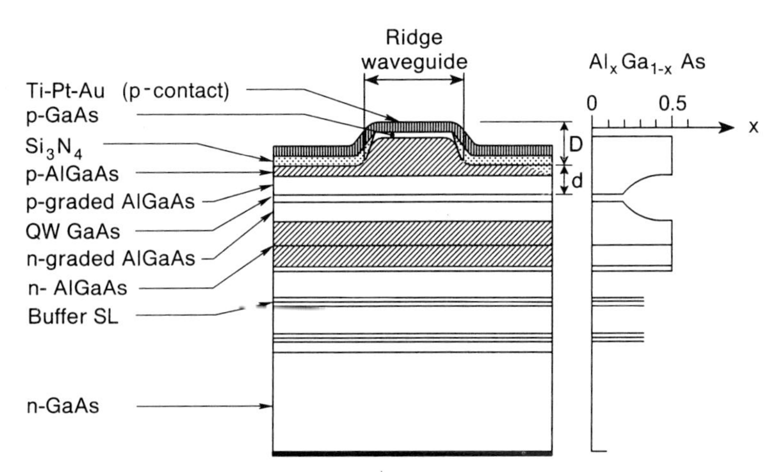

Fig. 1. Basic structure of AlGaAs single quantum well ridge laser.

[1]Parts of this text and selected figures have previously been published in Ref. 4.

the lasers are to be formed. The remaining ridges form the dielectric loading for the lateral optical waveguide, and since the current injection occurs only through the top of the ridge, a considerable degree of lateral current confinement is also obtained. We have found that lasers of this type are quite efficient and can have excellent beam properties even at high powers [10]. They are also relatively simple to construct, having for instance no regrown interfaces. Despite (or perhaps owing to) this simplicity, they have been shown to be very reliable provided the mirrors are suitably passivated. A 7-nm-thick GaAs quantum well is centered in a parabolically shaped AlGaAs graded-index region where the barrier height between the cladding and the quantum well corresponds to an aluminum content of 20%. The Al content of the 1.8-μm-thick cladding layers is 40%. The epitaxial growth is performed by molecular beam epitaxy (MBE), which allows good control and uniformity. The waveguide is constructed so that the optical mode is extended somewhat in the vertical direction. This reduces the optical power density at the facet. Higher-power operation is then made possible at the expense of a slightly higher threshold current.

The ridges are usually made by wet etching and, to obtain a single lateral mode, the etching is terminated 400 nm above the graded-index region. The total etch depth is then of the order of 1.5 μm micrometers for typical power laser structures. Ridge widths are of the order of 3 μm. To provide electrical isolation of the bonding pads, an insulating layer of silicon nitride (Si_3N_4) is deposited on the top surface of the wafer using plasma-enhanced chemical vapor deposition (PECVD). This is done immediately after the ridge etch before stripping the photoresist. The natural undercut of the etch process causes the conformality of the silicon nitride to be incomplete. The photoresist can thus be lifted off with organic solvents, leaving contact openings in the nitride which are self-aligned to the tops of the ridges. The fabrication is completed by patterning a *p*-contact metal of Ti/Pt/Au using lift-off, lapping the back of the wafer to a total thickness of about 0.15 mm, and applying a Ge/Au/Ni/Au *n*-contact to the back. This last layer is alloyed and then reinforced with a Ti/Ni/Au layer for soldering purposes. The end mirrors of the cavity are formed by cleaving pieces of appropriate length from the crystal so that the waveguides are perpendicular to the cleaved crystal facets. These facets have a reflectivity of about 30% by virtue of the refractive index mismatch between the laser crystal ($n \simeq 3.5$) and air ($n = 1$). Such mirrors must be protected against corrosion, which is caused by the very high optical power densities at the facet combined with atmospheric oxygen or moisture [11]. Aluminum oxide (Al_2O_3) is often used for this purpose and has demonstrated good reliability when deposited with the ion-beam sputter method [12]. The reflectivity of the mirrors can be raised or lowered as necessary by applying suitable coatings. Low reflectivity ($<30\%$) is obtained by simply choosing the appropriate optical thickness for the passivation layer. If the optical thickness is one half-wave, the reflectivity is unchanged. The lowest reflectivity that can be obtained with a single film depends on the refractive index of the passivation; for Al_2O_3 ($n = 1.66$) it is about 1%. For our applications the required reflectivity for the front laser mirror is between 5 and 15%, which corresponds to a thickness of about 152 to 181 nm. For high reflectivity, a multilayer stack of alternating high- and low-index material, each layer having an optical thickness of one-fourth of a wave, is employed [13]. For use with lasers, Al_2O_3 and Si ($n \simeq 4$) are often used as the high- and low-index materials, respectively [14]. The large difference in index means that few layers are needed to obtain a high reflectivity. For instance, two pairs (i.e., four layers) yield a reflectivity of over 90%. These layers have also been deposited by ion-beam sputtering on the lasers described here. In practice, bars, which are as long as the lasers (usually between 0.3 and 1 mm) and about 10 mm or so wide, are cleaved from the parent wafer. The bars have many lasers side by side. A number of bars are then placed in a suitable holder so that the top and bottom contacts are shielded from the deposition. After the coatings are deposited, the bars are removed from the holder and the individual lasers on the bar are tested. Then the bars are diced into chips containing one or more lasers and then mounted in a suitable package and wire bonded. The left-hand side of Fig. 2 illustrates the basic fabrication and testing steps required for lasers with cleaved (conventional) mirrors.

3. ETCHED MIRROR TECHNOLOGY

Despite the fact that cleaved crystal facets make excellent mirrors, they have a number of disadvantages, one of the most serious being that the portions of the wafer at either end of the laser must be cleaved away to form the mirrors. This makes monolithic integration of other devices that would collect and further manipulate the laser light impossible.

One technique to overcome this difficulty is microcleaving [15], in which small pieces of material are cleaved off at each end of the individual lasers. Although this method can potentially result in maximum mirror quality, it is limited to flat facets and to fixed orientation (the crystal cleavage planes). More freedom in design is possible when using distributed Bragg reflectors (DBRs) [16], but the processing is more complicated (submicrometer li-

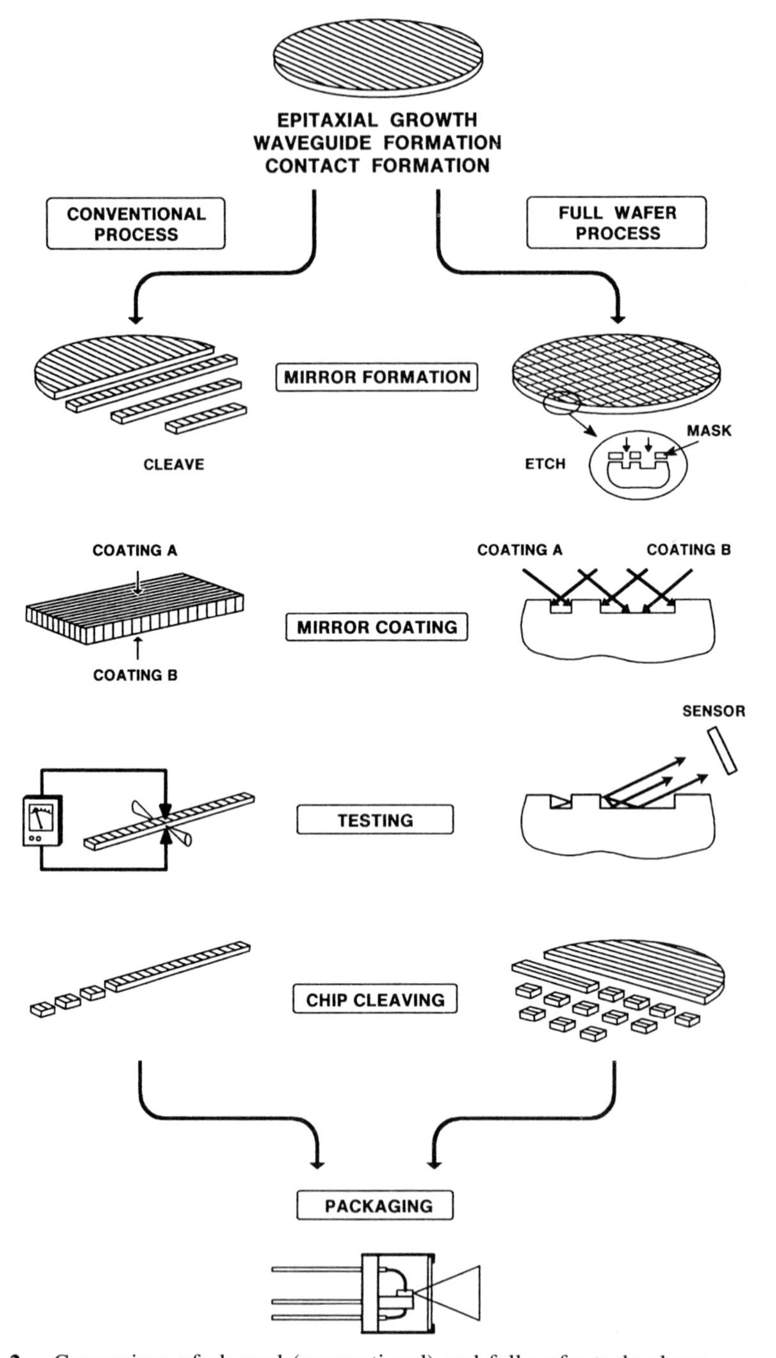

Fig. 2. Comparison of cleaved (conventional) and full-wafer technology.

thography, often with epitaxial overgrowth) and therefore costly to manufacture. The third method adopted here is the etching of trenches at each end of the laser waveguide down to the bottom of the lower cladding layer or deeper. The side walls of the etched trenches then act as the laser mirrors. Although considerable technical challenges are associated with etching and passivating these mirror facets, they offer a number of advantages. First, since they are defined by microlithographic techniques they can be made to practically any size, shape, and orientation. This allows a wide variety of cavities to be fabricated. It is in principle possible to develop effective output couplers to direct the light to other optoelectronic devices which could be integrated on the same wafer. Short-cavity lasers, groove-coupled-cavity lasers, lenses, staggered arrays, monitor photodiodes, beam deflectors, and beam-shaping devices are other applications of etched mirrors with future potential. Finally, etched mirrors simplify the fabrication and testing of lasers in general, since the lasers can be coated and tested before the wafer is diced into chips (full-wafer processing, testing, and screening). The amount of handling of fragile laser chips is therefore greatly reduced compared to more conventional fabrication methods. The basic structure of a laser with etched mirrors as it is employed in full wafer technology is shown in Fig. 3.

The goal of achieving etched mirrors of comparable quality to cleaved mirrors sets stringent requirements on the flatness, roughness, verticality, and degree of surface damage and makes it necessary to select an appropriate etching method carefully. In order not to affect the output beam properties, the facets have to be smooth within $\lambda/10$, which is 25 to 50 nm in III–V materials. This calls for a mask with extremely smooth edges. A nonvertical

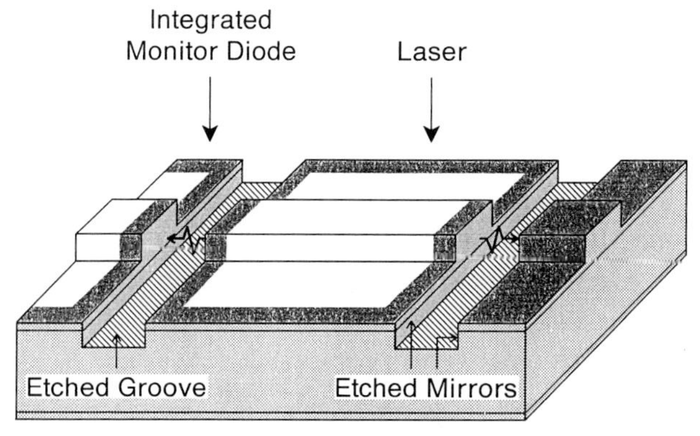

Fig. 3. Laser with etched mirrors and integrated monitor diode. (From Ref. 4. © 1991 IEEE.)

mirror with a tilt angle of 5° already reduces the facet reflectivity by 5 to 10%, depending on the vertical waveguide structure. The etching process, therefore, has to provide a high degree of anisotropy. As layers of different composition have to be etched through, a low etch rate selectivity is necessary; otherwise any lateral etching would result in a corrugated surface. High etch rates are preferable for depths of the mirror grooves between 4 and 8 μm.

Various techniques for the fabrication of laser mirrors in AlGaAs/GaAs material has been reported. We briefly review the wet etching, as a purely chemical process, followed by the sputter etching, as a purely physical process, and finally combinations of both physical and chemical processes.

Most wet-chemical etchants for AlGaAs show either isotropic etching or crystallographic preferential etching and are often selective for the aluminum concentration in the layers [17]. Despite these difficulties, it has been shown that vertical smooth facets and output powers of 30 mW in a single transverse mode can be achieved with crystallographic etching of specially tailored multiheterolayer combinations [18]. The limitations of this method regarding mirror orientation, heterolayer structure, and flat facets, however, make anisotropic dry-etching methods more attractive.

Sputter etching or ion-beam etching (IBE) with nonreactive ions from a discharge in a parallel-plate reactor or an ion-beam source have been used to etch laser mirrors and surface deflectors [19–21]. Nonreactive ion etching has detrimental effects such as high ion damage, redeposition of sputtered material, trenching, and strong mask erosion. The latter makes it necessary to tilt the substrate with respect to the ion beam, which imposes limits on the design of mirror shapes. Focused ion-beam etching (FIBE) has been used to fabricate high-quality mirror facets without the need for a mask [22]. By adding a reactive gas (chemically assisted FIBE) the etch rate can be enhanced [23], but the process is too slow to be applicable for the etching of thousands of facets and isolation trenches on a wafer.

Original work on reactive dry etching of mirror facets and other structures has been done in InP-based material [24, 25]. We will restrict our discussion to AlGaAs material here, but obviously the full wafer technology can be adapted to other III–V semiconductor materials.

On the one hand, all dry-etching techniques used for the fabrication of vertical facets in AlGaAs use ion bombardment to ensure anisotropy. On the other hand, ion bombardment is responsible for surface damage that may have an impact on device performance and long-term stability. This makes the reduction of both ion energy and beam current density as well as an increase of the chemical etching component highly desirable [26, 27]. Chlorine and bromine are the gases which form the most volatile halides with Al, Ga, and As at etching temperatures below 200°C, whereas fluorine-based chemistry [28] results in nonvolatile products with aluminum. Chlorine-based dry-etching methods are of special interest for AlGaAs material with arbitrarily varying Al concentration because of the small etching selectivity that can be achieved. Etched mirrors have been reported using Cl_2 reactive ion etching (RIE) [29]. Angled Cl_2 RIE has been used to fabricate 45° intracavity deflectors for surface emitting lasers [30]. In the RIE process, plasma interaction with chamber and electrode material has to be considered, and a load-lock system is usually required to achieve low oxygen and water pressures in the chamber. Reactive ion-beam etching (RIBE) using a reactive gas, which is fed into a Kaufman-type or electron cyclotron resonance (ECR) ion source, has the advantage that the plasma is confined in the ion source and chamber wall effects are avoided. An ultrahigh-vacuum system can be used, allowing etching of laser facets under very clean conditions. The combined chemical and physical action of chlorine ions has been used to etch facets for AlGaAs lasers by RIBE [31, 32]. The method allows an additional degree of freedom by tilting the substrate, whereby the angle of incidence and therefore the mirror or deflector profile can be controlled.

In contrast to other techniques, such as RIE and RIBE, chemically assisted ion-beam etching (CAIBE) has been shown to be a method that allows the decoupling of the chemical and physical etch components by using a nonreactive ion beam together with a separate feed for reactive gas introduced into the etching system near the sample surface [33]. The parameters that can be controlled independently are ion energy, ion current density, reactive gas flow or pressure, substrate temperature, substrate tilt angle, and rotation. Using modulated or pulsed ion-beam bombardment together with energy and mass analysis of the reaction products, it has been possible to gain insight into the etching mechanism of Cl_2./Ar-CAIBE of III–V compounds [34, 35]. Molecular chlorine is readily adsorbed on the AlGaAs surface, but the spontaneous chemical reaction rate at room temperature is negligibly low [36]. Ion bombardment enhances the dissociation into Cl radicals, which react spontaneously with Al, Ga, and As to form halides. The reaction itself can be accelerated by the ions. Arsenic chloride is removed from the surface by spontaneous desorption, whereas the less volatile chlorides of Al and Ga, which cover the surface at steady-state conditions, are removed mainly by ion-enhanced desorption.

An etching method that goes even further to decouple physical and chemical etching parameters is radical beam/ion beam etching (RBIBE), in which the chlorine is fed to the sample as a flux of radicals produced from molecular Cl_2. by cracking in an Evenson microwave cavity [37]. GaAs reacts spontaneously with Cl radicals at room temperature and etching, although at a low rate, occurs [38]. The addition of ion-beam bombardment allows the anisotropy of the etching to be controlled and the desorption rate of the reaction products to be enhanced.

Photochemical etching combining a laser beam with reactive gases is an entirely different approach to laser mirror fabrication [39]. However, there are limitations to side-wall angle control and the method is based on polymer film deposition.

Cl_2./Ar-CAIBE has been used for the facet etching of surface-emitting AlGaAs laser arrays. The shape of the vertical facets and the external planar or parabolic deflectors was controlled by the substrate tilt angle with respect to the incident ion beam [40, 41]. Broad-area lasers with high efficiency have been demonstrated using the same method for mirror etching [42]. In our study we have optimized the CAIBE parameters for the etching of mirror facets in single-transverse-mode AlGaAs/GaAs SQW-GRINSCH ridge lasers. Various mask materials were investigated for maximum smoothness of the mirror surface, and topography effects on the mirror flatness had to be avoided by adopting a new mirror design.

An outline of the full-wafer fabrication and testing procedure using etched mirrors and on-wafer facet coating is shown in Fig. 2. These processes are described in more detail in the following sections.

4. FULL-WAFER PROCESSING

4.1. Etching Process for Laser Mirrors

The CAIBE system [43] used in this work is shown schematically in Fig. 4. A broad-area Ar^+. ion beam is produced by a Kaufman-type ion source. Molecular chlorine is fed to the system as multiple jets from a ring feed. The geometry of the ring feed was optimized to achieve the most uniform chlorine concentration on a 2-inch wafer. Substrate rotation helps minimize shadowing effects caused by the directionality of the chlorine flux. The base pressure in the system is $<10^{-7}$ mbar. A cold trap provides a water partial pressure of $\simeq 5 \times 10^{-9}$ mbar and efficient pumping of excess chlorine. Even at a Cl_2. flow rate of 20 sccm the chamber pressure remains very low during the etching process ($<10^{-4}$ mbar). The substrate temperature does not exceed 50°C during etching. The native oxide layer on the surface of the AlGaAs samples was removed by sputter cleaning prior to etching.

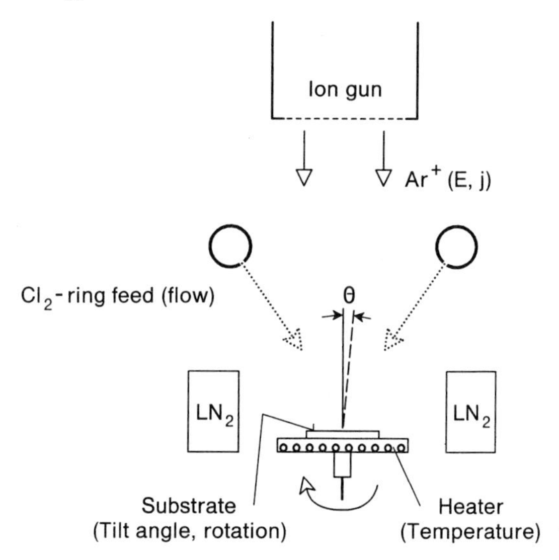

Fig. 4. Setup of the chemically assisted ion-beam etching (CAIBE) system. (From Ref. 4. © 1991 IEEE).

The etch rate of GaAs was investigated by varying the chemical etching component at a constant ion energy of 500 eV and a beam current density of 230 $\mu A/cm^2$. Figure 5 shows a linear increase of the etch rate as a function of chlorine flow rate. The physical sputter rate is enhanced by more than a factor of 100 at 15 sccm, which corresponds to a sputter yield of 30. Etching in this highly chemically dominated regime will reduce the surface damage considerably. An etch rate of 0.3 $\mu m/min$ is reasonable for an etch

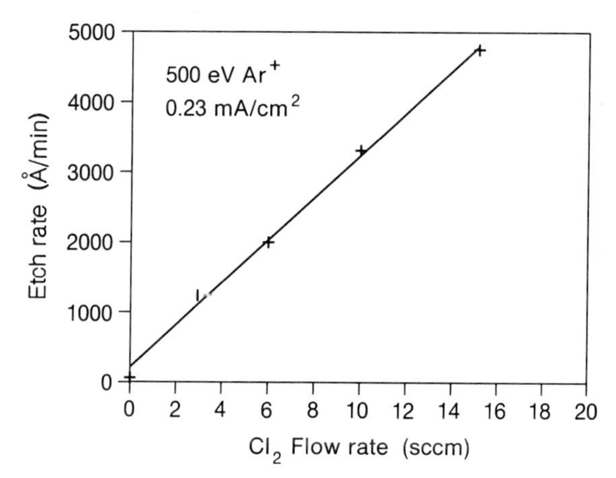

Fig. 5. Cl_2/Ar CAIBE etch rate versus Cl_2 flow rate for GaAs. (From Ref. 4. © 1991 IEEE).

depth of 6 μm, required in the case of ridge waveguide lasers. At an Ar^+ ion energy of 250 eV, etch rates of the same order were achieved by increasing the current density by a factor of 2. However, an ion energy of 500 eV was maintained as the typical mirror-etching condition owing to the more isotropic etching and the poor surface morphology at lower energies.

An important requirement for mirror etching is a low etch-rate selectivity between AlGaAs and GaAs. SQW-GRINSCH lasers have a complicated epilayer structure, in which the aluminum concentration may vary between 0 and 65% (for an emission wavelength of 720 nm). Because there is always some undercutting of the mask through etching by free radicals and molecules of chlorine, the side walls will show enhanced corrugation in the case of nonequal etch rates. The AlGaAs etch rates were measured on MBE-grown (100)-oriented GaAs/AlGaAs/GaAs samples as a function of Al content. In the same experiment, the physical component of the etching was varied by changing the ion-beam current density. The results obtained at constant ion energy and Cl_2 flow rate are shown in Fig. 6. Both Cl radical production and desorption of reaction products are enhanced by a higher current density. At ion-beam current densities below 150 $\mu A/cm^2$, a selectivity of up to 2 was found between GaAs and AlGaAs with a 50% Al mole fraction. This effect can be attributed to the formation of aluminum oxide on the AlGaAs surface due to the residual oxygen and water in the chamber. Below the transition point of 150 $\mu A/cm^2$, the etching process seems to be limited by the sputtering of aluminum oxide. At current densities above the

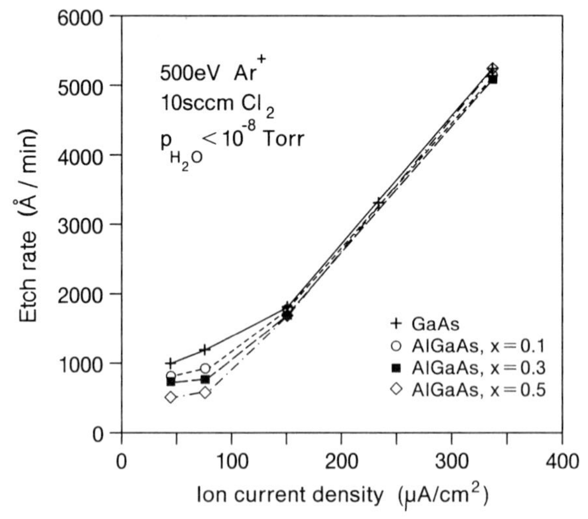

Fig. 6. CAIBE etch rates of AlGaAs as a function of ion-beam current density and Al concentration. (From Ref. 4. © 1991 IEEE).

transition point, the etching is limited by the amount of chlorine available on the surface. This result underlines the need for very low H_2O and O_2 pressure in a Cl_2/Ar-CAIBE system for equi-rate AlGaAs etching [44].

The anisotropy of the etching also depends on the current density at constant chlorine flow. At current densities of 300 $\mu A/cm^2$ or above, facet angles of less than 3° with respect to vertical can be achieved on a whole 2-inch wafer. At lower current densities the isotropic (chemical) etching is more pronounced, leading to mask undercutting and facet angles of up to 10 degrees. Good uniformity and orientation independence of the side-wall profiles as well as an excellent etched-surface morphology have been achieved using an Ar^+ ion energy of 500 eV, a current density of 300 $\mu A/cm^2$, and a chlorine flow of 15 sccm. The etch depth uniformity (which is not critical for etched mirrors) is 8% on 2-inch wafers.

The required smoothness of the etched mirrors (<30 nm to avoid scattering losses) is a challenge for the mask fabrication. In addition to smooth edges, the mask must have a vertical profile in order to avoid facet roughening by edge erosion and must be resistant to chlorine. An extensive investigation of mask materials showed that all single-layer metal masks produce an edge roughness of more than 100 nm, probably due to microcrystallinity. Dielectric layers such as SiO and SiO_2. showed partly very good results, but with poor reproducibility. The best edge smoothness was achieved with a multilayer resist (MLR) mask consisting of a hard-baked AZ-resist bottom layer, a thin SiO intermediate layer, and an AZ imaging resist layer. The mirror pattern is exposed and developed using contact printing. The pattern is then transferred into the SiO layer by CF_4-RIE and into the bottom AZ resist by O_2-RIE, whereby the top imaging resist was also removed. It is important that the intermediate layer is amorphous and that it is removed by another CF_4-RIE step. The remaining hard-baked resist stencil has extremely smooth edges with a roughness of the order of 10 to 20 nm. The etch rate selectivity for GaAs : hard-baked AZ is between 15 and 50 for typical CAIBE conditions. The mask is removable by oxygen ashing and rinsing in solvents. Using this mask and the described CAIBE process, a roughness of the order of 10 to 30 nm is achieved on the etched laser facets as seen in Fig. 7. The vertical striations on the facets seen at high contrast are produced by the edge roughness of the chromium mask plate used for contact lithography. Further improvement can be expected from optical reduction projection printing and from electron-beam direct writing of the mirror pattern.

Single-mode lasers used in optical storage and other applications require diffraction-limited focusing of the light, so the overall flatness of the mirror facets becomes important. Flatness of the mirror facets is not easy to achieve

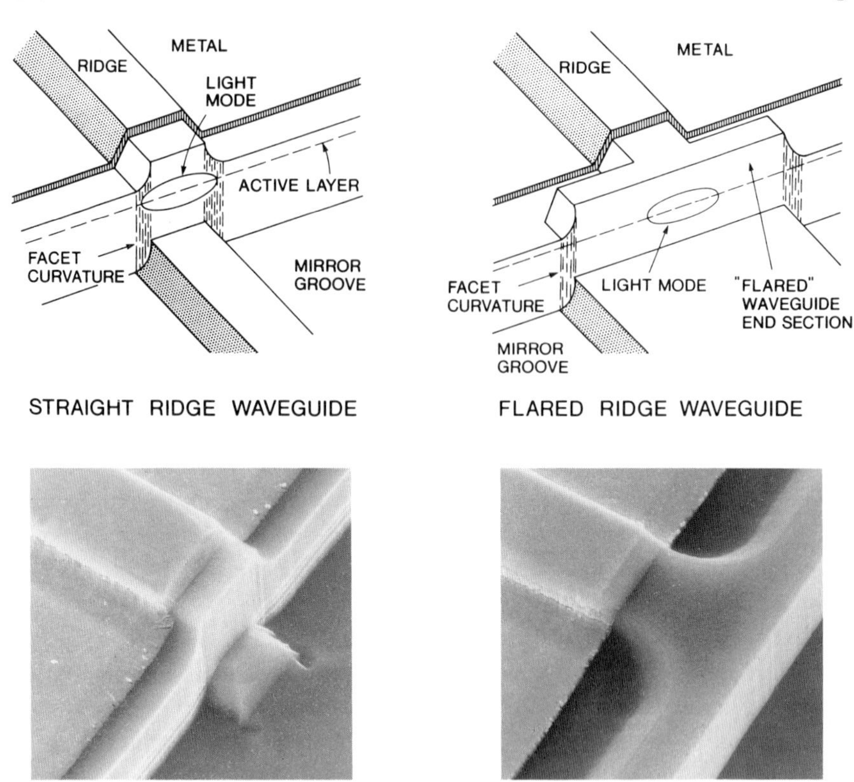

Fig. 7. Top row: schematic view of conventional straight ridge waveguide with etched facet and etched mirror with flared waveguide end section. Bottom row: SEM photographs of straight and flared waveguide etched-mirror facets. The waveguides are 3 μm wide. (From Ref. 4. © 1991 IEEE).

if the laser structure has a pronounced topography, as in the case of ridge waveguide and nonplanar substrate lasers. We found topography effects from lithography and from the CAIBE mirror-etching process that affect the mirror flatness of lasers with ridge waveguide structure. These result in a facet curvature near the ridge edges with a recess of the facet in the beam region of up to 300 nm and a highly distorted far-field intensity and phase distribution. By introducing broad ("flared") sections of the ridge near both mirror facets, all topography effects can be avoided (Fig. 7). The laterally nonguiding region has to be as short as possible to minimize coupling loss of the divergent beam to the back-traveling waveguide mode. We found that a length of 2 μm and a width of 24 μm for the flared section are enough to avoid any topography effect while retaining negligible mode coupling loss [45, 46]. The flared waveguide laser facets were flat within less than 20 nm. This improvement in overall flatness can clearly be seen in the scanning electron microscope (SEM) photographs in Fig. 7. The flared waveguide etched-mirror lasers have output characteristics and beam quality equivalent to those of cleaved-mirror lasers, whereas the far fields of straight waveguide lasers are multilobed or asymmetric. Using the widened waveguide approach, a similar improvement in etched-mirror quality can be expected for other laser structures with large topography, such as nonplanar substrate or certain buried-heterostructure lasers.

Beam distortion is also possible from light reflected at the bottom of the etched mirror groove in front of the laser facet. If the wafer is ultimately cleaved into individual laser chips, one has to provide enough clearance for the beam. The divergence angle of the output beam of SQW-GRINSCH lasers normal to the junction plane is usually quite large. In our case a clearance angle of 60° was required to avoid interference with reflected light. Taking into account the mirror etch depth and the thickness of the mirror coating on the bottom of the groove, the requirement for chip parting is that cleavage should take place within 9 μm of the front facet. Also, for laser arrays, it is required that the cleavage occur at 5 ± 4 μm from the facet over a distance of at least 2 mm. This was achieved by cleaving a flat to the 2-inch wafer right after MBE growth and high-precision mask alignment within 0.05° of the substrate axis for the first level of lithography. The positioning tolerance for the diamond scribing was ±1.5 μm. Two on-chip conductor loops, which cross the cleavage line at each tolerance boundary, allow a quick screening of in-spec chip parting after cleavage by electrical continuity testing.

4.2. Coating Process for Etched Mirrors

In the full-wafer laser fabrication technology, the etched laser mirrors are coated [47] for the same reasons cleaved mirrors are. The first is to passivate the mirrors, i.e., to eliminate or at least greatly slow down their degradation. The second is to modify their reflectivity as required by the application. For instance, lasers for optical storage usually have front mirrors with a slight antireflection coating (~10%) and back mirrors with a high (~90%) reflectivity coating. This ensures that most of the laser light exits through the front mirror, which raises the overall efficiency.

The material used to passivate the mirrors is aluminum oxide (Al_2O_3). It is deposited by ion-beam sputter deposition (IBD) [48] in a system shown schematically in Fig. 8. With this method, a directed beam of argon ions of a particular energy is used to sputter material from a solid target onto the

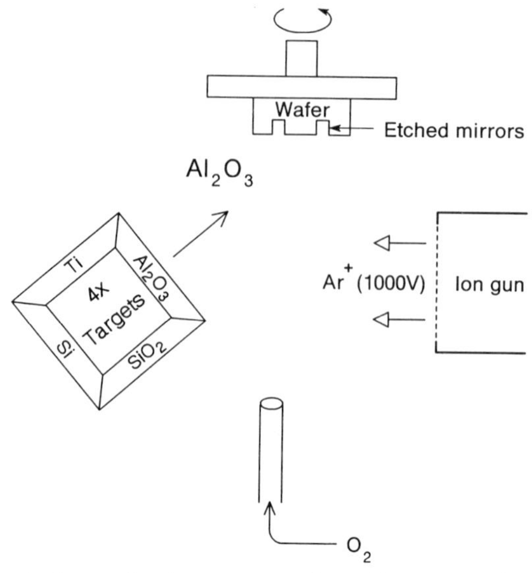

Fig. 8. Oblique incidence ion-beam sputter deposition system for on-wafer mirror coating. (From Ref. 4. © 1991 IEEE).

substrate. The advantage of the ion beam, compared to other sputtering techniques, is that the deposition is directional and thus allows the front and rear mirrors of the lasers to be coated differently as shown in Fig. 9. It is preferable to deposit the films by sputtering, since this usually results in denser films than those deposited by e-beam evaporation, which would otherwise be a possible directional deposition method. With IBD, the plasma is contained in the ion source; hence bombardment and possible damage of the substrate by energetic ions are minimized. Finally, relatively low gas pressures can be used in the chamber, which also improves the directionality of the deposition. The passivating properties of SiO_2 and Si_3N_4 deposited by this method were also studied and found to be inferior to those of Al_2O_3.

Briefly, the following process conditions are used: an argon ion-beam energy of 1000 eV, a beam current of about 70 mA, a chamber base pressure of $<10^{-7}$ mbar, and a chamber pressure during sputtering of about 8×10^{-4} mbar. In addition, when oxides are sputtered, oxygen is bled into the chamber at a rate of 10% of the total gas flow; otherwise the films tend to be metal-rich. Al_2O_3, SiO_2, and Si are sputtered from Al_2O_3, SiO_2, and Si targets, respectively. The substrate is not heated. Further details of the process and film properties can be found in Ref. 49.

As shown in Figs. 8 and 9, oblique-incidence deposition is employed to coat the side walls of the mirror trenches. One of the aims of the process

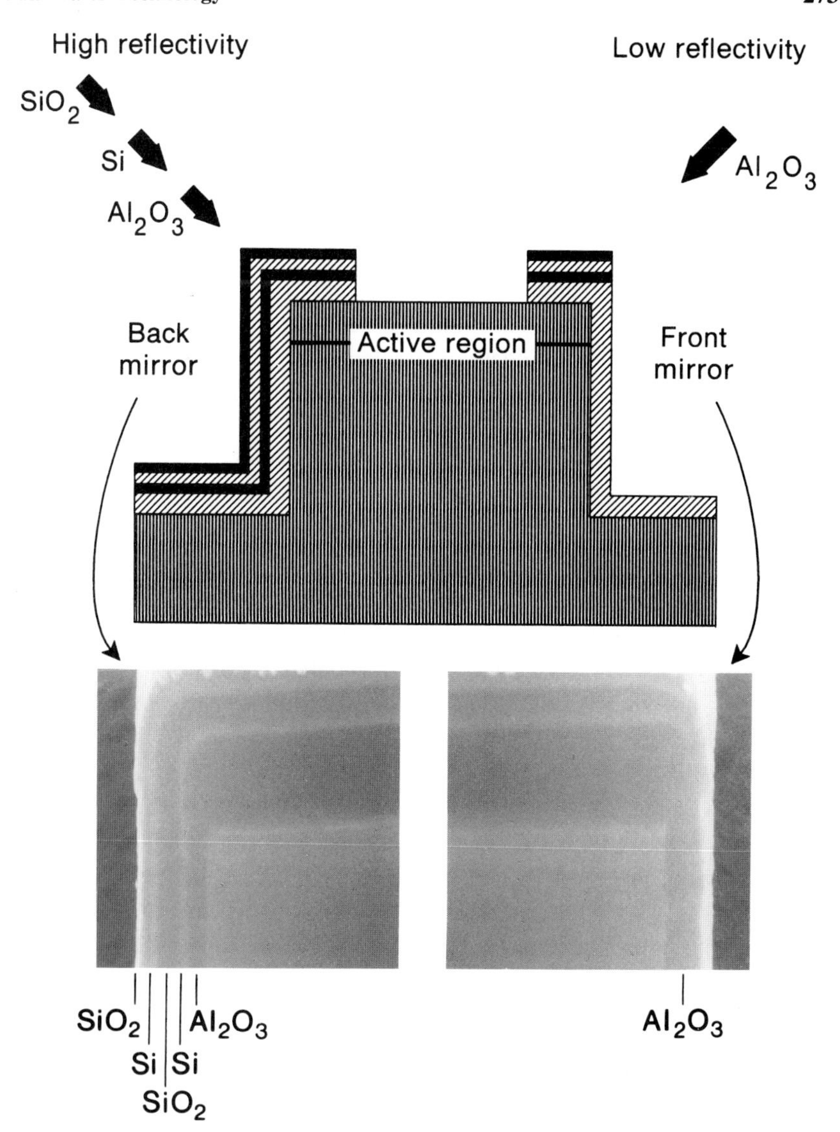

Fig. 9. Concept for high/low-reflectivity modification of etched mirrors. (From Ref. 4. © 1991 IEEE.)

optimization was to achieve a satisfactory uniformity of thickness and refractive index of the side-wall coatings over a wafer. This was done by optimizing the relative positions of the substrate and target as well as the tilt angle of the target. It is possible to coat the side walls of mirror trenches without shadowing from the opposite side wall of the trench. The refractive index of Al_2O_3 films obtained with this process is 1.66 (at 633 nm). This indicates that the film is dense compared to films prepared by e-beam evaporation [50]. Thickness and refractive index uniformities of 2.7% and $<0.3\%$ (1σ), respectively, have been obtained over a diameter of 40 mm.

Al_2O_3 films deposited at near-normal incidence by IBD have been shown to have excellent passivating properties for cleaved mirrors [12, 51]. Although no long-term stress tests have been conducted, we have attempted to estimate the impact that the oblique angle of deposition would have on the quality of the passivation. Such measurements were also made on lasers with cleaved mirrors, but using the full wafer deposition geometry. No significant difference between coatings applied at near-normal or oblique deposition angles has been found with respect to pulsed catastrophic optical damage (COD) or COD-limited lifetime at high output powers.

In this study, two types of coating have been applied to etched mirrors using the oblique-incidence IBD technique. The first was simply to passivate the front and back mirrors with Al_2O_3, leaving their natural reflectivity of about 30% unchanged. This provided Al_2O_3 films with an optical thickness of $\lambda/2$, where λ is the laser wavelength. Second, we applied coatings where the front-mirror reflectivity was about 10% and the back-mirror reflectivity about 90%. For this second type of coating, the front mirror coating was again Al_2O_3 but with the thickness reduced to produce the required reflectivity. For the back mirror, Al_2O_3 was used (because of its passivating properties) as the first low-index layer in a $\lambda/4$ stack of alternating high- and low-refractive-index materials. The high-index material was amorphous Si (a-Si), also deposited by IBD. A cross-sectional view of such mirror coatings is shown in Fig. 9. It is evident that the entire top surface of the wafer is coated along with the mirrors. Consequently, it is necessary to etch vias to the contact pads. This is done by RIE using CF_4. The etch rate of Al_2O_3 is quite low, so SiO_2 is used for the second and subsequent low-index layers in the $\lambda/4$ stack. From the SEM micrograph, it is evident that the coatings are dense, have good corner-covering properties and are uniform with depth into the mirror trench.

4.3. Process Sequence for Laser Fabrication

Fabrication starts with an MBE-grown AlGaAs SQW-GRINSCH structure followed by ridge waveguide fabrication and *p*-contact formation as described in Section 2. Then, the MLR mirror/trench mask is applied and patterned. The CAIBE process is performed at 500 eV, 300 $\mu A/cm^2$, and a Cl_2 flow of 15 sccm. An etch depth of more than 6 μm is required to make chip parting easier and to avoid beam reflection from the bottom of the etched groove. The resist mask is removed by O_2 plasma ashing. After transfer into the IBD chamber, the passivation and/or reflectivity modification layers are deposited separately on opposite facets. Vias are etched into the coating using CF_4-RIE to provide access to the contact pads. Following the wafer thinning, *n*-contact layers (Ge/Au/Ni/Au) are deposited on the back of the wafer and alloyed. After depositing a Ti/Ni/Au contact reinforcement layer, the lasers are then ready for full-wafer testing.

After FWT, the wafers are diamond scribed for chip parting. The chips have to be cleaved within 9 μm of the etched mirror in order to avoid far-field distortions by light reflection from the chute. This has been achieved by high-precision mask alignment within 0.05° of the substrate axis for the first level of lithography. Finally, the cleaved chips are soldered into a package and the top *p*-contacts of laser and monitor diodes are wire bonded to the package leads.

5. ETCHED MIRROR CHARACTERIZATION

The flared waveguide lasers with both mirror facets etched have power–current characteristics equivalent to those of cleaved-mirror lasers from the same chip. The continuous-wave (CW) light output characteristics of single-mode devices with a cavity length of 720 μm and uncoated facets are shown in Fig. 10. The threshold current is 16 mA and the differential external quantum efficiency is 74% for both cleaved and etched mirror lasers from the same chip. The output characteristics for two lasers with passivated etched

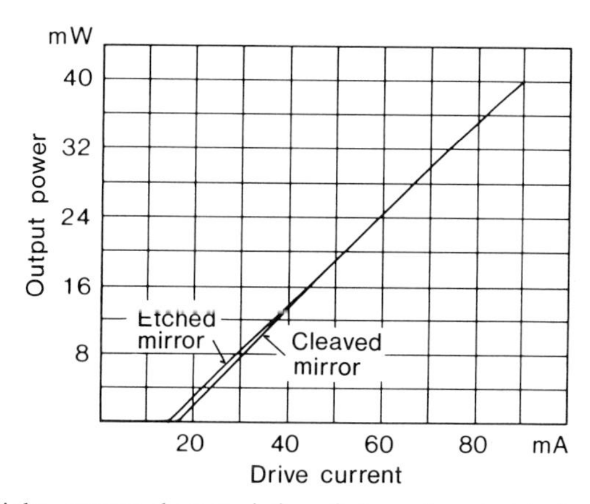

Fig. 10. CW light–current characteristics of cleaved- and etched-mirror lasers (uncoated).

mirrors are shown in in Fig. 11. The *P–I* curves are linear up to powers of more than 80 mW. Histograms of typical threshold current and front mirror efficiency on a 2-inch wafer with 1500 lasers are shown in Fig. 12.

Direct measures of the smoothness and verticality of the etched facets are the reflectivity (*R*) and scattering (*S*) coefficients that have been calculated from external quantum efficiencies of cleaved and etched mirror lasers [52]. Values of 0.33 and 0.02 for the reflectivity and scattering coefficients of flared waveguide etched mirrors are very close to those of ideally cleaved facets. As expected, the phase distortion caused by the irregular facets of straight waveguide devices results in poor (*S*, *R*) characteristics of these mirrors (Fig. 13).

The improvement of the mirror quality also shows up in the far-field intensity distribution. Straight waveguide etched-mirror lasers have multilobed or asymmetric far fields and Strehl ratios [53] that rarely exceed 0.85. Better beam quality is needed for applications requiring diffraction-limited focusing, such as optical storage. The lasers with the widened end sections have single-lobed and symmetric far fields up to high output power. The angular intensity distribution of the lateral (in junction plane) far field maintained the fundamental mode Gaussian beam profile up to power levels in excess of 50 mW. Vertical and horizontal beam angles $\theta_\perp$ and $\theta_\parallel$ of 25.5° and 9.6° full width at half-maximum (FWHM), respectively, were typically measured as shown in Fig. 14.

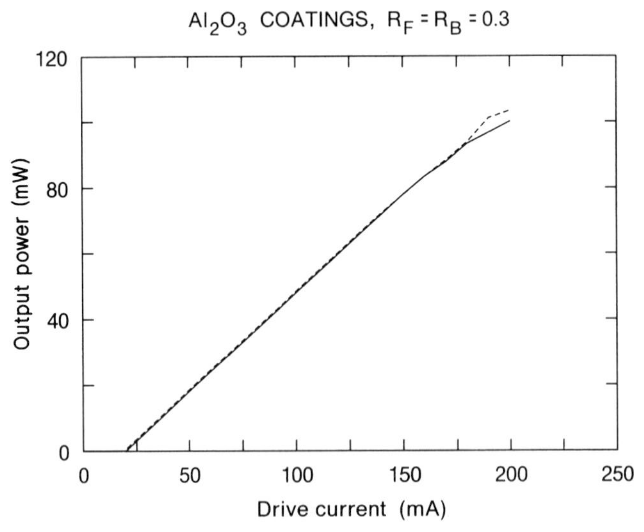

Fig. 11. CW light–current characteristics of two lasers with passivated etched mirrors. R_F and R_B are the reflectivities of the front and back mirrors, respectively.

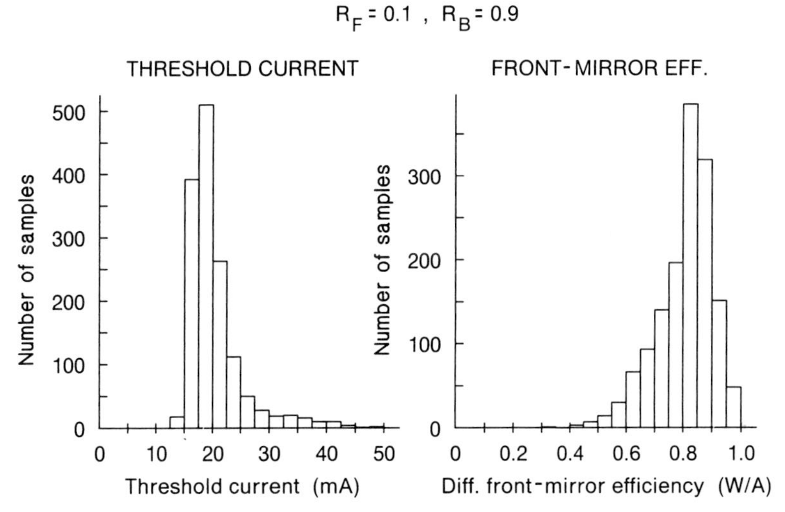

Fig. 12. Histograms of threshold and front mirror efficiency of reflectivity modified etched mirror lasers from a 2-inch wafer. (From Ref. 4. © 1991 IEEE)

For accurate measurements of the beam quality, the laser chips were cleaved close to the etched facets to avoid interference from light reflected at the bottom of the etched mirror trenches. To investigate the phase properties of the laser far field, the beam was collimated with a multielement lens of 6.5-mm focal length and a numerical aperture (NA) of 0.62 and then coupled into a Mach–Zehnder scanning interferometer with a phase resolution of $\simeq\lambda/100$. The phase distribution in the beam at 40 mW is shown in Fig. 15a. A very low root-mean-square (rms) value of the phase variation across

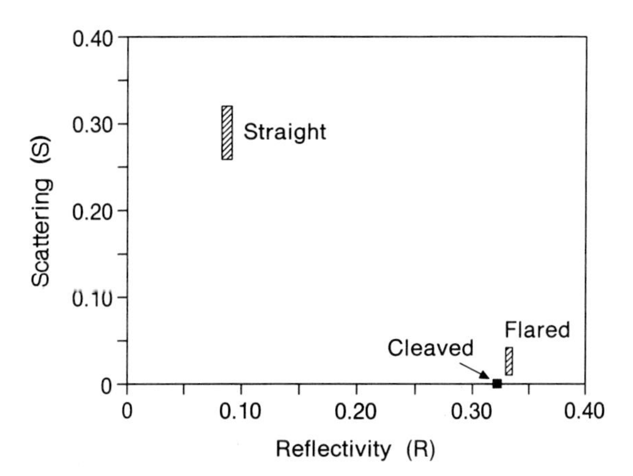

Fig. 13. Reflectivity and scattering coefficients of etched mirrors. (From Ref. 4. © 1991 IEEE.)

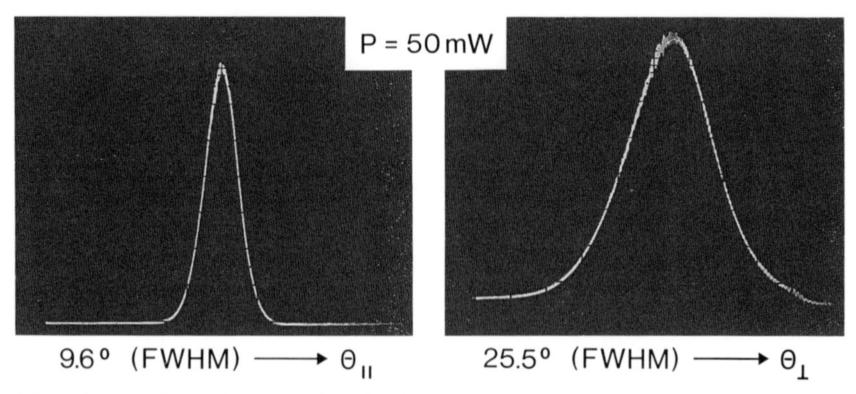

Fig. 14. Horizontal and vertical far fields of etched-mirror lasers. (From Ref. 4. © 1991 IEEE.)

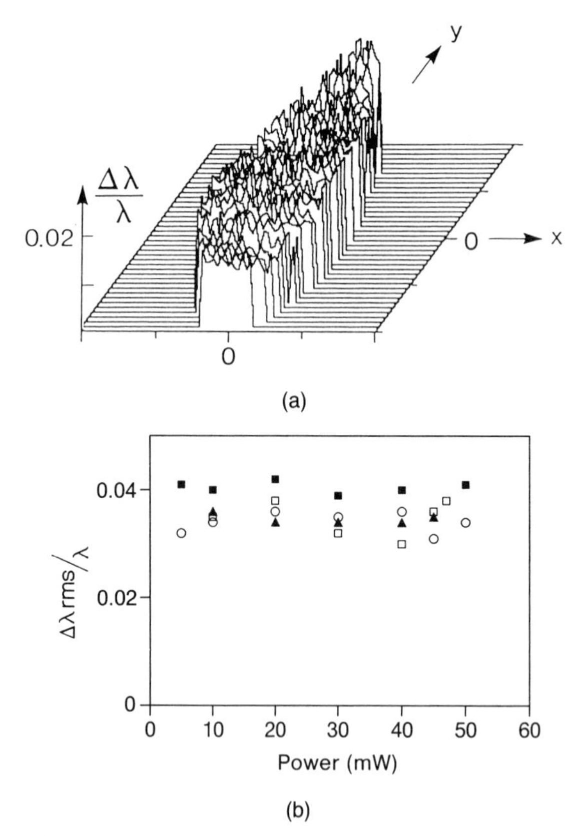

Fig. 15. (a) Two-dimensional representation of the wave front across the collimated beam of an etched-mirror laser at 40 mW. (b) Average spatial phase variation $\Delta\lambda_{rms}/\lambda$ versus optical output power for four arbitrarily selected lasers. (From Ref. 4. © 1991 IEEE.)

the beam (including contributions from the collimator lens) ranging between 0.03λ and 0.04λ was typically found. The observed very high beam quality as well as the low beam astigmatism changed little with output power, as demonstrated in Fig. 15b for a set of four lasers from different runs.

The measured average phase variations correspond to Strehl ratios higher than 0.94, which is the same as that measured for cleaved-mirror lasers, demonstrating that these etched-mirror lasers have the same beam quality as their cleaved counterparts.

Low degradation rates of the order of $\sim 2 \times 10^{-5}$ hr^{-1} have been reported for cleaved lasers passivated with Al_2O_3 as described in Section 2 [12]. By reducing contamination and ion damage on the facet surface before IBD passivation, we have achieved degradation rates between 10^{-5} and 10^{-6} h^{-1} for single-mode etched-mirror lasers at output power levels of 30 and 40 mW at 50°C. Several lasers have been tested at these conditions for more than 20,000 hours. Typical COD powers of these lasers are about 80 mW and values up to 120 mW have been obtained. Even better reliability is expected for *in situ* processing, with mirror etching and coating performed in the same vacuum system [54].

6. FULL-WAFER TESTING

6.1. Introduction

Full-wafer testing has been instrumental to the enormous progress made in VLSI electronic circuit technology. The same degree of progress is foreseen for FWT of optoelectronic devices, as demonstrated for laser diodes. Etched-mirror technology makes it possible to encompass part screening, wafer and process characterization, design evaluation, and process diagnostic functions with automated electrical and electro-optical measurements at the wafer level. This is done using a collection of integrated devices and components that are fabricated along with the production parts and designed using the same ground rules. This *monolithic integration* of all auxiliary test devices is important to the methodology of FWT as it precludes the introduction of hidden extraneous variables.

The major motivations behind the development of FWT are the process, design, and device characterization for rapid feedback of results by fast automated data acquisition. In addition, wafer and process characterization can substantially broaden the scope of testing, because FWP technology permits the integration of devices serving those specific tasks. Hence, the resulting documentation is more comprehensive but also more efficiently attainable. This is particularly valuable for process control and monitoring. The primary goal, however, is on-wafer part screening. The automated data acquisition saves time and costs and eliminates yield detractors caused by handling damage. It is particularly attractive when done after full-wafer burn-in.

6.2. Full-Wafer Chip Layout

The wafer layout for FWPT is governed by two objectives: to facilitate the automated data acquisition and to simplify the construction of new wafer designs. This is made possible by using a modular arrangement in which all devices with a particular function are collected on a module and all modules are of equal size and have the same footprint of contact pads. This allows a layout on a regular grid on which the elements of new wafer designs can easily be assembled from a library of existing modules.

The fully processed wafer in Fig. 16 illustrates this approach. It contains a collection of production-part modules arranged in vertical columns. They are interleaved by different test modules used for material and process characterization or diagnostic purposes. There are also auxiliary modules used for mask alignment, end-point detection, and device calibration. The combination of part modules and test modules reflects an early stage of development of this wafer design. As process yield improves, the combination of test and production modules will be changed in favor of the production parts to improve fabrication throughput.

We chose a 1.5-mm-square module having a column of 10 contact pads on each side. During testing, the wafer prober makes contact with the pads with a 20-pin probe card. After the wafer has been diced into modules, the same contact pads are used for wire bonding in the package. The peripheral pad configuration is particularly well suited for chips with several I/O connections, such as the characterization modules, or for laser arrays used in parallel optical recording as well as for designs that incorporate various degrees of optoelectronic integration.

6.3. Design of Production Parts

The SEM photograph in the upper left-hand corner of Fig. 16 shows a production part module carrying six diode lasers with integrated monitor diodes, which may be used to control the optical output power of the laser. The spacings between the lasers within the array are 140 μm. The module is configured to be packaged as chips with all six lasers and monitor diodes

Fig. 16. Fully processed 2-inch wafer comprising more than six hundred 1.5-mm^2 chips with over 5000 lasers. (From Ref. 4. © 1991 IEEE.)

connected to the package via bonding wires from the peripheral chip pads, as shown by the packaged laser array in Fig. 17. There is a second *p*-contact to the laser for measuring its $V(I)$ response unaffected by contact and lead resistance.

The monitor diode is configured like the laser but with a wider ridge and a cavity that is 100 μm long, sufficient to provide maximum sensitivity. Its front facet is tilted to eliminate adverse reflected-light interference, as shown in the SEM photograph in the upper right-hand corner of Fig. 16. Because

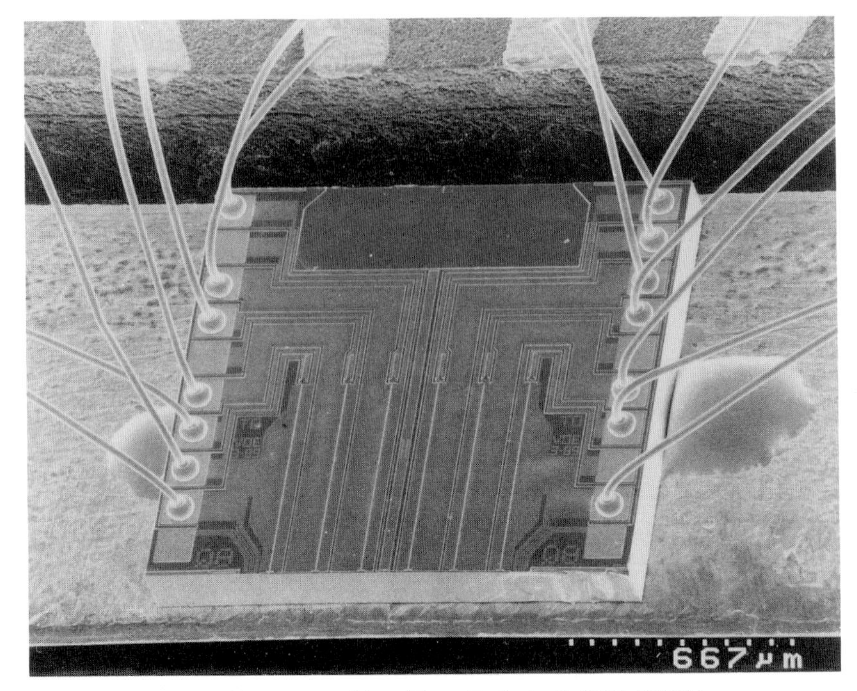

Fig. 17. Packaged laser array chip. (From Ref. 4. © 1991 IEEE.)

the monitor diode is reverse-biased in operation, whereas the the laser is forward-biased, it is necessary to insulate the monitor and laser diode circuits electrically from each other. Such insulation is provided by extending the mirror trenches so that they completely surround the devices, including their contact pads. This is necessary because the fabrication process does not allow leads to cross a mirror trench. Similar trenches are employed to insulate individual circuits as well as to provide thermal and optical isolation between adjacent components.

6.4. Testing of Laser Diodes

The integrated monitor diode allows the laser's $V(I)$ and the back-facet $P(I)$ response to be measured. This allows some characterization of the laser's impedance characteristics and optical output behavior, including a check of modal stability. The modal sensitivity is due to the tilted facet and limited detection volume of the monitor diode, which senses the diverging laser beam with a spatially nonuniform sensitivity. Changes of modal content are readily discernible from the differentiated $P(I)$ characteristics.

Such measurements, however, do not allow an evaluation of the spatial characteristics of the output beam such as beam divergence or astigmatism, nor do they permit sensitivity calibration of the monitor diode. Therefore, the scope of FWT would be rather limited if certain provisions were not added.

Figure 18 illustrates the basic concept of FWT, which makes use of integrated detector diodes for threshold and uniformity measurements and of the so-called reflecting chutes for optical beam characterization via an external linear diode array. This reflecting chute is a key element of comprehensive FWT and is simply a wide mirror trench which faces the laser's front facet. The bottom of this trench acts as a horizontal mirror which deflects the lower half of the diverging laser beam into a combined beam that projects at a shallow angle over the wafer surface. For typical beam divergences, the chutes must be about 200 μm long and 40 μm wide. They may be interleaved between adjacent lasers to conserve wafer space. A laser with a flared front mirror and a reflecting chute section is shown in Fig. 19.

The reflectivity of the chutes is quite high, given the shallow angle of incidence. This configuration is known as *Lloyd's mirror*, and it effectively images the laser's near field symmetrically below the mirror plane [55]. The chute mirror is about 4 μm below the beam axis, and the resulting far field looks as though it originated from two sources 8 μm apart. Figure 20 shows the horizontal and vertical beam profiles detected by the 2000-element linear diode array of the tester. While the horizontal profile is not affected by the reflecting chute, the vertical one exhibits the expected interference between the direct and the reflected beam segments [56]. The intrinsic beam profile is obtainable via a deconvolution of the measured profile.

6.5. Full-Wafer Part Screening

After completion of FWP, the wafers are tested with an automated procedure comprising a collection of parametric measurements. The results are then

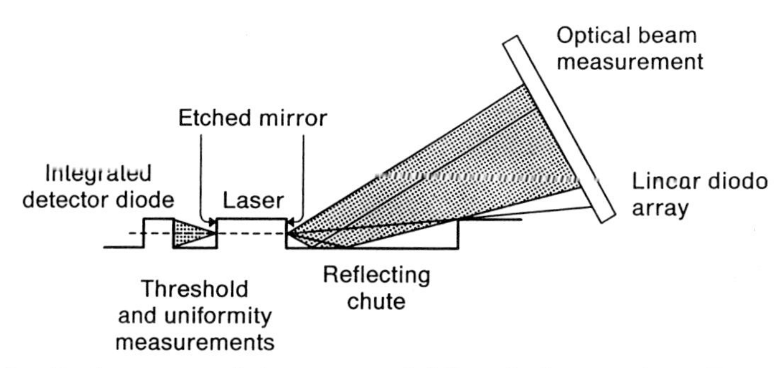

Fig. 18. Basic concept of the automated full-wafer laser testing. (From Ref. 4. © 1991 IEEE.)

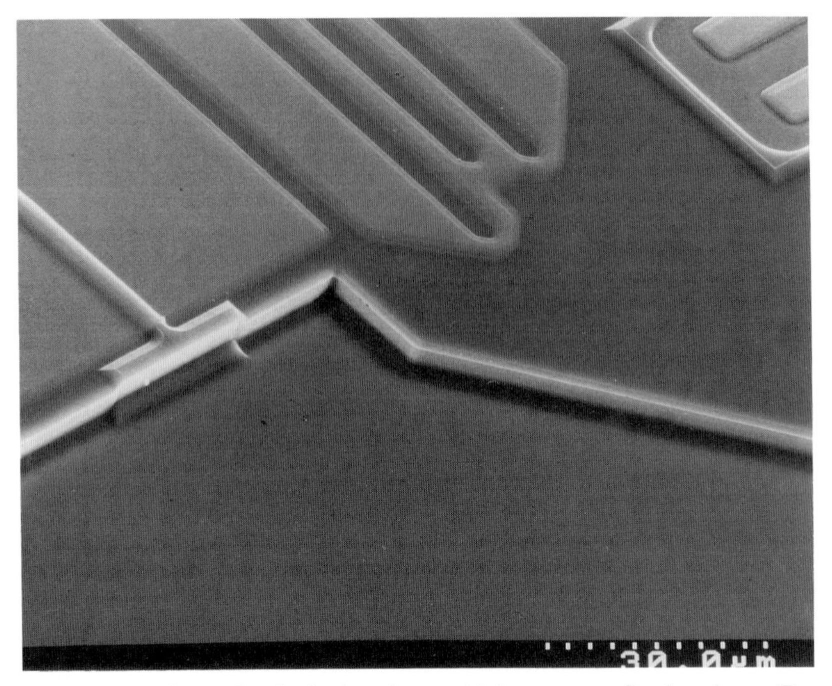

Fig. 19. Front facet of etched-mirror laser with large-area reflecting chute. (From Ref. 4. © 1991 IEEE.)

compared against the set of screening criteria and the parts labeled accordingly. A typical testing routine includes measurements of all important electro-optical parameters such as the $P(I)$ and $V(I)$ characteristics from which the threshold current, voltage, power, efficiency, and linearity are calculated. The $P(I)$ data are least-squares fitted to a polynomial of degree 2, and the results are used to control the subsequent testing routines. They comprise a far-field and spectral characterization made at several power levels. The far-field profiles, measured with an external linear diode array, are then fitted to a Gaussian and the FWHM divergence angle, ripple, and skew parameters are calculated.

6.6. Wafer and Process Characterization

Although the part-screening routine provides yield statistics, it does not reveal the causes of yield loss. Manufacturing often employs concurrently processed monitor wafers to evaluate critical fabrication steps. This approach

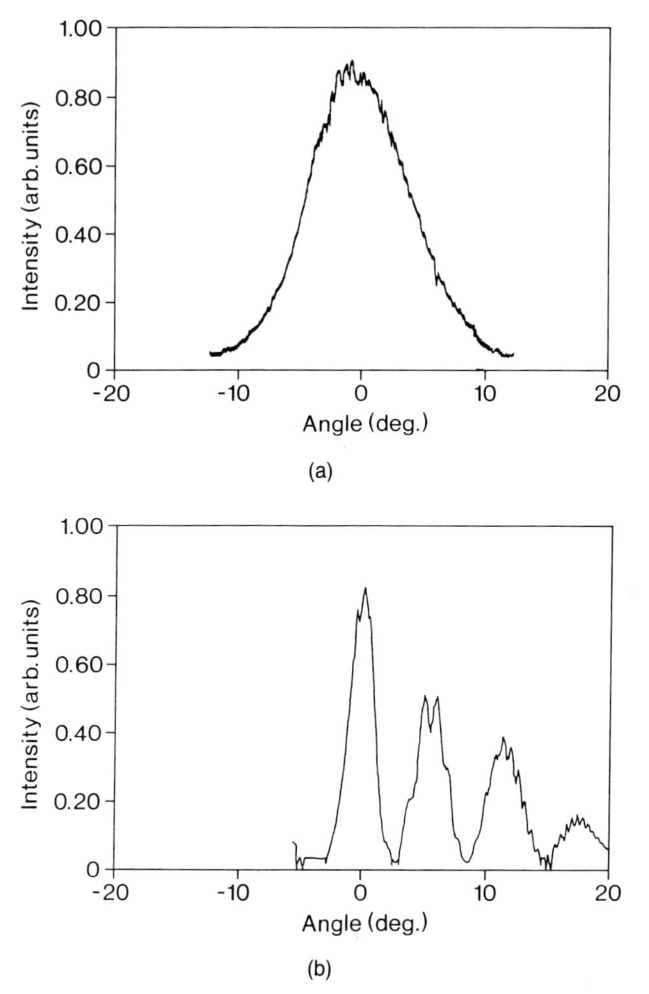

Fig. 20. On-wafer measurement of the far field intensity distribution measured via reflecting chute: (a) horizontal (parallel junction plane) and (b) vertical, showing interference effects.

frequently suffers from the unequal exposure history of the monitor wafers and from the problem of incorporating the nonroutine analytical results into the data base.

FWT avoids these shortcomings by integrating the process-monitor test sites onto the same wafer as the lasers. There are two types of test sites: P/N modules, with current flow across the junction, and P/P modules, where

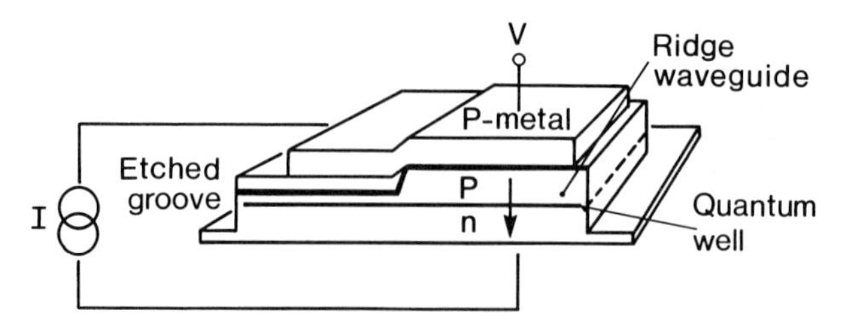

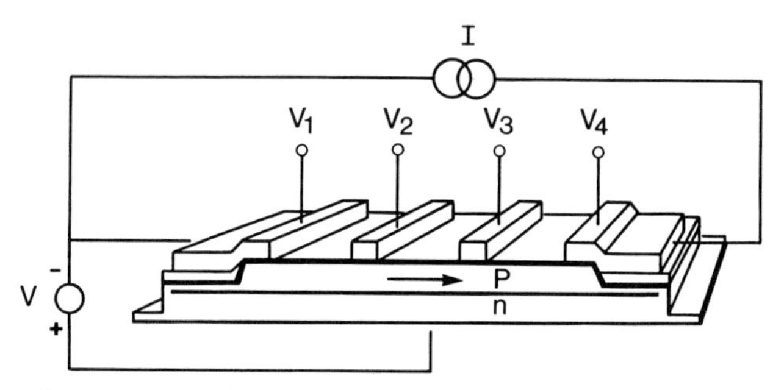

Fig. 21. P/N (top) and P/P (bottom) characterization modules. (From Ref. 4. © 1991 IEEE.)

the current flows parallel to the junction. Figure 21 shows the basic structures of the P/N (top) and P/P (bottom) modules.

6.7. P/N Characterization Modules

P/N modules are used in routine testing and for diagnostic purposes when a given critical design parameter is varied about the design point. Functional performance measurements then serve to evaluate process specifications versus a given design point, or vice versa.

A P/N module, as used in routine testing, contains a set of lasers whose lengths vary over one order of magnitude—an option uniquely provided by the FWP technology. It allows characterization of the electrical and optical properties of both the waveguide and the etched mirrors, such as internal efficiency, internal loss, and the mirror reflectivity ratio.

Because short lasers are more sensitive to the mirror characteristics, it has proved beneficial to include a module with a shortened cavity length to make an initial assessment of mirror-related issues (such as COD levels, burn-in dynamics, or long-term degradation) on a compressed time scale.

Very short lasers are also used to construct a device which permits mirror characterization via a simple electrical measurement. The device consists of two sites. One contains a large number of very short laser cavities (merely 8 μm long) that are electrically connected in parallel. The reference site contains only a few long lasers whose combined cavity length equals that on the first site. The measurement entails recording the forward and reverse-biased *I*(*V*) characteristics for each site. The difference between these currents is the additional current produced by the laser facets [27]. The results show that there is a marked increase of current density at the laser facet (which has been attributed to the bending of the laser's band structure near the facet and the presence of recombination sites at the mirror surface). It has been found that this facet current depends strongly on mirror roughness and surface contamination/oxidation, which can be characterized by a simple electrical measurement with this device. Figure 22 shows a section of the 8-μm-long cavity laser array, which has been used successfully to optimize etching and mirror passivation processes. It also demonstrates the potential of FWP to fabricate short-cavity lasers.

Fig. 22. Test site for mirror recombination current measurement. (From Ref. 4. © 1991 IEEE.)

6.8. P/P Characterization Modules

The P/P modules serve to characterize fabrication parameters such as resistivities, processing windages, and alignment errors. Their measuring principle generally takes advantage of the fact that the sheet resistances of the various topological levels (attainable via all masking permutations) differ by several orders of magnitude. They range from the mirror trench, which serves as an electrical insulator, to the metallization, which serves to interconnect the various resistive components constructed from suitable areas on the other two levels, namely the top and the base of the ridge.

Conceptually the simplest characterization circuit is a rectangle on some level. If the ratio of the sides of a rectangle is known, its sheet resistance can be determined—or, if the latter is known, the former can be derived. The example serves to show that the problem typically contains more than one unknown. One possible solution is to use the parameter obtained from one site as input in the evaluation of another parameter at the next site. This method suffers from propagation of errors, such as those caused by fluctuating probing resistances.

The problems of error propagation and contact resistance can be largely alleviated by constructing the measuring circuits in a bridge configuration so that the input impedance is a function of one parameter while the differential output is a function of the other, preferably the more critical one, since the differential measurement has much greater sensitivity and is impervious to contact resistance.

Such bridge circuits are used throughout the P/P modules. The construction of the branch circuits depends on the parameter to be measured, but the design objectives are identical: to devise a configuration of elements such that the counter branches are affected in an opposite way by the parameter of interest but are equally affected by all extraneous parameters. So far, this common-mode rejection goal has been achieved in most—but not all—designs.

Parameters routinely determined in this fashion include processing windages, sheet resistances of the fabricated layers, the ridge topography, the thinning of the conductor at the ridge slope, mask alignment errors, and the ridge top–metal contact resistance.

6.9. The Full-Wafer Test System

The present test system was designed to serve two somewhat different applications: production testing and part development. Whereas the former demands efficiency, the latter demands accuracy and flexibility. To meet these requirements, we chose a commercial semiconductor characterization system controlled by a workstation. It is interfaced to a 48-pin matrix switch connected to a standard automated wafer-probing station suitable for attaching the components of the optical characterization system. The $P(I)$ measurement is made with a photodiode large enough to collect the entire diverging laser beam.

For beam characterization, an optical spectrum analyzer and a 2000-element linear diode array were added to the parametric test system via an IEEE 488 interface. This allows simultaneous spectral and far-field measurements.

The optical spectrum analyzer is fiber-coupled to the laser. Physically, this fiber is located directly above the linear far-field array. The two measurements can then be overlapped, reducing test time. With the multiple power measurement capability, spectral shifts and mode content can be measured as a function of power. The overall block diagram of the full-wafer test system is shown in Fig. 23.

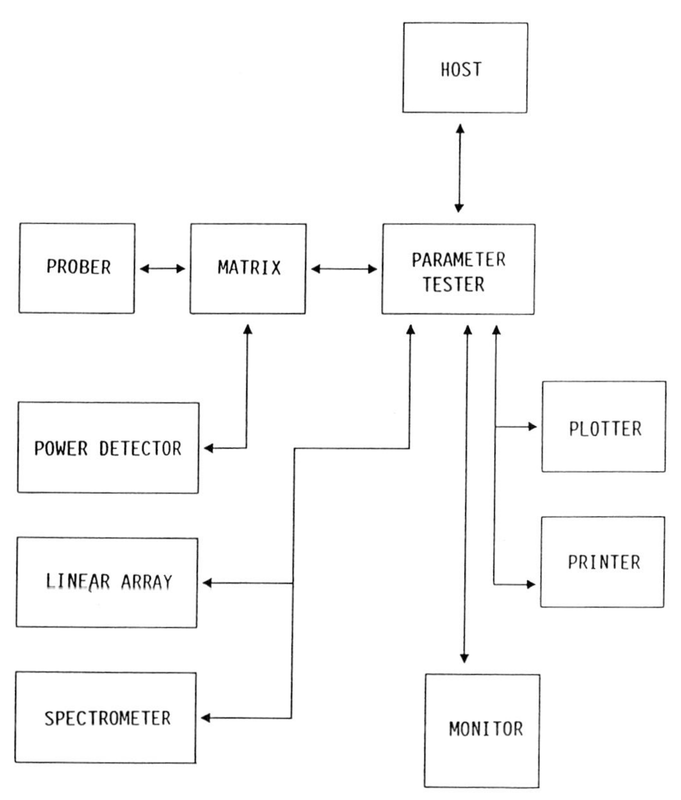

Fig. 23. Block diagram of full-wafer test system. (From Ref. 4. © 1991 IEEE.)

6.10. Data Analysis and Reports

Large volumes of data are generated for each wafer, which are analyzed for uniformity across the wafer, and plots such as the wafer map shown in Fig. 24 can be generated quickly on the workstation. If nonuniformities are detected, these changes can be correlated with the data from test sites and reported to process engineers for process modification or equipment maintenance.

Wafer histograms can be made and these data together with data from previous lots that have been added to control charts can be used to monitor production processes as a function of time. In the development areas, the data can be used to evaluate various process variations for their effectiveness.

7. SUMMARY

A new concept for full-wafer processing and testing for semiconductor laser fabrication in the AlGaAs/GaAs material system has been demonstrated. The approach is based on chemically assisted ion-beam etching for the laser-mirror formation. The technique routinely provides excellent mirror quality with mirror roughness of less than 20 nm, resulting in mirror reflectivities of about 30% and scattering losses of less than 2%. Lasers with two etched mirrors have been fabricated that have output power–drive current characteristics equivalent to those of lasers on the same wafer with both mirrors cleaved up to over 40 mW output power (CW) for uncoated mirrors. Single-mode operation exceeding 50 mW output power has been achieved for an SQW-GRINSCH ridge laser structure with coated, etched mirrors. The full-wafer technology is capable of producing AlGaAs/GaAs lasers for applications in optical storage (30–40 mW output power) with device performance close to that of standard cleaved lasers. This is mainly due to the CAIBE process and the flared waveguide concept, which provide smooth and flat mirrors. To our knowledge, record values for mirror scattering, optimum mirror reflectivity, and equivalence to cleaved mirrors in terms of laser threshold, efficiency, and maximum single-mode output power have been achieved. Promising results for uniformity and reproducibility of major laser diode properties on processed 2-inch wafers have been obtained.

Furthermore, the etching technique has been used to fabricate special devices and structures for on-wafer parametric laser and beam property characterization. The concept developed allows complete on-wafer testing of laser

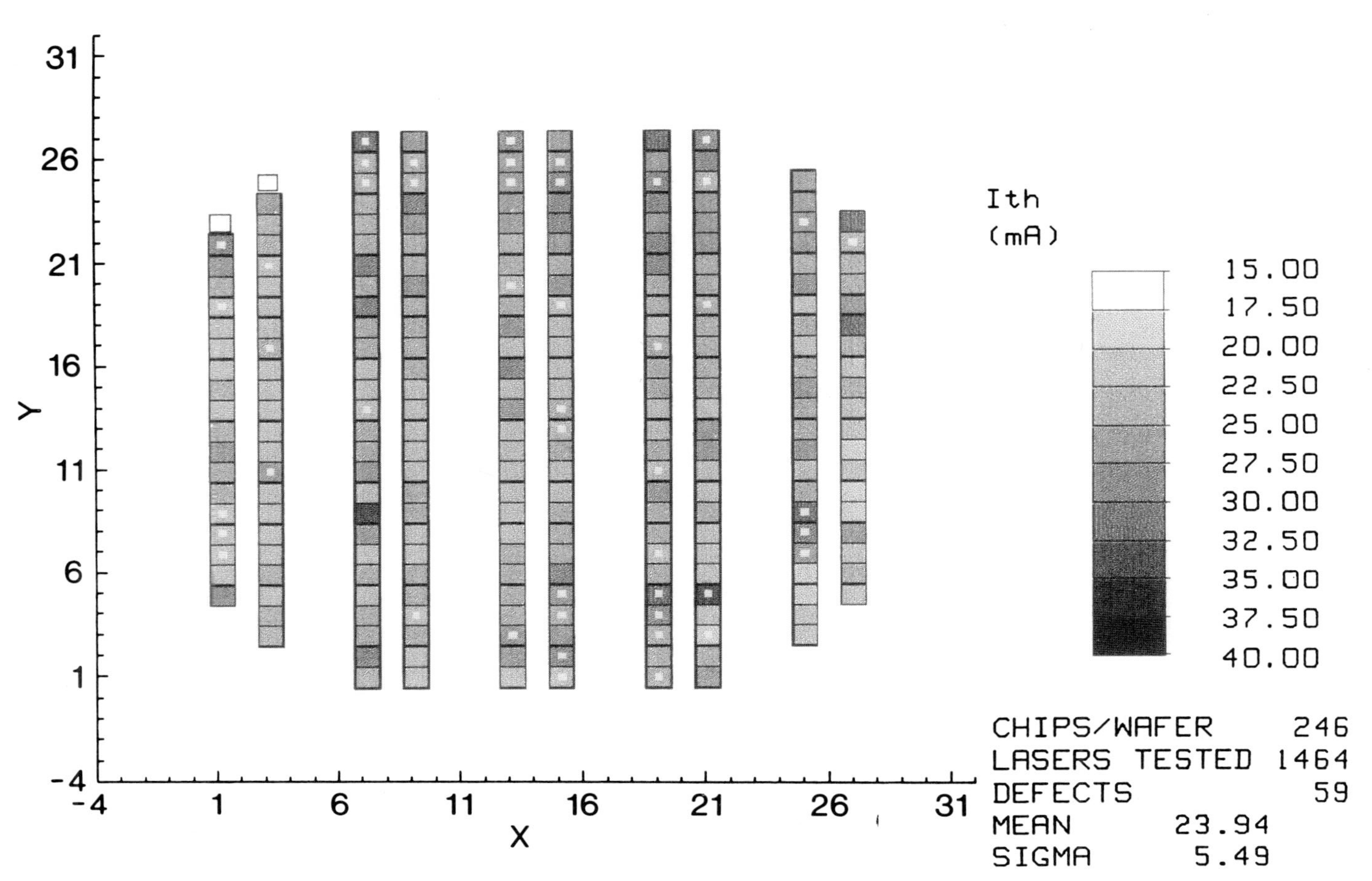

Fig. 24. Threshold current mapping of a 2-inch wafer with more than 1400 lasers tested.

diodes, including their beam properties, without any cleaving. The use of integrated detector diodes for uniformity measurements and reflecting chutes for beam diagnostic purposes provides efficient and fast laser testing. The concept also incorporates many test sites for process characterization which provide important feedback about process improvement/optimization such as sheet resistance, ridge dimensions, lithographic alignment errors, and mirror surface leakage. Considerable improvements in testing throughput in conjunction with a much broadened scope of testing make FWT a major achievement in large-scale laser fabrication. The approach described has been developed for lasers to be used in optical storage applications at wavelengths of 830 and 856 nm. However, the entire concept can be applied to semiconductor laser fabrication in any other material system, as well as for any other wavelength range and laser structure. The major difference will be the adaptation of the mirror-etching process to the composition of the material.

Figure 2 illustrates the major differences between cleaved (conventional) and full wafer technology. It demonstrates that many steps in processing and testing have to be performed on cleaved bars/chips with the conventional approach, whereas in the case of full-wafer technology the chips are cleaved just prior to packaging. Even on-wafer burn-in can be envisioned based on the full-wafer concept. FWP improves process yield and throughput by reducing bar and chip handling to an absolute minimum; it is now limited to the final packaging step. Full-wafer technology is changing semiconductor laser fabrication from a discrete-device to a VLSI-like technique.

Full-wafer technology has great potential for future integrated optoelectronics applications, since the use of high-quality etched mirrors eliminates previous restrictions on chip size and hence will allow the fabrication of optoelectronic integrated circuits. New functions will be made possible by shaping and tilting the mirrors to fulfill the specific requirements of a given application. Figure 25 summarizes the three major improvements offered by full-wafer technology, namely large-scale fabrication, testing, and its potential for future OEIC applications.

ACKNOWLEDGMENTS

The authors thank all members of the Laser Science and Technology department of the IBM Zurich Research Laboratory for their excellent and continuous support in epitaxy, processing, and testing. In particular, the contributions of K. Daetwyler, H.P. Dietrich, L. Perriard, and G. Sasso in processing; M. Benedict, N. Cahoon, H.K. Seitz, and P. Wolf in full-wafer

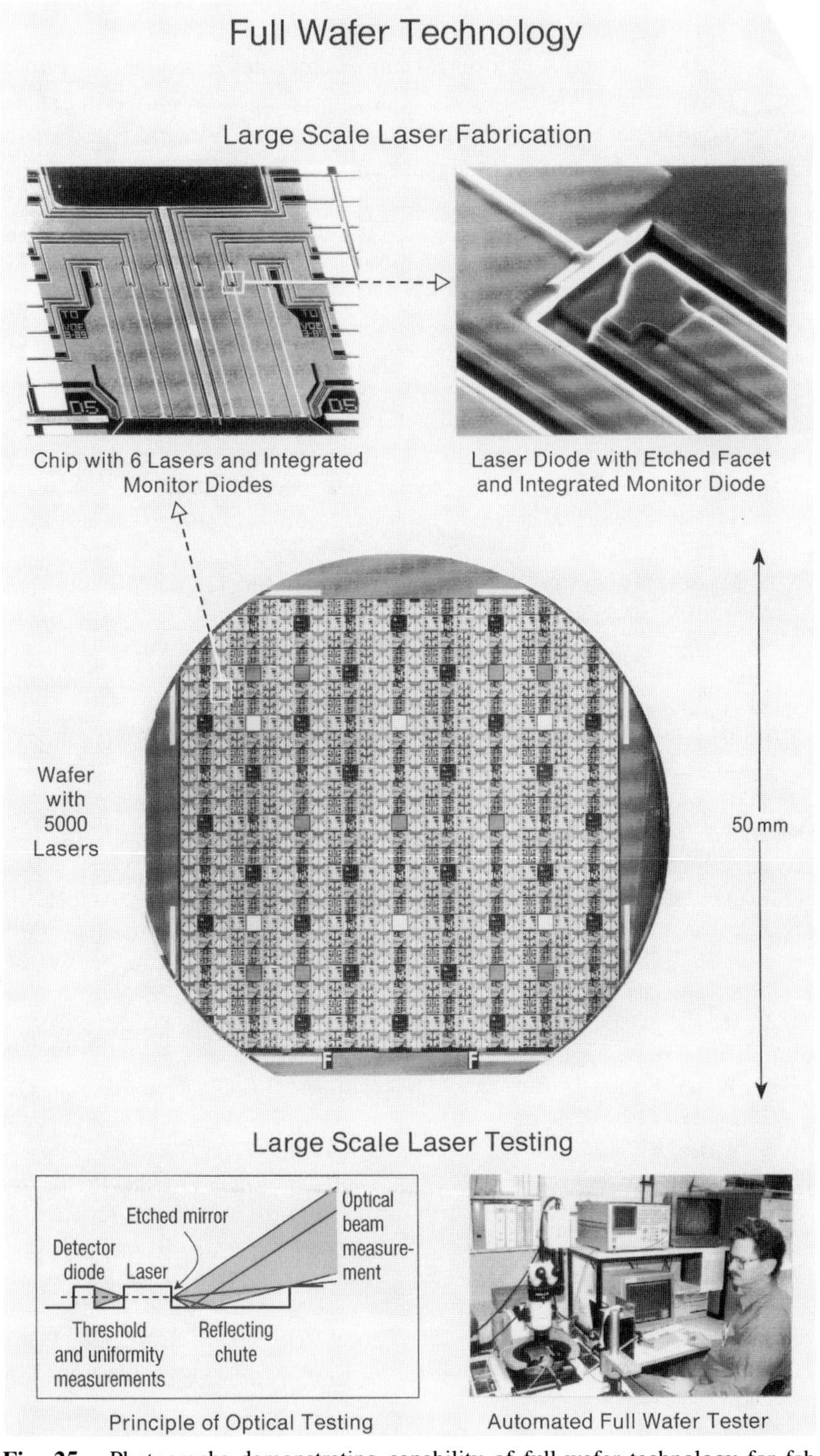

Fig. 25. Photographs demonstrating capability of full-wafer technology for fabrication, testing, and integration.

testing; G.L. Bona and A. Moser for mirror characterization; and T. Forster and A. Oosenbrug in life testing are highly appreciated. We are also grateful for the continuous encouragement and support from V. Graf, head of the Laser Science and Technology department.

REFERENCES

1. Y. Suematsu and S. Arai, *Proc. IEEE* **75**(11), 1472 (1987).
2. S. R. Forrest, *Proc. IEEE* **75**(11), 1488 (1987).
3. H. Matsueda, *J. Lightwave Technol.* **LT-5**(10), 1382 (1987).
4. P. Vettiger, M. K. Benedict, G. L. Bona, P. Buchmann, N. Cahoon, K. Dätwyler, H. P. Dietrich, A. Moser, H. K. Seitz, O. Voegeli, D.J. Webb, and P. Wolf, *IEEE J. Quantum Electron.* **27**(6), 1319 (1991); see also **28**(1), 387 (1992).
5. S. M. Sze, "Physics of Semiconductor Devices," Ch. 12, p. 681. Wiley, New York, 1981.
6. G. H. B. Thompson, "Physics of Semiconductor Lasers." Wiley, Chichester, 1980.
7. W. T. Tsang, *Appl. Phys. Lett.* **39**(2), 134 (1981).
8. H. Kawaguchi and T. Kawakami, *IEEE J. Quantum Electron.* **QE-13**, 556 (1977).
9. C. Harder, P. Buchmann, and H. P. Meier, *Electron. Lett.* **22**(20), 1081 (1986).
10. H. Jaeckel, G. L. Bona, P. Buchmann, H. P. Meier, P. Vettiger, W. J. Kozlovsky, and W. Lenth, *IEEE J. Quantum Electron.* **27**(6), 1560 (1991).
11. I. Ladany, M. Ettenberg, H. F. Lockwood, and H. Kressel, *Appl. Phys. Lett.* **30**(2), 87 (1977).
12. F. R. Gfeller and D. J. Webb, *J. Appl. Phys.* **68**(1), 14 (1990).
13. E. Hecht, "Optics." Addison-Wesley, Reading, MA, 1987.
14. M. Ettenberg, *Appl. Phys. Lett.* **32**(11), 724 (1978).
15. O. Wada, S. Yamakoshi, and T. Sakurai, *IEEE J. Quantum Electron.* **QE-20**(2), 126 (1984).
16. S. Wang, *IEEE J. Quantum Electron.* **QE-10**(4), 413 (1974).
17. D. J. Stirland and B. W. Strughan, *Thin Solid Films* **31**, 139 (1976).
18. M. Wada, K. Hamada, T. Shibutani, H. Shimizu, M. Kume, K. Itoh, G. Kano, and I. Teramoto, *IEEE J. Quantum Electron.* **QE-21**, 658 (1985).
19. N. Bouadma, S. Grosmaire, and F. Brillouet, *Electron. Lett.* **23**(16), 855 (1987).
20. J. J. Yang, M. Jansen, and M. Sergant, *Electron. Lett.* **22**(8), 438 (1986).
21. C. L. Shieh, J. Mantz, K. Alavi, and R. W. Engelmann, *Electron. Lett.* **24**(6), 343 (1988).

22. J. Puretz, R. K. DeFreez, R. A. Elliott, and J. Orloff, *Electron. Lett.* **22**(13), 700 (1986).
23. N. Takado, K. Asakawa, T. Yuasa, S. Sugata, E. Miyauchi, H. Hashimoto, and M. Ishii, *Appl. Phys. Lett.* **50**(26), 1891 (1987).
24. L.A. Coldren, K. Iga, B. I. Miller, and J. A. Rentschler, *Appl. Phys. Lett.* **37**(8), 681 (1980).
25. M. A. Bösch, L. A. Coldren, and E. Good, *Appl. Phys. Lett.* **38**(4), 264 (1981).
26. S. W. Pang, W. D. Goodhue, T. M. Lyszczarz, D. J. Ehrlich, R. B. Goodman, and G. D. Johnson, *J. Vac. Sci. Technol. B* **6**(6), 1916 (1988).
27. A. Scherer, H. G. Craighead, M. L. Roukes, and J. P. Harbison, *J. Vac. Sci. Technol. B* **6**(1), 277 (1988).
28. E. L. Hu and R. E. Howard, *Appl. Phys. Lett.* **37**(11), 1022 (1980).
29. G. A. Vawter, L. A. Coldren, J. L. Merz, and E. L. Hu, *Appl. Phys. Lett.* **51**(10), 719 (1987).
30. T. Takamori, L. A. Coldren, and J. L. Merz, *Appl. Phys. Lett.* **55**(11), 1053 (1989).
31. T. Yuasa, T. Yamada, K. Asakawa, S. Sugata, M. Ishii, and M. Uchida, *Appl. Phys. Lett.* **49**(16), 1007 (1986).
32. H. Nakano, S. Yamashita, T. P. Tanaka, M. Hirao, and M. Maeda, *J. Lightwave Technol.* **LT-4**(5), 574 (1986).
33. G. A. Lincoln, M. W. Geis, S. Pang, and N. N. Efremow, *J. Vac. Sci. Technol. B* **1**(4), 1043 (1983).
34. S. C. McNevin and G. E. Becker, *J. Appl. Phys.* **58**(12), 4670 (1985).
35. M. Balooch, D. R. Olander, and W. J. Siekhaus, *J. Vac. Sci. Technol.* **B4**, 794 (1986).
36. N. Furuhata, H. Miyamoto, A. Okamoto, and K. Ohata, *J. Electron. Mater.* **19**(2), 201 (1990).
37. J. A. Skidmore, L. A. Coldren, E. L. Hu, J. L. Merz, and K. Asakawa, *J. Vac. Sci. Technol.* **B6**, 1885 (1988).
38. D. G. Lishan and E. L. Hu, *Appl. Phys. Lett.* **56**(17), 1667 (1990).
39. K. Yamada, *Proc. Symp. GaAs and Related Compounds,* Japan (1989), Inst. Phys. Conf. Ser. No. 106, Ch. 8, p. 599.
40. T. H. Windhorn and W. D. Goodhue, *Appl. Phys. Lett.* **48**(24), 1675 (1986).
41. J. P. Donnelly, W. D. Goodhue, T. H. Windhorn, R. J. Bailey, and S. A. Lambert, *Appl. Phys. Lett.* **51**(15), 1138 (1987)
42. P. Tihanyi, D. K. Wagner, A. J. Roza, H. J. Vollmer, C. M. Harding, R. J. Davis, and E. D. Wolf, *Appl. Phys. Lett.* **50**(23), 1640 (1987).
43. P. Buchmann, H. P. Dietrich, G. Sasso, and P. Vettiger, *Proc. Int Conf. Microlithography,* Vienna (September 1988), p. 485.
44. K. Asakawa and S. Sugata, *J. Vac. Sci. Technol. B* **3**(1), 402 (1985).
45. P. Buchmann, G. L. Bona, N. Cahoon, K. Dätwyler, A. Moser, P. Vettiger, O. Vögeli, and D. Webb, *Proc. IEEE/LEOS Annual Meeting,* Orlando (October 1989), p. 480.

46. G. L. Bona, P. Buchmann, R. Clauberg, H. Jäckel, P. Vettiger, O. Vögeli, and D. J. Webb, *IEEE Photon. Technol. Lett.* **3**(5), 412 (1991); see also **3**(12), 1155 (1991).
47. K. Iga, Y. Mori, and Y. Kotaki, *Bull. P.M.E. (T.I.T.)* (58), 17 (1986).
48. M. Vasari, C. Misano, and L. Lasaponara, *Thin Solid Films* **117**, 163 (1984).
49. D. J. Webb, H.-P. Dietrich, F. R. Gfeller, A. Moser, and P. Vettiger, *Proc. Materials Research Society Fall Meeting,* Boston (1988) (L.E. Rehn, J. Greene, and F.A. Schmidt, eds.), Vol. 128, p. 507. Materials Research Society, Pittsburgh, 1989.
50. H. K. Pulker, *Appl. Optic.* **18**(12), 1969 (1979).
51. A. Moser, E.-E. Latta, and D. J. Webb, *Appl. Phys. Lett.* **55**(12), 1152 (1989).
52. H. Saito, Y. Noguchi, and H. Nagai, *Electron. Lett.* **22**(22), 1157 (1986).
53. M. Born and E. Wolf, "Principles of Optics," p. 461. Pergamon, New York, 1980.
54. M. Uchida, S. Ishikawa, N. Takado, and K. Asakawa, *IEEE J. Quantum Electron.* **QE-24**(11), 2170 (1988).
55. M. Born and E. Wolf, "Principles of Optics," p. 262. Pergamon, New York, 1980.
56. J. Salzman and A. Yariv , *Appl. Phys. Lett.* **49**(8), 440 (1986).

Chapter 8

EPITAXIAL LIFT-OFF AND RELATED TECHNIQUES

Winston K. Chan and Eli Yablonovitch*

Bellcore
Red Bank, New Jersey

1. INTRODUCTION

Optical systems inevitably need electronic as well as optical components. A major challenge is to combine electronic and optical functions in a way that will obtain the best performance (which includes reliability) for the lowest cost. There is a spectrum of methods to try to accomplish this (Fig. 1). At one end is the purely hybrid approach of assembling the various components with solders, wire bonds, epoxies, etc. This approach has been highly successful in achieving high performance, particularly at lower frequencies where parasitics are not a limiting factor, but it is costly because the assembly is done serially—one module at a time—and often the results depend on the

*Now at the Department of Electrical Engineering, University of California at Los Angeles, Los Angeles, California.

ISBN 0-12-200420-5

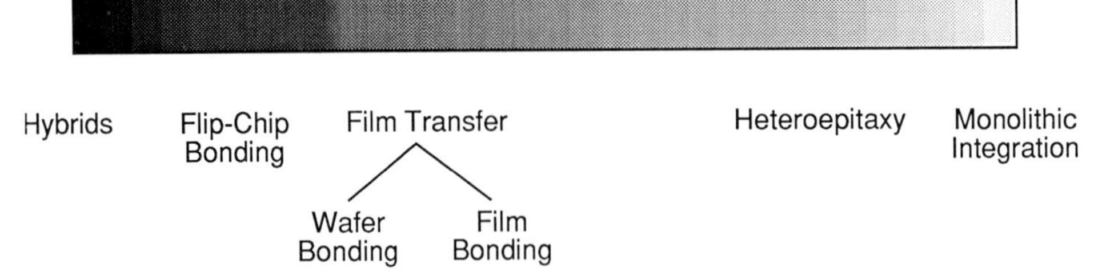

Fig. 1. Spectrum of the various approaches to optoelectronic integration. Monolithic integration is at one end and hybrids are at the opposite end. In the middle are different approaches that try to address the weaknesses of monolithic integration and hybrids.

skill of the assembler. At the other end of the spectrum, following the development of electronic integrated circuit technology, considerable effort has been expended in the monolithic integration of electronic and optical devices. The strength of this approach is that many circuits are made in parallel, in a highly reproducible manner, and with greatly reduced electrical parasitics by employing thin-film fabrication techniques commonly used in making integrated circuits. However, a drawback of monolithic integration is that one is forced to realize all the functions in a semiconductor material system that is dictated primarily by the wavelength of light to be used. Not every device made with the chosen material system would be the best possible for the application, and consequently the overall performance of the integrated circuit may be worse than that of the corresponding hybrid where one has the freedom to choose the best type of device for each function. Between these two extremes are various methods that address these fundamental problems. One of the goals of heteroepitaxy, such as the growth of GaAs on Si, is to enable the monolithic integration of GaAs optical devices with Si electronics so one can have the best of both technologies on a single substrate. In spite of the advances made in this area over the last decade, there remain serious questions on material quality and there are many material combinations of technological interest, such as InP on $LiNbO_3$, that cannot be realized with the current state of the art.

Closer to the hybrid end of the spectrum are several related methods referred to here collectively as film transfer. They have the common theme of bonding a thin device film to a host substrate to avoid growing the film directly on the host, but they differ in the details of how this is accomplished. The film is often less than a few micrometers thick, in some cases comprising only the necessary device layers, and is attached with at most a thin layer of solder or adhesive. The result is a structure that is sufficiently planar for thin-film fabrication techniques to delineate small features with accurate alignment and low electrical parasitics. The two essential steps of film transfer are separating the film from its growth substrate and bonding it to the host. If the film is separated before it is bonded, we have the film bonding techniques; but if the film is bonded before it is separated, we have the wafer bonding techniques. Device fabrication may occur before (preprocessing) or after (postprocessing) the transfer, resulting in four strategies that we will be discussing.

A final entry in the spectrum is flip-chip bonding [1], also known as solder bonding or C-4 (controlled collapse chip connection) processing. A completed chip is bonded to a substrate by small balls of solder, which usually double as electrical interconnects. The boundary between film transfer and flip-chip bonding is not sharp. Flip-chip bonding may be thought of as wafer bonding of preprocessed devices, with the bonds made only at selected areas. But with all of the interconnects made as an integral part of bonding, further processing is often unnecessary and the film separation step is superfluous.

2. TECHNIQUE

In this section, we will review film transfer techniques relevant to optoelectronics. We will then concentrate on epitaxial lift-off (ELO) because it is the most widely used method for optoelectronics. Rather than compare in detail variations that have evolved in different laboratories working on ELO, we will try give an overview and point out major variations.

2.1. Film Transfer Techniques

An early publication by Konagai *et al.* [2] on film bonding was the peeled film technology (PFT), in which a film was separated from its growth substrate by a highly selective etch of a sacrificial layer with hydrofluoric acid and attached to a host substrate in an unspecified way. This was based on a proposal by Milnes and Feucht [3]. Konagai *et al.* also proposed stacking many device and sacrificial layers to obtain many transferable films with a single growth. In ELO, which will described in greater detail in the following section, the device film also is separated (lifted off) from its growth substrate by highly selective chemical etches and then attached (grafted) to the host substrate by van der Waals forces [4, 5]. In another film bonding technique, cleavage of lateral epitaxial films for transfer (CLEFT) [6], the film is separated from the growth substrate by cleaving along the interface (parallel to the growth surface), which was weakened by narrow carbon stripes deposited before the start of growth. In film bonding, there is a point in which the fragile film is supported by neither the growth nor the host substrate. Film handling at this critical point is facilitated by a relatively thick wax or polymer film in PFT and ELO or by wafer bonding to a temporary substrate in CLEFT.

In the wafer bonding methods, wafers are bonded together before the device film is separated, bypassing the need to handle thin films. The device layer finishes in an upside-down position in the wafer bonding techniques. The first effort was by Stern and Woodall [7], who separated bonded GaAs laser diodes from their growth substrate by using hydrochloric acid to etch an aluminum-rich AlGaAs sacrificial layer. Now, the most widely used wafer bonding technique, known as direct bonding or bond-and-etchback silicon on insulator (BESOI), is the bonding of the thermal oxides on silicon wafers to obtain silicon-on-insulator (SOI) wafers [8–10]. The oxide layers of the mating wafers are fused together at temperatures as high as 1400°C, resulting in an interface as strong as fused quartz; lower temperatures can still yield strong bonds. One wafer is thinned either by mechanical lapping or chemical etching to a thickness suitable for device fabrication. This process has been adapted so that one of the wafers can be InP or GaAs [11, 12]. Low-temperature glass [13], metal [14, 15], and organic adhesive [16] have also been used to bond a growth wafer to a host substrate. Anodic or field-assisted bonding of GaAs to glass has been reported as well [17]. Two semiconductor wafers heated while in direct contact can fuse together by forming an alloy at the interface [18]. InP and GaAs wafers have been fused together with variations of this technique [19, 20].

2.2. Epitaxial Lift-off

The method used to remove the epitaxial films from their growth substrates depends on the material system. In the GaAs-AlAs system, a sacrificial layer of AlAs is undercut to separate the epitaxial film from the GaAs growth substrate. This method exploits the high selectivity of hydrofluoric acid in etching AlAs without etching GaAs. In the InP-$In_{0.53}Ga_{0.47}As$ system, the InP substrate selectively etches in hydrochloric acid with a layer of $In_{0.53}Ga_{0.47}As$ as an etch stop. Although a sacrificial layer of $In_{0.53}Ga_{0.47}As$ sandwiched between the InP substrate and an InP layer or a 5-nm-thick pseudomorphic AlAs layer could be used with a phosphoric acid–peroxide–water or hydrofluoric acid etch, respectively, it is more convenient to remove the substrate.

The most commonly employed ELO procedure begins by covering the top of the wafer with Apiezon W wax, making sure that no wax has run over the edge. If the device structure includes layers that will be attacked by the selective etch, they can be protected by etching the periphery of the wafer past these layers and covering the side walls with wax, or by etching mesas past these layers in the case of preprocessed devices. Next, the coated wafer is immersed in HF for GaAs or HCl for InP. Typical times are overnight for a centimeter-square GaAs sample and an hour for an InP sample of any size. Because of the high selectivity of these etches, the samples can be overetched considerably without any ill effects. When the film has separated from the substrate, it is rinsed in deionized water and transferred to the new substrate. It is important that the host substrate be flat for the film to adhere. When grafting to a silicon wafer on which circuits have been made, it is necessary to planarize the wafer first [21, 22]. In the basic ELO process, the film is free to slide on a thin layer of deionized water that was dragged out with it. When the film is correctly positioned, the water is gently squeezed out and the assembly is allowed to dry, often with a weight or spring load on top. As the water dries, surface tension from the water film pulls the film and the substrate closer together until short-range, attractive van der Waals forces hold them together. The wax can then be removed.

Cross-sectional transmission electron microscopy studies show there is an amorphous interfacial layer whose thickness can vary from zero to only a few tens of nanometers along the interface [5]. The composition of this layer probably includes the hydrated oxides of the film and its host.

2.3. Device Fabrication

There are two strategies for making devices on the transferred film. One is to preprocess the devices and transfer completed devices; the other is to postprocess the devices on a bare grafted film. Some processing, such as the deposition of an adhesion, contact, or passivation layer, may be done to the underside of the film in either case. The important distinction between pre- and postprocessing is that the film must be accurately aligned for bonding in preprocessing but need not be in postprocessing. Postprocessing is the only option in some wafer bonding methods because of the high temperatures needed for the bonding processes; moreover, in some cases, the preprocessed devices are upside-down after bonding and may require additional etch steps to be contacted from above.

Film transfer with preprocessed devices becomes very much like a hybrid technology, but with much thinner layers to enable thin-film technology to be used for low parasitic and high-density interconnects. Advantages of preprocessing over postprocessing are that one can use standard devices, perhaps from a foundry, that the processing of the grafted devices and the processing of the host devices do not interfere with each other, that one can graft devices tested to be good, and that little of the film is wasted if diced chips are grafted one at a time. Completed devices are thus lifted off and grafted to the new substrate, perhaps with devices already on it, and interconnected using one or more levels of metallization and interlevel dielectric. Because it is desirable to dice the film into chips and graft each chip independently, chip separation techniques compatible with ELO have been investigated [23, 24]. Without going to a direct-write lithography system that can be custom programmed for each grafted chip, the alignment of devices on the grafted film with those on the host substrate is of utmost importance. One may have large pads to ensure that interconnections can be made with a fixed mask in the presence of some misalignment error, but the penalty in excess capacitance or real estate consumption may offset any potential advantages of film transfer. Accurate placement of ELO chips is therefore currently being investigated: The lifted-off chips are temporarily grafted to a transparent polyimide film and their alignment to the host is done with a contact mask aligner to obtain 2-μm alignment accuracy [24]. Manipulators with motorized micrometers can give a few micrometers accuracy [23]. An ELO film has also been self-aligned to the substrate with 5-μm accuracy with the surface tension of water by pretreating the substrate so that it is hydrophilic in the region where the film is to attach and hydrophobic elsewhere [25].

Another concern is that stresses from materials deposited in making the preprocessed device may bow the ELO film to such an extent that it interferes with the film transfer [26]. Although low stress metallization and dielectric layers as well as stress-compensating layers can reduce the bowing, these may not always be practical. Moreover, stress can be present even in bare, unprocessed films because of slight changes in the lattice parameter through the film. For example, AlAs, often taken to be perfectly lattice matched to GaAs, is actually about 0.1% mismatched, so any asymmetric GaAs-AlGaAs heterostructure has an intrinsic stress.

The other approach is to postprocess the film, i.e., totally fabricate the devices on a bare grafted film. This obviously eliminates the need for precision alignment when bonding the film and has the added advantages that one may be able to compensate for changes arising from the different environment of the film (such as back surface depletion of a field-effect transistor) during the device fabrication and that one can intermix the processing steps of the host and ELO film devices to obtain a richer set of structures than by grafting preprocessed devices. As mentioned, postprocessing may be the only option for some of the wafer bonding methods. Although submicrometer device alignment is readily achieved with postprocessing, one faces the problems associated with device fabrication on grafted films, including film adhesion and elevated-temperature (>300°C) processing.

The succession of processing steps typical of optoelectronic device fabrication places a more stringent demand on film adhesion than that from the one or two steps required to interconnect preprocessed devices. In some applications, moreover, it is essential to remove all semiconductor between device mesas. With pure van der Waals bonded films, many mesas do not adhere once they are isolated [27]. The adhesion is enhanced by substituting the van der Waals bonds between the ELO film and substrate with a stronger bond. A thin layer of palladium [28] or indium [29] has been used to join the film to the host while making an *n*-type ohmic contact to the bottom of the film. Palladium bonding is particularly attractive because it occurs at room temperature. Gold [30] or a gold–zinc [24] alloy can likewise make a *p*-type contact to the bottom of the film. Organic glues and adhesives can be used but they need to be chosen carefully because, during cure, many of them shrink and release a reaction product, such as water, that can be trapped under the film. Also, most of them cannot tolerate high temperatures. Grafting to a thin, cured polyimide layer improves adhesion, particularly when the host is not flat [31]. Another method that has successfully enhanced adhesion of mesas is to "tape" them down by straddling the mesa with a material that adheres well both to it and to the substrate before completely separating the mesas [32].

The dominant effect of high-temperature processing is the formation of blisters and craters up to 0.5 mm in diameter that result from the vaporization of entrapped material, such as small particles or water [33, 34]. The density of trapped particles decreases when the processing is done in a clean-room environment. Most particles are introduced during transfer of the film from the etch solution to the rinse and from the rinse to the host substrate. An all-underwater process in which the etch solution is replaced by rinse water and all film handling is done completely within the rinse water greatly reduces the particle density [23]. A vacuum prebake is sometimes beneficial in removing trapped water, but a more practical solution is to etch the film into mesas before any high-temperature processing [35]. This uncovers the trapped material in area between mesas, limits any blistering to one mesa rather than to an extended area, and provides a short escape path (tens of micrometers) for vapors both during the high-temperature process itself and during a vacuum prebake.

3. APPLICATIONS

The types of applications reported for ELO and film transfer in general fall into three broad, although not mutually exclusive, categories. First are those in which ELO is a form of hybrid technology, and grafted devices are electrically interconnected with devices on the host substrate. Second are those in which some detrimental property of the growth substrate is eliminated by separating the growth substrate from the device layers and replacing the substrate with one that is more suitable. Finally, there are those in which the grafted film interacts with the host substrate or a device on the host substrate in a manner that cannot be mediated by an interconnect wire.

3.1. Interconnecting Devices

Much of the effort in using ELO as a hybrid technology has concentrated on grafting III–V optical devices to a silicon wafer where there can be complex electronics. Typical applications envisaged are optical interconnects between chips, displays with complex driver circuits, and signal processing by silicon electronics in an optical communication system. Lasers [15, 36], light-emitting diodes (LEDs) [24, 37], and optical modulators [30, 38, 39] have all been grafted to bare silicon. Making ELO lasers with cleaved facets introduced some complications to the fabrication [36]. Photodetectors [21] and LEDs [22] have been grafted to and interconnected with working circuits on a silicon wafer. Another active area is the grafting of GaAs electronics onto a new host substrate. GaAs metal–semiconductor field-effect transistors (MESFETs) [40], high-electron-mobility transistors (HEMTs) [41], and resonant tunneling diodes [29] have been grafted to glass as a demonstration of the ELO technique. III–V devices have been grafted to III–V hosts to combine GaAs MESFETs with long-wavelength optical devices [25, 42]. GaAs MESFETs have been grafted to an $LiNbO_3$ modulator to provide on-wafer drive electronics [43]. III–V electronic devices can operate at very high speeds, but their integrated circuit technology lags that of silicon. The integration of silicon VLSI electronics with high-speed III–V electronics in critical areas is an attractive alternative for obtaining both high speed and high complexity. With this as a motivation, both GaAs MESFETs and HEMTs grafted on silicon have been reported [44, 45]. ELO offers an advantage over heteroepitaxy in that the material of the layer between the silicon and GaAs devices need not be GaAs but can be chosen to minimize the parasitic capacitances that degrade the speed of the GaAs circuit [35].

3.2. Altering Substrate Properties

The growth substrate does not always have the best properties for device performance. For example, the dielectric constant of a GaAs or InP substrate is high, leading to high interconnect and bond pad capacitances that may degrade the speed. Grafting a high-speed device to a substrate with lower dielectric constant, such as beryllia or sapphire, reduces these parasitic capacitances [46, 47]. Similarly, grafting a laser to a substrate with a lower index of refraction alters its spontaneous emission spectrum [48]. ELO has also been used to avoid the traps present in semi-insulating GaAs substrates that lead to sidegating and leakage currents [35]. Optical absorption of the substrate can been avoided by ELO [49, 50]. The issue of high substrate cost for solar cells was addressed by lifting off the solar cells from the growth substrate and reusing the growth substrate [2].

There are times when the available growth substrates are not adequate even for growth. The alloy system InGaAsP spans a two-dimensional compositional space, but only the compositions lattice matched to a binary are routinely accessible. ELO can create new substrates with lattice constants different from those of the binaries. When a strained layer below its critical thickness is lifted off and grafted to a host, its lattice parameter relaxes to a value previously unavailable for growth [51]. A variation of this idea is to relieve the stress in a strained layer grown on a sacrificial layer [52–54]. The sacrificial layer underneath etched mesas is partially undercut-etched to form stress-relieved cantilevers that are kept in place by the unetched portion of the sacrificial layer.

ELO can also be an aid for analysis of optoelectronic materials. Without the substrate, a secondary ion mass spectrometry (SIMS) analysis can begin from the back of a layer to obtain better depth resolution there and to avoid possible interference from material near the top surface [55, 56].

3.3. Merging Devices

The third class of applications merges grafted devices with either the host substrate or devices on the host substrate so that they interact in ways other than through a metal wire. ELO presents an opportunity to do this because the glueless bonding of films consisting of only the necessary layers allows the active region of the film and substrate to be separated vertically by only a few tens of nanometers. Interactions with short characteristic lengths become a possibility. An analogy in electronics would better illustrate the point: Although a bipolar transistor consists of two *p–n* junction diodes, it is not possible to make a hybrid bipolar transistor from two discrete diodes because it is necessary that the diodes be less than a few minority-carrier diffusion lengths apart.

A simple form of optical interaction mediated by ELO is a waveguide formed with a grafted guiding layer and host substrate cladding layer [57]. Semiconductor quarter-wave stacks have been grafted to fibers to form optical resonators [58]. Grafted semiconductor films have also been vertically coupled to optical waveguides of ion-exchanged glass and proton-exchanged $LiNbO_3$ [59–61] for waveguide detectors. Enhanced optical coupling that is insensitive to the details of the bonded interface has been achieved by burying the semiconductor layer in the core of the waveguide for detector [62] and emitter [63]. A laser was grafted at the bottom of a well to butt-couple it to a glass waveguide [14]. A high-reflectivity bottom mirror that is insensitive to the wavelength and the incidence angle of light has been obtained by grafting a film to metal and used to reduce the threshold current of lasers [7] or to enhance the quantum efficiency of LEDs [64, 65] by recycling photons that would otherwise be lost in the substrate.

Grafting a GaAs film on a narrow metal finger results in a Schottky diode between them that can apply an electric field in the semiconductor to collect photogenerated carriers for an inverted metal–semiconductor–metal (MSM) photodetector [66] or to deplete and undeplete the GaAs for an inverted gate MESFET [67]. When a film is grafted to a substrate with ribs on its surface, the film bends as it tries to conform to the rib. The bending stress can alter the semiconductor band structure substantially in the vicinity of a rib, as demonstrated by the redshift of the exciton peak from an expediently placed quantum well in the grafted film [68]. Applying an electrostatic potential between the film and substrate modulates the amount of redshift [69].

4. UNANSWERED QUESTIONS

In spite of its successes, ELO is still at an early stage of development and many important questions have not been answered. Included are questions concerning the reliability of grafted devices and the chemical and physical nature of the grafted film and the bonded interface. These questions apply not just to ELO but to all film transfer techniques.

In addition to the usual reliability concerns about semiconductor devices, several issues are unique to ELO. First is trapped dust particles and how they affect yield. Second is the film adhesion as influenced by mechanical, thermal, and electrical stresses. Encapsulating the film is one way to prevent adhesion failure. However, devices will almost certainly be grafted to host substrates of different materials, so the effects of differential thermal expansion need to be studied. Under some conditions, there is slippage at the van der Waals bonded interface even during modest thermal cycling [70]. The effect of differential thermal expansion can be minimized by doing the bonding at a different temperature, such as the middle of the intended operating temperature range. Another concern is the bottom, exposed surface that is very close to the active layers. Top and bottom capping layers of either epitaxial semiconductor or separately deposited dielectric to protect the active layers would help, but these may interfere with device function and certainly would add to the step height that interconnects would have to cover. Again, an encapsulant should be beneficial. Another question is whether inhomogeneities in the bonding can cause local peaks in the stress or in the temperature rise and thereby accelerate failure. In the case of bonding with some sort of thick adhesive, stresses arising from the bonding process or from differential thermal expansion need to be investigated.

Many types of optical and electronic compound semiconductors have been made by ELO and related film transfer techniques. Performance comparisons that have been made against devices fabricated on the growth wafer are very favorable, quite often with differences within the normal variations seen among devices that are nominally the same. However, more stringent comparisons are required in some cases: lasers made by ELO have been broad-area lasers that are not too sensitive to material quality, and the only reported transferred film, single transverse mode lasers were made by wafer bonding rather than film bonding [14, 15]. Small changes in the dark current of InGaAs p-i-n photodetectors [61], but not of GaAs MSM photodetectors [21], have been reported; a more systematic study is required in this area. Except for one continuous test of an ELO MESFET that showed no change after 100 hours at room temperature [45], device degradation has not been studied.

A better understanding of the van der Waals bonding process and the resultant interface is required. The bonded interface can affect the device reliability, the $1/f$ noise properties, and the performance of merged devices, where some interaction takes place through the interface. With a fundamental understanding of the interface, it may even be possible to engineer it to the specific needs of the device. We also need to see if there are subtle changes in the ELO film that may be apparent only with more stringent testing.

Many possible ELO applications involve devices, such as lasers and power transistors, that have large power dissipation. Even assuming that the interfacial layer has the thermal conductivity of the worst solid conductor, the thermal resistance is extremely small because of the thinness of the layer. Although the ELO bond itself does not limit the heat dissipation, the host substrate in some instances would necessarily be a poor thermal conductor. Heat dissipation will probably be a limitation in such applications as driver electronics or lasers grafted on poor thermal conductors such as $LiNbO_3$ or glass. The heat can be removed by some other channel in such cases, but the overall complexity in the final packaged device may negate any advantages gain by ELO.

5. OUTLOOK

Assuming that all concerns about reliability and material quality prove to be unfounded or are adequately addressed, what is the outlook for ELO and other film transfer techniques? An assessment requires comparison with other methods capable of achieving similar performance in a particular application. For electrically interconnecting dissimilar devices with low parasitics, flip-chip bonding, already a well-established technique, is an alternative. High speeds (>20 GHz) have been demonstrated for flip-chip bonded detectors [71], and infrared focal plane arrays with more than 65,000 connections between the detector array and the silicon circuit have been made with 30-μm-square pixels [72]. A potential advantage of film transfer techniques is that of attaining even higher densities of interconnects. Optics and electronics are less densely interconnected to each other for most other optoelectronic applications than they are in focal plane arrays, so film transfer techniques do not offer any obvious advantages over flip-chip bonding here. Purely electronic applications will very likely require a high interconnect density between grafted and host substrate devices, so this may be an important area for film transfer. Another difference between film bonding and flip-chip bonding is in the orientation of the device; in applications where the device cannot be bonded upside-down, ELO and film bonding is certainly the more attractive option.

Success of film transfer for electrical interconnect applications, therefore, relies on finding applications in which flip-chip bonding either cannot provide adequate interconnect density or gives the wrong device orientation. In the other classes of applications just described, altering substrate properties and merging devices, no well-established competing method exists for accomplishing these goals and film transfer can become an important technology.

Heteroepitaxy, which is also in an early stage of development, is an alternative to film transfer for some of the applications considered. Film transfer, at this time, offers better material quality and a far larger variety of material combinations than heteroepitaxy does. Even compared to the most advanced heteroepitaxial system, GaAs on Si, film transfer offers more flexibility through lower processing temperatures and through the wider choice of buffer layers between the Si and GaAs.

Film transfer is manufacturable; silicon bipolar circuits made on direct bonded SOI wafers are commercially available [73]. But for the optoelectronics and compound semiconductors, it is not clear at this point exactly which variation of film transfer will be the most suitable for various purposes. The intended application will control the choices in many cases, so there may be different methods for different applications.

6. CONCLUSION

Film transfer techniques for optoelectronics in the last few years went to a stage of favorable initial demonstrations. Workers from laboratories around the world made grafted optoelectronic devices and were encouraged by the similarity of the performance of grafted devices to that of conventionally made devices. The time has already come to examine the performance of grafted devices critically in terms of reliability, performance, and uniformity and to consider where these techniques will have an impact in optoelectronics. In the next few years, we expect more commercial penetration of film transfer techniques to solve manufacturing, packaging, and systems problems.

REFERENCES

1. L. F. Miller, *IBM J. Res. Dev.* 239 (May 1969).
2. M. Konagai, M. Sugimoto, and K. Takahashi, *J. Cryst. Growth* **45**, 277 (1978).
3. A. G. Milnes and D. L. Feucht, *Eleventh Photovoltaic Specialists Conf.* 338 (May 1975).

4. E. Yablonovitch, T. Gmitter, J. P. Harbison, and R. Bhat, *Appl. Phys. Lett.* **51**, 2222 (1987).
5. E. Yablonovitch, D. M. Hwang, T. J. Gmitter, L. T. Florez, and J. P. Harbison, *Appl. Phys. Lett.* **56**, 2419 (1990).
6. R. W. McClelland, C. O. Bozler, and J. C. C. Fan, *Appl. Phys. Lett.* **37**, 560, (1980).
7. F. Stern and J. M. Woodall, *J. Appl. Phys.* **45**, 3904 (1974).
8. J. B. Lasky, S. R. Stiffler, F. R. White, and J. R. Abernathey, *1985 Int. Elect. Dev. Meeting Tech. Dig.* 684 (1985).
9. M. Shimbo, K. Furukawa, K. Fukuda, and K. Tanzawa, *J. Appl. Phys.* **60**, 2987 (1986).
10. W. P. Maszara, G. Goetz, A. Caviglia, and J. B. McKitterick, *J. Appl. Phys.* **64**, 4943 (1988).
11. G. G. Goetz and A. M. Fathimulla, *Proc. Electrochemical Soc.*, paper 309 (March 1990).
12. V. Lehmann, K. Mitani, R. Stengl, T. Mii, and U. Gösele, *Jpn. J. Appl. Phys.* **28**, L2141 (1989).
13. G. A. Antypus and J. Edgecumbe, *Appl. Phys. Lett.* **26**, 371 (1975).
14. M. Yanagisawa, H. Terui, K. Shuto, T. Miya, and M. Kobayashi, *IEEE Photon. Technol. Lett.* **4**, 21 (1992).
15. C. L. Shieh, J. Y. Chi, C. A. Armiento, P. O. Haugsjaa, A. Negri, and W. I. Wang, *Electron. Lett.* **27**, 850 (1991).
16. R. J. M. Griffiths, I. D. Blenkinsop, and D. R. Wight, *Electron. Lett.* **15**, 629 (1979).
17. B. Hök, C. Dubon, and C. Ovrén, *Appl. Phys. Lett.* **43**, 267 (1983).
18. R. H. Rediker, S. Stopek, and J. H. R. Ward, *Solid State Electron.* **7**, 621 (1964).
19. Z. L. Liau and D. E. Mull, *Appl. Phys. Lett.* **56**, 737 (1990).
20. Y. H. Lo, R. Bhat, D. M. Hwang, M. A. Koza, and T. P. Lee, *Appl. Phys. Lett.* **58**, 1961 (1991).
21. C. Camperi-Ginestet, Y. W. Kim, N. M. Jokerst, M. G. Allen, and M. A. Brooke, *IEEE Photon. Technol. Lett.* **4**, 1003 (1992).
22. A. Ersen, private communication.
23. P. Demeester, I. Pollentier, P. De Dobbelaere, C. Brys, and P. Van Daele, *Semicond. Sci. Technol.* **8**, 1124 (1993).
24. C. Camperi-Ginestet, M. Hargis, N. Jokerst, and M. Allen, *IEEE Photon. Technol. Lett.* **3**, 1123 (1991).
25. I. Pollentier, P. Demeester, P. Van Daele, D. Rondi, G. Glastre, A. Enard, and R. Blondeau, *Proc. Third Int. Conf. on Indium Phosphide and Related Materials*, 268 (1991).
26. I. Pollentier, Y. Zhu, B. DeMeulemeester, P. Van Daele, and P. Demeester, *Microelectron. Eng.* **15**, 153 (1991).
27. P. Demeester, I. Pollentier, L. Buydens, and P. Van Daele, *SPIE Vol.*

1361: Physical Concept of Materials for Novel Optoelectronic Device Applications part 2, 987 (1990).
28. E. Yablonovitch, T. Sands, D. M. Hwang, I. Schnitzer, T. J. Gmitter, S. K. Shastry, D. S. Hill, and J. C. C. Fan, *Appl. Phys. Lett.* **59**, 3159 (1991).
29. A. J. Tsao, V. K. Reddy, and D. P. Neikirk, *Electron. Lett.* **27**, 484 (1991).
30. G. W. Yoffe, *Electron. Lett.* **27**, 1579 (1991).
31. I. Pollentier, L. Buydens, P. Van Daele, and P. Demeester, *Proc. 22nd European Sol. State Device Res. Conf.-ESSDERC'92* (H. E. Maes, R. P. Mertens, and R. J. Van Overstraeten, eds.), p. 207. Elsevier, Amsterdam, 1992.
32. W. K. Chan, A. Yi-Yan, and T. J. Gmitter, *IEEE J. Quantum Electron.* **27**, 717 (1991).
33. E. Yablonovitch, K. Kash, T. J. Gmitter, L. T. Florez, J. P. Harbison, and E. Colas, *Electron. Lett.* **25**, 171 (1989).
34. J. F. Klem, E. D. Jones, D. R. Myers, and J. A. Lott, *Inst. Phys. Conf. Ser.*, No. 96, 387 (1988).
35. W. K. Chan, D. M. Shah, T. J. Gmitter, L. T. Florez, B. P. Van der Gaag, and J. P. Harbison, *Proc. SOTAPOCS XII*. Electrochemical Society, 1990.
36. I. Pollentier, L. Buydens, P. Van Daele, and P. Demeester, *IEEE Photon. Technol. Lett.* **3**, 115 (1991).
37. I. Pollentier, P. Demeester, A. Ackaiert, L. Buydens, P. Van Daele, and R. Baets, *Electron. Lett.* **26**, 193 (1990).
38. L. Buydens, P. De Dobbelaere, P. Demeester, I. Pollentier, and P. Van Daele, *Opt. Lett.* **16**, 916 (1991).
39. G. W. Yoffe and J. M. Dell, *Electron. Lett.* **27**, 558 (1991).
40. C. Van Hoof, W. De Raedt, M. Van Rossum, and G. Borghs, *Electron. Lett.* **25**, 136 (1989).
41. D. R. Myers, J. F. Klem, and J. A. Lott, *Proc. 1988 Int. Electron. Dev. Meeting* 704 (1988).
42. I. Pollentier, L. Buydens, A. Ackaert, P. Demeester, P. Van Daele, F. Depestel, D. Lootens, and R. Baets, *Electron. Lett.* **26**, 925 (1990).
43. A. C. O'Donnell, I. Pollentier, P. Demeester, P. Van Daele, and A. D. Carr, *Electron. Lett.* **26**, 1179 (1990).
44. D. M. Shah, W. K. Chan, T. J. Gmitter, L. T. Florez, H. Schumacher, and B. P. Van der Gaag, *Electron. Lett.* **26**, 1865 (1990).
45. D. M. Shah, Investigation of Epitaxial Lift-off GaAs and Langmuir-Blodgett Films for Optoelectronic Device Applications, Doctoral Dissertation, ECE Dept., New Jersey Institute of Technology (1992).
46. H. Schumacher, T. J. Gmitter, H. P. LeBlanc, R. Bhat, E. Yablonovitch, and M. A. Koza, *Electron. Lett.* **25**, 1653 (1989).
47. P. G. Young, S. A. Alterovitz, R. A. Mena, and E. D. Smith, *Proc. 1991 Int. Semicond. Device Res. Symp.* 689 (1991).
48. E. Yablonovitch, T. J. Gmitter, and R. Bhat, *Phys. Rev. Lett.* **61**, 2546 (1988).

49. G. Augustine, N. M. Jokerst, and A. Rohatgi, *Appl. Phys. Lett.* **61**, 1429 (1992).
50. F. Kobayashi and Y. Sekiguchi, *Jpn. J. Appl. Phys.* **31**, L850 (1992).
51. J. DeBoeck, C. Van Hoof, K. Deneffe, and G. Borghs, *Jpn. J. Appl. Phys.* **30**, L423 (1991).
52. K. Kawasaki, S. Sakai, N. Wada, and Y. Shintani, in "Gallium Arsenide and Related Compounds 1990" (K. E. Singer, ed.), *Inst. Phys. Conf. Ser.* No. 112, 269 (1990).
53. J. DeBoeck, G. Zou, M. Van Rossum, and G. Borghs, *Electron. Lett.* **27**, 22 (1991).
54. G. F. Burns and C. G. Fonstad, *IEEE Photon. Technol. Lett.* **4**, 18 (1992).
55. C. J. Palmstrøm, S. A. Schwarz, E. Yablonovitch, J. P. Harbison, C. L. Schwartz, L. T. Florez, T. J. Gmitter, E. D. Marshall, and S. S. Lau, *J. Appl. Phys.* **67**, 334 (1990).
56. S. A. Schwarz, C. J. Palmstrøm, C. L. Schwartz, T. Sands, L. G. Shantharama, J. P. Harbison, L. T. Florez, E. D. Marshall, C. C. Han, S. S. Lau, L. H. Allen, and J. W. Mayer, *J. Vac. Sci. Technol. A* **8**, 2079 (1990).
57. A. Yi-Yan, M. Seto, T. J. Gmitter, D. M. Hwang, and L. T. Florez, *Electron. Lett.* **26**, 1567 (1990).
58. J. M. Dell and G. W. Yoffe, *Electron. Lett.* **27**, 26 (1991).
59. A. Yi-Yan, W. K. Chan, T. J. Gmitter, L. T. Florez, J. L. Jackel, E. Yablonovitch, R. Bhat, and J. P. Harbison, *IEEE Photon. Technol. Lett.* **1**, 379 (1989).
60. W. K. Chan, A. Yi-Yan, T. J. Gmitter, L. T. Florez J. L. Jackel, D. M. Hwang, E. Yablonovitch, R. Bhat, and J. P. Harbison, *IEEE Photon. Technol. Lett.* **2**, 194 (1990).
61. A. Yi-Yan, W. K. Chan, C. K. Nguyen, T. J. Gmitter, R. Bhat, and J. L. Jackel, *Electron. Lett.* **27**, 87 (1991).
62. W. K. Chan, A. Yi-Yan, T. J. Gmitter, L. T. Florez, N. Andreadakis, and C. K. Nguyen, *Electron. Lett.* **27**, 410 (1991).
63. A. Yi-Yan, W. K. Chan, T. S. Ravi, T. J. Gmitter, R. Bhat, and K. H. Yoo, *Electron. Lett.* **28**, 341 (1992).
64. I. Pollentier, A. Ackaert, P. De Dobbelaere, L. Buydens, P. Van Daele, and P. Demeester, *Physical Concepts of Materials for Novel Optoelectronic Device Applications I: Materials Growth and Characterization* SPIE Vol. 1361, part 2, 1056 (1990).
65. I. Schnitzer, E. Yablonovitch, C. Caneau, and T. J. Gmitter, *Appl. Phys. Lett.* **62**, 131 (1993).
66. C. Camperi-Ginestet, N. M. Jokerst and S. Fike, *Opt. Soc. Am. Annu. Meet.*, paper FFF3 (September 1992).
67. W. K. Chan, D. M. Shah, T. J. Gmitter, and C. Caneau, *Electron. Lett.* **28**, 708 (1992).
68. W. K. Chan, T. S. Ravi, K. Kash, J. Christen, T. J. Gmitter, L. T. Florez, and J. P. Harbison, *Appl. Phys. Lett.* **61**, 1319 (1992).

69. J. A. Yater, K. Kash, W. K. Chan, T. Gmitter, T. S. Ravi, L. T. Florez, and J. P. Harbison, Meeting of the American Physical Society (March 1993).
70. J. F. Klem, E. D. Jones, D. R. Myers, and J. A. Lott, *J. Appl. Phys.* **66**, 459 (1989).
71. O. Wada, M. Makiuchi, H. Hamaguchi, T. Kumai, and T. Mikawa, *J. Lightwave Technol.* **9**, 1200 (1991).
72. A. Fowler, R. Joyce, I. Gatley, J. Gates, and J. Herring, *Infrared Detectors, Focal Plane Arrays and Imaging Sensors* SPIE Vol. 1107, 22 (1989).
73. F. Goodenough, *Electron. Design* 35 (December 19, 1991).

PART IV

State-of-the-Art Discrete Components

Chapter 9

ELECTRONICS FOR OPTOELECTRONIC INTEGRATED CIRCUITS

John R. Hayes

Bellcore
Red Bank, New Jersey

It is likely that future optoelectronic integrated circuits (OEICs) will comprise both active optical devices, such as photodetectors and/or lasers, and active electronic devices, such as field effect transistors (FETs) and/or bipolar junction transistors (BJTs). In addition, for such operations as wavelength selection in wavelength division multiplexing (WDM) applications, it may be beneficial to incorporate passive optical components (i.e., waveguides, gratings, and couplers) along with the active optoelectronic components, further complicating the integration but greatly simplify packaging complexities because only a single chip needs to be dealt with. It has also often been suggested that advantage can be obtained from the monolithic integration of lasers with laser driver circuitry, to form a monolithic OEIC transmitter, and photodiodes with amplifying electronics, to form a low-noise OEIC receiver. However, unlike the laser driver case, in which relatively low levels of electrical integration are typically required to interface the OEIC chip to more complex system electronics (presumably CMOS, BiCMOS, or GaAs MESFET), the receiver OEIC electronics typically require a higher level of integration. The reason is that, in addition to signal

ISBN 0-12-200420-5

amplification, a number of signal conditioning and related functions need to be accomplished at the full bandwidth of the system before the signal can be "truly detected." These additional circuits perform the function of automatic gain control, retiming, and decision and have the effect of conditioning the signal so that it can be digitally processed. For both the receiver and transmitter OEIC applications the component density requirements will typically be heavily on the side of the analog electrical components, with the number of optical devices incorporated on chip in a small minority. Based on current hybrid designs, the typical size of a receiver chip will approach several hundred electrical components, and for practical systems applications both the optical and electrical components should be of sufficiently high performance that overall system performance requirements can be met.

For the purpose of discussion in this chapter, we will make the assumption that an OEIC receiver can be expected to detect photons, thereby converting the optical signal to electronic form, and then amplify this detected signal to a sufficient level to allow a decision to be made about whether a 1 or a 0 has been received. Following this decision, the restored signal can be passed onto another chip or subsystem, presumable realized in a digital technology having orders of magnitude higher levels of integration (i.e., CMOS, BiCMOS, or GaAs MESFET) for subsequent digital signal processing. The hybrid integration of the OEIC with this signal processing electronics can be accomplished using any of a number of techniques such as multichip module or solder-bump technology. Clearly, given this scenario, high-performance electrical devices are required for the electronic component of an OEIC chip. However, because OEICs may require only medium- to large-scale levels of electronic integration, advanced device technologies such as high electron mobility transistor (HEMT) or heterojunction bipolar transistor (HBT) technologies can be considered for this application.

During the past decade the high-frequency performance of III–V semiconductor electronic devices has improved at an impressive rate, with high-frequency operation well in excess of 100 GHz being demonstrated [1, 2] for discrete devices realized with both GaAs and InP materials. This evolution in high performance can be attributed to the combination of advances in device fabrication technology and the incorporation of semiconductor heterojunctions into device designs. The development of epitaxial material growth—or more precisely, from a device design point of view, the degree of control of epitaxial growth—has enabled researchers to tailor device material structures to take advantage of energy band structures designed to achieve specific performance improvements.

To appreciate the advantages of heteroepitaxial material structures, consider the variety of field effect and bipolar transistors currently being investigated for various electronic applications as illustrated in Fig. 1. Each device type has its particular advantages and disadvantages, as does each device category. FETs exhibit high input impedance with low parasitic capacitance and do not suffer from charge storage when operated under saturation conditions. On the other hand, HBTs have a vertical geometry (allowing high packing density), a high transconductance (a small change in modulation voltage leads to a large change in signal current), and a threshold voltage that is determined by the intrinsic material band structure.

The advantages of III–V compound semiconductor materials for high-performance, high-speed electronics are widely recognized. These advantages stem in part from the versatile device structures made possible by advanced heteroepitaxy and in part from excellent electrical characteristics of lattice-matched and in some cases limited lattice-mismatched heterostructures. The superior electron velocity–electric field characteristic (v_e vs. E) of a number of technologically relevant semiconductor materials is illustrated in Fig. 2. The low electric field mobility (defined by the slope of the v_e–E curve close to the axis) for the III–V material shows values from a few thousand to 10,000 cm^2/V s and exceeds that of silicon in all cases. In addition, electrons in these III–V materials achieve their peak and saturated velocities at low electric fields relative to silicon, so that the total voltage drop across a device operated at the saturated carrier velocity is 3–10 times smaller. High electric field transport is important in all devices because transport in the region under the gate of an FET and in the collector region of an HBT takes place in the presence of high electric fields, and in high-performance devices carrier transit time through these regions of the device can be the principal source of internal delay. As shown in Fig. 2, the v_e–E characteristic of III–V materials also differs significantly from that of silicon in that there is typically a region of negative differential resistance resulting from details of

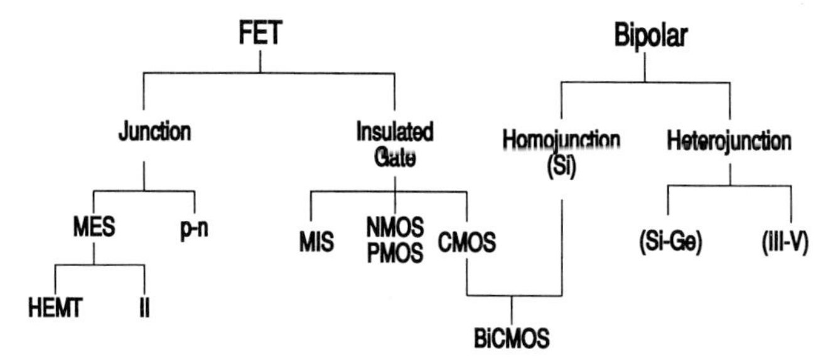

Fig. 1. Family tree of electronic devices including both field effect transistors and bipolar transistors.

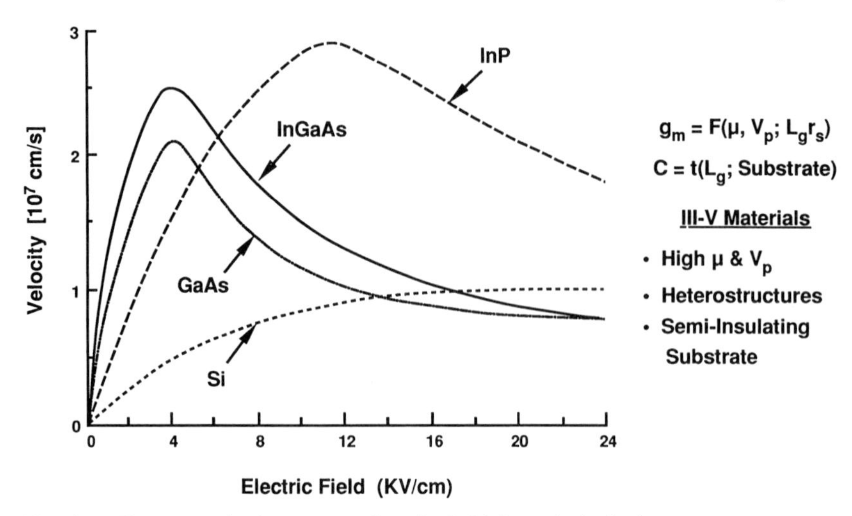

Fig. 2. Electron velocity versus electric field for relatively low or nominally undoped *n*-type InP, InGaAs (lattice matched to InP), GaAs, and silicon. As the figure indicates, silicon, due to its different band structure details and electron dynamics, does not exhibit the region of negative differential mobility that characterizes transport in III–V materials.

the electron dynamics and the band structure. In any practical device the carrier velocity varies significantly as carriers are transported through the device structure, but for a given class of devices, materials with high low-field mobility and high peak velocity (e.g., InGaAs) generally exhibit lower parasitic resistance and greater current-carrying capacity. In addition, the lower the electric field required to achieve saturated carrier transport, the lower will be the power dissipation to achieve high-speed operation.

Two figures of merit (FOM) characterize high-speed microwave transistors: the unity current gain cutoff frequency, f_T, and the maximum frequency of oscillation, f_{max}. These relatively easily determined figures of merit conveniently characterize the influence of a number of complex interactions determining the high-speed performance of a particular transistor type, but devices with comparable geometries and having active regions composed of materials with good transport properties, such as InGaAs, are expected to outperform those fabricated with other materials.

Over the past years a number of variations on material structures have been explored to optimize high-performance device operation. At present, the structures of choice are remarkably similar regardless of the material system used; this is illustrated in Fig. 3a for FETs and in Fig. 3b for HBTs. FET optimization is achieved in part by maximizing the channel carrier den-

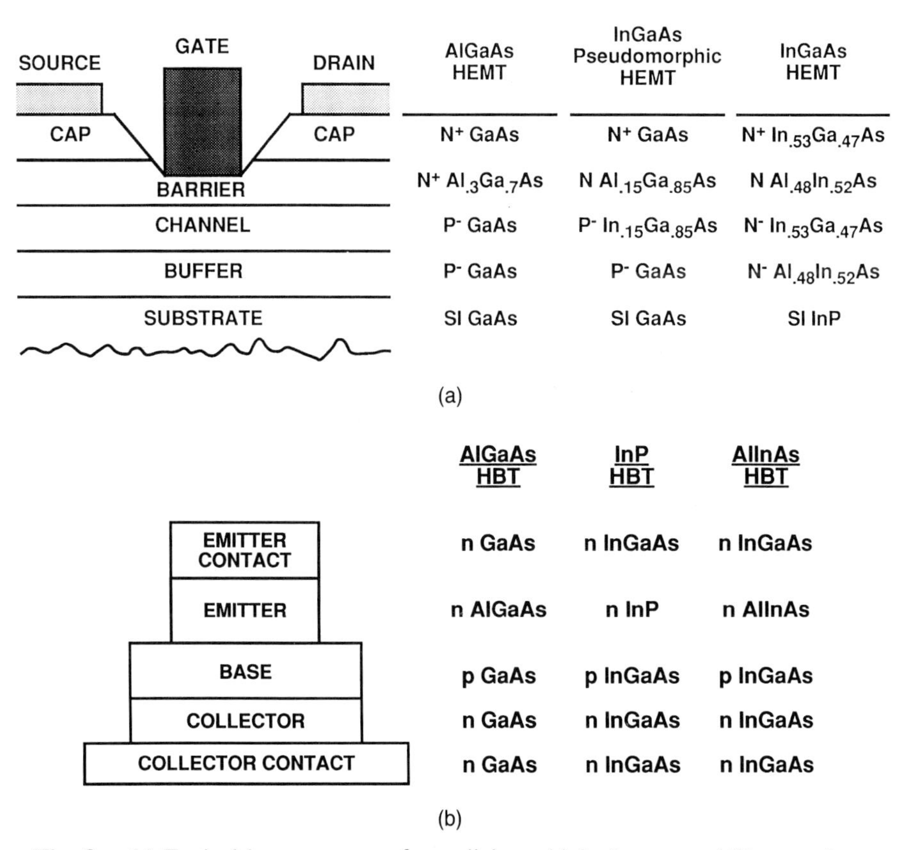

Fig. 3. (a) Typical layer sequence for realizing a high electron mobility transistor (HEMT). As indicated, apart from a change of constituent materials, the device structures for the three most common HEMT devices are identical. (b) Typical layer structure of a heterojunction bipolar transistor, comprising a wide energy bandgap emitter region and narrower energy bandgap base and collector regions. As indicated, the basic material layer strucuture is the same for the three types of devices shown.

sity while minimizing carrier scattering. This is best accomplished currently by using modulation doping, which, when combined with short gate lengths, results in reduced input capacitance and minimum carrier transit time [2]. A high doping level in the contact regions to reduce parasitic contact resistance is also critical. A typical HEMT [3] is shown in Fig. 4. The device structure incorporates a narrow-bandgap channel region and a wider-bandgap layer that is responsible for donating electrons to the channel. This wide-bandgap layer also forms a good Schottky barrier for the gate electrode with minimal current leakage at room temperature. A typical set of current–volt-

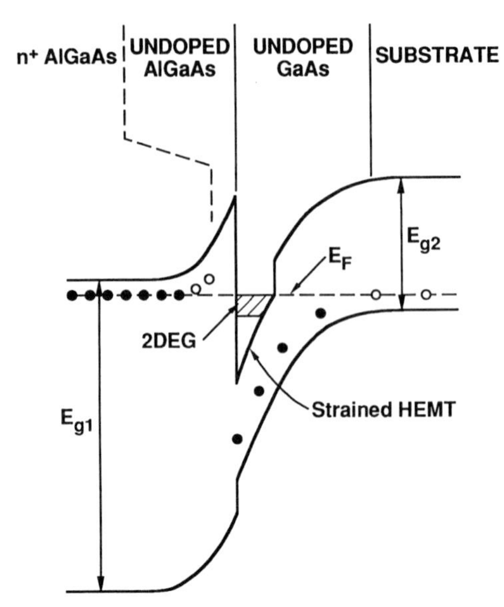

Fig. 4. Energy diagram of an HEMT. The conducting channel of the device resides at the interface between the wide- and narrow-bandgap energy materials.

age characteristics for such a device, with a gate width of 0.25 μm, are shown in Fig. 5 [3].

Enhanced HBT performance derives principally from the use of heterostructures to realize a valence bandgap difference between the emitter and base [4] and in some cases between base and collector junctions. These bandgap differences limit undesired carrier injection and allow optimizing the doping of the various layers to achieve high injection efficiency for minority carriers injected from the emitter into the base. Heterostructures can also minimize parasitic elements that degrade device performance. A schematic diagram of an Npn heterojunction bipolar transistor is shown in Fig. 6. The device comprises a narrow-bandgap collector that is moderately doped *n*-type, a heavily doped *p*-type narrow-bandgap base region, and a moderately doped *n*-type wider-bandgap emitter. The significant difference between an HBT and a silicon homojunction bipolar transistor is this use of the wide-bandgap emitter. The wide bandgap suppress hole injection from the base into the emitter, permitting a significant increase in base doping. The increase in base doping lowers the base resistance and allows one to operate the transistor at higher base currents and hence higher speed. It has been demonstrated that the advantages of the HBT approach pioneered for III–V transistors can also be used to realize higher-performance silicon tran-

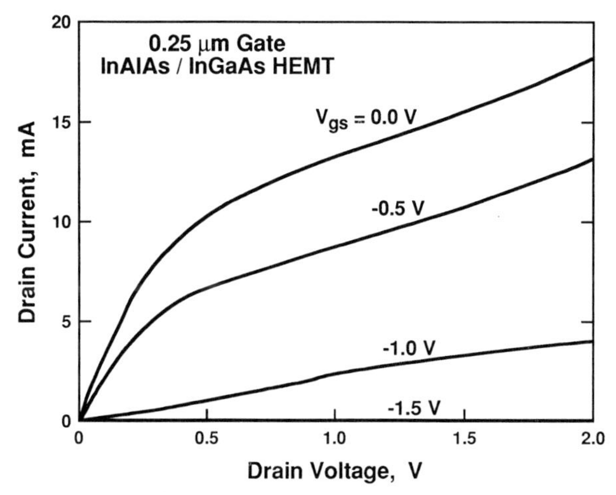

Fig. 5. A typical set of electrical characteristics for an InAlAs/InGaAs, 0.25-μm gate length HEMT. The electrical characteristics show the drain current versus source–drain voltage as a function of gate voltage. The transistor had a gate length of 0.25 μm, a gate width of 20 μm, and a maximum transconductance of 1100 mS/mm.

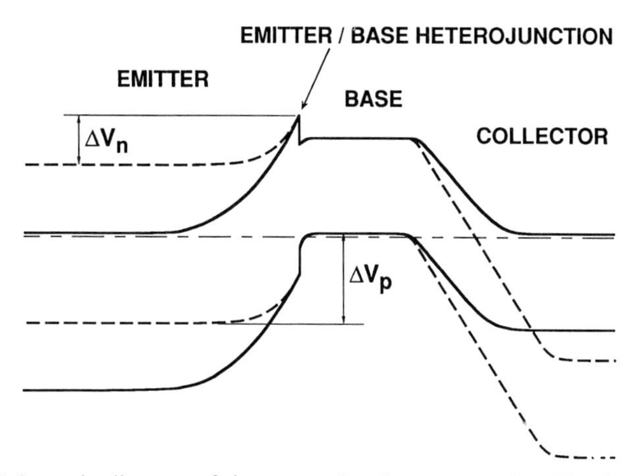

Fig. 6. Schematic diagram of the energy band structure of an Npn heterojunction bipolar transistor. The emitter comprises a wide-bandgap energy material doped *n*-type, whereas the base is heavily doped *p*-type for low base resistance. The solid lines represent the energy band structure of the device with no bias applied; the dashed lines represent the device under bias, with emitter–base junction forward biased and base–collector junction reverse biased.

sistors [5] by combining a silicon emitter with relatively narrower-bandgap SiGe base regions.

From the preceding discussion it can be inferred that the design of modern III–V devices is well characterized and that enhanced performance can be achieved by minimizing the influence of extrinsic parameters limiting device performance (extrinsic resistances and capacitances). However, the intrinsic material parameters that ultimately limit high field transport (e.g., the LO phonon energy and band structure) are unique to a particular material and, to a considerable degree, are outside the control of a device designer. Because it is not possible to change these basic material parameters, performance beyond that determined for a particular material system can be obtained only by translating a device design to another material system, one with superior intrinsic transport properties. The advantage of optimizing intrinsic material properties is illustrated in Fig. 7, which shows the dependence of the frequency performance of field effect transistors on gate length for GaAs and InGaAs materials. A similar performance comparison should

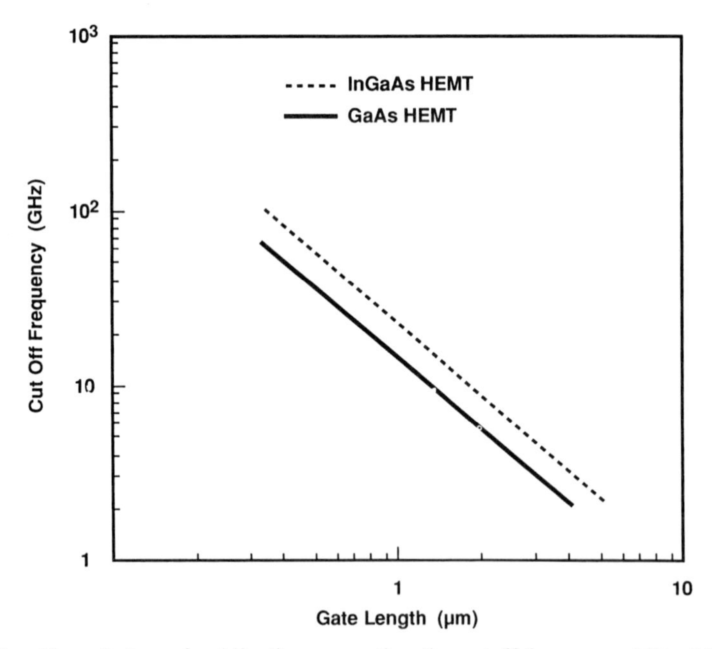

Fig. 7. Compilation of publications reporting the cutoff frequency (f_T) of InGaAs and GaAs channel field effect transistors as a function of gate length. The f_T for comparably doped channels having identical gate width is determined principally by the intrinsic transport properties of the channel material. The effective channel velocity that each line represents is shown in the figure.

be possible for HBTs, but there is not sufficient data to attempt this here, so our discussion will be restricted to the FET case. However, limited HBT data support the conclusion drawn for FETs. It can be shown that the gradient of f_T with respect to L_g in Fig. 4 represents an effective velocity that characterizes the transport of carriers under the gate region, which in turn is determined by the details of material parameters. As can be seen for the case of InP/InGaAs materials, this effective velocity is greater than that of AlGaAs/GaAs materials, consistent with the electron velocity–field characteristic illustrated in Fig. 2.

Careful design is required to ensure that individual transistors operate close to their peak performance when integrated into large circuits. However, in general, circuits incorporating large numbers of transistors are found to operate at speeds somewhat reduced from the highest speed achievable with a single device. For the case of silicon bipolar devices, digital circuits have been demonstrated to operate in excess of $0.5–0.8 f_t$, while for III–V devices comparable ICs are found to operate typically at only $\sim 0.1–0.25 f_t$. This translates to ~10–25 GHz for very high performance silicon and ~20–30 GHz for III–V devices. Nevertheless, properly designed GaAs ICs have been demonstrated to have as much as a factor of 10 advantage in reduced power dissipation at the same operating speed as silicon circuits. For the case of InP/InGaAs transistors, the relatively immature materials and processing technologies available mean that circuit design remains a research topic, with no circuits larger than a hundred transistors having been reported to date. However, research results obtained on substituting InP/InGaAs/InAlAs for GaAs/AlGaAs in HBT and FET IC designs look extremely promising [6].

Advanced epitaxial techniques for preparing material structures for InP/InGaAs transistors are only now emerging from research laboratories, almost a decade behind similar advances for GaAs. One significant advantage of this material system for HBT applications is the low surface recombination velocity characteristic of InP and InGaAs surfaces relative to GaAs/AlGaAs [7], which contributes to enhancing the injection efficiency of the HBT emitter–base junction. In fact, the threshold characteristics of InP/InGaAs HBTs are found to be quite similar to those of silicon bipolar devices which benefit from nearly ideal surface recombination characteristics [6].

In conclusion, discrete InP-based devices have been shown to have better performance than those of GaAs due to intrinsic material advantages. This fact, combined with the application of InP-based materials for fabricating optical components for use in the 1.3- to 1.5-μm optical fiber transmission windows, suggests that further development of InP-based electronics will result in a technology base for OEICs to meet the needs for fiber communication systems.

1. DEVICE NOISE PERFORMANCE

In order to realize the highest level of performance for an OEIC, it is necessary to select the appropriate device technology for integration. We consider first the case of an analog receiver. The question of whether a field effect transistor or a bipolar transistor is the superior electronic active device for fiber-optic receiver low-noise preamplifiers has been discussed by Smith and Personic [8], who pointed out that the sensitivity of a receiver using bipolar transistors for the preamplifier may be superior to that of an FET preamplifer if the base resistance, and the noise associated with it, can be sufficiently reduced.

2. RECEIVER NOISE IN FIELD EFFECT TRANSISTORS

In detail, following Smith and Personic, the noise current associated with an FET, assuming the gate current is negligible, is given by

$$\langle i^2 \rangle_{\text{circuit}} = (4kT/R_L)I_2B + 4kT\Gamma/g_m(2\pi C_T)^2 I_3 B^3, \tag{1}$$

where kT is the thermal energy, R_L is the load resistance, I_2 and I_3 are factors determined by the pulse shape, I_{gate} is the gate leakage current, C_T is the total capacitance (detector + parasitic + gate), g_m is the transconductance, B is the bit rate, and Γ relates the device's actual noise figure to that of an ideal resistor ($4kTB$) and is 1.75 for short gate lengths.

The absolute minimum noise level that a receiver preamplifer can obtain is given by allowing $R_L \to \infty$ to yield the following limiting sensitivity for FET receivers:

$$\langle i^2 \rangle_{\text{circuit-min}} = 4kT\Gamma/g_m(2\pi C_T)^2 I_3 B^3. \tag{2}$$

Thus the figure of merit for an FET preamplifer is given by

$$\text{FOM} = g_m/C_T^2. \tag{3}$$

In this limit, optimum performance can be obtained by maximizing the g_m while minimizing C_T. Note that the limiting value of g_m is given by v_s/L_g and that the optimum gate capacitance is given by setting $C_{\text{gate}} = C_{\text{photodiode}}$

or $C_T = 2C_g$. For these conditions, with gate capacitance proportional to L_g, the figure of merit for an FET can be related to the intrinsic carrier transport properties as FOM $\rightarrow v_s/L_g^2$; for equal gate length FET receivers the FOM $\propto v_s$ and the material with the highest v_s will in principle have the best perfromance.

3. NOISE IN BIPOLAR TRANSISTORS

The noise analysis for a bipolar transistor is a more complex function of material and device parameters than is the case for an FET. It will not be discussed in detail in this section because it can easily be found elsewhere [8], but as in the FET limiting case, as $R_L \rightarrow \infty$ the expression for the minimum circuit noise for a bipolar transistor simplifies to

$$\langle i^2 \rangle_{\text{circuit}} = (8\pi kT)(C_T/\beta^{1/2})(I_2 I_3)^{1/2} B^2 + 4kTr_b(2\pi)2(C_d + C_s)^2 I_3 B^3, \quad (4)$$

where r_b is the base resistance; β is the transistor gain; C_T is a combination of the photodetector, transistor, and photodiode capacitances; C_d is the photodetector capacitance; and C_s is the parasitic capacitance. Hence, neglecting the base resistance, which is a reasonable assumption for a heterojunction bipolar transistor, the figure of merit for the bipolar transistor is given by

$$\text{FOM} = \beta^{1/2}/C_T. \quad (5)$$

When the base resistance is small, which will be the case for heterojunction bipolar transistor, the bipolar devices have a slower bit-rate dependence than in the case of an FET (B^2 as opposed to B^3)—thus, in principle, making the bipolar a superior device for high-frequency communication receivers. However, the FET-based receiver is superior for low-bit-rate operation.

4. HETEROJUNCTION BIPOLAR TRANSISTOR DESIGNS

Heterojunction bipolar transistors are distinct from homojunction transistors in that they incorporate an emitter fabricated from material with a bandgap energy that is wider than that used for the base [3]. The bandgap energy variation through a typical HBT structure is shown in Fig 6. The bandgap difference is divided between the conduction and valance bands, with the fraction of the difference that appears in the conduction (valence) band determined by the difference in electron affinity (ionization potential) for the materials used to form the heterojunction. For the case of the most popular

HBT materials, AlGaAs/GaAs, InP/InGaAs, and AlInAs/InGaAs, the conduction band/valence band ratios are believed to be 65 : 35, 40 : 60, and 65 : 35, respectively. The advantages of a heterojunction bipolar transistor over a homojunction bipolar transistor can be appreciated by examining the electron and hole current flow in the device. For this discussion the usual expressions for these currents have been simplified by neglecting recombination currents and the difference in the density of states between the emitter and base materials. The inclusion of recombination current in general affects the device performance only when the emitter dimensions or the injected current densities are extremely small.

For this simplified case, the electron current flow from the emitter to the base is given by

$$I_{eb} = n_e v_e \exp(-qV_n/\kappa T), \tag{6}$$

where n_e is the emitter free electron concentration and v_e is the effective electron velocity, given by

$$v_e = D_n/(L_b \sinh(W_b/L_b)). \tag{7}$$

It is typical for a transistor to be designed with a base width (W_b) at least an order of magnitude less than the minority carrier diffusion length (L_b), so this expression can be approximated by

$$v_e = D_n/W_b \tag{8}$$

For the holes injected from the base into the emitter, a similar expression can be obtained:

$$I_{be} = p_b v_p \exp(-qV_p/\kappa T), \tag{9}$$

where v_b can be approximated by D_h/L_e. With this approximation the current gain is given by

$$h_{fe} = (n_e v_e/p_b v_p) \exp(\Delta E_g). \tag{10}$$

Clearly, for an Npn transistor it is essential that the valence band discontinuity exceed the conduction band discontinuity.

For the case in which the emitter and base materials are identical, such as silicon, ΔE_g is zero. In this case, to achieve appreciable current gain it is necessary to dope the emitter significantly higher than the base region, and then the gain can than be approximated by

$$h_{fe} = (n_e v_e/p_b v_p). \tag{11}$$

Typical doping levels for homojunction devices are of the order of 10^{21} cm^{-3} for the emitter and 10^{17} cm^{-3} for the base. In the case of a heterojunction, the bandgap difference allows the doping level required in the emitter to be relaxed while the base doping level can be increased, permitting the transistor to operate with a significant number of advantages over the homojunction case.

The high-frequency performance for analog applications is often characterized by f_t (the frequency at which the current gain goes to unity) and f_{max} (the frequency at which the power gain goes to unity). For an HBT, f_t is given by the sum of the internal device delays (emitter charging, base transit, collector transit, and collector charging). Designs to maximize f_t and f_{max} yield the best device performance that can be obtained.

Consider

$$f_T = (1/2\pi)(\tau_E + \tau_B + \tau_C + \tau_C'), \tag{12}$$

where the base transit time is given by

$$\tau_B = W^2/2D_B \tag{13}$$

with W the base width and D_B the diffusion coefficient. The emitter depletion layer transit time is given by

$$\tau_E = r_e\,(C_e + C_c + C_p) \approx (kT/qI_E)(C_e + C_c + C_p), \tag{14}$$

where r_e is the emitter resistance, C_e the emitter capacitance, C_c the collector capacitance, and C_p the parasitic capacitance.

The collector transit time is determined by the carrier transport across the collector depletion region and is given by

$$\tau_C = x_c/2v_s. \tag{15}$$

The collector charging time is defined as

$$\tau_C' = r_c C_c, \tag{16}$$

where r_c is the collector resistance. The maximum frequency of oscillation is related to f_T and is given by

$$f_{max} = \{f_T/(1/2\pi(R_B C_c))\}^{1/2}. \tag{17}$$

Thus it can be seen that perfromance is enhanced by minimizing parasitic resistances and capacitances and by taking advantage of materials with good transport properties.

Although we have discussed HBTs in terms of abrupt heterojunction interfaces, it is usually the case for these devices that abrupt junctions are intentionally replaced with compositionally graded structures that enhance the device perfromance by aiding carrier transport through elimination of detrimental interface fields. This is particularly helpful for for the emitter–

base junction, where the grading achieves a low collector–emitter offset voltage and also improves base hole confinement [9]. In the case of the base, compositional grading can also be used to advantage to introduce a quasi-electric field that aids the transport of minority carriers [10]. For example, if there is a graded bandgap difference of 0.3 eV across a 100-nm-thick base region, the minority carriers experience a quasi-electric field of 3 kV/cm, which, even in highly doped material, will dramatically increase carrier transport beyond diffusive and reduce the base transit time. In the case of devices using InGaAs as the material for the HBT collector region, the low reverse bias junction breakdown voltage that characterizes this material is such that these devices typically operate close to the point where the performance of receivers using this single-heterojunction technology can be compromised. In applications where this presents a problem, it is attractive to use a wider-bandgap energy material, such as InP, in the collector region. However, this can complicate the design unless the base–collector junction is appropriately graded, since a significant built-in potential will exist in the case of an abrupt junction that will impede efficient electron collection.

5. HETEROJUNCTION BIPOLAR TRANSISTOR DEVICE GEOMETRY

The typical layer structure of an InP/InGaAs heterojunction bipolar transistor is shown in Fig. 8. Starting at the top epitaxial layer, there is a 2000-Å-thick InGaAs heavily doped *n*-type contact layer, followed by the 2000-Å InP emitter layer which is about 5000 Å thick and doped in the neighborhood of 5×10^{17} cm^{-3}. The next layer is InGaAs, forming the base region, which is heavily doped *p*-type to as high a level as possible. The level of base doping is determined to some degree by the rate of diffusion of the dopant species during growth. In the case of GaAlAs/GaAs HBTs, molecular beam epitaxy (MBE) growth is widely used, and until recently beryllium [12] has been most popular *p*-type dopant. Beryllium has a low diffusion coefficient under the quiescent conditions occurring during growth. However, during operation, out-diffusion of beryllium from the base into the collector has been found to result from the strong bias-induced stress, and this has created long-term reliability problems. The use of carbon [13] has been shown to provide a satisfactory alternative base dopant. In the case of OMCVD growth, Zn is often used [14] because it can be relatively easily

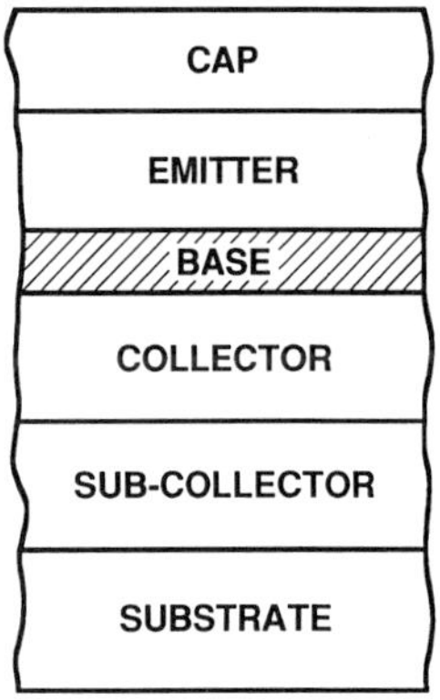

Layer	Doping	Thickness	Composition
CAP	1×10^{19}	1000 Å	n^+ InGaAs
EMITTER	5×10^{17}	3000 Å	n^- InP or InAlAs
BASE	$\sim 5 \times 10^{19}$	~1000 Å	p^+ InGaAs
COLLECTOR	$\sim 7 \times 10^{16}$	~5000 Å	n^- InGaAs or InP
SUB-COLLECTOR	$\sim 4 \times 10^{18}$	~5000 Å	n^+ InGaAs
SUBSTRATE	Semi-Insulating <100> InP Substrate		

Fig. 8. Typical layer structure used to realize an InP-based heterojunction bipolar transistor. The doping level, layer thickness, and semiconductor composition are indicated on the diagram. A typical *p*-type dopant for the base region could be Be or Zn, and a typical *n*-type dopant for the other layers could be Si, Sn, or S.

incorporated in large amounts. However, in this case the diffusion coefficient is found to be large enough that it is necessary to take into account the out-diffusion from the base into the emitter that can occur during growth. Excessive out-diffusion results in a potential barrier to electron injection and, in particularly extreme cases, results in forming a homojunction; i.e., the base–emitter junction will be formed in the emitter layer. When designing a transistor with a diffusing dopant it is possible to make allowances for the out-diffusion during growth by setting back the dopant from the interface. This set back is typically 100 Å [15], but the exact details depend on the doping level, temperature during growth, and vacancy concentration. Preceding the growth of the base layer, the collector and collector contact layers are grown. These *n*-type layers are typically doped in the high 10^{16} cm^{-3} to low 10^{17} cm^{-3} range for the collector, with the contact layer being very heavily doped.

Parasitic base resistance arising from the lateral space between the physical base–emitter junction and the base contact metal can be minimized if self-aligned techniques are used for the base contact metallization. A schematic diagram of a self-aligned process [15] is shown in Fig. 9. A typical structure will have an emitter finger dimension of around 30 mm^2. The self-aligned process is achieved by defining the emitter by etching and then simultaneously depositing the emitter and base contacts.

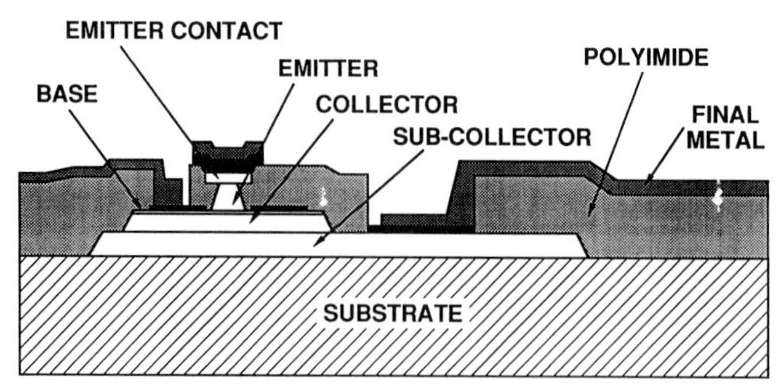

Fig. 9. Cross section of a typical self-aligned InP/InGaAs heterojunction bipolar transistor. The self-aligned base contact is obtained by undercutting the InP emitter using the InGaAs emitter contact layer as the etch mask. Polyimide is used as the insulator to isolate the conductiong metal from the semiconductor regions of the device.

6. HIGH-PERFORMANCE FET DESIGNS

It can be shown that high-performance, high-speed FETs are best realized by having conducting channels formed by thin sheets of high-density, highly mobile charge. A key advance in the design of FETs was the invention of modulation doping, which allowed all these conditions to be satisfied simultaneously in the type of FET referred to as a HEMT (high electron mobility transistor). The material structure used in HEMT devices is shown in Fig. 10 and consists of a heavily doped wide-bandgap energy material grown on a nominally undoped narrower-bandgap material. This structure allows the free electrons in the wide-bandgap material to transfer to the lower-bandgap material, forming a thin sheet of charge at the material interface. This sheet of charge produces a conduction channel in which the electrons are separated from their donor atoms and therefore not subjected to the mobility limitations resulting from ionized donor scattering. Both a large sheet charge density (typical values can be in excess of 10^{12} cm^{-2}) and a high mobility due to the absence of ionized impurities in the channel region result. At low electric fields, the two-dimensional electron gas sheet is confined close to the interface over an approximate distance of 100 Å.

The high-performance FET's fabricated using modulation-doped channels have come to be known by many acronyms: modulation-doped field effect transistor (MODFET), selectively doped heterostructure transistor (SDHT), two-dimensional electron gas field effect transistor (TEGFET), and high electron mobility transistor (HEMT). The devices were first demonstrated in the AlGaAs/GaAs semiconductor alloy system in 1978 by Momuri *et al.* [16].

InGaAs	Si-doped	5 nm
InAlAs	undoped	15 nm
InAlAs	Si-doped	10 nm
InAlAs	undoped	3 nm
InGaAs	undoped	30 nm
InAlAs	undoped	0.5 μm
	S. I. InP Substrate	

Fig. 10. Typical layer structure of a high-performance InAlAs/InGaAs HEMT. The nominally undoped InGaAs channel region is typically *n*-type and the material structure can be deposited by either molecular beam epitaxy or organometallic chemical vapor deposition.

The HEMT device is structurally similar to a metal-oxide-semiconductor field effect transistor (MOSFET). For this comparison, the oxide layer is replaced by the wide-bandgap semiconductor material, and when the charge is transferred to the narrow-gap material this layer is left positively charged, as is the oxide metal in an MOS device. The material structure of a typical InAlAs/InGaAs HEMT is shown in Fig. 10. It consists of a semi-insulating Fe-doped substrate and typically incorporates a thick InAlAs buffer layer (0.5 μm), a thin InGaAs channel layer (30 nm), and a wide-bandgap energy InAlAs layer (28 nm). This layer is heavily doped beginning about 3 nm from the channel interface and supplies the electrons to the channel region. The 3-nm spacing provides additional separation of the ionized donors from the conducting electrons by physically offsetting the donors from the channel. Finally, a thin 5-nm InGaAs cap is grown to assist in ohmic contact formation. The energy band structure of a typical HEMT was shown in Fig. 4.

The approximate expression for the f_T of a field effect device is given by

$$f_T = g_m/(1/2\pi C_{gs}) \approx v_s/2\pi L_g, \tag{18}$$

where g_m is the transconductance (change in drain current with applied gate bias) and C_{gs} is the gate source capacitance. This expression can be related to the intrinsic velocity field characteristic, where v_s is the average velocity of carriers under the gate region and L_g is the gate length. Since the electric field accelerating the channel electrons varies as a function of position in the channel, the electron velocity varies throughout the device structure with v_s representing some kind of average velocity. Hence, as discussed earlier, the material with superior high electric field transport properties will yield the highest performance for a given gate length.

The maximum frequency of oscillation for an FET is given by

$$f_{max} = f_T/\{2(r_1 + f_T \tau_3)\}^{1/2} \tag{19}$$

where

$$r_1 = (R_g + R_i + R_s)/R_{ds} \tag{20}$$

and

$$\tau_3 = 2\pi R_g C_{gd}. \tag{21}$$

In this expression R_g is the gate resistance, R_i is the charging resistance, R_s is the source resistance, R_{ds} is the channel resistance and the intrinsic resistance under the gate, and C_{gd} is the gate–drain capacitance.

Since the HEMT channel layer is thin (typically 10 nm), some recent research efforts have been aimed at incorporating lattice-mismatched layers into the channel of the device to take advantage of the resulting strain. This strain is accommodated elastically, so no dislocations occur and the strained material can be designed to have superior electron transport properties. Lattice-mismatched layers using InGaAs on InP, with the indium mole fraction as high as 0.75, have been investigated. This large indium fraction differs significantly from the latticed-matched value of 0.53 and results in a mismatch of 1.5–2.0% over a distance of 10 nm [17]. Such a mismatch is within the limits of what can be tolerated without defect or dislocation creation, and the high field transport properties of these high-indium layers are found to be superior to those of the lattice-matched layers, in part because the increased indium content lowers the bandgap and reduces the effective mass of the electrons. Beyond a point, the strain-induced change in the band structure leads to a degradation of transport properties compared with the bulk material.

7. OPTOELECTRONIC INTEGRATED CIRCUITS

7.1. Receivers

In the past decade a large number of GaAs- and InP-based OEICs have been realized. These circuits comprise a number of combinations of device structures. In the case of InP-based receivers, InAlAs/InGaAs HEMTs have been integrated with pin and MSM photodiodes [18, 19] to realize both single-channel and multichannel OEICs. Device arrays have been investigated because for array applications OEICs allow reduced packaging constraints. Receiver arrays with as many as eight nearly identical channels have been realized which operate at up to 1.2 GHz with a sensitivity of −23.8 dBm at 1.0 Gbit/s and a bit error rate of 10^{-9}. These circuits exhibit an adjacent channel crosstalk of less than −28 dB [20].

InP/InGaAs Npn heterojunction bipolar transistors have been integrated with pin photodiodes to realize OEIC receivers [21]. This approach shows significant promise, with performance at 10 Gbits/s with sensitivity of −15.5 dBm at a bit error rate of 10^{-9} having been reported [22].

7.2. Transmitter

The transmitter laser driver circuit is a key analog component of a lightwave system. Driver circuit design is simple to understand, but it is challenging to achieve high performance because of the many constraints this circuit has to satisfy. It must be able to deliver a large modulation current to the laser diode with extremely sharp rise and fall times, and it should add minimal jitter to the laser response to obtain systems capable of operating with large dynamic range.

Modern lasers used for optical communications require drive currents of the order of 50 mA and operate at voltages around 2 V at bit rates of many GHz. When used for high-bit-rate, direct detection systems, the laser diode should be prebiased to threshold in order to minimize the delay between the input modulation current pulse and the output light pulse. For analog modulation operation the drive circuit requirements may be more complex than for digital applications because a highly linear reponse is typically required. In many cases the laser driver for an analog communication system has additional circuitry to compensate for the nonlinear response of the laser light output with drive current.

Lasers have a low impedance, typically around a few ohms, that is determined principally by the bulk resistance of the diode and the associated resistance and inductance of the bond wire. If the laser is monolithically integrated or is in extremely close contact with the drive circuitry, one has only the laser impedance plus the bond wire inductance of the interconnect to drive to consider, but the likelihood of laser junction heating from the transmitter circuit must be taken into account in determining overall performance. If the driver is remote from the laser, then for high-speed operation an impedance controlled line, typically 50 Ω, should be used, and this results in a more complex transmitter package.

Since the laser exhibits low impedance, it is best driven by a high-impedance current source. This is best accomplished using a differential amplifier approach. The circuits that are typically used for laser drivers comprise few components and are often simple in concept, but to drive the desired current level at multi-Gbit speeds they require components that have a high level of performance. Bipolar devices can easily satisfy both the wide bandwidth and high current drive capability requirements. In particular, InP-based HBTs may be nearly optimal for this application because their low turn-on voltage allows the circuit to operate at lower power levels than GaAs-based counterparts. In addition, there is the advantage of the ability to integrate the laser monolithically on a common substrate to help minimize some of the packaging difficulties.

Laser drivers realized using both GaAs- and InP-based HBTs have been demonstrated with operation in the multigigabit regime. The laser driver demonstrated by Runge *et al.* [23] was based on the cherry hopper design principle of cascading amplifier stages with mismatched input and output impedances and featured a cascode output buffer. This circuit had a simulated bandwidth of 12 GHz and was tested up to 11 Gbit\s. When the circuit was used to drive 100 mA into 50 Ω, which is adequate for operation at SONET OC-192 bit rates, an excellent (open) digital modulation output eye was obtained. The chip was found to dissipate only 1.6 W. InP-based HBT laser drivers have also been demonstrated by Banu *et al.* [24], who used a conventional differential emitter-coupled pair design and achieved 10-Gbit/s operation with a peak-to-peak output current of 100 mA. For this demonstration, rather than driving a 50-Ω line, a 2.8-Ω load resistor was fabricated on chip to mimic the incorporation of a monolithic laser. It should be noted that to date neither of these very high performance laser driver circuit results has actually been used to drive a laser source.

REFERENCES

1. P. M. Asbeck, M. F. Chang, J. A. Higgins, N. K. Sheng, G. J. Sullivan. and K. C. Wang, *IEEE Trans. Electron. Devices* **36**, 10, 2032 (1989).
2. L. D. Nguyen, A. S. Brown, M. A. Tompson, and L. M. Jelloian, *IEEE Trans. Electron. Devices* **39**(9), 2007 (1992).
3. P. S. D. Lin, W-P. Hong, O-H. Kim, R. Bhat, B. van der Gaag, and H. Schumacher, *Electron. Lett.* **26**(11), 763 (199?).
4. H. Kroemer, *Proc. IEEE* **70**(1), 13 (1982).

5. G. L. Patton, J. H. Comfort, B. S. Meyerson, E. F. Crabbe, G. J. Scilla, E. De Fresart, J. M. C. Stork, J, Y-C. Sun, D. L. Harame, and J. N. Burhghartz, *IEEE Electron. Device Lett.* **11**(4), 171, (1990).
6. C. W. Farley, K. C. Wang, M. F. Chang P. M. Asbeck, R. B. Nuling, N. H. Sheng, R. Pierson, and G. J. Sullivan *Electron. Device Lett.* **10**, 377 (1989).
7. H. C. Casey, Jr. and E. Buehler, *Appl. Phys. Lett.* **30**(5), 247, (1977).
8. R. G. Smith and S. D. Personic, "Receiver Design for Optical Fiber Communication Systems," Ch. 4. Semiconductor Devices for Optical Communications. Springer-Verlag, Berlin, 1982.
9. J. R. Hayes, F. Capasso, R. J. Malik, A. C. Gossard, and W. Wiegmann, *Appl. Phys. Lett.* (1983).
10. J. R. Hayes, F. Capasso, A. C. Gossard, R. J. Malik, and W. Wiegmann, *Electron. Lett.* **19**(11), 410, (1983).
11. B. Jalali, R. N. Nottenburg, Y-K. Chen, D. Sivco, D. A. Humphrey, and A. Y. Cho, *IEEE. Electron. Lett.* **10**, 391 (1989).
12. M. Illgems, *J. Appl. Phys.* **48**, 1278 (1977).
13. R. Bhat, J. R. Hayes, E. Colas, and R. Esagui, *IEEE Electron. Device Lett.* **9**(9), 442 (1988).
14. J. R. Hayes, R. Bhat, H. Schumacher, and M. Koza, *Electron. Lett.* **23**(24), 1299 (1987).
15. L. G. Shantharama, H. Schumacher, J. R. Hayes, R. Bhat, M. Koza, and R. Esagui, *Electron. Lett.* **25**, 127 (1989).
16. T. Mimura, S. Hiyamizu, T. Fujii, and K. Nanbu, *Jpn. J. Appl. Phys* **19**, L225 (1980).
17. P. S. D. Lin, W. P. Hong, O-H. Kim, R. Bhat, B. Van der Gaag, and H. Schumacher, *Electron. Lett.* **26**, 763 (1990).
18. H. Yano, G. Sasaki, N. Nishiyana, M. Murata, K. Kamiyama, and H. Hayashi. *IEE Electron. Lett.* **28**, 503 (1992).
19. W-P. Hong, G. K. Chang, R. Bhat, J. L. Gimlett, C. Nguyen, G. Sasaki, and M. Koza, *IEE Electron. Device Lett.* **25**, 1561 (1989).
20. H. Yano, M. Murata, G. Sasaki, and H. Hayashi. *J. Lightwave Technol.* **10**(7), 933 (1992).
21. S. Chandrasekhar, A. G. Dentai, C. H. Joyner, B. C. Johnson, A. H. Gnauck, and G. J. Qua, *IEE Electron. Device Lett.* **26**, 1880 (1990).
22. S. Chandrasekhar, L. M. Lunardi, A. H. Gnauck, D. Ritter, R. A. Hamm, M. B. Panish, and G. J. Qua, *IEE Electron. Lett.* **28**(5), 466 (1992).
23. K. Runge, R. Standley, D. Danial, J. Gimlett, R. B. Nubling, R. L. Pierson, S. M. Beccue, K.-C. Wang, N-H. Sheng, M-C. F. Chang, D. M. Chen, and P. M. Asbeck, *IEEE J. Solid State Circuits* **27**, 1332 (1992).
24. M. Banu, D. Jalali, R. Nottrnburg, D. A. Humphrey, R. K. Montgomery, R. A. Hamm, and M. B. Panish, *IEE Electron. Lett.* **27**, 278 (1991).

Chapter 10

LASERS AND MODULATORS FOR OEICS

Larry A. Coldren

University of California at Santa Barbara
Santa Barbara, California 93106

1. INTRODUCTION

This chapter deals with the issue of converting signals from electrical currents or voltages into discernible signals on lightwaves, given the context of optoelectronic integration. This transduction process is clearly central to the viability of any optoelectronic integrated circuit (OEIC) that has optical outputs. In this section OEIC is interpreted to apply to a wide range of possible circuits, ranging from complex integrated transmitter chips to simple arrays of lasers that might be flip-chip bonded to an electronic IC to facilitate output coupling [1]. Thus, problems of compatibility with electronics may or may not be an important issue in all cases. In fact, the primary reasons for manufacturing lasers in integrated form may be to facilitate wafer-scale testing and multielement packaging.

The primary emphasis of the discussion will be on diode lasers. However, a summary of electro-optic modulators, which can also be used for the desired electro-optic transduction, will be given for comparison. The dual problem of optical-to-electrical transduction is the subject of Chapter 11.

ISBN 0-12-200420-5

Also, some of the issues involved in fabricating the laser structures to be discussed were introduced in Part III.

Diode lasers have gone through two decades of evolution (and occasionally revolution) as discrete components. During this process their properties have steadily improved as better technology and more stringent criteria have emerged. However, now that we have become convinced that integrated lasers are desired, if only to provide wafer-scale manufacture and testing economies, it is not clear that the evolution has led us to optimal device geometries. Thus, we need to consider carefully all the necessary criteria for a given OEIC before trying to optimize our lightwave sources. Clearly, such issues as manufacturability, reliability, and ultimate cost should be considered strongly, because these are some of the principal reasons integration is being considered in the first place.

Even with the added boundary conditions imposed by integration, diode lasers still have a few generic elements that need to be provided. There must be a gain region in which carriers can be confined, a means of effectively injecting carriers into the gain region from an electrical bias, a means of laterally confining the photons in the vicinity of the gain region for low optical propagation loss and good interaction with the carriers, and mirrors for optical feedback and desired output coupling.

Historically, diode lasers have used cleaved facets for the mirrors. The first major departure for integrated lasers is that the substrate cannot be cleaved to form mirrors. And, although some excellent work has been done with "microcleaving," a process in which a miniature cantilever is first formed by etching and then broken off to form a mirror facet [2], it seems clear that other more IC-compatible processes will gain favor in OEIC manufacture. Integrated mirrors for conventional in-plane lasers can be made either by careful etching of facets perpendicular to the laser waveguide or by forming gratings in the waveguide for Bragg reflection. Also, one type of surface-emitting laser uses a vertical cavity with multilayer Bragg mirrors grown above and below the active region during the initial epitaxial growth, so that no additional processing is necessary for the mirrors.

Figure 1 illustrates the two basic classes of diode lasers that will be considered in this work. The optical output of an edge-emitting laser is directed parallel to the plane of the substrate, and that of a surface-emitting laser is directed normal to the plane of the substrate. The internal optical propagation axis of the edge emitter is always in-plane also, whereas the internal propagation axis of a surface emitter may be either in-plane or "vertical cavity," as illustrated in the figure. The in-plane-cavity, edge-emitting laser is most appropriate for coupling into integrated planar waveguides. Such

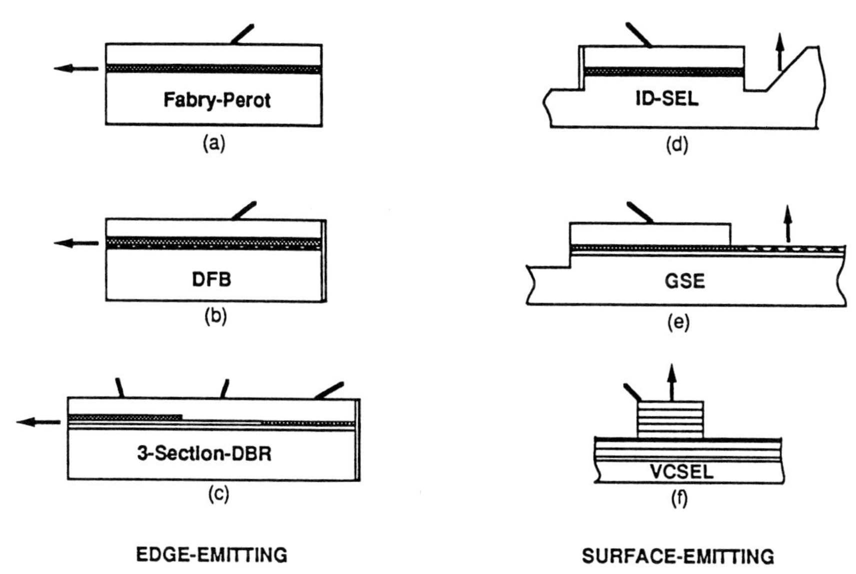

Fig. 1. Edge-emitting (a–c) and surface-emitting (d–f) emitting laser schematic cross sections.

integration is especially desirable for photonic integrated circuits, such as multisection tunable lasers, heterodyne receivers, and wavelength division multiplexing circuits primarily found in telecommunications applications. The distributed Bragg reflector (DBR) types also provide single-frequency operation, desirable in applications where dispersion may be a problem or where coherent or some other narrowband detection may be desired.

The surface-emitting laser geometries seem to be most desirable for high-density optical interconnection from circuits, although they would also seem to offer advantages for the mass production of individual devices. Wafer-scale testing is facilitated, and, in the vertical-cavity cases, it is also argued that the output beams are easier to couple to fibers or through some other optical network because of their relatively low divergence. Two-dimensional arrays are possible for the ultimate in interconnection density or compatibility with free-space interconnection, often associated with "optical computing" architectures. Such separately addressable, and perhaps multiwavelength, arrays have potential for applications in telecommunications and control of multielement systems (e.g., phased-array radars) as well as in computer interconnection. Finally, both edge-emitting and surface-emitting arrays have been coherently coupled to provide larger diffraction-limited powers than are normally possible from a single element [3–5]; however, we consider this type of integration outside the bounds of the current discussion.

To complete the discussion of transducers to convert electrical signals to optical ones, we also briefly consider electro-optic modulators. Of course, as indicated in Fig. 2, modulators require an optical bias as well as the electrical data signal in order to provide an encoded optical output. However, this complication has a few potential benefits. For example, electro-optic gain is possible if the optical bias is high enough, and a common optimized source can be used to bias numerous devices, permitting the use of coherent effects as well. It is also argued that reliable, uniform arrays of modulators may be easier to manufacture; that heteroepitaxy, such as GaAs/Si, can be used; that higher modulation speed may be possible; and that on-chip power dissipation may be less than for a laser array. The latter points have become less clear with advances in low-threshold and high-speed lasers.

Figure 2 illustrates both waveguide and "surface-normal" modulators. The waveguide types are most relevant for photonic integrated circuits (PICs), in which the input and/or output is to be propagated in a waveguide to another device along the surface of the substrate. For example, it may be desirable to externally modulate an in-plane laser for reduced wavelength chirp. In the surface-normal modulator, lightwaves arrive and leave the device nearly perpendicular to the wafer surface. This kind of structure is

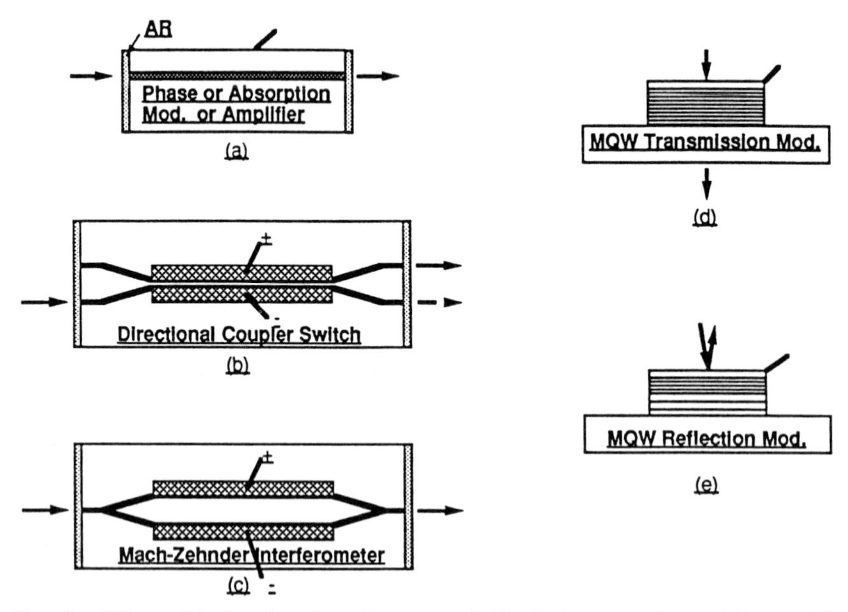

Fig. 2. Waveguide (a–c) and surface-normal (d–e) electro-optic modulators: (a), (d), and (e) are cross sections; (b) and (c) are top views.

somewhat similar to the vertical-cavity surface-emitting lasers, and, not too surprisingly, it has some of the same desirable attributes. The main additional issue with modulators as compared to lasers for optical interconnection is coupling a lightwave in as well as out. In many modulators, it turns out that there is a trade-off between optical bandwidth and modulation efficiency, so the most efficient devices can work only over a relatively narrow optical wavelength window. This may also lead to a serious limitation on the temperature range of operation for such devices, since the usable window shifts with temperature.

2. LASERS FOR OEICS

2.1. The Enabling Technologies

Any diode laser requires a gain region to compensate losses and mirrors for feedback as well as output coupling. In this subsection we shall first consider gain regions that may be desirable for OEIC compatible lasers, and then we shall review possible integrable mirror configurations.

For the purposes of this book, we must also distinguish between a manufacturing process for discrete lasers and one that leads to some level of integration. Perhaps the lowest level of integration that could qualify as relevant to OEICs is an integrated one-dimensional array of lasers. However, even this seems to be stretching the point, since even the most primitive of manufacturing techniques generates such arrays by default. Thus, we shall focus on the issues related to integrating lasers with a more complex OEIC or PIC, in which numerous other components are included. Again, we shall use the definitions that OEIC refers primarily to the integration of optical devices with electronic integrated circuits and PIC refers primarily to the integration of photonic devices with minimal electronics involved. Although our main emphasis here will be on OEICs, most of the discussion related to in-plane devices is relevant to PICs as well. With such an emphasis, then, the application of optical interconnection is given more weight than optical communication. However, in most cases the design criteria for the laser and modulator components in an OEIC will probably fit both applications.

2.1.1. Gain Regions and Simple Relationships for OEIC Lasers

A diode laser gain region amplifies lightwaves by using the stimulated recombination of injected carriers to generate additional photons. For OEICs it is desirable for this process to be very efficient, so that the drive current and thermal dissipation can be kept to a minimum. For overall maximum efficiency, the threshold current, I_{th}, must be low, and the external differential efficiency, η_d, above threshold must be high. For both, it is important that all of the diode current contribute to holes and electrons that recombine in the active region. (The fraction of the input current contributing to such carriers is called the laser internal quantum efficiency, η_i. Leakage currents and poor carrier confinement to the active region tend to reduce η_i.) For low threshold current, the active material must exhibit significant gain at low carrier densities, the cavity losses (equal to threshold gain) must be low, and the volume of the active region should be as low as possible, limited by the need for a reasonable overlap with the optical mode and the ability to dissipate heat. For high differential efficiency, the cavity losses should be dominated by output coupling rather than internal losses.

Low threshold in modern lasers is facilitated by the use of quantum well or strained-layer quantum well active regions [6–9]. Quantum wells are formed simply by growing very thin active layers ~100 Å thick. In this case the states carriers can occupy are reduced in number and are more bunched in energy compared with thicker bulk active regions [10]. Figure 3 compares the density of states functions, $\rho(E)$, for these cases. For a given injection current, it is possible to fill a larger proportion of the electron or hole states at the lowest allowed energies in quantum wells, and since a lasing transition involves only a very narrow band of electron and hole energies, the gain can be higher. Therefore, cavity losses can be compensated to reach threshold at a lower current.

Equally important is the more obvious fact that the thinner quantum well provides less active material to pump, so that the density of injected carriers (without consideration of their energy distribution) is already higher for a given current. Similarly, if the length and width of the active region are also reduced, then the current required to sustain a desired carrier density is further reduced. Thus, it should be obvious that small active volume is good, provided that a sufficient overlap with the optical mode exists. This factor is important in considering the merits of low-active-volume, vertical-cavity, surface-emitting lasers or in-plane lasers with highly reflecting facet coatings.

The addition of strain in these quantum wells allows even lower laser threshold currents for a given cavity loss [9, 11, 12]. This occurs because the highest-lying valence band can be distorted to more closely resemble the shape of the conduction band. That is, the energy density of states for holes is reduced. As a result, a greater proportion of the available hole states are filled for a given injection current, and again the gain is increased.

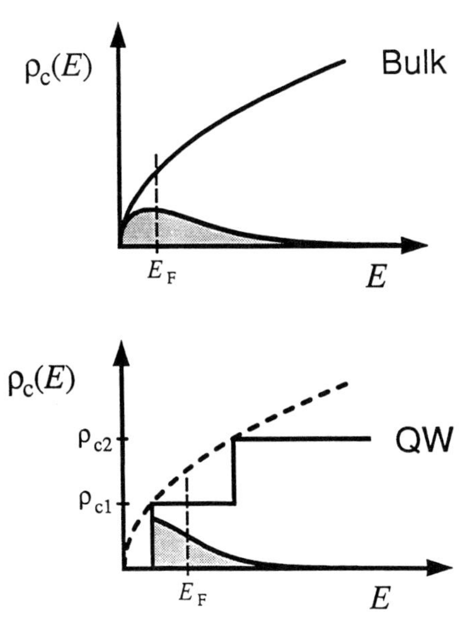

Fig. 3. Schematic density of states function $\rho_c(E)$ vs energy. E, relative to the conduction band edge for bulk and quantum-well active regions. Shaded areas represent density of filled electron states for the indicated Fermi levels, E_F. In a quantum well, each quantized level steps up the magnitude of the density of states function by a constant amount. Note that the edge of each step lines up with the bulk limit of the density of states function (indicated by the dashed line).

Typically, a carrier density of $\sim 2 \times 10^{18}$ cm^{-3} is required in III–V semiconductors to reach transparency [13–17], i.e, the point that divides gain from loss on the carrier density scale. This number is nearly the same for bulk or quantum well active regions, but the gain slope or differential gain is higher in quantum wells. The transparency value can be considerably lower for strained-layer quantum wells, and the differential gain can be even higher.

With this qualitative introduction, let us now proceed to summarize the physics of semiconductor lasers in mathematical terms, so that the dependences on various factors can be made more explicit. Again, the threshold gain, g_{th}, is determined by the optical cavity losses, which are composed of internal losses, α_i, and mirror losses, α_m. If the optical cavity is designed properly, so that most of the optical power generated above threshold is coupled out through the mirrors, most of the additional carrier flux injected appears as a useful photon flux at the laser's output. These physical concepts are embodied in the mathematical expressions for threshold modal gain, Γg_{th}, which is the net gain that must be added to the mode for threshold, and differential quantum efficiency, η_d, which is the number of photons out of the mode per electron recombining in the device above threshold.

Referring to Fig. 4a, the threshold condition is $r_1 r_2 \exp(-2j\beta_{ath} l_a) \exp(-2j\beta_{pth} l_p) = 1$, where r_1 and r_2 are the amplitude reflection coefficients of the two laser mirrors, l_a and l_p are the active and passive region lengths, $\boldsymbol{\beta}_a = \beta_a - j(\alpha_{ia} - \Gamma g)/2$ and $\boldsymbol{\beta}_p = \beta_p - j\alpha_{ip}/2$ are the active and passive region complex propagation constants, respectively, and Γ is the transverse mode confinement factor. Then, solving for the threshold gain and wavelength, we find that [13–17]

$$\Gamma g_{th} = \alpha_i + \alpha_m \tag{1a}$$

and

$$\lambda_{th} = 2(\mu_a l_a + \mu_p l_p)m, \tag{1b}$$

where $\alpha_i = \alpha_{ia} + (l_p/l_a)\alpha_{ip}$, $\alpha_m = (1/l_a)\ln(1/R)$, $R = |r_1 r_2|$ is the mean mirror power reflectivity magnitude, μ_a and μ_p are the active and passive waveguide effective indicies, and m is the longitudinal mode number. (We have assumed the reflector phases to be zero at the lasing wavelength; this can usually be accomplished by proper location of the mirror reference plane.) Threshold "modal" gain refers to the net incremental gain given to the transverse optical mode that is propagating back and forth between the cavity mirrors. Strictly speaking, it is the average transverse gain, weighted by the normalized mode cross section, $U(x,y)$, or, $\langle g \rangle = \langle U(x,y)| g(x,y) | U(x,y) \rangle$, but if $g(x,y)$ is constant over some active region (A) and zero elsewhere, it can be removed from the integral, leaving $\langle g \rangle = \Gamma g$, where $\Gamma = \int_A |U|^2 \, dx \, dy$. Figure 4 illustrates several different laser waveguide–active region configurations that have been used to confine the optical mode and the carriers transversely. For the carriers' in-plane single-quantum-well case, $\Gamma \sim 3\%$, whereas for a vertical-cavity laser it is possible for Γ to be $\sim 80\%$.

Since α_m is the output coupling cavity loss above threshold, $(\alpha_i + \alpha_m)$ is the total cavity loss, and η_i is the fraction of injected carriers recombining in the active region, the differential efficiency can be derived as [16]

$$\eta_d = \eta_i \, \alpha_m/(\alpha_i + \alpha_m). \tag{2}$$

Since the output power increases above the threshold current according to η_d, we can express the output power from mirror 1 as

$$P_{o1} = F_1 \eta_d (I - I_{th}) \, h\nu/q, \tag{3}$$

where $h\nu$ is the photon energy, q is the electronic charge, and F_1 is the

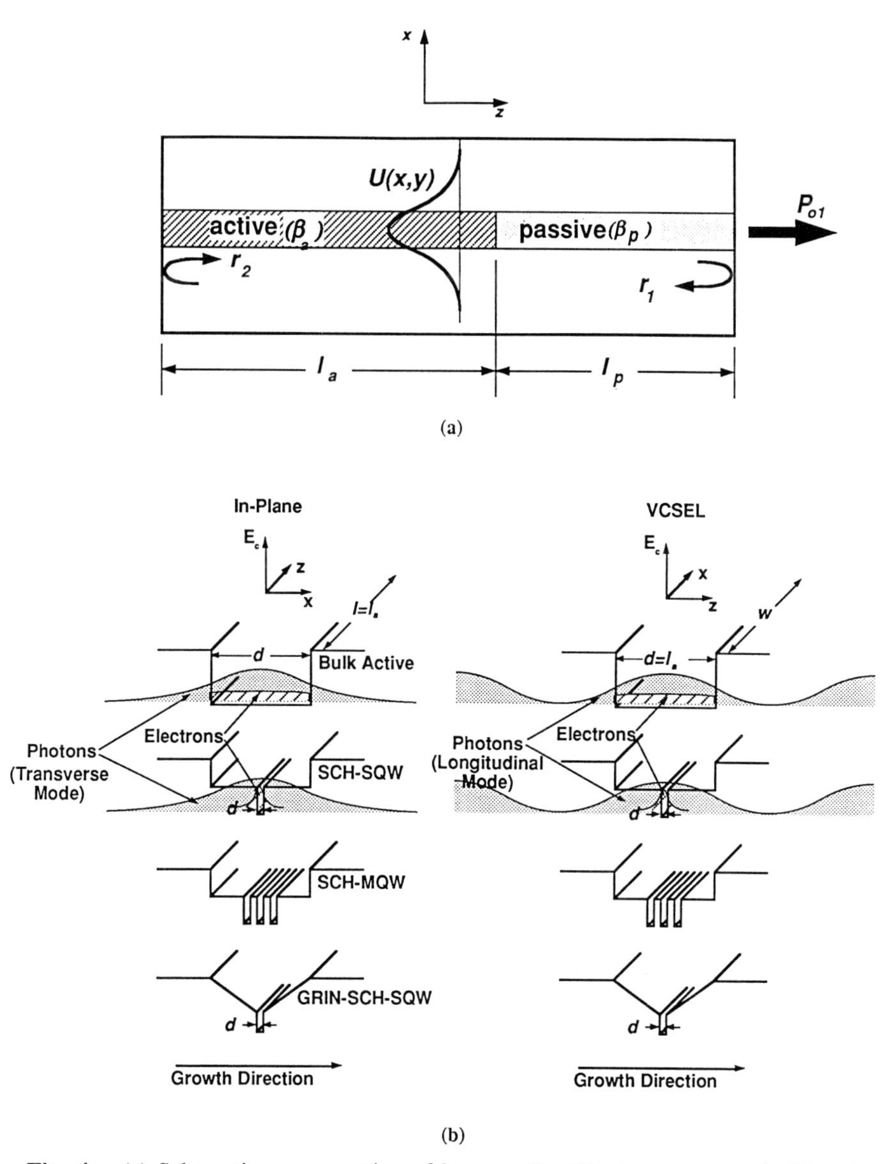

Fig. 4. (a) Schematics cross section of laser cavity. The transverse mode $U(x, y)$ is assumed to propagate in the z direction along the active and passive waveguides. (b) Schematics of conduction band edge energy E_c vs. x and z, the distances normal or perpendicular to the substrate's surface for bulk and quantum well active regions. In the quantum well cases, separate-confinement heterostructures (SCHs) are used. Photon and electron confinements are indicated by dotted and hatched regions, respectively, for both in-plane and vertical-cavity lasers. For the in-plane case, the optical wave propagates along z.

fraction of the power lost by the cavity at the mirrors that is coupled to the output. It can be shown that

$$F_1 = T_1/[(1 - R_1) + (R_1/R_2)^{1/2}(1 - R_2)],$$

where T_1 is the intensity transmission coefficient of output mirror 1, and R_1 and R_2 are the intensity reflection coefficients of mirrors 1 and 2, respectively.

Actually, Eqs. (1) to (3) have little to do with the physics of carrier injection and gain. They are primarily derived from the characteristics of the optical cavity. In this simple model, only the factor η_i, the internal quantum efficiency, is needed to describe the relevant carrier collection and confinement effects. The rest of the problem is contained in I_{th}.

In determining the threshold current, we must relate the gain to the injected current. For this, we must know a great deal about the physics of carrier–photon interactions as well as the radiative versus nonradiative recombination rates in the active region. From rather involved theoretical calculations [16–18], it can be shown that the gain for a given photon energy, E, takes on the general form

$$g(E) = K(E)\, \rho_{red}(E)[\, f_c(E_2) - f_\nu(E_1)], \tag{4}$$

where the "interaction coefficient" $K(E) = q^2h|M_T|^2/[2\varepsilon_0 c m_0^2 \mu E]$; $|M_T|^2$ is the transition matrix element; ε_0 is the free-space dielectric constant; μ is the refractive index; c is the free-space light speed; m_0 is the free-electron mass; $\rho_{red}(E)$ is a "reduced" density of states representing the density of transition state pairs, obtained under momentum conservation by using a "reduced" effective mass in $\rho_c(E_2)$; and f_c and f_ν are the conduction and valence band quasi-Fermi functions, which are determined by the electron and hole carrier densities and the initial and final electron state energies E_2 and E_1 ($E = E_2 - E_1$). The primary differences between bulk, quantum well, and strained-layer quantum well lasers are due to changes in the density of states functions, illustrated in Fig. 3, and the resulting different values of f_c and f_ν for a given injection level [10, 18].

In all cases for no current injection in undoped material, the probability of electron states being filled in the conduction band, $f_c \sim 0$, and the probability of such states being filled in the valence band, $f_\nu \sim 1$, so the material absorbs maximally. As carriers are injected, f_c increases and f_ν decreases. When sufficient carriers are injected to make f_c and f_ν equal, the material has reached transparency, and for larger current injection, $f_c > f_\nu$, so that g is positive. The Fermi functions f_c and f_ν are indirectly determined by the known number of injected carriers filling the lowest available states from

$$n = \int \rho_c(E_2)\, f_c(E_2)\, dE_2 = p = \int \rho_\nu(E_1)\, [1 - f_\nu(E_1)]\, dE_1, \tag{5}$$

where we have assumed undoped material or high injection, so that $n = p$. The shaded areas in Fig. 3 represent this integral.

In common III–V materials, such as GaAs and InP, the valence band density of states is much larger than in the conduction band. Thus, as equal numbers of holes and electrons are injected, a larger fraction of the conduction band states must be filled. In fact, for transparency just above the bandgap energy in bulk GaAs, $f_c = f_v \sim 0.8$, indicating that most of the conduction band electron states are filled, while only a small fraction of the valence band states are empty. For four times this injection, $f_c \sim 1$, while $f_v \sim 0.6$ at this same energy. Thus, according to Eq. (3), only about 40% of the maximum absorption has been turned over into gain.

Recently, there has been a lot of excitement over strained active regions in diode lasers. The primary reason is that the density of states in the highest-lying valence band can be significantly reduced, so that $f_c - f_v$ can approach unity under strong injection. Thus, higher maximum gain is possible and, perhaps more important, transparency is reached for $f_c = f_v \sim 0.5$, which requires much less carrier injection than for the unstrained case.

Now, to relate the carrier density n, to the current I, we equate the supply rate of injected carriers J/qd to the recombination rate $\mathcal{R} = n/\tau_e$ below threshold. Since the current density contributing carriers to the active region, $J = \eta_i I/wl$,

$$\eta_i I/(qdwl) = \mathcal{R}(n) = n/\tau_e, \tag{6}$$

where d is the active region thickness measured perpendicular to the current flow, w and l are the width and length of the active region measured in the plane, the active region volume $V = dwl$, and τ_e is the carrier lifetime. (Note that the earlier defined active cavity "length" l_a, may be either l or d, dependent on whether an in-plane or vertical-cavity structure is used, respectively.) The recombination rate $\mathcal{R}$ can be split into radiative $\mathcal{R}_{rad}$ and non-radiative $\mathcal{R}_{nr}$ terms, and it is also sometimes expanded into a powers series $(An + Bn^2 + Cn^3)$, where the various terms approximate defect (An), spontaneous (Bn^2), and Auger (Cn^3) recombination, respectively [14, 19]. Since the carrier density is clamped above threshold for uniform active region gain, $\mathcal{R}$ does not change above threshold.

With Eqs. (1 to 6) we have completely specified the power out versus current into some laser structure. However, the relationships are parametric and/or transcendental, so a closed-form analytical expression is not possible. The primary problem is with the gain expression. For an accurate appraisal of a laser's characteristics, it is best to use a computer calculation of Eq. (4) [18, 20–22]. Figure 5 gives the results of such a calculation for

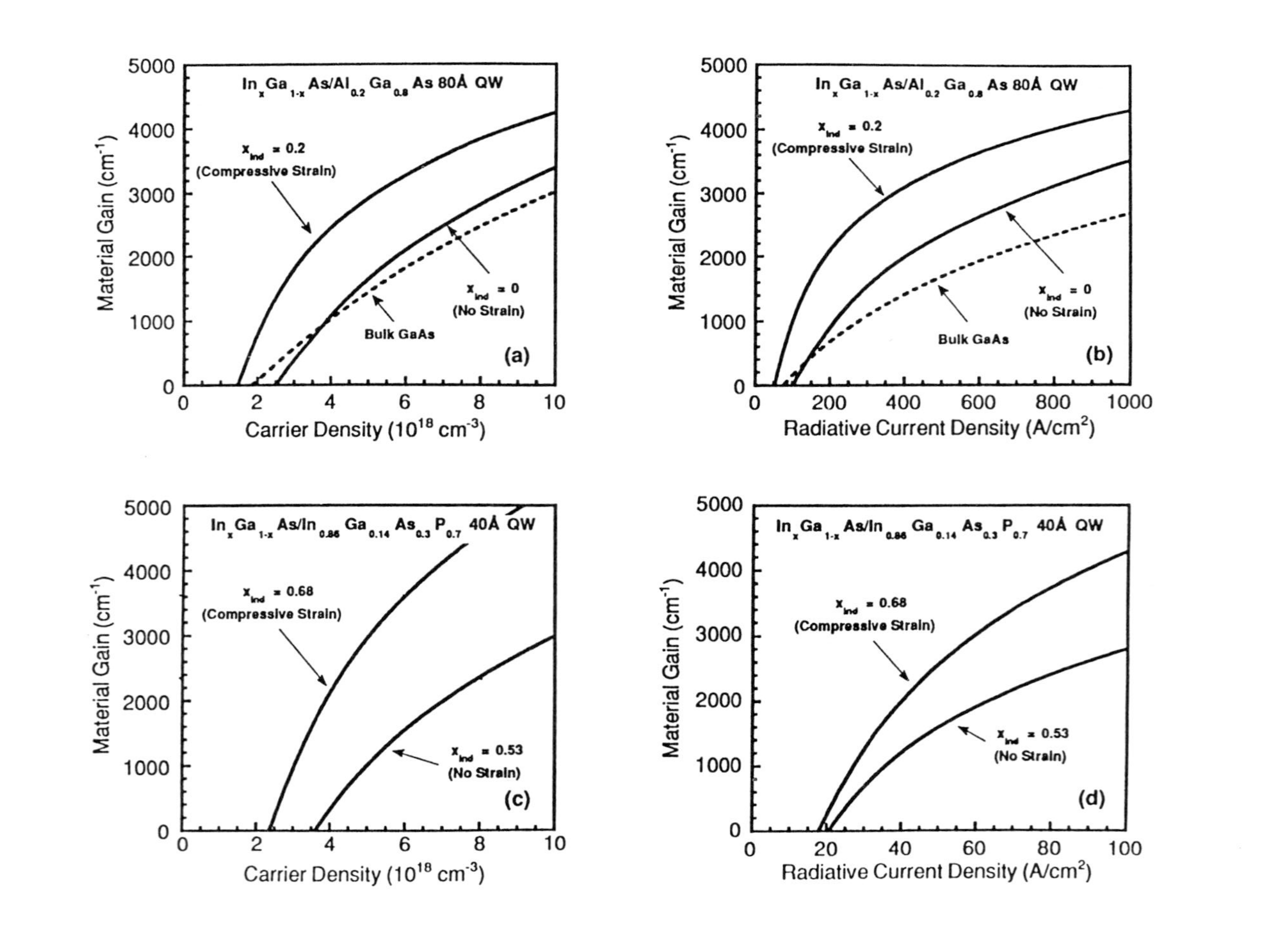
In x Ga 1-x As/Al 0.2 Ga 0.8 As 80Å QW
x Ind = 0.2
(Compressive Strain)
x Ind = 0
(No Strain)
Bulk GaAs
(a)
Material Gain (cm-1)
Carrier Density (10^18 cm-3)
In x Ga 1-x As/Al 0.2 Ga 0.8 As 80Å QW
x Ind = 0.2
(Compressive Strain)
x Ind = 0
(No Strain)
Bulk GaAs
(b)
Material Gain (cm-1)
Radiative Current Density (A/cm2)
In x Ga 1-x As/In 0.86 Ga 0.14 As 0.3 P 0.7 40Å QW
x Ind = 0.68
(Compressive Strain)
x Ind = 0.53
(No Strain)
(c)
Material Gain (cm-1)
Carrier Density (10^18 cm-3)
In x Ga 1-x As/In 0.86 Ga 0.14 As 0.3 P 0.7 40Å QW
x Ind = 0.68
(Compressive Strain)
x Ind = 0.53
(No Strain)
(d)
Material Gain (cm-1)
Radiative Current Density (A/cm2)

several cases [18]. Here, active layers of a certain thickness were assumed, and the effects of band mixing in the valence band were also included. For comparison, examples of quantum wells and strained-layer quantum wells are given. A reference "bulk" curve that neglects quantum effects is also given. Only radiative recombination is included in the radiative current density, $J_{rad} = qd\mathcal{R}_{rad}$. If other nonradiative recombination is important, an additional current density $J_{nr} = qd\mathcal{R}_{nr}$, necessary to sustain these effects, must be added.

In order to derive a closed-form expression for the output power vs. current in, the gain vs. carrier density in Fig. 5 can be approximated by a simple analytic formula,

$$g \cong g_0' \ln[(n + n_s)/(n_o + n_s)]. \tag{7a}$$

In this approximation, g_0' is an empirical gain coefficient, n_0 is the transparency carrier density, and n_s is a shift to force the natural logarithm to be finite at $n = 0$ so that the gain equals the unpumped absorption, $K(E)\,\rho_{red}(E)$. However, if we restrict our attention to positive gains, $g \geq 0$, Eq. (7a) can be further approximated as

$$g \cong g_0 \ln(n/n_0), \qquad (g \geq 0), \tag{7b}$$

provided that we use a new gain coefficient g_0. In this case the differential gain, $\partial g/\partial n = g_0/n$. As shown by Fig. 5, n_0 and $\partial g/\partial n$ will be quite different for bulk, quantum well, and strained-layer quantum well active regions, and this is the basis for many of the arguments for and against certain of these structures. Fitting Eq. (7b) to these plots, the strained 80-Å InGaAs/GaAs quantum well yields $g_0 \sim 2500$ cm^{-1} and $n_0 \sim 1.5 \times 10^{18}$ cm^{-3}, and the 80-Å GaAs quantum well gives $g_0 \sim 2400$ cm^{-1} and $n_0 \sim 2.5 \times 10^{18}$ cm^{-3}. For the InP substrate cases, the strained 40-Å InGaAs/InP gives $g_0 \sim 4000$ cm^{-1} and $n_0 \sim 2.4 \times 10^{18}$ cm^{-3}, and the unstrained 40-Å InGaAs quantum well gives $g_0 \sim 3000$ cm^{-1} and $n_0 \sim 3.6 \times 10^{18}$ cm^{-3}.

Fig. 5. Material gain vs. carrier density, n, and radiative current density, J_{rad}, for strained and unstrained quantum wells on GaAs (top) and InP (bottom). Dashed (bulk) curves represent gain for identical thickness GaAs layer ignoring quantum effects. (From Ref. 18.) When grown on a GaAs substrate, InGaAs is compressively strained when the indium mole fraction, $x_{Ind} > 0$. On an InP substrate, InGaAs is lattice matched when $x_{Ind} = 0.5321$; it is compressively strained when $x_{Ind} > 0.5321$; and tensile strain is produced when $x_{Ind} < 0.5321$. Notice the marked improvement in gain characteristics in both material systems when compressive strain is used.

Now we can combine Eqs. (1) and (7b) to get the threshold carrier density,

$$n_{th} \cong n_0 \exp[g_{th}/g_0] = n_0 \exp[(\alpha_i + \alpha_m)/\Gamma g_0]. \tag{8}$$

From Eq. (6), $I_{th} = qV\mathcal{R}(n_{th})/\eta_i$, and for the best laser material the recombination at threshold is dominated by spontaneous recombination, or $\mathcal{R} = \mathcal{R}_{rad} \cong Bn^2$. Thus,

$$I_{th} \cong qVBn_0^2 \exp[2(\alpha_i + \alpha_m)/\Gamma g_0]/\eta_i \tag{9}$$

where for most III–Vs of interest the bimolecular recombination coefficient $B \sim 10^{-10}$ cm^3/s. Equation (3) together with (9) can now be used for a closed-form expression of output power vs. applied current. However, since we are usually trying to minimize the current needed for a given required power from one mirror, P_{01}, we solve for I:

$$I \cong qP_{01} (\alpha_i + \alpha_m)/(F_1\eta_i h\nu\, \alpha_m) + qVBn_0^2 \exp[2(\alpha_i + \alpha_m)/\Gamma g_0]/\eta_i, \tag{10}$$

where the first term is the additional current required above threshold to obtain power P_{01}, and the second term is the threshold current, or Eq. (9).

For best accuracy, one is better off using Eq. (1) in combination with Fig. 5 for the threshold current term. Alternatively, if one starts with the three-parameter fit of Eq. (7a) and follows through with the same substitutions, a more accurate threshold current expression can be obtained. Nevertheless, Eqs. (9) and (10) give reasonable accuracy in simple analytic expressions which correctly show that it is always desirable to reduce the transparency value and increase the differential gain of the active material. Both points argue in favor of using quantum well, especially strained-layer quantum well, active regions. Figure 6 compares experimental threshold currents on a variety of strained and unstrained quantum well lasers on GaAs and InP [23, 24]. Relative to the cavity design, the equations also indicate that it is desirable to reduce the cavity loss $(\alpha_i + \alpha_m)$ and volume V, subject to retaining a reasonably large transverse confinement factor, Γ. (Note that $\Gamma \propto d$ for in-plane SCH lasers, but it depends only weakly on w and l in vertical-cavity lasers.) Thus, the merits of using vertical-cavity surface emitters or short-cavity in-plane lasers with coated facets are also suggested.

Due to the exponential dependence on g_{th}/g_0 in Eqs. (9) and (10), it may be beneficial to use more than one well to increase Γ as indicated in Fig. 4. Again, this dependence is a result of the saturation of the gain as the carrier density is increased to nearly fill the lowest set of states indicated in Fig. 3. Thus, by distributing the carriers over N_w wells, the gain per well is reduced by less than N_w times, but the modal gain is still multiplied by

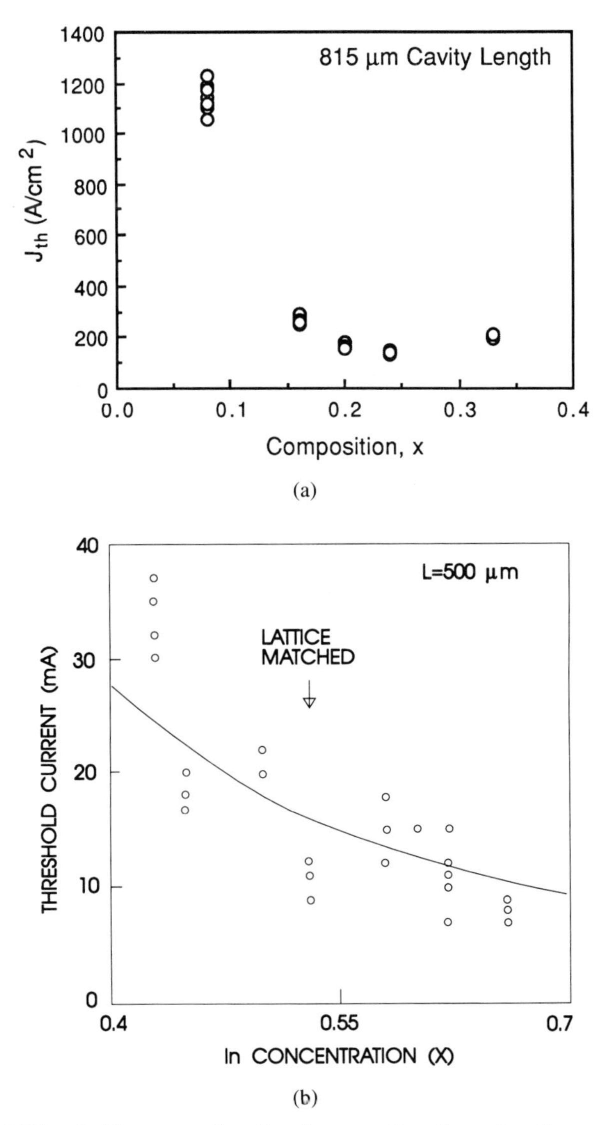

Fig. 6. (a) Threshold current density, J_{th}, as a function of well composition for 150 × 815 μm cleaved lasers with 70-Å $In_xGa_{1-x}As$ quantum wells on a GaAs substrate. (From Ref. 23.) (b) Threshold current of buried-heterostructure $In_xGa_{1-x}As$/InP multiple-quantum-well lasers vs. well composition. (From Ref. 24.)

nearly N_w times this value. For such multiple-quantum-well (MQW) lasers Eqs. (9) and (10) are still valid, but one must be sure to multiply the single-well confinement factor Γ_1 and volume V_1 by the number of wells N_w. That is, for an MQW laser, from Eq. (9) or the second term in (10), one can explicitly write

$$I_{thMQW} \cong qN_wV_1Bn_o^2 \exp[2(\alpha_i + \alpha_m)/(N_w\Gamma_1 g_0)]/\eta_i. \quad (11)$$

Here, we have assumed a separate confinement waveguide, so that the optical mode does not change significantly as more wells are added. Also, the number of wells is limited to the number that can be placed near the maximum of the optical mode. Of course, the optimum number of wells is the number that minimizes Eq. (11), neglecting nonradiative recombination.

Another factor that has not been discussed so far is modulation bandwidth. The 3-dB bandwidth of a diode laser is about 30% higher than the resonance frequency, f_r, if parasitics can be neglected [25]. Figure 7 illustrates an experimental small-signal response of a strained-layer SQW laser, which illustrates the rapid roll-off above resonance and the relatively large resonance peak at low biases [26]. To avoid excess frequency chirping and pattern effects due to this intensity ringing, systems designers sometimes prefer to use a modulation bandwidth less than f_r. The resonance frequency can be approximated by

$$f_r = (1/2\pi)\,[\Gamma g_0 l_a \eta_i v_g (I - I_{th})/(qVl_t n_{th})]^{1/2}, \quad (12)$$

where v_g is the mode group velocity, $(\partial g/\partial n)|_{n_{th}} = g_0/n_{th}$, and $l_t = l_a + l_p$.

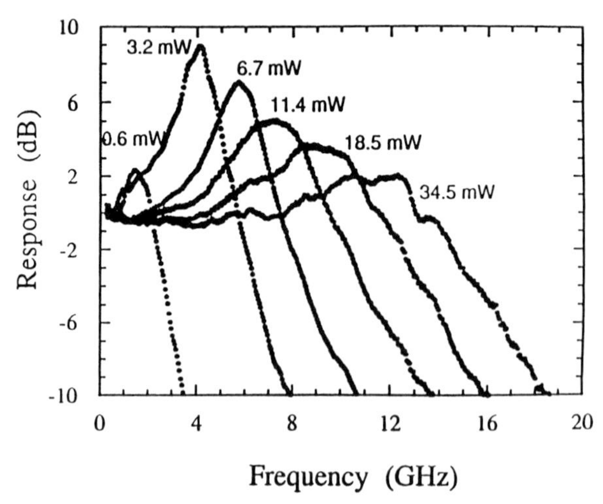

Fig. 7. Modulation response of a strained SQW laser at various power levels for a device with a 400-μm-long cavity and a 2-μm-wide ridge. (From Ref. 26.)

Equation (12) also suggests the desirability of having a large differential gain and a small active region volume, again provided that the transverse confinement factor is not compromised. This argues in exactly the same direction as the preceding arguments for low current. That is, a short-length in-plane laser with facet coatings or a vertical-cavity laser should be fast. Another important factor in Eq. (12) relevant to the balance between the two terms in Eq. (10) is $(I - I_{th})^{1/2}$. Figure 7 illustrates this square-root dependence on current. Thus, if the device is to be fast, it may be necessary to reduce the threshold term at the expense of increasing the above-threshold term, so that the operating point for a given power out is placed well above the threshold value. This may be an important consideration, since one is generally considering relatively high-speed links when optical interconnection is to be employed.

If nonradiative recombination is important at threshold, an additional nonradiative threshold current component must be added as outlined earlier. For the long-wavelength InGaAsP/InP materials, nonradiative recombination is known to be very important [14, 27, 28]. In fact, were it not for such recombination, the threshold current densities of lasers using such materials would be lower than those using GaAs quantum wells, as indicated in Fig. 5, in which only J_{rad} is plotted. If such higher-order nonradiative carrier recombination is important at threshold, one must add another component to the threshold current due to $\Re_{nr}(n_{th}) \cong Cn_{th}^3$. Then, Eqs. (9) and (10) should be increased by

$$I_{nr_{th}} \cong CqVn_0^3 \exp[3(\alpha_i + \alpha_m)/\Gamma g_0]/\eta_i, \tag{13}$$

where for 1.3-μm InGaAsP material, the Auger coefficient $C \sim 3 \times 10^{-29}$ cm^6/s, and for 1.55-μm material it is about two or three times larger [29–31]. The cubic dependence on n_{th} places more importance on reducing the threshold carrier density in this material system. In fact, this additive Auger term dominates Eq. (9) for carrier densities above n_{th} 3×10^{18} or 1.5×10^{18} cm^{-3} at 1.3 and 1.55 μm, respectively. This fact focuses more attention on reducing cavity losses, $(\alpha_i + \alpha_m)$, and maintaining a large confinement factor Γ. Primarily for these reasons, quantum wells have not been too successful in long-wavelength lasers, since quantum wells usually imply higher threshold carrier densities and smaller confinement. This problem is somewhat mitigated by the use of MQW structures, but the improvement over a single bulk layer is marginal, especially in the InGaAsP system, where the conduction band offset is relatively small. Nevertheless, the internal losses do tend to be smaller with an MQW structure, which leads to some clear advantages. However, with the advent of strained-layer InGaAs/InGaAlAs quantum wells on InP, the improvement may be more dramatic, since all the parameters affecting n_{th} move in the right direction. In fact, the Auger coefficient C, itself, may also be reduced due to the splitting of the valence bands [32, 33]. Thus, even at long wavelengths we may eventually be able to approximate the threshold currents by Eq. (9) without the additive term of Eq. (13).

2.1.2 Mirrors and Output Couplers for OEIC Lasers

As may be evident from the discussion of required laser drive current for a given power out, the effective mirror loss $\alpha_m = (1/l_a)\ln(1/R)$ should be kept low for low laser threshold, but it must be larger than the internal incremental losses α_i for a large fraction of the generated light to be coupled out in a useful form. [Don't forget that $\alpha_i = \alpha_{ia} + (l_p/l_a)\alpha_{ip}$, so that losses in any passive section are effectively "amplified"; this is especially important in vertical-cavity lasers.] Basically, α_m is a useful output coupling loss, but it is still a loss. The explicit mathematical statement of the associated tradeoff is given by Eq. (10), in which the second term is the laser threshold and the first term is the additional current required for a given power out above threshold. To make the second term small, it is desired to make α_m small; to make the first term small, it is desired to make α_m large. Figure 8 gives a simple schematic illustrating this trade-off between the first and second terms of Eq. (10). As α_m is increased, the threshold goes up, but so does the slope. Thus, the curves cross at some point providing the same operating point. However, as also indicated, for minimum current at a given power out, all else being constant, there is an optimum α_m.

As discussed, the argument for high speed also argues in favor of reducing

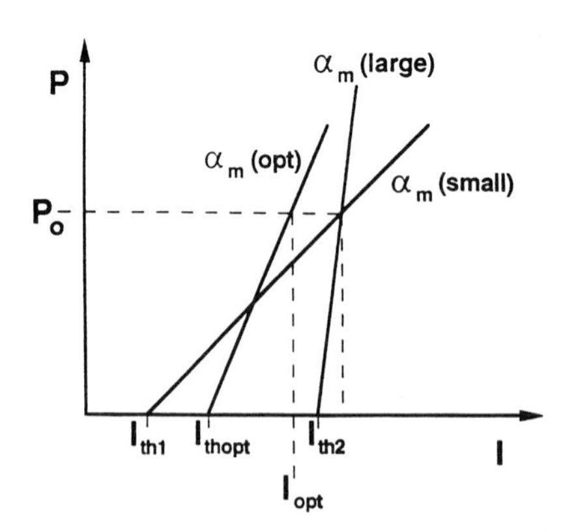

Fig. 8. Schematic output power vs. injection current for different mirror reflectivities; α_m (opt) gives lowest current for given power out.

the mirror loss, perhaps to even below the optimum value for minimum current indicated in Fig. 8. Again, the important factor in Eq. (12) relevant to this discussion is $(I - I_{th})^{1/2}$. That is, if the device is to be fast, it may be necessary to reduce the threshold below the optimum value for minimum total current so that the operating point is at several times the threshold value.

Provided sufficient speed can be accomplished, a counter argument is that only the slope efficiency matters in determining the *modulation* current amplitude. That is, if we suppose that the laser is operated by dc biasing it to some cw point and then adding a certain modulation signal, the magnitude of the modulated power is determined only by the slope dP/dI about the bias point. This argues for a larger than optimum α_m, as also indicated in Fig. 8. Thus, it may be difficult to make any global decision about the desired tradeoff until all aspects of the particular application are clear. In what follows, we shall continue to assume that the net bias current should be kept small.

Etched Facets. For integrable in-plane-cavity lasers, whether surface emitting or not, etched mirrors have emerged as one of the preferred technologies [34]. These mirrors mimic cleaves in that they are fabricated after all growth and most processing is complete, and they form a simple discrete reflection plane that should have very predictable properties. They are particularly interesting if the laser's output is not to be coupled to an integrated waveguide structure. For such in-plane active–passive waveguide coupling, a strong argument can be made for grating mirrors [35], although the technology of growing lasers over predefined gratings still frightens many potential OEIC manufacturers. As discussed earlier the use of in-plane grating couplers will also be reviewed in the PIC section of Part V.

Attempts to make high-quality, reproducible etched facets have taken several paths over the past decade or so. Initial attempts to make etched facet lasers used wet chemical etching [36], and over the years some relatively good results have been obtained with this technique [37, 38]. However, primarily because of the difficulty in reproducing these good results in a multilayer laser structure, anisotropic dry etching has become more important [39–41]. Of course, dry etching also has had a long history of less than perfect results, which date back over a decade to the first work of the author [41]. Nevertheless, high-yield wafer-scale processing of dry etched-facet lasers has been demonstrated [42]. Although some reliability issues still are unclear, it would appear that such processing can be used for the large-scale manufacture of integrated lasers.

Some of best recent results were discussed in Chapter 7. As outlined there, it is currently possible to create an entire wafer full of etched-facet lasers with high performance and good yield. Numerous other researchers have obtained similar quality facets on a somewhat smaller scale [43, 44]. For surface emission, 45° angled mirrors have also been used either internal [45–47] or external [48–50] to the cavity. Figure 9 shows shows a sampling of scanning electron micrographs indicating the appearance of some of these results. In the sections to follow we shall review the device results in some detail. Here, we would like to discuss the dry etching fabrication technologies.

As mentioned, the first dry-etched facet lasers were formed on InGaAsP/InP lasers by reactive ion etching (RIE). By RIE we refer to the use of a reactive gas in a parallel-plate RF sputter etching geometry [39, 41, 51–53]. Generally, higher etch rates can be obtained even with reduced power density and cathode voltage as compared to sputter etching or ion milling. Also, since the etch products are volatile (a criterion for an RIE gas–substrate combination) and no redeposition occurs, very deep and narrow features can be obtained. The etched wall follows the direction of the ions, which impinge normal to the substrate due to the parallel-plate geometry. RIE is the most widely used reactive etching process. The system is simple and robust, and it can easily be scaled to a large size for production. Even though the ion voltage can be reduced to values of ~200 V while retaining a very planar anisotropic etch, there continues to be some difficulty with etching damage (especially for ions above 300 V), which may limit device reliability.

Over the years a number of other reactive dry etching processes have emerged. Figure 10 is an attempt to summarize some of these. Reactive ion beam etching (RIBE) describes the process of using a reactive gas in an ion-beam etching (IBE) gun to do reactive etching. Although the first chlorine RIBE of GaAs was carried out by the author and co-workers [54], most of the successful laser facet work has been carried out by Asakawa and co-workers [55, 56]. His work has also focused on the use of a microwave electron cyclotron resonance (ECR) excited plasma in the gun. One of the main advantages of the RIBE process is that the substrate can be accurately angled with respect to the ion beam to create angled facets. (Actually, it is possible to mimic such etching in an RIE system if the substrates are placed in a field-free region either behind a grid at the cathode potential or in a recess within the cathode [57–58].) Asakawa has also demonstrated radical beam etching (RBE), in which unaccelerated chlorine radicals, generated in a microwave ECR plasma, are used to do purely chemical dry etching [59].

(a)

Fig. 9. Scanning electron micrographs of facet and surface morphologies for various etching procedures. (a) Nonselective RIE of GaAs/AlGaAs double heterostructures. The Cl_2 pressure is 0.5 mtorr, the self-bias voltage is −350 V, and the etching time is 45 min. The ~1-μm-thick dark layer on top is hard-baked photoresist. (From Ref. 64.) (b) Angled RIE etched GaAs substrates using recessed substrate

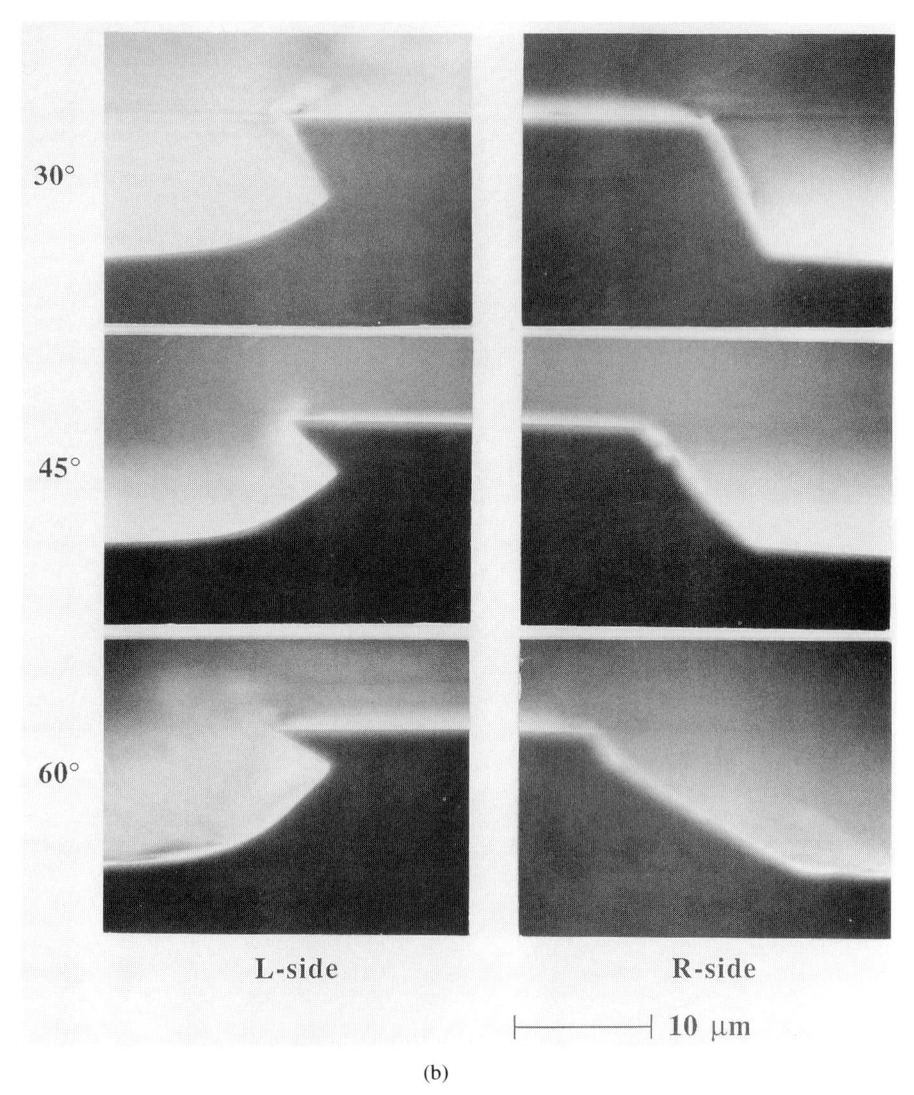

(b)

Fig. 9. (*continued*) holder. The Cl_2 pressure is 0.5 mtorr, the self-bias voltage is −450 V, and the etching time is 60 min. (From Ref. 58.) (c) Ar^+ ion-milled (IBE) facet of a GaAs/AlGaAs heterostructure laser. (From Ref. 49.) (d) Anisotropic RIBE etched profile in an AlGaAs/GaAs heterostructure wafer with corresponding schematic of heteroepitaxial layers. The pressure is 0.6 mtorr, the $Cl_2{}^+$ ion voltage is 300 V, and the etching time is 30 min. (From Ref. 55.) (e) RBIBE etched facet on GaAs. The Cl_2 pressure is 0.8 mtorr, the Ar^+ ion voltage is 200 V, and the etching time is 7 min. (From Ref. 61.)

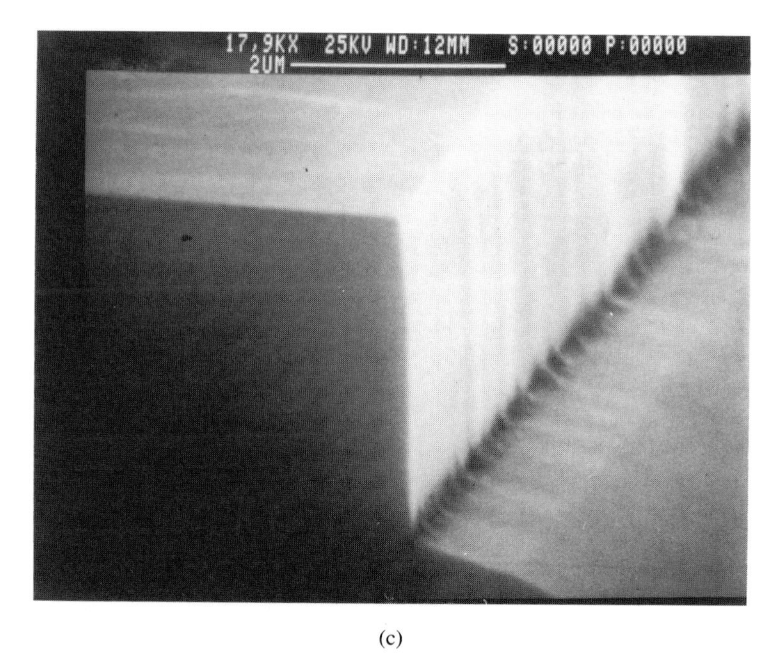

(c)

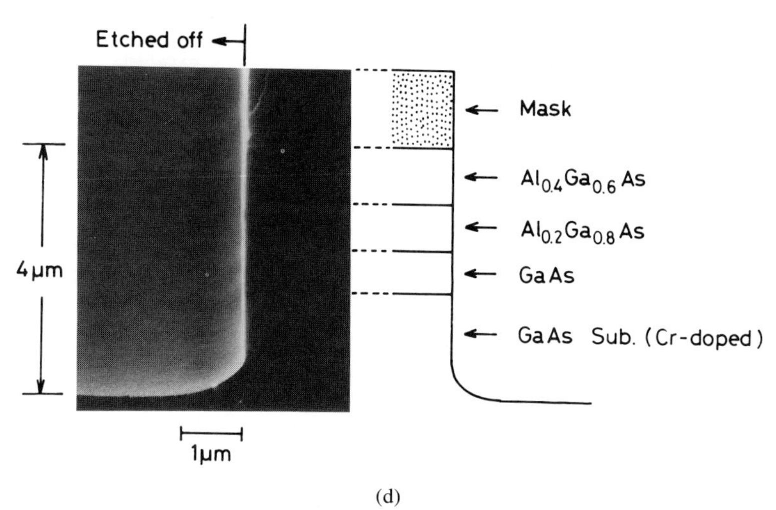

(d)

Fig. 9. (*continued*)

This results in a very low damage etch, but the etching is not anisotropic as desired for planar laser facets.

In order to keep the chlorine out of their ion guns but still gain some of the advantages of reactive etching, some workers have used a chemically assisted ion-beam etching (CAIBE). In this case a more conventional argon

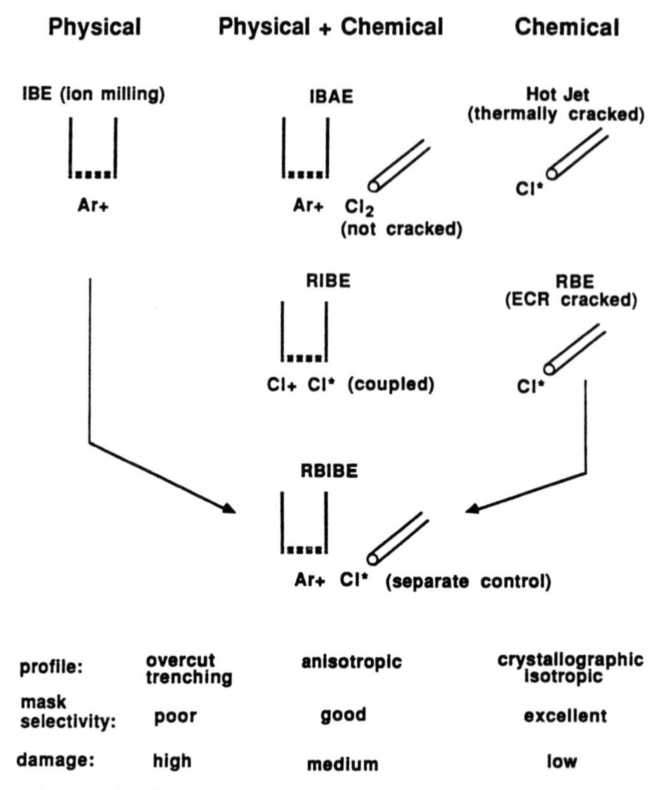

Fig. 10. History of existing dry etching technologies. RBIBE combines RBE with IBE to allow the entire etching spectrum to be spanned. The etched facet profile, mask selectivity, and damage are varied as the chemical and physical etching components are varied. Ion-beam assisted etching (IBAE) is also referred to as CAIBE.

ion-beam etching system is used together with chlorine gas metered into the etch chamber simultaneously [60]. This again allows the use of relatively low energy (~200 V) Ar ions for low damage, but it provides for an increased etch rate from the chemical enhancement. A further refinement on this process has been to combine RBE with CAIBE to provide radical-beam ion-beam etching (RBIBE), in which the chlorine gas is first cracked in a microwave cavity before being directed to the substrate being bombarded by an Ar ion beam [61]. Uniquely, this system allows a complete range of etching, from purely chemical (RBE), to purely physical (IBE). It has been shown to provide nearly damage-free anisotropic etching [62].

After the first RIE work, which used a chlorine–oxygen mixture on InGaAsP/InP lasers, there has been quite a bit of work, mostly on AlGaAs/GaAs, using pure chlorine gas [63, 64]. Several different Freon gases have also been used in an RIE process [39, 65], and in spite of their attractiveness from a safety standpoint, the use of such carbon-containing gases has some drawbacks. First, if subsequent processing is going to be carried out, there may be a problem with carbon contamination of the device, which will compromise its properties. Second, carbon-containing gases tend to form polymers that coat the insides of the etching systems. Thus, the systems develop a "memory" which results in a dependence of the current etching properties on past history. Poor etch rate and wall slope control can result. Finally, the etching chemistry can get complicated, leading to difficulties in extending existing experience to a new situation. Nevertheless, some workers still favor a Freon-based process. On the other hand, pure chlorine RIE is very clean and the chemistry is relatively simple. As with the Freons, simple photoresist can be used as an etch mask, although for the best side-wall control either a trilevel resist process or a deposited metal or dielectric masking pattern is used.

For AlGsAs/GaAs etching, chlorine RIE, RIBE, or RBIBE can be carried out at room temperature, since the AlCl and GaCl etch products have a fairly high vapor pressure. In a clean load-locked system the etch rates for AlGaAs and GaAs appear to be about the same. In systems with residual water vapor and oxygen the Al-containing alloys etch at a much lower rate. Thus, this fact can also be used to obtain a high degree ($\sim 35:1$) of etch selectivity of GaAs over AlGaAs if oxygen is added [66]. Selective etching of GaAs on AlGaAs has also been demonstrated in Freon 12 [67]. There are no known good selective etches of AlGaAs on GaAs.

Room-temperature chlorine reactive etching of InGaAsP/InP is difficult because the etch product InCl is less volatile. This problem can be addressed by using more of a physical etch component to sputter away the InCl, or the temperature of the substrate can be raised [52, 68]. Increasing the physical etch component can be accomplished by increasing the ion beam voltage. Unfortunately, it is well known that the threshold for damaging InP is lower than for GaAs, so inevitably more damage is created. Fortunately, the damage seems to be less of a problem in InGaAsP/InP lasers because the surface, interface, and defect recombination velocity is about two orders of magnitude lower than in AlGaAs/GaAs lasers. Addition of oxygen seems to increase the etch rate of InP and InGaAsP at lower temperatures. Although this is not fully understood, it is possible that some higher-volatility perchlorates are formed. Adding oxygen is desired if one wishes to oxidize a metal mask material, such as titanium, to prevent it from etching. Of course, for the AlGaInAs/InP system, the addition of oxygen can result in a selective etching of the InP over the AlGaInAs. Also, one must keep in mind that to use the more exotic RIBE and RBIBE processes it is important to use gases that are compatible with the ion gun. Oxygen will certainly attack hot filaments. Generally, an ECR excited ion gun is best whenever reactive gases are used, although it has been found that hot tungsten filaments can survive chlorine environments if the filament is kept very hot during exposure to the gas [69]. (That is, the filament is most vulnerable when only warm.)

InP and InGaAsP have also been successfully etched in a methane-containing plasma [70]. The etching chemistry here can form metal organics similar to those used in MOCVD deposition of these compounds. Some very straight-walled and smooth etching has been reported. However, the author is not aware of any determined effort to form laser facets with this process. Also, the etch products can be rather messy, so only RIE is advised here.

Grating Mirror Formation. As indicated in Fig. 1, grating mirrors can be used for laser feedback and output coupling in several different geometries. One of their key distinguishing features is that their net reflection coefficient is large only over a small wavelength range, because it is due to the summation of small reflection components distributed over many wavelengths of propagation distance. Thus, the relative reflection bandwidth of a grating, $\Delta f/f_0$, is approximately equal to the reciprocal of the number of reflection elements, or $\lambda/2l_g$, since the reflection components from the end of the grating must still be nearly in phase with those from the first components. For values of the grating reflection, r_g, approaching unity, the bandwidth gets wider as the effective penetration depth, l_{ge}, of the optical power into the grating is reduced, so that l_g must be replaced by l_{ge}. Figure 11 gives example normalized plots of grating reflection spectra for different discontinuity strengths. Because of their limited bandwidth, grating reflectors can be used to make single-frequency lasers, since only one longitudinal mode would have a large reflection coefficient and a low mirror loss.

The complex reflection coefficient, $\boldsymbol{r}_g(\lambda)$, and the penetration depth, l_{ge}, are related to the strength of the grating discontinuities, usually defined by a constant κ, which is actually the coupling constant between forward and backward waves [71–74]. For a grating composed of a square-wave index discontinuity, $\pm\Delta\mu$; $\kappa = (2/\Lambda)(\Delta\mu/\mu)$, where the grating period $\Lambda = \lambda/2\mu$ at the Bragg condition [72, 73]. For other grating shapes κ is scaled according to the relative Fourier coefficients. Of course, one must realize that the effective index grating will have a very different shape from the physical grating etched along one side of a waveguide. Therefore, κ is usu-

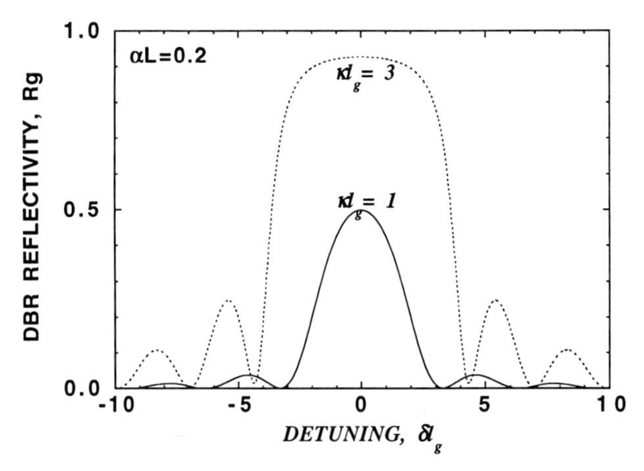

Fig. 11. Power reflectivity for a distributed Bragg reflector as a function of relative detuning δL from the Bragg wavelength for two values of κl_g. $\alpha_g l_g$ represents the power propagation loss through the grating.

ally calculated from some effective index perturbation theory [74]. In any case, once κ is derived the complex reflection coefficient is [74]

$$r_g(\lambda) = \kappa \tanh \gamma l_g / [\gamma + j\delta \tanh \gamma l_g], \tag{14}$$

where $\delta = \beta - \beta_0$, β_0 is the propagation constant at the Bragg frequency, and $\gamma = (\kappa^2 - \delta^2)^{1/2}$. (The grating reflection reference plane is assumed to be at the first effective index down-step, so that the reflection phase at the Bragg wavelength is zero.) The maximum magnitude of $\boldsymbol{r}_g(\lambda)$ is

$$|r_g|_{max} = \tanh \kappa l_g, \tag{15}$$

and the effective penetration depth, defined as half of the reflection phase slope near the reflection maximum, $(\partial\phi_R/\partial\beta)/2$, is

$$l_{ge} = (1/2\kappa) \tanh \kappa l_g. \tag{16}$$

The maximum reflection magnitude, $|\boldsymbol{r}_g|_{max}$, can also be obtained directly from the effective indicies in a square-wave grating (or quarter-wave stack) [75]. Assuming j quarter-wave layers ($j/2$ grating periods) and an effective index that switches between a low value of μ_l and a high value of μ_h,

$$|r_g|_{max} = (1 - b)/(1 + b), \tag{17}$$

where

$$b = [\mu_l/\mu_h]^{j+1}.$$

In a laser with one discrete mirror and one grating mirror, the grating reflector phase and energy storage can be approximated by using a total cavity length, $l_t = l_a + l_{ge}$, equal to the mirror spacing plus the penetration depth into the grating, for calculating either mode spacing or internal loss, α_i. (By using l_a for the mirror spacing and $l_{ge} = l_p$, we have assumed that the cavity is filled with active material; if not, of course, the mirror spacing must also include any additional passive length.) The mirror loss for a particular mode is still given by the expression after Eq. (1) with $R = |r_1 \boldsymbol{r}_g(\lambda)|$.

The use of gratings as passive end mirrors for a laser cavity is conceptually simple, due to the analogy with cleaved mirrors. Although simple in concept, the fabrication of these so-called distributed Bragg reflector (DBR) lasers can be very complex, since an active–passive waveguide transition must exist, and several material regrowths may be necessary [76, 77]. Nevertheless, as discussed in Part V, the in-plane DBR laser is an excellent candidate for integration with passive waveguides as in PICs or for forming multisection tunable lasers [35]. In fact, these two latter applications will probably be the primary ones for in-plane DBRs in the future.

Distributed feedback (DFB) lasers use gratings in the same region as their active regions. In their simplest form the grating is uniform, and there is no change in transverse structure along the laser's entire length [71–74, 78]. Thus, there is no active–passive waveguide transition, and in-plane lasers can be fabricated with fewer steps than DBRs. Primarily for these reasons, DFB lasers are commercially available but DBRs are not.

The fabrication of in-plane DFB or DBR lasers can involve one of several possible procedures as outlined by the cross sections in Fig. 12. For the more simple DFB, the first basic issue is whether to place the grating below or above the active layer in the growth sequence. Forming the grating first may remove the need for one regrowth step, but it means that the critical active layer must be grown on top of the corrugated surface, perhaps separated only by a very thin buffer/waveguide layer. For the DBR, this may not actually save a regrowth step because one still needs to remove the active layer over the grating and then regrow. The second case is forming the grating above the active layer, after it and a thin covering waveguide layer are grown. For the DFB, the grating is then etched in the cover waveguide, and finally the top cladding layer is regrown. For the DBR, the active layer is first etched away over the grating region, then the grating is formed in the lower waveguide layer, and finally the top cladding layer is regrown. If a tunable DBR structure is desired, a phase shift section can also be formed at this time by removing the active layer from an additional length between the grating and active regions [35, 77, 79]. For either the DFB or DBR case, the next step is to form the lateral waveguide. This might involve

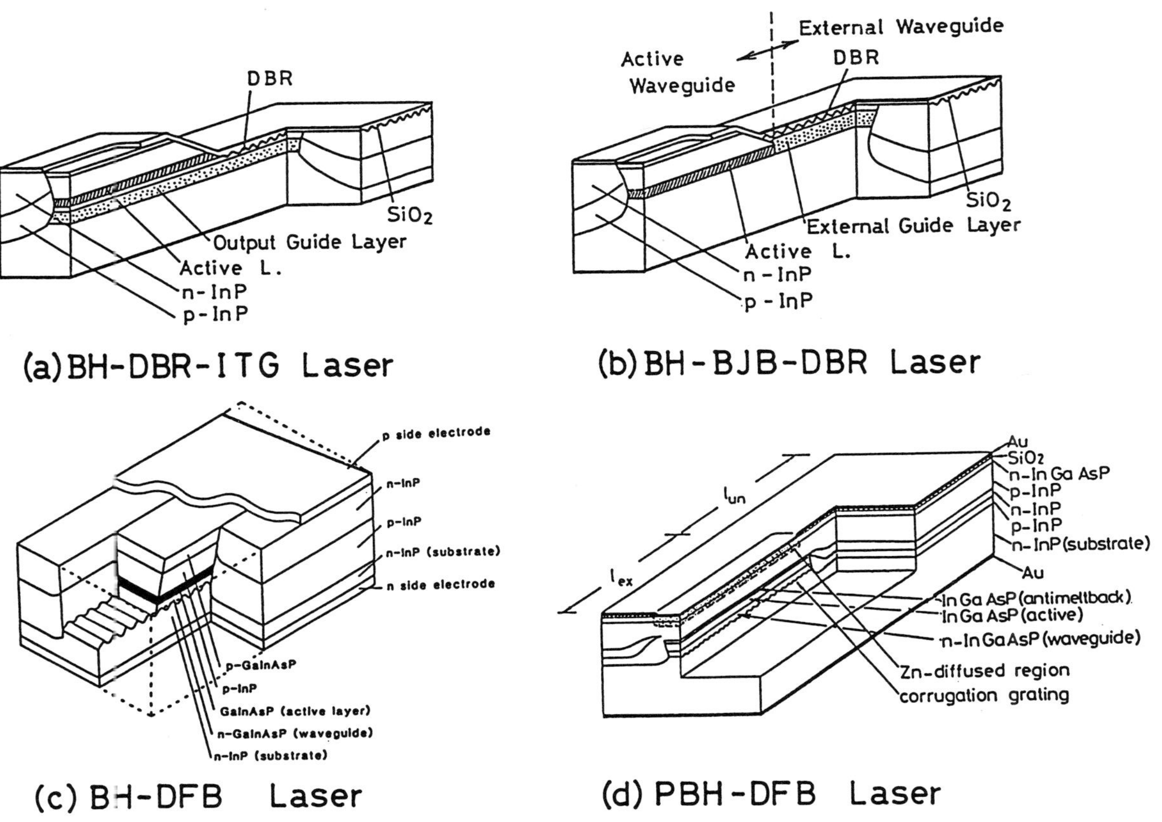

Fig. 12. Schematic representation of GaInAsP/InP BH-DBR and DFB lasers. (From Ref. 77. © 1983 IEEE.) (a) BH-ITH-DBR laser, (b) BH-BJB-DBR laser, (c) BH-DFB laser, (d) PBH-DFB laser.

etching away the guiding and active layers in the regions beside the active channel, followed by a regrowth of lateral cladding materials. It might also be possible to combine this step with the transverse cladding regrowth in the devices with the grating above the active layer.

The patterning and etching of the grating structure can also be accomplished in a variety of ways. The most common patterning technique is via holographic optical exposure of photoresist [80]. Figure 13 shows schematics of two commonly used systems. After exposure and development, the samples are wet or dry etched, cleaned, and finally reinserted into the episystem for regrowth. In order to increase the contrast of the resist patterns, oblique evaporation of a metal followed by oxygen reactive etching to remove the uncoated photoresist has occasionally been used before the semiconductor etching step [81]. In combination with RIE, this can lead to gratings with very high aspect ratios. Although the holographic exposure technique can provide long uniform grating patterns, it is not flexible for creating gratings with such variations as occasional quarter-wave shifts. Thus, there is an increasing use of e-beam direct writing for this purpose [82–84].

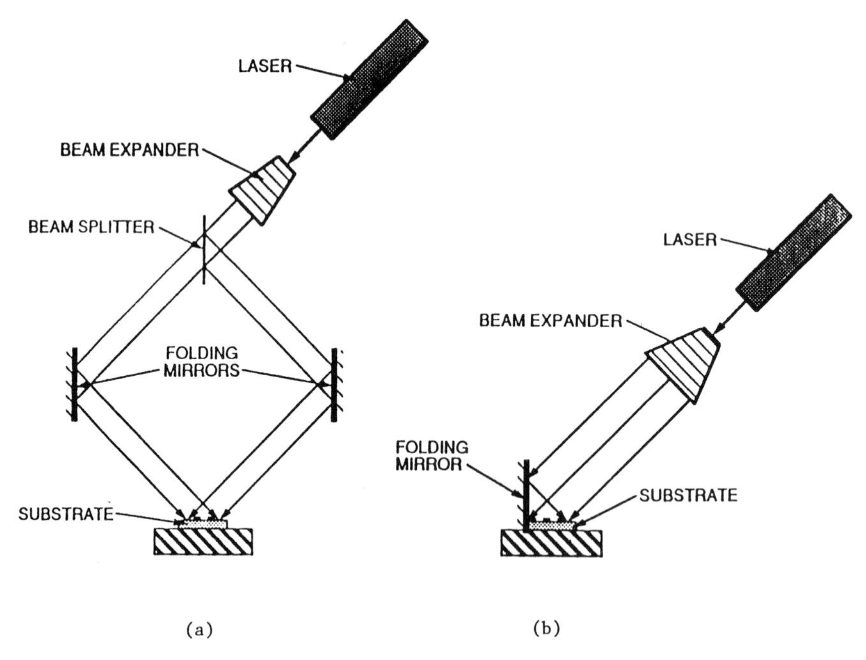

Fig. 13. Holographic grating exposure systems. Grating period is found by (a) rotating both holding mirrors and translating substrate upward or (b) by rotating substrate/folding mirror assembly. Technique (a) is more susceptible to vibration and air currents.

The e-beam technique can provide complex patterns with good contrast in the resist. However, only the best systems have sufficiently small aberations to write coherent first-order gratings over millimeter areas, and the exposure time is long and costly compared with the optical holographic technique. Also, it should be realized that quarter-wave shifts can be incorporated in a holographic technique if additional dielectric or resist layers are first defined.

The fabrication of vertical-cavity DBR lasers has quite a different set of associated problems. Because the propagation direction is along the crystal growth direction, it is relatively easy to create the active–passive interfaces as well as the passive grating mirrors in a single growth step. Changing the materials during growth is a common and straightforward procedure. The more difficult problems are to make electrical contacts with the various buried layers and to form the lateral waveguiding structure. These problems are discussed later in the device section.

Because of this different geometry, the DBR mirrors in a vertical cavity surface-emitting laser (VCSEL) are typically much shorter and have a much higher κ compared with in-plane lasers. For example, some devices that incorporated an InGaAs strained-layer active region have used pure GaAs and AlAs quarter-wave layers for the mirrors [85–87]. In this case $\Delta\mu \sim \pm 0.28$, and 15.5 mirror periods gives $R_{gmax} = 0.992$. In fact, the coupled mode formula Eq. (14) is only approximately valid for such strong discontinuities. Equation (17) is useful for the peak reflectivity, but it is more common to use a transmission matrix approach [88] for the whole wavelength response. This method also allows one to use slightly different periods along the length of the mirror, and the important effects of different terminations can be included. For example, in top-surface emitters it is common to make use of the air–semiconductor interface in combination with several mirror periods. Of course, the back mirror is terminated by the GaAs substrate. Figure 14 shows $\ln[1/|\boldsymbol{R}_g|_{max}]$ versus number of mirror periods for AlAs/GaAs DBR mirrors with both air and GaAs terminations.

In some VCSEL geometries, it is desirable to make electrical contact with the active layer through the mirrors. Unfortunately, the same large hetero-barriers that lead to a large κ also cause large series resistance and voltage drops. Some researchers have used graded interfaces to reduce these undesirable effects [86, 89]. Fortunately, this has only a small effect on κ, since it is proportional to the fundamental Fourier coefficient, and even if one went to the extreme of creating a sinusoidal grating, κ would only be reduced to $\pi/4$ of its value for a square-wave grating. Of course, such a

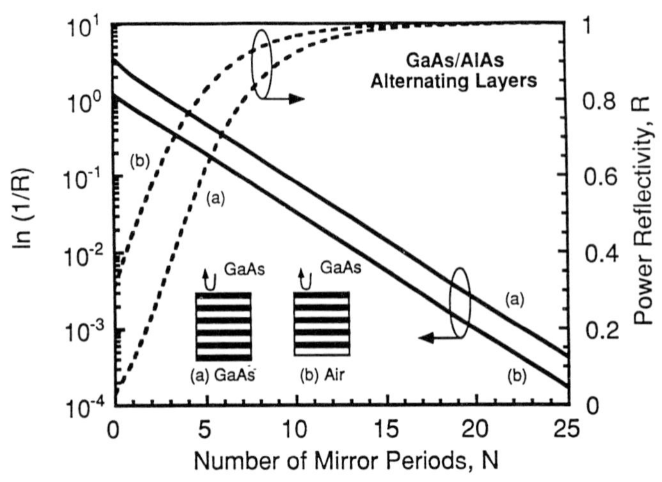

Fig. 14. Reflectivity of typical surface-emitting laser (SEL) GaAs/AlAs alternating quarter-wavelength mirror stacks as a function of the number of mirror pairs. Case (a) corresponds to a stack terminated on GaAs (the bottom mirror of an SEL). Case (b) corresponds to a stack terminated on air (the top mirror of some types of SELs). The solid curves, corresponding to the left axis, are proportional to the mirror loss experienced by the cavity. The dashed curves, corresponding to the right axis, indicate power reflectivity (field reflectivity squared). In the inserts, black corresponds to AlAs and white corresponds to GaAs. Note that for zero number of mirror pairs, case (a) is left with a residual layer of AlAs (for proper matching to the GaAs termination). As a result, $N = 15$ (for example) should be interpreted as $N = 15.5$ for case (a).

reduction can easily be compensated by increasing the grating length, l_g, by the same factor as κ is reduced, since only κl_g appears in $\boldsymbol{r}_g(\lambda)$.

Second-order grating reflectors can be used for surface-emitting output coupling from an in-plane laser as shown in Fig. 1. In this grating surface-emitting (GSE) laser case [90, 91], part of the guided input light is reflected back down the input waveguide, part is transmitted in the same guide, and part is coupled out perpendicularly to the waveguide in both the upward and downward directions. This dual coupling is made possible by using both the first- and second-order Fourier components of the effective index perturbation due to the grating. With this "second-order" grating, the grating period is equal to the guided wavelength. Thus, the fundamental Fourier component has this same period, and it can only couple light into the perpendicular directions because there is no component to scatter the light backward. On the other hand, the second Fourier component is at twice the grating's spatial frequency, and thus it satisfies the Bragg condition for coupling the forward to the backward guided wave, as would a grating having this period.

As the foregoing suggests, one must calculate the Fourier components of the effective index grating in order to obtain the coupling coefficients for the backward and perpendicularly scattered waves. Unfortunately, because coupling to the vertical waves and the backward waves occurs simultaneously, one can not simply use the second-order coupling constant, κ_2, with Eqs. (14) and (16) for the laser cavity's reflection coefficient and effective length. In effect, each coupling direction appears as a loss to the other. Certainly, the use of κ_2 in Eq. (15) gives an upper limit on the grating's feedback for the in-plane laser cavity. Generally, the maximum reflectivity would be less than half of this value [92–95].

2.2. Integrable Laser Configurations, Applications, and Characteristics

In this section we shall review three categories of diode lasers that may find use in OEICs. They are (1) in-plane cavities with in-plane emission, (2) in-plane cavities with surface emission; and (3) vertical cavities with surface emission. In each case specific configurations which appear to have potential for use in OEICs will be identified, the more specific areas of application will be discussed, and the characteristics of state-of-the-art devices will be reviewed. Whenever possible, the limitations on performance and yield will be identified, and issues related to their compatibility with ICs as well as their ease of coupling to output media will be discussed.

2.2.1 In-Plane Cavities—In-Plane Emission

This category is clearly the one on which the most data exists, since it has been the mainstream of research, development, and manufacture over the history of diode lasers. Our purpose here is not to summarize all of this history but rather to focus on configurations that are relevant for OEICs. There are two basic OEIC applications in which edge emission is important: (1) direct coupling to an output medium which lies in the same plane and (2) excitation of a passive optical waveguide integrated on the same chip.

In the first case, a linear array of lasers can be integrated along the edge of a chip for output coupling into an array of waveguides or fibers which provide an optical interconnection to some other circuit, board, or machine [1, 96, 97]. The lasers may be integrated directly on the electronic IC, or they may be integrated only with each other, or they may be integrated with monitoring detectors and, perhaps, a driving or interfacing circuit. Integration directly on an electronic IC to provide optical outputs is the most difficult form of integration for lasers. Compatible processing is necessary, the laser yield must be perfect or the entire circuit is lost, and the heat generated by the electronic IC degrades the performance of the lasers. The previous sentence applies for any laser–electronic IC integration, regardless of the laser type. However, this most advanced form of integration is probably least attractive for edge-emitting lasers, because they must still be located along the chip edge, and little advantage over some hybrid integration with a separate laser array is realized.

At the other extreme is a hybrid integration or packaging, in which the output laser array is separate from the electronic IC [1, 98]. If the laser chip contains only an array of lasers, the primary benefit would probably be the

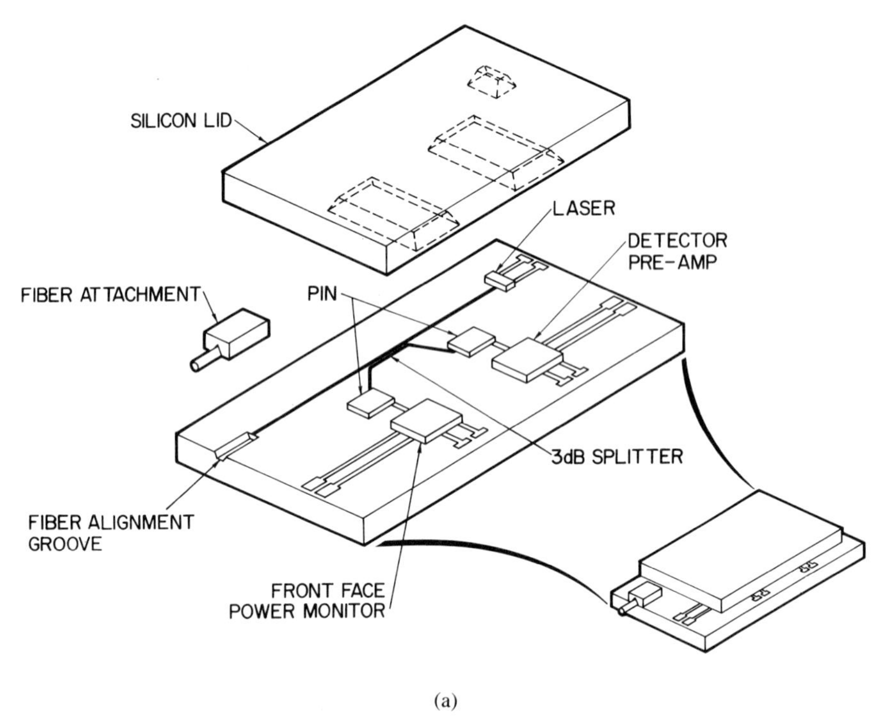

(a)

Fig. 15. (a) Packaging options for combining compound semiconductor optical devices with integrated electronic circuits, such as silicon CMOS. (b) Light output versus injection pulsed current (*L–I*) characteristics on dry etched and cleaved facet lasers. (A) corresponds to a conventional cleaved laser, (B) corresponds to an as-etched laser without facet passivation, and (C) corresponds to an *in situ* Al_2O_3-passivated etched mirror laser. Dry etching was via RIBE. (From Ref. 43. © 1988 IEEE.)

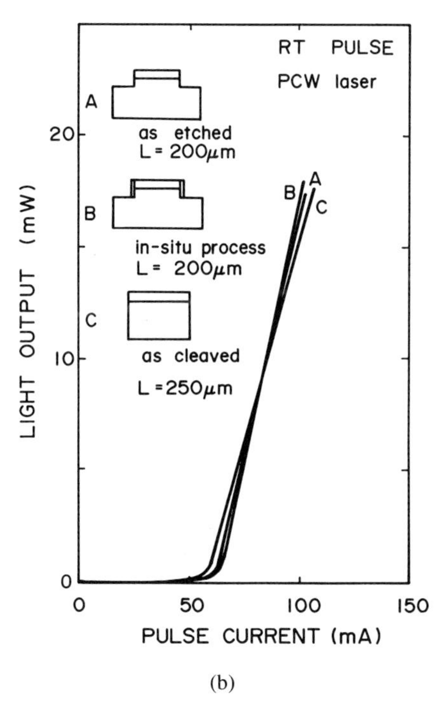

(b)

Fig. 15. (*continued*)

ability to mass couple to an output waveguide array with a single alignment step. After all, packaging/coupling of single lasers presently dominates their net cost, and one of the key goals of integration is to reduce cost. If 50 lasers could be packaged with the same effort it presently takes to package 1, perhaps the packaging cost could become a less dominant factor. For this kind of integration conventional cleaving could work, but wafer-scale manufacturing and testing will probably demand the use of an etched facet (or grating) technology, followed by conventional wafer dicing close to the facet. The etched facet (grating) process also facilitates the incorporation of a monitoring detector with no additional steps. For such an application the most important issues are obtaining high yield, reliability, and for gratings low-cost manufacture. All lasers in the array must be of high quality, or the entire array is rejected. Etched facet technology still has unproven reliability. Because the use of grating mirrors still requires an etched or cleaved facet for good beam quality, they can be contemplated only if a single-frequency output is necessary. Figure 15 shows simple schematics of the type of monolithic or hybrid integration for IC interconnection suggested here, and it shows some recent etched facet laser results.

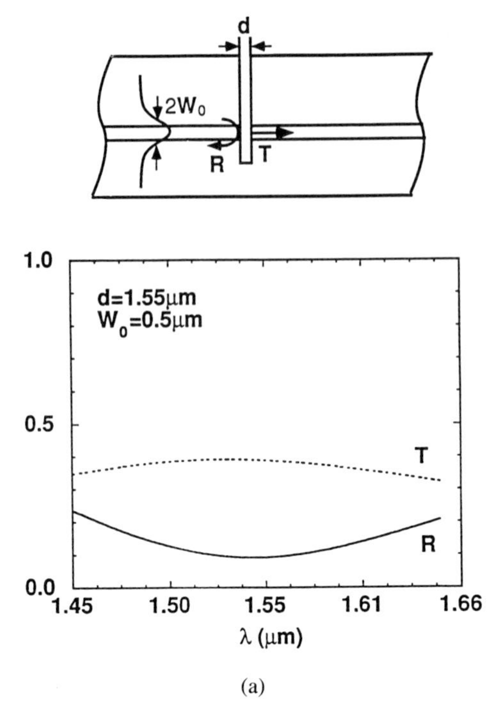

(a)

Fig. 16. (a) Schematic for a narrow etched groove coupler and its power reflection and transmission spectrum. Gaussian beam with half-width W_0 is used to model the diffraction through the groove. (b) Schematic for a grating coupler and its power reflection and transmission spectra designed at wavelength 1.55 μm with corresponding parameters indicated.

The second important application of edge-emitting lasers is in coupling to on-chip passive waveguides which are part of some integrated optical circuit (or PIC). In this case, there seem to be two choices for the necessary partially transmissive waveguide mirror component—either a narrow etched groove or a grating. The two are illustrated in Fig. 16, along with a schematic of their reflective and transmissive properties. The narrow etched groove concept was actually the first to be explored [99], even though most recent work has focused on the passive grating mirror [35, 100]. With modern etched facet techniques, the groove is viable, and it certainly requires less substrate area than the grating. Its early use was related to two-section, coupled-cavity laser work, where its properties were modeled as those of a low-finesse parallel-plate Fabry–Perot [101]. More recently, it has also been explored for waveguide crossing beamsplitters [102].

The use of grating mirrors in PICs is discussed at greater length in Part

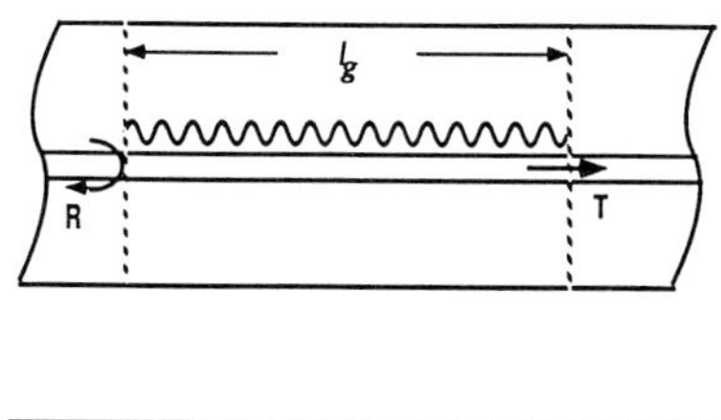

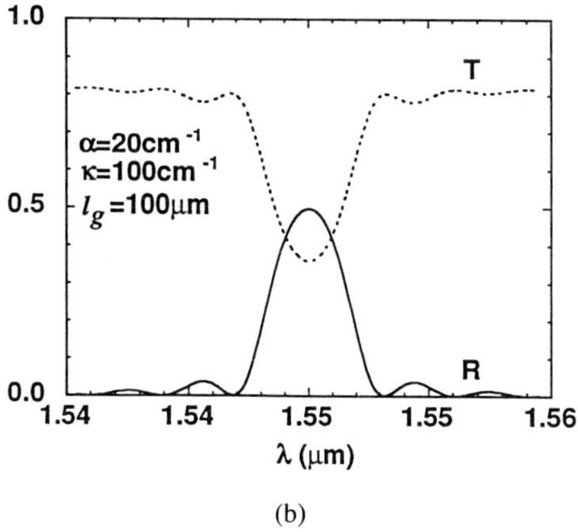

(b)

Fig. 16. (*continued*)

V. As discussed there, one important example is an integrated coherent receiver, in which a local oscillator is integrated. For this and other such applications, it seems that a single-frequency, or tunable single-frequency, laser is desired. Thus, for the necessary partially transmissive waveguide mirror, a grating is probably the best choice. Therefore, either a DFB or DBR laser is desired. As discussed earlier and outlined in Fig. 12, the DBR requires an active–passive waveguide transition between the active region and grating and the DFB does not. However, for integrated DFBs this transition is required at the output end of the grating, where the electrical pumping is ended. Thus, the associated regrowth technology is required in either case.

Because of the DBR laser's conceptual similarity to discrete reflector lasers, the equations of the laser theory section can also be applied to DBRs with little reinterpretation. The main issues are defining an effective mirror loss, α_m, and an effective total cavity length, l_t. As discussed in the grating section, the definition of α_m given after Eq. (1) is valid if the magnitude of Eq. (14) is used in $R = |r_1 r_g(\lambda)|$. Also, with the output taken from the grating power transmission, T_g, and since $R_g + T_g \neq 1$, due to some additional loss, the output power P_{01}, given by Eq. (3), must properly include $T_1 = T_g$ in F_1, which is the fraction of power lost by the cavity that is effectively coupled out. The total laser cavity length, l_t, is given by the separation between the gratings plus the effective grating penetrations, l_{ge}, into each grating as given by Eq. (16).

An important problem with the narrow-bandwidth DBR mirrors is that there may not be a longitudinal laser mode near the reflection maximum. In this case, the laser threshold would occur for the lowest loss mode, as determined by $|r_g(\lambda)|$ at the wavelength of the mode. Also, the exact wavelength of the mode is determined by the nonzero grating reflection phase. In other words, the threshold gain and current of a DBR laser can vary significantly as a function of the total round-trip cavity phase for the lowest loss mode. For an in-plane laser this phase is generally measured in the thousands of radians, so a small cavity phase change (caused by a temperature change, carrier density change or some electro-optic phase shifter) can move the mode to be in or out of alignment with the DBR maximum. However, in vertical-cavity DBRs the cavity is so short that the layers must have a very well-defined thickness for the mode to be centered on the reflection maximum, but this alignment is then relatively stable.

The DFB laser has some inherent advantages over the DBR. For example, because the grating fills the entire cavity, anything that affects the round-trip phase of the optical mode will also affect the grating bandpass in the same way. As a result, the longitudinal optical mode is fixed relative to the grating pass band, and if the mode is positioned for low mirror loss at one temperature and carrier density, it will always be. However, with DFB lasers aligning the mode with the best feedback wavelength is not trivial. In fact, if a uniform grating is used, two degenerate longitudinal modes spaced on opposite sides of the lowest loss point exist [71, 78]. If the device is cleaved at the end of one of the gratings for output coupling, the resulting reflection must be added to that of the grating. This yields an unknown threshold and spectrum, since the reflection phase from a cleave is generally random relative to the grating. However, in the uniform-grating DFB this end reflection will tend to have the beneficial effect of shifting the mode spectrum so that only a single mode can lase [100]. To provide reproducibly a mode at the Bragg wavelength, where the loss is lowest, a quarter-wavelength shift can be added in the center of the grating. This shift provides the necessary half-wavelength-wide cavity, measured between the grating index down-step reference planes, where the reflection phase is zero at the Bragg wavelength. In this case end reflections must be suppressed.

Although the principles outlined in the theory section still apply, it is more difficult to identify the parameters in that theory for the DFB laser. The quarter-wave-shifted DFB is basically a laser composed of two active mirrors with reflectivities that can be increased by pumping and an extremely small intermediate spacing that functions only to position the mode properly. Thus, for each active mirror, $(R_g + T_g)$ generally exceeds unity. However, because the gain and distributed reflection are intertwined, the threshold gain equation is more complex than Eq. (1). The threshold condition is simply $r_{g1}r_{g2} = 1$, but the complex r_g values are calculated from grating expressions that include the gain in a complex propagation constant, $\boldsymbol{\beta}$, in Eq. (14) [103, 104]. Generally, numerical techniques are necessary to determine the threshold gain as a function of the grating discontinuity strength [103–107]. Figure 17 gives the results of such a calculation for standard ($\phi = 0$) and quarter-

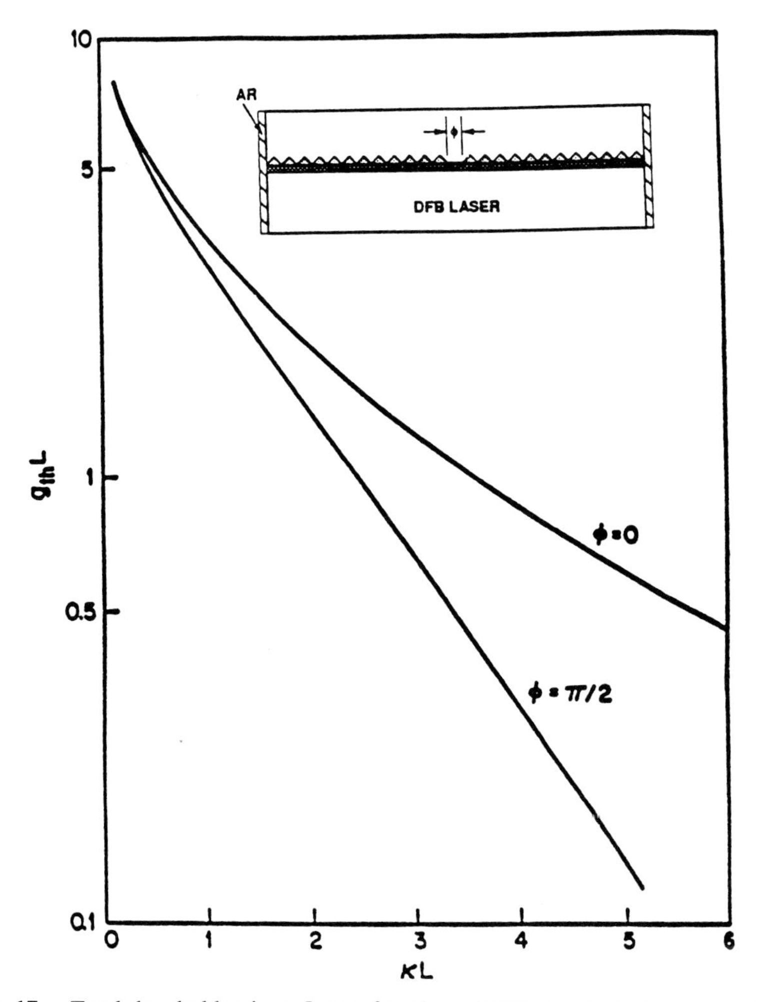

Fig. 17. Total threshold gain $g_{th}L$ as a function of DFB length L, for zero and π/2 grating phase shift in the center of the laser. (From Ref. 106. © 1985 IEEE).

wave-shifted ($\phi = \pi/2$) DFBs, in which no end reflections exist [106]. This result can then be used in conjunction with Fig. 5 or some approximate analytical expression for the gain as a function of current.

Multisection tunable DFB and DBR lasers are desirable in any sort of multiwavelength system, whether or not coherent detection is used. They may also find use in sensing or spectroscopy. In such systems it is necessary either to set the wavelength to some particular value or to sweep it more or less continuously. In either case, one requires electronic tunability to be able

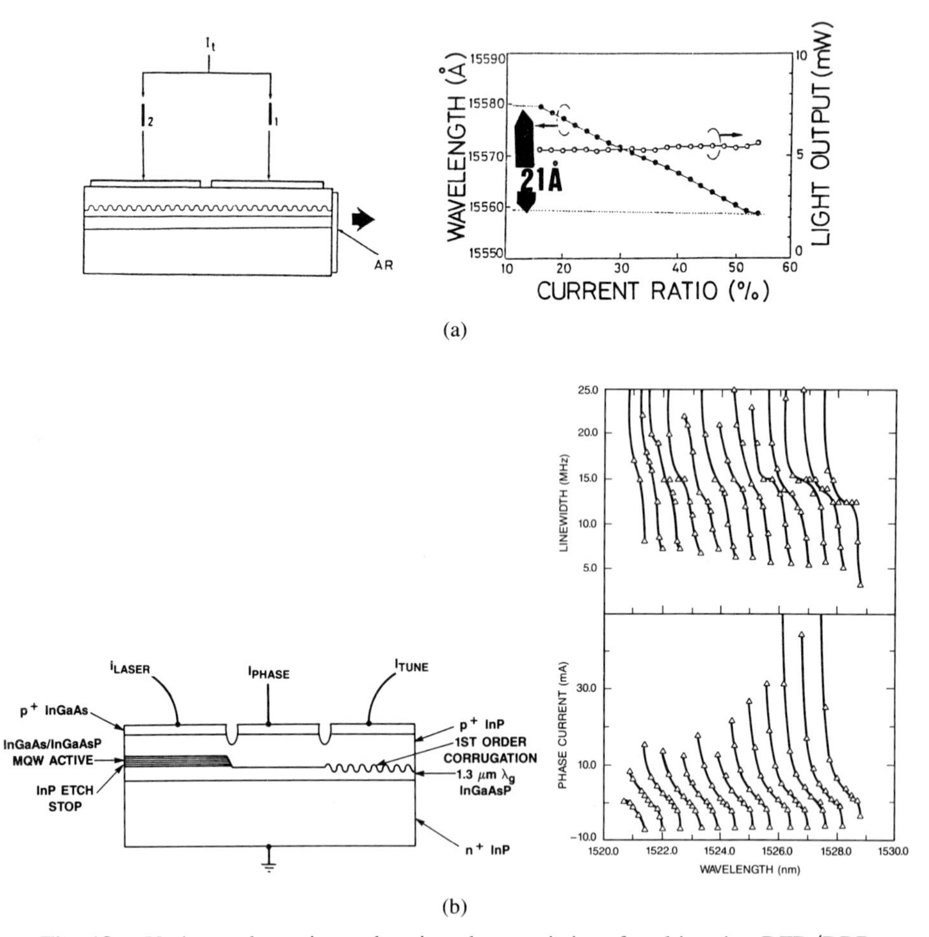

Fig. 18. Various schematics and tuning characteristics of multisection DFB/DBR lasers (a) Schematic and lasing spectra at various current ratios for a 1.5μm multielectrode DFB laser. (From Ref. 108.) (b) Schematic and tuning characteristics of a three-section MQW-DBR. (From Ref. 35.) (c) Schematic and wavelength vs. tuning current I_t for 400-μm TTG laser. (From Ref. 116).

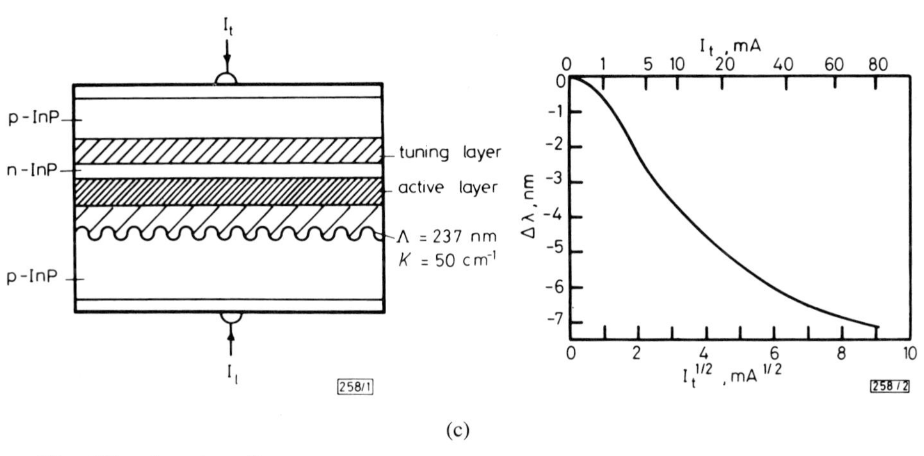

(c)

Fig. 18. (*continued*)

to make the wavelength adjustment and hold it over some range of temperature. In the multisection DBR laser, the wavelength is tuned by changing the grating mirror's index of refraction, which shifts its Bragg wavelength to select different longitudinal modes of the cavity and thereby changes the mode number m in Eq. (1b), and/or by changing the index μ_p of the passive phase-shift section within the cavity to change the net round-trip phase observed by the laser modes, so that they move in wavelength, as given again by Eq. (1b). In the DFB, the grating and the cavity are superimposed, so a shift in index shifts both the mode-selective grating filter and the cavity phase simultaneously. Thus, in the DBR it is generally necessary to adjust two indices to tune the wavelength and keep the mode aligned with the minimum in cavity loss [35, 79, 108–110], but in the DFB tuning the cavity phase tunes the mode and the minimum loss point together. However, since all of the DFB cavity is active, it is generally also necessary to adjust the current into two sections to maintain constant output while tuning [79, 111–113]. Thus, for tunable DFBs at least two separately pumped sections are necessary along the active region, and for DBRs at least three separated contacted sections are necessary—one active, one grating mirror, and one intracavity phase shifter—in order to obtain complete wavelength coverage over some range. Figure 18 gives schematics of tunable DFB and DBR lasers along with some results for each.

Also shown in Fig. 18 are results for a new tunable twin-guide (TTG) configuration [114], which has a passive, separately biased tuning layer above the active layer in a DFB laser. Provided this layer is sufficiently close to intercept a reasonable portion of the transverse waveguide mode, the laser

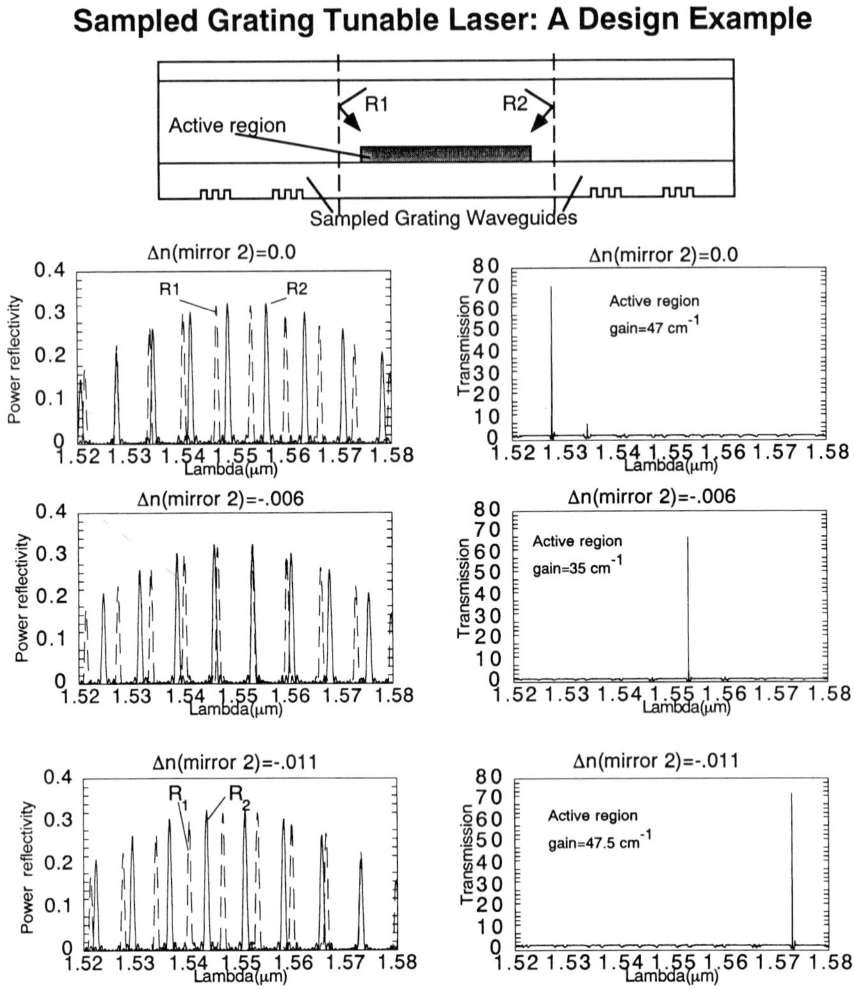

(a)

Fig. 19. (a) Schematic and illustrative tuning of sampled-grating DBR laser. Nominal lengths are: active region, 300 μm; mirror 1, 550 μm; mirror 2, 500 μm. Mirror 1 uses a sampling window of 5 μm and sampling period of 50 μm, mirror 2 a sampling window of 5 μm with a sampling period of 45 μm. Both mirrors have a nominal grating pitch of 235 nm, kappa (before sampling) of 125 cm^{-1}, and passive loss of 5 cm^{-1}. The left column shows mirror reflection spectra for three different index changes induced in mirror 2; the right column shows the corresponding device transmission response with gain in the active region (From Ref. 117.) (b) Wavelength tuning using a mismatched-waveguide directional-coupler filter (MWDCF) with grating-assisted codirectional coupling. Index change Δn_{1f} in the coupled section tunes the filter and index change Δn_{1p} in the uncoupled section tunes the cavity mode for continuous tuning. (From Ref. 118. © 1991 IEEE.)

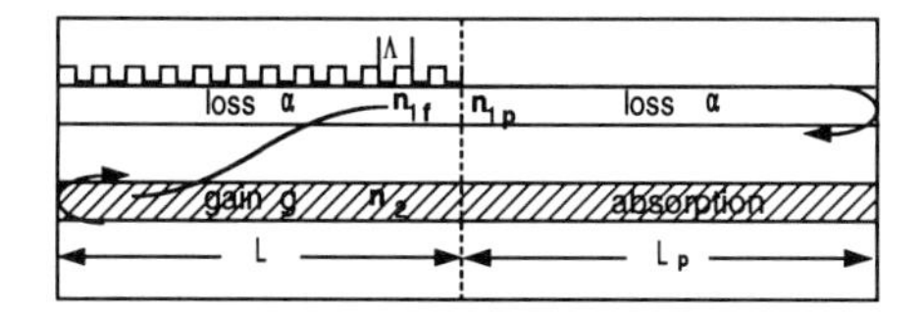

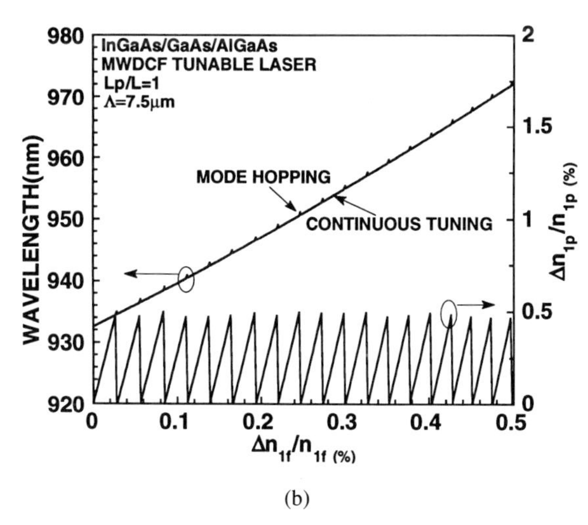

(b)

Fig. 19. (*continued*)

can be tuned by varying its index. As shown by the results, this laser can also give continuous tuning over about the same range as the multisection DFB and DBR configurations. Because the tuning layer fills the entire cavity length, it should be possible to tune the lasing mode without shifting it relative to the grating stopband. The main problem with the TTG configuration is in providing the required separate biases to the two closely spaced layers without compromising other properties of the laser. Spreading resistance in the spacing layer between the two layers must be kept low, but with high doping the internal loss of the laser suffers. The resistive spacer layer may also provide laterally nonuniform pumping of the active region, leading to further compromises in the laser's efficiency.

For the configurations considered in Fig. 18, the net relative tuning of the wavelength is limited by the available effective index change that can be obtained, i.e., $\Delta\lambda/\lambda < \Delta\mu/\mu$. For carrier injection into a passive region, the useful $\Delta\mu/\mu$ is limited to less than 1% due to associated free carrier losses. For an active section the index shift is similarly limited due to partial carrier clamping above threshold. This gives a limit of ~10 nm in the 1 μm wavelength range [79]. Some practical experimental devices have approached this limit [115, 116].

Because some systems require a considerably wider tuning range and the gain material can have a sufficient bandwidth to accommodate a wider range, there have been some proposals for new configurations to exceed this $\Delta\mu/\mu$ limit. Figure 19 shows schematics of two ideas being pursued in the author's laboratory [117, 118]. The first is a modified DBR concept in which the two end mirrors are "sampled" gratings. This sampling provides a grating reflection spectrum that has multiple reflection bands spaced by ~5–10 nm as determined by the sampling period. However, if the sampling period of the two mirrors is different, only one pair of reflection maxima will be aligned for some pair of grating indices, thus selecting only a single longitudinal laser mode. Also, if the sampling periods of the two gratings are not very different, it is possible to align an adjacent pair of reflection maxima by only a slight index change. Thus, as with a vernier scale it is possible to shift the wavelength of the net cavity loss minimum, where two reflection maxima align, by a much larger amount than the shift in index, $\Delta\mu/\mu$, as also indicated in Fig. 19.

Using this sampled grating concept, ~30 nm of discontinuous tuning has been demonstrated in a two-section DFB scheme [117]. The main problem with this geometry is that it requires four electroded sections to perform the function just outlined. As in the three-section DBR, there must be a gain and phase shift section, but here there are two mirror sections. Tuning the mirrors together results in a sweeping of the effective mode selection filter across a range of $\Delta\mu/\mu$; tuning one relative to the other "changes channels" to a new band [117]. For this type of complex control the device probably needs to be packaged with a microprocessor! A decade ago this would have been a joke; today it describes a fairly standard configuration. In fact, the three-section DBR probably also needs this kind of control circuit for best reliability.

The second scheme indicated in Fig. 19 uses codirectionally coupled waveguides with different propagation constants to create an intracavity tunable mode-selection filter [118–120]. A coarse grating is used along one waveguide to provide an additional **k**-vector component to compensate for the different propagation constants. Since the propagation constants are wavelength dependent, this provides **k**-vector matching only at one wavelength, so that a mode selection function is realized. Now, if the propagation constants are relatively close together, a slight change in the effective index of one guide gives a large change relative to the index difference between the two. Consequently, the tuning is proportional to $\Delta\mu/(\mu_2 - \mu_1)$, and this can be much larger than $\Delta\mu/\mu$ in one guide. In this case the control of the device is somewhat simpler than in the first example. However, the bandwidth of the tunable mode selection filter is also proportional to $\Delta\mu/(\mu_2 - \mu_1)$, and simulations indicate that it may be difficult to obtain good single-frequency characteristics without some additional frequency-selective element in the cavity. Thus, the control algorithm for this structure could be as complex as for the other. Tuning as large as 57 nm has been reported [120].

2.2.2 In-Plane Cavities—Surface Emission

Lasers that emit vertically, or perpendicularly to the surface, offer a new degree of freedom for use in OEICs. With such devices the geometry of the OEIC and the placement of the optical outputs become less constrained. Thus, there has been a significant amount of research over the past several years on this topic. The prospects of higher speed, lower crosstalk, and greater connectivity density offered by using optical interconnection between electronic ICs and boards are at the core of much of the activity. The interconnection might use optical waveguides, such as fibers, or free-space optics, or some combination of both to provide the desired lightwave path between the outputs and inputs of two subsystem elements. Also, some sort of optical branching and/or switching within the lightwave path can be used to fan out and/or vary the routing. Surface emission is particularly well suited for generating two-dimensional arrays of lasers, which in turn are desirable in a highly connected system. The use of free-space interconnection provides the ultimate in interconnection density, but the mechanical positioning of the various elements then becomes critical [121, 122].

In this section we will specifically consider three types of surface emitters. All are in-plane lasers, but the outputs are either coupled out directly in the vertical direction or deflected into the vertical direction after emitting from an in-plane etched facet. As indicated in Fig. 1, etched facets angled at 45° or second-order gratings have been used for these purposes. Figures 20–22 give more explicit schematics of the three configurations, together with representative characteristics. The first two use the 45° angled facet, but in one it is outside the cavity (integrated deflector SEL) [49, 50] and in the other it is internal to the cavity (folded-cavity SEL) [47, 48]. The third uses the vertical output coupling from a second-order grating mirror (grating SEL) [91, 123, 124], as already mentioned.

The IDSEL is the oldest and probably most simple to understand of these structures. As indicated in Fig. 20, it consists of an in-plane etched-facet laser with an external integrated mirror formed at 45° to the plane. The 45°

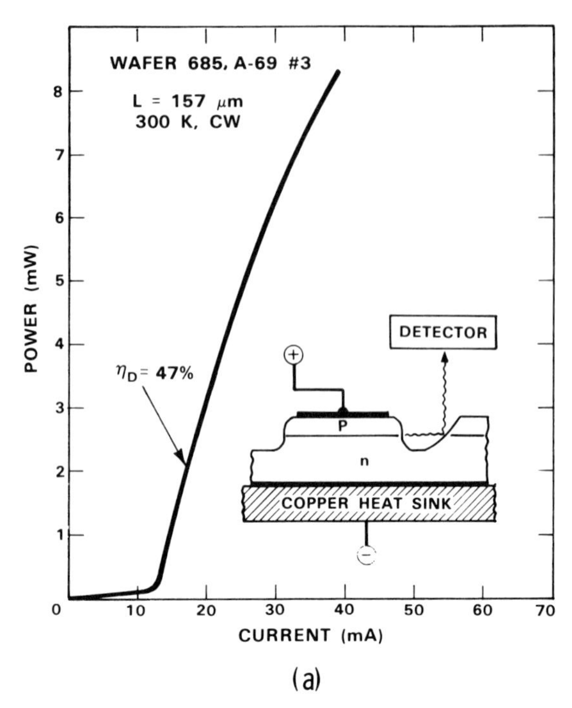

Fig. 20. Characteristics of an integrated deflector surface-emitting laser (IDSEL). (From Ref. 50.) (a) *L–I* curve for cw operation of the laser shown by the schematic inset; (b) far-field pattern. Lasers are mounted *p* side up on a a copper heat sink as indicated in the inset.

mirror has been formed by reactive dry etching with AlGaAs/GaAs [125–127] and by mass transport of InP with InGaAsP/InP [50]. The external mirror has also been curved to recollimate the laser's output into a low diffraction angle beam [50, 125]. This is an important attribute of this structure, but it requires extremely good etching control. Its other primary advantages derive from its similarity to other in-plane lasers. That is, other than the issues already mentioned for etched facets, one can be reasonably sure that there will be no catastrophic surprises once an OEIC manufacturing plan is developed. As indicated by the experimental results in Fig. 20, good IDSEL lasers with high yield have been fabricated [50]. Aside from far-field emission, its theoretical characteristics are similar to those of other in-plane lasers, and one can apply the earlier theoretical formalism by only including the proper output mirror transmission factor T_1, in P_{01}, as in the DBR case. Here, T_1 represents the net transmission of the output mode, which is partially transmitted through the vertical laser mirror facet and then reflected

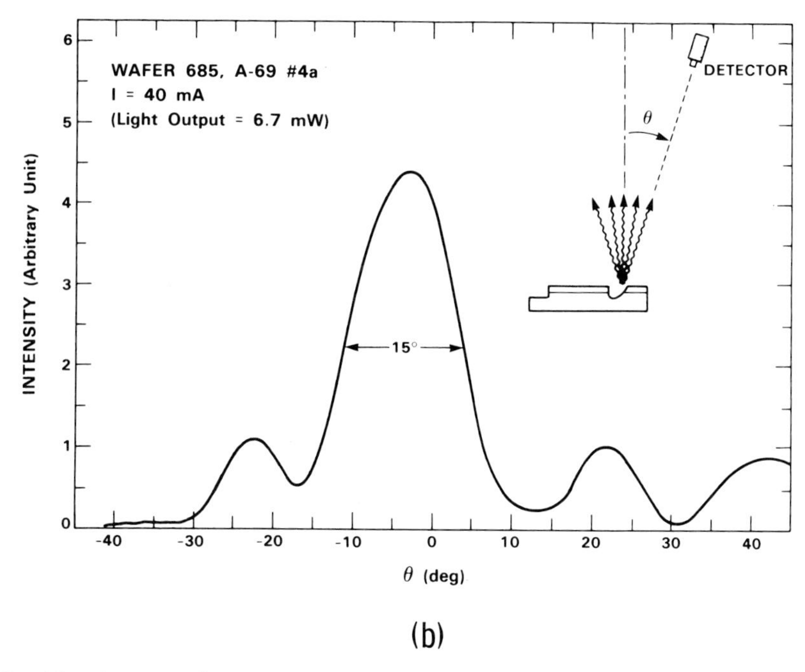

Fig. 20. (*continued*)

off the 45° mirror; and R, as always, is the net reflectivity back into the cavity mode from the laser mirror.

The FCSEL is a newer structure that is emerging as a promising configuration. The embodiment shown in Fig. 21 uses 45° mirror facets etched at both ends of the in-plane cavity to deflect the mode internally either upward or downward [128]. For ease of manufacture, the facets can be etched at the same time in this case. The downward lightwave is reflected from a grown-in DBR mirror, and the output is taken from the transmission at the surface after being deflected upward. In earlier embodiments [47], the rear facet was a simple vertical etched facet, which required another etch step but which did not have the additional diffraction loss in the DBR mirror. The primary problem in this structure is reducing the mode diffraction and other losses encountered in propagating from the in-plane waveguide to the in-plane mirrors and back. As indicated in Fig. 21, the characteristics of this configuration can be modeled by considering an equivalent in-plane laser with unguided passive sections near each mirror. The complex round-trip coupling factor C includes all scattering, misalignment, and diffraction losses encountered by lightwaves emerging from and then reexciting the desired transverse waveguide mode. Thus, for a mirror reference plane at the end

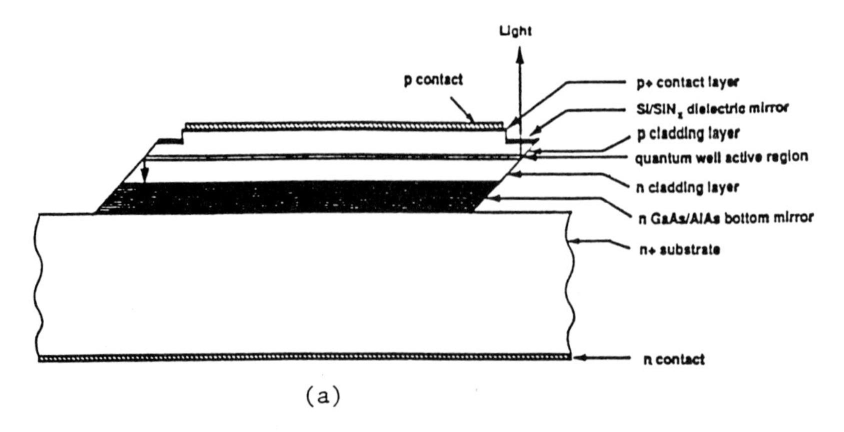

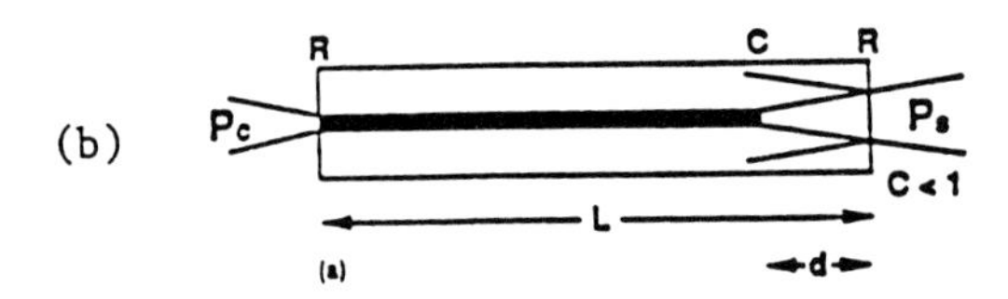

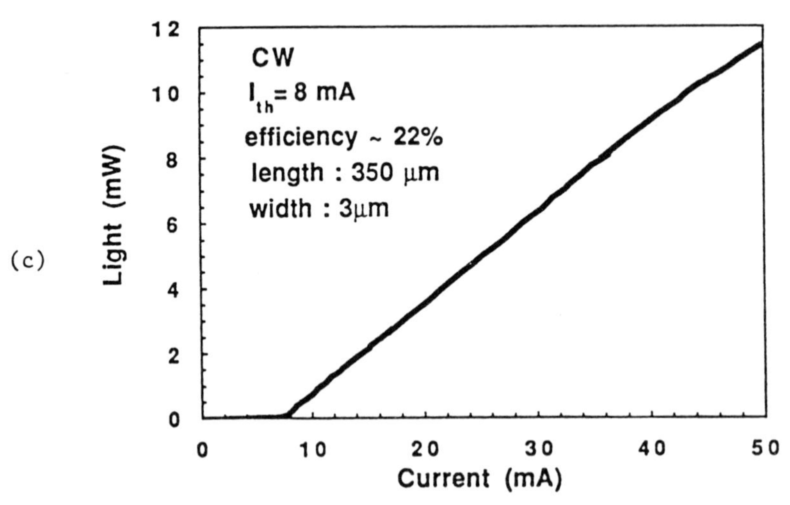

Fig. 21. Folded-cavity surface-emitting laser (FCSEL). (a) Schematic diagram of the device structure. (From Ref. 128. © 1991 IEEE.) (b) Model of the device as an in-plane laser with unguided sections to account for the loss through the angled facets and coupling back into the cavity. (From Ref. 47.) (c) *L–I* curve of a 5-mm-wide, 400-mm-long device under CW operation. (From Ref. 128. © 1991 IEEE.)

of the active layer, $\mathbf{R}_j = C_j R'_j$, where $j = 1, 2$ for the output and back mirrors, respectively, and the boldface type indicates that the complex $\mathbf{R}_j$ includes the roundtrip propagation phase delay. R'_j is the reflection from the surface ($j = 1$) or the back mirror ($j = 2$). For mode shapes and distances encountered in the experiments, $|C| \sim 0.8$ has been observed [47, 128]. Again, the output power, P_{01}, must use the proper T_1 in F_1 to account for these losses. In this case, $T_1 \cong (1 - R'_1)$, but scattering losses from the reflection off the 45° mirror must also be added if it is not collected by the receiving optics; thus, the approximation.

Figure 22 shows results for the GSE laser. As suggested by the discussion in the grating section, this is a considerably more complex structure to analyze, since three-way simultaneous coupling occurs in the grating coupler. That is, the input waveguide mode is partially reflected (via the second spatial harmonic of the grating), partially deflected upward and downward (via the fundamental spatial harmonic), and partially transmitted. Thus, three simultaneous coupled-mode equations must be considered in the analysis. Figure 22 summarizes the results of such a calculation for a particular configuration [129]. One problem is to redirect the upward or downward lightwave so that both can be used as the output. A buried grating or surface metallization has been used to perform this function [92]. One of the key attributes of this configuration is that the "extra output" which continues down the waveguide can be used to phase lock another integrated GSE laser. Thus, arrays of these lasers with coherently related outputs are relatively easy to obtain [5, 130, 131], and high-power diffraction-limited outputs seem possible, because the in-plane active regions are relatively easy to heat sink. However, in such arrays it is difficult to phase the output from the separate grating couplers properly to form the diffraction-limited output, unless the bias to the gain regions in between are separately adjusted. Also, it is desirable to weight the output levels from the various array elements to obtain reasonable far-field side-lobe levels. Thus, the control of these arrays becomes very complex. For most OEIC applications, in which some sort of optical interconnection is sought, it is unlikely that such large-area arrays will find much use. In fact, one of the chief concerns about the use of the GSE laser in OEICs is the relatively large substrate area it requires.

2.2.3 Vertical-Cavities—Surface-Emission

Although the structures discussed in the preceding section provided surface emission, the optical mode still propagated along the length l of their active regions in the plane of the substrate. For vertical-cavity SELs (VCSELs) that are still pumped vertically across dimension d, we need to exchange

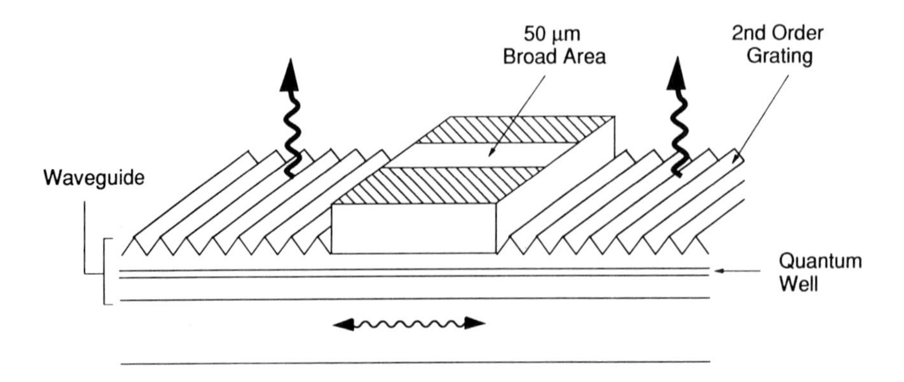
50 μm
Broad Area
2nd Order
Grating
Waveguide
Quantum
Well
SEE DETAIL VIEW
500 μm
500 μm
U
R
T
S

(a)

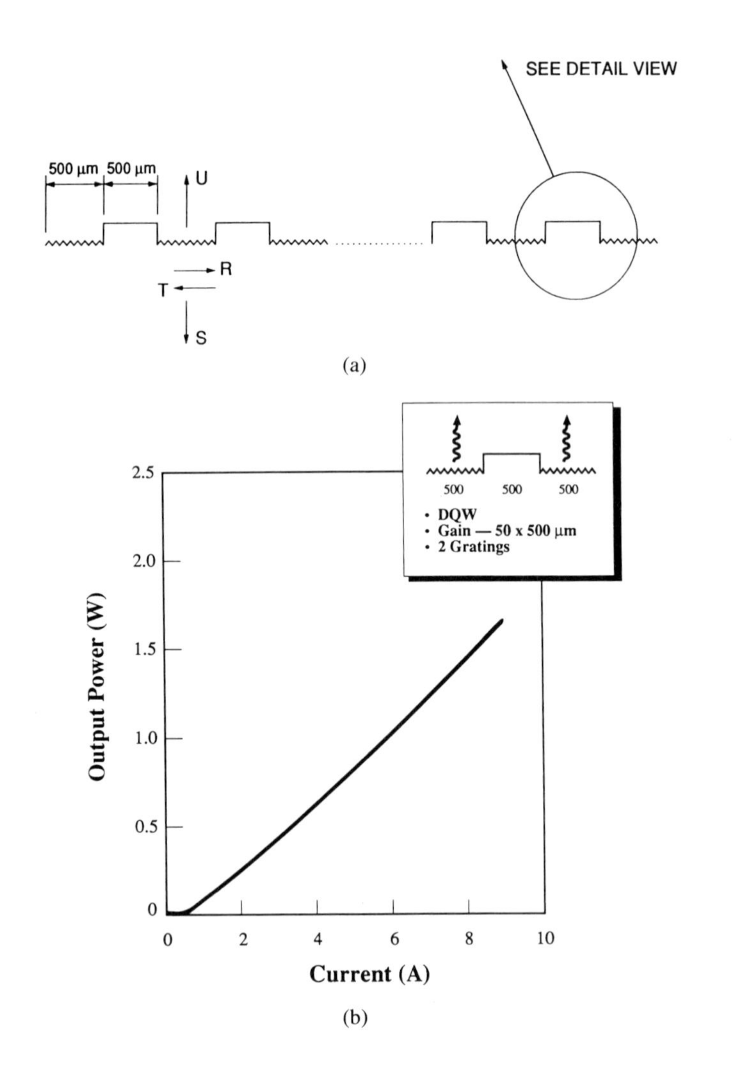
500
500
500
• DQW
• Gain — 50 x 500 μm
• 2 Gratings
Output Power (W)
2.5
2.0
1.5
1.0
0.5
0
0
2
4
6
8
10
Current (A)

(b)

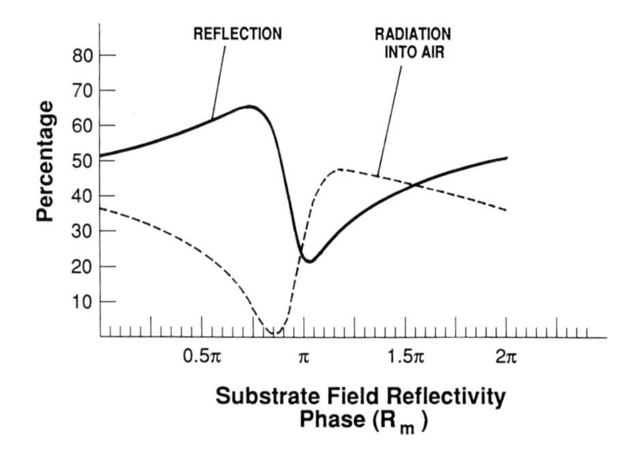

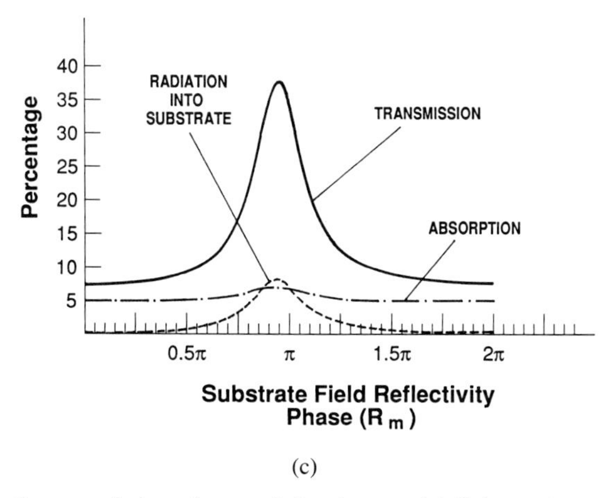

(c)

Fig. 22. Grating coupled surface-emitting lasers. (a) Schematic representation of grating coupled surface-emitting laser array. (b) *L–I* curve for a single gain element GSE. (From Ref. 123.) (c) Calculated percentage reflection, transmission, radiation, and absorption as a function of the phase of the substrate field reflectivity $\{|R_m| = 0.95\}$ for a second-order rectangular grating of amplitude 100 nm and length 50 μm (From Ref. 29.)

our optical cavity "length" and "thickness" dimensions to fit the analysis. As indicated by Fig. 23, the optical mode propagates normal to the surface for the entire cavity length, and it crosses the active region along its dimension, d, in VCSELs. The confinement factor, Γ, is determined by the mode overlap with the active area cross section, which is wl in this case. Thus, d and l have effectively changed roles. Therefore, in our laser cavity analysis, the active region "length" $l_a = d$, and the total cavity length l_t adds

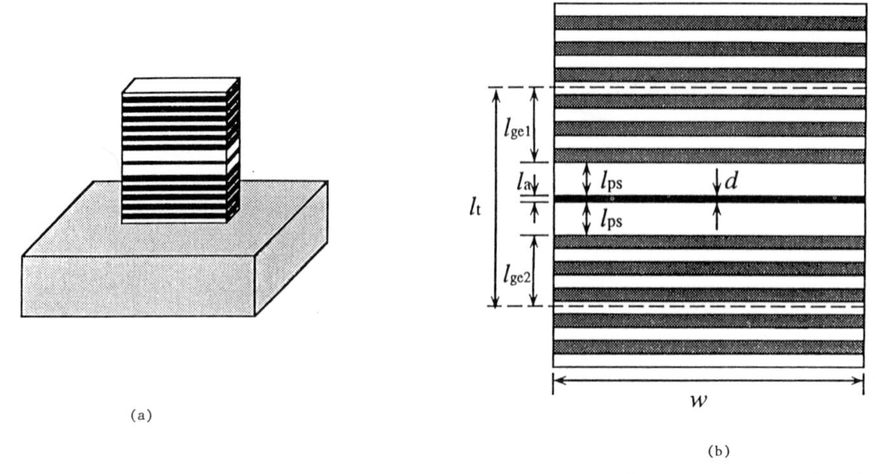

Fig. 23. (a) Schematic view of a dry etched VCSEL device. (b) Cross-sectional view showing relevant lengths and widths.

the net passive length l_p, comprised of cavity spacer layers and grating penetrations measured in the vertical directions, i.e., $l_p = l_{ps} + l_{ge1} + l_{ge2}$. With these interpretations, the equations in the preceding sections are all valid. (However, if a laterally pumped structure is to be analysed, d and w must be interchanged in the gain equations.)

The result of these dimension interpretations is that the confinement factor for a VCSEL is typically close to unity, and the longitudinal fill factor, l_a/l_t is very small. In fact, for dimensions taken from recent low threshold VCSELs on GaAs, $\Gamma \sim 0.8$ and $l_a/l_t \sim 0.01$ [85, 86]. Thus, the "volume confinement factor," $\Gamma l_a/l_t \sim 1\%$, which is only slightly smaller than in typical in-plane quantum well lasers. However, because it is possible to place the gain only at peaks of the electric field standing wave in a VCSEL, the effective modal gain can be doubled compared with a case where the gain is averaged over the peaks and nulls of the standing wave [132–135]. Therefore, in comparing the required "volume modal gain" for threshold, we multiply Eq. (1a) by (l_a/l_t) for both in-plane and VCSEL lasers, but we replace g by 2g for the VCSELs which have the proper active region design. If we further suppose that $l_a \cong l_t$ for the in-plane (IP) case, and that (l_a/l_t) is small for the VCSEL (VC) case, we find that

$$\Gamma g_{th} \cong \alpha_{ia} + (1/l_t)\ln(1/R) \qquad \text{[IP];} \tag{18a}$$

$$2(l_a/l_t)\Gamma g_{th} \cong \alpha_{ip} + (1/l_t)\ln(1/R) \qquad \text{[VC].} \tag{18b}$$

Now, since the coefficients of the threshold gain are nearly the same, if the passive losses are nearly the same we observe that similar threshold gains will be obtained if $[(1/l_t)\ln(1/R)]_{VC} \sim [(1/l_t)\ln(1/R)]_{IP}$. Since $l_{tIP} \sim 200 l_{tVC}$, this requires that $\ln(1/R_{IP}) \sim 200\ln(1/R_{VC})$; or for $R_{IP} = 0.32$, this suggests that $R_{VC} \sim 0.994$, which is about what researchers have found from more careful analyses. The results of such an analysis are illustrated in Fig. 24. Here, radiative VCSEL currents for threshold and 1 mW of power out for one to three InGaAs quantum wells versus the mean mirror reflectivity are plotted [136]. The quantum well theory summarized in Fig. 5 was used, and an internal loss of 30 cm^{-1} was assumed.

The net result of this discussion is that one should be able to obtain similar threshold gains and, therefore, current densities in VCSELs. Also, it should be clear that the external differential efficiencies should be similar, since the internal and mirror losses were assumed to be in the same ratio. To many

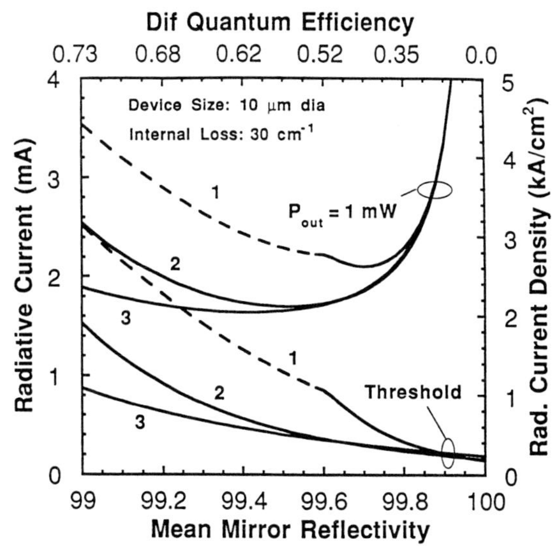

Fig. 24. The threshold current and current needed to reach 1 mW of output power for a circular 10-μm-diameter surface-emitting laser is shown for active regions with one, two, and three InGaAs/GaAs 80-Å QWs. For the assumed internal loss, α_i = 30 cm^{-1}, the two- and three-quantum-well cases provide better overall performance in terms of both threshold current and current required to reach a given output power. For the single-quantum-well case shown, the reflectivity must be kept larger than 99.6% to avoid the second quantized level. In this regime (dashed curves), we can no longer neglect effects such as barrier leakage and nonradiative Auger currents; hence, the radiative current shown in the figure no longer approximates the total current needed to achieve the given material gain. For the two- and three-quantum-well cases, this transition point does not occur until the reflectivity drops below 99%, providing much more tolerance in the mirror design

this result is unexpected, some would argue revolutionary. To date, these similar characteristics have not been realized, primarily because the internal losses of current primitive VCSELs are much higher than those obtained in state-of-the-art in-plane lasers [137]. At this writing, the lowest measured threshold current density is ~600 A/cm^2 for a single InGaAs/GaAs quantum well [86], which is several times larger than in comparable in-plane lasers, but this was obtained in a VCSEL with an estimated internal loss of ~30 cm^{-1}. Also, differential efficiencies >30% have been obtained with threshold current densities <1 kA/cm^2 for double quantum well structures [137]. Once cavity losses and lifetimes become similar to those of in-plane lasers, it is believed that many of the VCSEL's characteristics will be similar to those of in-plane lasers. Of course, one should not lose sight of the very low active region volume in VCSELs, which naturally leads to very small threshold currents as indicated by Eq. (9) in the active region section. Thus, even if the current densities are somewhat higher than for in-plane lasers, the net current tends to be smaller in the VCSEL. However, it may be possible to make very short active region in-plane lasers with HR mirror coatings to approach the same active volume as VCSELs, but since surface emission is desirable for OEICs, perhaps the VCSEL will have an inherent geometric advantage.

Figure 25 summarizes some recent results for VCSELs [87, 137–138]. Most results so far have been for GaAs substrates. The strained-layer InGaAs active regions are particularly useful, because high gain and efficiency are desired. Progress with InGaAsP/InP devices has been slow, probably again due to high internal losses, but this is also related to the difficulties in obtaining high gains and the necessarily high mirror reflectivities in this system. Strain may be even more useful than in the GaAs system, because a reduction in the threshold carrier density is more significant here. That is, the additive nonradiative threshold current term, Eq. (13), due to Auger recombination (Cn^3), can be reduced significantly. Thus, strained quantum wells may enable the creation of practical VCSELs on InP.

Because of their small footprint and inherently low threshold current, VCSELs seem suitable for OEIC applications. They are wafer-scale man-

Fig. 25. (a) Room-temperature cw light vs. current for different sizes indicated with three quantum wells. (From Ref. 139). (b) Light vs. current curve and lasing spectrum of the indicated VCSEL with four pairs of electron beam-deposited SiO_2/Si. The device active reason is 10 μm in diameter. (From Ref. 87.) (c) Emission spectra at ~0.1 mA below and above threshold for a 10-μm-diameter buried-implant laser for which the *L-I* curve is shown in the inset. (From Ref. 138).

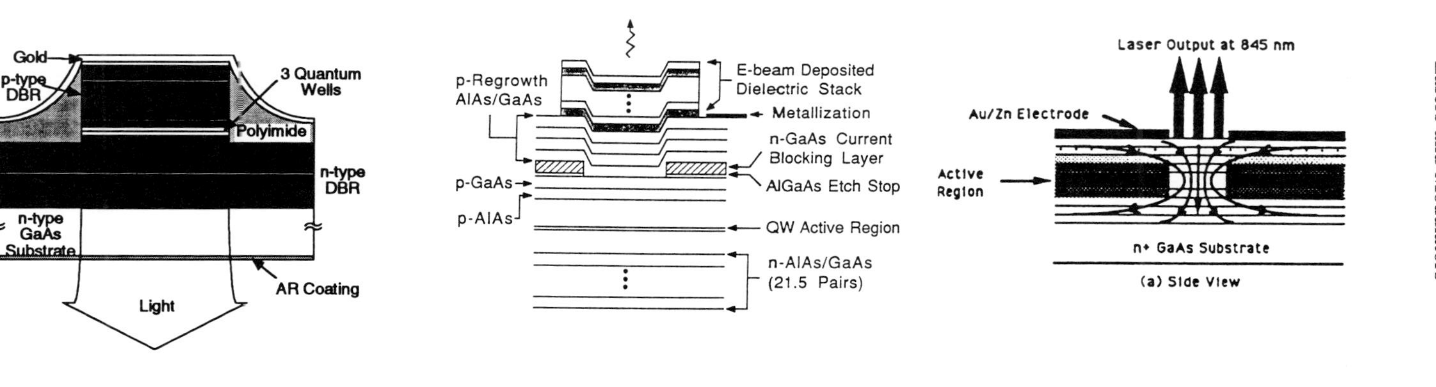
Gold
p-type DBR
3 Quantum Wells
Polyimide
n-type DBR
n-type GaAs Substrate
AR Coating
Light
p-Regrowth AlAs/GaAs
E-beam Deposited Dielectric Stack
Metallization
n-GaAs Current Blocking Layer
AlGaAs Etch Stop
p-GaAs
p-AlAs
QW Active Region
n-AlAs/GaAs (21.5 Pairs)
Laser Output at 845 nm
Au/Zn Electrode
Active Region
n+ GaAs Substrate
(a) Side View

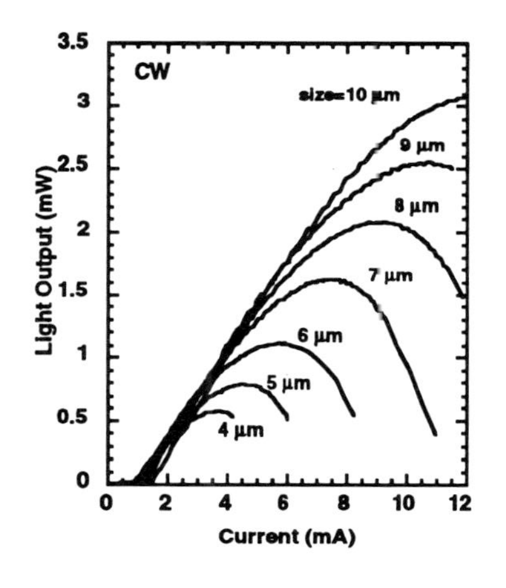
CW
size=10 μm
9 μm
8 μm
7 μm
6 μm
5 μm
4 μm
Light Output (mW)
Current (mA)

(a)

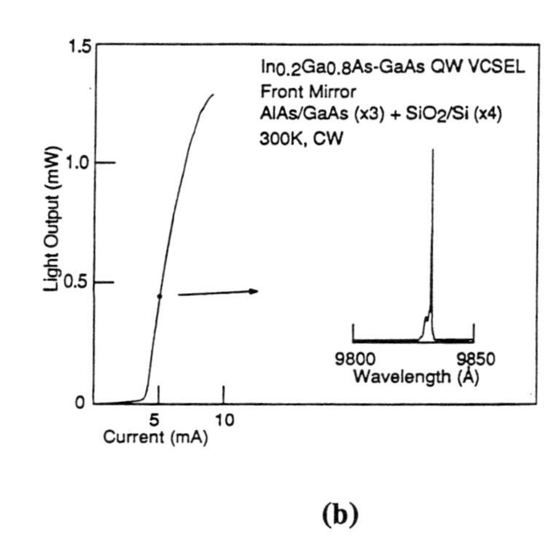
In0.2Ga0.8As-GaAs QW VCSEL
Front Mirror
AlAs/GaAs (x3) + SiO2/Si (x4)
300K, CW
9800
9850
Wavelength (Å)
Light Output (mW)
Current (mA)

(b)

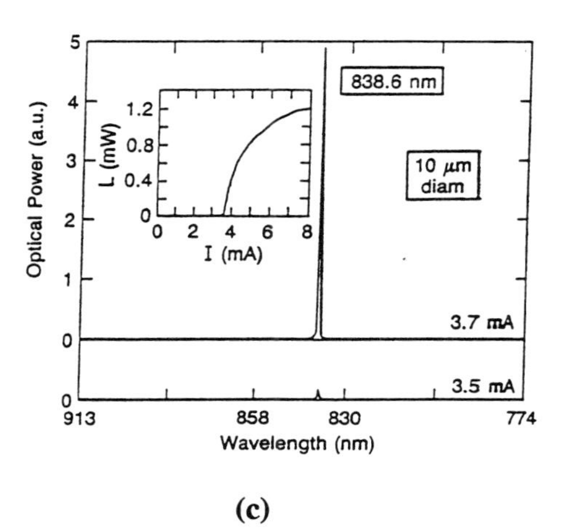
838.6 nm
10 μm diam
3.7 mA
3.5 mA
L (mW)
I (mA)
Optical Power (a.u.)
Wavelength (nm)

(c)

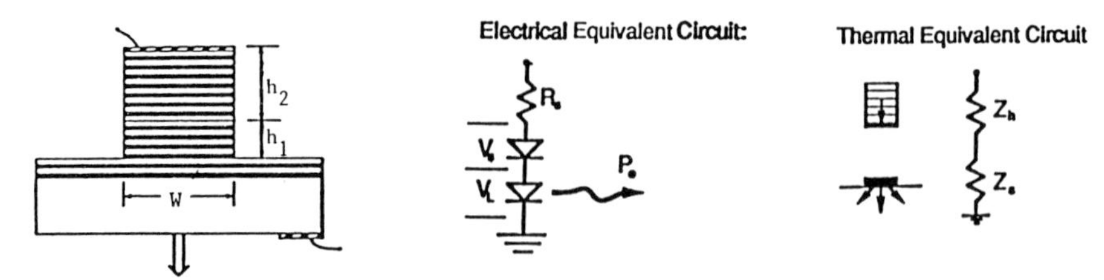

Fig. 26. Electrical and thermal equivalent circuits for the generic mesa-etched laser structure indicated.

ufacturable and testable, and if made with diameters of ~6 μm they also seem to have a significant fiber coupling advantage over in-plane SELs. In fact, the low numerical aperture of their output beams is one of the most striking attributes of VCSELs [139, 140]. That is, whether coupling into a waveguide, such as a fiber, or through some free-space optical network, the low divergent beams are highly desirable. Potential problems still exist in compatibility with electronics and in obtaining sufficient output power. The compatibility issues derive from the need to grow the thick multilayer mirror stacks; the power limitation is fundamentally related to the need to dissipate heat in structures with less than 100% efficiency.

Because the device is so small, the heat dissipation issue is more significant than in typical in-plane lasers. However, cw outputs of ~1 mW should be no problem, as may be seen from the following discussion. Referring to the electrical and thermal equivalent circuits in Fig. 26, the power dissipation is

$$P_{disp} = I^2R_s + IV_s + I_{th}V_L + (1 - \eta_{ex})(I - I_{th})V_L. \tag{19}$$

If we assume a VCSEL with a series resistance of 100 Ω, a threshold current of 1 mA, a differential efficiency of 50%, a threshold voltage of 1.5 V, and a series voltage drop $V_s = 0$, then, for 1 mW of light out, a power of 4 mW is dissipated. (Interestingly, only 0.9 mW of this is spent in the series resistance.) Assuming a worst-case etched mesa geometry, where the heat must flow a distance h to the bottom of the circular mesa through an AlAs/GaAs mirror stack via a thermal impedance Z_h and then spread out in the GaAs substrate via a thermal impedance Z_s, we obtain a net series thermal impedance of [141]

$$Z = 4h/(\pi w^2\xi_h) + 1/(2w\xi_s), \tag{20}$$

where ξ_h is the thermal conductivity of the mirror stack, taken as 0.45 W/(cm- K) for AlAs, and ξ_s is the thermal conductivity of the substrate, also taken as 0.45 W/(cm- K). Thus, if we assume 4 mW of heating 1 μm above the substrate and $w = 7\ \mu m$, the active region temperature rise is ~10°C for this example. If the mesa etching stops at the active region, so that $h = 0$, about 7 mW can be dissipated, or approximately 2 mW of optical output can be generated, for the same junction temperature rise in our example. From Eqs. (19) and (20) it is clear that a better heat sink geometry would have to be used to obtain much higher outputs from VCSELs. The buried active region designs, such as the proton-implanted configuration, can have thermal impedances nearly half that of etched mesa designs. However, in principle the etched mesas can be "buried" in some even better heat-conducting material. Also, if back emission is used, these can be flip-mounted onto good heat sinks.

3. MODULATORS FOR OEICS

3.1. Introduction

As mentioned earlier, electro-optical modulators may also be used to transduce electrical signals onto lightwaves. Also, phase modulators may be useful in tunable lasers, frequency modulators, or photonic switching. Here, however, we shall consider only the creation of intensity modulators using either index or absorption (gain) modulation in their active regions. As indicated in Fig. 2, with absorption or gain modulation, an intensity modulator can be constructed by simply passing the lightwave through a waveguide or bulk region containing the absorptive material. With index modulation some sort of switch or interferometer must be formed to provide the desired intensity modulation. Figure 2 shows a Mach–Zehnder interferometer as well as a directional coupler switch as examples. In the case of index modulation, actual electro-optical signal gain is possible, because the "valving action" involved in these devices provides a certain modulation depth on the input optical bias, regardless of its level. In the case of absorption modulation the induced current flowing across the active region requires a drive power proportional to the input optical power, and the electrical signal input power is always greater than the optical signal output power. A semiconductor optical amplifier is one kind of absorption modulator, in which the absorption is negative. Of course, it does not have electro-optical signal gain, but the optical output can be larger than the optical input, so that other system losses can be compensated. Thus, complex photonic switching networks with no insertion loss can be designed. In fact, insertion loss is one of the key problems of most modulators.

The modulator has a number of other attributes that might make it desirable for electro-optic transduction. For example, if operated in a reverse-biased configuration, it is tolerant to material defects, so that heteroepitaxy of different materials, e.g., GaAs on Si, can be used for reliable devices [142, 143]. The same factor also leads to a projection of high yield on large-scale arrays, although uniformity may be more of a problem than defects in this case. In addition, a reverse-biased index modulation configuration suggests a zero bias or holding power. In the case of absorption modulation, absorption of the optical input bias must be considered, so device heating may be a major issue.

In the case of photonic integrated circuits, waveguide modulators offer several desirable alternatives to directly modulating lasers. Among them is the ability to work on foreign substrates, such as Si. Also, one of the key interests is for "chirp-free" modulation. Generally, when a laser is modulated, the output wavelength chirps as the index is modulated along with the gain. Although similar effects occur in both absorption and index modulators, it is possible to create designs in which chirp is very small. For example, in the Mach–Zehnder interferometer modulator, chirp-free modulation is possible if equal out-of-phase signals are applied to both legs [144].

Surface-normal modulators are most useful in applications in which a fairly high density of outputs is desired. Index modulators are desired to reduce absorbed power and the associated heating effects, but the inherently short interaction length and the need to construct an interferometer or switch to provide intensity modulation create difficult boundary conditions for such devices. Thus, absorption modulation is usually used.

Finally, since an electro-optical modulator really multiplies the optical input by the electrical drive signal, it can be used to operate on information already in optical form. Thus, with two-dimensional arrays of surface-normal modulators, it is possible to form "spatial light modulators," which can be used for many interesting switching and signal processing applications [145, 146]. This multiplication capability might be considered as one aspect of "optical computing." Of course, this overused term also includes everything from free-space interconnection of electronic ICs [147, 148] to bistable optical latches, which are formed by series connecting modulators to create self-electro-optic-effect devices (SEEDs) [149–151]. Fortunately, we do not have the charter in this book to decipher the many possible applications of modulators to optical computing.

3.2 Waveguide Modulators

As already outlined, both index and absorption effects can be used for intensity modulation of waveguided lightwaves. Also, the absorption can be negative, as in optical amplifiers. In this section, properties of specific types of waveguide modulators which use one of these effects are considered.

Figure 27 shows three different configurations that use index modulation

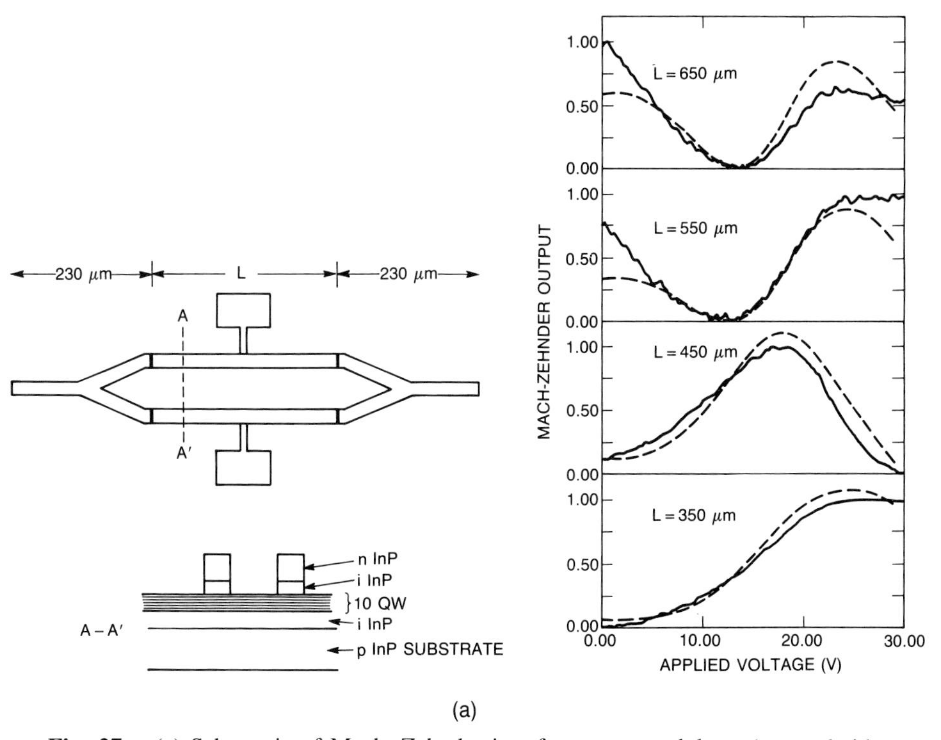

Fig. 27. (a) Schematic of Mach–Zehnder interferometer modulator (top and side views) and relative output intensity as a function of voltage for reverse bias applied between a single interferometer arm and the grounded substrate. Characteristics for four different lengths are shown. Solid curves are measured and dashed curves are calculated as described in the text. (From Ref. 152.) (b) Schematic of strip-loaded directional coupler and relative power output for coupled channel and input channel as a function of voltage. Circles represent data; solid line was calculated for the InGaAsP QW, dashed line for bulk InGaAsP. (From Ref. 153.) (c) Schematic diagram of guide–antiguide modulator index and optical field profiles for on and off states; on/off ratio vs. wavelength for different active region lengths for TE and TM modes. (From Ref. 154.)

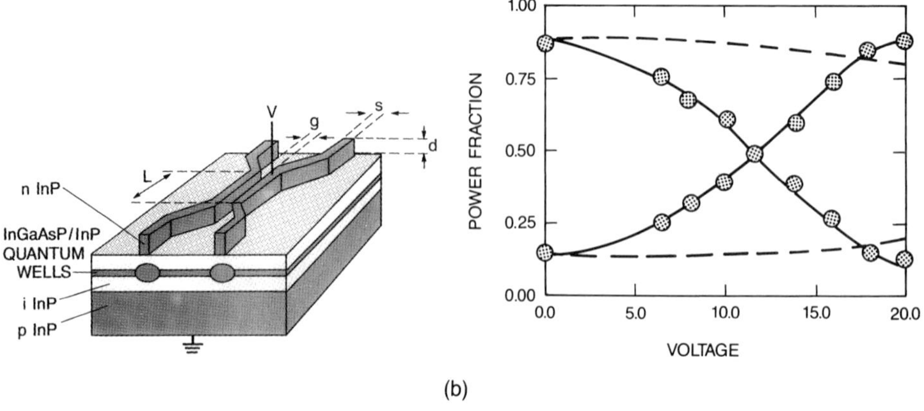

(b)

Fig. 27. (*continued*)

along with example results. The first uses interferometry [152] and the second two use optical switching to provide intensity modulation [153–155]. Waveguide crossing switches have also been explored [156, 157]. If the linear electro-optic effect is used, the effective optical bandwidth of operation can be relatively large. However, if quadratic electro-optic effects, which are associated with the absorption edge, are used, an optical bandwidth of only ~10 nm results. The electrical bandwidth is generally limited by the *RC* time constant in reverse-biased *pin* designs, and bandwidths of ~5–10 GHz tend to result for active lengths of ~1 mm in semiconductor waveguides. For traveling-wave designs, in which the electrical signal is propagated on a microstrip line along with the optical wave, bandwidths >100 GHz should be possible [158]. However, if the index modulation is due to free-carrier injection, then the free-carrier lifetime dominates and the bandwidth can be reduced to ~100 MHz.

In the Mach–Zehnder interferometer the input is split by a Y-junction into two legs containing separate phase shifters, and the output is formed by recombining these two legs in another Y-junction. Thus, when the phase shifters are adjusted for equal phase delays (modulo 2π), the two legs recombine to replicate the input. But when the two branches are out of phase, odd symmetry higher-order modes, which radiate away in the single-mode output waveguide, are formed at the second Y-junction. If the two legs are driven simultaneously in a push–pull fashion, the frequency modulation (or wavelength chirp) associated with the phase-advanced and -retarded legs cancels.

The directional coupler switch can be used as an intensity modulator if only one output is used. However, it has the additional feature of providing

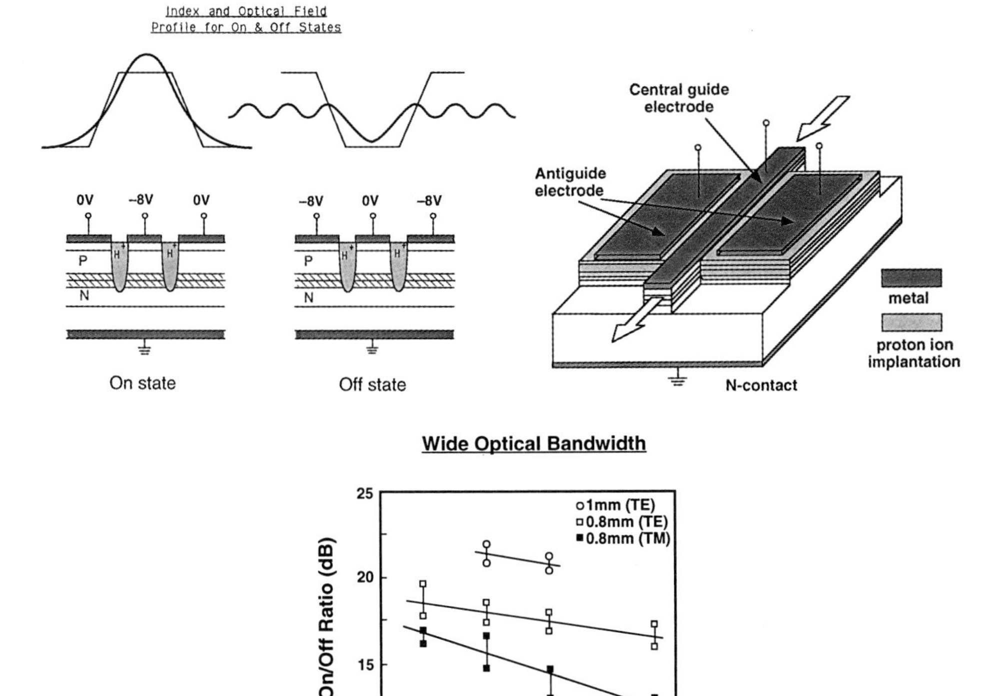

(c)

Fig. 27. (*continued*)

the complementary output, which is useful in some applications. In this case, the directional coupler is typically designed for 100% coupling of lightwaves from one waveguide to the other when no index shifts are added. Then, electrodes over one or both of the guides along the coupling region can be used to shift the index of one guide relative to the other, so that the phase coherent coupling is spoiled and the optical energy stays in the original guide. Again, chirp-free operation is possible if simultaneous positive and negative index shifts are created in the two guides. However, totally eliminating the chirp is somewhat more difficult than in the Mach–Zehnder case.

The third device is a "guide–antiguide" modulator in which the input is either simply transmitted through to the output or switched to a set of radiation modes, which are not coupled to the single-mode output guide [155]. Thus, high-extinction modulation is possible over a wide range of optical wavelengths, because no critical wavelength-dependent balance is required, as in the first two devices. The operation of the device may be viewed as due to a confinement-factor modulation, since it involves a change in the effective lateral index profile as the voltages on the central guide and the two surrounding antiguide electrodes are changed. Because the entire lateral index profile is due to electro-optical index changes, it can be changed over the entire range from a strongly guiding to a strongly antiguiding profile. The strongly guided output guide acts as a spatial mode filter to discriminate against unwanted radiation or weakly guided modes. Besides its wide optical bandwidth, the main desirable feature of this structure is its simplicity. Also, by operating the guiding and antiguiding electrodes in a push–pull fashion, it is possible to obtain a first-order chirp cancellation, since part of the energy propagates in both regions.

Figure 28 shows a schematic of an absorptive intensity modulator along with illustrative results for both loss [159, 160] and gain [161, 162] modulation. As shown, the design of the loss or gain modulator seems to be much simpler than that of the index modulators outlined in Fig. 27, and as indicated in the results, it is possible to obtain very nice on/off ratios. With the amplifier, the output can actually be larger than the input in the "on" state. These are all desirable features, and it is especially likely that the use of modulated amplifiers will find considerable use in photonic integrated circuits, in which loss compensation is important. The negative qualities of these absorptive modulators include (1) limited optical bandwidth, (2) limited electrical bandwidth, (3) power consumption/heat generation, (4) high chirp, and (5) for amplifiers, problems with reflections and crosstalk.

For either gain or loss modulation, the optical bandwidth is small, because both effects can only be modulated near the absorption edge of the active semiconductor material. For both, the effective optical wavelength bandwidth is ~10 nm in regions slightly above (gain) or below (loss) the absorption edge for bulk active material. For quantum well material the bandwidth is generally even smaller, although the efficiency can be correspondingly higher. For quantum well absorption modulation, a quantum-confined Stark shift of the electron–hole exciton absorption line by an applied electric field is usually used [163, 164]. In this case, the optical bandwidth for absorption modulation is ~4 nm. For gain modulation, the more narrow quantum well spectral linewidth can result in a similar optical bandwidth restriction. However, if the device is operated so as to fill portions of two quantum states in the on state, the bandwidth of amplification can actually be higher than in the bulk case [165].

The electrical bandwidth of reverse-biased absorption modulators can be

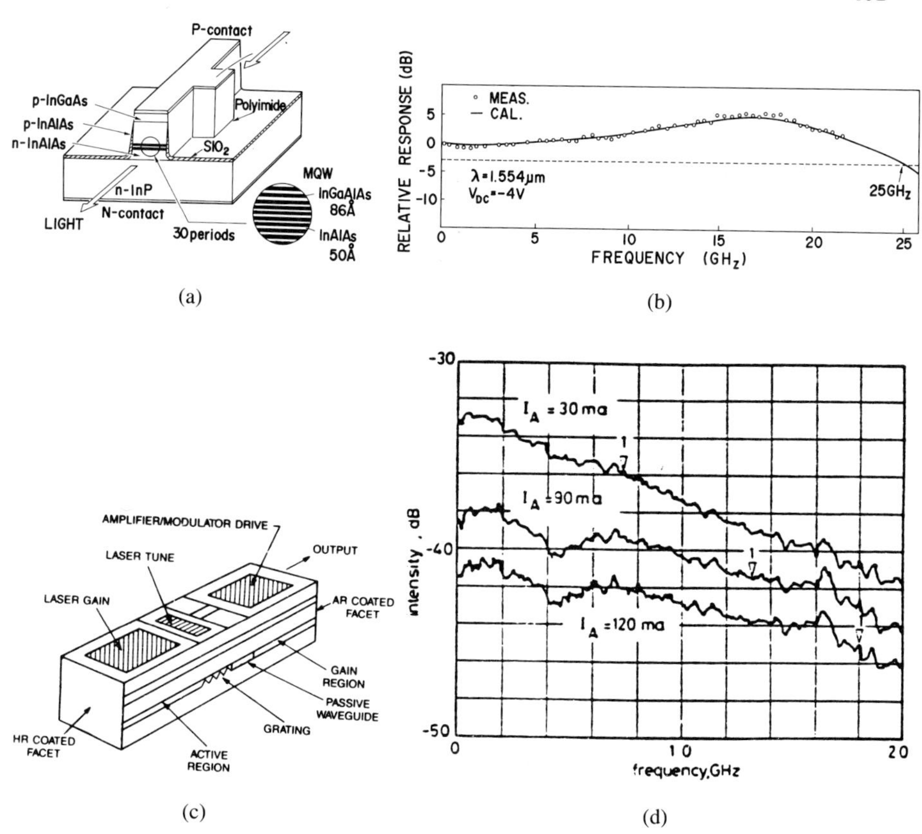

Fig. 28. (a) Schematic view of InGaAlAs/InAlAs MQW waveguide modulator. (b) Frequency response of MQW modulator,driven by a 5-dBm RF signal and a dc bias of 4 V (From Ref. 160.) The extrapolated 3-dB bandwidth is estimated to be 25 GHz. (c) Schematic of integrated strained-layer MQW laser–amplifier which is used as a gain modulator. (d) Small-signal frequency response for different amplifier bias currents. (From Ref. 170.)

limited by the creation of carriers that must be swept out of the active region as in photodetectors. However, this sweep-out time still generally allows bandwidths of ~20–30 GHz, which is usually larger than the *RC*-limited bandwidth [166]. This *RC* bandwidth also tends to be a little higher than in index modulators, simply because the length of the devices tends to be shorter. Of course, traveling-wave geometries are possible here as well. In multiple-quantum-well designs there can also be a problem with hole trapping in the active region. In AlGaAs/GaAs or InGaAlAs/InP this effect can be dealt with, but for InGaAsP/InP, in which the MQW hole barriers are higher, hole trapping appears to be the dominant speed limitation [167, 168]. In gain modulators, the bandwidth is again limited by the injected carrier lifetime, but if operated in the high-gain region, where stimulated effects dominate the carrier recombination, this can still be ~1 GHz [169].

For gain or absorption modulation chirp is a major problem, and for small-signal modulation the amount of chirping tends to be similar to that from directly modulated lasers. But, for large-signal modulation, the wavelength chirping effect tends to be much better than with lasers, since the relaxation ringing effects associated with the resonant laser cavity are absent [170].

3.3. Surface-Normal Modulators

As in the case of lasers, efficiency is one of the most important criteria for modulator devices in OEICs. Thus, we focus on designs which can provide efficient electro-optical conversion. In most surface-normal, or transverse, modulators there is a major difficulty in obtaining high modulation efficiency [171, 172] because the interaction length is limited. For the same reasons, it is difficult to obtain high contrast (on/off ratio). These problems are especially severe if drive voltages are limited (≤5 V) to be compatible with those of ICs. However, as shown by the schematics in Fig. 29, there have been several attempts to address this interaction length problem by using reflectors at one or both ends of the modulator's active region [173–

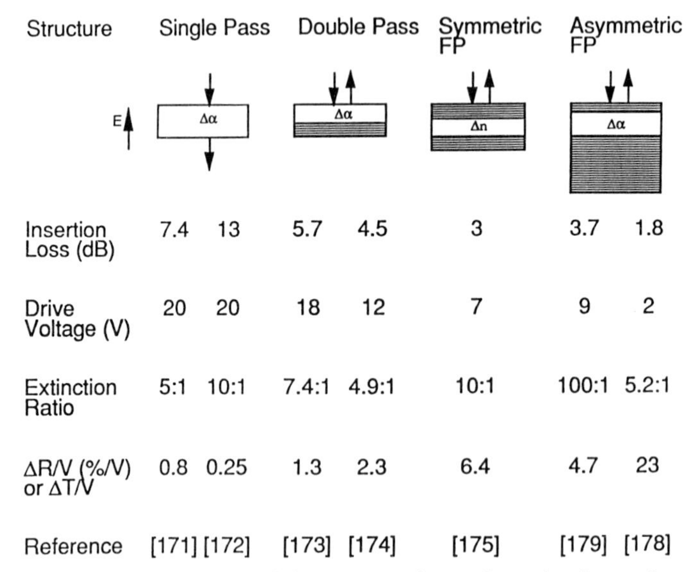

Fig. 29. Comparison of state-of-the-art experimental results for surface-normal electro-optic modulators. All devices have MQW active regions.

179]. For a single back-side reflector the interaction length is doubled. In the case of two reflectors, the effective interaction length is multiplied by the finesse, F, of the Fabry–Perot cavity that is formed, where $F = \pi\sqrt{R/(1 - R)}$, and R is the mean mirror reflectivity. Using this latter approach, high-contrast reflection modulation has been obtained with only 2 V of drive [178] in an asymmetric unbalanced resonator configuration [176]. Unfortunately, the use of such Fabry–Perot cavities may also lead to some reduction in the optical bandwidth as well as tight tolerances on the layer thicknesses within the structure. Nevertheless, if low voltage and high contrast are desired, the vertical Fabry–Perot reflection modulator may be the best solution.

Either absorption or index modulation (or both) may be used in surface-normal modulators. For simple transverse modulators, absorption modulation is most straightforward. But obtaining sufficiently high absorption changes, $\Delta\alpha/\alpha$, is always a problem. In fact, near the strongly absorbing region, $\Delta\alpha/\alpha \leq 3$, typically. For electro-optic modulation, $\Delta\alpha$ is induced by a voltage change across some active region. The electric field so induced can be modulated as fast as the RC time constant will allow, but if significant numbers of carriers are generated in the absorption process, the speed can be limited by a sweep-out time or minority carrier lifetime. This effect typically shows up as a reduction in speed as the optical input intensity is increased. The electric field–induced absorption change is generally obtained by using the change in the semiconductor absorption edge, which usually shifts to lower energies as the electric field is increased. In bulk regions, this phenomenon is associated with the Franz–Keldish effect [180], and in MQW regions it may be due to several effects, but the quantum-confined Stark effect (QCSE) [163] associated with changes in an excitonic absorption line is generally largest. As a result of the association with these slight changes in some absorption edge, the wavelength range over which $\Delta\alpha/\alpha$ is large is limited to ~4 nm for MQW and ~8 nm for bulk active regions. For a given field change, $\Delta\alpha/\alpha$ is largest for the QCSE in MQWs, but the maximum $\Delta\alpha/\alpha$ obtainable can be almost as large in bulk regions, because the field can be varied over a wider range. Thus, the wider optical bandwidth associated with bulk layers can be obtained if a larger drive voltage can be used. However, the contrast is always quite limited if a simple transmission modulator is constructed from this $\Delta\alpha/\alpha$ effect.

In order to increase the contrast and at the same time reduce the drive voltage, an asymmetric Fabry–Perot (ASFP) reflection modulator has been explored [176–179], as mentioned earlier. In this case, a high back mirror reflectivity $R_b \sim 100\%$ is combined with a moderate front mirror reflectivity

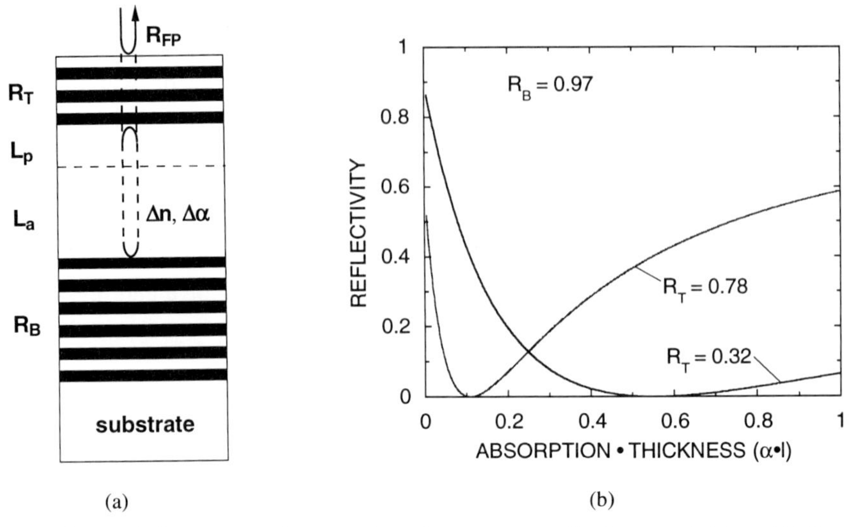

Fig. 30. (a) Schematic of a typical surface-normal reflection modulator. (b) Reflectivity (R_{FP}) of the ASFP modulator as a function of the cavity loss and cavity length product for top mirror reflectivity of 0.32 and 0.78. (c) Layer structure with $R_T = 0.32$. The topmost layer is doped with Be to 10^{18} cm^{-3}, the MQW region is undoped, and the bottom grating is doped with Si to 10^{18} cm^{-3}. (d) Narrowband reflectivity spectra for the ASFP modulator (shown in c) at different bias voltages. (From Ref. 181.) Reflectivity is measured relative to a gold mirror ($R \approx 0.95$). (e) Layer structure with $R_T = 0.78$. (f) Narrowband reflectivity spectra for the ASFP modulator (shown in e) at different bias voltages.

$R_f \sim 30$–80%. As shown in the schematic in Fig. 30, the mirrors are doped n and p, respectively, and the active cavity of length l_a is undoped, so that the field across it can be varied with reverse bias. For this configuration the net reflection from the top surface is given by [183]

$$R_1(\lambda) = [(r_f - r_b t^2)/(1 - r_f r_b t^2)]^2, \qquad (21)$$

where $R_f = r_f^2$, $R_b = r_b^2$, $t^2 = \exp(-2j\boldsymbol{\beta}_a l_a)\exp(-2j\boldsymbol{\beta}_p l_p)$, l_p accounts for any passive cavity region, such as mirror penetrations, and $\boldsymbol{\beta}_a = \beta_a - j\alpha_a/2$ and $\boldsymbol{\beta}_p = \beta_p - j\alpha_p/2$. The magnitude of Eq. (21) is plotted in Fig. 30 versus the cavity attenuation ($\alpha_t l_t = p\alpha_a l_a + \alpha_p l_p$) for two different front mirror reflectivities. If the active material in the cavity has low absorption for no voltage applied, the net reflection is dominated by the back reflector, so the reflection is high. Now, if a voltage is applied to increase the absorption, the reflection component from the rear mirror is attenuated to eventually equal that from the front, so the net reflection is lowered to zero. This condition is that $R_f = R_b \exp(-2\alpha_t l_t)$. Of course, if the cavity absorption is

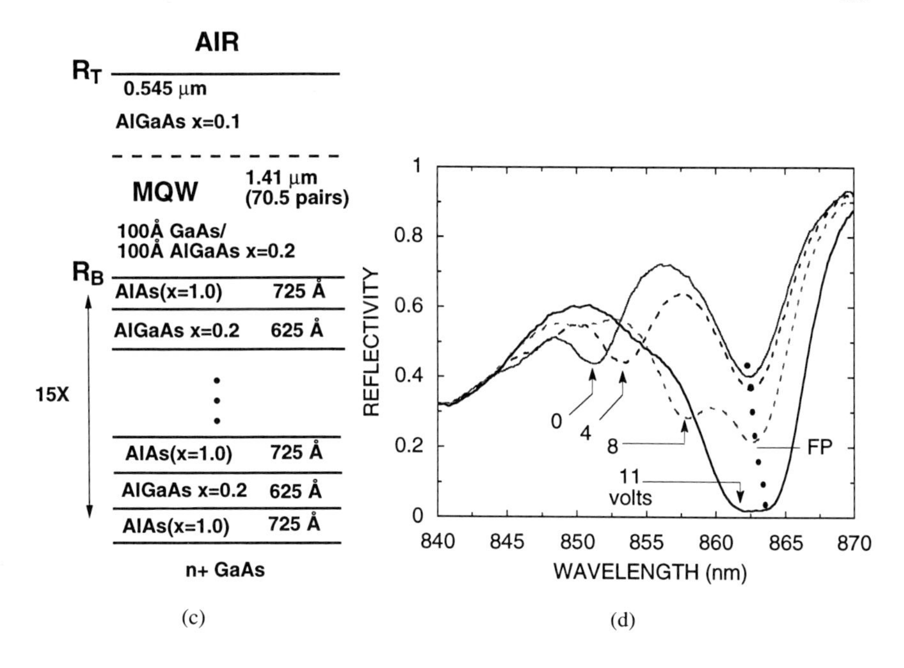
AIR
R_T
0.545 μm
AlGaAs x=0.1
MQW
1.41 μm (70.5 pairs)
100Å GaAs/
100Å AlGaAs x=0.2
R_B
AlAs(x=1.0) 725 Å
AlGaAs x=0.2 625 Å
15X
AlAs(x=1.0) 725 Å
AlGaAs x=0.2 625 Å
AlAs(x=1.0) 725 Å
n+ GaAs
REFLECTIVITY
WAVELENGTH (nm)
0
4
8
11 volts
FP

(c) (d)

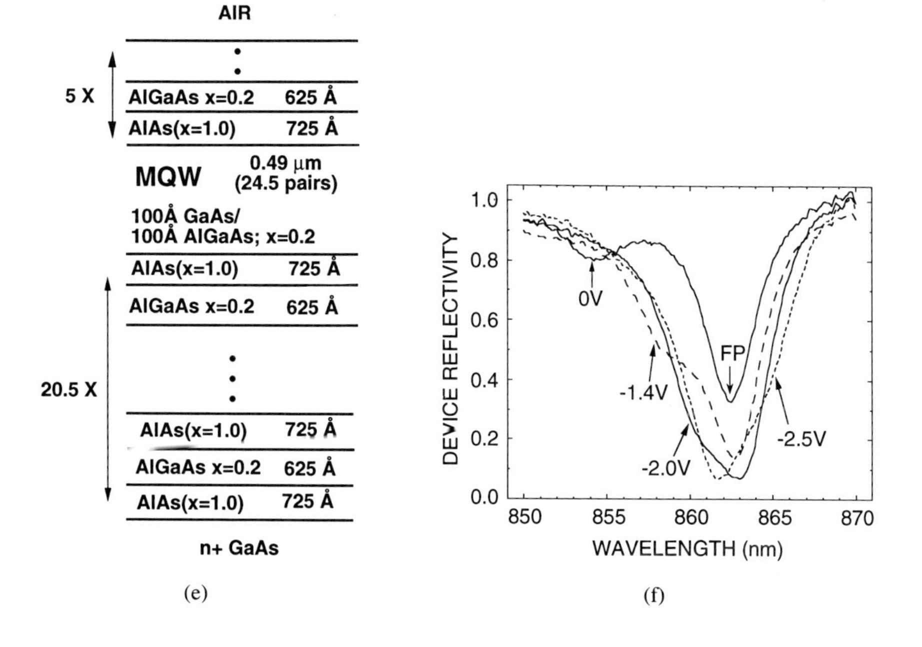
AIR
5 X
AlGaAs x=0.2 625 Å
AlAs(x=1.0) 725 Å
MQW
0.49 μm (24.5 pairs)
100Å GaAs/
100Å AlGaAs; x=0.2
AlAs(x=1.0) 725 Å
AlGaAs x=0.2 625 Å
20.5 X
AlAs(x=1.0) 725 Å
AlGaAs x=0.2 625 Å
AlAs(x=1.0) 725 Å
n+ GaAs
DEVICE REFLECTIVITY
WAVELENGTH (nm)
0V
-1.4V
FP
-2.0V
-2.5V

(e) (f)

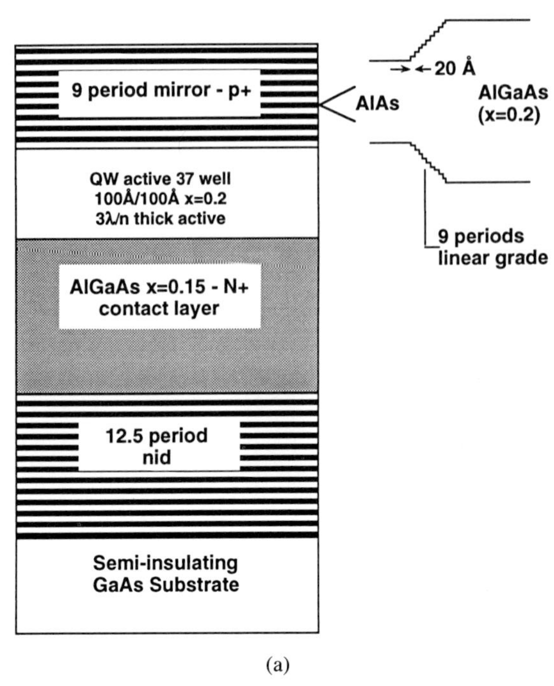

(a)

Fig. 31. (a) Layer structure for an SFP modulator designed to operate at high frequency. (b) Narrowband reflectivity spectra for the SFP modulator at different bias voltages. (c) Measured modulation efficiency as a function of frequency for a device of 42 × 42μm. Dashed line represents calculated performance based on measured values of resistance and capacitance. (From Ref. 186.)

increased further, the reflection will then become dominated by the component from the front mirror. Figure 30 also shows some experimental spectra for MQW ASFP modulators, which illustrate the desired modulation [174, 177]. As expected, the experiments show that the residual absorption (at $V = 0$) results in a significant insertion loss in the on state. In addition, we observe that little sacrifice in optical bandwidth occurs for the low-Q cavity (~4 nm for 10-dB contrast) but that the bandwidth is somewhat reduced for the higher-Q cavity (~2 nm). Fortunately, the material index dispersion helps broaden the useful optical bandwidth slightly, if the exciton absorption line is near the Fabry–Perot resonance [177]. Also, it is worth noting that the voltage for the off state can have a large tolerance if the cavity loss is sat-

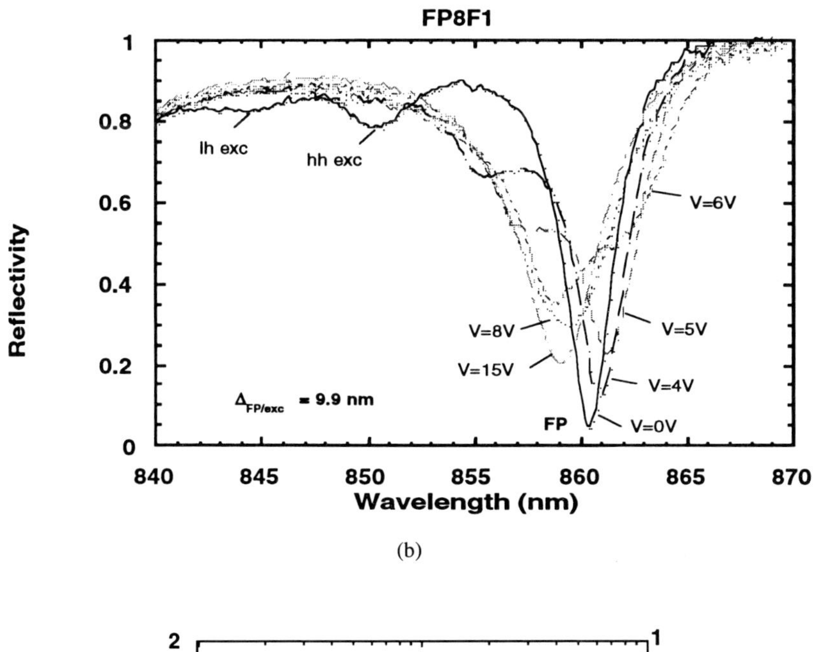

(b)

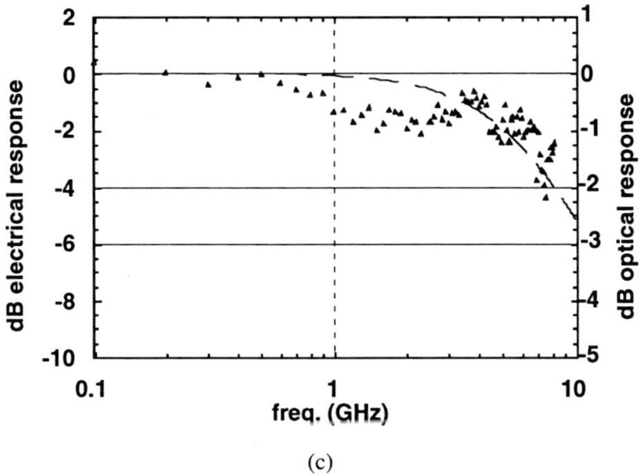

(c)

Fig. 31. (*continued*)

urating near the optimum value. That is, the R vs. V curve can stay near zero for a wide range of voltages. Nevertheless, the layer thickness tolerances on growing such a structure are still very tight (~0.5%), because the Fabry–Perot resonance must be accurately positioned from the absorption edge [182, 183].

Absorption-type modulators have two basic problems, as already mentioned. First, the absorbed energy results in local heating, which will shift the properties of a particular modulator and those of nearest neighbors. Thus, array nonuniformities and crosstalk result. Second, the absorption generates free carriers, which will result in nonuniform fields and a reduction in modulation bandwidth if allowed to accumulate. Thus, for many applications, it would be more desirable to use index modulation together with some interferometric effect to convert it to intensity modulation. Actually, the first Fabry–Perot electro-optical experiments were aimed at fulfilling this goal [184]. Figure 31 summarizes some results for a symmetric Fabry–Perot (SFP) configuration, in which intensity modulation of some wavelength is accomplished by shifting the resonance line back and forth in wavelength [185]. The modulator bandwidth is found to be RC limited. The cavity is balanced so that the net reflection on resonance is approximately zero. For this case, the index modulation is also associated with the field-induced shifts in an absorption edge, since these quadratic electro-optical effects are larger than the less wavelength-dependent linear electro-optic effect. Thus, again the device layer thicknesses are very critical. Also, in order to obtain a reasonable modulation depth, it is necessary to use a relatively high-finesse cavity so that the Fabry–Perot resonance can be shifted an amount $\Delta\lambda$, which is a significant fraction of its linewidth, $\delta\lambda = \lambda^2/(2nl_t F)$. The relative shift in the resonant wavelength $\Delta\lambda/\lambda = (l_a/l_t)\,\Delta n/n$ in the cavity, and $\Delta n/n <<$ 1%, if the resonance is separated from the absorption edge sufficiently to neglect absorption. (The length fill factor results from the fact that the index cannot be varied over the entire effective cavity length due to the penetration into the mirrors, i.e., $l_t = l_a + l_{geb} + l_{get}$.) Thus, to shift the resonance by half of its linewidth ($\Delta\lambda = \delta\lambda/2$) in order to get ~50% modulation depth, $\Delta n l_a F = \lambda/4$. As shown in Fig. 31, this is difficult to accomplish with reasonable finesses. For example, for $\Delta n = 0.003$, $l_a = 0.5$ μm (for 100 kV/cm at 5 V), and $\lambda = 0.9$ μm, $F = 150$ to satisfy the condition for 50% modulation depth. In Fig. 31, $F \sim 40$. The requirement for high finesse results in a very small optical bandwidth, which must be $< 0.1\delta\lambda$ for reasonable contrast. In our example with $F = 150$, the optical bandwidth would be <1 Å for $l_t = 1$ μm; in Fig. 31 it is ~3 Å. The slight increase in loss tends to improve the contrast if the contrast is low; however, it tends to reduce the contrast if a larger shift is used.

REFERENCES

1. T. Horimatsu and M. Sasaki, *J. Lightwave Technol.* **7**, 1612 (1989) and references therein.
2. H. Blauvelt, N. Bar-Chaim, D. Fekete, S. Margalit, and A. Yariv, *Appl. Phys. Lett.* **40**, 289 (1982).
3. D. R. Scifres, W. Streifer, and R. D. Burnham, *IEEE J. Quantum Electron.* **QE-15**, 917 (1979).
4. D. Botez and J. C. Connolly, *Appl. Phys. Lett.* **43**, 1096 (1983).
5. G. A. Evans, N. W. Carlson, J. M. Hammer, M. Lurie, J. K. Butler, S. L. Palfrey, R. Amantea, L. A. Carr, F. Z. Hawrylo, E. A. James, C. J. Kaiser, J. B. Kirk, and W. F. Reichert, *IEEE J. Quantum Electron.* **25**, 1525 (1989).
6. A. Yariv, *IEEE Circuits Devices* **5**, 25 (1989).
7. M. Asada, A. Kameyama, and Y. Suematsu, *IEEE J. Quantum Electron.* **QE-20**, 745 (1984).
8. Y. Arakawa and A. Yariv, *IEEE J. Quantum Electron.* **QE-21**, 1666 (1985).
9. E. Yablonovitch and E. O. Kane, *J. Lightwave Technol.* **LT-4**, 504 (1986).
10. Y. Arakawa and A. Yariv, *IEEE J. Quantum Electron.* **QE-22**, 1887 (1986).
11. J. I. Hashimoto, T. Katsuyama, J. Skinkai, I. Yoshida, and H. Hayashi, *Appl. Phys. Lett.* **58**, 879 (1991).
12. T. R. Chen, L. Eng, B. Zhao, Y. H. Zhuang, S. Sanders, H. Morkoc, and A. Yariv, *IEEE J. Quantum Electron.* **26**, 1183 (1990).
13. H. Ghafoori-Shiraz, *Opt. Quantum Electron.* **20**, 153 (1988).
14. G. P. Agrawal and N. K. Dutta, "Long-Wavelength Semiconductor Lasers," Ch. 2 and 3, Van Nostrand Reinhold, New York, 1986.
15. C. H. Henry, R. A. Logan, and F. R. Merritt, *J. Appl. Phys.* **51**, 3042 (1980).
16. G. H. B. Thompson, "Physics of Semiconductor Laser Devices," Ch. 2. Pitman Press, Great Britain, 1980.
17. A. Yariv, "Optical Electronics," 3rd ed., Ch. 15. Holt, Rinehart & Winston, New York, 1985.
18. S. W. Corzine, R. H. Yan, and L. A. Coldren, in "Quantum Well Lasers," Ch. 1 (P. S. Zory, ed.). Academic Press, Boston, 1992.
19. E. Wintner and E. P. Ippen, *Appl. Phys. Lett.* **44**, 999 (1984).
20. S. Colak, R. Eppenga, and M. F. H. Schuurmans, *IEEE J. Quantum Electron.* **QE-23**, 960 (1987).
21. R. Nagarajan, T. Kamiya, and A. Kurobe, *IEEE J. Quantum Electron.* **25**, 1161 (1989).

22. D. Ahn and S. L. Chuang, *IEEE J. Quantum Electron.* **24**, 2400 (1988).
23. K. J. Beernink, P. K. York, and J. J. Coleman, *Appl. Phys. Lett.* **55**, 2585 (1989).
24. H. Temkin, T. Tanbun-Ek, R. A. Logan, D. A. Cebula, and A. M. Sergent, *IEEE Photon. Technol. Lett.*, **3**, 100 (1991).
25. J. E. Bowers, *Electron. Lett.* **22**, 1119 (1986).
26. R. Nagarajan, T. Fukushima, J. E. Bowers, R. S. Geels, and L. A. Coldren, *Electron. Lett.* **27**, 1058 (1991).
27. M. Asada and Y. Suematsu, *IEEE J. Quantum Electron.* **QE-19**, 917 (1983).
28. A. Sugimura, *IEEE J. Quantum Electron.* **QE-17**, 627 (1981).
29. C. B. Su, J. Schlafer, J. Manning, and R. Olshansky, *Electron. Lett.* **18**, 1108 (1982).
30. M. Asada and Y. Suematsu, *Appl. Phys. Lett.* **41**, 353 (1982).
31. C. B. Su, J. Schlafer, J. Manning, and R. Olshansky, *Electron. Lett.* **18**, 595 (1982).
32. A. R. Adams, *Electron. Lett.* **22**, 249 (1986).
33. E. P. O'REILLY and A. GHITI, in "Quantum Well Lasers," Ch. 2 (P. S. Zory, ed.). Academic Press, Boston, 1992.
34. S. Harder, B. J. Van Zeghbroeck, M. P. Kesler, H. P. Meier, P. Vettiger, D. J. Webb, and P. Wolf, *IBM J. Res. Dev.* **34**, 568 (1990).
35. T. L. Koch and U. Koren, *J. Lightwave Technol.* **8**, 274 (1990).
36. J. L. Merz and R. A. Logan, *J. Appl. Phys.* **47**, 3503 (1976).
37. K. Iga, T. Kambayashi, K. Wakao, K. Moriki, and C. Kitahara, *IEEE J. Quantum Electron.* **QE-15**, 72 (1979).
38. K. Iga and B. I. Miller, *Electron. Lett.* **16**, 342 (1980).
39. E. L. Hu and R. E. Howard, *Appl. Phys. Lett.* **37**, 1022 (1980).
40. G. J. Sonek and J. M. Ballantyne, *J. Vac. Sci. Technol.* **B2**, 653 (1984).
41. L. A. Coldren, K. Iga, B. I. Miller, and J. A. Rentschler, *Appl. Phys. Lett.* **37**, 681 (1980).
42. P. Buchmann, H. P. Dietrich, G. Sasso, and P. Vettiger, *Microelectron. Eng.* **9**, 485 (1989).
43. M. Uchida, S. Ishikawa, N. Takado and K. Asakawa, *IEEE J. Quantum Electron.* **24**, 2170 (1988).
44. A. Kasukawa, M. Iwase, Y. Hiratani, N. Matsumoto, Y. Ikegami, M. Irikawa, and S. Kashiwa, *Appl. Phys. Lett.* **51**, 1774 (1987).
45. T. Yuasa, N. Hamao, M. Sugimoto, N. Takado, M. Ueno, H. Iwata, Y. Tashiro, K. Onabe, and K. Asakawa, in "Conference on Lasers & Electro-Optics 1988," Anaheim, CA (Optical Society of America, Washington, DC), paper W06, 258 (May 1988).
46. A. J. SpringThorpe, *Appl. Phys. Lett.* **31**, 524 (1977).
47. T. Takamori, L. A. Coldren, and J. L. Merz, *Appl. Phys. Lett.* **55**, 1053 (1989).
48. W. D. Goodhue, K. Rauschenbach, C. A. Wang, J. P. Donnelly, R. J. Bailey, and G. D. Johnson, *J. Electron. Mater.* **19**, 463 (1990).

49. J. J. Yang, M. Sergant, M. Jansen, S. S. Ou, L. Eaton, and W. W. Simmons, *Appl. Phys. Lett.* **49**, 1138 (1986).
50. Z. L. Liau and J. N. Walpole, *Appl. Phys. Lett.* **46**, 115 (1985).
51. J. L. Vossen, *J. Electrochem. Soc.* **126**, 319 (1979).
52. L. A. Coldren and J. A. Rentschler, *J. Vac. Sci. Technol.* **19**, 225 (1981).
53. N. Vodjdani and P. Parrens, *J. Vac. Sci. Technol.* **B**, 1591 (1987).
54. M. A. Bosch, L. A. Coldren, and E. Good, *Appl. Phys. Lett.* **38**, 264 (1981).
55. K. Asakawa and S. Sugata, *J. Vac. Sci. Technol.* **B3**, 402 (1985).
56. T. Yuasa, T. Yamada, K. Asakawa, S. Sugata, and M. Ishii, *Appl. Phys. Lett.* **49**, 1007 (1986).
57. G. D. Boyd, L. A. Coldren, and F. G. Storz, *Appl. Phys. Lett.* **36**, 583 (1980).
58. T. Takamori, L. A. Coldren, and J. L. Merz, *Appl. Phys. Lett.* **53**, 2549 (1988).
59. S. Sugata and K. Asakawa, *J. Vac. Sci. Technol.* **B5**, 894 (1987).
60. M. W. Geis, G. A. Lincoln, N. Efremow, and W. J. Piacentini, *J. Vac. Sci. Technol.* **19**, 1390 (1981).
61. J. A. Skidmore, L. A. Coldren, J. L. Merz, E. L. Hu, and K. Asakawa, *Appl. Phys. Lett.,* **53**, 2308 (1988).
62. J. A. Skidmore, D. L. Green, J. A. Olsen, D. B. Young, E. L. Hu, L. A. Coldren, and P. M. Petroff, *Int. Symp. on Electron, Ion, and Photon Beams,* Seattle, (May 1991).
63. E. L. Hu and R. E. Howard, *J. Vac. Sci. Technol.* **B**, 85 (1984).
64. G. A. Vawter, L. A. Coldren, J. L. Merz, and E. L. Hu, *Appl. Phys. Lett.* **51**, 719 (1987).
65. S. J. Pearton, W. S. Hobson, U. K. Chakrabarti, G. E. Derkitis, Jr., and A. P. Kinsells, *J. Electrochem. Soc.* **137**, 3892 (1990).
66. E. L. Hu and L. A. Coldren, *Proc. SPIE Symp. Adv. Process. Semicond. Devices* **797**, 8 (1987).
67. A. Seabaugh, *J. Vac. Sci. Technol.* **B**, 77 (1988).
68. V. M. Donnelly, D. L. Flamm, C. W. Tu, and D. E. Ibbotson, *J. Electrochem. Soc.* **129**, 2533 (1982).
69. M. W. Geis, N. N. Efremow, and G. A. Lincoln, *J. Vac. Sci. Technol.* **B4**, 315 (1986).
70. T. R. Hayes, M. A. Dreisbach, P. M. Thomas, W. C. Dautremont-Smith, and L. A. Heimbrook, *J. Vac. Sci. Technol.* **B7**, 1130 (1989).
71. H. Kogelnik and C. V. Shank, *J. Appl. Phys.* **43**, 2327 (1972).
72. S. Wang, *IEEE J. Quantum Electron.* **QE-10**, 413 (1974).
73. M. C. Amann, S. Illek, C. Schanen, and W. Thulke, *Appl. Phys. Lett.* **54**, 2532 (1989).
74. H. A. Haus, "Waves and Fields in Optoelectronics," Ch. 8, Prentice Hall, Englewood Cliffs, NJ, 1984.
75. S.W. Corzine, R. H. Yan, and L. A. Coldren, *IEEE J. Quantum Electron.* **27**(6), 1359–1367, (June, 1991).

76. K. Komori, S. Arai, Y. Suematsu, I. Arima, and M. Aoki, *IEEE J. Quantum Electron.* **25**, 1235 (1989).
77. Y. Suematsu, S. Arai, and K. Kishino, *J. Lightwave Technol.* **LT-1**, 161 (1983).
78. A. Yariv and M. Nakamura, *IEEE J. Quantum Electron.* **QE-13**, 233 (1977).
79. L. A. Coldren and S. W. Corzine, *IEEE J. Quantum Electron.* **QE-23**, 903 (1987).
80. C. V. Shank and R. V. Schmidt, *Appl. Phys. Lett.* **23**, 154 (1973).
81. N. Tsumita, J. Melngailis, A. M. Hawryluk, and H. I. Smith, *J. Vac. Sci. Technol.* **19**, 1211 (1981).
82. M. Suehiro, T. Hirata, M. Maeda, and H. Hosomatsu, *Jpn. J. Appl. Phys.* **29**, L1217 (1990).
83. C. F. Zah, C. Caneau, S. G. Menocal, A. S. Gozdz, P. S. D. Lin, F. Favire, A. Yi-Yan, T. P. Lee, A. G. Dentai, and C. H. Joyner, *Electron. Lett.* **25**, 650–651 (1989).
84. T. Katoh, Y. Nagamune, G. P. Li, S. Fukatsu, Y. Shiraki, and R. Ito, *Appl. Phys. Lett.* **57**, 1212 (1990).
85. Y. H. Lee, J. L. Jewell, A. Scherer, S. L. McCall, S. J. Walker, J. P. Harbison, and L. T. Florez, *Electron. Lett.* **25**, 1377 (1989).
86. R. S. Geels, S. W. Corzine, J. W. Scott, D. B. Young, and L. A. Coldren, *IEEE Photon. Technol. Lett.* **2**, 234 (1990); and M. G. Peters, B. J. Thibeault, D. B. Young, J. W. Scott, F. H. Peters, A. C. Gossard, and L. A. Coldren, *Appl. Phys. Lett.* **63** (25), 3411–3413, (1993).
87. C. Lei, T. J. Rogers, D. G. Deppe, and B. G. Streetman, *Appl. Phys. Lett.* **58**, 1122 (1991).
88. S. Ramo, J. R. Whinnery, and T. VanDuzer, "Fields and Waves in Communication Electronics," Ch. 11. Wiley, New York, 1965.
89. K. Tai, L. Yang, Y. H. Wang, J. D. Wynn, and A. Y. Cho, *Appl. Phys. Lett.* **56**, 2496 (1990).
90. P. Zory, *Appl. Phys. Lett.* **22**, 125 (1973).
91. Zh. I. Alferov, V. M. Andreev, S. A. Gurevich, R. F. Kazarinov, V. R. Larionov, M. N. Mizerov, and E. L. Portnoy, *IEEE J. Quantum Electron.* **QE-11**, 449 (1975).
92. A. Hardy, D. F. Welch, and W. Streifer, *IEEE J. Quantum Electron.* **26**, 50 (1990).
93. R. J. Noll and S. H. Macomber, *IEEE J. Quantum Electron.* **26,** 456 (1990).
94. S. H. Macomber, *IEEE J. Quantum Electron.* **26**, 2065 (1990).
95. A. Hardy, R. G. Waarts, D. F. Welch, and W. Streifer, *IEEE J. Quantum Electron.* **26**, 843 (1990).
96. J. W. Goodman, F. J. Leonberger, S. Kany, and R. A. Athale, *Proc. IEEE,* **72**, 850 (1984).
97. O. Wada, H. Nobuhara, T. Sanada, M. Kuno, M. Makiuchi, T. Fujii, and T. Sakurai, *J. Lightwave Technol.* **7**, 186 (1989).

98. C. H. Henry, G. E. Blonder, R. F. Kazarinow, *J. Lightwave Technol.* **7**, 1530 (1989).
99. L. A. Coldren, B. I. Miller, K. Iga, and J. A. Rentschler, *Appl. Phys. Lett.* **38**, 315 (1981).
100. W. Streifer, D. R. Scifres, and R. D. Burnham, *IEEE J. Quantum Electron.* **QE-12**, 422 (1976).
101. L. A. Coldren, T. L. Koch, T. J. Bridges, E. G. Burkhardt, B. I. Miller, and J. A. Rentschler, *Appl. Phys. Lett.* **46**, 5 (1985).
102. J. S. Osinski, C. E. Zah, R. Bhat, R. J. Contolini, E. D. Beebe, T. P. Lee, K. D. Cummings, and L. R. Harriott, *Electron. Lett.* **23**, 1156 (1987).
103. H. Sato and Y. Hori, *IEEE J. Quantum Electron.* **26**, 467 (1990).
104. J. E. A. Whiteaway, G. H. B. Thompson, A. J. Collar, and C. J. Armistead, *IEEE J. Quantum Electron.* **25**, 1261 (1989).
105. W. Streifer, D. R. Scifres, and R. D. Burnham, *IEEE J. Quantum Electron.* **QE-13**, 134 (1977).
106. S. L. McCall and P. M. Platzman, *IEEE J. Quantum Electron.* **QE-21**, 1899 (1985).
107. W. Streifer, R. D. Burnham, and D. R. Scifres, *IEEE J. Quantum Electron.* **QE-11**, 154 (1975).
108. K. Kobayashi and I. Mito, *J. Lightwave Technol.* **6**, 1623 (1988).
109. Y. Kotaki and H. Ishikawa, *IEEE J. Quantum Electron.* **25**, 1340 (1989).
110. S. L. Woodward, I. M. Habbab, T. L. Koch, and U. Koren, *IEEE Photon. Technol. Lett.* **2**, 854 (1990).
111. M. Kuznetsov, L. W. Stulz, T. L. Koch, U. Koren, and B. Tell, *Electron. Lett.* **25**, 686 (1989).
112. X. Pan, H. Olesen, and B. Tromborg, *IEEE Photon. Technol. Lett.* **2**, 312 (1990).
113. M. Kuznetsov, *IEEE J. Quantum Electron.* **24**, 1837 (1988).
114. C. F. J. Schanen, S. Illek, H. Lang, K. W. Thulke, and M. C. Amann, *Proc. IEEE* **137**, 69 (1990).
115. T. L. Koch, U. Koren, and B. I. Miller, *Appl. Phys. Lett.* **53**, 1036 (1988).
116. S. Illek, W. Thulke, C. Schanen, H. Lang, and M. C. Amann, *Electron. Lett.* **26**, 46 (1990).
117. V. Jayaraman, L. A. Coldren, S. Denbaars A. Mathur, and P. D. Dapkus, *Integrated Photonics Research '92*, New Orleans, paper no. WF1-1 (April 13–16, 1992); V. Jayaraman, A. Mathur, L. A. Coldren, and P. D. Dapkus, *IEEE Phononics Technology Lett.* **5**(5), 489 (1993).
118. Z. M. Chuang and L. A. Coldren, *IEEE—LEOS Annual Meeting*, San Jose, CA, paper no. SDL13.1 (November 4–7, 1991); and Z. M. Chuang and L. A. Coldren, *Photonic Technology Lett.* **5**(10) (1993).
119. M. Mittelstein, D. Mehuys, A. Yariv, J. E. Ungar, and R. Sarfaty, *Appl. Phys. Lett.* **54**, 1092 (1989).
120. R. C. Alferness, U. Koren, L. L. Buhl, B. I. Miller, M. G. Young, T.

L. Koch, G. Raybon, and M. C. A. Burrus, *Integrated Photonics Research 1992*, New Orleans, LA (April 13–16, 1992).

121. F. B. McCormick and M. E. Prise, *Appl. Opt.* **29**, 2013 (1990).
122. A. Dickinson and M. E. Prise, *Appl. Opt.* **29**, 2001 (1990).
123. D. F. Welch, R. Parke, A. Hardy, R. Waarts, W. Streifer, D. R. Scifres, *Electron. Lett.* **25**, 1038 (1989).
124. G. A. Evans, N. W. Carlson, J. M. Hammer, M. Lurie, J. K. Butler, S. L. Palfrey, R. Amantea, L. A. Carr, F. Z. Hawrylo, E. A. James, C. J. Kaiser, J. B. Kirk, W. F. Reichart, S. R. Chinn, J. R. Shealy, and P. S. Zory, *Appl. Phys. Lett.* **53**, 2123 (1988).
125. T. H. Windhorn and W. D. Goodhue, *Appl. Phys. Lett.* **48**, 1675 (1986).
126. J. J. Yang, M. Jansen, and M. Sergant, *Electron. Lett.* **22**, 438 (1986).
127. W. D. Goodhue, K. Rauschenbach, C. A. Wang, J. P. Donnely, R. J. Bailey, and G. D. Johnson, *J. Electron. Mater.* **19**, 463 (1990).
128. C. P. Chao, K-K. Law, and J. L. Merz, *Device Research Conf.* IIA-3 (1991).
129. A. Hardy, D. F. Welch, W. Streifer, *IEEE J. Quantum Electron.* **25**, 2096 (1989).
130. N. W. Carlson, G. A. Evans, J. M. Hammer, M. Lurie, L. A. Carr, F. Z. Hawrylo, E. A. James, C. J. Kaiser, J. B. Kirk, W. F. Reichert, D. A. Truxal, J. R. Shealy, S. R. Chinn, and P. S. Zory, *Appl. Phys. Lett.* **52**, 939 (1988).
131. D. J. Bossert, R. K. DeFreez, H. Ximen, R. A. Elliot, J. M. Hunt, G. A. Wilson, J. Orloff, G. A. Evans, N. W. Carlson, M. Lurie, J. M. Hammer, D. P. Bour, S. L. Palfrey, and R. Amantea, *Appl. Phys. Lett.* **56**, 2068 (1990).
132. R. Geels, R. H. Yan, J. W. Scott, and S. W. Corzine, R. J. Simes, and L. A. Coldren, *Conf. on Lasers & Electro-Optics 1988*, Anaheim, CA (Optical Society of America, Washington, DC), paper WM1, 206 (May 1988).
133. S. Corzine, R. S. Geels, R. H. Yan, J. W. Scott, and L. A. Coldren, *IEEE Photon. Technol. Lett.* **1**, 52 (1989).
134. S. Corzine, R. S. Geels, J. W. Scott, R. H. Yan, and L. A. Coldren, *IEEE J. Quantum Electron.* **25**, 1513 (1989).
135. M. Y. A. Raja, S. R. J. Brueck, M. Osinski, C. F. Schaus, J. G. McInerney, T. M. Brennan, and B. E. Hammons, *IEEE J. Quantum Electron.* **25**, 1500 (1989).
136. L. A. Coldren, S. W. Corzine, R. S. Geels, A. C. Gossard, K-K. Law, J. L. Merz, J. W. Scott, R. J. Simes, and R. H. Yan, *SPIE 1990*, Aachen, Germany, **1362**, 24 (1990).
137. R. S. Geels, S. W. Corzine, and L. A. Coldren, *IEEE J. Quantum Electron.* **27**(6) 1359–1367 (1991); and M. G. Peters, J. W. Scott, B. J. Thibeault, and L. A. Coldren, submitted *Phontonics Technology Lett.* (1993).
138. B. Tell, Y. H. Lee, K. F. Brown-Goebeler, J. L. Jewell, R. E. Leibeguth, M. T. Asom, G. Livescu, L. Luther, and V. D. Mattera, *Appl. Phys. Lett.* **57**, 1855 (1990).

139. R. S. Geels and L. A. Coldren, *Electron. Lett.* **27**(21), 1984 (1991), and F. J. Peters, M. G. Peters, D. B. Young, J. W. Scott, B. J. Thibeault, S. W. Corzine, and L. A. Coldren, *Electronics Letters* **29**(2), 200 (1993).
140. P. Gourley, T. M. Brennan, B. E. Hammons, S. W. Corzine, R. S. Geels, R. H. Yan, J. W. Scott, and L. A. Coldren, *Appl. Phys. Lett.* **54**, 1209 (1989).
141. L. A. Coldren, S. W. Corzine, R. S. Geels, and J. W. Scott, *Conf. Lasers & Electro-Optics 1991*, Baltimore, MD (Optical Society of America, Washington, DC), paper JThA1, 338 (May 1991). Also, J. W. Scott, R. S. Geels, S. W. Corzine, and L. A. Coldren, *J. Quantum Electron.* **29**(5), 1295 (1993).
142. P. Barnes, P. Zouganeli, A. Rivers, M. Whitehead, and G. Parry, *Electron. Lett.* **25**, 995 (1989).
143. K. W. Goossen, G. D. Boyd, J. E. Cunningham, W. Y. Jan, D. A. B. Miller, D. S. Chemla, and R. M. Lum, *IEEE Photon. Technol. Lett.* **1**, 304 (1989).
144. F. Koyama and K. Iga, *J. Lightwave Technol.*, *6*, 87, (1988).
145. H. S. Hinton, *IEEE J. Selected Areas in Commun.* **6**, 1209 (1988).
146. A. L. Lentine, D. A. B. Miller, J. E. Henry, J. E. Cunningham, L. M. F. Chirovsky, and L. A. D'Asaro, *Appl. Opt.* **29**, 2153 (1990).
147. D. A. B. Miller, *Opt. Lett.* **14**, 146 (1989).
148. J. W. Goodman, H. E. Midwinter, H. S. Hinton, *OSA Proc. Photon.* (Optical Society of America, Washington, D.C.), **3**, 164 (1989).
149. D. A. B. Miller, D. S. Chemla, T. C. Damen, A. C. Gossard, W. Wiegmann, T. H. Wood, and C. A. Burrus, *Appl. Phys. Lett.* **45**, 13 (1984).
150. A. L. Lentine, H. S. Hinton, D. A. B. Miller, J. E. Henry, J. E. Cunningham, and L. M. F. Chirovsky, *IEEE J. Quantum Electron.* **25**, 1928 (1989).
151. K-K. Law, R. H. Yan, and L. A. Coldren, *Appl. Phys. Lett.* **57**, 1345 (1990).
152. J. E. Zucker, K. L. Jones, B. I. Miller, and U. Koren, *IEEE Photon. Technol. Lett.* **2**, 32 (1990).
153. J. E. Zucker, K. L. Jones, M. G. Young, B. I. Miller, and U. Koren, *Appl. Phys. Lett.* **55**, 2280 (1989).
154. M. Cada, B. P. Keyworth, J. M. Glinski, A. J. SpringThorpe, C. Rolland, and K. O. Hill, *J. Appl. Phys.* **69**, 1760 (1991).
155. T. C. Huang, Y. Chung, N. Dagli, and L. A. Coldren, *Appl. Phys. Lett.* **58**, 2211 (1991).
156. T. C. Huang, T. Hausken, K. Lee, N. Dagli, L. A. Coldren, and D. R. Meyers, *IEEE Photon. Technol. Lett.* **1**, 168 (1989).
157. K. Ishida, H. Nakamura, H. Matsumura, T. Kadoi, and H. Inoue, *Appl. Phys. Lett.* **50**, 141 (1987).
158. J. Nees, S. Williamson, and G. Mourou, *Appl. Phys. Lett.* **54**, 1962 (1989).
159. T. H. Wood, *J. Lightwave Technol.* **6**, 743 (1988).
160. K. Wakita, I. Kotaka, O. Mitoni, H. Asai, Y. Kawamura, and M. Naganuma, *J. Lightwave Technol.* **8**, 1027 (1990).

161. G. Eisenstein, R. S. Tucker, J. M. Wiesenfeld, P. B. Hansen, G. Raybon, B. C. Johnson, T. J. Bridges, F. G. Storz, and C. A. Burrus, *Appl. Phys. Lett.* **54**, 454 (1989).
162. U. Koren, T. L. Koch, B. I. Miller, G. Eisenstein, and R. H. Bosworth, *Appl. Phys. Lett.* **54**, 2056 (1989).
163. D. A. B. Miller and D. S. Chemla, in "Optical Nonlinearities and Instabilites in Semiconductors," Ch. 13 (H. Haug, ed.). Academic Press, Boston, 1988.
164. W. H. Knox, D. S. Chemla, D. A. B. Miller, J. B. Stark, and S. Schmitt-Rink, *Phys. Rev. Lett.* **62**, 1189 (1989).
165. M. Mittelstein, D. Mehuys, A. Yariv, J. E. Ungar, and R. Sarfaty, *Appl. Phys. Lett.* **54**, 1092 (1989).
166. G. D. Boyd, A. M. Fox, D. A. B. Miller, L. M. F. Chirovsky, L. A. D'Asaro, J. M. Kuo, R. F. Kopf, and A. L. Lentine, *Appl. Phys. Lett.* **57**, 1843 (1989).
167. T. H. Wood, J. Z. Pastalan, C. A. Burns, Jr., T. Y. Chang, N. J. Sauer, and B. C. Johnson, in *Conf. on Lasers & Electro-Optics 1991*, Baltimore, MD (Optical Society of America, Washington, DC), paper CMA5, 10 (May 1991).
168. M. Suzuki, H. Tanaka, and S. Akiba, *Electron. Letter.* **25**, 88 (1989).
169. G. Eisenstein, U. Koren, G. Raybon, T. L. Koch, J. M. Wiesenfeld, M. Wegener, R. S. Tucker, and B. I. Miller, *Appl. Phys. Lett.* **56**, 1201 (1990).
170. U. Koren, B. I. Miller, M. G. Young, T. L. Koch, R. M. Jopson, A. H. Gnauck, J. D. Evankow, and M. Chien, *Electron. Lett.* **27**, 62 (1991).
171. T. H. Wood, C. A. Burrus, D. A. B. Miller, D. S. Chemla, T. C. Damen, A. C. Gossard, and W. Wiegmann, *Appl. Phys. Lett.* **44**, 16 (1984).
172. T. Y. Hsu, U. Efron, W. Y. Wu, J. N. Schulman, I. J. D'Haenens, and Y. C. Chang, *Opt. Eng.* **27**, 372 (1988).
173. G. D. Boyd, D. A. B. Miller, D. S. Chemla, S. L. McCall, A. C. Gossard, and J. H. English, *Appl. Phys. Lett.* **50**, 1119 (1987).
174. R. B. Bailey, R. Sahai, C. Lastufka, and K. Vural, *J. Appl. Phys.* **66**, 3445 (1989).
175. R. J. Simes, R. H. Yan, D. G. Lishan, and L. A. Coldren, *15th European Conf. on Optical Comm.*, Sweden (1989).
176. M. Whitehead, G. Parry, and P. Wheatley, *IEE Proc. Part J* **136**, 52 (1989).
177. R. H. Yan, R. J. Simes, and L. A. Coldren, *IEEE J. Quantum Electron.* **25**, 2272 (1989).
178. R. H. Yan, R. J. Simes, and L. A. Coldren, *IEEE Photon. Technol. Lett.* **2**, 118 (1990).
179. M. Whitehead, A. Rivers, G. Parry, and J. S. Roberts, *Electron. Lett.* **25**, 984 (1989).
180. H. Shen and F. H. Pollak, *Phys. Rev. B—Condensed Matter* **42**, 7097 (1990).

181. R. H. Yan, R. J. Simes, and L. A. Coldren, *IEEE Photon. Technol. Lett.* **1**, 273 (1989).
182. R. H. Yan, R. J. Simes, L. A. Coldren, and A. C. Gossard, *Appl. Phys. Lett.* **56**, 1626 (1990).
183. R. H. Yan, R. J. Simes, and L. A. Coldren, *IEEE J. Quantum Electron.* **27**(7), (1991).
184. R. J. Simes, R. H. Yan, R. S. Geels, L. A. Coldren, J. H. English, A. C. Gossard, and D. G. Lishan, *Appl. Phys. Lett.* **53**, 637 (1988).
185. R. J. Simes, R. H. Yan, C. C. Barron, D. Derrickson, D. G. Lishan, J. Karin, L. A. Coldren, and M. Rodwell, *IEEE Photon. Technol. Lett.* **3**(6), 513–515 (1991).
186. R. J. Simes, dissertation, UCSB-ECE Technical Report #90-03 (1990).

Chapter 11

PHOTODETECTORS FOR OPTOELECTRONIC INTEGRATED CIRCUITS

Joe C. Campbell

Microelectronics Research Center
University of Texas
Austin, Texas

1. INTRODUCTION

One of the essential functions of an optical communications system is the conversion between electrical and optical signal formats in the transmitter and receiver. With an eye toward reducing costs while improving yield, performance, and functionality, an early thrust of optoelectronic integrated circuits has been to integrate the source or photodetector onto the same chip as the driver or amplifier electronics, respectively. To date, this has been more successful for receivers, primarily because photodetectors tend to be somewhat more flexible (although not insensitive) than lasers with regard to materials and fabrication techniques. The materials systems used in the fab-

ISBN 0-12-200420-5

rication of receiver OEICs have primarily been III–V compounds in spite of the excellent characteristics of Si photodetectors. For short-wavelength ($\lambda < 1.0\ \mu m$) applications such as local area networks and computer interconnects, where the transmission distances are short (<3 km) and low cost is essential, GaAs-based photodetector structures have frequently been utilized because they can easily be integrated with low-noise GaAs MESFET preamplifier circuits. At longer wavelength, where the primary application is long-haul, high-bit-rate optical communications, InP/InGaAsP sturctures have been used because GaAs and Si are transparent for $\lambda > 1\ \mu m$.

In order to achieve high sensitivities in receiver OEICs, it has been necessary to modify conventional photodetectors or to develop new structures and fabrication techniques that are compatible with transistor circuits without degrading the critical characteristics of the photodetectors. The performance of photodetectors can be evaluated in terms of three primary characteristics: responsivity, noise, and bandwidth. The responsivity is important in that it is a measure of how well the photodiode converts the incident optical signal into current. The noise of the photodetector can be a limitation on the minimum signal level needed to achieve a specified bit error rate, and the bandwidth must be sufficient to accommodate the transmission rate. The dark current and capacitance of a photodetector can also be significant attributes in that they influence the noise and the bandwidth. The dark current is a source of noise because it contributes "signal" in the absence of light and the capacitance is a factor in the frequency response through the *RC* time constant, but more significantly the capacitance of the photodetector contributes to the total front-end capacitance, which is inversely proportional to the receiver figure of merit [1].

Several types of photodetectors have been employed, to varying degrees, for lightwave applications. In roughly ascending order of structural complexity, the lightwave photodetectors are photoconductors, metal–semiconductor–metal (MSM) photodiodes, p-i-n photodiodes, phototransistors, and avalanche photodiodes. Photoconductors and phototranisitors have not been widely utilized for hybrid lightwave receivers, and it does not appear that this will change for OEIC receivers. Photoconductors are limited by their relatively slow speed of response [2], and phototransistors have not exhibited a performance advantage compared to the p-i-n/field effect transistor (FET) combination [3, 4]. Avalanche photodiodes, on the other hand, have been successfully implemented in many receiver circuits, particularly for high bit rates where the performance advantage of the APD is maximized [5, 6]. However, APDs have not been incorporated into receiver OEICs. One reason for this is the complexity of the APD structure. The APDs that have achieved the best performance, such as the separate absorption and multiplication (SAM) APD, are multilayer structures and their performance is very sensitive to the parameters of the constituent layers. In addition, the higher bias voltages, temperature stabilization, and precise bias control required by APDs have made them unattractive for the initial development stage of OEICs. As crystal growth procedures improve and as OEICs become widely used components, APDs may be utilized in high-performance receiver OEICs. At present, however, receiver OEICs rely almost entirely on p-i-n photodiodes or MSM photodetectors. Accordingly, the rest of this chapter will concentrate on integration techniques that have been employed for these two photodetectors and on their performance characteristics.

2. p-i-n PHOTODIODES

The p-i-n photodiode has become the most widely deployed photodetector for all lightwave applications including OEICs. The performance characteristics of p-i-n photodiodes, such as noise, bandwidth, dark current, and photoresponse, are well understood and documented, but their integration into OEICs has introduced significant fabrication challenges. As a result, this section will concentration on the techniques that have been employed for integration after a brief review of p-i-n photodiode fundamentals.

2.1. Responsivity

The responsivity of a photodetector is simply a measure of how efficiently it converts light into current. Ideally, each incident photon would result in the charge of one electron flowing in the external circuit. In practice, there are several physical effects, such as incomplete absorption, recombination, reflection from the semiconductor surface, and contact shadowing, that tend to reduce the responsivity. Once the photons have entered the semiconductor, the photodetection process consists of (1) absorption and (2) collection of the photogenerated electron–hole pairs. The primary absorption processes are free carrier absorption, band-to-band absorption, and band-to-impurity absorption. To the extent that band-to-band absorption dominates, the internal quantum efficiency depends only on the absorption coefficient α and the length of the absorbing region, l, such that $\eta_i = (1 - e^{-\alpha l})$. For a well designed p-i-n photodiode, absorption occurs in or near the space-charge region where the electric field can quickly separate the electron–hole pairs before they can recombine.

The external quantum efficiency is proportional the internal efficiency but also takes reflection and contact shadowing into account. Typically, owing to their relatively high index of refraction, semiconductors reflect approximately 30% of the incident light. Antireflection (AR) coatings consisting of single or multiple dielectrics such as SiN, Al_2O_3, Pb_2O_3, ZnS, or MgF_2 can routinely reduce this optical power loss to less than 5%. Contact shadowing occurs when an electrical contact blocks a portion of the active absorbing region. The net effect is to reduce the efficiency because the portion of the optical signal that is reflected by the contact is lost. The external quantum efficiency can be written as

$$\eta_e = \eta_i(1 - R)(S/A), \tag{1}$$

where R is the reflectivity, S is the optically sensitive portion of the photodetector, and A is the total photodetector area. For most p-i-n photodiodes, $S/A = 1$ and typical quantum efficiencies are 60 to 70% for devices without AR coatings and >90% for those with coatings.

2.2. Noise

The sensitivity of a lightwave receiver is frequently determined by the noise generated in the front end. In most cases, the noise of the preamplifier dominates that of the photodetector [1], particularly for the types of photodetectors that have been most successfully incorporated into OEICs. There are conditions, however, under which the photodetector noise can become significant. The so-called $1/f$ noise is important only at low frequencies and for almost all lightwave applications it is of little consequence. For the most part, the thermal noise component of these photodetectors can also be neglected compared to the shot noise of the dark current and the signal-induced photocurrent. The mean square shot-noise current can be written as

$$\langle i_n^2 \rangle = 2q(I_d + I_{ph})I_2 B, \tag{2}$$

where q is the charge on an electron, I_d is the dark current, I_{ph} is the photocurrent, I_2 is a Personick integral [1], and B is the bit rate.

The components of dark current in a p-i-n photodiode are diffusion current, generation–recombination current, and tunneling [7]. The diffusion part of the dark current is due to minority carriers that are thermally gererated in the p and n neutral regions and that subsequently diffuse into the depletion region. The diffusion current density is given by the Schockley equation [3]

$$J_{diff} = J_s[\exp(qV/kT) - 1] \tag{3}$$

where k is Boltzmann's constant, T is the junction temperature, V is the applied voltage, and I_s is the saturation current. Assuming complete ionization of impurities, the saturation current can be written as [8]

$$J_s = qn_i^2[D_p/L_pN_d + D_n/L_nN_a] \quad (4)$$

where n_i is the intrinsic carrier concentration, D_n and D_p are the minority carrier diffusion lengths, L_n and L_p are the minority diffusion lengths, and N_d and N_a are the densities of ionized donors and acceptors, respectively.

The generation–recombination current arises from the generation of carriers through midgap traps. In a depletion region of width W, an approximate expression for the generation–recombination current density is

$$J_{gr} = (qn_iAW/\tau_e)[\exp(qV/2kT) - 1] \quad (5)$$

where τ_e is the effective carrier lifetime [9]. At low bias voltages, the generation–recombination current is usually the primary source of dark current in a pin photodiode.

At high reverse bias, the dark current of narrow-bandgap pin photodiodes will be dominated by tunneling. In this bias regime, empty conduction band states on the n side of the junction will line up directly across from filled valence band states on the p side of the junction. If the barrier is low enough, quantum tunneling of electrons across the barrier can occur. The barrier will be narrower, and thus tunneling will be enhanced, when the bandgap is small and when the field is high. For direct bandgap semiconductors, the tunneling current is given by [10]

$$J_t = \gamma \exp[-2\pi\Theta m_0^{1/2} E_g^{3/2}/qhE_m] \quad (6)$$

where m_0 is the free electron mass, h is Planck's constant, E_m is the maximum electric field, and E_g is the bandgap energy. The parameter Θ is a dimensionless quantity given by $\Theta = \alpha(m_c^*/m_o)^{1/2}$, where m_c^* is the electron effective mass. For band-to-band tunneling the constant α is approximately unity and the prefactor $\gamma = (2m_c^*/E_g)^{1/2}(q^3E_mV/h^2)$. It has been shown by deep-level transient spectroscopy that the dominant tunneling mechanism may involve thermal activation of deep trap states and subsequent tunneling of carriers from the traps into band states [11]. In that case, the parameters α and γ will differ from their band-to-band values.

2.3. Frequency Response

Three physical effects determine the bandwidth of photodetectors for OEICs: (1) diffusion of carriers generated outside the space-charge region, (2) the RC time constant, and (3) the transist time. Since diffusion times are typically very large (~ns/μm), consideration is given in the device design to minimizing the amount of light absorbed outside the space-charge region. If either the resistance or capacitance is too high, the *RC* time constant can limit the speed of response. For most devices, however, the *RC* time constant is less than 10 ps. The ultimate limitation on the bandwidth of a well-designed photodetector for OEICs is the transit time, i.e., the time it takes the photogenerated carriers to drift through the space-charge region. For a transit-time-limited response, the frequency-dependent photocurrent is given by [12, 13]

$$\frac{i(\omega)}{i(0)} = \frac{1}{(1 - e^{-\alpha d})}\left[\frac{e^{j\omega\tau_n - \alpha d} - 1)}{(j\omega\tau_n - \alpha d)} - e^{-\alpha d}\frac{(e^{j\omega\tau_n} - 1)}{j\omega\tau_n} + e^{-\alpha d}\frac{(e^{j\omega\tau_p} - 1)}{j\omega\tau_p} + e^{-\alpha d}\frac{(1 - e^{j\omega\tau_p + \alpha d})}{(j\omega\tau_n + \alpha d)}\right] \quad (7)$$

where $\tau_n = W/v_n$ and $\tau_p = W/v_p$ are the electron and hole transit times, respectively, and v_n and v_p are the electron- and hole-saturated drift velocities.

The relative contributions of the transit time and the *RC* time constant to the 3-dB bandwidth are summarized in Fig. 1 [12]. The bandwidth is plotted versus the absorbing layer thickness, *l*, for photodiode radii of 5, 10, 20, and 50 μm. The transit-time limit increases rapidly with decreasing absorb-

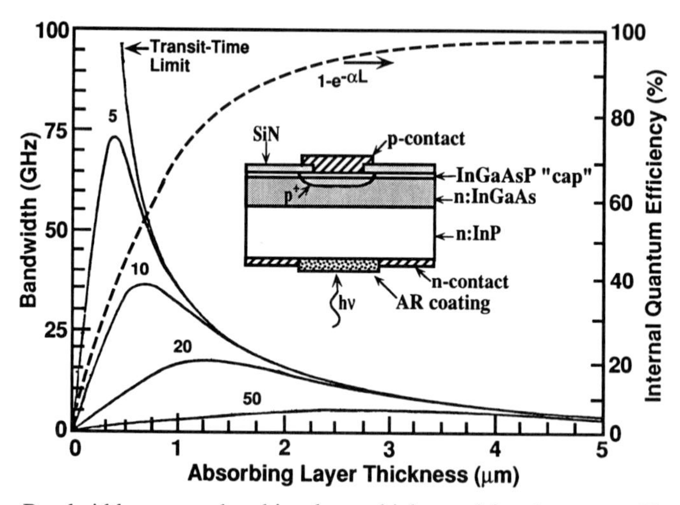

Fig. 1. Bandwidth versus absorbing layer thickness *l* for detector radii of 5, 10, 20, and 50 μm. The dashed line is the internal quantum efficiency $(1 - e^{-\alpha l})$. These curves assume electron and hole saturation velocities of $v_n = 7 \times 10^6$ cm/s and $v_p = 6.5 \times 10^6$ cm/s, respectively, and $\alpha = 1.15 \times 10^4$ cm^{-1} which corresponds to an incident wavelength of 1.3 μm. (From Ref. 12.)

ing layer thickness, reaching a value in excess of 75 GHz for $l < 0.5$ μm. It is clear from the figure that optimum performance is achieved, at a given value of l, when the diameter is small enough to give an RC time constant less than the transit time. Consequently, the highest bandwidths can be achieved with device diameters < 10 μm. Figure 1 also illustrates the trade-off between bandwidth and quantum efficiency. The dashed line shows the dependence of the internal quantum efficiency on l. To produce high quantum efficiency, a relatively thick absorption layer is required to attenuate the optical signal. For InGaAs a thickness of approximately 2 μm is needed to absorb > 85% of the light and the coresponding transit time–limited bandwidth is near 15 GHz.

2.4. Integration Techniques

The design of any receiver OEIC requires the simultaneous optimization of materials parameters for both the photodetector and the transistors in the preamplifier circuit. This can become difficult and necessitate performance trade-offs if the photodetector and the transistors have substantiallly different sturctures. The disparity in the materials requirements has been most acute for p-i-n photodiodes. For example, the most successful hybrid circuits have combined p-i-n photodiodes with FET preamplifier circuits. The p-i-n structure usually requires a thick (2 mm), depleted absorbing layer on a heavily doped substrate. Depletion is achieved by keeping the carrier concentration in the absorbing layer low (typically $< 10^{15}$ cm^{-3}). FETs, on the other hand, utilize thin (~0.5 mm) heavily doped ($>10^{17}$ cm^{-3}) channel layers on semi-insulating substrates. One of the primary thrusts of p-i-n/OEIC design has been to develop crystal growth and fabrication techniques to circumvent this problem.

The structures for p-i-n receiver OEICs can be broadly characterized as planar or nonplanar and each can be either horizontally or vertically integrated. Nonplanar structures present difficulties because steps and ridges limit the minimum feature size. With increasing bit rates and the resulting requirement for wider bandwidths, scaling some device dimensions such as gate length to thc submicrometer level will be necessary. The nonplanar steps and dips will make this difficult and can ultimately lead to degradation in performance, lower yields, and reduced reliability. On the other hand, planarization techniques can add to the cost and to the complexity of fabrication. The horizontal and vertical integration approaches are illustrated in Fig. 2. For horizontal integration (Fig. 2a), the photodetector and the electrical devices are laterally separated. The materials incompatibility neces-

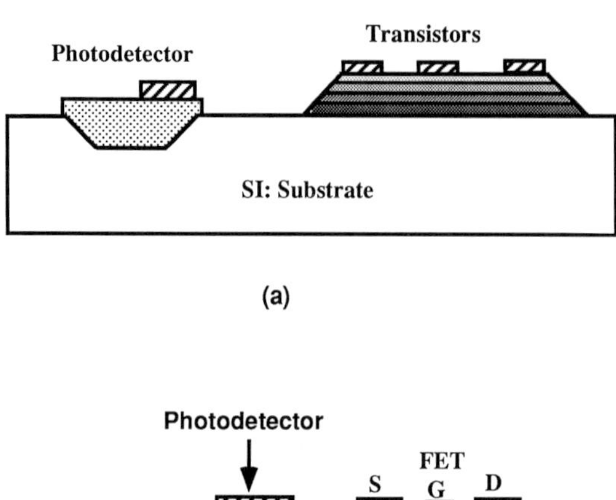

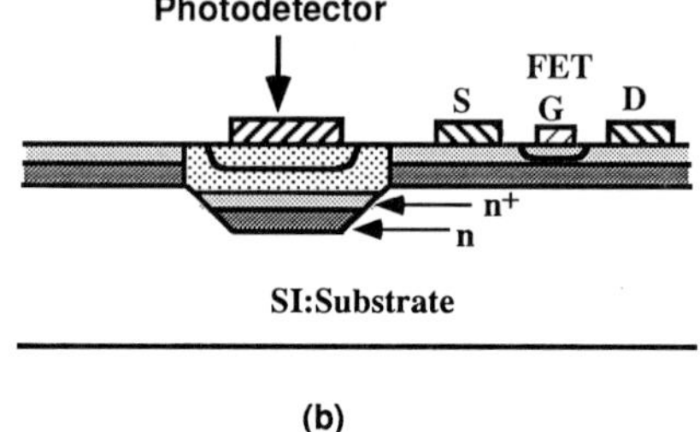

Fig. 2. Examples of (a) horizontal and (b) vertical integration schemes.

sitates different epitaxial layers in the photodetector and electronic regions. A disadvantage of this approach is that it often results in nonplanar structures as a result of steps at the photodetector–electronic interface. In the vertical approach, the electronic layers are grown on top of or beneath (Fig. 2b) the photodetector layers. Contacts to the photodetector are formed by selective diffusion or by selectively etching away the electronic layers. The advantages of this approach are its relative simplicity of fabrication and that it can be an excellent way to achieve planarity. On the other hand, the existence of conducting epitaxial layers below the electronic devices tends to increase the parasitic capacitance of the electronic devices and the interconnects.

2.4.1. *Planar Integration*

Planarization of p-i-n receiver OEICs has been achieved by ion implantation [14, 15], diffusion [15], and embedding the p-i-n in an etched well [16–18]. Many of the techniques that have been developed for integrating p-i-n photodiodes utilize patterned substrates combined with selective removal of epitaxial layers, an approach first developed by Kolbas *et al.* [16]. Typically, the photodetector epitaxial layers are grown on substrates that have been patterened with recesses or "wells" (Fig. 3a and b). These well regions are formed by selective chemical etching. The subsequent removal

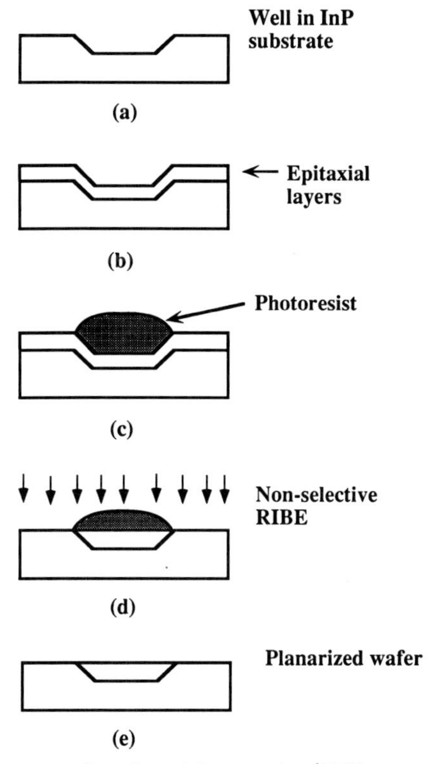

Fig. 3. Fabrication steps for planarizing a p-i-n/FET structure. (a) Formation of the recessed region in which the photodiode will be fabricated. (b) Growth of the p-i-n epitaxial layers. (c) Deposition of the photoresist etch mask. (d) Nonselective reactive ion beam etching. (e) Completed planar structure. (From Ref. 18. © 1990 IEEE.)

of unwanted epitaxial material outside the well regions can be accomplished by chemical etching [16] or by reactive ion etching (RIE) [17, 18]. The RIE method developed by Shimizu *et al.* [18] is illustrated in Fig. 3c and d. The depression left after the epitaxial layer growth is filled with photoresist and then the wafer is planarized by RIE. Finally, the photodetector is formed in the well region by selective diffusion. The characteristics of the photodiodes fabricated in this manner are as follows: dark current = 5 nA at −5 V, capacitance = 77 fF at −5 V, quantum efficiency (with antireflection coating) = 77%, and bandwidth = 13 GHz.

An alternative to planarization by etching has been achieved by Hata *et al.* [14] with a fully ion-implanted $In_{0.53}Ga_{0.47}As$ horizontal structure and by Kim *et al.* [15] with an $InP/In_{0.53}Ga_{0.47}As$ vertically integrated configuration formed by selective diffusion. A schematic cross section of the vertically

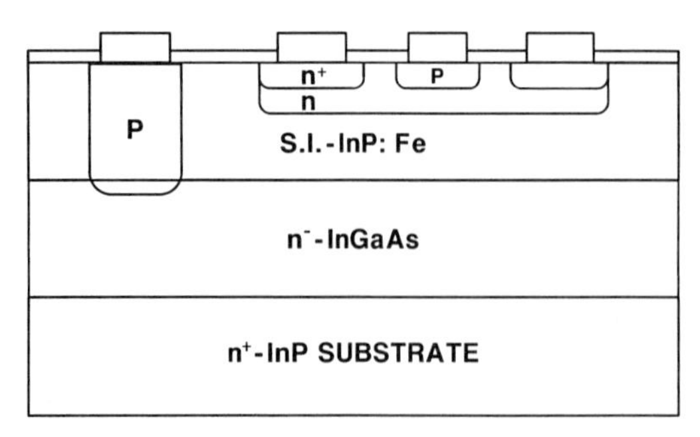

Fig. 4. Schematic cross section of an $In_{0.53}Ga_{0.47}As$ p-i-n/InP JFET. Planarization is achieved by a combination of Zn diffusion and ion implantation. (From Ref. 15. © 1988 IEEE.)

integrated p-i-n along with the companion ion-implanted JFET is shown in Fig. 4. An Fe-doped semi-insulating layer for the transistors was grown over the $In_{0.53}Ga_{0.47}As$ absorbing layer. The photodiode was fabricated into a 75-μm-diameter circle by Zn diffusion through the Fe:InP to the $In_{0.53}Ga_{0.47}As$ followed by contact formation. The dark current was 10 nA and the capacitance was 0.25 pF at −5 V.

2.4.2. Nonplanar Integration

Many of the nonplanar approaches to integration utilize mesa structures to provide isolation between the photodetector and the electronics. As with the planar structures, growth in recessed regions is often utilized for the photodetector region and the integration can be either vertical and horizontal. One distinction among nonplanar devices is the procedure used to fabricate the contact between the p-i-n and the electronics. This can be done by placing the contact along the surface, which usually has bumps and ridges between the two regions, or by using an air bridge. An example of horizontal integration with the growth of an $In_{0.53}Ga_{0.47}As$ p-i-n in a well is shown in Fig. 5 [19, 20]. An interesting aspect of this horizontal approach is the use of a GaAs epitaxial layer for the electronics. The $In_{0.53}Ga_{0.47}As$ photodiodes are similar in structure and performance to discrete, nonintegrated p-i-ns: the dark current and capacitance at −5 V are 20 nA and 55 fF, respectively, and the quantum efficiency is 95% at $\lambda = 1.3$ μm.

An example of a vertically integrated, nonplanar structure is shown in Fig. 6 along with an outline of the fabrication procedure [21]. Note that the $Al_{0.52}Ga_{0.47}As/In_{0.53}Ga_{0.47}As$ layers are grown over the whole wafer and remain underneath the p-i-n after the isolation step. The dark current and capacitance at −5 V are 6 nA and 80 fF, respectively, and the quantum ef-

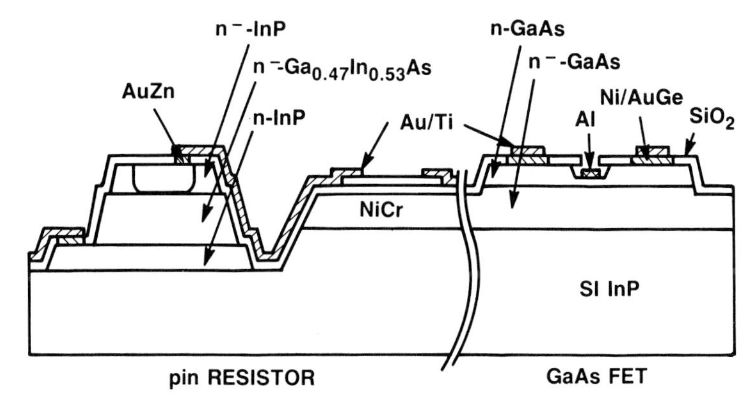

Fig. 5. Schematic cross section of an $In_{0.53}Ga_{0.47}As$ p-i-n horizontally integrated with a GaAs MESFET preamplifier. (From Ref. 19.)

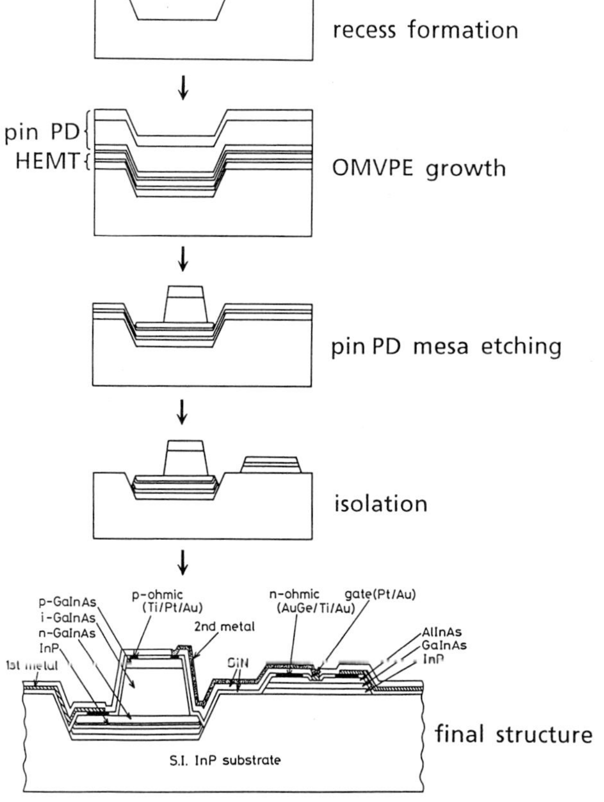

Fig. 6. Fabrication procedure for an $In_{0.53}Ga_{0.47}As$ p-i-n vertically integrated with an AlInAs/InGaAs HEMT. (From Ref. 21. © 1990 IEEE.)

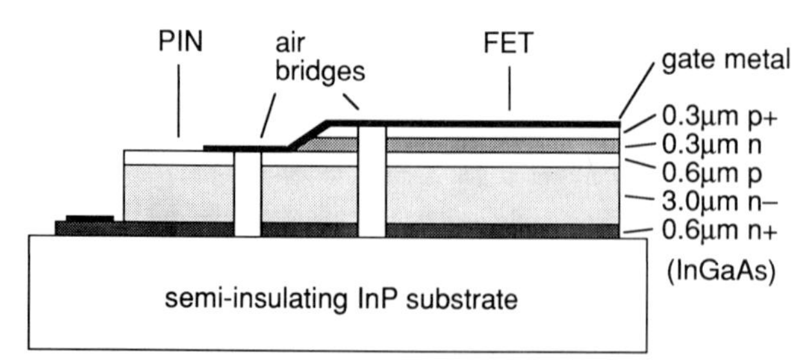

Fig. 7. Schematic cross section of a vertically integrated $In_{0.53}Ga_{0.47}As$ p-i-n with an $In_{0.53}Ga_{0.47}As$ JFET. Electrical isolation between the p-i-n and the FET is achieved through the use of air bridges. (From Ref. 22.)

ficiency is 76% at $\lambda = 1.3\ \mu m$. The bandwidth of the p-i-n is approximately 10 GHz. Figure 7 shows another vertically integrated structure in which the FET layers are grown on top of the p-i-n. The FET layers are removed from the p-i-n region during processing. A novel feature of this structure is the use of air bridges for interconnections [22]. The air bridge is formed by etching away some of the semiconductor underneath the metal contact. This has the advantages that the p-i-n can be made very small with low capacitance and the isolated contact pad can have a large area to facilitate packaging. This isolation technique has produced $In_{0.53}Ga_{0.47}As$ p-i-n photodiodes with dark currents as low as 4 pA and bandwidths as high as 25 GHz. In addition, the quantum efficiency is greater than 80% at $\lambda = 1.55\ \mu m$ [23]. Most integrated receivers have utilized FETs for the preamplifier circuits. An alternative configuration consisting of an $In_{0.53}Ga_{0.47}As$ p-i-n photodiode with an $InP/In_{0.53}Ga_{0.47}As$ bipolar preamplifier circuit has been reported by Chandrasekhar *et al.* [24, 25] (Fig. 8). The photodiode characteristics of this p-i-n are comparable to those of the p-i-n/FET structures: dark current = 35 nA at -10 V, quantum efficiency = 63%, and capacitance = 275 fF.

3. MSM PHOTODIODES

Much of the development of OEICs has emphasized the integration of already developed discrete devices. It is clear, however, that the realization of high-performance OEICs will also require the development of novel device elements and new combinations. The metal–semiconductor–metal (MSM) photodiode is an example of such a device. Although the MSM is not new [26], it did not play a significant role in the evolution of state-of-the-art hybrid lightwave receivers. It has, however, developed into one of the most promising photodetectors for receiver OEICs. As illustrated in Fig. 9, the

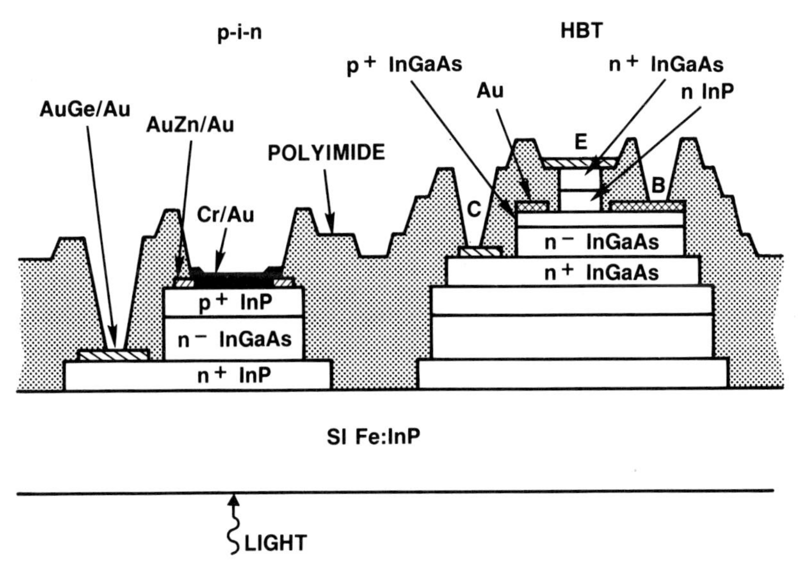

Fig. 8. Schematic cross section of an $In_{0.53}Ga_{0.47}As$ p-i-n vertically integrated with an $InP/In_{0.53}Ga_{0.47}As$ heterojunction bipolar transistor. (From Ref. 24. © 1990 IEEE.)

MSM is a planar device consisting of two interdigitated electrodes on a semiconductor surface. These electrodes are deposited so as to form back-to-back Schottky diodes. Absorption occurs near the semiconductor surface between the electrodes. The first OEICs to utilize MSMs were GaAs based. The development of similar circuits in InP-based compounds for long-wavelength telecommunications applications has been hampered by the low Schottky barrier heights in InP and InGaAs. This prevents the direct deposition of Schottky contacts on the InP or InGaAs surface. Various schemes to enhance

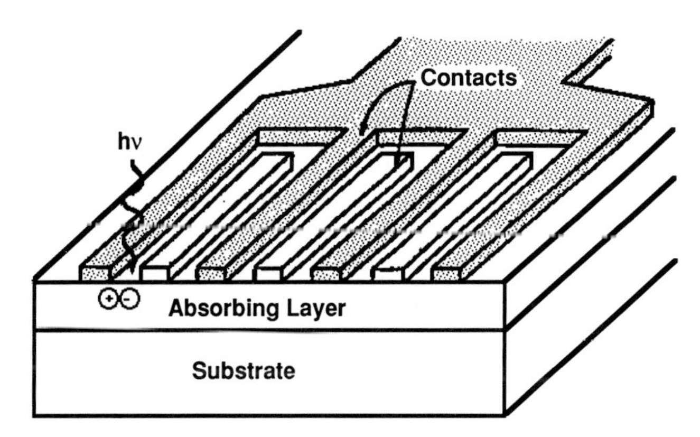

Fig. 9. Schematic cross section of an MSM photodiode.

the barrier height have been successfully demonstrated, and this has allowed the fabrication of long-wavelength OEICs.

The positive attributes of MSMs are numerous. The principal advantage, which derives from the simplicity of the planar contact configuration, is its compatibility with MESFET and HEMT circuitry. In addition, MSM photodiodes are relatively easy to fabricate, they do not add significantly to the complexity of the circuit, and, typically, they exhibit very low capacitance per unit area. The low capacitance is important because (1) it permits relatively large-area devices, which facilitates coupling to optical fibers, and (2) its contribution to the total front-end capacitance is negligible. If there is no photoconductive gain, the bandwidth can be very high for sufficiently narrow electrode spacing. The two limiting factors in the performance of MSMs are dark current and quantum efficiency. It has frequently been difficult to obtain low dark current in an MSM because the lateral current flow can be significantly influenced by the semiconductor surface or heterojunction interfaces. Nevertheless, improvements in device design have resulted in GaAs and InGaAs devices with low dark current. The degradation in quantum efficiency caused by electrode shadowing cannot be completely eliminated, and this is probably the primary disadvantage of MSM photodiodes. There are, however, fabrication techniques such as the formation of submicrometer electrodes that can reduce the effect of contact shadowing.

3.1. Responsivity

There are two aspects of the photoresponse of an MSM photodetector: (1) the internal quantum efficiency may reflect the presence of an internal gain mechanism and (2) the ultimate limit on the external quantum efficiency is fixed by electrode shadowing. Figure 10 shows the photocurrent versus bias voltage. The low-voltage knee occurs because the electric field is insufficient to separate the photogenerated carriers until the flat-band condition is achieved at the anode. Above the knee, the photoresponse is relatively flat. There is a slight rise in signal with bias as the depletion width increases. The internal quantum efficiency η_i can be estimated from the fractional surface area exposed and the penetration depth of the light [27]. At 7 V, η_i approaches 100%. The further increase in the photocurrent at higher bias is therefore evidence of an internal gain mechanism. The origin of this gain may be operative at lower voltages as well but is masked by incomplete depletion of the absorption layer. Gain has been reported in GaAs and $In_{0.53}Ga_{0.47}As$ MSM structures [28, 29]. There is still some uncertainty with regard to the physical origin of the gain, but it is most probably due to deep-

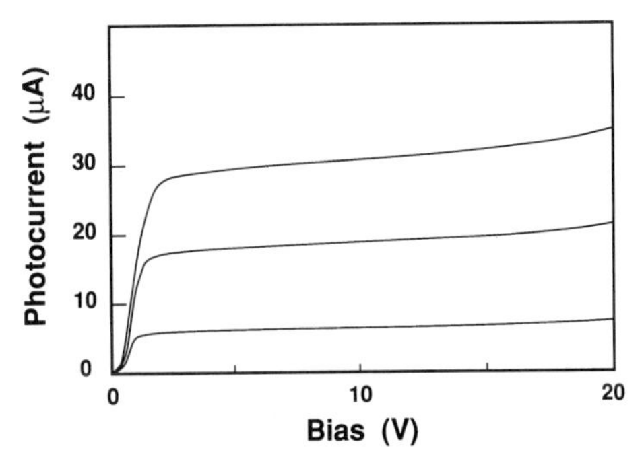

Fig. 10. Photocurrent versus voltage for an $In_{0.52}Al_{0.48}As/In_{0.53}Ga_{0.47}As$ MSM photodiode. The electrode width was 1 μm and the interelectrode spacing was 2 μm. (From Ref. 30. © 1990 IEEE.)

level traps. Although this gain can boost the dc signal, in general it is undesirable because of its adverse effect on the frequency response.

The MSM is illuminated through the top electrode structure, which is opaque to a fraction of the incident signal. This blocking effect can be accounted for by the relation

$$\frac{S}{A} = \frac{s}{s + w} \tag{8}$$

in the expression for the external quantum efficiency as given by Eq. (1). The width of the contact fingers is w, and s is the interdigital spacing. Figure 11 shows the calculated quantum efficiency at $\lambda = 1.55$ μm for a 1-μm electrode width and electrode spacings of 1, 2, and 3 μm versus the thickness of the absorbing layer. For reference, the efficiency of a p-i-n photodiode is also shown. Although contact shadowing cannot be completely eliminated, it is clear from Fig. 11 that it is desirable to reduce the contact width to the extent possible. This could be done with "submicron" photolithography techniques. It is also possible to use transparent electrodes or back illumination, but the low electric field strength underneath the electrodes gives rise to poor frequency response.

3.2. Noise

As discussed, the primary noise source for a p-i-n photodiode is the thermal noise associated with its series resistance and the shot noise of the photocurrent and the dark current. In most instances, the shot noise dominates. Similar considerations apply to the MSM, particularly at high frequencies. At lower frequencies ($\leq$10 MHz) a $1/f$-type noise similar to that of FETs

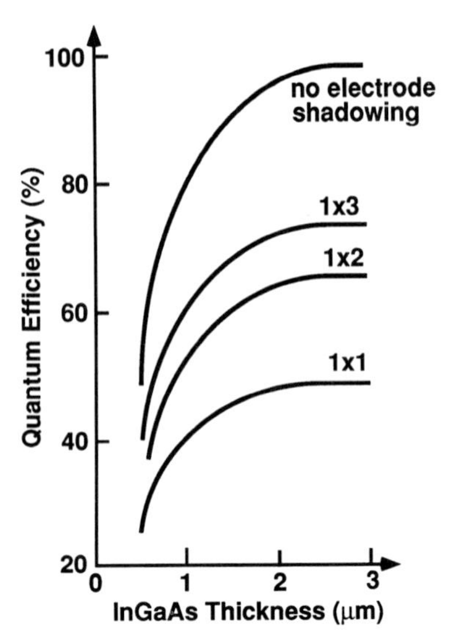

Fig. 11. Calculated external quantum efficiency of an $In_{0.53}Ga_{0.47}As$ MSM photodiode at $\lambda = 1.55$ μm versus absorbing layer thickness. The electrode width is 1 μm and the spacing is 1, 2, and 3 μm. (From Ref. 30. © 1990 IEEE.)

is often observed. This might be expected, since the MSM and the FET employ lateral current flow, which can involve strong interactions with the surface or a heterojunction interface. The origin of this noise is thought to be traps and it has been greatly reduced by improving the quality of the epitaxial layers [30].

Several physical effects contribute to the dark current of an MSM photodetector. A model that provides a complete and accurate description of the dark *I–V* curve has yet to be developed. A full treatment would involve an accurate description of the potential barriers under the electrodes, inclusion of different current transport mechanisms including thermionic emission and tunneling, and a two-dimensional analysis of the electrode electric field profile. Nevertheless, the overall behavior of the dark current can be adequately described by a one-dimensional representation of back-to-back Schottky diodes.

In an MSM photodiode at low bias voltage, the dark current is adequately described by a thermionic emission–diffusion model [31]. The current density is similar in form to the diffusion current in Eq. (3) except that the saturation current J_s is replaced by a thermionic emission term J_{st}, which is given by

$$J_{st} = A_n^* T^2 \exp\{-q(\varphi_n - \Delta\varphi_n)/kT\} + A_p^* T^2 \exp\{-q(\varphi_p - \Delta\varphi_p)/kT\} \quad (9)$$

where $A^*_{n,p}$ are the effective Richardson constants for the electrons and holes, respectively, $\varphi_{n,p}$ are the respective barrier heights, and $\Delta\varphi_{n,p}$ are the changes in the barrier heights due to image charge effects.

Ito and Wada [32] have shown that the dark current of GaAs MSM photodiodes can be accurately described in terms of a thermionic emission model and that it is sensitive to the type of metal used for the Schottky barrier. When the electron barrier height φ_n is greater than 0.7 eV, the dark current increases exponentially with φ_n. This can be explained by considering the trade-off between electron injection at the reverse-biased contact and hole injection in the forward-biased contact. At low bias voltage, the main source of dark current is electron injection. As the bias is increased, however, hole injection, for which the barrier height is $\varphi_p = E_g - \varphi_n$, can become significant. It follows from the necessity to minimize both hole and electron injection simultaneously that the lowest dark current is achieved when the electron and hole barrier heights are approximately equal, i.e., when $\varphi_p \approx \varphi_n \approx E_g/2$. This condition can be satisfied with WSi_x contacts by varying the alloy composition. Dark current values less than 1 nA at 10 V have been obtained with $x = 0.64$.

For long-wavelength MSM photodiodes the low Schottky barrier height in InGaAs would result in unacceptably high dark current were it not for barrier enhancement techniques such as strained layers of GaAs [33] and $Al_xGa_{1-x}As$ [34, 35] and Langmuir–Blodgett films [36]. To date, the most successful approach has utilized the growth of a thin, lattice-matched layer of $In_{0.52}Al_{0.48}As$ on the $In_{0.53}Ga_{0.47}As$ absorbing layer [37–40]. The dark current of a typical $In_{0.52}Al_{0.48}As/In_{0.53}Ga_{0.47}As$ MSM is shown in Fig. 12. The leakage current densities are 2.5×10^{-3} A/cm^2 and 3×10^{-2} A/cm^2 at 15 V and 30 V, respectively, and the breakdown voltage is $\approx$40 V. In the low-bias regime (<1 V) the dark current is negligible. A “knee” in the dark IV curve near the flat-band voltage ($V_{fb} \approx 1$ V) occurs as the anode contact becomes forward biased. At intermediate bias levels ($1\ V < V < 20\ V$) the dark current can be represented by a thermionic emission model. The details of the dark current in this bias range will depend critically on the characteristics of the $In_{0.52}Al_{0.48}As$ enhancement layer, such as its thickness and compositional profile and the abruptness of the heterojunction. In general it has been shown that holes dominate the carrier transport and that the effective barriers for holes and electrons are 0.65 and 0.7 eV, respectively [30]. At higher bias, tunneling becomes significant. This component of the dark current increases rapidly above $\approx$ 20 V as the bias voltage reduces the effective barrier height, and then further increase of the bias yields triangular barriers of diminishing width. In this bias regime, a tunneling current cal-

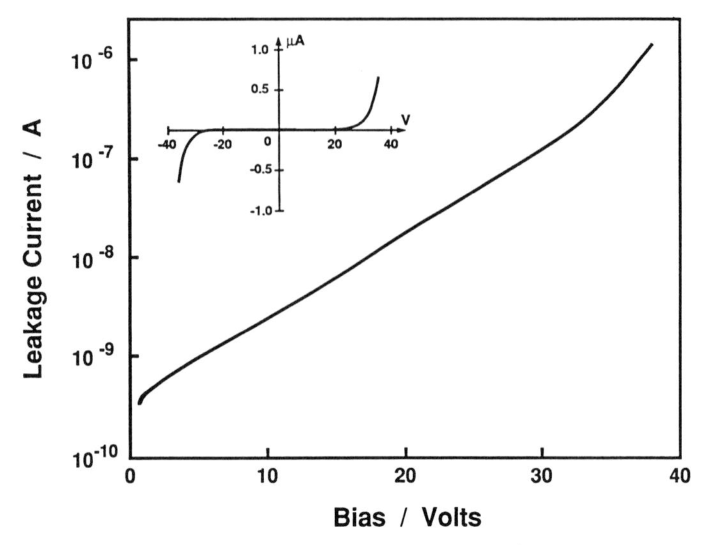

Fig. 12. Dark current versus voltage of an $In_{0.52}Ga_{0.48}As/In_{0.53}Ga_{0.47}As$ MSM photodiode. The electrode width was 1 μm and the spacing was 2 μm. The total device area was 350 μm^2. (From Ref. 30. © 1990 IEEE.)

culation using the potential dropped across the barrier layer predicts a slope for the log(I) versus V curve similar to that observed in Fig. 12 [30]. Near breakdown, the importance of impact ionization rises, but it is difficult to determine the relative magnitudes of the avalanching and tunneling components. Nevertheless, the essential point is that for intermediate biases (10 V $< V_b <$ 20 V) the shot noise of the dark current is much lower than the noise of state-of-the-art preamplifiers. Consequently, the dark current of MSM photodiodes can usually be ignored compared with that of the transistor amplifier circuits.

3.3. Capacitance

The capacitance of an MSM photodiode depends on the total device area and the width and separation of the electrode fingers. An approximate expression for the capacitance can be obtained from Lim and Moore's [41] calculation for an undoped, infinitely thick semiconductor:

$$C = \frac{\varepsilon_0(1 + \varepsilon_r)A}{(s + w)} \times \frac{F[(^{\pi}/_{2})\backslash\kappa]}{F[(^{\pi}/_{2})\backslash\kappa']} \tag{10}$$

where ε_r is the relative dielectric constant, A is the total photodetector area, and $F[(\pi/2)\backslash\kappa]$ is a complete elliptic integral of the first kind,

$$F[\varphi\backslash\kappa] = \int_0^{\varphi} [1 - \kappa^2 \sin^2 \theta]^{-1/2}\, d\theta \tag{11}$$

where

$$\kappa = \tan^2[\pi w/4(s + w)] \quad \text{and} \quad \kappa' = \sqrt{1 - \kappa^2}$$

Figure 13 shows the calculated capacitance of an MSM structure for electrode spacings of 1, 2, and 3 μm. The device is assumed to be square, with L being the length of one of its sides. The solid and dashed lines are for electrode widths of 1 and 0.5 μm, respectively. The capacitance of an InGaAs p-i-n photodiode with absorbing layer thicknesses of 0.5 and 2 μm is shown for comparison. It is clear that the MSM photodiode exhibits significantly lower capacitance per unit area.

3.4. Frequency Response

The frequency response or bandwidth of MSM photodiodes will be a primary consideration for high-bit-rate applications. As discussed previously, low capacitance per unit area is one of the attractive attributes of the MSM photodiode. Capacitances of less than 100 fF are typical and the associated RC bandwidth (assuming an impedance of 50 Ω) is > 32 GHz. The effect of the transit time, on the other hand, can be substantial, depending on the

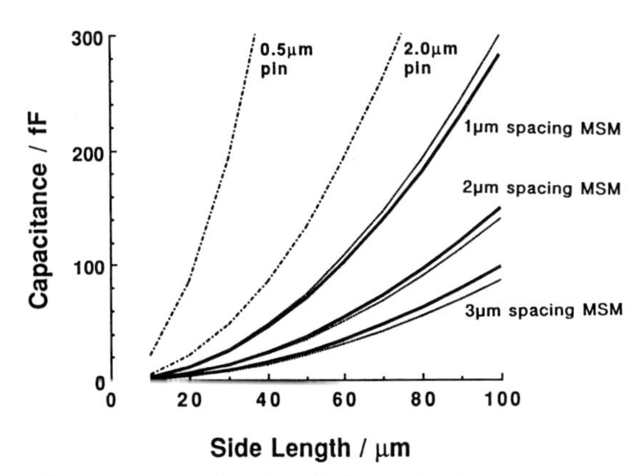

Fig. 13. Capacitance versus edge length for an MSM photodiode. The interelectrode spacing was 1, 2, and 3 μm. The solid and dashed lines are for 1-μm- and 0.5 μm-wide electrodes, respectively. The capacitance of a p-i-n photodiode with 0.5-μm and 1.0-μm-thick absorption layers has been plotted for comparison. (From Ref. 30. © 1990 IEEE.]

electrode spacing, which is typically 1 to 3 μm corresponding to approximate transit-time bandwidths of 18 and 12 GHz, respectively [30]. If the bias is not sufficient to deplete the absorption region, diffusion can degrade the frequency response. Also, the field must be adequate for the carriers to achieve saturated drift velocities. Consequently, a relatively high bias (10 to 20 V) is usually required to achieve fast response. To date, essentially all of the MSM photodetectors that have been reported have been transit time limited in their frequency response. It follows that the electrode spacing is the primary factor in determining the bandwidth. Initial analysis of the transit time of MSMs has utilized one-dimensional models developed for p-i-n photodiodes [12,13] with the electrode spacing being substituted for the thickness of the p-i-n absorbing layer. This approach is appropriate for very thin absorption layers but can only provide an upper limit for the relatively thick layers required for optical communications. The reason is that the equipotential lines in a layer of finite thickness are curved, which leads to longer paths and hence slower response for the photogenerated carriers. The two-dimensional electric field profile also influences the wavelength dependence of the response. Longer wavelength signals are absorbed farther from the surface (and the electrodes), where the field is weaker and the curvature of the equipotential lines is more pronounced. From these considerations it is clear that the bandwidth will peak for short interelectrode spacing, thin absorbing layers, and shorter wavelengths. It should be noted, however, that the first two of these conditions lead to lower quantum efficiency.

Many of the device results that have been reported to date have been pulse response measurements. For example, Wada *et al.* [37] have reported a full width at half-maximum (FWHM) pulse width of 14.7 ps for an $In_{0.53}Ga_{0.47}As$ MSM with an $Al_{0.52}In_{0.48}As/In_{0.53}Ga_{0.47}As$ graded superlattice barrier enhancement layer. MSMs with abrupt [30] and graded [40] InAlAs surface layers have yielded response times of 40 and 50 ps, respectively. Most of the measured impulse responses have exhibited a long tail on the trailing edge, which makes it difficult to determine the actual temporal performance. In theory, the bandwidth can be obtained by deconvolving the system response and performing a Fourier transform into the frequency domain. In practice, it is difficult to achieve an accurate estimation by this technique, particularly when the system response is a significant fraction of the pulse width. To date, the fastest MSM photodetectors have been fabricated using electron beam lithography to define ≈25-nm finger spacing and width [42]. Figure 14a shows a scanning electron micrograph of one of these MSMs. A recombination-limited response of 0.87 ps was obtained with an MBE-

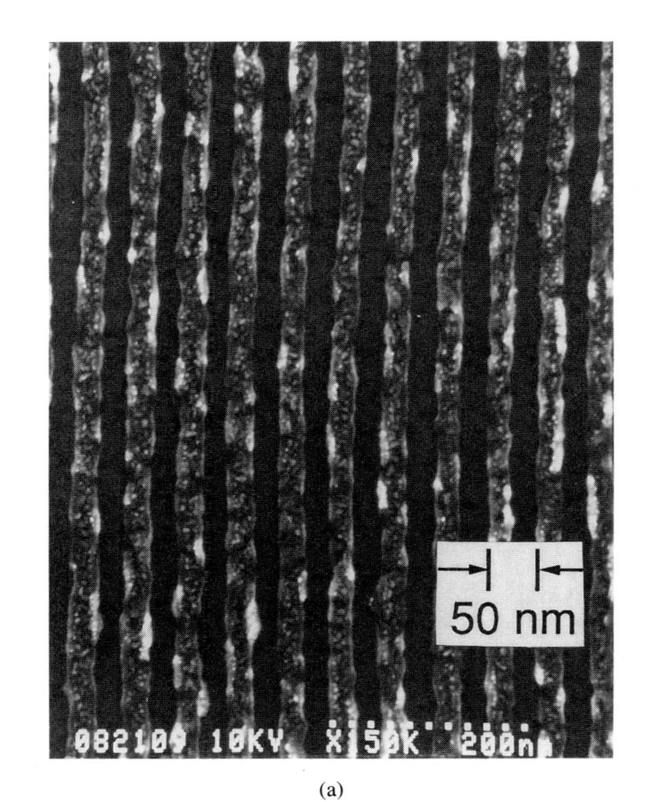

(a)

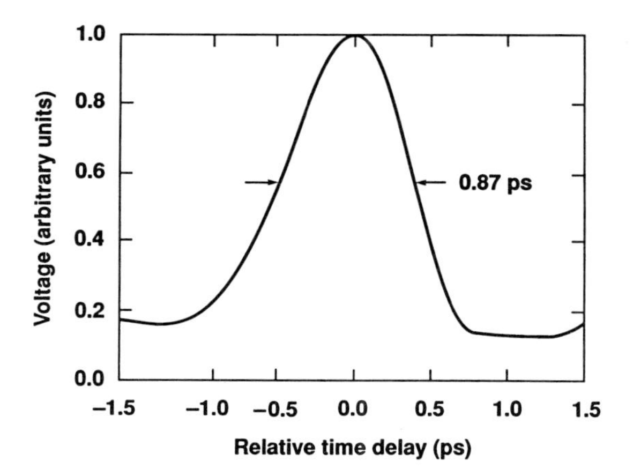

(b)

Fig. 14. (a) Scanning electron micrograph of an MSM photodetector with finger spacing and width of 25 nm. (b) Impulse response of a recombination time–limited MSM on low-temperature GaAs with 300-nm spacing and width. (From Ref. 42.)

grown, low-temperature GaAs epitaxial layer (Fig. 14b) and transit time–limited pulse widths of 1.5 and 10.7 ps were achieved on bulk GaAs and Si, respectively. Comparable results have not been achieved by the long-wavelength MSM photodiodes, but bandwidths sufficient for today's highest-bit-rate systems have been reported. A bandwidth of 10 GHz has been reported for an $In_{0.42}Ga_{0.58}As$ MSM with a graded cap layer [43]. With further improvements in the device structure of the long-wavelength MSMs, particularly in the design of the barrier-enhancement cap layers, bandwidths in excess of 18 GHz are projected [30].

3.5. Integration

As mentioned, one of the primary attributes of the MSM photodetector is its compatibility with FET circuits. As a result, there have been numerous reports of MSM/FET monolithic receivers. The circuits can be classified by photodetector material, the type of FET structure, and whether the integration scheme is vertical or horizontal. The MSMs are usually either InGaAs with barrier enhancement or GaAs, as dictated by the operating wavelength of the application, and the FETs are primarily MESFETs [44–49] and HEMTs [50–55].

Wada *et al.* [44] have fabricated 1 × 4 arrays of GaAs MSMs with MESFET preamplifiers for operation near 850 nm. These circuits demonstrated a small MSM capacitance of 0.1 pF and a bit rate of 1.5 Gbit/s, nonreturn to zero (NRZ). No optical crosstalk was observed and the electrical crosstalk between circuits in the array was below −20 dB. In order to facilitate interfacing with other circuits, a two-stage amplifier circuit was subsequently incorporated [45]. This resulted in a sensitivity of 400 V/W with sufficient bandwidth to permit operation at 2 Gbit/s. Higher speed was reported by Harder *et al.* [46] using enhancement/depletion 0.35-μm recessed-gate MESFET technology. The MSM photodetector had a dark current of 0.8 nA, a responsivity of 0.2 A/W, and a capacitance of 12 fF. The short gate length of this circuit yielded an $f_T \approx 40$ GHz and resulted in a receiver bandwidth of 5.2 GHz.

The most complex receiver reported to date has also used the GaAs MSM/MESFET combination. Crow *et al.* [49] successfully combined the MSM with GaAs self-aligned, refractory-gate MESFET processing to achieve a chip that included photodetection, amplification, clock recovery, and deserialization. This circuit contained over 2000 active devices and operated at 1 Gbit/s with a sensitivity of −22 dBm at a bit error rate of 10^{-9}. Contributing to this high performance were the low capacitance of the MSM

(150 fF) and the ability to locate it near the preamplifier to reduce the parasitic capacitance (20 fF).

By incorporating HEMTs into the preamplifer circuits, higher f_T values can be obtained, which leads to improved circuit performance. A bandwidth of 8.2 GHz was achieved using an enhancement/depletion 0.5 μm recessed-gate AlGaAs/GaAs HEMT preamplifier circuit [54], and Ketterson *et al.* [55] have achieved a 3-dB transimpedance bandwidth of 5.6 GHz using AlGaAs/InGaAs/GaAs pseudomorphic HEMTs. For long-wavelength applications the most complex circuit reported to date (14 components) utilized the InAlAs/InGaAs material system [52]. This circuit included the MSM photodetector, an HEMT preamplifier and cascode amplifier, Schottky level-shifting diodes, and an HEMT output impedance driver.

4. CONCLUSION

Of all the photodetectors that have been used for lightwave applications, the p-i-n photodiode and the MSM photodiode appear to be the most suitable for OEIC applications. The facts that both structures are relatively easy to produce and that their performance is not easily degraded by small variations in materials or fabrication procedures are strong advantages for OEICs, where materials incompatibility between the optical and electronic components can add significant complexity to the integration process. The key performance parameters for these devices are noise (usually dark current), capacitance, bandwidth, and quantum efficiency. As a result of contact shadowing in the MSM, the p-i-n has the advantage of higher quantum efficiency but tends to have higher capacitance per unit area. The MSM is easier to integrate, at least for GaAs devices, and can have somewhat higher dark current due to trap-related gain and the fact that current flow is lateral, which can also introduce excess noise at low frequencies. These differences aside, however, the characteristics of both types of photodiodes are sufficient to provide excellent performance in multigigabit receiver OEICs.

REFERENCES

1. R. G. Smith and S. D. Personick, Receiver design for optical communications, in "Semiconductor Devices for Optical Communications" (H. Kressel, ed.), p. 89. Springer-Verlag, New York, 1980.
2. S. R. Forrest, *IEEE J. Lightwave Technol.* **LT-3**, 347 (1985).

3. M. C. Brain and D. R. Smith, *IEEE Trans. Electron Devices* **Ed-30**, 390 (1983).
4. J. C. Campbell and K. Ogawa, *J. Appl. Phys.* **53**, 1203 (1982).
5. B. L. Kasper and J. C. Campbell, *J. Lightwave Technol.* **LT-5**, 1351 (1987).
6. J. N. Hollenhorst, D. T. Ekholm, J. M. Geary, V. D. Mattera, Jr., and R. Pawelek, *Proc. SPIE Conf. on High Frequency Analog Communications*, Vol. 995, p. 53 (1988).
7. S. R. Forrest, O. K. Kim, and R. G. Smith, *Solid-State Electron.* **26**, 951 (1983).
8. W. Schockley, *Bell Syst. Tech J.* **28**, 435 (1949).
9. S. M. Sze, "Physics of Semiconductor Devices." Wiley, New York, 1969.
10. J. L. Moll, "Physics of Semiconductors." McGraw-Hill, New York, 1964.
11. R. Trommer and H. Albrecht, *Jpn. J. Appl. Phys.* **22**, L364–L366 (1983).
12. J. E. Bowers, C. A. Burrus, and R. J. McCoy, *Electron. Lett.* **21**, 812 (1985).
13. J. C. Campbell, B. C. Johnson, G. J. Qua, and W. T. Tsang, *J. Lightwave Technol.* **7**, 778 (1989).
14. S. Hata, M. Ikeda, T. Amano, G. Motosugi, and K. Kurumada, *Electron. Lett.* **20**, 947 (1984).
15. S. J. Kim, G. Guth, G. P. Vella-Coleiro, C. W. Seabury, W. A. Sponsler, and B. J. Rhoades, *IEEE Electron Devices Lett.* **9**, 447 (1988).
16. R. M. Kolbas, J. Abrokwah, J. K. Carney, D. H. Bradshaw, B. R. Elmer, and J. R. Biard, *Appl. Phys. Lett.* **43**, 821 (1983).
17. S. Miura, O. Wada, M. Makiuchi, and K. Nakai, *Appl. Phys. Lett.* **48**, 1461 (1986).
18. J. Shimizu, Y. Inomoto, N. Kida, T. Terakado, and A. Suzuki, *IEEE Photon. Technol. Lett.* **2**, 721 (1990).
19. A. Suzuki, T. Itoh, T. Terakado, K. Kasahara, K. Asano, Y. Inomoto, H. Ishihara, T. Torikai, and S. Fujita, *Electron. Lett.* **23**, 954 (1987).
20. J. Shimizu, T. Suzaki, T. Terakado, S. Fujita, K. Kasahara, T. Itoh, and A. Suzuki, *Electron. Lett.* **26**, 824 (1990).
21. H. Yano, K. Aga, H. Kamei, G. Sasaki, and H. Hayashi, *J. Lightwave Technol.* **8**, 1328 (1990).
22. D. Wake, E. G. Scott, and I. D. Henning, *Electron. Lett.* **22**, 719 (1986).
23. D. Wake, R. H. Walling, I. D. Henning, and D. G. Parker, *Electron. Lett.* **25**, 967 (1989).
24. S. Chandrasekhar, B. C. Johnson, M. Bonnemason, E. Tokimitsu, A. H. Gnauck, A. G. Dentai, C. H. Joyner, J. S. Perino, G. J. Qua, and E. M. Monberg, *IEEE Photon. Technol. Lett.* **2**, 505 (1990).
25. S. Chandrasekhar, A. G. Dentai, C. H. Joyner, B. C. Johnson, A. H. Gnauck, and G. J. Qua, *Electron. Lett.* **26**, 1881 (1990).
26. T. Sugeta, T. Urisu, S. Sakata, and Y. Mizushima, *Jpn. J. Appl. Phys.* **19** (Suppl. 19–1), 459 (1980).
27. D. A. Humphreys, R. J. King, D. Jenkins, and A. J. Moseley, *Electron. Lett.* **21**, 1187 (1985).

28. M. Ito and O. Wada, *IEEE J. Quantum Electron.* **QE-22**, 1073 (1986).
29. D. L. Rogers, J. M. Woodall, G. D. Pettit, and D. McIntiff, *IEEE Electron. Devices Lett.* **9**, 515 (1988).
30. J. B. D. Soole and H. Schumacher, *IEEE J. Quantum Electron.* **27**, 737 (1991).
31. S. M. Sze, D. J. Coleman, Jr., and A. Loya, Current transport in metal-semiconductor-metal (MSM) structures, *Solid-State Electron.* **14**, 1209–1218 (1971).
32. M. igo and O. Wada, *IEEE J. Quantum Electron.* **QE-22**, 1073 (1986).
33. H. Schumacher, H. P. LeBlanc, J. B. D. Soole, and R. Bhat, *IEEE Electron Devices Lett.* **9**, 607 (1988).
34. T. Kikuchi, H. Ohno, and H. Hasegawa, *Electron. Lett.* **24**, 1208 (1988).
35. W. P. Hong, G. K. Chang, and R. Bhat, *IEEE Trans. Elect. Devices* **ED-4**, 659 (1988).
36. W. K. Chan, G. K. Chang, R. Bhat, N. E. Schlotter, and C. K. Nguyen, *IEEE Electron. Devices Lett.* **EDL-10**, 417 (1989).
37. O. Wada, H. Nobuhara, H. Hamguchi, T. Mikawa, A. Tackeuchi, and T. Fujii, *Appl. Phys. Lett.* **54**, 16 (1989).
38. J. B. D. Soole, H. Schumacher, H. P. LeBlanc, R. Bhat, and M. A. Koza, *IEEE Photonics Technol. Lett.* **1**, 250 (1989).
39. D. H. Lee, S. L. Sheng, N. J. Sauer, and T. Y. Chang, *Appl. Phys. Lett.* **55**, 1863 (1989).
40. H. T. Griem, S. Ray, J. L. Freeman, and D. L. West, *Appl. Phys. Lett.* **56**, 1067 (1990).
41. Y. C. Lim and R. A. Moore, *IEEE Trans. Electron. Devices* **ED-15**, 173 (1968).
42. Y. Liu, W. Khalil, P. B. Fisher, S. Y. Chou, T. Y. Hsiang, S. Alexandrou, and R. Sobolewski, *IEEE Device Research Conf.,* Cambridge, MA (1992).
43. C. Jagannath, A. N. M. Masum Choudhury, A. Negri, B. Elamn, and P. Haugsjaa, *Appl. Phys. Lett.* **58**, 325 (1991).
44. O. Wada, H. Hamaguchi, M. Makiuchi, T. Kumai, M. Ito, K. Nakai, T. Horimatsu, and T. Sakurai, *J. Lightwave Technol.* **LT-4**, 1694 (1986).
45. H. Hamaguchi, M. Makiuchi, T. Kumai, and O. Wada, *IEEE Elect. Devices Lett.* **EDL-8**, 39 (1987).
46. C. S. Harder, B. Van Zeghbroeck, H. Meier, W. Patrick, and P. Vetteger, *IEEE Electron Devices Lett.* **9**, 171 (1988).
47. W. S. Lee, G. R. Adams, J. Mun, and J. Smith, *Electron. Lett.* **22**, 147 (1986).
48. S. J. Wajtczuk, J. M. Ballantyne, Y. K. Chen, and S. Wanuga, *Electron. Lett.* **23**, 574 (1987).
49. J. D. Crow, C. J. Anderson, S. Bermon, A. Callegari, J. F. Ewen, J. D. Feder, J. H. Greiner, E. P. Harris, P. C. Hoh, H. J. Hovel, J. H. Magerlein, T. E. McKoy, A. T. S. Pomerene, D. L. Rogers, G. J. Scott, M. Thomas, G. W. Mulvey, B. K. Ko, T. Ohashi, M. Scontras, and D. Widiger, *IEEE Trans. Electron Devices* **36**, 263 (1989).

50. W.-P. Hong, G.-K. Chang, R. Bhat, J. L. Gimlett, G. K. Nguyen, G. Sasaki, and M. Koza, *Electron. Lett.* **25**, 1561 (1989).
51. L. Yang, A. S. Sudbo, W. T. Tsang, P. A. GARBINSKI, and R. M. Camarda, *IEEE Photonics Technol. Lett.* **2**, 59 (1990).
52. G.-K. Chang, W.-P. Hong, J. L. Gimlett, R. Bhat, G. K. Nguyen, G. Sasaki, and J. C. Young, *IEEE Photon. Technol. Lett.* **2**, 197 (1990).
53. H. S. Fuji, S. Ray, T. J. Williams, H. T. Griem, J. P. Harrang, R. R. Daniels, M. J. LaGasse, and D. L. West, *IEEE J. Quantum Electron.* **27**, 769 (1991).
54. V. Hurm, J. Rosenzweig, M. Ludwig, W. Benz, M. Berroth, A. Huelsmann, G. Kaufel, K. Koehler, B. Raynor, and J. Schneider, *Electron. Lett.* **27**, 734 (1991).
55. A. A. Ketterson, M. Tong, J.-W. Seo, K. Nummila, J. J. Morikuni, S.-M. Kang, and I. Adesida, *IEEE Photon. Technol. Lett.* **4**, 73 (1992).

PART V

Optoelectronic Integrated Circuits (OEICs)

Chapter 12

CURRENT STATUS OF OPTOELECTRONIC INTEGRATED CIRCUITS

O. Wada

Fujitsu Laboratories Ltd.
Atsugi 243-01, Japan

John Crow

IBM, T. J. Watson Research Center
Yorktown Heights, New York 10598

1. INTRODUCTION

Discrete optoelectronic devices using III–V semiconductors are one of the keys to optical telecommunication, data processing, and sensoring systems.

ISBN 0-12-200420-5

But they cannot be utilized without being linked with other components such as electronic ICs and optical guided-wave elements. Integrated optoelectronics which organize all necessary devices into an integrated form can realize compact, realiable components exhibiting high-performance functions. During this decade, many advances have been made in different areas of integrated optoelectronics. Not only have optoelectronic integrated circuits (OEICs)[1–4] been developed as transmitters and receivers for telecommunications, but also novel optical functional circuits for advanced signal processing have been created by the use of integrated optoelectronic device structures.

The purpose of this section is to summarize the present technology of integrated optoelectronics. We first explain various categories, advantages, and potential applications of integrated optoelectronics. Recent developments are reviewed with a special emphasis on transmitter and receiver OEICs based on GaAs and InP materials, and technological challenges for further improving performance and functions are discussed.

2. CATEGORIES AND ADVANTAGES OF INTEGRATED OPTOELECTRONICS

In this section, we will cover all integrated devices and circuits that incorporate semiconductor optoelectronic elements. Figure 1 shows the three major areas with reference to three basic characteristics: optoelectronic conversion and electronic and optical signal processing [5, 5a]. Lasers and photodetectors are the best representation of optoelectronic conversion, while standard OEICs combine these devices with electronic signal processing elements. The mature III–V semiconductor electronics and optoelectronics field can offer its strong resources to accomplish this. The integration of optoelectronic devices with other optical devices such as waveguides can provide smart chips which have advanced optical processing features. These semiconductor-based circuits, referred to as photonic integrated circuits (PICs), will be treated in detail in this chapter, in contrast to the dielectric material–based "integrated optics" [5b]. These two integration schemes fully encompass existing device technology. Revolutionary optical devices can be designed by more efficiently utilizing the interaction between photons and electrons within the integrated optoelectronic device structure. One such device is the bistable laser, which uses saturable absorption structures. This type of integration requires drastic alteration of device and system designs but opens up possibilities for optical functions that do not presently exist.

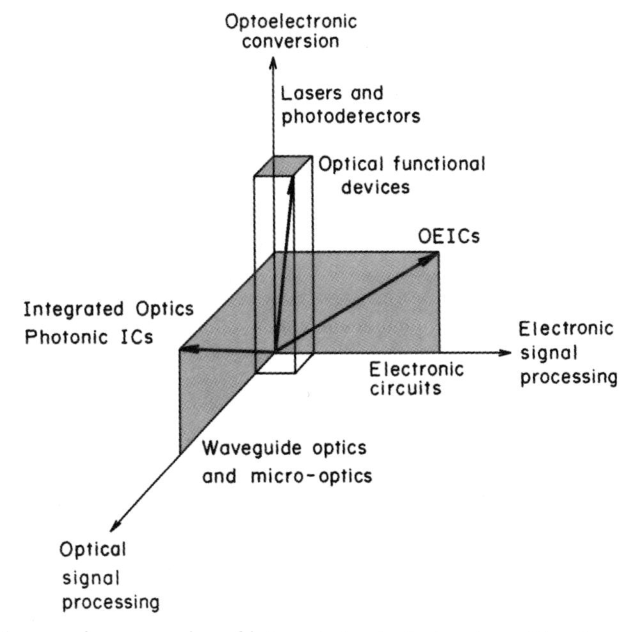

Fig. 1. Three major categories of integrated optoelectronics with reference to three basic functions.

OEIC chips have also been made using silicon technology. The main advantage of Si OEIC is the mature IC technology base, and this has led to their early use in the consumer and industrial application markets. By using an IC technology base, these OEICs offer a limited set of optical components and performance. For example, Si OEICs have currently been developed only for speeds less than 100 Mbps and only for the receiver function. To enhance Si OEIC potential, many researchers postulate fabricating III–V materials for the optoelectronic device on an Si chip containing the electronics. Although products using Si OEICs will be mentioned in covering the status of the field, this section will concentrate on OEICs and hybrid integration in modules using III–V materials.

The advantages of integration can be represented by those of the OEICs, which are shown in Fig. 2 [6]. The three axes are the important divisions of advantage, performance, manufacturability, and function, with some potential application systems given for example. Reduction of parasitics by integration provides high-speed and low-noise circuit performance [6a]. This is useful for developing telecommunications systems which can operate at ultrahigh speed, such as 10 Gbps, and coherent systems for which wide-bandwidth receivers are required. The integration leads to system compo-

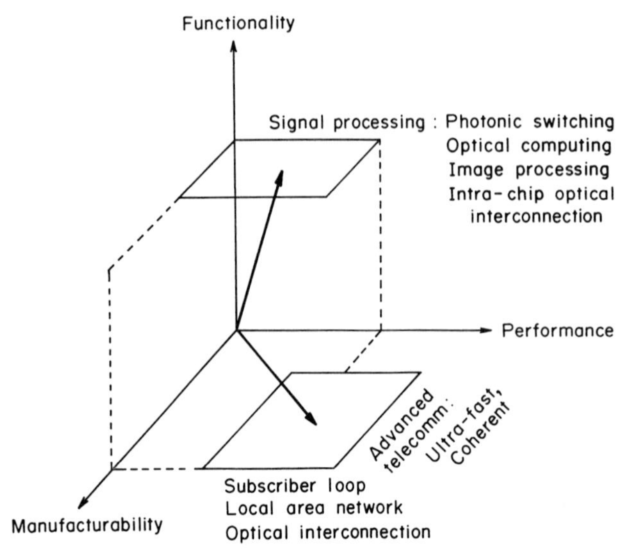

Fig. 2. Application fields of OEICs shown in space defined by three axes of important advantages. One is to improve both the performance and manufacturability, and the other is to improve functions. A similar view applies to waveguide-based integration and integrated optoelectronic functional devices.

nent-count reduction, assembly simplification, and reliability enhancement, all of which lead to lower cost. This advantage has been used by the consumer audio industry in developing hybrid integrated read/write heads for the compact disc player, reducing the product's thickness as well as eliminating a number of optical assembly steps in head fabrication [6b]. Also, such improvements in manufacturability can widen the field of lightwave components into broadband telecommunication networks and optical interconnections in data processing systems [4]. Optical interconnections can eliminate electromagnetic noise, crosstalk, and grounding problems, as well as the RC-limited delay time of existing electrical interconnection [6c]. By integrating electronic circuits with optical elements and by arraying them, either one- or two-dimensionally, lightwave systems with advanced functions can be constructed to make it possible to manage massive quantities of information at high speed. A variety of novel optoelectronic functions, including logic operation, wavelength control, and light beam steering, can also be generated within integrated structures, opening future architectures of photonic switching and optical computing systems. In the same direction, optical interconnection at the chip level should ultimately become practical as the technology develops, breaking the limits of speed and integration scale of existing Si electronics.

The focus of technological areas and advantages will depend on the target applications. Most of the published reports have been in the area of transmitter and receiver OEICs aimed at telecommunications applications. The first OEIC experimetal study was reported in 1978–79 by Yariv's group at Caltech on the integration of an AlGaAs/GaAs laser with a Gunn diode [7] and with GaAs MESFETs [7a] on GaAs substrates. Since then, researchers have pursued many different approaches to integrating devices in both GaAs-based [8, 8a] and InP-based [9, 9a] material systems, utilizing device technologies developed for lasers and transistors [9b–9d]. Laboratory demonstrations have already shown various high-performance circuits, and now manufacturing systems are being mobilized to develop practical applications.

3. STATUS OF OPTOELECTRONIC INTEGRATED CIRCUIT DEVELOPMENT

3.1. Integrated Structures and Processing

The material systems which have been most energetically investigated are the lattice-matched epitaxial systems such as AlGaAs/GaAs with light emitters and detectors near the 0.8-μm wavelength and GaInAsP/InP and AlInGaAs/InP for the 1.3- and 1.55-μm wavelengths. Non-lattice-matched heteroepitaxial structures are starting to be used in OEIC fabrication. Thus, material selection for optoelectronic and electronic device elements can be broadened to cover various combinations of III–V [10] and even II–VI compound semiconductors [11] and Si [12]. Recent growth techniques such as molecular beam epitaxy (MBE) and metalorganic chemical vapor deposition (MOCVD) are extensively used instead of conventional liquid-phase epitaxy (LPE) due to their high-quality crystal, uniformity, and throughput.

In terms of fabrication, the most significant difference between an OEIC and a conventional electronic IC is that the OEIC integrates significantly different functions and structures onto one chip. Enormous efforts have been made to solve the incompatibility problem between device structures and develop reproducible fabrication technology. Figure 3 illustrates representative integrated structures for a GaAs-based OEIC receivers in which thick photodetector structures must be integrated with thin transistor structures [3]. The vertical structure can be fabricated by a single growth procedure but suffers from performance degradation due to capacitive coupling among transistors on the conductive substrate. This is the key reason for the limi-

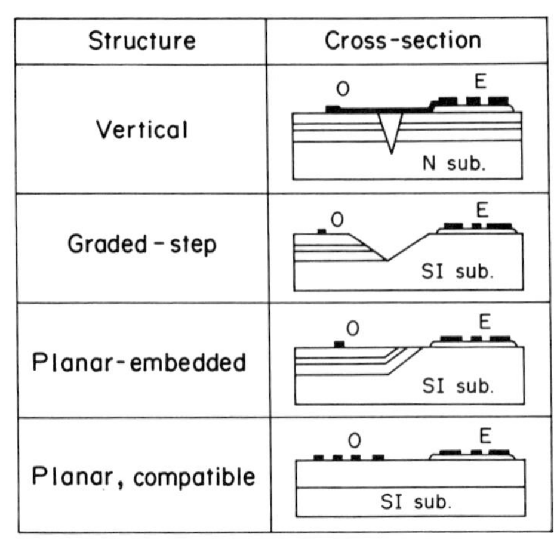

Fig. 3. Various integrated structures used for OEIC receivers.

tation of the speed performance of Si OEICs to around 100 Mbps. Efforts have been devoted to accommodating elements horizontally on a semi-insulating substrate to preserve the advantage of high-speed transistor circuits. Key areas of development in this field are the planarity of the structure and the process compatibity among elements. In graded-slope and planar-embedded structures, engineering of the dry etching technique has been able to provide surfaces with sufficient planarity for device processing. Process compatibility has been achieved by introducing metal–semiconductor–metal (MSM) photodiodes [13, 14] instead of PIN photodiodes.

In laser integration, the structure design is more difficult than for photodetectors because built-in waveguide cavities and mirrors are necessary. For InP-based materials, no standard electronic IC technology has yet emerged regardless of the material's advantages in high electron saturation velocity and mobility in GaInAs and the tight carrier confinement in AlInAs/GaInAs heterojunctions. However, various obstacles are being overcome in these fields to realize planar, compatible OEIC structures, which we discuss in the following sections.

3.2. GaAs-Based OEICs

3.2.1. GaAs-Based Transmitters

In fabricating OEIC transmitters utilizing lasers, low laser threshold current is necessary to avoid heating and a fabrication process for forming monolithic cavity mirrors is indispensable. Quantum well laser structures grown by either MBE [15–17] or MOCVD [18, 19] have been used extensively to reduce the threshold current. Mirror formation has been carried out by applying either a microcleavage technique [20–22] or improved wet/dry etching, and smooth vertical facets have been formed by non-material-selective reactive ion beam etching (RIBE) with Cl_2 gas [19, 23, 24]. Integrated transmitters have been made using serial fabrication of the lasers and transistors due to structural variations.

Figure 4 shows the circuit diagram and photomicrograph of a four-channel OEIC transmitter array fabricated by the Fujitsu group by using a planar,

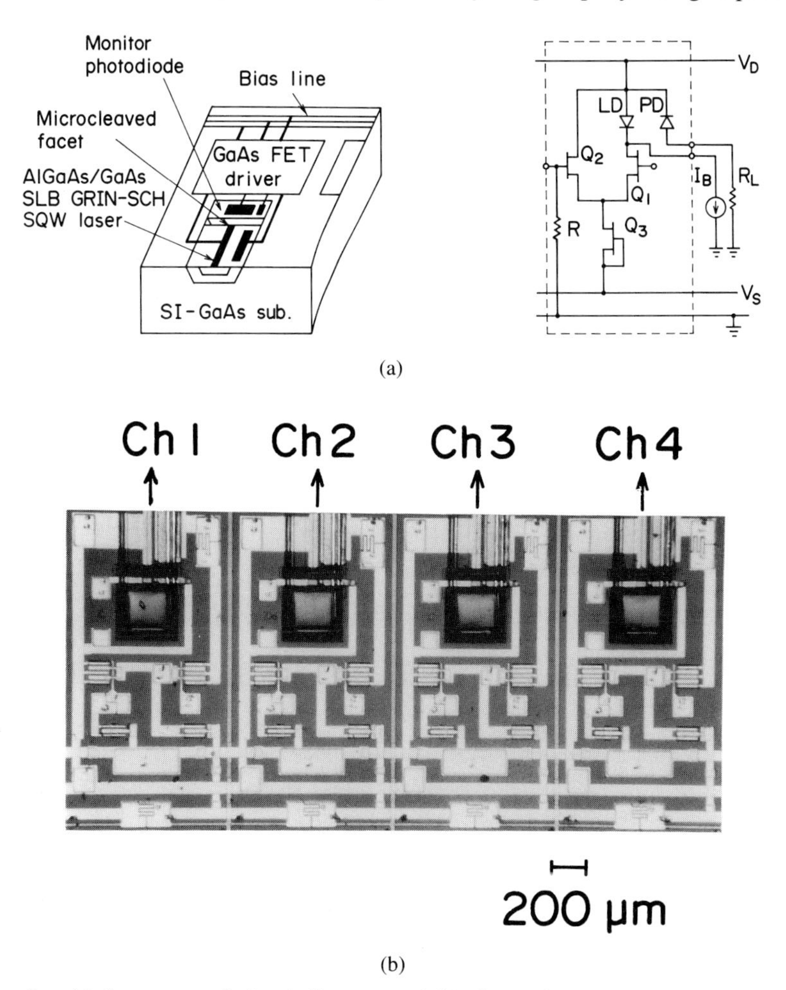

Fig. 4. (a) Structure and circuit diagram and (b) photomicrograph of a four-channel OEIC transmitter array. AlGaAs/GaAs superlattice-buffered (SLB) graded-index waveguide separate confinement (GRIN-SCH) single quantum well (SQW) lasers have been grown by MBE and embedded planar in an SI-GaAs substrate.

embedded structure as reported by Kuno *et al.* [25, 26] Each circuit channel consists of an MBE-grown AlGaAs/GaAs GRIN-SCH single-quantum-well (SQW) laser having a ridge waveguide with a microcleaved facet, a monitor photodiode formed using the same heterojunction structure as the laser, and three MBE-grown GaAs MESFETs with 2-μm gates as a driver circuit. A reasonably low and uniform threshold current of 17.5 ± 3.5 mA was observed even for an eight-channel array, although it was higher than that for discrete lasers similarly fabicated on a flat substrate (10 mA). The conversion ratio of input voltage to optical power was 6 mW/V, large enough to be used for the ECL interface. The response speed characteristics are shown in Fig. 5, along with the crosstalk to the other circuit channels [26]. The useful device speed is limited primarily by the relaxation oscillation frequency near 1 GHz, but four-channel array operation has been demonstrated at a bit rate of 1.5–2 Gbps, (NRZ) non return to zero code. Crosstalk can be caused by electromagnetic, optical, and thermal interference among circuit channels. The crosstalk observed in Fig. 5 is due primarily to electromagnetic coupling but also partly to the optical feedback to the FETs in neighboring channels at high frequencies. However, the crosstalk was still less than −20 dB in the useful frequency range below 0.6 GHz. No thermal crosstalk was observed, in this case because of the low threshold current and sufficient separation between lasers (1 mm). Preliminary aging tests car-

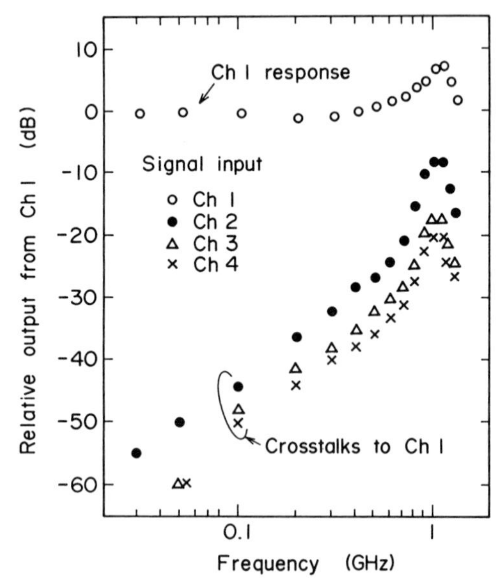

Fig. 5. Modulation and crosstalk characteristics versus frequency in a four-channel OEIC transmitter array.

ried out on these integrated lasers showed a threshold current degradation rate of 8%/kh for 3-mW output power at 40°C. This result is encouraging for an integrated laser, but it still indicates needs for high-reliability lasers for OEICs.

Matsueda *et al.* [27, 19] at Hitachi fabricated a transmitter consisting of 14 active elements including an MQW laser, a monitor photodiode, and a GaAs MESFET input amplifier in 1985. They used a standard Si ion-implantation technique for forming FET channels and RIBE with Cl_2 gas for etching interior facets of the laser and the photodiode, aiming for high reproducibility. The laser structure was grown by MOCVD in a well before the FET processing. The laser threshold current was 31 mA even with one facet etched by RIBE. A conversion ratio as high as 8 mW/V was obtained by use of a two-stage differential circuit for the amplifier, and dynamic response was demonstrated at 2 Gbps, NRZ. Although reasonable operation of the whole circuit was achieved, the performance obtained seems to have been limited by the degradation of device characteristics such as the laser threshold current and the FET's *k*-value, caused by restrictions in device structure design and processing needed for integration.

The advantage of OEICs in incorporating useful electronic signal processing function within a chip has been examined by several groups. Hamada *et al.* [28] at Matsushita demonstrated a light source for optical disc readers consisting of 38 elements, including a GaAs MESFET ring oscillator circuit for high-frequency laser modulation, which reduces the reflection-induced laser noise. A threshold current of 30 mA was obtained for their LPE-grown lasers and modulation operation was achieved at 800 MHz. A transmitter with the largest integration scale has been reported by Kerney *et al.* [22] at Rockwell, who integrated a transverse junction stripe laser with a microcleaved facet and a circuit consisting of more than 200 ion-implanted GaAs MESFETs for a 1:4 multiplexer and laser driver. Their LPE-grown laser showed 40-mA threshold current and multiplexer operation was obtained for the 160-MHz clock. The most severe problem in their fabrication to incorporate a laser into a standard MESFET IC chip was incompatibility in structure and processing between the laser and MESFETs, leading to yield problems.

The application of low-threshold-current MQW lasers has enabled monolithic integration of transmitters, and the fabrication process has been simplified and stabilized significantly by using ion-implantation and dry etching techniques. However, the fabrication yield so far achieved is not sufficient for large-volume production. This has been caused by incompatibility of device structures as well as insufficient reproducibility of the fabrication process, which is essentially based on serial fabrication of standard laser and MESFET ICs. These problems have often resulted in degradation of device characteristics such as the laser threshold current, differential quantum efficiency, FET transconductance, and cutoff frequency. There have been very few reports on the reliability of OEIC transmitters. Whether monolithic integration enhances the reliability is not yet shown.

3.2.2. GaAs-Based Receivers

Early OEIC receivers were fabricated by using vertical structures involving an AlGaAs/GaAs PIN photodiode and GaAs MESFET amplifier [29], but their overall speed was degraded because of capacitive coupling between the FETs. The PIN/FET-amplifier receivers were integrated on SI (semi-insulating)-GaAs substrates by a few research groups using either graded steps with similar structures [30] or planar embedded structures [31, 31a]. Transimpedance amplifiers have been used in most work and have attained fairly high speeds with reasonable sensitivity (1 Gbps, −18.6 dBm) [32].

Planar, compatible structures incorporating a GaAs MSM photodiode have provided complex MESFET circuits on SI-GaAs substrates. Figure 6 shows the circuit diagram and the photomicrograph of a four-channel MSM/amplifier receiver array incorporating 52 elements, which was developed at Fujitsu in 1985 [32, 33]. With a 100-μm-square area and 3-μm interdigitated lines and spaces, the MSM photodiodes were integrated with 2-μm gate MESFETs using an MOCVD-grown epitaxial structure. Highly uniform characteristics were observed over the entire array; the receiver sensitivity determined for the eight-channel array was 105 ± 10 V/W. Adding to the structural merit, the MSM photodiodes have the advantage of reducing the capacitance to one-fourth that of PIN photodiodes with the same area, enabling high-speed and low-noise performance. The frequency dependences of output response, input equivalent noise current, and crosstalk to neighboring channels are shown in Fig. 7. The cutoff frequency 1.1 GHz and the noise floor of 5 $pA/Hz^{1/2}$ are consistent with the circuit parameters used. The crosstalk was dominated by electromagnetic interference between neighboring amplifiers and was below −37 dB for up to 0.6 GHz.

Essentially the same MSM/MESFET structure has been utilized in much recent work on switching system and computer applications. The fabrication process has been simplified further by the use of an SI-GaAs substrate for both the photoabsorption layer and the MESFET channel layer together with a standard ion-implantation technique. Pedrotti *et al.* [34] at Rockwell have developed MSM/amplifier receivers for the 16 × 16 crossbar switch. Yamanaka *et al.* [35] at NTT have reported a time/space-division switch con-

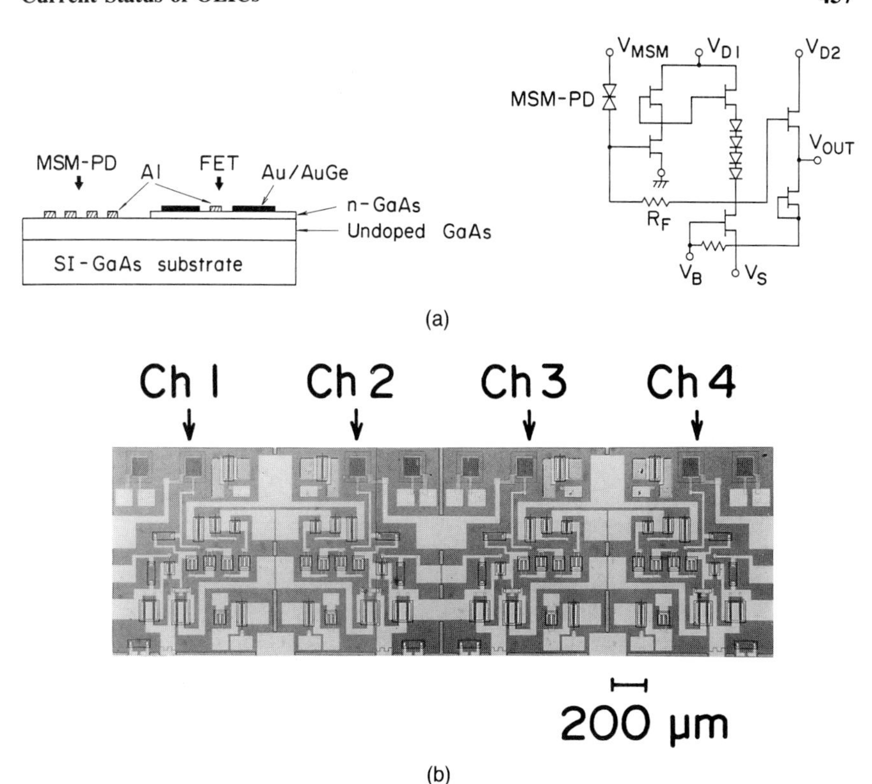

Fig. 6. (a) Structure and circuit diagram and (b) photomicrograph of a four-channel OEIC receiver array. A planar, process-compatible structure has been achieved using GaAs MSM photodiodes.

sisting of 231 MESFET gates. An LSI-level GaAs OEIC receiver integrating four MSM/preamplifier channels with deserializing and clock recovery circuits, composed of 8000 MESFETs, has been fabricated by Crow *et al.* at IBM for optical interconnections within computer networks [36]. Operation at 1 Gbps was demonstrated and will be described later. The advantage of the fast response of MSM photodiodes has also been demonstrated by Zeghbroeck *et al.* at IBM for a single MSM photodiode (105 GHz) [37] as well as for MSM/MESFET amplifier receivers (5.2 GHz) [38]. Takano *et al.* [38a] of Sony have reported a small-size MSM/MESFET receiver block involving a four-stage, 36-dB-gain postamplifier compatible with a DCFL circuit, for chip- and board-level data communication. They have shown 5-Gbps operation with an optical input power of 200 μW and a power dissipation as low as 8.2 mW.

Another approach to realizing planar, compatible structures is the use of

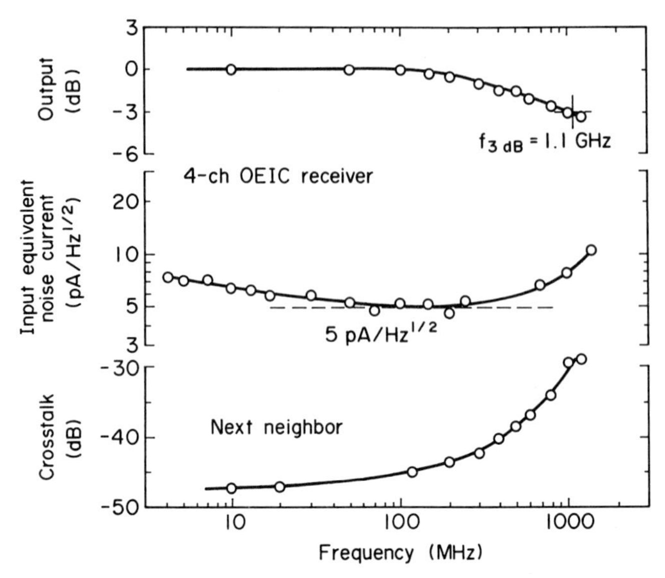

Fig. 7. Output response, input equivalent noise current, and crosstalk versus frequency of received signal for a four-channel OEIC receiver array.

photoconductors. Bulk GaAs photoconductor/MESFET integration was studied by Matsuo *et al.* [39] In order to obtain high receiver sensitivity, circuit design for minimizing the effect of high dark current in a photoconductor is important. An AlGaAs/GaAs modulation-doped (MOD) photoconductor has been integrated with a MODFET high-impedance amplifier and associated with an equalizing circuit by Chen *et al.* [40] at AT&T to achieve a sensitivity of −32.6 dBm at 90 Mbps.

Because there is no strict requirement for a waveguiding structure in a photodetector, planar and compatible OEIC structures have been successfully developed for receiver OEICs. MSM-based planar, compatible structure is one of the most significant breakthroughs achieved in OEIC fabrication technology, generating the most developed OEIC chip. Its process compatibility with standard GaAs MESFET IC technology is extremely important for developing useful OEIC receivers cost-effectively. Performance of such receivers has been shown by many research groups to be approaching that of hybrid circuits, as will be described in detail later. Although no explicit life test data have been reported, they should be able to exhibit reliability as high as that of standard GaAs MESFET ICs.

3.3. InP-Based OEICs

Excellent matching of GaInAsP and AlGaInAs lattices on InP substrates to long-wavelength optical transmission systems motivated the development of

InP-based OEICs. In 1980, Leheny *et al.* [41] at AT&T reported the first InP-based PIN/FET receiver OEIC. Immature transistor technology in this material system, however, has somewhat limited the integration scale compared with GaAs-based OEICs. MESFETs are not as useful as in GaAs-based system due to the low Schottky barriers, so other transistor structures such as metal–insulator–semiconductor (MIS) FETs, junction (J) FETs, and heterobipolar transistors (HBTs) are used.

A GaInAsP/InP transmitter was fabricated by Shibata *et al.* [41a] at Matsushita in 1985, in which a buried-heterostructure (BH) laser and a driver circuit consisting of three HBTs were integrated on an n^+-type InP substrate. The chip width was designed to be equal to the laser cavity length, so that conventional substrate cleavage was used for forming laser facets. Also, the fabrication process was simplified by using a second LPE growth to form both the burying layers for lasers and the HBT structures. Compact, pigtailed modules have been developed, and functional modulation of the entire module was demonstrated at 0.4 Gb/s. This performance, however, was no better than that of well-designed hybrid modules, and this is speculated to have been the largest obstacle in their earliest effort to commercialize this OEIC. The chip structure employed is basically a vertical integrated structure in which the transistor operation speed can be limited due to interdevice coupling. The performance limit observed in this OEIC is attributed to this reason.

Efforts have been devoted to improving performance by introducing horizontal integrated structures and refining device structures and fabrication processes. Suzuki *et al.* [42] of Toshiba fabricated a laser/FET OEIC using MOCVD to grow an GaInAsP/InP structure on an SI-InP substrate. They introduced an InP MESFET with a self-aligned, slightly oxidized Schottky barrier gate and also a self-aligned constricted mesa (SA-CM) laser with a mass-transported BH structure. The transmitter exhibited high-speed operation with −3 dB frequency of 6.6 GHz. Lo *et al.* [43] of Bellcore reported the integration of a GaInAsP/InP λ/4-shift SA-CM distributed feedback (DFB) laser and an AlInAs/GaInAs MODFET on an SI-InP substrate as shown in Fig. 8. This showed ultrahigh-speed operation of up to 10 Gbps.

There have been a number of studies of InP-based OEIC receivers, most of which considered PIN photodiodes because of the difficulties of obtaining low- leakage Schottky barriers. A vertically integrated structure combined with ion implantation was used by Kim *et al.* [44] to simplify the PIN/JFET-amplifier receiver structure for medium-speed (0.4 Gbps) operations. Horizontal integration of PIN photodiodes and amplifiers on SI-InP substrates has been implemented by several groups using InP MISFETs [45],

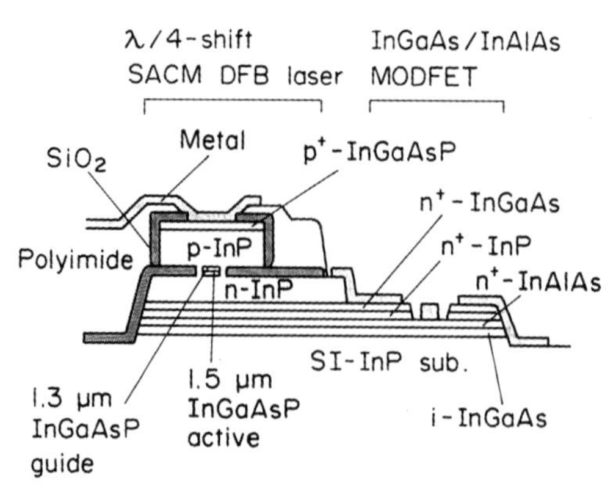

Fig. 8. Structure of DFB-laser/MODFET transmitter fabricated on InP substrate. Modulation of a self-aligned constricted-mesa (SA-CM) DFB laser has been shown up to 10 Gbps. (From Lo *et al.*, Ref. 43.)

JFETs [46–48], and HBTs [49]. A simple fabrication process consisting of a single MOCVD growth and selective etching technique was developed by Spear *et al.* [46] of STC in 1989. Figure 9 shows the structure of their PIN/HJFET amplifier receiver, in which heterojunction (HJ) FET layers and PIN layers are sequentially grown on an SI-InP substrate and the unwanted PIN layers are then removed by selective InP/GaInAs etching in the amplifier area. Such a receiver containing six HJFETs has exhibited a sensitivity of −27.8 dBm at 1.2 Gbps [46a]. Also, Lee *et al.* [47] have succeeded in integrating four channels of this receiver circuit using an identical process

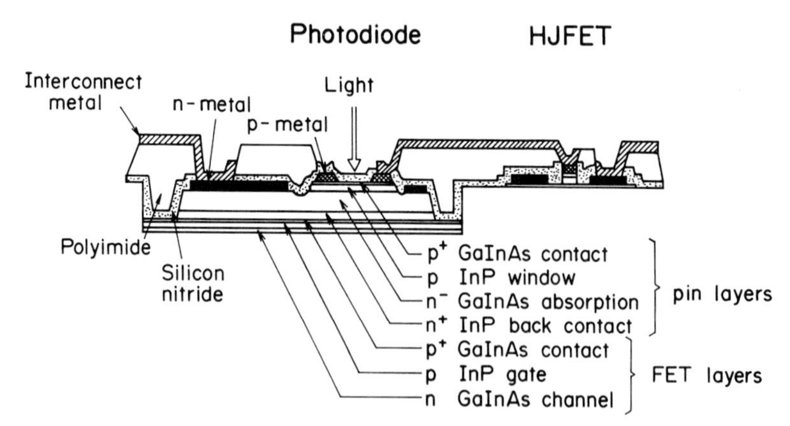

Fig. 9. Structure of InP-based PIN/HJFET OEIC receiver fabricated by a single MOCVD growth run on an SI-InP substrate. A patterned substrate and material-selective etching technique have been used to simplify the fabrication process. (From Spear *et al.*, Ref. 46.)

[47]. Such simple fabrication processes are extremely important for high-yield manufacture of OEICs.

To improve the receiver sensitivity, especially at high bit rates, it is important to use a front-end transistor with a high cutoff frequency, f_T. Nobuhara *et al.* at Fujitsu [50] and Yano *et al.* at Sumitomo [51] have introduced AlInAs/GaInAs high-electron-mobility transistors (HEMTs) in PIN/amplifier receivers to attain high performance around 2 Gbps. Chandrasekhar *et al.* [52] at AT&T have integrated AlInAs/GaInAs HBTs in their PIN/amplifier receiver to achieve an excellent sensitivity, for example, -21 dBm at 4 Gbps. Operation at 10 Gbps has been reported on receivers using HEMTs [52a] and HBTs [52b].

The possibility of using planar MSM photodiodes in this material system has been pursued. Low leakage of less than 100 nA has been achieved in an MSM photodiode having an AlInAs barrier enhancement layer [53]. This enhancement layer is associated with an AlGaInAs graded bandgap layer grown on top of a GaInAs photoabsorption layer to eliminate the hole pile-up effect. Barrier enhancement has also been shown when non-lattice-matched GaAs [54] and AlGaAs [55] cap layers have been used. Hong *et al.* [56] at Bellcore have fabricated receivers using an AlInAs-capped MSM photodiode with AlInAs/GaInAs HEMTs, which should create a simple, reproducible OEIC fabrication process.

Due to the strong technological background of InP-based optoelectronic devices, the performance of InP-based OEICs has improved rapidly. Demonstration of 10-Gbps operation achieved in InP-based transmitters and receivers is very encouraging in terms of realizing the advantage of OEICs in enhancing high-speed performance. However, the fabrication technique so far reported is basically a serial processing of optoelectronic and electronic devices applying existing processes, leaving reproducibility and uniformity issues to be resolved for large-volume and low-cost manufacturing.

3.4. Si-Based OEICs

OEICs have been fabricated using Si IC technology as a base, but the function has been limited to an optical receiver. The indirect gap of Si makes it unattractive, if not impossible in view of the recent observation of light emission from porous Si [56a], to consider this material for light generation. However, it has been used as an optical waveguide material and is extensively used as a packaging substrate for optical components [56b] (discussed later).

In 1985, an Si-based OEIC receiver was fabricated for data links at Motorola by Hartman *et al.* [56c], using a PIN photodiode and a bipolar IC amplifier. A special 5-μm-deep implant was added to the ECL IC process and used for the photodiode to improve its responsivity. A 4-dB penalty in responsivity (compared to the dc response) for the shallow photoabsorption region was incurred. Using an 80-μm-diameter photodiode and a 1-μm bipolar device, a receiver sensitivity of about −10 to −15 dBm was achieved in the 125–500-Mbps data rate range.

The fundamental problem of Si photodetectors for high-speed data link applications is their mismatch to the low-loss wavelengths of optical fiber. Even at the 850-nm fiber transparency window, the absorption depth necessary for efficient generation of photocarriers is over 10 μm. This distance must be depleted if the device speed is not to be limited by photocarrier diffusion current. This magnitude of depleted depth is incompatible with fast transit times, low bias voltages, and other IC processes. However, low efficiency and slower photodiodes may be acceptable for some applications, as mentioned later. The development of silicon-on-insulator (SOI) technology [56d], including SIMOX (separation by implantation of oxygen) process [56e], in the CMOS IC industry has generated renewed interest in improving the performance of Si OEIC receivers, but it is still doubtful that they will ever compete in speed or sensitivity against the III–V receivers. It should be mentioned that there are trends in the data link industry (mentioned in Part I) to migrate to 780- and 670-nm lasers, developed for the high-volume audio disk, bar code reader, and optical storage markets. The Si OEIC can offer improved performance at these shorter wavelengths.

The main feature of the Si OEIC receiver is its potential for cost savings and size reduction compared with hybrid receivers. This is achieved because the Si IC process is well developed and high yielding and because single-chip modules are lower in cost and higher in reliability than multichip modules. These features have led to a product use of Si OEICs in the optical data storage and audio industries. In 1973 Sharp started to develop an Si OEIC called the OPIC and have used it in industrial components such as circuit isolators and office data links [56f, 56g]. The Si IC process includes a 10–20-μm-thick epitaxial, intrinsic (100 Ω-cm) buffer layer, which is used to improve the efficiency of the photodiode. Performance has been limited to 8–16 Mbps in the current products, but extensions to 50–100 Mbps should be possible. The chip technology cross section is shown in Fig. 10. Sony has also developed an Si OEIC called the photodiode IC (PDIC) for use in the low-profile audio disk and minidisk products [6b, 56h]. A diagram of this PDIC chip, on which a microprism and a laser are aligned to form an extremely compact (7 mm thick) read/write head, is shown in Fig. 11a.

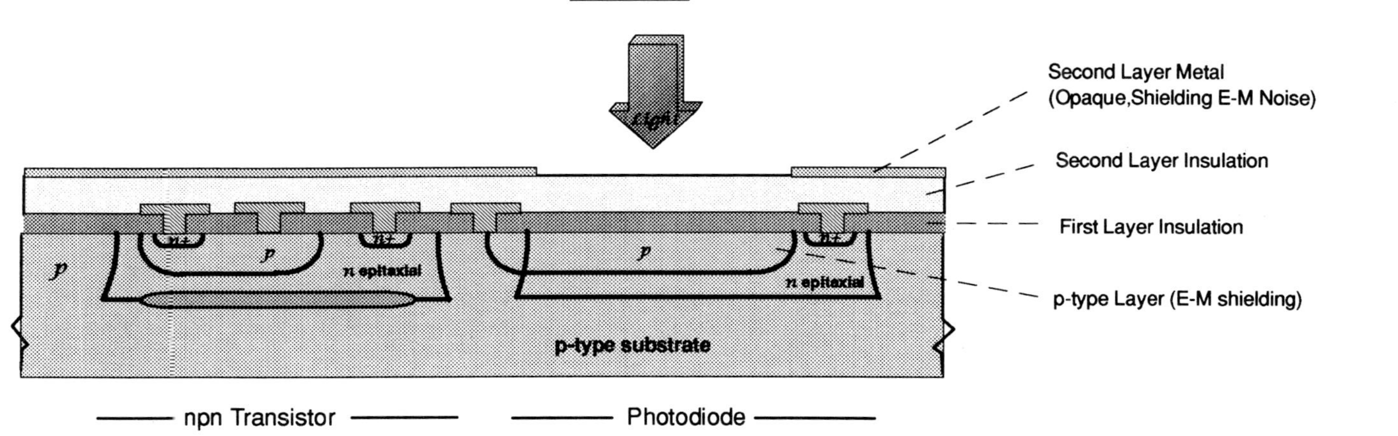

Fig. 10. Cross section of Si-based OEIC receiver chip. (From Nagao and Yamamoto, Refs. 56f and 56g.)

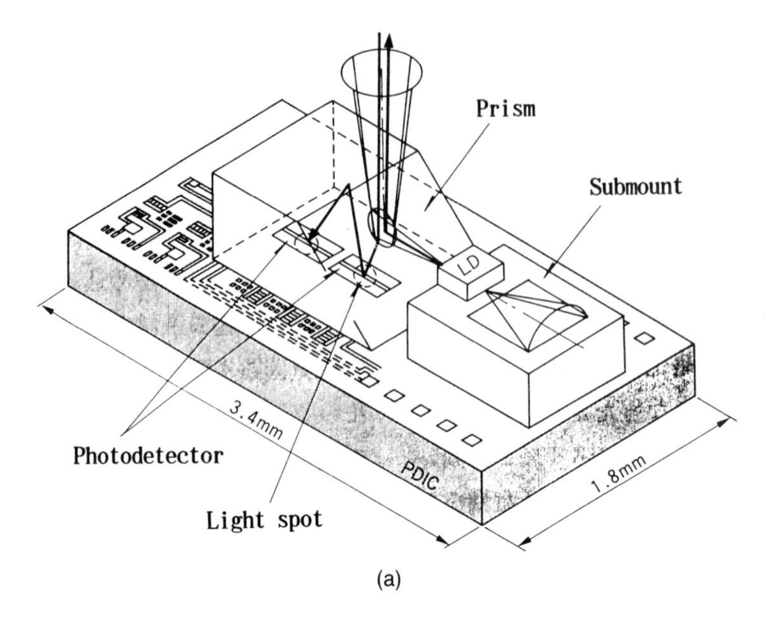

(a)

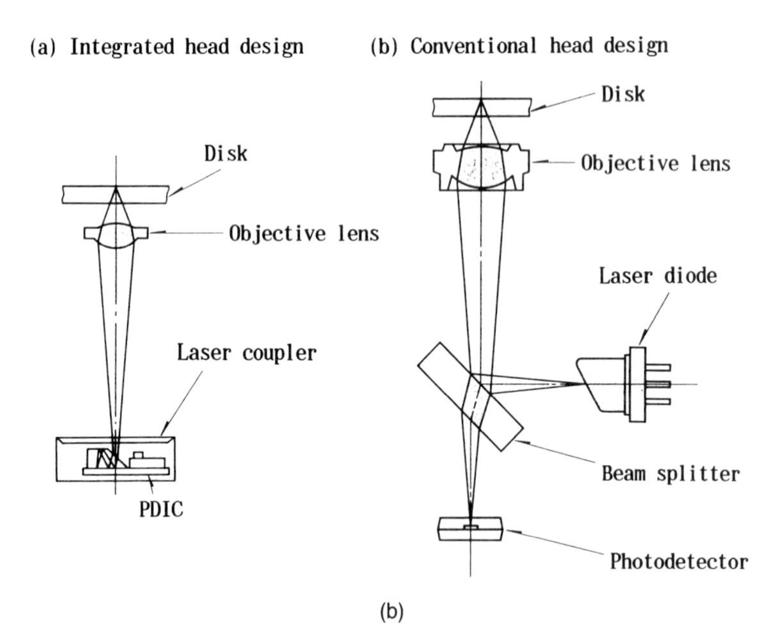

(b)

Fig. 11. (a) Structure of read/write head consisting of an Si-based OEIC receiver chip with other components mounted on top. (b) Diagrams showing optical path in read/write head assembly for both OEIC-based compact design and conventional design with discrete components. (From Refs. 6b and 56h.)

Figure 11b shows the part count and size reduction possible when integrated optoelectronics is used in read/write head modules.

3.5. Mixed Semiconductor Integration

Independent selection of materials for each optoelectronic and electronic device would enhance OEIC's performance, manufacturability, and functionality. In this direction, proposals have included a monolithic approach using non-lattice-matched heteroepitaxy as well as a hybrid approach using flip-chip integration, grafting, and lift-off techniques.

3.5.1. Monolithic

Remarkable advances have been made in non-lattice-matched heteroepitaxy for various combinations of III–V compounds and substrates including silicon. This has been reviewed in Part II. Key issues involve solving the problems of crystal defects and stress which are induced by differences in lattice constants as significant as 4% as well as thermal expansion coefficients at the interface. The defect density in GaAs/Si has been lowered to 10^6 cm^{-2}, and continuous wave (CW) laser operation has been achieved by optimizing the growth conditions, including buffering and annealing [57, 58]. Figure 12 shows an OEIC fabricated by Choi *et al.* [12] at MIT with an AlGaAs/GaAs LED grown by MBE on an Si MOSFET circuit. Basic light modulation at 27 Mbps (MOSFET limited) has been shown.

Suzuki *et al.* [9] at NEC have fabricated long-wavelength OEICs by using MBE-grown GaAs/InP heteroepitaxial structures. High-reliability, lattice-

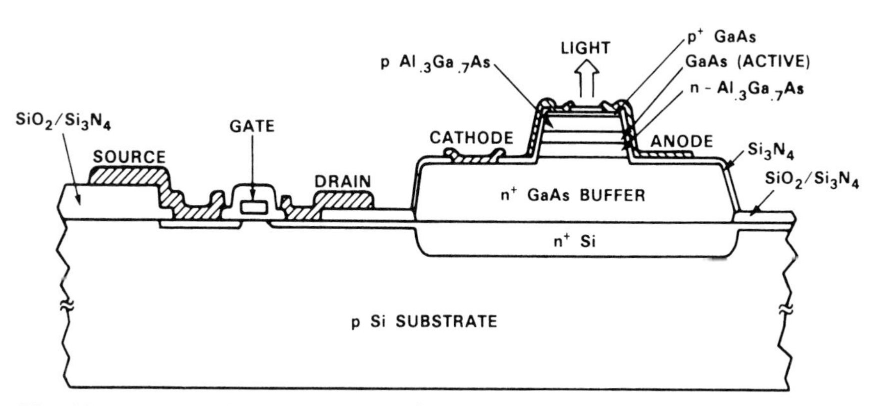

Fig. 12. Cross section of an AlGaAs/GaAs LED integrated with an Si MOSFET circuit. The III–V layer has been grown by MBE on an Si substrate after finishing major processes for MOSFETs. (From Choi *et al.*, Ref. 12.)

matched GaInAsP/InP optoelectronic devices and standard GaAs-MESFET technology have been combined. The interface quality problem is less severe in the FET because it is a majority carrier device with a current direction parallel to the interface. Both the laser/driver transmitter and PIN/amplifier receiver have performed excellently at a bit rate of 1.2 Gbps. MOCVD has also been used to grow GaAs/InP heterostructures in the fabrication of transmitters consisting of GaInAsP/InP lasers and GaAs MESFETs [59]. Efforts in non-lattice-matched heteroepitaxy have also been made with other combinations of materials. Fallahi *et al.* [11] have reported a CdTe interdigitated photoconductor with an AlGaAs/GaAs FET on a SI-GaAs substrate, implying a perspective on HgCdTe/GaAs OEICs for far-infrared sensing.

3.5.2. Hybrid

Although fabricating transmitters and receivers from discrete optoelectronic and IC devices allows the optimization of each device for its associated function, historically the resulting modules have often been degraded by uncontrollable parasitics and the size and cost of the modules have often limited their widespread use. Hybrid techniques in packaging are being developed which may allow mixing of optimized devices into compact structures. Flip-chip integration, one of the advanced hybrid techniques, can provide most of the OEIC advantages without suffering from material and processing difficulties. Sussmann *et al.* [60] at Plessey first demonstrated the flip-chip integration of a GaInAs/InP PIN photodiode on a GaAs amplifier chip to realize receiver sensitivity (−36 dBm at 650 Mbps) better than that of monolithic receivers. Hamaguchi *et al.* [61] at Fujitsu extended this technique to fabricate a receiver using a monolithic-lens, ultrasmall-junction PIN photodiode, shown in Fig. 13, to achieve easy fiber alignment, low capacitance, and high receiver sensitivity at high speed (−29.8 dBm at 2 Gbps). An additional advantage of flip-chip integration is the integration of avalanche photodiodes (APDs), which have so far been difficult to combine in OEICs because of their sensitive structures. Hamano *et al.* [61a] at Fujitsu fabricated a receiver module using a GaInAs/InP APD and an Si bipolar amplifier exhibiting a sensitivity of −23 dBm at 10 Gbps, NRZ. Although drawbacks exist, such as the complexity of the assembly process and the thermal expansion limit in integration scale, the flip-chip technique remains useful for simple optoelectronic integration for advanced telecommunications systems.

Another approach would be to combine different materials by a grafting or lift-off technique, in which a thin semiconductor epitaxial layer is transferred onto a different substrate and bonded by van der Waals forces. Po-

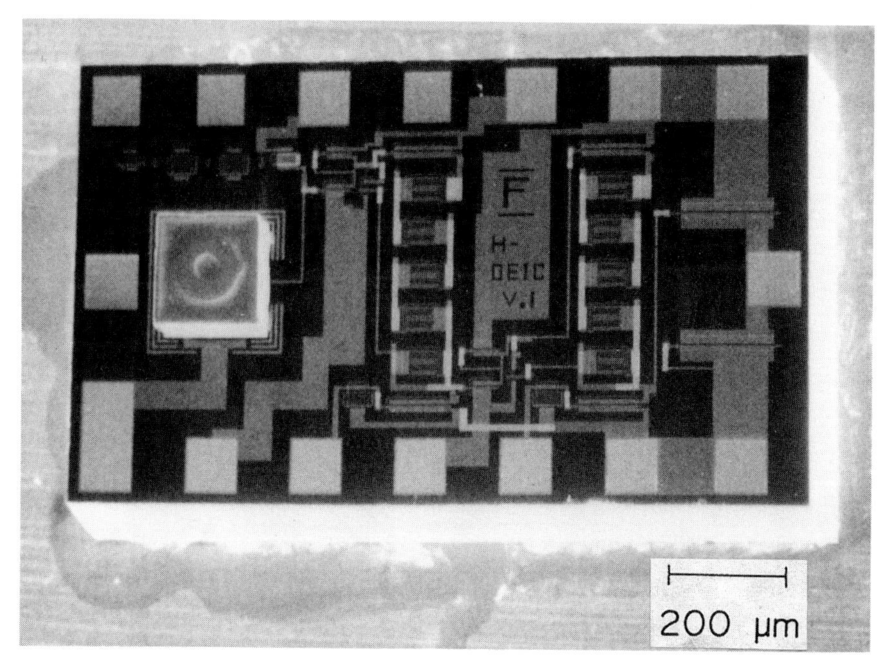

Fig. 13. Scanning electron micrograph of flip-chip integrated GaInAs-PIN/GaAs-amplifier receiver. The PIN photodiode has a back-illuminated structure with a monolithic microlens on the light entry surface, which allows easy fiber alignment as well as small junction area.

lentier *et al.* [62] reported an LED/FET using a lift-off GaAs/InP structure. Yi-Yan *et al.* [63] used lift-off to integrate $LiNbO_3$ waveguides and GaAs devices. Reliability in both high-temperature processing and long-term operation needs to be demonstrated.

3.6. Present Status of OEICs

3.6.1. OEIC Chip Performance

Significant advances in both performance and integration complexity have been achieved for OEIC chips. Figure 14 shows the receiver sensitivity versus the bit rate for previously reported GaAs- and InP-based OEICs and the best hybrids using GaInAs PIN photodiodes and APDs [6]. Since this plot includes a mix of receiver designs and operating wavelengths, an accurate comparison cannot be made, but it still illustrates the trend of performance improvement. The degradation in early OEIC performance was due to insufficient capacitance reduction and poor transistor characteristics, but these

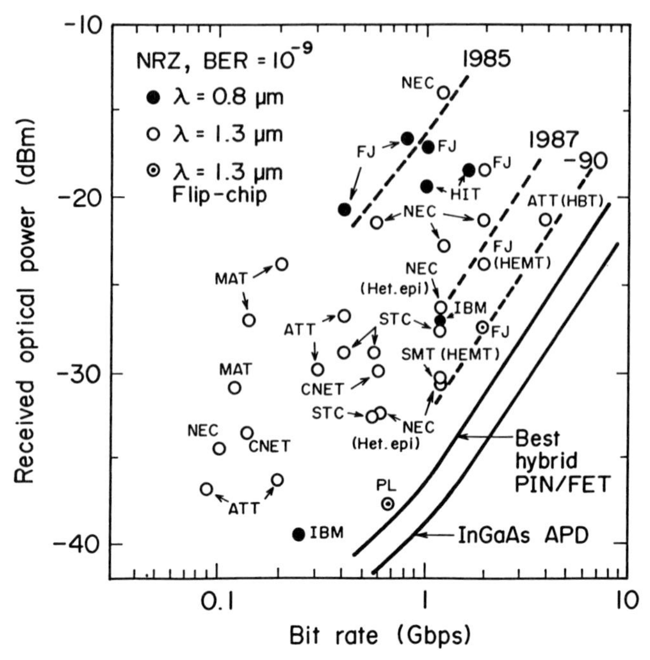

Fig. 14. Receiver sensitivity plotted versus operation bit rate for previously reported GaAs- and InP-based OEIC receivers, together with best hybrid data for both PIN photodiodes and APDs for reference. The dashed curve referring to average performance in a particular year indicates significant progress in recent years. The improvement is due to high-performance transistors such as heteroepitaxial GaAs MESFETs and InP-based HEMTs and HBTs.

problems have nearly been solved. In particular, the introduction of low-capacitance photodiodes and high-cutoff-frequency transistors, such as HEMTs, HBTs, and heteroepitaxial GaAs MESFETs, has improved sensitivities to almost the hybrid level. Flip-chip integration has already been shown to be comparable to the best hybrids. The remaining degradation in OEICs is attributed to unoptimized designs of low-noise amplifier circuits and to device degradation due to the complicated OEIC process.

The advantages of integration should be confirmed by systems application. Optical link experiments have been carried out in both short- and long-wavelength regions. GaAs-based OEIC link modules have shown transmissions through a multimode fiber at a bit rate of 800 Mbps, NRZ over a dispersion-limited distance of 2 km as reported by the Fujitsu group [64]. The IBM group has shown a bit error rate of 10^{-15} at 1 G bps for unpackaged OEICs, indicating that the OEIC's reliability is sufficiently high for application to data communication even within computers. An experiment using a pair of GaAs/InP heteroepitaxial OEICs and a single-mode fiber has been conducted by the NEC group, and transmission at 1.2 Gbps over a distance of 52.5 km has been achieved [65]. These results have demonstrated the advantages of OEICs in implementing compact modules with reduced part count and fairly high performance. Function and performance of current OEIC chips thus seem to be rapidly approaching practicality for conventional data communications systems. Recently, IBM has reported on high yields (greater than 80%) of GaAs MESFET based receivers, fabricated in 32 channel monolithic arrays. Over 5000 receivers from 6 different wafers were fabricated at a commercial GaAs foundry and characterized.[1] If this level of yield can be maintained in commercial runs, and maintained following packaging, 500–1000 Mb/sec receiver chips costing an order of magnitude less than hybrid counterparts (i.e., less than $10 each) should be producible.

3.6.2. Packaging Technology

Integrated optoelectronics offers the potential of over an magnitude reduction in module size for the same function, but this cannot be achieved without advanced multichip optoelectronic packaging. Also, manufacturable optoelectronic packaging developments are prerequisite for practical system applications. Low-loss optical coupling, high-speed electrical connections, and efficient heat removal are key issues. Figure 15 shows the structure of a 4 × 4 optical switch fabricated at Fujitsu using GaAs-based OEIC reçeiver and transmitter arrays and a GaAs-IC electronic cross-point switch [66]. Slant-end fiber arrays enable coupling to MSM photodiodes with horizontal access, low loss ($<$0.2 dB), and large alignment tolerance ($>$40 μm). This is extremely useful for developing compact, planar, and manufacturable packaging. Taper-end fiber arrays exhibited laser coupling with a reasonable loss (1.2 dB). Microstriplines have been utilized for interchip connections. This has resulted in a compact optical switch exhibiting an optical gain slightly above 0 dB and a crosstalk of −20 dB for full optical switch operation at 560 Mbps, NRZ.

If self-alignment of optical components can be developed, packaging manufacturability would be much improved. Wale *et al.* [67] GEC Marconi have proposed the application of flip-chip bonding using solder reflow, originally developed for electronic IC chip attachment, to align an optical waveguide to a fiber with micrometer accuracy. Jackson *et al.* [68] of IBM have

[1]J. D. Crow, "Technology Challenges for Optical Interconnects in Multichip Module Environment," pg. 489, *Proceedings of the 6th LEOS Annual Meeting,* November 15–18, 1993, San Jose, CA.

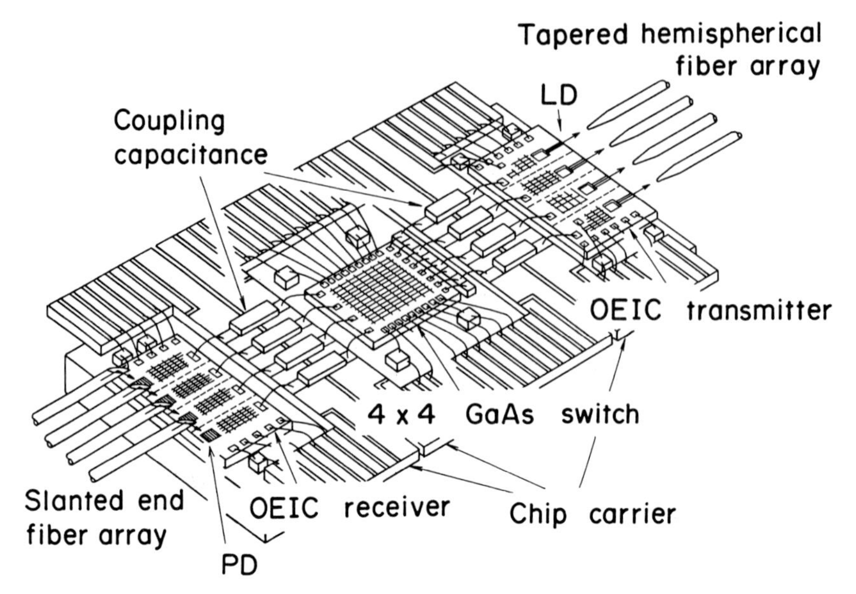

Fig. 15. Structure of a packaged 4 × 4 optical switch module composed of OEIC receiver and transmitter arrays and a GaAs switch circuit. Horizontal access, low-loss fiber coupling, and high-speed electrical connections have been achieved.

advanced this approach further by fabricating a 7 mm by 15 mm module with GaAs four-channel transmitter and receiver OEIC chips flip-chip mounted onto a Si carrier, self-aligned to optical waveguides. The eight multimode waveguides were self-aligned to the two chips within 2 μm, simultaneously with making the 20 electrical I/O connections between chips and carrier. The module shown in Fig. 16, was demonstrated in a 1-Gbps link.

4. TECHNOLOGICAL CHALLENGES

4.1. OEIC Fabrication Technology

To manufacture OEICs with useful functions and high performance, efficient and reproducible material growth, device processing, and packaging techniques must be developed. Whether this is possible depends on what direction OEIC structures take. As is evident in recent OEIC experiments, receivers are closer than transmitters to this goal, because of more complicated structures for lasers. Quantum well lasers are useful in simplifying OEIC structures. Planar, compatible GaAs-based transmitter structures have been proposed using a lateral current-injection (LCI) laser [69] with MQW struc-

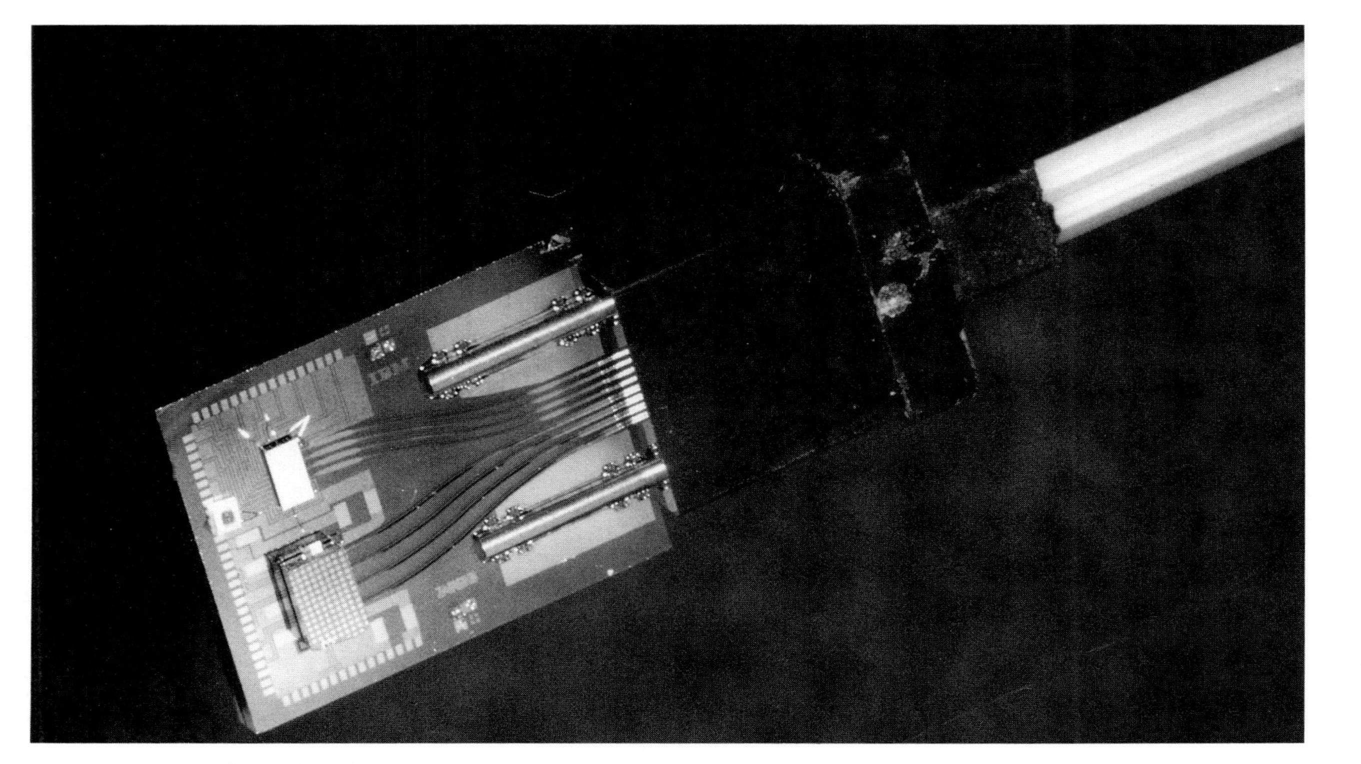

Fig. 16. Photograph of a packaged optoelectronic multichip module for a parallel data link. A four-channel transmitter and receiver are coupled with waveguides on an Si substrate. (From Jackson *et al.*, Ref. 68.)

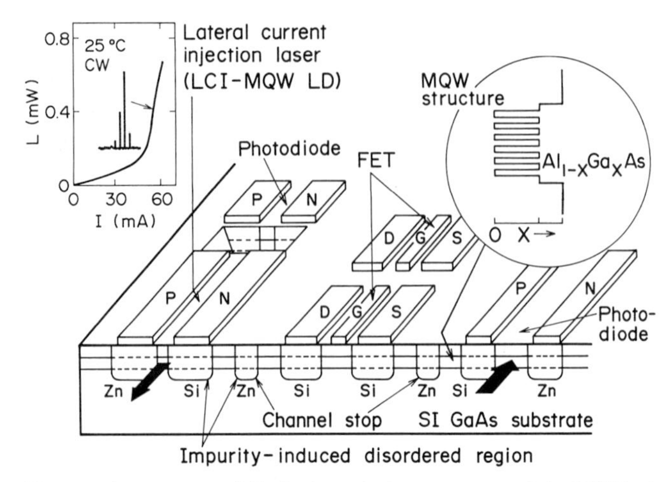

Fig. 17. Proposed structure of GaAs-based planar, compatible OEIC. A common MQW structure is used for lateral current injection (LCI) lasers, MQW-FETs, and photodetectors. Planar, index-guide lasers can be formed by impurity-induced disordering of MQW.

tures as shown in Fig. 17 [70]. The impurity-induced lattice disordering of MQW occurring during Zn and Si diffusion is used to form lateral contacts as well as a built-in index guided cavity. The same MQW structure can also be used for FET channels and photoabsorbers, making the whole OEIC structure planar and process compatible. Preliminary experiments have verified CW operation of the LCI laser with a threshold current of 47 mA [71] and basic operation of MQW-FETs [70]. Recent MQW laser performance has shown submilliampere thresholds [72] and response speed on the order of tens of picoseconds [73]. This indicates the possibility of introducing laser OEICs in optical interconnections at the intrachip level without power consumption and delay time problems [6b]. Further developments in quantum size effects, including quantum wires and boxes [74], are expected to broaden the field of devices and functions for OEICs.

In material growth, the controllability of the interface is key. GaInAsP/InP MQWs have been advanced less than AlGaAs/GaAs MQWs because of the less stable interface of four elements. However, recent MOCVD-grown materials have shown clear exciton absorption spectra [75]. Excellent laser characteristics, including a nearly ideal quantum efficiency, have also been shown by GaInAsP/InP strained-layer MQWs [76]. Improvements are also expected to continue in non-lattice-matched heterostructures. III–V/Si heteroepitaxy will be particularly important for applications in which OEICs are needed for optical interconnections within Si-based electronic systems.

The demonstration of room-temperature, CW operation of GaInAsP lasers grown on Si for more than 6000 h is very encouraging for non-lattice-matched heteroepitaxy [76a]. Possibilities include intrachip interconnections using waveguides formed over the wafer surface [6c]. Another possibility is in interwafer communication using parallel optical signal transfer in free space. A three-dimensional optically coupled common (3D-OCC) memory as shown in Fig. 18a has been proposed by Koyanagi *et al.* [77], in which fast, optical data transfer in the vertical direction is combined with the lateral electrical interconnections to provide intelligent memory for real-time parallel processor systems. Figure 18b depicts a proposed structure for 3D-OCC memory which consists of stacked OEIC wafers prepared by SOI technology.

In processing OEICs, selective-area growth remains a prerequisite for integrating different device structures. Single-step growth on a patterned substrate and consequent selective etch can be an excellent practical technique, as pointed out in regard to Fig. 9. Another direction is an *in situ* growth technique which is described in more detail in another chapter. Hashimoto *et al.* [78, 79] at OJRL have developed a system which combines an MBE system and a focused ion beam implanter, as well as other chambers for etching, cleaning, and annealing, and connected through a wafer transfer chamber. Temkin *et al.* [80] at AT&T have introduced a high-uniformity gas source MBE system for the growth. These techniques will be extremely important as circuits become more complex with more precise structural requirements. It is interesting to note that a planar, compatible OEIC structure as proposed in Fig. 17 can be efficiently produced using *in situ* processing.

4.2. Enhancement of Function

4.2.1. Waveguide-Based Integration

Integration of active and passive waveguide devices with optoelectronic conversion devices can provide various optical processing capabilities which could not be attained with simple electronic circuits. Compact and rigid integration of elements is most suited for optical processing circuits because they need precise alignment between elements. Many interesting features of this type of integration have been shown: a monolithic light source using a DFB laser and an absorption modulator to provide ultralow chirp (0.01 nm) and ultrahigh speeds (10 Gbps) [81]; a directional coupler switch having a DFB laser source at the input port and a photodetector at the output port, which is useful as optical nodes in a network [82]; a multiwavelength light source incorporating an array laser, a waveguide combiner, and an output laser amplifier for wavelength division multiplexing (WDM) systems [83];

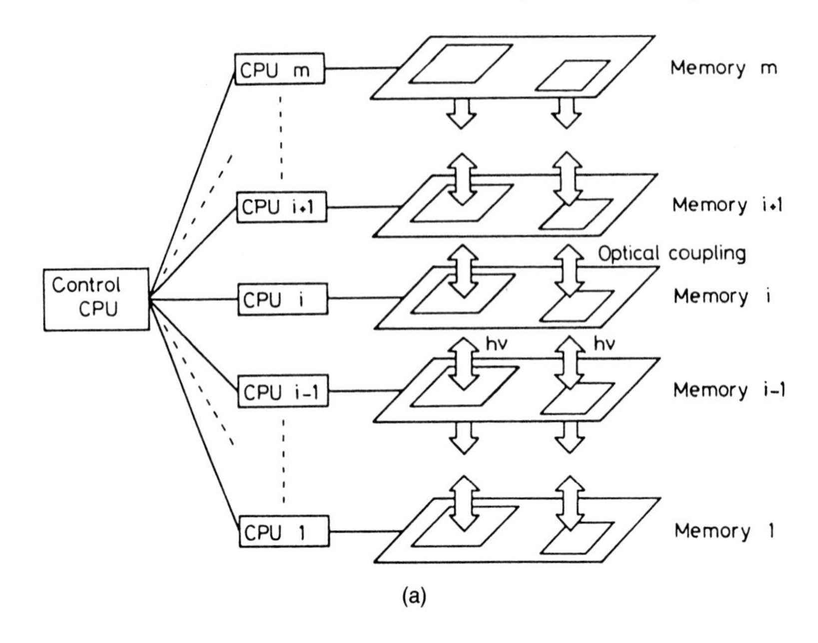

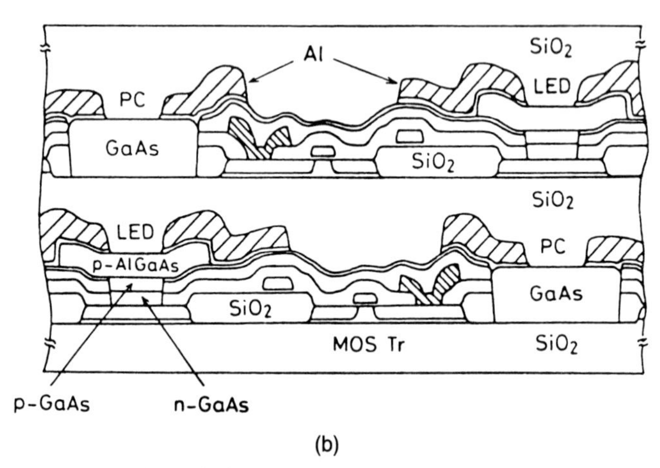

Fig. 18. (a) Proposed parallel processor system and (b) multilayer structure for three-dimensional optically coupled common memory. Horizontal data transfer is carried out electrically within the wafer and vertical, interwafer data transfer can be done optically. Heteroepitaxial GaAs-based LEDs and photoconductors (PCs) are used in each memory cell. (From Koyanagi *et al.*, Ref. 77.)

and a coherent optical receiver consisting of a local oscillator laser, a directional coupler, and photodiodes [84, 85].

Most of the technological issues in this area, such as the fabrication of different element structures on one chip, are common throughout integrated optoelectronics. Likewise, the reduction of optical coupling losses among elements and external optical fibers is characteristic in this area. Butt-joint coupling can join elements for low-loss operation, but fabrication is complicated. Gradual coupling using either a multilayer structure, which incorporates an intermediate refractive index layer for impedance matching between the device and the waveguide [86], or a grating coupler with second-order corrugation [87] is useful for device implementation. The impedance matching structure has been used in a short-size photodetector/waveguide exhibiting an 11-GHz cutoff frequency without loss of quantum efficiency [88]. One possibility for coupling a fiber to waveguides at low loss is to use large-core waveguides. To provide good mode matching between the fiber and waveguides, a diluted MQW waveguide has been proposed in which very thin (3.5 nm) GaInAsP layers are inserted into an InP core layer, creating a low coupling loss of 0.2 dB [89]. Figure 19a shows the structure of a coherent optical receiver integrating a diluted MQW waveguide directional coupler and a balanced pair of photodiodes [90]. The coupling between the waveguide and the photodiode is provided through the

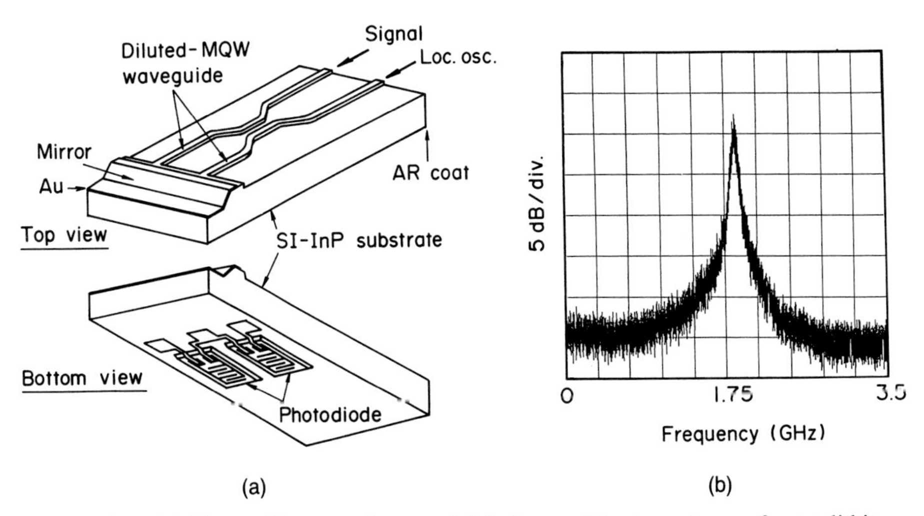

Fig. 19. (a) Top and bottom views and (b) observed beat spectrum of monolithic coherent receiver. The output lights from the directional coupler are coupled into balanced planar PIN photodiodes through the InP substrate. Heterodyne operation has been shown at 1.8 GHz.

transparent InP substrate by using a 45° mirror. This structure has demonstrated heterodyne detection with a sufficient (>20 dB) signal-to-noise ratio at 1.8 GHz as shown in Fig. 19b [91].

There is no doubt that the merging of such waveguide-based photonic integration with OEICs will generate very useful components. One example is coherent receivers involving electronic amplifiers. Another interesting area encompasses optical interconnections within the chip which can, for example, eliminate the skew problem in clock signal distribution [6c]. The fabrication processes are expected to develop to enable such integration, and the reduction of the propagation loss in non-lattice-matched waveguides (1 dB/cm for GaAs guides on InP) is encouraging [92].

4.2.2. Optical Functional Devices with Integrated Structures

In the future, optoelectronic devices and circuits are expected to find applications in optical interconnections, photonic switching, and optical computing systems. Toward these goals, high parellelism of data transmission and processing and digital optical logic functions are considered to be the most important research areas. Functions can be enhanced by the use of integrated optoelectronic device structures, as mentioned previously. There are two important directions in this view. Two-dimensional integration of optoelectronic devices, including surface-emitting lasers [92a], is an extremely important technique for realizing highly parallel data transmission and processing. In addition, various novel optical logic functions can be generated by integrating device structures so that the interaction between electron and photon systems is controlled within their structures.

Among many desirable optical logic functions, such as switching and memory, optical bistability is basically required and investigated most with a variety of devices proposed so far [5, 93]. Self-electro-optic effect devices (SEEDs) proposed by Miller *et al.* [94] at AT&T have a PIN photodiode structure with an AlGaAs/GaAs MQW photoabsorption layer. Bias voltage–dependent exciton absorption in MQW, the quantum confined Stark effect, and an external feedback circuit generate bistability in optical transmission. Tight wavelength control is necessary and, even though no optical gain is expected, the structure is suitable for two-dimensional integration. Lentine *et al.* [94a] have demonstrated a two-dimensional array of 64×32 symmetric SEEDs (S-SEEDs) with an optical switching energy of 0.8–2.5 pJ and an operation speed as fast as 1 ns.

Other types of bistable device structures include PNN and PNPN diodes. In double-heterostructure optoelectronic switches (DOES)[95], vertical to surface transmission electrophotonic (VSTEP) devices [96], and integrated

LED/phototransistor(HPT) devices [97, 98], the phototransistor action and external feedback are a mechanism for generating optical bistability. Built-in light-emitting structures can provide optical gain, which offers an important advantage of cascadability and large fan-out for optical processor applications. Kasahara *et al.* [99] at NEC have reported VSTEP devices, which operate as dynamic memory with a turn-off time of 2.5 ns and a low holding power (20 μW) [96], and a two-dimensional array with 1000 devices. They have just introduced a vertical cavity laser [92a] into this structure to improve light power and directionality [100]. Two-dimensional arrays of functional devices were reported by several groups. Matsuda *et al.* [101] at Matsushita fabricated a photonic parallel memory (PPM), integrating 1000 GaInAsP/InP LED/HPT devices on an InP substrate. They have shown a turn-on energy of 1.5 pJ and demonstrated image storage using this array. Figure 20 shows the structure of PPM, which includes reset HPTs for optical erase. A 10 × 10 array with a minimum reset power of 1.6 μW has been demonstrated. The integration level of two-dimensional arrays so far fabricated is at least three orders of magnitude lower than in electronic ICs such as DRAM chips. The switching power reported is only slightly worse than that of conventional transistors. Future progress is expected, pointing out what advantage of optical devices should be targeted by these devices.

Optical bistability can also be provided by a tandem laser structure in which a low-biased section acts as the saturable absorber. Since Lasher's proposal [102], many experimental studies have tried to attain the advantages of both optical gain and a response speed as fast as that of conventional lasers [103]. A 1-Gbps optical-set, electrical-reset memory utilizing a GaInAsP/InP bistable laser has been fabricated by optimizing the size of

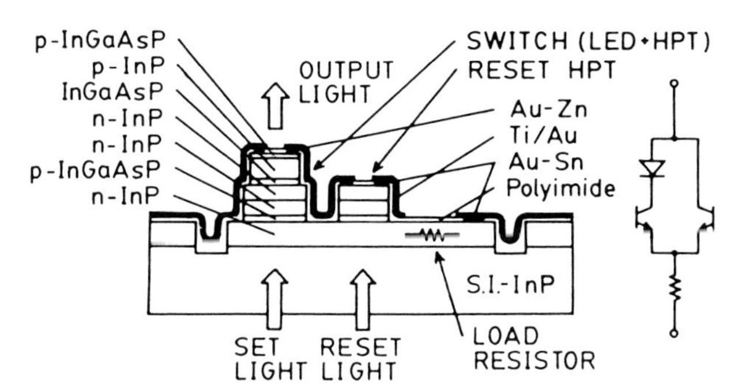

Fig. 20. Cross section and equivalent circuit of unit cell integrated in optically erasable photonic parallel momory. Each cell consists of an LED/HPT bistable switch and a reset HPT. (From Matsuda *et al.*, Ref. 101.)

two laser sections so that the carrier density fluctuation is minimized [104]. An optical reset operation has also been demonstrated by using beat vibration between the laser and injected light [104a] and a gain quenching effect [105]. Gain quenching occurs when light with a wavelength longer than that of lasing is injected into and amplified within the device. The switching energy needed for a bistable laser is in a range of 10 to 100 fJ, which is no greater than those of other optical bistable devices and transistors [5a]. The major drawback of bistable lasers is rather large dc power consumption, but reducing the laser threshold current is expected to solve this in the future. Two-dimensional integration of these devices can be achieved by integrating 45° mirrors or second-order corrugations for vertical emission.

One important advantage of using optical devices in switching and computing is the large data capacity which can be provided by using the WDM technique. Wavelength conversion and filtering devices are required for such systems. Figure 21a shows a GaInAsP/InP wavelength conversion laser fabricated at Fujitsu by integrating a bistable laser and a distributed Bragg re-

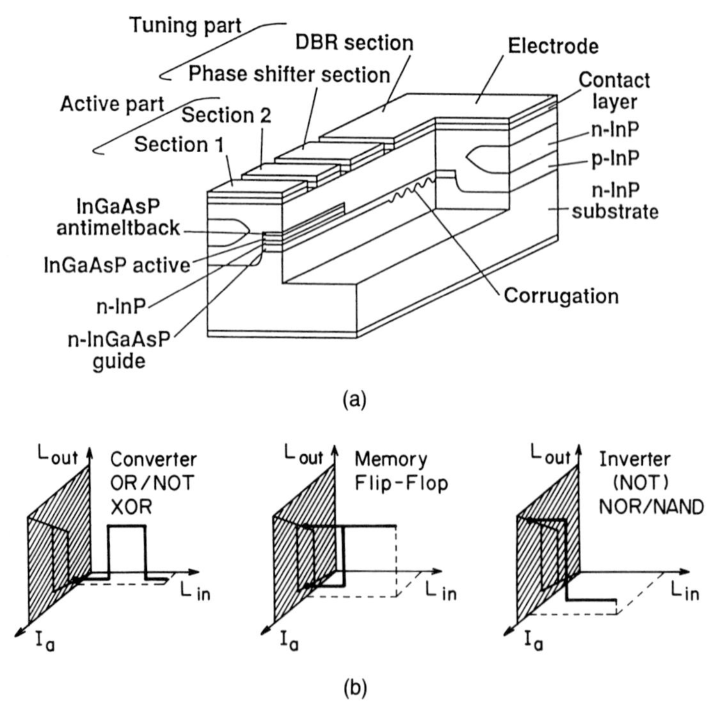

Fig. 21. (a) Structure and (b) various functions achieved by integrated wavelength conversion laser. A variety of optical functions can be achieved by choosing the active part current with respect to the light–current hysteresis loop.

flector (DBR) wavelength tuner [106]. In addition to wavelength tunability over 4.5 nm, this device has a variety of optical logic functions based on the saturable absorber and gain quenching [107]. By adjusting the active laser current (I) with respect to the light–current hysteresis loop, the operation mode can be selected, as indicated in Fig. 21b. Optical logic operation at 1 Gbps has been demonstrated [108].

Other optical functions so far demonstrated include lasers with beam steering functions which use either the current-induced refractive index change in the DBR[109] or the beam-waist location control in twin-gain-guided lasers [110]. Whether these novel devices can be practically implemented in the future will be determined not only by the device performance but also by their ability to be incorporated into system architecture and packaging technology.

4.3. Demonstration of Manufacturability

Much of the previous work on fabricating integrated structures suffered from reproducibility and yield problems. MSM photodiode-based GaAs receiver OEICs are an exceptional example; integration scale exceeding 8000 devices has been achieved due to their planar, compatible structure. Receiver OEIC functional yield comparable to GaAs IC yield has been demonstrated. Taking into account that GaAs LSIs such as 30-kbit gate array chips have already been developed, large-scale OEICs can be fabricated as far as compatibility with standard process is established. The yield problem is more severe in OEIC transmitters due to lack of process-compatibility between the laser and electronic devices in most of previous OEICs. As for the fabrication of InP-based OEICs, even the receiver structure is complex, including substrate etch and refill growth, resulting in less uniformity and reproducibility. According to Lee *et al.* [111] of BNR Europe, fabrication of an eight-channel PIN/JFET receiver array on 2-inch InP wafers has shown an yield of fully functional arrays better than 34%. This result is encouraging for this complexity and element count (104 elements), but further improvement in structure and processing is required for low-cost production. Two-dimensional arrays of integrated devices such as SEEDs and VSTEP devices have shown an integration scale exceeding 1000 devices in one chip, although no explicit yield data are available. This supports the assertion that simplicity and stability of the fabrication process are crucial for improving yield as well as uniformity. Uniformity of the substrate and epitaxial layer has nearly been masked by process reproducibility and uniformity in previous OEIC fabrication, but it should be investigated in more detail on increasing the integration scale.

Once the manufacturing process is established for an integrated chip, reliability of the whole chip is expected to increase. However, no clear demonstration of this has been made on any integrated optoelectronic circuits or devices. An early result on aging of GaAs-based OEIC transmitters showed a nonnegligible threshold current degradation, indicating that a stable fabrication technique needed to be developed. The recent result by IBM mentioned earlier indicates great progress over the last 5 years. Integration using lasers is most sensitive to the quality of substrates and epitaxial structures, and high quality must be achieved in laser structures after necessary processing steps including regrowth and facet etching and coating. The defect density in a substrate required for discrete laser fabrication is of the order of 10^4 cm^{-2}, but a density of 10^2 cm^{-2} or less is definitely required for two-dimensional integration at the 1k device level. Low-defect substrates as well as growth and processing techniques that do not induce defects are a challenge for the future.

5. CONCLUDING REMARKS

The development of discrete optoelectronic devices has been pursued to find solutions to specific applications such as high-performance optical communications systems. In contrast, integrated optoelectronics can produce a variety of advantages and applications which have previously been difficult. For their realization, development is needed to cover and unite many different technologies, in both circuit fabrication and system application. Since their initiation, OEICs have been primarily investigated for telecommunication applications. Laboratory prototypes of basic transmitters and receivers on both GaAs- and InP-based material systems have shown performance comparable to or, in some cases, even better than that of commercial hybrid products available today. Manufacturable fabrication technologies are now strongly required so that they can be used in practical systems.

Generation of new optical functions using waveguide-based integration as well as integrated optoelectronic device structures opens future avenues for lightwave components toward, not only advanced telecommunication systems such as WDM and coherent systems but also better optical interconnections, photonic switching systems, and optical computers. Research in those areas has started only recently but is advancing rapidly, as indicated by many encouraging circuit demonstrations including waveguide integrated circuits, optical logic devices, and their two-dimensional integration. In those areas, the requirements for devices are strongly linked with the architectures and design specifications of overall systems, most parts of which are constructed with electronics. Therefore, it is important to plan where and how these integrated optoelectronics will be used in a target system, so that the most effective application of optoelectronics and electronics can be determined.

In order to drive research efforts toward system-scale implementation of integrated optoelectronics, an emphasis should still be placed on the development of the fabrication technology. In previous work on integrated optoelectronics, device uniformity has been a general problem, resulting in low yield and restriction for arraying. Development of planar, compatible structures and fabrication techniques is still a key. The reliability of integrated optoelectronic circuits and devices has not yet been sufficient for practical application. In particular, even conventional discrete lasers have not shown reliability equivalent to that of electronic devices and ICs. Realization of high-reliability integrated lasers is indispensable for their practical application. A broader application area is opened by mixed material integration using non-lattice-matched heteroepitaxy, where more research targeting high reliability is needed.

Integrated optoelectronic circuits must be designed properly for their target applications. In previous work, however, it was not possible to use sufficient design tools, including simulation techniques. CAD techniques are less developed for optoelectronic devices than for electronic devices and ICs. In OEIC transmitters and receivers, an approach mixing analog and digital electronic functions is very important and a new design technique is required for this purpose; this will be discussed in the next chapter. No design technique has been developed for complicated devices such as integrated optoelectronic functional devices. More work should be devoted to developing device and circuit models enabling the design of the whole optoelectronic system.

System implementation of integrated optoelectronics needs a practical optoelectronic packaging technique. Techniques so far used have been basically an extension of technology existing in the field of telecommunication link modules. Optoelectronic packaging more suitable for high-density integration is seriously required, because manufacturability, reliability, and cost issues depend critically on this step of manufacturing process. This subject will be reviewed in Chapter 16. The range of technology to be covered is thus extremely wide, but ongoing research shows encouraging progress in many fields, as discussed in the following chapters, and great contributions are expected to generate novel, useful optical components.

REFERENCES

1. A. Yariv, *IEEE Trans. Electron Devices* **ED-31**, 1656 (1984).
2. I. Hayashi, *Tech. Dig. Int. Conf. Integrated Optics and Optical Fiber Commun. (IOOC'83)*, Tokyo, p. 170 (1983).
3. O. Wada, T. Sakurai, and T. Nakagami, *IEEE J. Quantum Electron.* **QE-22**, 805 (1986).
4. R. F. Leheny, *IEEE Circuits Devices Mag.* **5**, 38 (1989).
5. O. Wada, and S. Yamakoshi, *SPIE Proc.* **1215**, 28 (1990).

5a. O. Wada, *SPIE Proc.* **1462**, to be published.

5b. T. Tamir, "Integrated Optics." Springer-Verlag, Berlin, 1975 (last edition: 1982).

6. O. Wada, *Int. J. High Speed Electron.* **1**, 47 (1990).

6a. S. Margarit and A. Yariv, in "Semiconductors and Semimetals," Vol. 22(W. T. Tsang, ed.), Part E, Ch. 2, Academic Press, Orlando, FL, 1985.

6b. Nikkei Mechanical, September 16, 1991 issue, no. 358, p. 66.

6c. J. W. Goodman, F. I. Leonberger, S. Y. Kung, and R. A. Athale, *Proc. IEEE* **72**, 850 (1984).

7. C. P. Lee, S. Margalit, I. Ury, and A. Yariv, *Appl. Phys. Lett.* **32**, 806 (1978).

7a. M. Yust, N. Bar-Chaim, S. H. Izadpanah, S. Margalit, I. Ury, D. Wilt, and A. Yariv, *Appl. Phys. Lett.* **35**, 795 (1979).

8. O. Wada, *Opt. Quantum Electron.* **20**, 441 (1988).

8a. O. Wada, in "Properties of Gallium Arsenide," 2nd ed., EMIS Datareviews Ser. No. 2, p. 732. INSPEC, Stevenage, UK, 1990.

9. A. Suzuki, K. Kasahara, and M. Shikada, *IEEE J. Lightwave Technol.* **10**, 1479 (1987).

9a. J. Shibata and T. Kajiwara, *Opt. Quantum Electron.* **20**, 363 (1988).

9b. M. Dagenais, R. F. Leheny, H. Temkin, and P. Bhattacharya, *IEEE J. Lightwave Technol.* **8**, 846 (1990).

9c. M. Hirano, *Optoelectron Devices Technol.* **2**, 137 (1987).

9d. S. R. Forrest, *IEEE Trans. Electron Devices* **ED-32**, 2640 (1985).

10. A. Suzuki, T. Itoh, T. Terakado, K. Kasahara, K. Asano, Y. Inomoto, H. Ishihara, T. Torikai, and S. Fujita, *Electron. Lett.* **23**, 954 (1987).
11. M. Fallahi, F. Therez, D. Esteve, D. Kendil, M. Barbe, and G. Cohen-Solal, *Electron. Lett.* **24**, 1245 (1988).
12. H. K. Choi, G. W. Turner, T. H. Windhorn, and B. Y. Tsaur, *IEEE Electron Device Lett.* **EDL-7**, 500 (1986).

13. M. Ito, O. Wada, K. Nakai, and T. Sakurai, *IEEE Electron Device Lett.* **EDL-5**, 531 (1984).
14. T. Sugeta, T. Urisu, S. Sakata, and Y. Mizushima, *Jpn. J. Appl. Phys.* **19**, (Suppl. 19–1) 459 (1980).
15. T. Fujii, S. Hiyamizu, S. Yamakoshi, and T. Ishikawa, *J. Vac. Sci. Technol.* **B3**, 776 (1985).
16. T. Sanada, S. Yamakoshi, O. Wada, T. Fujii, T. Sakurai, and M. Sasaki, *Appl. Phys. Lett.* **44**, 325 (1984).
17. F. Brillouet, A. Clei, A. Kampfer, S. Biblemont, R. Azoulay, and N. Duhamel, *Electron. Lett.* **22**, 1259 (1986).
18. C. S. Hong, D. Kasemset, M. E. Kim, and R. A. Molano, *Electron. Lett.* **20**, 733 (1984).
19. H. Nakano, S. Yamashita, T. P. Tanaka, M. Hirao, and M. Maeda, *IEEE J. Lightwave Technol.* **LT-4**, 574 (1986).
20. H. Blauvelt, N. Bar-Chaim, D. Fekete, S. Margalit, and A. Yariv, *Appl. Phys. Lett.* **40**, 289 (1982).
21. O. Wada, S. Yamakoshi, T. Fujii, S. Hiyamizu, and T. Sakurai, *Electron. Lett.* **18**, 189 (1982).
22. J. K. Kerney, M. J. Helix, R. M. Kolbas, S. A. Jamison, and S. Ray, *Tech. Dig. GaAs IC Symp.* New Orleans (1982), p. 38.
23. K. Asakawa, and S. Sugata, *Jpn. J. Appl. Phys.* **22**, L653 (1983).
24. T. Yuasa, M. Mannoh, K. Asakawa, K. Shinozaki, and M. Ishii, *Appl. Phys. Lett.* **48**, 748 (1986).
25. M. Kuno, T. Sanada, H. Nobuhara, M. Makiuchi, T. Fujii, O. Wada, and T. Sakurai, *Appl. Phys. Lett.* **49**, 1575 (1986).
26. O. Wada, H. Nobuhara, T. Sanada, M. Kuno, M. Makiuchi, T. Fujii, and T. Sakurai, *IEEE J. Lightwave Technol.* **LT-7**, 186 (1989).
27. H. Matsueda, M. Hirao, T. P. Tanaka, H. Kodera, and M. Nakamura, *Tech. Dig. Int. Symp. GaAs and Related Compounds*, Karuizawa, Japan (1985), Inst. Phys. Conf. Ser. No. 79, p. 655.
28. K. Hamada, N. Yoshikawa, H. Shimizu, T. Otsuki, A. Shimano, K. Itoh, G. Kano, and I. Teramoto, *Extended Abstracts, 18th Conf. Solid State Devices and Materials*, Tokyo (1986), p. 181.
29. O. Wada, H. Hamaguchi, S. Mirua, M. Makiuchi, K. Nakai, T. Horimatsu, and T. Sakurai, *Electron. Lett.* **19**, 1031 (1983).
30. O. Wada, H. Hamaguchi, S. Miura, M. Makiuchi, K. Nakai, H. Horimatsu, and T. Sakurai, *Appl. Phys. Lett.* **46**, 981 (1985).
31. R. M. Kolbas, J. Abrokwah, J. K. Carney, D. H. Bradshaw, B. R. Elmer, and J. R. Biard, *Appl. Phys. Lett.* **43**, 821 (1983).
31a. S. Miura, O. Wada, M. Makiuchi, and K. Nakai, *Appl. Phys. Lett.* **48**, 1461 (1986).
32. O. Wada, H. Hamaguchi, M. Makiuchi, T. Kumai, M. Ito, K. Nakai, T. Horimatsu, and T. Sakurai, *IEEE J. Lightwave Technol.* **LT-4**, 1694 (1986).

33. M. Makiuchi, H. Hamaguchi, T. Kumai, M. Ito, O. Wada, and T. Sakurai, *IEEE Electron Device Lett.* **EDL-6**, 634 (1985).
34. K. D. Pedrotti, S. Beccue, W. J. Haber, B. P. Brar, G. Robinson, and M. K. Kilcoyne, *IEEE J. Lightwave Technol.* **8**, 1334 (1990).
35. N. Yamanaka, S. Kikuchi, and T. Takada, *Tech. Dig. European Conf. Optical Commun. (ECOC)*, Gothenberg, Sweden (1989), Vol. 1, p. 280.
36. J. D. Crow, *Tech. Dig. Conf. Optical Fiber Commun. (OFC'89)*, p. 83 and *Tech. Dig. Int. Conf. Integrated Optics and Optical Commun. (IOOC'89)*, Kobe, Japan (1989), vol. 4, p. 86.
37. B. J. V. Zeghbroeck, W. Patrick, J.-M. Halbout, and P. Vettiger, *IEEE Electron Device Lett.* **9**, 527 (1988).
38. C. S. Harder, B. V. Zeghbroeck, H. Meier, W. Patrick, and P. Vettiger, *IEEE Electron Device Lett.* **EDL-9**, 171 (1988).
38a. C. Takano, K. Tanaka, A. Okubora, and J. Kasahara, *Tech. Dig. IEEE GaAs IC Symp.* (1991), p. 209.
39. N. Matsuo, H. Ohno, and H. Hasegawa, *Jpn. J. Appl. Phys.* **23**, L648 (1984).
40. C. Y. Chen, N. A. Olsson, C. W. Tu, and P. A. Garbinski, *Appl. Phys. Lett.* **46**, 681 (1985).
41. R. F. Lehney, R. E. Nahory, M. A. Pollack, A. A. Ballman, E. D. Beebe, J. C. Dewinter, and R. J. Martin, *Electron. Lett.* **16**, 353 (1980).
41a. J. Shibata, I. Nakano, Y. Sasai, S. Kimura, N. Hase, and H. Serizawa, *Appl. Phys. Lett.* **45**, 191 (1984).
42. N. Suzuki, H. Furuyama, Y. Hirayama, M. Morinaga, K. Eguchi, M. Fushibe, M. Funamizu, and M. Nakamura, *Electron. Lett.* **24**, 467 (1988).
43. Y. H. Lo, P. Grabbe, M. Z. Iqbal, R. Bhat, J. L. Gimlett, J. C. Young, P. S. D. Lin, A. S. Gozdz, M. A. Koza, and T. P. Lee, *IEEE Photonics Technol. Lett.* **2**, 673 (1990).
44. S. J. Kim, G. Guth, G. P. Vella-Coleiro, C. W. Seabury, W. A. Sponsler, and B. J. Rhodes, *IEEE Electron Device Lett.* **9**, 447 (1988).
45. A. Antreasyan, P. A. Garbinski, V. D. Mattera, Jr., H. Temkin, N. A. Olsson, and J. Filipe, *IEEE Photon. Technol. Lett.* **1**, 123 (1989).
46. D. A. H. Spear, P. J. G. Dawe, G. R. Antell, W. S. Lee, and S. W. Bland, *Electron. Lett.* **25**, 156 (1989).
46a. W. S. Lee, D. A. H. Spear, M. J. Agnew, and S. W. Bland, *Tech. Dig. Conf. Optical Fiber Commun. (OFC'90)*, San Francisco (1990), PD-28.
47. W. S. Lee, D. A. H. Spear, P. J. G., Dawe, and S. W. Bland, *Tech. Dig. European Conf. Optical Commun. (ECOC'90)*, Brighton, U.K. (1990), p. 465.
48. R. N. L. Nguyen, M. Allovon, P. Blanconnier, S. Vuye, and A. Scavennec, *IEEE J. Lightwave Technol.* **6**, 1507 (1988).
49. S. Chandrasekhar, B. C. Johnson, M. Bonnemason, E. Tokumitsu, A. H. Gnauck, A. G. Dentai, C. H. Joyner, J. S. Perino, G. J. Qua, and E. M. Monberg, *IEEE Photon. Technol. Lett.* **2**, 505 (1990).

50. H. NOBUHARA, H. HAMAGUCHI, T. FUJII, O. AOKI, M. MAKIUCHI, AND O. WADA, *Electron. Lett.* **19**, 1246 (1988).
51. H. YANO, K. AGA, H. KAMEI, G. SASAKI, AND H. HAYASHI, *IEEE J. Lightwave Technol.* **8**, 1328 (1990).
52. S. CHANDRASEKHAR, A. G. DENTAI, C. H. JOYER, B. C. JOHNSON, A. H. GNAUCK, AND G. J. QUA, *Electron. Lett.* **26**, 1880 (1990).
52a. S. CHANDRASEKHAR, *Tech. Dig. Int. Conf. Indium Phosphide and Related Materials*, Newport, RI (April 1992), p. 67.
52b. Y. AKAHORI, Y. AKATSU, A. KOHZEN, AND J. YOSHIDA, *IEEE Photon. Technol. Lett.* **4**, 754 (1992).
53. O. WADA, H. NOBUHARA, H. HAMAGUCHI, T. MIKAWA, A. TACKEUCHI, AND T. FUJII, *Appl. Phys. Lett.* **54**, 16 (1989).
54. H. SCHUMACHER, H. LEBLANC, J. SOOLE, AND R. BHAT, *IEEE Electron Device Lett.* **36**, 659 (1989).
55. T. KIKUCHI, H. OHNO, AND H. HASEGAWA, *Electron. Lett.* **24**, 1209 (1988).
56. W.-P. HONG, G.-K. CHANG, R. BHAT, J. L. GIMLETT, C. K. NGUYEN, G. SASAKI, AND M. KOZA, *Electron. Lett.* **25**, 1562 (1989).
56a. L. T. CANHAM, *Appl. Phys. Lett.* **57**, 1046 (1990).
56b. T. MIYASHITA, M. KAWACHI, AND M. KOBAYASHI, *Proc. Optical Fiber Commun. Conf.*, vol. 4 (1988), p. 173.
56c. D. H. HARTMAN, M. K. GRACE, AND C. R. RYAN, *IEEE J. Lightwave Technol.* **LT-3**, 729 (1985).
56d. J.-P. COLINGE, "Silicon-on-Insulator Technology, Materials to LSIs." Kluwer Academic Publishers, Boston, 1991, p.162.
56e. M. A. GUERRA, *Solid State Technol.* (November 1990), p. 75.
56f. H. NAGAO AND M. YAMAMOTO, *IEICE Tech. Rep.* **89**, (288), CPM89–70, 7 (1989). (In Japanese)
56g. SHARP ELECTRONIC COMPONENTS CATALOG, OPIC (August 1991).
56h. S. MIYAOKA, presented at IEEE/LEOS Summer Topical Meeting on Integrated Optoelectronics (August 1992), Santa Barbara, CA.
57. S. SAKAI, T. SOGA, M. TAKEYASU, AND M. UMENO, *Appl. Phys. Lett.* **48**, 413 (1986).
58. D. G. DEPPE, N. HOLONYAK, JR., D. W. NAM, K. C. HSIEH, AND G. S. JACKSON, *Appl. Phys. Lett.* **51**, 637 (1987).
59. Y. H. LO, C. CANEAU, R. BHAT, L. T. FLOREZ, G. K. CHANG, J. P. HARBISON, AND T. P. LEE, *Electron. Lett.* **25**, 666 (1989).
60. R. S. SUSSMANN, R. M. ASH, A. J. MOSELEY, AND R. C. GOODFELLOW, *Electron Lett.* **21**, 593 (1985).
61. H. HAMAGUCHI, M. MAKIUCHI, T. KUMAI, O. AOKI, AND O. WADA, *Tech. Dig. Optoelectronics Conf. (OEC'89)*, Tokyo (1989), p. 194.
61a. H. HAMANO, T. YAMAMOTO, Y. NISHIZAWA, Y. OIKAWA, H. KUWATSUKA, A. TAHARA, K. SUZUKI, AND A. NISHIMURA, *Electron. Lett.* **27**, 1602 (1991).
62. I. POLENTIER, L. BUYDENS, A. ACKAERT, P. DEMEESTER, P. VAN DAELE, F. DEPESTEL, D. LOOTENS, AND R. BAETS, *Electron. Lett.* **26**, 925 (1990).

63. A. Yi-Yan, W. K. Chan, T. J. Gmitter, L. T. Florez, J. L. Jackel, E. Yablonovitch, R. Bhat, and J. P. Harbison, *IEEE Photon. Technol.* **1**, 379 (1989).
64. T. Horimatsu, T. Iwama, Y. Oikawa, T. Touge, M. Makiuchi, O. Wada, and T. Nakagami, *IEEE J. Lightwave Technol.* **LT-4**, 680 (1986).
65. T. Suzuki, S. Fujita, Y. Inomoto, T. Terakado, K. Kasahara, K. Asano, T. Torikai, T. Itoh, M. Shikada, and A. Suzuki, *Electron. Lett.* **24**, 1283 (1988).
66. T. Iwama, T. Horimatsu, Y. Oikawa, K. Yamaguchi, M. Sasaki, T. Touge, M. Makiuchi, H. Hamaguchi, and O. Wada, *IEEE J. Lightwave Technol.* **LT-6**, 772 (1988).
67. M. J. Wale, C. Edge, F. A. Randle, and D. J. Pedder, *Tech. Dig. European Conf. Optical Commun. (ECOC'89)*, Gothenburg, Sweden (1989), vol. 2, p. 368, ThA19–7.
68. K. P. Jackson, E. B. Flint, M. F. Cina, D. Lacey, J. M. Trewhella, T. Caulfield, and S. Sibley, *Tech. Dig. EIA/IEEE Electronic Components & Technology Conf.* (May 1992), San Diego, p. 93.
69. A. Furuya, M. Makiuchi, O. Wada, and T. Fujii, and H. Nobuhara, *Jpn. J. Appl. Phys.* **26**, L134 (1987).
70. O. Wada, A. Furuya, and M. Makiuchi, *IEEE Photon. Technol. Lett.* **1**, 16 (1989).
71. A. Furuya, M. Makiuchi, and O. Wada, *Electron. Lett.* **24**, 1282 (1988).
72. E. Kapon, S. Shimhony, J. P. Harbison, L. T. Florez, and P. Worland, *Appl. Phys. Lett.* **56**, 1825 (1990).
73. R. Nagarajan, T. Kamiya, A. Kasukawa, and H. Okanoto, *Appl. Phys. Lett.* **55**, 1273 (1989).
74. Y. Arakawa and A. Yariv, *IEEE J. Quantum Electron.* **QE-22**, 1887 (1986).
75. M. Sugawara, T. Fujii, S. Yamazaki, and K. Nakajima, *Appl. Phys. Lett.* **54**, 1353 (1989).
76. P. J. A. Thijs and T. Van Dongen, *Electron. Lett.* **25**, 1735 (1989).
76a. M. Sugo, H. Mori, and Y. Itoh, *Tech. Dig. Int. Conf. Indium Phosphide and Related Materials* (April 1992), Newport, RI, p. 642.
77. M. Koyanagi, H. Takata, H. Mori, and J. Iba, *IEEE J. Solid-State Circuits* **25**, 109 (1990).
78. H. Hashimoto and E. Miyauchi, *Extended Abstracts 16th Conf. Solid State Materials and Devices*, Kobe, Japan (1984), p. 121.
79. A. Takamori, S. Sugata, K. Asakawa, E. Miyauchi, and H. Hashimoto, *Jpn. J. Appl. Phys.* **26**, L142 (1987).
80. H. Temkin, L. R. Harriott, R. A. Hamm, J. Weiner, and M. B. Panish, *Appl. Phys. Lett.* **54**, 1463 (1989).
81. H. Soda, M. Furutsu, K. Sato, N. Okazaki, S. Yamazaki, H. Nishinoto, and H. Ishikawa, *Electron. Lett.* **26**, 9 (1990).
82. S. Sakano, H. Inoue, H. Nakamura, T. Katsuyama, and H. Matsumura, *Electron. Lett.* **22**, 594 (1986).

83. U. Koren, T. L. Koch, B. I. Miller, G. Eisenstein, and R. H. Bosworth, *Appl. Phys. Lett.* **54**, 2056 (1989).
84. H. Takeuchi, K. Kasaya, Y. Kondo, H. Yasaka, K. Oe, and Y. Imamura, *IEEE Photon. Technol. Lett.* **1**, 398 (1989).
85. T. L. Koch, U. Koren, R. P. Gnall, F. S. Choa, F. Hernandez-Gil, C. A. Burrus, M. G. Young, M. Oron, and G. I. Miller, *Electron. Lett.* **25**, 1621 (1989).
86. R. J. Deri and O. Wada, *Appl. Phys. Lett.* **55**, 2712 (1989).
87. T. L. Koch, P. J. Corvini, W. T. Tsang, U. Koren, and B. I. Miller, *Appl. Phys. Lett.* **51**, 1060 (1987).
88. R. J. Deri, N. Yasuoka, M. Makiuchi, O. Wada, A. Kuramata, and H. Hamaguchi, *Appl. Phys. Lett.* **56**, 1737 (1990).
89. R. J. Deri, N. Yasuoka, M. Makiuchi, A. Kuramata, and O. Wada, *Appl. Phys, Lett.* **55**, 1495 (1989).
90. R. J. Deri, T. Sanada, N. Yasuoka, M. Makiuchi, A. Kuramato, H. Hamaguchi, O. Wada, and S. Yamakoshi, *IEEE Photon. Technol. Lett.* **2**, 581 (1990).
91. N. Yasuoka, T. Sanada, M. Makiuchi, A. Kuramata, H. Hamaguchi, O. Wada, and R. J. Deri, *Tech. Dig. Int. Conf. InP and Related Mat.*, Cardiff, U.K. (1991) to be published.
92. R. J. Deri, R. Bhat, J. P. Harbison, M. Seto, A. Yi-Yan, L. T. Florez, M. Koza, and A. Y. Lo, *IEEE Photon. Technol. Lett.* **2**, (1890).
92a. K. Iga and F. Koyama, "Surface Emitting Diode Lasers and Arrays." Academic Press, New York, to be published.
93. N. Streibel, K.-H. Brenner, A. Huang, J. Jahns, J. Jewell, A. W. Lohmann, D. A. B. Miller, M. Murdocca, M. E. Prise, and T. Sizer, *Proc. IEEE* **77**, 1954 (1989).
94. D. B. A. Miller, D. S. Chemla, T. C. Damen, T. H. Wood, C. A. Burrus, A. C. Gossard, and W. Wiegmann, *IEEE J. Quantum Electron.* **QE-21**, 1462 (1985).
94a. A. L. Lentine, F. B. McCormick, R. A. Novotny, L. M. F. Chirovsky, L. A. D'Asaro, R. F. Kopf, J. M. Kuo, and G. D. Boyd, *IEEE Photon. Technol. Lett.* **2**, 51 (1990).
95. G. W. Taylor, R. S. Mand, J. G. Simmons, and A. Y. Cho, *Appl. Phys. Lett.* **49**, 1406 (1986).
96. K. Kasahara, Y. Tashiro, N. Hamano, M. Sugimoto, and T. Yanase, *Appl. Phys. Lett.* **52**, 679 (1988).
97. A. Sasaki, K. Matsuda, Y. Kimura, and S. Fujita, *IEEE Trans. Electron Devices* **ED-29**, 1382 (1982).
98. K. Matsuda, K. Takimoto, D. H. Lee, and J. Shibata, *IEEE Trans. Electron Devices* **37**, 1630 (1990).
99. K. Kasahara, Y. Tashiro, I. Ogura, M. Sugimoto, S. Kawai, and K. Kubota, *Tech. Dig. Int. Conf. Integrated Optics and Optical Fiber Commun. (IOOC'89)*, Kobe, Japan (1989), vol. 3, p. 158.

100. M. Sugimoto, T. Numai, I. Ogura, H. Kosaka, and K. Kasahara, *Tech. Dig. IEEE/LEOS Annual Meeting*, Boston (1990), PD8.
101. K. Matsuda, H. Adachi, T. Chino, and J. Shibata, *IEEE Electron Device Lett.* **11**, 442 (1990).
102. G. J. Lasher, Solid State Electron. *7*, 707 (1964).
103. H. Kawaguchi, *Appl. Phys. Lett.* **45**, 1264 (1984).
104. T. Sanada, T. Odagawa, T. Machida, K. Wakao, and S. Yamakoshi, *Extended Abstracts Conf. Solid State Devices and Materials*, Tokyo (1989), p. 321.
104a. K. Inoue and K. Oe, *Electron. Lett.* **24**, 512 (1988).
105. T. Odagawa, T. Sanada, and S. Yamakoshi, *Extended Abstracts Conf. Solid State Devices and Materials*, Tokyo (1988), p. 331.
106. S. Yamakoshi, K. Kondo, M. Kuno, K. Kotaki, and H. Imai, *Tech. Dig. Conf. Optical Fiber Commun. (OFC'88)*, New Orleans (1988), PD10.
107. H. Nobuhara, K. Kondo, S. Yamakoshi, and K, Wakao, *Electron. Lett.* **25**, 1485 (1989).
108. K. Kondo, H. Nobuhara, S. Yamakoshi, and K. Wakao, *Tech. Dig. Int. Topical Meeting Photonic Switching*, Kobe, Japan (1990), 13D-9.
109. K. Yamashita, H. Nagata, Y. Kubota, and K. Tone, *Tech. Dig. Int. Topical Meeting Photonic Switching*, Kobe, Japan (1990), p. 125.
110. S. Mukai, M. Watanabe, H. Itoh, H. Yajima, Y. Hosoi, and S. Uekusa, *Opt. Quantum Electron.* **17**, 431 (1985).
111. W. S. Lee, D. A. H. Spear, A. D. Smith, S. A. Wheeler, and S. W. Bland, *Electron. Lett.* **28**, 612 (1992).

Chapter 13

SCALING AND SYSTEM ISSUES OF OEIC DESIGN

R. Bates, John Crow, J. Ewen, and D. Rogers

IBM, T.J. Watson Research Center
Yorktown Heights, New York 10598

1. INTRODUCTION

The previous chapter has reviewed the status of OEIC integration today, summarizing chip performance and identifying some remaining problems. This status has also been reviewed by Dagenais *et al.* [1]. Much of this early work is characterized by the combination of optimized designs for discrete devices to demonstrate the feasibility of integration. A general accomplishment of OEIC development to date is that integrated functions are approaching—and even comparable to, in the case of receivers—the performance of discrete device chip hybrid modules.

ISBN 0-12-200420-5

Much larger levels of integration have been demonstrated in the GaAs material system [2, 3], in which many thousands of electronic devices have been integrated along with optical components. The advantage of larger scales of integration is the ability to add functions at the chip level which can make the overall optical subsystem easier to design and use, while simultaneously maintaining the card density required in modern electronic systems. If the technology is sufficiently mature, the integration also offers higher reliability and lower cost at the subsystem level. Even with a medium scale of integration, all the analog functions associated with an optoelectronic interface can be combined on a single chip, as in the low-profile read/write head or the integrated receiver chips described in the previous chapter. Also, subsystem fault isolation and failure recovery can be handled at the chip level, making the module more robust and consequently easier to use. Products using automatic optical subsystem shutdown techniques are available in audio disks (Philips blanking circuits) and data links. (Finisar Gb/s modules). The large-scale OEIC chip can offer complete light-to-logic function (even decoding into data words), so that the digital system designer can treat optical and analog operations as "black box" functions. In addition, arrays of these functions on a single chip can be offered to greatly reduce the card size occupied by the O–E interface.

With large levels of integration come a number of new design issues which must be addressed. For the optoelectronic components, the major issues are uniformity of device characteristics, yield, and reliability (especially a concern for OEIC transmitters). Noise and crosstalk become more important as the level of integration increases. For OEICs, electrical, optical, and thermal crosstalk must be controlled to ensure reliable link operation at the system level. These concerns are only magnified with the addition of large amounts of analog and digital electronics. Crosstalk between the various analog, digital, and optoelectronic functional blocks can seriously degrade the overall system performance. The OEIC complexity at this point stresses the ability of both the tools and the models to predict these crosstalk effects.

The following sections address these issues from the perspective of increasing levels of integration within the OEIC. First, the questions of reliability, noise, and crosstalk in OEIC transmitters are examined. Next, the design issues of integrated optical receivers and receiver arrays are addressed. The challenges of combining these tranmitter and receiver elements along with LSI levels of electronics are examined.

2. MEDIUM-SCALE OEIC TRANSMITTERS

The advantages cited for OEIC transmitters include improvements in size, cost, ruggedness, and overall reliability [4–6]. Nakamura *et al.* [7] also claim speed as a major advantage, showing the serious eye degradation that results at 6.3 Gbit/s with a 0.3-nH, 0.5-mm-length wirebond between the laser and drive electronics; however, results at 10 Gbit/s with an improved hybrid approach [8] suggest that the implication that OEIC transmitters are unavoidable for very high-speed single-channel applications is, as yet, premature.

However, there are optical subsystem operation issues in laser transmitters in data link applications which might serve to drive the development of an integrated transmitter technology. These include (1) enhanced laser reliability and recovery from device failure, (2) enhanced laser link robustness against noise and reflections, (3) laser safety, and (4) arrays of efficient laser O–E converters with support circuits for optical buses. An integrated transmitter module, ultimately with an OEIC, could offer these additional link performance functions while maintaining low cost and compact size. As no OEICs have been developed which demonstrate these features, the following sections will discuss the system issues motivating such an OEIC. Topics include laser yield, reliability, coherence, and crosstalk.

2.1. High-Availability Transmitters

The present proven reliability of semiconductor lasers, at either ~0.8 or ~1.3 μm, appears inadequate to satisfy the demanding requirements of many data communication applications [9], despite the encouraging improvements being reported, [10, 11]. One well-known way to circumvent this type of component problem is to improve the overall system availability by duplicating the critical parts; for example, this technique has been applied to discrete laser diodes used in various undersea fiber-optic digital transmission systems [12] to meet the stringent performance required in such environments; specifically, the duplicated components are operated in a "cold standby" mode and powered on individually in turn as required.

With discrete lasers, the optical signals may be coupled to the main fibers using a number of techniques, e.g., mechanical fiber switches, polarizing beamsplitters, or polarization-maintaining fibers and couplers. With a monolithic array of redundant lasers, e.g., *n* sets of *m* devices, the *m* devices in each set can be spaced close enough to couple directly into a multimode fiber; for example, with 12-μm spacing between a set of three devices, the variation in power butt-coupled into a flat-cleaved 50/125 μm fiber is less than 3 dB.

The improvement in availability from cold standby laser redundancy has been analyzed for both discrete devices [13, 14] and monolithic arrays [15]. In general, there will be both some correlation between the intrinsic failure probability density functions (pdf) of the individual devices and some interaction between them while operating, i.e., some wear-out during the cold standby period. Figure 1 shows an example of the computed variation in median lifetime required to achieve an effective average failure rate (AFR) of 50 FIT (failures in 10^9 hours) with two devices, assuming the individual failure pdfs are lognormal with a standard deviation of 1.

The minimum lifetime requirement arises when there is neither correlation between the lifetimes of the devices nor interaction during operation; the discrete device implementation approaches this case. At the other extreme, with complete correlation and interaction, the two devices behave as one and, for this example, necessitate a factor of 4 improvement in lifetime to achieve the same FIT rate. In a monolithic array, one expects correlation coefficients between devices in a set to approach unity, due to their close proximity but, hopefully, low interaction coefficients; for sets of two devices, this amounts to approximately a factor of 2 reduction in required lifetime compared to a single device with no redundancy.

This type of analysis can be extended to investigate the effectiveness of various maintenance strategies when the redundant components form a field-replaceable unit, For example, in a two-device array, the system would be designed to turn on the second device immediately after the failure of the

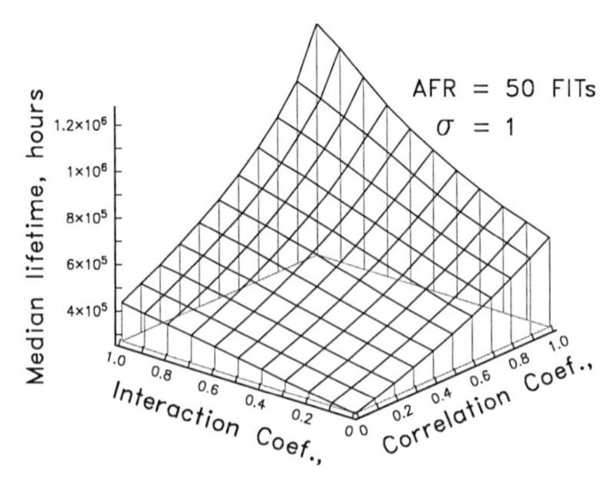

Fig. 1. Example of lifetime requirements for a general two-device array.

first device to ensure uninterrupted availability; however, the maintenance plan may require that at a convenient time within some later period, the whole unit is replaced. Figure 2 shows an example of the computed variation in available time Δt to schedule maintenance and replacement of the partially failed unit after 1 and 10 years of system life, expressed in thousands of hours (accepting a 1% probability that the second device may already have failed before the first). The median intrinsic lifetime and standard deviation of the individual devices are 100 kh and 1, respectively. The dashed part of the surface lies below $\Delta t = 0$; the solid part is positive and represents useful available time.

Evidently, for such a maintenance strategy to be practical for this example, it would be important that the interaction between the devices be very low, although high correlation between the lifetimes of the two devices would be desirable. If the interaction coefficient was significant, it would be advantageous to alternate the "power-on" and "standby" lasers periodically to reduce the likelihood of turning on the standby only to discover it had already failed.

2.2. Low Optical Noise Transmitters

A major consideration in optical data links is how to offer high-performance (Gb/s), robust (low BER) links at low cost—comparable to copper wire

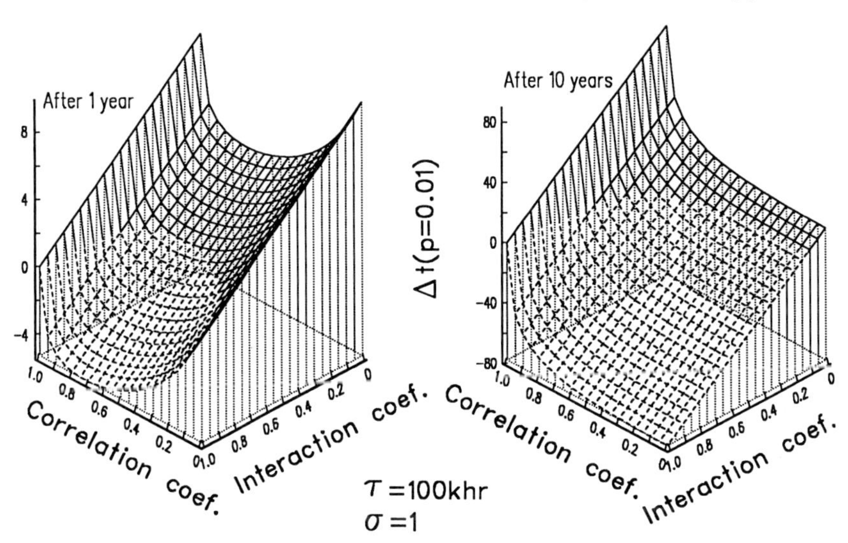

Fig. 2. Examples of available time to schedule maintenance during the system lifetime.

links. This generally excludes the use of precision alignment components associated with single-mode fibers or of external optical components (e.g., optical isolators) to suppress noise-inducing reflections. Light-emitting diode (LED) and multimode fiber links are used for a few hundred Mb/s links in products today (e.g., FDDI (fiber distributed data interface) and FCS (fiber channel standard) networks).

However, a low-coherence laser transmitter in a multimode fiber link promises more efficient E–O conversion, more optical power, and higher speed (Gb/s). Laser structures which offer low coherence were discussed in Part IV. A transmitter with coherence-reducing driver circuits integrated into the chip or module is another solution.

First, consider the effects of a laser coupled to a multimode fiber link. The combination of regular high-coherence lasers and multimode guides with mode-selective loss is well known to result in modal noise [16] and the possibility of unacceptable bit error ratio "floors." Rather than trying to minimize mode-selective loss through the use of tight-tolerance connectors and so forth and hence lose the original reason for choosing multimode guides, the best solution to this problem is to use low-coherence lasers.

Low-coherence lasers can be considered as effectively having poor frequency stability; their effectiveness in avoiding modal noise can be illustrated with the use of Fig. 3. This shows a single-frequency laser coupled to a cylindrical waveguide A, length l, supporting two modes E_1 and E_2; a similar waveguide B is coupled to A, but is offset by the waveguide radius a. With the phase difference shown between the two modes, no energy is coupled from A to B; if the phase difference between the two modes changes by 180°, the coupled energy rises to a maximum; a further 180° change and it drops to zero again.

Using this laser/waveguide link for data transmission, the received signal

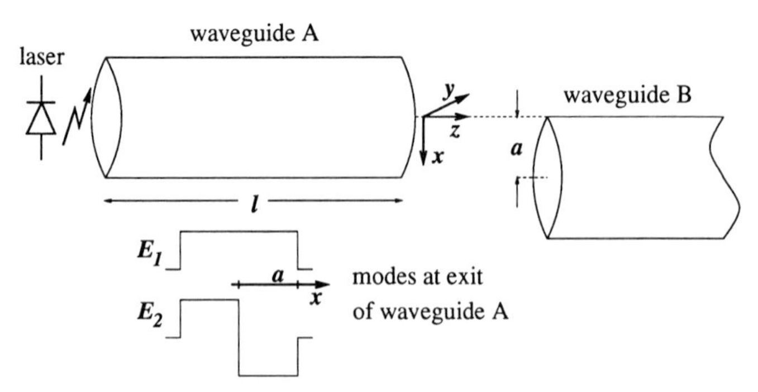

Fig. 3. Two-mode waveguides misaligned to give mode-selective loss.

will be corrupted if the fluctuations in coupled energy—modal noise—occur at a rate less than or comparable to the rate of data transmission. If, however, the energy fluctuations are forced to occur at a much higher rate, they will be averaged out by the low-pass filter response of the receiver, to allow error-free transmission. Specifically, if the average and differential mode velocities are v and Δv, respectively, then it will take an optical frequency change $\Delta f = v^2/l\Delta v$ for a 360° phase change between the two modes; this change must occur, on average, many times each bit period.

A rigorous analysis of this problem by Dändliker [17] relates the necessary waveguide dispersion and frequency instability or source coherence to ensure minimal modal noise within the receiver bandwidth; this analysis assumes a Lorentzian linewidth for the spectral lines of the laser, arising from random phase noise. Similar results can be derived for self-pulsating lasers [18, 19], where the broad linewidths, or low coherence, arise from high-index frequency modulation of the optical carrier, coupled from the pulsations in the intensity; minimal modal noise is generated by any mode-selective loss, as long as it occurs beyond some distance that is proportional to the product of the source coherence and the waveguide bandwidth. Figure 4, from Thompson *et al.* [20], compares the performance of regular, high-coherence lasers with that of self-pulsating, low-coherence, lasers in a data link, with and without mode-selective loss; with the fiber discontinuity after 5 m—a typical fiber jumper length—no evidence of modal noise or an error ratio floor is evident for the low-coherence laser.

Self-pulsating lasers have the additional desirable property of maintaining a relatively low level of intensity noise in the presence of optical reflections [21, 22]; for this reason and because of their compatibility with multimode guides, they are now being widely applied for robust, low-cost computer data links [23].

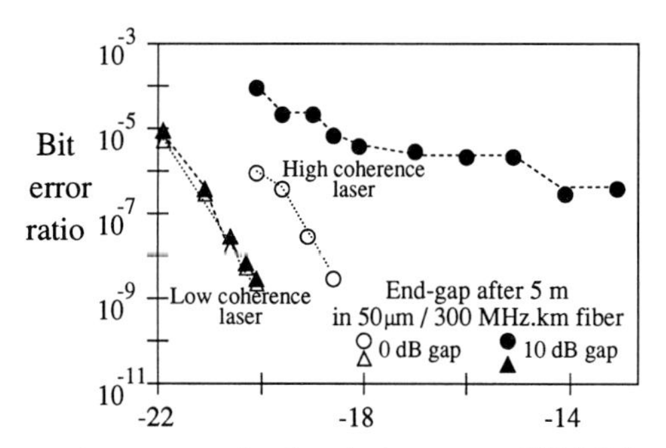

Fig. 4. Demonstration of efficacy of low-coherence lasers with mode-selective loss. (From Ref. 20.)

The volume of self-pulsating lasers manufactured annually dwarfs the volume of regular, high-coherence telecommunications lasers by approximately two orders of magnitude [24]; however, in contrast to the latter type, details of the designs of commercial self-pulsating laser diodes have not been published in the open literature. Petermann [25] includes a general description of their operation and some aspects of the design and characteristics of these important devices are given by Poh [26] and Harder *et al.* [27].

As an alternative to using self-pulsating lasers, the coherence of regular lasers may be reduced by adding external modulation; this was proposed by Bosch *et al.* [28, 29] and has been further studied by Hügli and Bates [30]. This approach is particularly suitable for InP-based devices, where it is more difficult to make high-reliability, high self-pulsating frequency lasers.

External modulation circuits generating a frequency spectrum with components at multiples of the data rate can be a source of noise for the rest of the link, especially the link receiver if it is packaged close to the transmitter. Integration of such a modulator into a laser module along with the laser driver circuit would be preferable. This could be either a hybrid or monolithic integration with the laser, depending on factors such as chip yield and cost.

With both self-pulsating and externally modulated lasers, it is necessary that the frequency of the coherence-reducing signal is much greater than the frequency of the fundamental modulation; for example, for binary baseband transmission at a rate of B bit/s, the self-pulsating or external modulation frequency should be $>\sim 2.5 \times B$ Hz for the link penalty to be less than 1 dB [19, 30]; this will ensure that the intermodulation products between the baseband signal and the coherence-reducing signal lie sufficiently outside the receiver bandwidth. For example, Fig. 5 shows the computed received

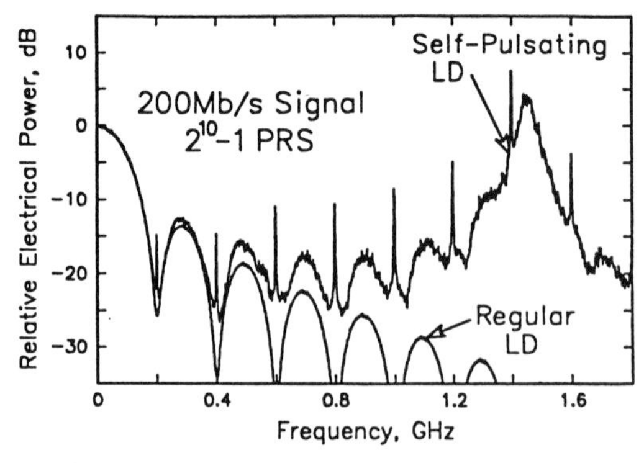

Fig. 5. Examples of computed electrical spectra from a regular and self-pulsating laser with 200 Mbit/s baseband data modulation.

electrical spectra from a regular and a self-pulsating laser, both being modulated at 200 Mb/s with a $2^{10} - 1$ pseudorandom signal (PRS); whereas the spectral envelope for the regular laser falls off with the characteristic $sinc^2$ shape, that for the self-pulsating laser peaks up again around harmonics of the self-pulsating frequency, with lower sidebands encroaching back on the data modulation signal.

The exact penalty will depend on a number of factors, including the modulation depth, the details of the receiver bandwidth, and the characteristics of the timing recovery circuit.

2.3. Optical Transmitter Arrays with Low Optical, Electrical, and Thermal Crosstalk

An array of devices in an optical transmitter offers the opportunity for reduced packaging costs in multichannel applications; however, it brings with it the additional system design issues of acceptable optical, electrical, and thermal crosstalk between the channels. Some discussion of these system issues and the necessary component requirements follows.

Figure 6 illustrates a schematic model of part of an n-channel transmitter array, made up of electrical circuits, lasers, and guides; this can be used to make a first-order analysis of the optical and electrical crosstalk requirements for a data transmission system, using non-return-to-zero (NRZ) modulation. Let each laser couple a power P into its own guide and $P\chi_o$ into its immediate neighboring guides. Assuming that the coupled optical power decreases with distance according to a "fall-off" factor δ_o, then, in general, $P\chi_o\delta_o^i$ will be coupled into its $(i + 1)$th nearest neighbor. With this model, the middle guide in the array will receive the worst mean-square optical crosstalk σ_{ox}^2, given by

$$\sigma_{ox}^2 = \frac{(P\chi_o)^2}{2}\left(\frac{1 - \delta^{n-1}}{1 - \delta^2}\right). \tag{1}$$

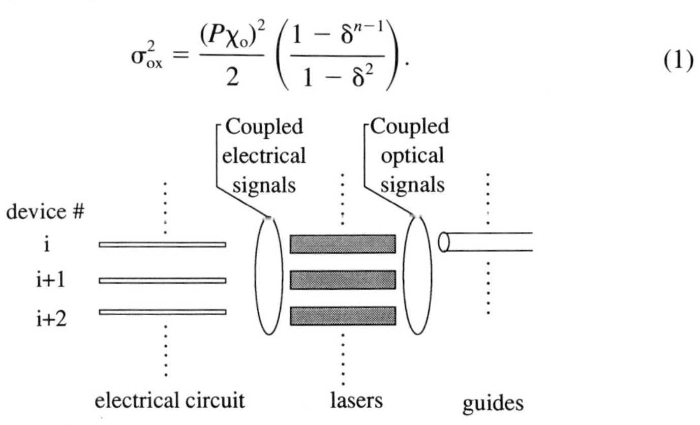

Fig. 6. Model for electrical and optical crosstalk in transmitter array.

Modeling this crosstalk as a random Gaussian process, the change in optical power, $M_{o/o}$ required to maintain the bit error ratio BER in an ideal binary data transmission system, due to optical crosstalk, is readily given by

$$M_{o/o} = -5 \log_{10}\left(1 - 2\chi_o^2 A^2 \frac{1 - \delta_o^{n-1}}{1 - \delta_o^2}\right) \text{dB} \tag{2}$$

where BER $= \frac{1}{2}$ erfc $A/\sqrt{2}$.

Figure 7 shows the computed optical crosstalk requirements, using this simple model for various fall-off factors; the main figure is computed for an optical link penalty $M_{o/o} = 0.1$ dB; the inset figure shows how the requirements vary with penalty for a 3-dB fall-off factor in a five-channel array. Evidently, for any reasonable fall-off factor, i.e., >2 dB, the crosstalk requirement quickly reaches an asymptotic value that is independent of the size of the array.

For a practical system, some additional crosstalk margin must be included to account for variations in laser output and coupling efficiency; ~2 dB should also be included if the channels are operated synchronously, to account conservatively for the non-Gaussian nature of the interference. The published data on optical crosstalk measurements indicate that, for laser spacing

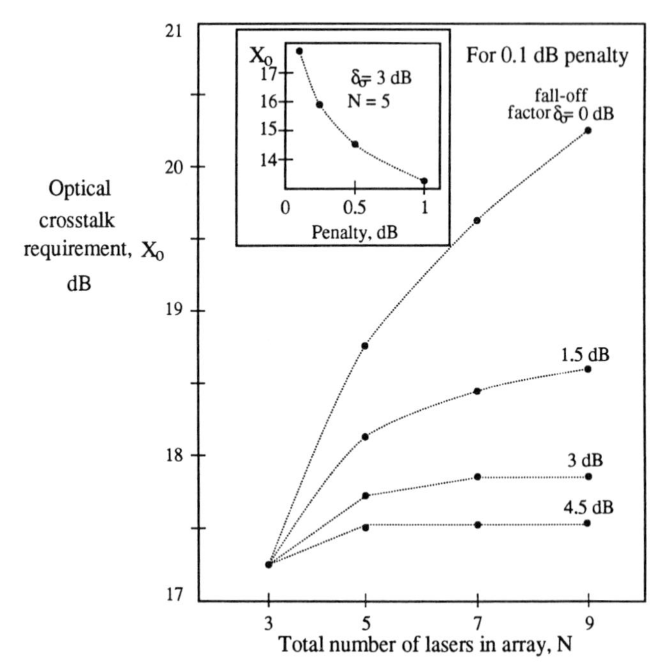

Fig. 7. Optical crosstalk requirements for transmitter arrays.

~250 μm, these requirements are easily satisfied; e.g., both Ewen *et al.* [31] for edge-emitting lasers and Maeda *et al.* [32] for surface-emitting lasers give optical crosstalk values better than 50 dB.

The system requirements for electrical crosstalk can be analyzed in a similar way, but now the crosstalk can arise within the electrical circuits, as well as within the laser array; typically it will have some frequency-dependent characteristic, for example, due to inductive coupling in the wire-bonding of the integrated circuits. The analysis is best illustrated with an example in which the "resistive" and "reactive" crosstalk components have equal magnitude at the data rate, *B*; thus the crosstalk coefficient between the drive currents, *I*, of the *i*th and $(i + 1)$th laser, $\chi_e(f)$, is given by

$$\chi_e(f) = \frac{I_i}{I_{i+1}} = \frac{\chi_E}{\sqrt{2}}\left(\frac{1 + jf}{B}\right). \tag{3}$$

The coupled crosstalk power can be computed by numerically integrating the products of the sinc^2 power spectral density of transmitted NRZ data pulses, the squared modulus of the crosstalk coefficient, and some factor to account for finite rise and fall times of the pulses. For example, with the rise and fall times modeled as a single-pole response with $t_{r/f}B = 0.15$, the change in optical power, $M_{o/e}$ required to maintain the bit error ratio BER in an ideal binary data transmission system, due to electrical crosstalk, is given by an expression similar to that for the optical case, but with a factor $1.34\chi_E^2$ in place of $2\chi_o^2$. The requirements can be derived from Figure 7 with the adjustment, for this example,

$$\chi_E = 2 \times \chi_o - 1.74 \text{ dB}. \tag{4}$$

Figure 8 illustrates some measurements of electrical crosstalk in a transmitter array, in both the time and frequency domains; laser *i* is modulated with a 1-Gb/s NRZ pseudorandom signal; the accompanying traces show the signal coupled to lasers $i + 1$ and $i + 2$. In the time domain, the trace for laser *i* shows the expected eye diagram; the coupled signals on lasers $i + 1$ and $i + 2$ show expanded versions of this; since the coupling is predominantly inductive, these coupled signals are virtually derivatives of the original signal; the inductive coupling effect is more apparent in the frequency domain. For this example, the crosstalk coefficient is approximately 25 dB and the fall-off factor 8 dB, both measured at the bit rate; hence, using the preceding results, the optical link penalty from electrical crosstalk in a five-channel array of this type would be less than 1 dB at 1 Gb/s.

The published data on electrical crosstalk measurements give results that are much closer to the requirements than those just given for optical cross-

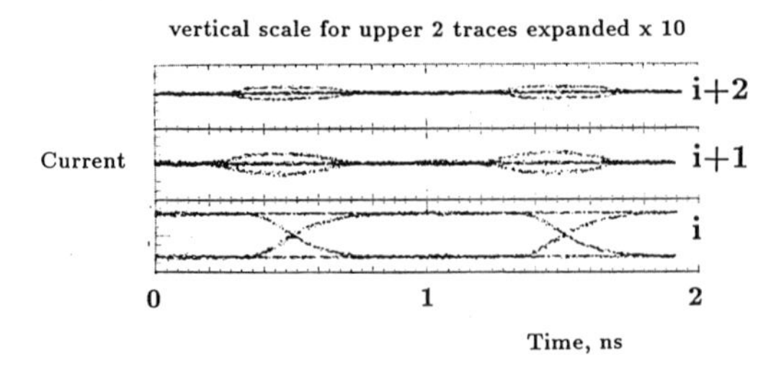

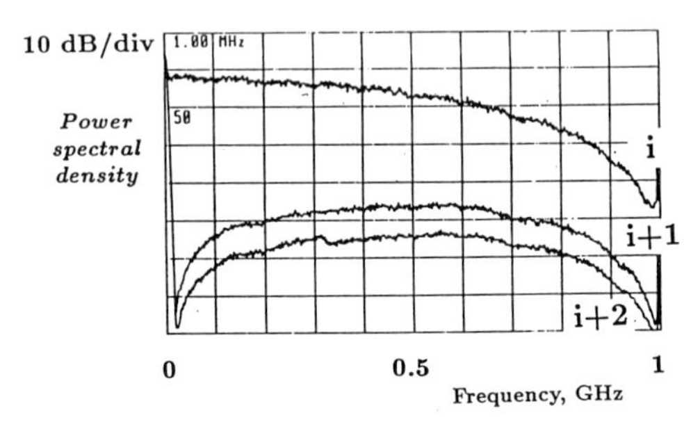

Fig. 8. Some time- and frequency-domain measurements of electrical crosstalk in a transmitter array.

talk, for optical crosstalk, indicating that it is potentially a more difficult problem to solve. For example, Ewen *et al.* [31] cite values better than 30 dB up to 3 GHz; Maeda *et al.* [32] give values better than 35 dB below 1 GHz but as poor as 10 dB at 5 GHz; Wada *et al.* [33] cite 28 dB at 500 MHz but as poor as 14 dB at 1 GHz.

The thermal issues associated with transmitter arrays involve the independent issues of crosstalk and resistance. Sometimes these terms are confused: thermal crosstalk concerns the heating effect of each device on the others; thermal resistance concerns the heating of the devices above the ambient temperature. Typically, thermal resistance, rather than thermal crosstalk, is the more serious system issue in transmitter arrays; for completeness, both will be discussed in the following as they pertain to system requirements for transmitter arrays.

For data transmission systems with line coding to control pattern run lengths, thermal crosstalk does not directly contribute to the link penalty, since its time constant is approximately milliseconds and much longer than those associated with the laser modulation. Even for low-duty-cycle applications it is unlikely to be a problem, since the actual crosstalk values are typically relatively low; e.g., Sato and Murakami [34] cite a value of 15°C/W for an InGaAsP/InP array on a silicon heat sink, with 250-μm device spacing.

Thermal resistance is the more serious system issue in transmitter arrays because the relative cross-sectional area of each device's heat path is less, due to integration (i.e., there will be less opportunity for lateral heat spreading); also, the devices may have higher electrical series resistance, due to the limited size and structure of the contact and the associated interconnection [33]. The system issue is simply an accentuated version of the one facing single-device transmitters, where if the device threshold current rises too high through aging and increases in ambient temperature, thermal runaway will occur as the bias control circuit attempt to maintain constant optical output power.

Laff *et al.* [35] have made a thorough three-dimensional analysis of the heat flow with a GaAs transmitter array mounted on a silicon block, in turn mounted on a large copper heat sink; this showed that, with high threshold current devices, some form of active cooling was necessary to avoid thermal runaway. Similar trends were demonstrated by Bouley [36] with a simpler analysis; this is outlined here to illustrate the sensitivity of the problem and the need for low threshold currents in transmitter arrays.

The rise in device junction temperature above the ambient, ΔT, is a function of the thermal resistance to the ambient, R_{th}, the voltage drop across the device, V, its series resistance, R_s, its CW threshold current, I_{th}, and its average modulation current, $\bar{I}_m$:

$$\Delta T = R_{th}(I_{th} + \bar{I}_m)[V + R_s(I_{th} + \bar{I}_m)] \tag{5}$$

with the CW threshold current related to the pulsed threshold current, I'_{th}, according to $I_{th} = I'_{th} \exp \Delta T/T_0$, where T_0 is the characteristic temperature. We'll consider the worst-case values of junction temperature rise ΔT when the devices have aged, resulting in a *50%* increase in original threshold current, and are operating at an ambient case temperature of 50°C.

For example, a typical single-channel GaAs-type transmitter may have $V = 2$ V, $R_{th} = 50$°C/W, $R_s = 2.5\ \Omega$, $T_0 = 150$ K, and $\bar{I}_m = 5$ mA; for a beginning-of-life pulsed threshold current of 23 mA at 25°C, the worst-case scenario yields a ΔT of 5°C. However, in an multichannel array, where both thermal and electrical resistances may double, to 100°C/W and $R_s = 5\ \Omega$, respectively, ΔT now rises to 11°C in the worst case; to maintain the same ΔT rise of only 5°C requires more than halving the original pulsed threshold current, to 10 mA. Similar requirements can be computed for InGaAsP/InP-type transmitters. Hence, for both types of array transmitters, there is a strong incentive to use quantum well–based devices for their advantage of low threshold current.

3. MEDIUM-SCALE OEIC RECEIVERS

The design of an optical receiver for an OEIC is very similar to a corresponding hybrid design. The principal differences are the potentially low parasitic capacitances possible in the integrated design and the possible reduction in electronic component performance due to compromises made in implementing the integrated detector. In what follows, the issues that arise in optimizing the sensitivity of an optical receiver will be reviewed with special attention to those most affected by the integrated design. The relative merits of the different types of integrated detectors possible will be evaluated with special attention being given to the lateral type of detectors (MSM, lateral PIN) as opposed to vertical devices such as the conventional PIN diode.

What encompasses an optical receiver depends on point of view. From a high-level system point of view, it might include all of the optoelectronics and circuitry needed to present a parallel bus interface to the computer. From a device-level point of view, it might include only the detector and a single transistor preamplifier stage. The following discussion is confined to just those circuits and electro-optic components needed to recover a single line digital version of the coded signal sent by the optical transmitter.

The design of a receiver circuit for optical communications involves a number of factors, the most important of which are the sensitivity and dynamic range. The basic goal is to achieve the lowest possible noise at a given bandwidth or data rate. The OEIC design differs mostly in the techniques needed to achieve these goals.

In an OEIC design, the variety of components is generally limited compared with a hybrid design. Capacitors are available, but only in values less than a few picofarads, and inductors with usable values are in most cases impractical except at the very highest frequencies. Usually, only one polarity of transistor is available. For bipolar this is usually the NPN, while for MESFETs it is usually the *n*-channel device. The lack of these components can be compensated for by using innovative circuit techniques such as active filters, using feedback circuits, and using depletion mode transistors as load devices. Examples of such techniques will be discussed. In most cases these alternative techniques prove to be as effective as the ones they replace.

The OEIC design is sometimes hampered by the compromises in component performance inevitably caused by the diverse materials requirements of the transistors and detectors. For GaAs MESFET OEICs using MSM detectors and operating at 0.8-μm wavelengths, these compromises have been small or nonexistent because of the compatibility of the detector with the electronic IC process. For other types of OEICs, the compromise can be more severe. For OEICs fabricated on InP using independent layer growths for the electronic and electro-optic devices, the compromise is often small, resulting in larger parasitic capacitances than the ideal. Nevertheless, these parasitic capacitances can still be lower than those attainable with hybrid designs. For long-wavelength OEICs fabricated using GaAs MESFET layers grown on InP the compromise can involve lower-performance electronics and low-frequency effects due to trapped charge.

OEIC design can benefit from the low parasitic capacitances possible in an integrated approach. This can be particularly important with preamplifiers designed to operate with the extremely low capacitance detectors used with single-mode fibers. As mentioned earlier, in designing optical receivers for fiber-optic communication systems one of the most important considerations is maximizing the receiver's sensitivity under a set of constraints dictated by the application. The most important of these constraints is generally the bandwidth of the linear portion of the amplification system, although other constraints such as the dynamic range and power consumption may also affect the design. For this reason, in what follows, it is assumed that the bandwidth is the primary constraint.

Another area of application for OEIC receivers that is closely related to that of optical communications is optical storage. In this application the receiver is required to recover a low-level amplitude- or polarization-modulated optical signal which represents the data stored on the disk. An OEIC receiver can be of particular advantage in this application because of its ability to reduce the size and weight of the read–write head, which reduces disk access time. Also, the small feedback loop between detector and preamplifier can increase noise immunity to other electronic components.

As with optical communication, the lowest possible noise at a given bandwidth is important. This is particularly important for magneto-optic detection systems, since with sufficiently low noise powers (1 μW shot noise–limited power) it becomes possible to dispense with the complicated differential detection technique currently used.

Figure 9a shows a typical magneto-optic differential detection system. During a read operation a fraction of the light reflected from the disk is passed through a polarizing beamsplitter set at an 45° angle to the initial polarization direction. This results in two amplitude-modulated signals 180° out of phase. This signal is then passed to a pair of detectors and electronics that measure the difference of the two signals. By measuring this difference, the common-mode noise due to the disk is subtracted out. At the low optical power levels, made possible by a low-noise OEIC receiver, the disk noise

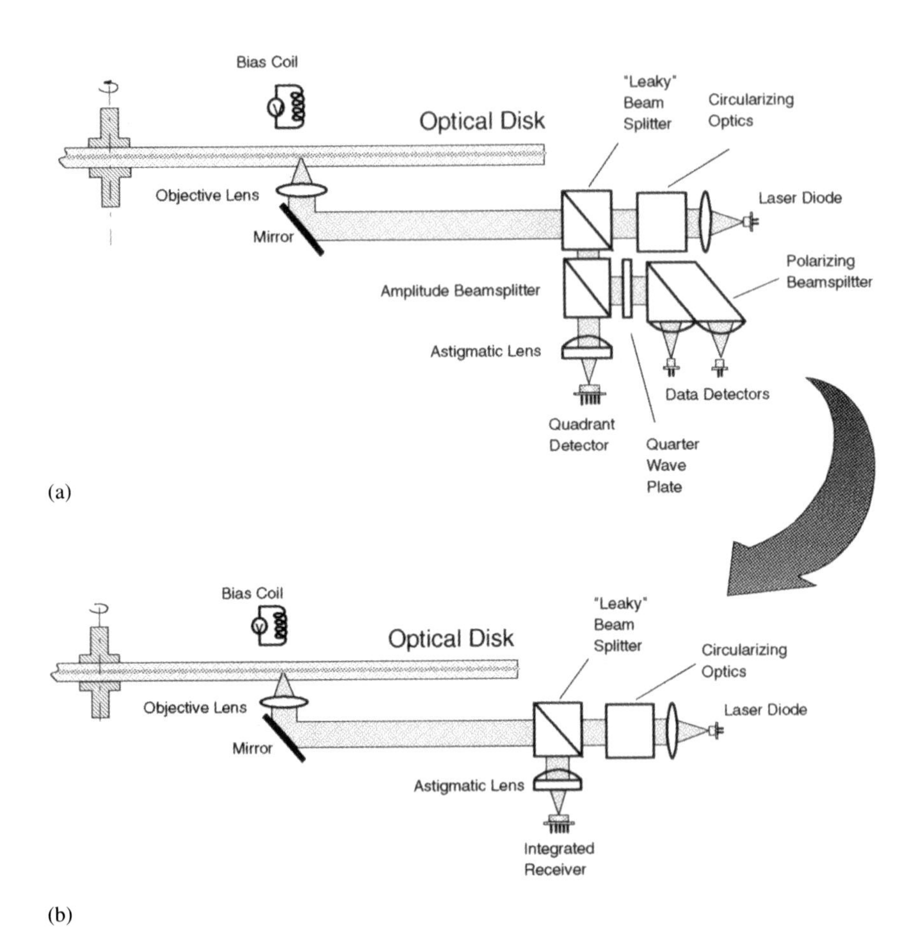

Fig. 9. Magneto-optic receiver chip.

can be made small relative to the data signal, allowing a simpler single-beam detection technique. Figure 9b shows an example of how the use of an OEIC and a single-ended detection technique can reduce the number of components needed. In this design the two data detectors, three beamsplitters, and the quadrant detector used for focus and tracking control are combined into a single OEIC with a integrated quadrant detector.

Another important feature of integrated detectors is that it is easy to make multisegment detectors that, in addition to recovering the data signal, recover focus and tracking information. Figure 10 shows such an integrated receiver chip for a differential detection head, where quadrant detectors have been integrated with sense and servo amplifiers. This chip is 1 × 2 mm in

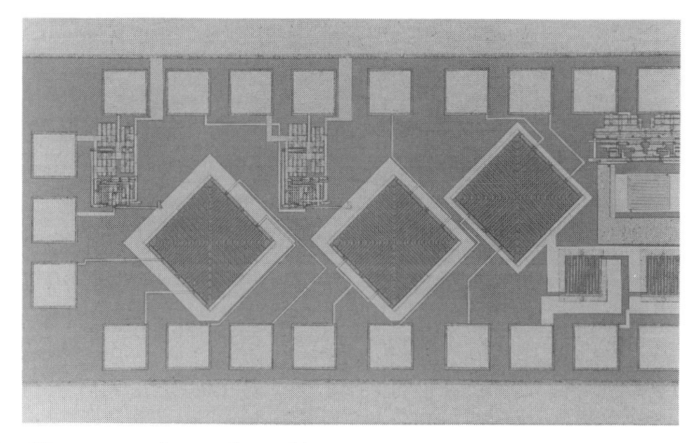

Fig. 10. Magneto-optic receiver chip.

size. For multibeam storage systems it is easy to see how this chip can be enlarged to include multiple read channels.

3.1. Detectors

As mentioned in Part IV on detectors, two types of detectors are commonly used in OEICs. They are the vertical PIN diode and the lateral MSM detector. In reality, there are also lateral PIN detectors and, in principle, vertical MSMs. In what follows, because issues of photodiode integration are being reviewed, the expression *PIN detector* will be used to refer to the vertical detectors and the expression *MSM* to refer to lateral types of detectors.

The performance achievable in a fiber-optic receiver designed with these types of detectors is determined primarily by their responsivity, bandwidth, dark current, and capacitance, with the speed of the system governed primarily by the bandwidth and capacitance and the sensitivity governed by the capacitance, dark current, and responsivity. As mentioned in Part IV, the noise due to the detector itself can usually be neglected compared with other circuit noise sources.

In a properly designed preamplifier, the circuit noise is limited by the capacitance of the detector. For a given bandwidth target, a smaller detector capacitance allows the design of a preamplifier with smaller transistors and higher impedances resulting in less circuit noise. For this reason, it is not the dark current but the capacitance and responsivity that most often determine the maximum sensitivity achievable in an OEIC receiver. The capacitance determines the circuit noise level and the responsivity determines how this noise relates to the optical signal.

Of the two types of detectors, the MSM and the PIN, which is capable of higher sensitivity at a given bandwidth is a complicated issue. In general, the PIN detector is capable of larger responsivity but has a much larger capacitance than the MSM detector. On the other hand, the MSM detector is capable of low capacitance but also has lower responsivity because light is blocked by its interdigitated finger structure. To determine which detector is better in a given application, one must look more quantitatively at the relation between detector capacitance and noise. In general, it may be shown by scaling arguments [36] that for a given preamplifier design the rms value of the noise current is proportional to the square root of the detector capacitance. The sensitivity, which is defined here as the minimum optical power needed to achieve given bit error rate, is given by

$$P_{\mathrm{m}} = \frac{Q\langle i_{\mathrm{N}}^2\rangle^{1/2}}{\eta} \tag{6}$$

where P_{m} is the minimum detectable mean power needed to achieve a given error rate; Q is the signal-to-noise ratio, or quality factor, needed for that error rate; $\langle i_{\mathrm{N}}^2\rangle$ is the mean square noise current referred to the photocurrent; and η is the detector responsivity. For an error rate of 10^{-9}, Q is approximately 6.2. Notice that the minimum detectable power is inversely proportional to the responsivity but proportional to the rms noise, which in turn is proportional to the square root of the detector capacitance. Using a detailed model of the PIN and MSM detector capacitances and using experimental bandwidth versus finger spacing results for the MSM detector, it has been shown that for data rates below about 10 GHz the MSM detector can, in principle, outperform its PIN counterpart [37]. Figure 11 shows the ratio of the best sensitivity possible at a given bandwidth in a receiver optimized for

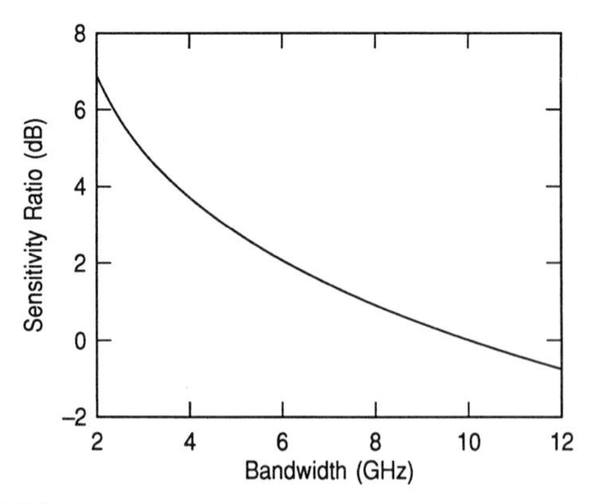

Fig. 11. MSM/PIN receiver sensitivity ratio.

an MSM photodiode versus one optimized for a PIN photodiode. Performance of MSM-based preamplifiers comparable to that of state-of-the-art hybrid designs has been demonstrated [38], although the full performance advantage predicted in Fig. 11 has not yet been experimentally demonstrated.

3.2. Preamplifier Design

The choice of a preamplifier design for an OEIC can often be motivated by factors other than the maximum sensitivity. OEICs are particularly attractive for use in computer communications in which a large number of high-speed communications links are needed. This places an emphasis on low cost and high packing density. In these applications performance in terms of sensitivity may not be as important as speed, dynamic range, power consumption, and cost.

Two types of preamplifiers have commonly been used in fiber-optic systems: the integrating amplifier and the transimpedance amplifier (Fig. 12). Although the integrating amplifier is capable of delivering higher sensivity, it requires a precise equalization circuit which can be difficult to fabricate without the precise components available in hybrid designs. It is also difficult to achieve a large dynamic range using this type of preamplifier. Since the sensitivity improvement using an integration type of design is not large, a transimpedance amplifier has been most commonly used. Preamplifiers

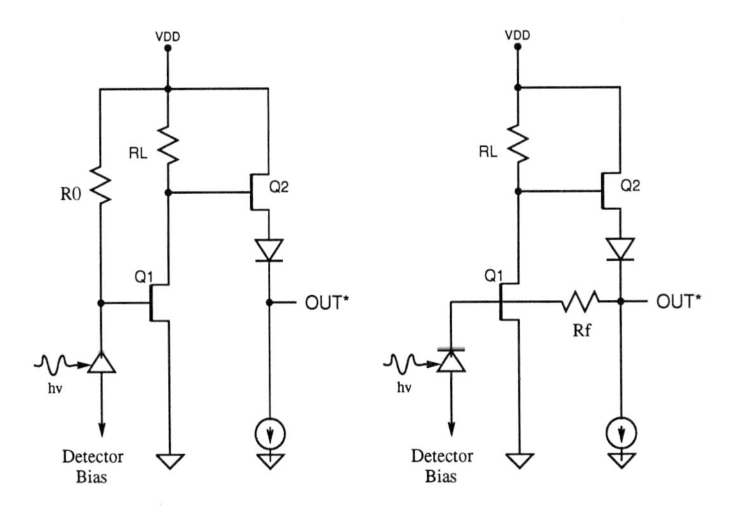

Fig. 12. Transimpedance preamplifier circuits.

with integrating designs have, however, been successfully implemented [37, 38] with good performance.

There are variations in the type of transimpedance amplifier that is appropriate for a given application. Two types of modifications are often employed. One type uses variations on the gain block, the other variations on the feedback element.

Variations on the gain block include using multiple gain stages and using cascode stages to improve the open-loop gain and to reduce the effects of Miller capacitance. Multiple gain stages are usually implemented at lower frequencies where the additional phase change introduced by the added stages has less effect on the stability of the amplifier and the added open-loop gain can improve the sensitivity–bandwidth trade-off.

The use of cascode gain stages is especially effective in FET designs, where the open-loop gain can be limited by the relatively high drain conductance of these devices. Again, at higher bandwidths the effectiveness of this technique is limited. The higher gain results in a lower open-loop bandwidth, which can result in excessive overshoot and ringing in the response of the transimpedance amplifier.

Variations on the feedback element are usually directed at making the preamplifier operate over a larger dynamic range. This may involve the use of a passive element such as a simple diode circuit to create a clamping effect at high signal levels or the use of active elements such as a FET connected in parallel across the transimpedance resistor whose channel resistance is controlled by an AGC circuit.

Aside from such modifications, the primary goal in designing a preamplifier is to achieve the lowest possible noise while retaining a bandwidth adequate for the needed bit rate. In general, the bandwidth needed in the receiving system will be constrained on the lower end by the need to keep the data pattern "eye" open vertically and on the upper end by need to limit the noise bandwidth. In a particular design the optimum bandwidth is a complicated function of the code spectrum [39] and cannot be generally specified, although bandwidths greater than the baud rate usually result in excess noise.

Once the type of preamplifier and the required bandwidth have been chosen, the size of the circuit components must be matched to the detector capacitance in order to minimize the noise. The most important of these components is the amplifying transistor (Q1 in Fig. 12). Its size, which generally determines the size of the other components in the preamplifier, is chosen to minimize the noise. For an integrating type of preamplifier the size of this transistor is chosen so that the input capacitance of the preamplifier equals the photodiode capacitance. For a transimpedance amplifier the size of this transistor is a more complicated issue. In order to estimate the size of this transistor, consider the circuit noise current referred to the input of the preamplifier. This may be written as

$$\langle i_n^2 \rangle = \frac{4kT}{R_f} B + 2qI_L B + \frac{4kT\Gamma[2\pi(C_D + C_{in})]^2 B^3}{3g_m} \tag{7}$$

where R_f is the transimpedance, C_D is the detector capacitance, and C_{in} is the input capacitance of the preamplifier. Γ is the FET channel noise factor, which is unique to the FET type. It is approximately 1.8 for GaAs and about 0.7 for silicon FETs. For simplicity, it was assumed that the frequency response of the system was flat out to a bandwidth limit B. The first term is the thermal noise due to the transimpedance resistor, the second is the shot noise due to any leakage currents in the FET and the detector, and the third term is the thermal noise in the FET channel. Note that the transconductance, g_m, the input capacitance, C_{in}, and the FET leakage currents are proportional to the width of the transistor, w. Using this fact and assuming that the transimpedance, R_f, is inversely proportional to the transistor width, we can rewrite Eq. (7) as

$$\langle i_n^2 \rangle = \frac{4kTBw}{R_{f0}} + 2qI_{L0}Bw + \frac{4kT\Gamma w^2(C_D + C_{in0}w)^2 B^3}{3g_{m0}w} \tag{8}$$

This expression is a function of two independent variables, R_{f0} and w. The variables I_{L0}, C_{in0}, and g_{m0} are characteristic of the IC process, and C_D, the detector capacitance, is determined by the application. w is determined in terms of R_f by the requirement that the bandwidth be fixed.

For a given preamplifier design there will exist a relationship between the bandwidth of the preamplifier, the normalized transimpedance R_{f0}, and the detector capacitance. For the simple case in which the open-loop bandwidth of the preamplifier gain stage is large and the effects of the postamplifier are neglected, this relation can be written as

$$B = \frac{A_0}{2\pi R_f(C_D + C_{in})} = \frac{A_0 w}{2\pi R_{f0}(C_D + C_{in0})w} \tag{9}$$

where A_0 is the open-loop gain of the preamplifier. This relation between the bandwidth, the FET width w, and the normalized transimpedance R_{f0} may then be inverted to give R_f in terms of w and B:

$$R_{f0} = \frac{A_0 w}{2\pi B(C_D + C_{in0}w)} \tag{10}$$

Substituting this expression in Eq. (8), we arrive at an expression for the noise in terms of the one parameter, w.

$$\langle i_n^2 \rangle = 2I_{L0}qw + \frac{8B^2\pi kT(C_D + C_{in0}w)}{A_0} + \frac{4B^3GkT(C_D + C_{in0}w)^2}{3g_{m0}w} \tag{11}$$

Figure 13 shows the noise versus w for typical MESFET values for g_{m0}, and A_0 for several different bandwidths B. A clear minimum appears at a particular value of w which determines the optimum design point. Notice that above a certain value for w the noise varies relatively slowly. This optimum FET width is the most important parameter of the design. Once the size of the amplifying transistor has been determined, the sizes of the other transistors in the circuit follow from biasing and output loading considerations.

The increase in noise at small values of w is due FET channel noise [the third term in Eq. (8)]. This is a general feature of these types of curves. At large values of w the noise due to the transimpedance resistor and leakage current dominates because the larger FET sizes require a lower transimpedance to achieve the same bandwidth.

The preceding calculation involved many simplifying assumptions. In a realistic system the bandwidth is not flat up to a given frequency but falls off much more slowly. The result is that the FET channel noise, which increases with frequency, is emphasized more. Also, the relationship between the bandwidth and the FET width given by Eq. (10) ignores the contribution of the postamplifier in reducing the bandwidth. This can have an important effect in reducing the FET channel noise, which increases rapidly with frequency. In general, though, for a given design a more realistic

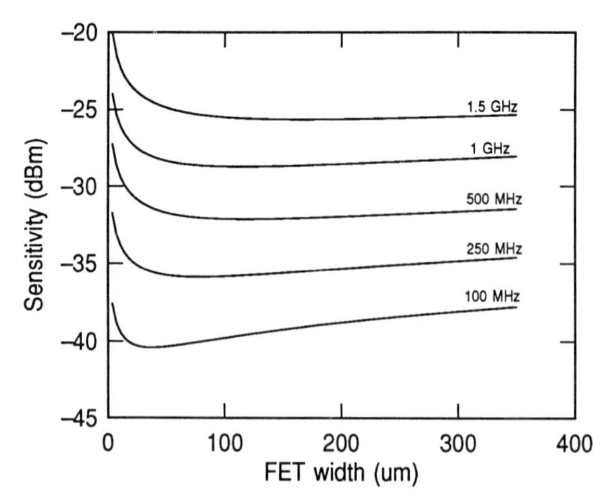

Fig. 13. Noise current vs. normalized transimpedance.

expression for the frequency response of the preamplifier–postamplifier combination may be used as above to arrive at a plot of the noise versus FET width, and the transistor size may be optimized. In hybrid designs it is often impossible to affect such an optimization because the transistor sizes and detector sizes are governed by what is available for the individual components. With an OEIC design, greater freedom exists in the choice of these components, and due to the lower parasitic capacitances it is possible to achieve very high performance designs.

This optimization can be especially effective when using MSM detectors, since not only the size of the detector can be varied but also the finger spacing and thus its capacitance. Using the largest possible finger spacing, commensurate with the required detector bandwidth, the detector capacitance can be minimized to match the application.

3.3. Monolithic Receiver Design

For larger OEICs in which the preamplifier is combined with a postamplifier and decision circuits, and perhaps even with additional circuitry such as clock recovery circuits and deserializers, noise immunity also becomes an important consideration. This has led to the use of a differential form of transimpedance amplifier [40] that is much more immune to supply noise. As shown in Fig. 14, the differential transimpedance amplifier consists of two identical transimpedance amplifiers connected in a differential arrangement. Note that the second detector electrode is capacitively coupled to the second half of the preamplifier to increase the signal level at the output.

Differential circuits are useful not only in the preamplifier to improve noise immunity but also in subsequent amplifier stages, for noise immunity and to minimize coupling of these stages' signal back to the preamplifier through the supply, causing instability. This is especially important in monolithic designs, where it is difficult to bypass the power supplies at as many points as in hybrid designs.

The differential amplifier shown in Fig. 15 is an example of how the output level of each gain stage can be controlled using feedback rather than capacitive coupling between stages. In this circuit common-mode feedback is used to control the average output level. The bias current for the differential pair, which controls the output signal levels, is supplied by the transistor Q1. The drain current of this transistor is, however, determined such that its gate voltage is two diode drops below the mean of the two output voltages, OUT and OUT*. This type of common-mode feedback circuit de-

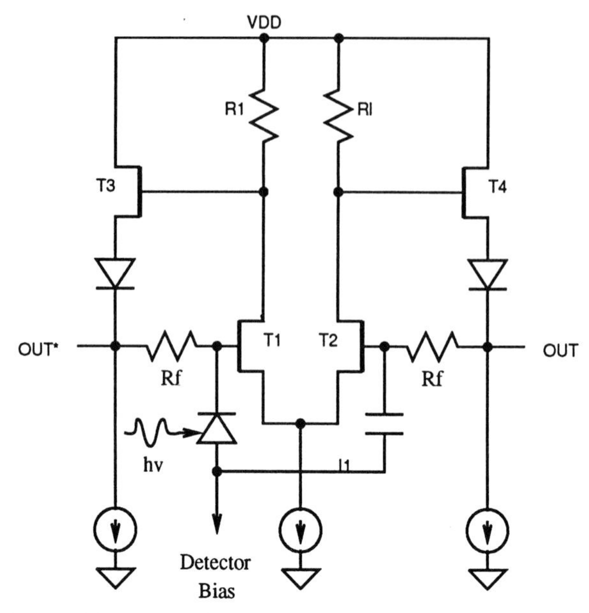

Fig. 14. Differential preamplifier circuit.

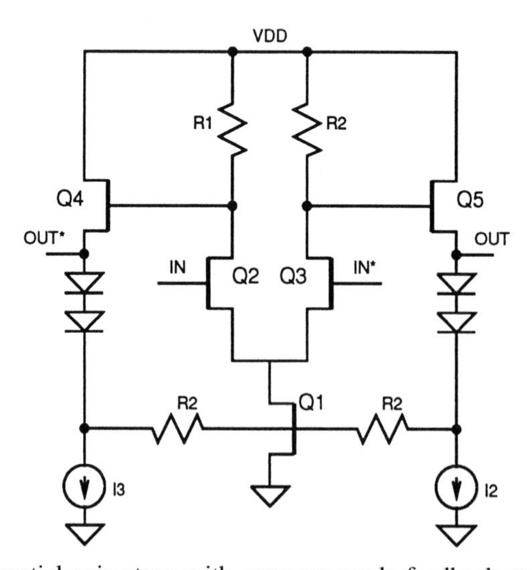

Fig. 15. Differential gain stage with common-mode feedback stabilization.

termines the mean of the two output voltages with immunity to variations in device transconductance, supply voltage, and temperature effects.

In large OEIC designs, noise coupling from the supply is not the only consideration. Noise coupled capacitively into the preamplifier circuit can also be a problem. In particular, care must be exercised with the differential preamplifier to minimize capacitance between high-speed signal or supply lines and the detector coupling capacitor, which may be physically large. Again, differential circuits are useful in this context in that the signal lines come in pairs, which tends to cancel capacitive coupling.

4. CHALLENGES FOR LARGE-SCALE OEICS

With increasing levels of integration come a number of new design issues which must be addressed: mixed analog–digital circuits, optical components and link elements, noise and crosstalk, modeling techniques, speed/power/density, and so on. From the point of view of large-scale IC design, the presence or absence of optics per se is of relatively minor importance. It is the mixed analog and digital circuits which distinguish typical OEICs from either microprocessors or MMICs. Furthermore, the wide range of signal levels present in typical OEICs provide unique challenges as compared with other mixed-mode ICs. As an example, consider the receiver OEIC of ref. 2, which contains a photodetector and preamplifier, a retiming circuit, and a deserializer circuit (demultiplexer and synchronization) all operating at 1 Gb/s. The front-end receiver circuits must deal with input signals of a few microamperes, while the deserializer provides 10 outputs at ECL levels (0.8 V) switching at 100 MHz. Not only is the potential crosstalk a serious concern, but also the models and modeling techniques to predict noise and distortion are nontrivial. These issues associated with designing large-scale OEICs are discussed in more detail in the following sections.

4.1. Noise and Crosstalk

Noise and crosstalk are serious concerns in designing OEICs. Both optical and electrical noise mechanisms exist which can adversely affect the overall performance. These noise issues can be classified in three broad categories: noise-sensitive components, noise-generating elements, and optical noise.

The noise-sensitive components are usually associated with receiver OEIC elements because of the smaller signal levels present. Receiver circuits are high-gain, wide-bandwidth amplifiers, which are sensitive to noise over a wide frequency spectrum. Noise injected into the receiver front end will directly degrade the overall link performance. The retiming circuits (phase-lock loop) are also high-gain circuits which are sensitive to noise; however, in this case only noise close to the operating frequency is important. Since any retiming circuit contains some nonlinear element in the front end to regenerate the clock component [41], noise which is harmonically related to the operating frequency must also be considered. In the transmitter, any bias control circuits must also be considered noise sensitive, since these feedback circuits contain high-gain amplifiers, although the problem is usually not as severe as for the receiver because the bandwidths are orders of magnitude smaller.

Noise-generating circuits are primarily the various driver circuits in both transmitter and receiver OEICs. A laser or LED driver requires large switching currents relative to the other circuits on the chip and is also among the fastest circuits. This gives rise to very fast edges, which can generate broadband noise components. The electrical off-chip drivers also generate large switching currents. Although the edge rates are usually not as high as for a laser driver, there are typically many such drivers on a single chip (10–100) and the aggregate noise is significant.

As noted, the fast edges present in the driver circuits generate broadband noise which can degrade receiver sensitivity. It is also important to note that in many applications, the clocks within the chip are derived from the data signal itself, and these clocks are used throughout the chip, including the off-chip drivers. The net effect is that much of the switching noise will contain strong components at the data frequency and its harmonics. Since this switching noise will fall within the passband of the retiming filter and yet have an effectively random phase relationship to the incoming data signal, the retiming circuit performance can be degraded by the resultant increase in the jitter in the recovered clock. Because the noise sources are driven by the data signal, it is possible for the performance degradation to be pattern dependent: certain patterns produce phase relationships which affect the circuit performance more than others. This is especially important in data communication links, where block coding is typically preferred over scrambling the data (e.g. 8B/10B coding for FCS and 4B/5B coding for FDDI applications). This situation is made even worse if multiple channels are included on the same chip. In this case, adjacent channels will not operate at exactly the same frequency, because they are derived from different data signals. However, the frequencies will be close enough to lie within the passband of the retiming filter and degrade the performance as before.

Because the frequencies are not identical, the relative phases will change continually, further exacerbating the problem.

A number of potential crosstalk paths exist between the noise-generating circuits and those sensitive to the noise. There is direct electromagnetic interference, either capacitative or inductive coupling between different circuit components. Crosstalk through the substrate can be either resistive or capacitative or a combination of the two. Finally, crosstalk in the power supply is a ubiquitous problem in any IC design and even more so in OEICs because of the sensitive receiver circuits.

4.2. Circuit Design

As noted in the previous section, noise and crosstalk are serious concerns when designing an OEIC; however, numerous circuit techniques exist to address these issues.

Circuits can be made less sensitive to both power supply noise and EMI by using a fully differential design wherever possible. This is especially important in the preamplifier and receiver circuits. The only penalty associated with using differential designs is the larger chip area required for the additional devices and wiring. The basic differential amplifier circuit is shown in Fig. 16. Although it is shown using FETs as the active devices, completely analogous circuits exist for bipolar technologies. A current steering stage is formed by the differential pair T1–T2 and current source I1. The source-follower stages, T3 and T4, buffer the switching stage from the output load. When biased in the linear regime, it forms the basis for the receiver circuits. When operated in a switching mode, it forms the basis for a differential current switch logic family which is useful in the retiming circuit.

An ideal differential amplifier will reject all common-mode noise and is

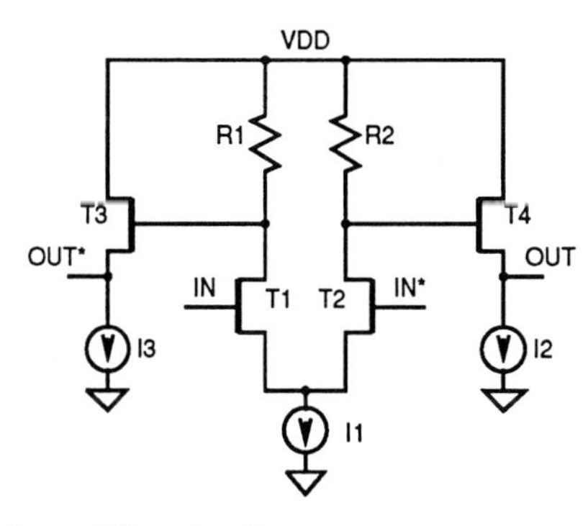

Fig. 16. Differential amplifier circuit.

particularly good for minimizing sensitivity to power supply variations. In practice, the amount of common-mode noise rejection depends on how symmetrical the differential circuit can be made, and this depends critically on the parasitic elements in the circuit as well as the device parameters.

To illustrate the importance of symmetry, consider the simplified representation of the differential current switch shown in Fig. 17. Current sources I1 and I2 represent the result of the current switch, T1, T2, and I1, of Fig. 16. In practice, these elements will have some parasitic capacitance associated with them which has been represented by the capacitors, C, in Fig. 17. Ideally, the output voltage, V_{out}, is independent of the supply voltage, V_{dd}, and this holds true at low frequencies even when the components are mismatched. However, the full expression for the small-signal output voltage as a function of the power supply voltage is

$$\left|\frac{V_{out}}{V_{dd}}\right| = \frac{\sqrt{(R1 - R2)(R1 + R2)}\,\omega C}{\sqrt{(1 + C^2 R1^2 \omega^2)(1 + C^2 R2^2 \omega^2)}} \tag{12}$$

Assuming that $R1 - R2 = \delta$, that δ is small so that $R1 \approx R2 = R$, and $\tau = RC$, this equation reduces to

$$\left|\frac{V_{out}}{V_{dd}}\right| = \frac{\sqrt{2\delta/R}\,\omega\tau}{(1 + \omega^2\tau^2)} \tag{13}$$

The sensitivity of the output to power supply noise is then directly proportional to the fractional mismatch between the load resistors. Similar expressions can be calculated for the other elements in the circuit as well. To achieve good noise immunity, it is important to match the devices as closely as possible through the choice of devices (i.e., considering local statistical variations), device orientation and layout, and proximity effects (e.g., backgating in GaAs MESFET technologies). Careful physical layout is also required to ensure that asymmetric parasitic loading does not degrade the common-mode rejection of the circuit.

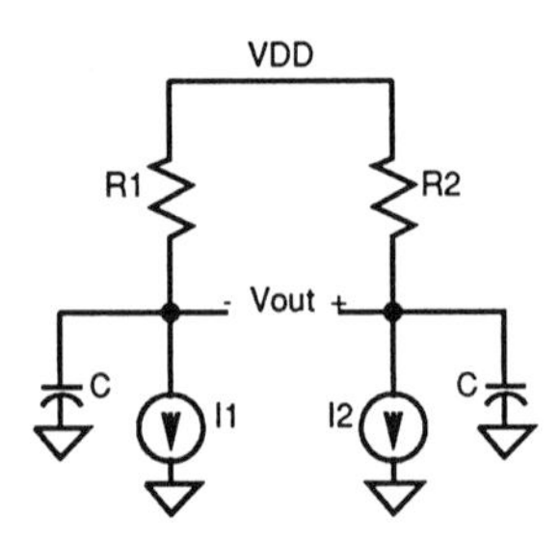

Fig. 17. Sample common-mode rejection calculation.

Differential circuits have other advantages as well. The overall delay through the circuit is effectively averaged over both the rising and falling edges of the signal, since both true and complement signals are used throughout. This results in extremely small delay skew, the difference in delay between rising and falling edges, and minimizes pulse distortion through the circuit. Again, in a perfectly symmetrical differential circuit, the delay skew is zero. In practice, the delay skew for a differential circuit can be an order of magnitude less than that of a comparable single-ended circuit. This is especially important in the retiming circuit, where pulse distortion and the ability to align signals relative to one another are critical.

Differential circuits can also be used to minimize the generated noise as well. As shown in Fig. 16, the total supply current is independent of the logic state of the circuit; therefore, the switching noise induced on the power supply is reduced compared with a single-ended circuit. Differential driver circuits are usually limited to the laser or LED driver, or perhaps a handful of special-purpose high-speed electrical interfaces. The extra area consumed by these circuits is not a major concern; however, doubling the number of chip I/O pads to accommodate differential drivers is usually prohibitive in large-scale OEIC designs. When single-ended electrical off-chip drivers are used, care must be taken to control the edge rates, $\Delta I/\Delta t$ and $\Delta V/\Delta t$, in conjunction with the package parasitics in order to limit the generated noise.

Figure 18 illustrates a typical chip with the important packaging parasitics included. In practice, there will always be some parasitic inductance associated with the power supply connections between the chip and the first decoupling capacitor which provides a low-impedance return path for the supply. When the various circuits on the chip switch, a noise is induced on the power supply, $\Delta V = L\,\Delta I/\Delta t$, depending directly on the parasitic inductances and edge rates. Typically the internal circuits (those which don't interface directly to chip I/O pads) switch relatively small amounts of cur-

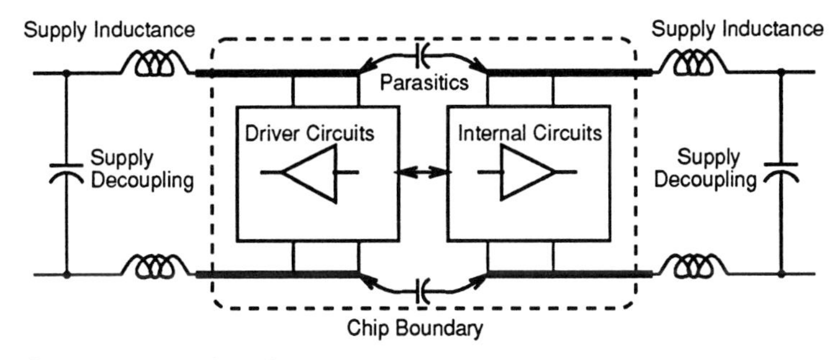

Fig. 18. Power supply noise.

rents, and the overall noise will be averaged over many circuits all switching at slightly different times. The off-chip drivers are required to switch much larger currents, especially when driving high-speed terminated transmission lines, and in synchronous systems the outputs will all tend to switch at nearly the same time. These drivers then tend to be the worst noise generators on the chip.

One way to minimize the effect of this noise on the sensitive circuits is to use separate power supplies for the internal circuits and the off-chip drivers as shown in Fig. 18. Parasitic coupling is still possible between the different supplies, especially capacitative coupling, since the supply buses tend to be physically large to handle the required currents. Care must be taken in the physical layout to minimize these potential crosstalk paths. Another potential concern is the electrical interface between the internal circuits and the off-chip drivers. Careful design is required to insure adequate logic noise margins across this interface.

In addition, it is possible to minimize noise sensitivity in the retiming circuits by choosing the appropriate design strategy. Any clock recovery circuit consists of a nonlinear element followed by a narrowband filter. This filter can be implemented using passive techniques such as SAW filters [42], or dielectric resonators [43], or active techniques such as phaselock loops (PLLs) [41]. The passive techniques tend to be less sensitive to noise and crosstalk than PLLs because they are relatively low-gain, open-loop systems. A PLL, on the other hand, is a high-gain, closed-loop system, in which the overall loop stability and dynamic behavior must be controlled. PLLs also tend to be more complicated circuits with more potential crosstalk mechanisms. However, the passive filters typically require special materials ($LiNbO_3$, quartz, barium titanate) which are not amenable to integration with the electronic components. These filters are also relatively bulky compared with the OEIC chip itself. This presents even more packaging problems when the levels of integration rise to the point that multiple channels are integrated on a single OEIC.

It is possible to integrate a high-performance PLL on an OEIC along with other logic [44]. The block diagram of a basic PLL is shown in Fig. 19 and consists of a nonlinear element (transition detector), a phase detector, a narrowband loop filter, and a voltage-controlled oscillator (VCO). For the purposes of this discussion, the loop filter and the VCO are the most critical components.

Because of the narrow bandwidth of the loop filter required for good performance, it is difficult to integrate this filter on chip using standard analog techniques. Large-value capacitors must be placed off chip, and the design

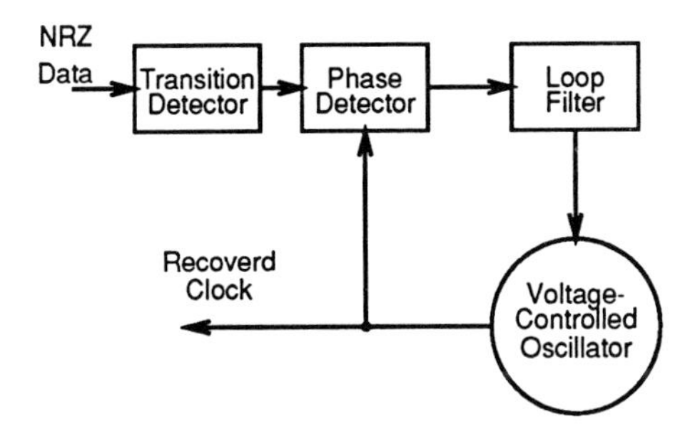

Fig. 19. Basic phaselock loop block diagram.

then suffers from the same density problems as with SAW filters or dielectric resonators. Active filters can minimize the size of the capacitors required, but it is still very difficult to contain the design on chip with typical processes available (e.g., capacitors $\lesssim$ 100 pF). Alternatively, it is possible to use digital filtering techniques [45]. Reasonable filters can be implemented using standard logic blocks (counters), which require no off-chip components, no high-performance, high-precision analog components, and are easily integrated with other logic on the OEIC [46]. More devices are required using the digital filter approach, but in an LSI environment devices are cheap and the circuits are digital, which means they are very insensitive to noise and crosstalk.

The VCO can be implemented in a number of ways. It is possible, although difficult, to integrate inductors so that standard oscillators using *LC* tank circuits can be constructed. Low-noise relaxation VCOs [46] can easily be integrated in a variety of technologies. The timing capacitors required are on the order of a few picofarads, which can easily be contained on chip in a reasonable area. These circuits can dissipate very little power and are useful from 10 MHz to 5 GHz and above. Ring oscillators are also useful as integrated VCOs [44, 47]. These are very simple to integrate, are compatible with other digital logic, and can be used into the multi-GHz range. A ring oscillator requires no analog components (inductors or capacitors) and so is suitable for integration in a wide variety of processes. Because the frequency is set by the delay of the individual stages, it is possible to span extremely wide frequency ranges with a single oscillator by simple, digital control of the number of stages [47].

4.3. Modeling

To say that good modeling is critical is perhaps the ultimate tautology, but that makes it no less important and OEICs present a number of challenges along these lines. With large-scale OEICs, as with any LSI circuit, accurate simulation is paramount, because of the large amount of resource required to design and fabricate a chip. It has to work the first time (or at least the second). Key is the mixed analog and digital nature of these circuits. This affects not only the types of simulators used but also models required. The digital logic depends on the switching characteristics of a device, typically how much current it provides when it is "on" and how much it leaks when it is "off". For an analog amplifier it is much more important, for example, to know the bias point for maximum f_T rather than the maximum current per unit device width or area. This complicates both the model development and parametric testing because many more parameters need to be accurately included to adequately predict the circuit behavior.

True mixed-mode simulators (1s and 0s freely mixed with analog voltages and currents) are in their infancy. Some circuits, receivers and drivers, can be adequately modeled using a circuit-level simulator such as SPICE. The digital logic can be handled by a variety of commercially available logic simulators. The interface between these two worlds is rather fuzzy and usually requires some simulation tricks or custom programming by the designer in order to bridge the gap. The retiming circuit, especially for PLLs, typically falls in this gap. It can contain significant amounts of both analog and digital circuits, and at the very least it involves widely varying time constants (high-frequency VCO and narrowband filter), which make standard circuit simulators difficult to use (CPU limited). Usually one must code behavioral models for the PLL functional blocks (Fig. 19) and then use either a circuit or a logic simulator. Alternatively, one can write a custom simulator based on analytic models of the PLL to evaluate the overall performance. Both of these alternatives have the disadvantage that the one-to-one correspondence between the logical design and the physical design is lost, making the verification process (ensuring that the circuit fabricated is the same one simulated) much more difficult and time consuming.

It is also necessary to model the optical components and the optical link. For example, the dispersion in the fiber will distort the signal as it propagates, and this must be taken into account when designing the receiver circuits, trading signal distortion for noise bandwidth and sensitivity. As another example, mode-partition noise in a laser, combined with the dispersion of the fiber, can lead to excess jitter in the incoming data signal, which must be taken into account when designing the effective Q of the retiming circuit. This in turn affects a number of elements in the PLL such as the VCO stability. Again, the tools that would allow a designer to weigh one component or functional block against another in order to optimize the overall system performance are lacking.

Finally, it is necessary to be able to model accurately the various parasitic elements, both on chip and those associated with the package. This is important not only from a raw performance point of view but also in order to minimize potential crosstalk and noise problems as discussed previously.

4.4. Example: GaAs OEIC

A number of general concepts have been discussed in the previous sections. To better illustrate these ideas, consider a specific OEIC [2] implemented using GaAs MESFETs as an example. A chip photograph of this OEIC is shown in Fig. 20.

This chip contains four receivers consisting of an integrated MSM photodetector, amplifier, and level restore circuit; a PLL retiming circuit; and a demultiplexer along with byte synchronization circuits. Containing approximately 8000 devices, it qualifies as a large-scale OEIC. Here the focus is on the receiver circuit and its interaction with other circuits on the chip

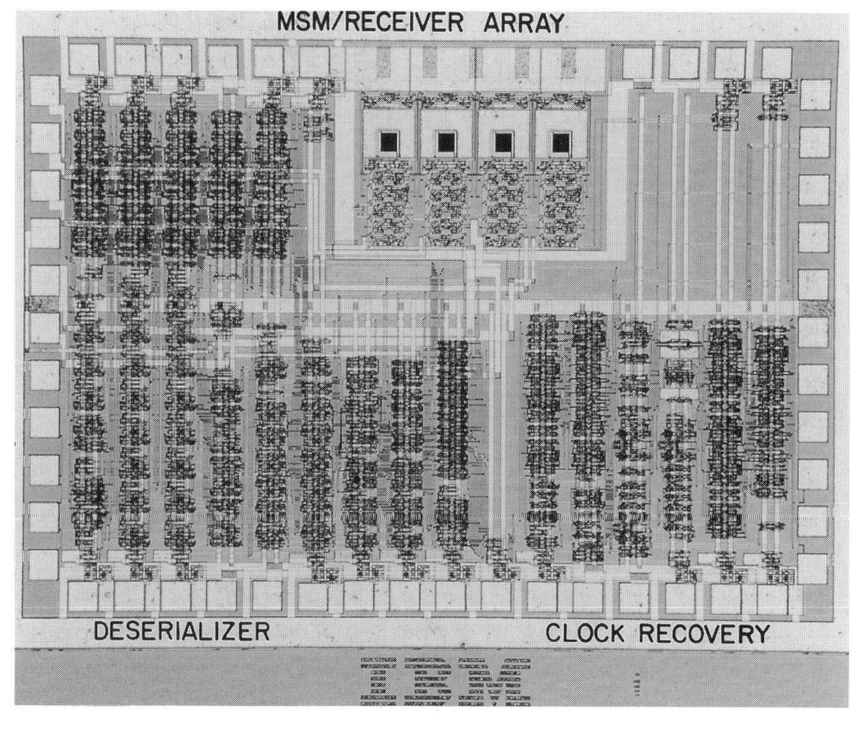

Fig. 20. GaAs OEIC.

as well as the package. The block diagram and equivalent circuit for this portion of the chip is shown in Fig. 21.

The circuit consists of an MSM photodetector, a decoupling capacitor which is integrated on the chip, a transimpedance preamplifier, a number of postamplifier stages, the off-chip driver, and the parasitic inductance, L_P, from the chip to the power supply connection on the module. In practice, there is also a parasitic capacitance, C_P, between the power supply bus for the off-chip driver and the decoupling capacitor, C, because both of these structures are relatively large (Fig. 20). A 3-D capacitance modeling program has been used to extract the parasitic capacitance using the dimensions from the physical layout of the chip, and the value is approximately 18 fF. The parasitic inductance due to a 2-mm-long, 1-mil gold wirebond is about 2 nH. These parasitics have been included along with the circuit models in a SPICE-like simulator to evaluate the overall receiver response.

The eye diagram in response to a random data pattern is shown in Fig. 22. The input signal level is equivalent to a -21 dBm optical input, which is about 2 dB above the predicted noise sensitivity limit for a 10^{-9} bit error rate. Also, the off-chip driver circuits have been biased so as to have an output transition rate of approximately 1 V/ns. The resulting response is well behaved, with only a slight ringing on some transitions, and one would expect this circuit to achieve the predicted sensitivity.

Figure 23 illustrates the receiver response when the off-chip drivers are biased to produce an edge transition rate of 3 V/ns. All other parameters are the same as in the previous figure, but the response has changed dramatically. There is significant ringing in the response due to the feedback

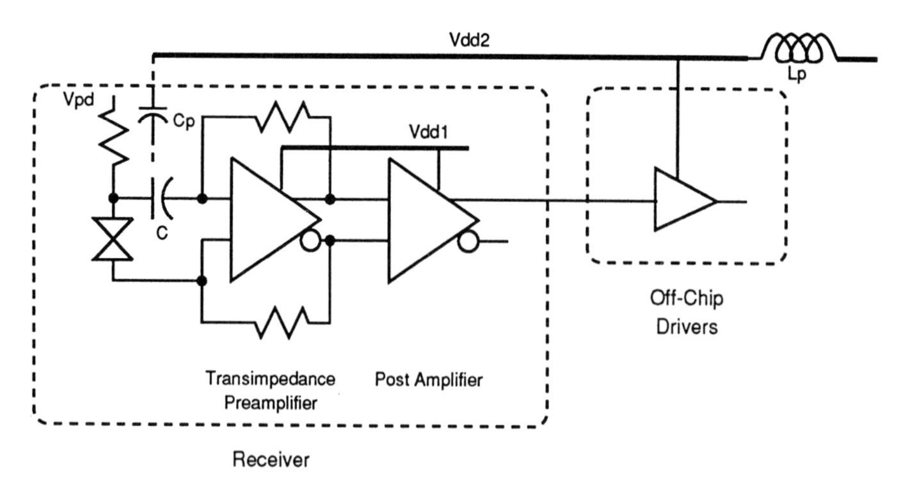

Fig. 21. Receiver block diagram.

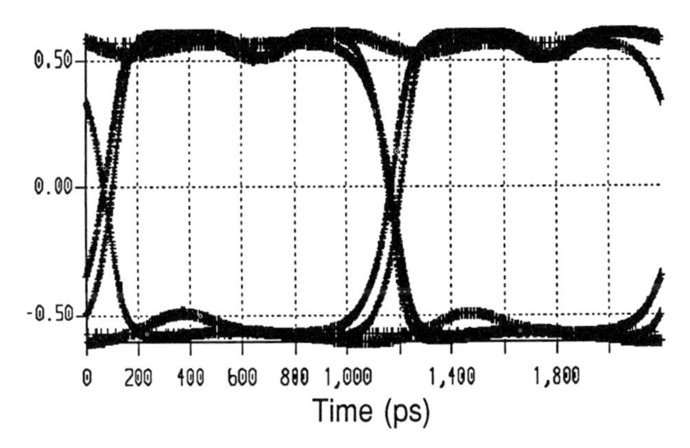

Fig. 22. Simulated receiver response at 1 V/ns transitions.

via the parasitic capacitance, C_p, and the eye diagram is significantly degraded. Simulations with smaller input signal levels indicate that the circuit will oscillate. In this case one would expect the error rate performance to be significantly degraded.

The bit error rate performance of this chip at 1 Gb/s has been measured for both of these operating conditions and the results are shown in Fig. 24. The sensitivity at a 10^{-9} error rate is -20 dBm, and the error rate response is well behaved as the optical power is decreased. This situation corresponds to output transitions of 1 V/ns (Fig. 22). The second curve in Fig. 24 corresponds to output transitions of 3V/ns. The sensitivity has been degraded by 1.7 dB as compared with the previous case, in agreement with the simulations of Figs. 22 and 23. Also notice that the measured error rate (bullets) increases much more rapidly as the optical power is decreased than one would expect from Gaussian noise alone (solid line). This indicates that the

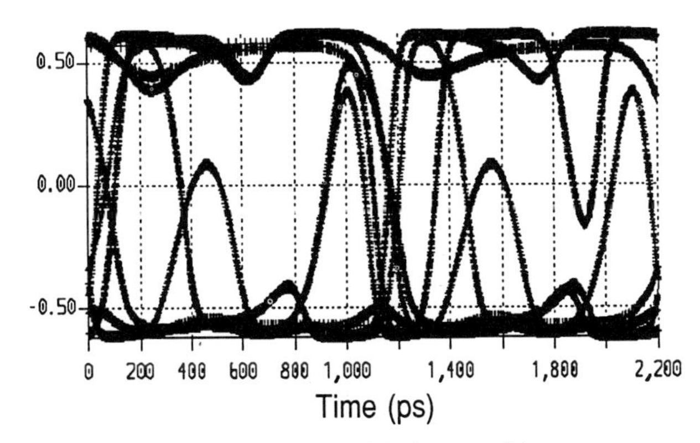

Fig. 23. Simulated receiver response at 3 V/ns transitions.

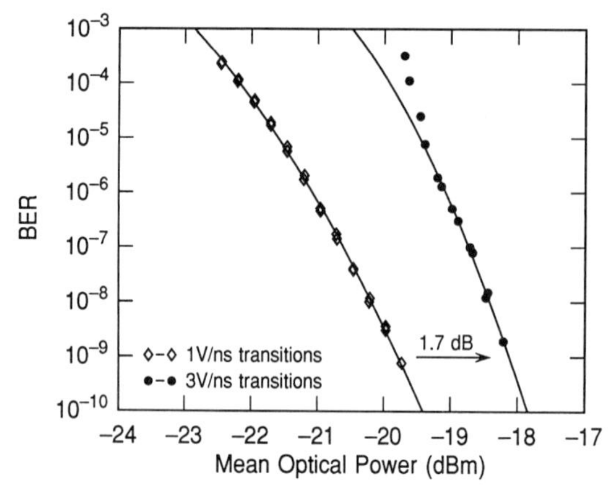

Fig. 24. Receiver bit error rate performance.

crosstalk is sufficient in this regime that the receiver is beginning to oscillate.

The previous discussion indicates the importance of modeling not only the circuits but also the parasitic elements of the chip and package layout. It also demonstrates the importance of controlling the characteristics of the miscellaneous circuits on the chip, the off-chip drivers, in order to achieve the expected performance. Once the models are in place, strategies to improve the performance can be evaluated. In the previous example, consider decreasing the parasitic power supply inductance from 2 nH to 1 nH. The simulated eye diagram is shown in Fig. 25.

The response is significantly improved, even though the off-chip drivers are still operating at 3 V/ns edge rates. By simply shortening a wirebond or adding a second wirebond in parallel with the first, this crosstalk "prob-

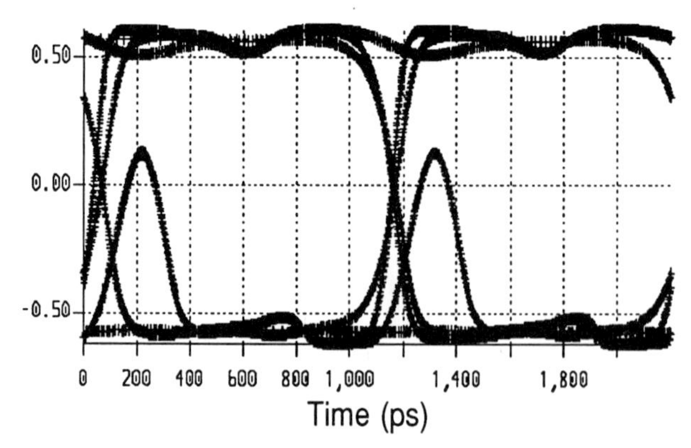

Fig. 25. Simulated receiver response with 1-nH supply inductance.

lem" can be significantly reduced. The crosstalk and noise issues of large-scale OEICs can be readily handled, and simple, cost-effective solutions can be found, provided that sufficient care is taken to model the OEIC within the whole optical system or link of which it is part.

5. SUMMARY

Large-scale OEICs can be designed and fabricated with raw performance (speed and sensitivity) rivaling that of comparable discrete alternatives. It is necessary to address the unique challenges presented by large-scale OEICs, noise and crosstalk, by extensive modeling of the circuits, including the effects of chip and package parasitics as well as optical noise sources and distortion. In this way it is possible to realize the density and cost advantages associated with integration.

REFERENCES

1. M. Dagenais, R. Leheny, H. Temkin, and P. Bhattacharya, Applications and challenges of OEIC technology: a report on the 1989 Hilton Head workshop. *IEEE J. Lightwave Technol.* **LT-8**, 846–862 (June 1990).
2. J. F. Ewen *et al.*, High speed GaAs fiber optic link. *Proc. Int. Conf. on Int. Optics and Fiber Optic Comm.* (1989).
3. N. Yamanaka *et al.*, A GaAs 1.8 Gb/s opto-electronic universal switch LSI with monolithically integrated photo-detector and laser driver for b-isdn. *Proc. ECOC '89* Gothenburg, Sweden (September 1989).
4. K. Kobayashi, Integrated optical and electronic devices, in "Optical Fiber Telecommunications II" (S. E. Miller and I. P. Kaminow, eds.), ch. 15, pp. 601–630. Academic Press, Orlando, FL.
5. T. Horimatsu and M. Sasaki, OEIC technology and its application to subscriber loops. *IEEE J. Lightwave Technol.* **LT-7**, 1612–1622 (November 1989).
6. Y. H. Lo, P. Grabbe, M. Z. Iqbal, R. Bhat, J. L. Gimlett, J. C. Young, P. S. D. Lin, A. S. Gozdz, M. A. Koza, AND T. P. Lee, Multigigabits/s 1.5 μm $\lambda/4$ shifted DFB OEIC transmitter and its use in transmission experiments. *IEEE Photon. Technol. Lett.* **2**, 673–674 (September 1990).
7. M. Nakamura, N. Suzuki, and T. Ozeki, The superiority of opto-electronic integration for high-speed laser diode modulation. *IEEE J. Quantum Electron.* **QE-22**, 822–826 (1986).
8. T. Suzaki, Y. Suzuki, H. Yamada, S. Fujita, H. Hida, M. Kitamura, and M. Shikada, 10 Gb/s optical transmitter module with MQW DFB-LD and DMT driver IC. *Electron. Lett.* **26**, 151–152 (January 1990).

9. J. D. Crow, Optical interconnect technology for multiprocessor networks (a critical review), in "Digital Optical Computing" (R. A. Athale, ed.), pp. 77–99. SPIE Optical Engineering Press, 1990.
10. A. R. Goodwin, R. G. Plumb, M. J. Robertson, and D. James, GaAlAs lasers with aging rates of less than 0.1% per thousand hours. *8th European Conf. on Optical Communications*, pp. 107–109, (1982).
11. O. Fujita, Y. Nakano, and G. Iwane, Reliability of semiconductor lasers for undersea optical transmission systems. *IEEE J. Lightwave Technol.* **LT-3**, 1211–1216 (December 1985).
12. P. K. Runge and P. R. Trischitta, The SL undersea lightwave system. *IEEE J. Selected Areas Commun.* **SAC-2**, 784–793 (November 1984).
13. K. Aida and M. Amemiya, Undersea transmission-system reliability with laser-diode standby-redundant optical repeaters. *IEEE Trans. Reliability* **R-33**, 439–441 (December 1984).
14. K. Amemiya, K. Aida, and M. Aiki, Reliability analysis of submarine transmission systems using repeaters with redundant optical sources. *Trans. Inst. Electron. Commun. Eng. Jpn.* **J68-A**, 1053–1060 (October 1985).
15. R. J. S. Bates, Using monolithic laser arrays for improved transmitter availability in computer data links, in "Fiber Optics Reliability: Benign and Adverse Environments," pp. 80–85. SPIE Vol. 842, 1987.
16. R. E. Epworth, The phenomenon of model noise in analog and digital optical fiber systems. *4th European Conf. on Optical Communications*, pp. 492–501 (1978).
17. R. Dændliker, A. Bertholds, and F. Maystre, How model noise in multimode fibers depends on source spectrum and fiber dispersion. *IEEE J. Lightwave Technol.* **LT-3**, 7–15 (February 1985).
18. R. J. S. Bates, Multimode waveguide computer data links with self-pulsating laser diodes. *Proc. ICO/OITDA Int. Topical Meeting on Optical Computing*, p. 9E3 (April 1990).
19. R. J. S. Bates, Low coherence laser diodes for computer data-links, in "Lasers and Electro-Optic Society Annual Meeting," Boston, MA (November 1990).
20. J. E. Thompson, T. H. Baird, and R. D. Hoffe, Modal noise in laser-based multimode optical interconnects, in First International Workshop on Photonic Networks, Components and Applications (October 1990), "Optics and Photonics" vol. 12, pp. 254–259. World Scientific Publ., 1991.
21. T. Tanaka, T. Kawano, and T. Kajimura, High-power operation in self-sustained pulsating AlGaAs semiconductor lasers with multiquantum well active layer. *Appl. Phys. Lett.* **53**(25), 2471–2473 (1988).
22. S. Matsui, H. Takiguchi, H. Hayashi, S. Yamamoto, S. yano, and T. Hijikata, Suppression of feedback-induced noise in short-cavity V-channeled substrate inner stripe lasers with self-oscillation. *Appl. Phys. Lett.* **43**(3), 219–221 (1993).
23. R. L. Soderstrom, T. R. Block, D. L. Karst, and T. Lu, Low cost high

performance components for computer optical data links, *IEEE Lasers and Electro-Optics Society Annual Meeting*, Orlando, FL. (October 1989).

24. Optoelectronics Industry Technology and Development Association, Japan, Present and Projected Market for Optoelectronic Sources (March 1989).
25. K. PETERMAN, Laser diode modulation and noise, in "Advances in Optoelectronics," pp. 145–151. Kluwer Academic, Boston (1988). ISBN 90-277-2672-8.
26. B. S. POH, Prediction of self-sustained oscillations in buried heterostructure stripe lasers. *IEE Proc*. **132**(pt. 1, no. 1), 29–33 (1985).
27. C. H. HARDER, K. Y. LAU, AND A. YARIV, Bistability and pulsations in cw semiconductor lasers with a saturable absorption. *Appl. Phys. Lett*. **39**, 382–384 (September 1981).
28. F. BOSCH, G. L. DYBWAD, AND C. B. SWAN, Laser fiber-optic digital system performance improvements with superimposed microwave modulation, *Conf. Lasers and Electro-Optic*, p. TuDD7. IEEE, February 1980.
29. Laser digital transmitter, U.S. Patent 4,317,236.
30. R. HÜGLI AND R. J. S. BATES, Reduction of modal noise in high speed short distance computer data links by an electrical signal. *IEEE J. Lightwave Technol*. **LT-10**, 1788–1793 (December 1992).
31. J. F. EWEN K. P. JACKSON, R. J. S. BATES, AND E. B. FLINT, GaAs fiber optic modules for optical data processing networks. *IEEE J. Lightwave Technol*. **LT-10**, 1755–1763 (December 1992).
32. M. W. MAEDA *et al*., Multigigabit/s operation of 16-wavelength vertical-cavity surface-emitting laser array. *IEEE Photon. Technol. Lett*. **3**, 863–865 (October 1991).
33. O. WADA, H. NOBUHARA, T. SANADA, M. MAKIUCHI, T. FUJII, AND T. SAKURAI, Optoelectronic integrated four-channel transmitter array incorporating AlGaAs/GaAs quantum-well lasers. *IEEE J. Lightwave Technol*. **LT-7**, 186–197 (February 1989).
34. K. SATA AND M. MURAKAMI, Experimental investigation of thermal crosstalk in a distributed feedback laser array. *IEEE Photon. Technol. Lett*. **3**, 501–503 (June 1991).
35. R. A. LAFF, L. D. COMERFORD, J. D. CROW, AND M. J. BRADY, Thermal performance and limitations of silicon-substrate packaged GaAs laser arrays. *Appl. Opt*. **17**, 778–784 (March 1978).
36. D. ROGERS, Integrated optical receivers using MSM detectors. *J. Lightwave Technol*. **9**(12), (1991).
37. D. ROGERS AND R. BATES, A fully integrated high sensitivity PIN/FET optical receiver at 250 mbaud. *Proc. Eur. Conf. on Opt. Commun*. (1988).
38. H. YANO *et al*., Low Noise Current Optoelectronic Integrated Receiver with an internal Equalizer for Gbit/s long wavelength optical communications. *Proc. Opt. Fiber Commun. Conf*. (1990).
39. T. V. MUOI, Receiver design for high-speed optical-fiber systems. *J. Lightwave Technol*. **LT-2**, 243–267 (1984).

40. D. L. ROGERS, Monolithic integration of a 3-GHz detector/preamplifier using a refractory-gate, ion-implanted MESFET process. *Electron Device Lett.* **EDL-7**, (November 1986)
41. F. M. GARDNER, "Phaselock Techniques," 2nd ed. Wiley, New York, 1979.
42. R. ROSENBERG, D. ROSS, P. TRISCHITTA, D. FISHMAN AND C. ARMITAGE, Optical fiber repeatered transmission systems utilizing SAW filters. *IEEE Trans. Sonics Ultrason.* **30**, 119–126 (May 1983).
43. P. WALLACE, R. BAYRUNS, J. SMITH, T. LAVERICK, AND R. SHUSTER, A GaAs 1.5 Gb/s clock recovery and data retiming circuit. *ISSCC Digest of Tech. Papers,* pp. 192–193, IEEE (1990).
44. J. EWEN, D. ROGERS, A. WIDMER, F. GFELLER, AND C. ANDERSON, Gb/s fiber optic link adapter chip set. *GaAs IC Symposium,* Nashville, TN, pp. 11–14 (November 1988).
45. W. LINDSEY AND C. CHIE, A survey of digital phase-locked loops. *Proc. IEEE* **69**, 410–431 (April 1981).
46. M. WAKAYAMA AND A. ABIDI, A 30-MHz low-jitter high-linearity CMOS voltage-controlled oscillator. *IEEE J. Solid-State Circuits* **22**, 1074–1081 (December 1987).
47. R. BAUMERT, P. METZ, M. PEDERSEN, R. PRITCHETT, AND J. YOUNG, A monolithic 50–200 MHz GMOS clock recovery and retiming circuit. *IEEE Custom Int. Circuits Conf.,* pp. 14.5.1–14.5.4 (1989).

Chapter 14

MODELING FOR OPTOELECTRONIC INTEGRATED CIRCUITS

R. Baets, D. Botteldooren, G. Morthier, F. Libbrecht, and P. Lagasse

University of Gent—IMEC
Department of Information Technology
Gent, Belgium

1. INTRODUCTION

In optoelectronic systems component integration is widely considered to be a major condition for optimal performance and/or cost-effective manufacturing. Integration helps to reduce the parasitics and cost of packaging and brings the economics of wafer-scale production to optoelectronics.

There are different kinds of optoelectronic circuits:

ISBN 0-12-200420-5

- Integration of optoelectronic devices with electronic circuits
- Integration of optoelectronic devices with passive optical devices (waveguides, lenses, etc.)
- Integration of optoelectronic devices into arrays (e.g., laser arrays)

In the two first cases a problem is encountered in that very dissimilar devices need to be integrated. In general, this leads to very difficult technological problems which have slowed down the impact of optoelectronic ICs on practical applications.

Modeling has a role to play in this development. First, integration often implies that one has to search for a compromise in the device design due to the presence of other devices on the same chip. This leads to difficult tradeoffs and modeling can be useful in this, particularly because it enables quick time to market, cost-effective, and reliable designs. A second reason to use modeling is that integration, with close distances between devices, can lead to undesired *crosstalk effects* (electrical, optical, thermal, or even elastic). Again, modeling can help to establish the strength of the coupling effects and allows design of structures with optimized performance. Finally, modeling also allows us to gain a deep understanding of complex device characteristics (to "look inside the devices") and verify the device operation within a system environment.

In the process of designing an optoelectronic integrated circuit in detail, one may face a diverse range of questions, from basic questions about material properties to questions about the overall performance of the circuit. This implies that one may want to use a range of different models: quantum mechanical semiconductor models, macroscopic semiconductor models, electromagnetic models (for optical or electrical RF problems), circuit models, and system models. Ideally, one would like different levels of modeling to be decoupled from each other as much as possible, in such a way that each higher level uses only parameters provided by lower levels. This is a common situation in electronic integrated circuits. In optoelectronic devices and integrated circuits, however, one may need to combine different types of models into one model. This is the case in *semiconductor lasers,* where one may need to combine a semiconductor model with an electromagnetic model. In an integrated optic circuit with waveguide devices, it is difficult to do a circuit-like analysis because of the problem of defining a circuit description for the *radiation from waveguide discontinuities*. In a system-oriented model it is usual to make use of simple "blackbox" descriptions of the constituent components. This may be difficult in a situation in which the system performance is critically dependent on interactions between the components (e.g., optical feedback into a laser diode). In such a situation it may be necessary to combine a system model with a detailed device model.

In this chapter we describe a wide range of models useful in the analysis and design of optoelectronic ICs. We cover material models, waveguide device models, laser device models, and systems models. Some special attention is given to coupling problems. Purely electronic devices and their modeling are not discussed in this chapter. We refer ref. 1 for more details about this subject.

2. MODELING OF MATERIAL STRUCTURES

When designing optoelectronic devices and OEICs, the choice of an appropriate material structure is of primary importance. Optical properties of the various useful semiconductors have been measured and catalogued. The variety of semiconductor alloys used calls for some theoretical background to interpolate the properties of ternary and quaternary alloys from the known properties of the composing binary semiconductors. Moreover, in active devices, the influence of electric fields and carrier concentrations becomes important. These extra parameters can be taken into account only by making a large number of experiments or by using a suitable theoretical model. Finally, when low-dimensional semiconductor structures, such as quantum wells (QWs), quantum well wires (QWWs), and quantum dots (QDs) are introduced, the number of parameters that influence the optical properties increases further. For these material combinations, optimum design parameters can be obtained only by using adequate models.

The physical effects that make up the optical and especially the optoelectronic properties of bulk semiconductors on one hand and of quantum wells on the other hand are essentially the same. However, the relative contributions of different effects to the macroscopic properties and the approximations one can make differ sufficiently to make us consider both topics separately.

2.1. Models for Bulk Material

The physics underlying models for the evaluation of optical properties of semiconductors are mostly split into different contributions rather than trying to solve microscopic equations taking all effects into account. Depending on the problem studied, only the relevant contributions are taken into account.

2.1.1. Gain and Refractive Index Change

In semiconductor lasers high carrier concentrations are injected in the active layer. Moreover the optical field strength is high. Under these conditions the optical gain has to be calculated. Different approaches are reported in literature. They all predict the gain quite accurately for frequencies near the gain maximum. Most divergence between the models exists near the low energy side of the bandgap, the so called *band-tails* [2].

Ab initio calculations of the semiconductor band structure followed by a gain calculation require too much computer power and are never used. Furthermore a lot of models neglect carrier interactions or limit the effect of carrier interactions to a rigid band shift. In this case, gain as a function of pulsation is found as

$$g(\omega) = \frac{-4\pi^2 e^2}{ncm^2\omega} \sum_{c,v} \int_{BZ} \frac{2\,d\mathbf{k}}{(2\pi)^3} |\varepsilon.\mathbf{M}_{cv}(k)|^2\, \delta(E_c(\mathbf{k}) - E_v(\mathbf{k}) - \hbar\omega), \quad (1)$$

where n and m are the background refractive index and the electron mass, $\mathbf{M}_{cv}$ is the *matrix element* for optical transitions between conduction band c and valence band v. The summation is over all contributing conduction and valence bands; the integration is over the first Brillouin zone.

As the optical properties of semiconductors are influenced primarily by the band edges, a **k.p** *description* [3] of conduction band minima and valence band maxima, using measured masses, bandgaps, and transition probabilities, is often preferred to a rigorous calculation of the energy band structure of the semiconductor. Gain and refractive index change due to carrier injection can be obtained directly by combining the results of the **k.p** method with relaxation broadening [4]. A constant relaxation time results in an inaccurate value for the gain at low photon energies. To improve the results, the scattering mechanisms that introduce the low-energy optical transitions were studied and a better line form was introduced [5].

More recently, Haug *et al.* [6, 7] developed a gain model including *many-body effects*. The screened electron–hole Coulomb interaction that they introduce seems to enhance the gain. Details can be found in the references.

The direct influence of the strong optical field on the stimulated emission is insignificant and to our knowledge has never been taken into account in semiconductor gain calculations. However, the change of the gain due to the effect of the optical intensity on the carrier populations can become quite significant in laser modeling and may be calculated as a third-order gain contribution using the relaxation theory [2, 8]. Although no consensus has been reached on the exact nature of the underlying physics, the term is mostly referred to as *spectral hole burning* and attributed to thermal relaxation of the holes giving the optical transition.

2.1.2. Conduction Band–Valence Band Absorption

Conduction band–valence band absorption is modeled in exactly the same way as gain. Modulation of absorption and refractive index change by injecting carriers can be modeled using the gain models described earlier [9]. However, problems may arise if band tails are not taken into account accurately enough, as the energy region of interest for applications lies just below the bandgap.

2.1.3. Inter-Valence-Band Absorption

In devices operating at long wavelengths, photons can be absorbed by transitions of holes from one valence band (e.g., HH band) to another (e.g., SO band). To model this type of absorption, an adapted form of (1) has to be evaluated. It is important, however, to describe the valence bands accurately in this calculation because transitions occur at high $|k|$ values. In particular, nonparabolicity and dependence on k direction have to be taken into account.

2.1.4. Free-Carrier Effects

High carrier concentrations in semiconductors behave to a first-order approximation as a free carrier cloud with a well-chosen mass. Photons with low energy interact with the quasi-free particles, giving a specific absorption and refractive index variation. Qualitative values can be calculated theoretically [10], but quantitative free-carrier absorption and refractive index data can only be derived semiempirically.

2.1.5. Linear Electro-optic Effect (LEO)

When an electric field is applied to a semiconductor, both absorption and refractive index will change. For device applications one is interested primarily in the refractive index change. A linear (LEO) and a quadratic (Franz–Keldish) dependence on the applied electric field can be distinguished. In ref. 11 a semiempirical model is developed to describe the LEO effect. The dispersion is estimated theoretically, while the magnitude is found from comparison with experimental data.

2.1.6. *Franz–Keldish Effect*

Although the quadratic electro-optic effect can be studied in exactly the same way as the linear effect [12], an interesting full-theoretical calculation of the effect close to the bandgap can be found in ref. 13.

2.2 Models for Quantum Wells, Wires, and Dots

Models for the optical properties of low-dimensional semiconductor structures can essentially be the same as the ones used for bulk semiconductors. Quantization of the conduction band (CB) and the valence band (VB) into subbands and the existence of excitons at room temperature are the two most important features that determine the typical QW, QWW, and QD behavior.

2.2.1. *Gain and Refractive Index Change*

In contrast to the bulk calculations described earlier, QW gain calculations often start with evaluation of the subbands contributing to the gain. Simpler models for the subband structure solve the quantum mechanical particle-in-a-box problem for electrons and holes separately [14]. The calculation of the optical property itself is then a simple extension of the bulk calculation.

It has been pointed out, however, that coupling between light and heavy holes caused by the reduced symmetry in the QW has a significant influence on the calculated gain [15].The subband calculation now requires diagonalization of a **k.p** interaction matrix for all *k* vectors in the plane of the QW [16] or an approach based on a full microscopic description of the semiconductor materials in the QW, for example, a tight-binding approximation [17]. Both are quite time-consuming calculations.

Material structures with lattice-mismatched layers introduce *strain*. Extremely thin layers can accommodate a considerable amount of strain without relaxing. This strain alters the subband structure in the QW. It can easily be incorporated in the more sophisticated subband structure calculation [18].

2.2.2. *Conduction Subband–Valence Subband Absorption—Excitons*

Direct absorption between valence and conduction subbands can be modeled in exactly the same way as the gain, described before. However, the confinement of an electron and a hole in the small QW enhances the Coulomb attraction and allows *excitons* to be generated, even at room temperature. For each CB–VB subband pair that contributes to the absorption, exciton states have to be calculated. The overall problem is far from trivial. Generally one introduces a well-chosen form for the exciton wave function that makes the problem split into the known QW problem for the holes and the electrons and a two-dimensional problem in the plane of the QW that can be solved, for example, with variational techniques. The exciton can be studied neglecting hole coupling [19] or taking it into account [20]. The latter case is much more CPU-time consuming, especially if, as stated in the reference, it makes little sense to perform the calculation with the assumption of circular symmetry for the exciton wave function in the plane of the QW.

2.2.3. *Stark Shift*

The change of the absorption under influence of a static electric field is one of the most useful properties of low-dimensional semiconductor structures. Once the models for absorption are developed, it is a straightforward exercise to examine the influence of a static electric field [21, 22]. A few remarks have to be made, however:

- To model high-field Stark shifts, the model must be able to handle electrons and holes that tunnel out of the QW under influence of the field.
- The influence of the electric field on the line broadening must be taken into account to get a valid prediction of the absorption change.

2.2.4. *Carrier-Induced Effects*

The study of absorption and refractive index modulation introduced by injecting carriers in the QW region is considerably more complicated than the study of the equivalent modulation in bulk materials because one has to take into account many-body effects. The first-order effects are a filling of k-space that reduces the absorption at low photon energies and a screening of the Coulomb attraction that reduces exciton strengths. Detailed descriptions can be found in refs. 23 and 24.

2.3 Example: QW for Optimum Intensity Modulator Performance

As stated in the introduction, theoretical models for the optical properties can be used to find the best material and QW sizes for a certain application. In this example our target will be to obtain a maximum change of absorption under the influence of an electric field. We must choose how rigorous the model will be and to what extent it will rely on experimental parameters. An almost completely experimental approach starts from absorption spectra measured under various electric field conditions and fits Lorentz peaks (for the excitons) and step functions (for the band–band absorption) to them. For fields, QW sizes, etc. in between measurements the amplitudes, positions, and linewidths of the different contributing absorption peaks are found by interpolation.

A semiempirical model [25] using a single calculated exciton peak with amplitude and linewidth extracted from experimental data and a full theoretical calculation of the influence of the electric field gives remarkably good coincidence with experimental data. The model can be used as a fast design tool. As an example, the relative absorption changes as a function of photon energy and QW sizes for InGaAs QWs are shown in Fig. 1.

A last approach can be to make a full theoretical model as accurate as possible, taking into account all the relevant effects mentioned [26]. Although it cannot be expected that this type of modeling will yield quantitatively accurate absorption data, it has the advantage of allowing the study of types of QWs that were never actually grown.

3. MODELING OF WAVEGUIDE DEVICES

Many optoelectronic or photonic integrated circuits contain *waveguide devices,* as either passive or electrically driven structures. Modeling has an important role in the analysis of these waveguide devices. There are several reasons for this, one being that the characteristics of very few waveguide structures can be derived adequately from simple analytic expressions and that intuition is often misleading. Simple structures like *curved waveguides* or *junctions* can be analyzed only by numerical models. The high number

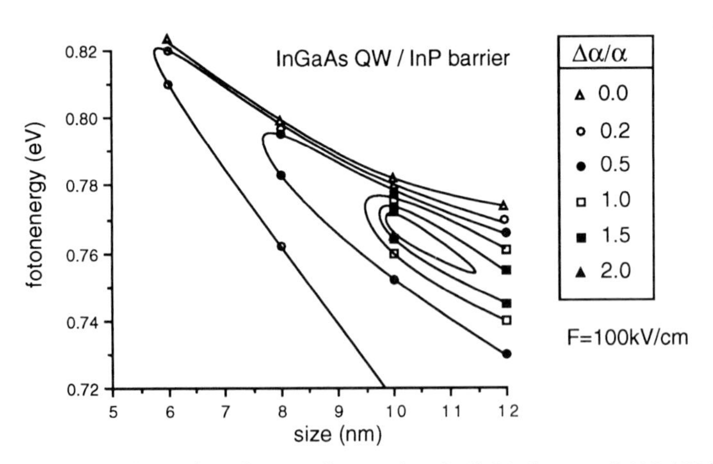

Fig. 1. Relative absorption changes for an electric field change of 100 kV/cm as a function of photon energy and QW size for InGaAs/InP QWs.

of structural parameters in these waveguide structures makes a purely experimental optimization by trial and error impractical.

The designer of a waveguide device or circuit faces a number of questions. The waveguide devices in the circuit will in general contain straight sections and transitory sections. Straight sections can be handled by modal methods, implying the reduction of a three-dimensional problem to a two-dimensional problem. However, this is practical only in waveguides with a low number of modes and under the condition that the radiation modes in the input and output plane have little impact on the overall performance. This latter condition is certainly not always fulfilled in integrated structures. Real multimode waveguide circuits are not easily analyzed. Often *ray tracing* is used in this case. This poses two problems. A fundamental problem is that any diffraction effects are neglected, although they may well be relevant. A practical problem is that it is cumbersome to deal with the entire three-dimensional configuration. Nevertheless, the behavior of the structure is often critically dependent on details of the geometric structure and the excitation. These are some reasons why multimode structures are not used much in integrated optical or optoelectronic circuits.

Transitory sections are more difficult to analyze. In fact, one now has to deal with a scattering problem, the rigorous solution of which is very difficult in most situations. Even a simple problem like the reflection and transmission at the interface between a waveguide and uniform space is relatively complex if worked out rigorously. This has led to the development of a number of approximate techniques, which are accurate only under a number of conditions such as low refractive index contrast (= the difference in the refractive index of neighboring points) or paraxial wave propagation. The most well-known example of these methods is the *beam propagation method* (BPM). In fact, BPM-like models form the only method that combines reasonable complexity with relatively general applicability. Yet, the basic BPM algorithm imposes many restrictions on the structure to be analyzed and there is intensive work to loosen these restrictions.

A special class of models is based on the *coupled mode theory*. This is a perturbation method that can be used for both modal analysis and propagative calculations. It uses knowledge about a waveguide structure to analyze a more complex structure that deviates to a small extent only from the simpler (unperturbed) structure. Coupled mode theory has been used widely to analyze coupled waveguides and waveguides with distributed Bragg reflection. In a sense, the basic BPM is also a coupled mode model with the unperturbed structure being uniform space and its solutions plane waves. In this section we focus attention on modal and propagative techniques.

3.1. Modal Analysis

A wide range of models and numerical techniques can be used to find the propagation constants and field profiles of a *stripe waveguide* [27]: finite-difference methods [e.g., 28], finite-element methods [e.g., 29], integral equation methods [e.g., 30], and mode-matching methods [e.g., 31]. Apart from being numerically different, all these methods differ in the following aspects:

- Vectorial solution or scalar approximation
- Occurrence of spurious modes
- Capability of handling graded-index structures
- Capability of handling high index contrast and/or very thin layers
- Capability of handling complex refractive indices (structures with gain or loss)
- Capability of handling anisotropic structures

Although all these methods claim to be rather rigorous, an extensive comparison [25] shows that the results for the propagation constant can differ substantially between the various models. These differences are due mainly to errors introduced by the numerical procedure, such as rounding-off errors, boundary box effects, and mesh discretization effects. Those errors normally become larger as the mode number increases. This shows that most implementations of these models do not contain much intelligence to find a good compromise, depending on the specific structure, between numerical accuracy and computation time.

The methods mentioned so far are relatively rigorous. A more approximate and simpler model is the well-known *effective index method* (EIM) [32]. It basically reduces the two-dimensional problem to a set of one-dimensional slablike problems in the transverse direction and a single one-dimensional slablike problem in the lateral dimension. The approximation is accurate only if the lateral variations in the waveguide structure are weak. This implies immediately that the method is adequate only for stripe waveguides, which can be considered to be laterally variable slab waveguides. If the lowest-order slab mode goes into cutoff at the left and right of the stripe, one faces an ambiguity. Nevertheless, the effective index method rapidly provides information about waveguide structures. With some experience from the user side concerning its validity for a given structure, it is a very useful method for a quick design. Because of the numerical efficiency of the method, it is possible to build it into more complex models such as those used for semiconductor lasers. There are a number of variations of the effective index method, such as the *weighted index method* [33], the *corrected EIM* [34], the *azimuthal EIM* [35], and the *spectral index method* [36]. Most of these variations target specific waveguide geometries, for which they perform better than the conventional EIM.

Figure 2 shows an example of the lateral electric field component of the two lowest-order modes in a polymeric waveguide, which was experimentally integrated with GaAs/AlGaAs lasers [37]. Both the EIM and the finite-difference method were employed for this particular configuration. It is seen

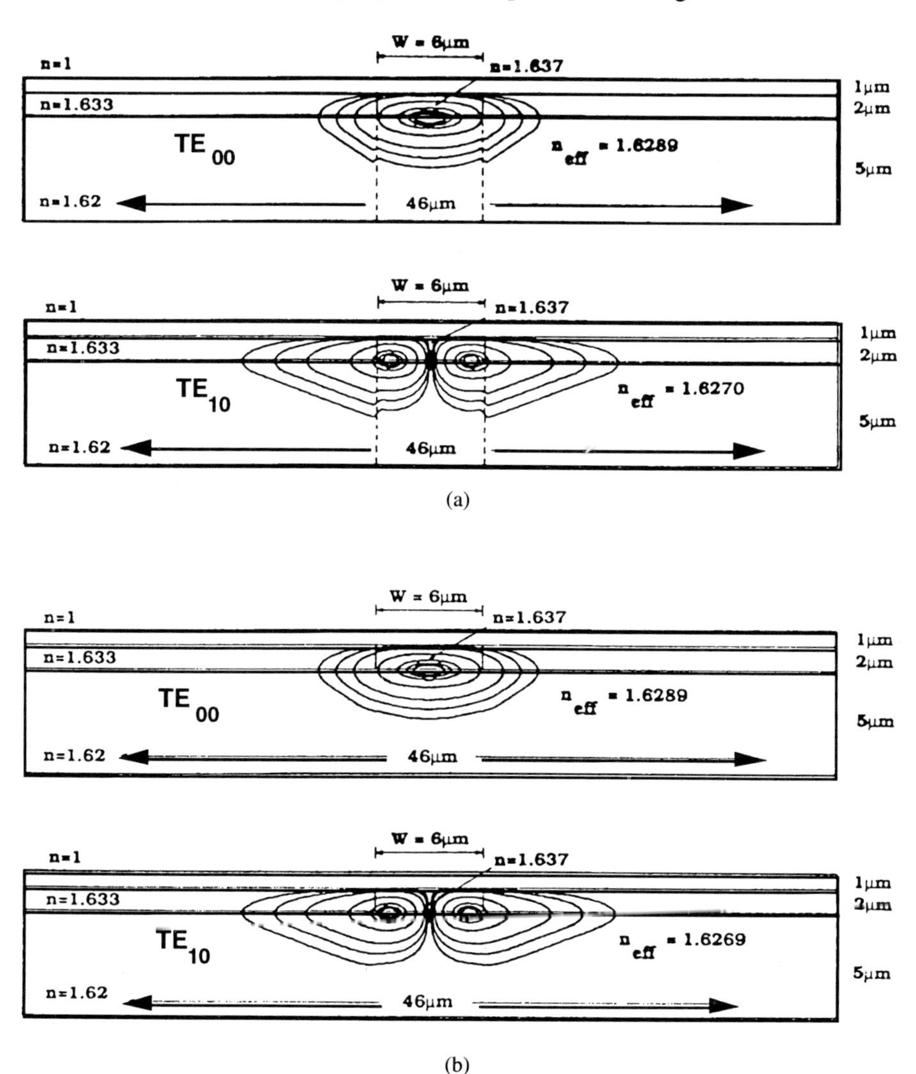

Fig. 2. Intensity contours for the two lowest-order modes in a polymeric stripe waveguide: (a) efective index calculation; (b) finite-difference calculation.

that the field distributions are similar for both methods, except for the small (unphysical) discontinuities in the EIM solution at the stripe boundaries. This is indeed an example where the EIM can be expected to behave well.

3.2. Propagative Models

In transitory, i.e., longitudinally variant, structures the guided modes are usually coupled to the *radiation modes* and, in consequence, part of the power guided by the waveguide structures is lost. One of the most commonly used methods that takes into account radiation losses is the *beam propagation method* (BPM) [38–41].

This method has been used widely for the analysis of *curved waveguides, couplers, junctions,* etc. In its original form, however, it is restricted to paraxial scalar fields and structures with a relatively small refractive index contrast. This implies that a surface waveguide at a material interface to air cannot easily be handled by BPM. Even for a thin single-mode III–V waveguide with a refractive index contrast of 0.2 to upper and lower cladding layer, the BPM is working at its limits.

The BPM consists of a stepwise propagation of a field through a structure. For each step the field is first propagated as if the structure were homogeneous and then a correction is applied. The method can be used for both three-dimensional and two-dimensional problems. Often a three-dimensional structure is reduced to a two-dimensional problem through the use of EIM before running the BPM (in two dimensions). In this way the computing times can be largely reduced and the common high index contrast in the transverse direction is no longer a problem.

The BPM has a number of important advantages. The algorithm is remarkably simple and numerically stable. This is mainly due to the fact that it is based on the fast Fourier transform algorithm. Although it is a noniterative algorithm, high accuracy can be obtained with an adequate choice of the numerical parameters (lateral mesh size, propagation step, etc.) Furthermore, the method can be readily used for nonlinear waveguides [42] and waveguides with loss or gain. This means that the BPM can be used in a laser model [43].

A great deal of effort has been spent to extend the applicability range of the BPM. One extension is the bidirectional BPM [44, 45], which is capable of analyzing structures with large reflections. Another extension is to include the vectorial electromagnetic effects and to allow for anisotropic materials. A different approach to beam propagation is to use finite-difference techniques rather than Fourier transform–based techniques [46]. This can offer either increased accuracy relative to the paraxial approximation or higher speed. Moreover, it allows one to implement alternative boundary conditions and to use a nonuniform mesh. The tranverse derivatives of the electric field are evaluated less accurately, however, than in the standard BPM.

4. MODELING OF LASER DIODES

Owing to their high efficiency, small size, and easy electronic modulation, *laser diodes* are the ideal light sources in optoelectronic circuits. However, their behavior is not always easily understood and depends on wave propagation in an electronically excited semiconductor, wherein various electronic, optical, and electro-optical interactions can take place. The basic structure of most semiconductor lasers, the *double heterojunction,* assures the confinement of excited carriers (electrons and holes) as well as the confinement of light (photons). It forms a potential well for the excited carriers and, at the same time, a waveguide for the light. The resulting high carrier and photon densities in the active layer then often make the semiconductor an optically nonlinear and nonuniform material. It must be emphasized that the complex refractive index in a laser diode depends not only on the injected currents but also on the optical field intensities. The complex behavior of laser diodes and the variety of output characteristics on which their performance can be judged have led to the development of different types of laser models. These models differ by the degree of complexity and the approximations applied, but also by the objectives of the modeling, i.e., what needs to be optimized.

Models for laser diodes can be based on the previously discussed material and waveguide models or at least make use of their outcome. Depending on the complexity of a laser model, material and/or waveguide models can be included entirely in the laser model. Alternatively, simple relations expressing the dependence of gain and refractive index on carrier density, wavelength, etc. or the dependence of the mode confinement factor on the transverse/lateral waveguide structure can find their way from material and waveguide models into a laser model.

4.1. Overview of Different Types of Laser Models

In a laser diode one is interested in finding the modes and the *resonance wavelengths* of a three-dimensional cavity; i.e., one is looking for a field distribution at a certain wavelength that reproduces itself after one round trip in the cavity. The stimulated emission depends on the field intensities and is therefore not uniform in the cross section or in the longitudinal direction. Because it depletes the carrier density, it follows that the spatial variation of carrier density and fields needs to be determined iteratively.

Models taking into account the detailed cross-sectional geometry are usually restricted to the analysis of the static behavior [47]. Such models serve well in the study of the emitted *near- and far-field pattern* or to investigate the influence of changes in waveguide dimensions or composition. They are also ideally suited for the static analysis of lasers in which the cross section varies in the longitudinal direction [48] or the optical field is not tightly confined by the waveguide (as in *gain-guided lasers* [49]). In the analysis, it is generally assumed that the carrier density is uniform over the active layer thickness. The electronic problem then consists of solving the *diffusion equation* in the active layer and the *Poisson equations* in the passive layers. The electromagnetic problem is often reduced to a two-dimensional problem by application of the effective index method to the transverse direction. This two-dimensional problem can be solved, e.g., with the BPM [43, 46, 47].

If the spectral or the dynamic behavior of a laser diode is to be analyzed, one can further simplify the spatial complexity of the problem by averaging over both the lateral and transverse directions. Each uniform waveguide can then be characterized by an effective index n_e and by *optical confinement factor* Γ (fraction of the optical power that is confined to the active layer) [50]. Nevertheless, the refractive index in the active layer depends on the carrier density or the field intensity and this must also be taken into account in the effective index. The variation of the effective index n_e caused by the carrier-induced refractive index change Δn_a can be approximated as

$$n_e^2 = n_{e,0}^2 + 2\Gamma n_{a,0}\Delta n_a \tag{2}$$

with $n_{a,0}$ the refractive index of the unpumped active layer. The reduced spatial complexity allows one to consider variations in the spectral or in the time domain. By including the spontaneous emission that couples into each mode and by taking into account the phase resonance conditions, the distribution of the optical power over the different modes and the wavelength of the latter can be determined. For a structure with *distributed feedback* the wave propagation for the longitudinal mode q can be described by the *coupled wave equations* Eqs. (3a) and (3b), while the carrier density in the active layer obeys the *continuity equation* Eq. (4) [51]:

$$\frac{\partial R_q^+}{\partial z} + \frac{1}{v_g}\frac{\partial R_q^+}{\partial t} + \left(j\,\Delta\beta_q - \frac{J_{sp}}{|R_q^+|^2}\right)R_q^+ = \kappa^+ R_q^- + F_q^+, \tag{3a}$$

$$-\frac{\partial R_q^-}{\partial z} + \frac{1}{v_g}\frac{\partial R_q^-}{\partial t} + \left(j\,\Delta\beta_q - \frac{J_{sp}}{|R_q^-|^2}\right)R_q^- = \kappa^- R_q^+ + F_q^-, \tag{3b}$$

$$\frac{\partial N}{\partial t} = \frac{J}{qd} - \frac{N}{\tau(N)} - \sum_q \frac{\Gamma g(N,\omega_q)}{h\omega_q wd}\,(|R_q^+|^2 + |R_q^-|^2) + F_N. \tag{4}$$

$R_q^{\pm}$ are the slowly varying complex amplitudes of the forward- and backward-propagating fields $E_q^{\pm}$ $[E_q^{\pm} = R_q^{\pm}\ \exp(j\omega_q t\ \pm\ \beta_0 z)]$ at the frequency $\omega_q \cdot \Delta\beta_q$ denotes the difference between the actual propagation constant of the waveguide and the reference propagation constant, and J_{sp} represents the average *spontaneous emission* that couples in the mode. The reference propagation constant in waveguides with a grating is usually chosen as the Bragg constant (= $m\pi/\Lambda$ for an mth order grating of period Λ). $\kappa^{\pm}$ are the *coupling coefficients* that describe the coupling between forward- and backward-propagating fields if a grating exists in the waveguide. The Langevin functions $F_q^{\pm}$ and F_N represent the spontaneous emission noise [52]. w and d are the width and thickness of the active layer and g is the optical gain.

Longitudinal variations of the carrier density, an effect known as *spatial hole burning* and of great importance in distributed feedback (DFB) lasers, can be accounted for by dividing the laser into small sections in which a uniform carrier density can be assumed. The static solution of the resulting multisection laser can then be determined iteratively. The dynamic behavior can be analyzed, e.g., by decomposing the coupled wave equations into equations for the amplitudes and the phases, by applying a Fourier transform to remove the time derivatives, and by linearizing the equations, which is justified for small-signal ac excitations such as small ac currents or noise sources. Discretization in the longitudinal direction then results in a linear set of algebraic equations that can be solved with standard numerical methods.

Most longitudinal models, developed since multisection lasers and DFB or distributed Bragg reflector (DBR) lasers started attracting a lot of attention, are still restricted to analysis of the static behavior [e.g., 53], although a few models also include the noise [54] or even an analysis of the dynamic behavior [49, 55]. One such modeling tool is *CLADISS* (Compound Laser Diode Simulation Software), which solves the longitudinal equations governing a DFB laser diode in a self-consistent way in both the static regime and small-signal dynamic operation [49]. It can handle the analysis of multisection lasers, with different sections being active or passive, with or without a grating and/or discrete reflections between different sections (e.g., due to discontinuities in the waveguide geometry). It allows one to calculate, e.g., the side mode suppression, the linewidth, and the RIN and small-signal ac responses. The small-signal analysis has also been extended with a calculation of second- and third-order harmonic distortion. A higher-order perturbation approach is used for this calculation.

The analysis of the dynamic and the noise properties of laser diodes is still often based on *rate equations* [46]. Such equations, obtained by averaging the longitudinal equations over the longitudinal coordinate, describe the variation in time of the number of photons in each mode, the frequency of each mode, and the number of carriers. Analytical solutions for the dynamic and noise characteristics can be obtained from these simple equations. The equations can easily be solved numerically for a large-signal regime. Alternatively, the rate equations are translated into an electrical equivalent circuit that can be handled by a circuit simulator such as *SPICE* [56, 57, 58].

It must finally be noticed that several models for the study of *thermal effects* in laser diodes have been developed. The detailed transverse/lateral geometry has a large influence on the thermal behavior and the (static or dynamic) heat equation is solved in these two dimensions in most models [59].

4.2. Lasers in OEICs

Distributed feedback and distributed Bragg reflector lasers are ideally suited for optoelectronic integration, since the feedback needed for resonance is provided by a grating rather than by a facet in such lasers. Facet reflectivities (which constitute the only feedback in Fabry–Perot lasers) between laser and, e.g., waveguide or amplifier are no longer needed in DFB lasers, and one facet (at the side of the grating section) is not required in DBR lasers. Up to now, DBR lasers have been used mainly as tunable, *local oscillator* lasers for coherent receivers, but the recent developments seem to indicate that similar or even better tuning can be achieved with AR (anti-reflection)-coated, multisection DFB lasers. Such lasers are to be preferred for integration purposes and could be integrated, for example, with a waveguide on one side and *a monitoring photodiode* on the other side.

Multisection DFB lasers also have the advantage that the currents in the different sections can be adjusted to obtain a certain wavelength, a large side mode suppression, a small linewidth, or a uniform FM response. For example, ordinary AR-coated and uniformly pumped lasers are not single mode at low power levels and nonuniform injection is one method, besides others like the inclusion of a phase shift in the grating or the use of non-

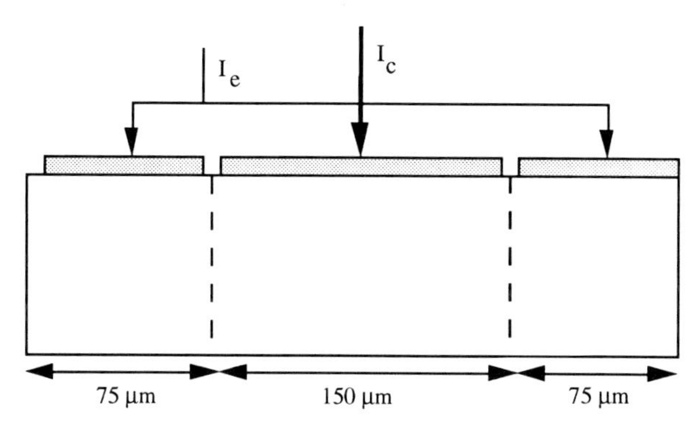

Fig. 3. Schematic drawing of a multielectrode λ/4-shifted laser, indicating the injected currents.

uniform waveguide or grating parameters, to overcome this. A three-section, λ/4-shifted laser (in which the grating phase is shifted by λ/4 at the center) is shown schematically in Fig. 3. Figure 4 illustrates how the laser becomes multimode under uniform injection and how this can be prevented by non-uniform injection.

Integrated structures often consist of a laser, a waveguide, and an *amplifier* or a detector [60, 61]. The waveguide and the amplifier and sometimes even the *detector* can then, for modeling purposes, be considered as sections of a more complicated multisection laser. In fact, a waveguide is nothing more than a passive, gratingless section, while an amplifier corresponds to an isolated active, gratingless section. Longitudinal models, such

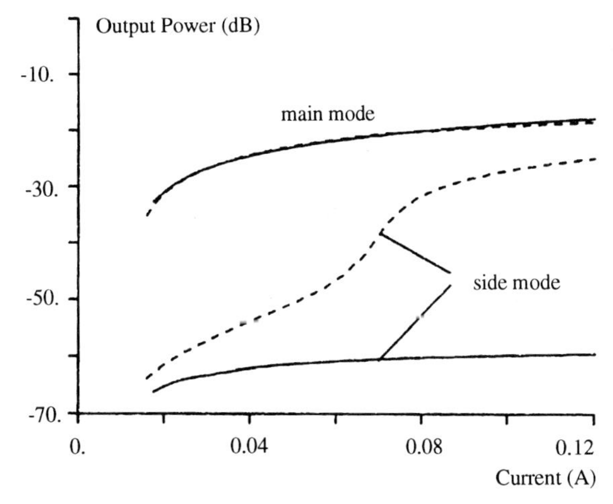

Fig. 4. Output power in main and side mode vs. the total injected current for $\kappa L = 3$; (---) $I_c = I_e$, (——) $I_c = 1.2 I_e$.

as the ones described before, can then be regarded as tools for the investigation of integrated structures and in particular for the investigation of the influence of discontinuities between the different sections. These *discontinuities* cause the light to reflect and eventually to couple back into the laser, where they can have a detrimental influence on the performance, especially on the spectral and tuning properties [62]. The influence of these reflections can be analyzed accurately only with the help of a large-signal dynamic model. Laser models can furthermore be used to determine optimum structures for the integrated structure formed by the laser, the waveguide, and the amplifier and to gain insight in the role of each component.

5. COUPLING PROBLEMS

The discussion so far has mainly concentrated on discrete devices within an OEIC. However, an integrated circuit contains a number of devices and the designer is interested in its overall performance. One of the advantages of integration is that the *parasitic effects* of *hybrid assembly* are avoided. However, integration may introduce new parasitic effects because with the close spacing the devices may interact in an undesired way; i.e., the operation state of one device may change as a function of the operation state of adjacent devices. Such coupling effects are not easily recognized experimentally and are generally neglected in the design or modeling phase. They lead to lower and more reproducible parasitic effects than hybrid assemblies anyway.

In fact, the modeling of such effects is often very difficult. Ideally one would like to limit detailed device modeling to single devices and treat the coupling effects by circuit analysis. This, however, may well be impossible for many types of coupling.

A first important type of coupling is optical coupling. Light scattered from one device may couple into another device and thereby cause crosstalk. Such coupling can be a problem in OEICs that contain both very weak and very strong optical signals (e.g., a coherent receiver) or in an optoelectronic array chip in which many optical signals pass through a dense array of optoelectronic devices (e.g., an $N \times N$ optical switch). With the increasing use of semiconductor optical amplifiers in OEICs, the potential problems of coupling effects may also increase. In an integrated optical circuit the overall modeling of these coupling effects would be very difficult because one has to handle a large area or volume. The pragmatic, but conservative, approach is to design the constituent devices in such a way that very little optical power leaks in or out of them except at well-defined input or output ports. In this context it may be useful to model the response of a waveguide device to an incident radiation field.

Electrical (or rather electromagnetic) coupling is the second source of concern. Many OEICs handle very high data rates and therefore *electromagnetic interference* effects between metallic signal paths may be relevant. Generally speaking, it is probably simpler to analyze these effects than it is to analyze the optical coupling effects, in view of the ratio between chip size and wavelength. Because of this the electromagnetic coupling effects can in many cases be handled by a set of lumped elements, implying that the overall performance can be analyzed with a circuit model. At very high data rates this is no longer true and one faces a design task similar to that of a *MMIC*.

Another form of interaction is *thermal coupling*. Many devices in OEICs dissipate a considerable amount of power. At the same time, many optoelectronic devices, especially those that operate near a band edge or are based on the principle of optical interference, are critically temperature sensitive. Even if the heat sink of the chip is kept at constant temperature, there may still be relatively important temperature differences across the chip surface. The operation point of the devices can therefore depend on the neighborhood of heat dissipation sources. Direct signal crosstalk will not occur, however, if the thermal time constants are slow compared with the lowest signal frequency.

Another important aspect of OEICs is the efficient coupling of signals, electrical and optical, into and out of the chip. This is related to *packaging* and is generally considered to be a difficult task. One has to couple optical and high-speed electrical signals from relatively large *connectors* into the small chip, often under fairly demanding thermal conditions. For the modeling of these passive electromagnetic problems, which are dielectric for the optical interfaces and metallic/dielectric for the electrical interfaces, a whole range of modeling tools exist with varying complexity and accuracy.

Although there is a lot of similarity between the optical and electrical coupling problems—both are electromagnetic problems—it is perhaps worth emphasizing that there is an important difference caused by the relative bandwidth. In the optical case the relative bandwidth of the optical signal is either extremely small or unrelated to the information content. This generally implies that the coupling needs to be analyzed only for one wavelength and the coupling is transparant for the information content. The electrical signals, however, have a wide relative bandwidth and therefore the coupling needs to be analyzed for a whole range of wavelengths or frequencies. Now the coupling is very much dependent on information content.

6. SYSTEM NOISE ANALYSIS

When OEIC's are used in systems with weak or complex signals, the noise behavior of the OEICs becomes important, because it will ultimately limit the system performance. This will be demonstrated in this section in the special case of an optical communication system. In digital communication systems the performance is normally expressed as the *bit error rate* (BER), whereas for analog systems the *signal-to-noise ratio (SNR)* is used. The performance of an entire circuit or system can be analyzed with either theoretical analytical models or numerical simulation tools.

6.1. Analytical Models

The analytical model for the BER of a digital communication system is given in a closed form as

$$\mathrm{BER} = 0.5\left(\int_{-\infty}^{d} p_1(x)\,dx + \int_{d}^{+\infty} p_0(x)\,dx\right) \tag{5}$$

if

$$\int_{-\infty}^{+\infty} x p_0(x)\,dx < \int_{-\infty}^{+\infty} x p_1(x)\,dx$$

and if the binary data 0 and 1 have the same probability of being transmitted. Here $p_0(x)$ and $p_1(x)$ are the probability density functions of the output signal at the decision time if a zero and a one, respectively, are detected, and d is the *decision threshold*. The minimum BER is obtained if both terms in Eq. (5) are equal, which also determines the optimum value for the threshold d. To derive the probability density functions $p_0(x)$ and $p_1(x)$, various techniques are available [63, 64]. They all require that the communication system be treated in its entirety. This leads rapidly to a complex problem which can be solved only by using approximations, both in system topology and in device characteristics.

As an example we will focus on the performance of a digital coherent optical communication system in which the probability density functions are computed by means of a simplified receiver model. The model for an *ASK system* is given in Fig. 5, where the *photocurrent* is modeled as

$$i(t) = A_d \cos(2\pi f_{IF} t + \phi(t)) + n(t) \tag{6}$$

where

A_d = amplitude of the carrier, containing the data

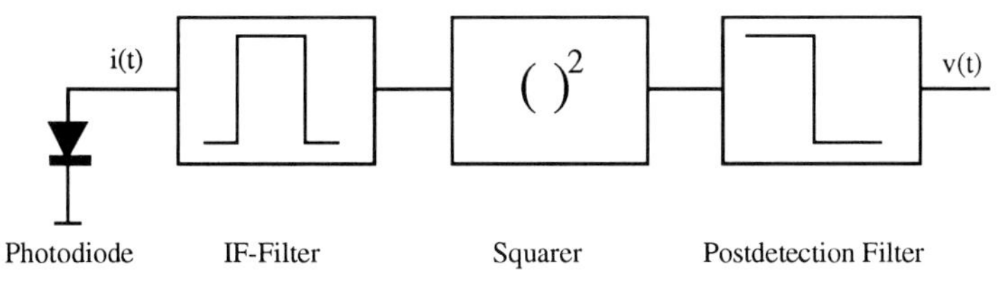

Fig. 5. Schematic drawing of an ASK coherent optical communication system.

f_{IF} = intermediate frequency
$\phi(t)$ = phase noise of the laser diode due to the nonzero linewidth
$n(t)$ = shot noise

Even in this simple case the probability density functions cannot be calculated exactly because of the phase noise. Various approximations, which can also be used for FSK or DPSK systems, were reported:

- Kazovsky (K) [65, 66] calculated the autocorrelation function of the output voltage and then assumed that the distribution of the output voltage was Gaussian.
- Garrett and Jacobsen (GJ) [67, 68] modeled the phase as a random intermediate frequency process and derived the probability density function considering a single sample of that process. This method has also been used by others [69].
- Foschini, Greenstein, and Vannucci (FGV) [70, 71] simulated the effect of phase noise numerically using the *Radon Nikodym derivative* and derived an analytical approximation to the moment-generating function (i.e., the Laplace transform of the probability density function) of the output voltage.

The FGV approach is believed to be the most accurate one, but it is valid only for an idealized receiver with high IF frequency, white Gaussian shot noise and receiver noise, and description of the filters as finite time integrators. The GJ and K approaches can also include real filter responses and receiver front-end parasitics. The K approach is the easiest to perform but the assumption of a Gaussian distribution of the output samples is crude and should be validated by a more rigorous method before usage. It has already been shown that for an ASK system the GJ and K approaches lead to similar results concerning the influence of the phase noise. For larger *IF linewidths* ($\Delta\nu >$ bit rate), however, there is a growing difference between the GJ and K results on the one hand and the FGV results on the other hand for the same receiver structure [72].

For the analytical analysis of analog optical communication systems one is often restricted to the computation of the autocorrelation function leading to the SNR [73–75].

6.2. Numerical Models

In the numerical system simulation model, the calculation of the performance is based on a *Monte Carlo analysis*. The circuit or system is broken down into a chain of several modules. For each module an appropriate model—simplified or rigorous—is constructed which describes the phenomena to be included. The performance is then evaluated by passing a sampled data version of the input signal through the various computer models of the modules. The resulting output signals can then be used to view the time evolution or to calculate the power spectrum just as if the system analyzer were performing measurements on a real system. The techniques used for modeling linear or nonlinear systems, discrete time signals, signal processing, and generation of random numbers [76–78] are therefore of vital importance.

To obtain the performance (BER) of a digital system, one can count the errors by comparing the output signal with the input signal. Unfortunately, because one needs the occurrence of approximately 10 errors before having a confident approximation, this method takes a long, mostly unpractical, time for high-performance systems (BER $< 10^{-9}$). In order to obtain an estimation of the performance in a more realistic time interval one has to use other BER estimation techniques. Some possibilities are :

- *Importance sampling* where the variance or the mean of the noise sources is increased to cause more errors. Naturally, after the simulation run the results should be multiplied by the appropriate normalization coefficient to give the real results [79–81].
- A semianalytical method where the unknown distribution is fitted to an analytical class of distributions.

Decreasing the simulation time by using importance sampling as in Monte Carlo simulations for radio frequency, satellite communication, or optical IM/DD (intensity modulated/direct detection) systems has not yet been reported for optical coherent communication systems. The use of the semianalytical method seems to be very interesting [55, 82]. Using the moments up to order six to fit the distribution $p(x)$ defined by the differential equation

$$\frac{dp}{dx} = \frac{a_2x^2 + a_1x + a_0}{b_2x^2 + b_1x + b_0} \tag{7}$$

to the unknown distribution of the output signal, only 5×10^5 data bits are needed in the data signal for the Monte Carlo simulation to obtain a 90% confidence interval of 0.5 dBm for the sensitivity at BER of 10^{-9}. The probability density function defined by Eq. (7) is a second-order approximation to the exact Fokker–Planck equation.

The performance of an analog communication system can easily be obtained by evaluating the output spectrum.

These numerical simulations are therefore a useful alternative to direct analytical evaluation. Furthermore, they provide more insight into the system performance through the evaluation of the simulation and the experience gained from simulation experiments. And because Monte Carlo simulations introduce no limitations to system topology and to the characteristics of the components, they can be used to study a wide variety of optical communication systems and the influence of component deficiencies. As an example, in Fig. 6 [83] the influence of the imbalance in the directional coupler of a balanced receiver for a 140 Mbit/s FSK system is given.

7. CONCLUSION

In this chapter we have described a number of modeling methods that can be useful in the design of OEICs. Contrary to the case of electronic ICs, the entire design of OEICs, from the details of the device level to the overall

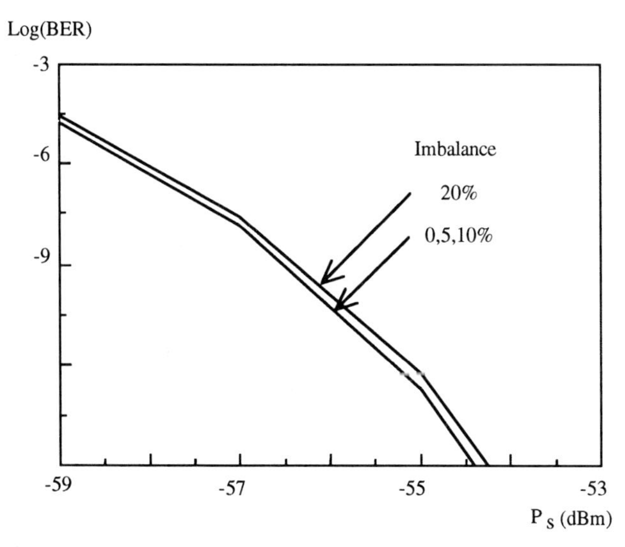

Fig. 6. Influence of imbalance in the directional coupler on the bit error rate as a function of the received power.

performance level including the package, is often done by a few people in the same group, at least at present. These people have to become acquainted with a very diverse range of problems—semiconductor, electromagnetic, electronic, systems—and have to use a wide range of different modeling tools. This diversity is certainly a problem and an impediment to rapid progress. It would not be desirable, however, to construct one modeling package that can handle all the problems encountered in the design of OEICs. It is important that the modeling tools be developed in user-friendly packages that can be readily interfaced to each other and used effectively by people who are not necessarily specialized in the physics of the underlying model.

At the end of this chapter it is appropriate to try to assess the merits that modeling has had so far in the design and understanding of optoelectronic devices and systems. One should probably make a distinction between cases in which modeling is meant to provide a better qualitative understanding and those in which it is expected to provide accurate quantitative results. As far as the first is concerned, one may think that sophisticated modeling does not give much more insight than simple intuitive reasoning or simple analytic calculation. This may indeed be true in many situations but certainly not in all. There are passive waveguide structures for which intuition or simple calculation is quite unable to predict the principle of operation. In complex laser structures the strong nonlinearity of the equations that describe the operation is such that it is difficult to predict even major features in the device performance. It should be emphasized, however, that even in these cases intuition and simplified analytic work are always of great help in deciding which structures are worth modeling.

For the second aspect, the quantitative aspect, we believe that the field of optoelectronic device modeling is on a learning curve. At present, in many situations it is inconceivable that the performance predicted by modeling fits nicely to experimental results. This can be due to many factors such as uncertainty about the structural parameters (geometry, compositions, etc.) and the material properties, but also effects not recognized as important and not included in the modeling. There is continuous and rapid improvement, however, and we believe that computer-aided design tools for accurate design will become available in the near future.

REFERENCES

1. C. M. Snowden (ed.), "Semiconducor Device Modelling." Springer-Verlag, Heidelberg, 1989.

2. M. H. COHEN, M. Y. CHOU, E. N. ECONOMOU, S. JOHN, AND C. M. SOUKOULIS, *IBM J. Res. Dev.* **23**, 82 (1988).
3. E. O. KANE, *J. Phys. Chem. Solids* **1**, 249 (1957).
4. M. ASADA AND Y. SUEMATSU, *IEEE J. Quantum Electron.* **QE-21**, 5 (1985).
5. M. ASADA, *IEEE J. Quantum Electron.* **25**, 2019 (September 1989).
6. H. HAUG AND S. KOCH, *Phys. Rev. A,* **39**, 356 (1989).
7. C. ELL, H. HAUG, AND S. W. KOCH, *Opt. Lett.* **14**, 356 (1989).
8. G. P. AGRAWAL, *J. Appl. Phys.* **63**, 1332 (1988).
9. D. BOTTELDOOREN AND R. BAETS, *Appl. Phys. Lett.* **54**, 1989 (1989).
10. SEEGER, "Semiconductor Physics." Springer-Verlag, Berlin, 1985.
11. S. ADACHI AND K. OE, *J. Appl. Phys.* **56**, 74 (1984).
12. S. ADACHI AND K. OE, *J. Appl. Phys.* **58**, 1499 (1984).
13. A. ALPING AND L. A. COLDREN, *J.Appl. Phys.* **61**, 2430 (1987).
14. M. ASADA, A. KAMEYAMA, AND Y. SUEMATSU, *IEEE J. Quantum Electron.* **QE-20**, 745 (1987).
15. DOYEOL AHN AND SHUN-LIEN CHUANG, *IEEE J. Quantum Electron.* **QE-26**, 13 (1990).
16. R. EPPENGA, M. F. SCHUURMANS, AND S. COLAK, *Phys. Rev. B* **36**, 1554 (1987).
17. J. N. SCHULMAN AND YIA-CHUNG CHANG, *Phys. Rev. B* **31**, 2056 (1985).
18. S. C. HONG, M. JAFFE, AND J. SINGH, *IEEE J. Quantum Electron.* **QE-23**, 2181 (1987).
19. R. L. GREENE, K. K. BAJAJ, AND D. E. PHELPS, *Phys. Rev. B* **29**, 1807 (1984).
20. KUN HUANG, JIANBAI XIA, BANGFEN ZHU, AND HUI TANG, *J. Luminescence* **40–41**, 88 (1988).
21. D. A. B. MILLER, J. S. WEINER, AND D. S. CHEMLA, *IEEE J. Quantum Electron.* **22**, 1816 (1986).
22. P. J. STEVENS, M. WHITEHEAD, G. PARRY, AND K. WOODBRIDGE, *IEEE J. Quantum Electron.* **24**, 2007 (1988).
23. S. SMITH-RINK, C. ELL, AND H. HAUG, *Phys. Rev. B* **33**, 1183 (1986).
24. G. LIVESCU, D. A. B. MILLER, AND D. S. CHEMLA, *IEEE J. Quantum Electron.* **24**, 1677 (1988).
25. G. LENGYEL, K. W. JELLEY, AND R. W. H. ENGELMANN, *IEEE J. Quantum Electron.* **26**, 296 (1990).
26. D. BOTTELDOOREN AND R. BAETS, *Proc. OSA Topical Meeting on Quantum Wells for Optics* **AND** *Optoelectronics,* Salt Lake City (March 1989).
27. D. DE ZUTTER e.a., *IEEE Proc.* **136**(pt. J), 273 (1989).
28. M. S. STERN, *IEE Proc.* **135**(pt. J), 56 (1988).
29. B. M. A. RAHMAN AND J. B. DAVIES, *J. Lichtwave Technol,* **LT2**, 682 (1984).
30. J. M. VAN SPLUNTER, H. BLOK, N. H. G. BAKEN, AND M. F. DANE, *URSI Int. Symp. on Electromagn. Theory*, Budapest, 321 (1986).
31. M. C. AMANN, *J. Lightwave Technol.* **LT4**, 689 (1986).
32. W. STREIFER AND E. KAPON, *Appl. Opt.* **18**, 3724 (1979).
33. P. C. KENDALL e.a., *IEE Proc. A 134*, 669 (1987).

34. J. J. G. M. Van der Tol and N. H. G. Baken, *Electron. Lett.* **24**, 207 (1988).
35. A. J. Marcatili and A. A. Hardy, *IEEE J. Quantum Electron.* **24**, 766 (1988).
36. B. V. Burke, *IEE Proc.* **137**(pt. J), 7 (1990).
37. M. Van Ackere, Ph.D. thesis (in Dutch), University of Gent (1990).
38. M. D. Feit and J. A. Fleck, *Appl. Opt.* **17**, 3990 (1978).
39. J. Van Roey, J. Van der Donk, and P. E. Lagasse, *J. Opt. Soc. Am.* **71**, 803 (1981).
40. P. E. Lagasse and R. Baets, *Radio Sci.* **22**, 1225 (1987).
41. L. Thylen, *Opt. Quantum Electron.* **15**, 433 (1983).
42. L. Thylen, E. M. Wright, G. I. Stegeman, and C. T. Seaton, *Opt. Lett.* **11**, 590 (1986).
43. R. Baets and P. E. Lagasse, *Electron. Lett.* **20**, 41 (1984).
44. P. Kaczmarski, R. Baets, G. Franssens, and P. E. Lagasse, *Electron. Lett.* **25**, 716 (1989).
45. P. Kaczmarski, Ph.D. thesis, University of Gent (1990).
46. D. Yevick and B. Hermansson, *IEEE J. Quantum Electron.* **QE-26**, 109 (1990).
47. J. Buus, *IEE Proc.* **132**(pt. J), 42 (1985).
48. R. Baets, J. Van de Capelle, and P. Vankwikelberge, *Ann. Télécommun.* **43**, 423 (1988).
49. G. Agrawal and N. Dutta, "Long-Wavelength Semiconductor Lasers." Van Nostrand Reinhold, New York, 1986.
50. J. Buus, *IEEE J. Quantum Electron.* **QE-18**, 1083 (1982).
51. P. Vankwikelberge, G. Morthier, and R. Baets, *IEEE J. Quantum Electron.* **QE-26**, 1728 (1990).
52. C. Henry, *J. Lightwave Technol.* **LT-4**, 288 (1986).
53. G. Agrawal and A. Bobeck, *IEEE J. Quantum Electron.* **QE-24**, 2407 (1988).
54. J. Whiteaway, G. Thompson, A. Collar, and C. Armistead, *IEEE J. Quantum Electron.* **QE-25**, 1261 (1989).
55. F. Libbrecht, G. Morthier, R. Baets, and P. Lagasse, *Proc. OFC'91*, 120, San Diego (1991).
56. I. Habermayer, *Opt. Quantum Electron.* **13**, 461 (1981).
57. D. M. Byrne, *Proc. OFC'92*, 301, San Jose, CA (1992).
58. T. Zhang, R. Hicks, and R. S. Tucker, *Microwaves & RF* **30**, 114 (1991).
59. M. Ito and T. Kimura, *IEEE J. Quantum Electron.* **QE-17**, 787 (1981).
60. U. Koren, B. Miller, G. Raybon, M. Oron, M. Young, T. Koch, J. DeMiguel, M. Chien, B. Tell, K. Brown-Guebler, and C. Burrus, *Proc. 12th Int. Semiconductor Laser Conf.*, 162, Davos, Switzerland (1990).
61. T. Koch, U. Koren, A. Gnauck, K. Reichman, H. Kogelnik, G. Raybon, R. Gnall, M. Oron, M. Young, J. DeMiguel, and B. Miller, *Proc. 12th Int. Semiconductor Laser Conf.*, 166, Davos, Switzerland (1990).
62. K. Petermann, "Laser Diode Modulation and Noise." KTK Publishers, Tokyo, 1988.
63. M. Schwartz, W. R. Bennett, and S. Stein, "Communication Systems and Techniques." McGraw-Hill, New York, 1966.

64. D. Middleton, "An Introduction to Statistical Communication Theory." McGraw-Hill, New York, 1960.
65. L. G. Kazovsky, P. Meisner, and E. Patzak, *J. Lightwave Technol.* **LT-5**, 770 (1987).
66. O. Tonguz and L. G. Kazovsky, *IEEE Photon. Technol. Lett.* **2**, 72 (1990).
67. I. Garrett and G. Jacobsen, *J. Lightwave Technol.* **LT-4**, 323–334 (1986).
68. G. Jacobsen and I. Garrett, *J. Lightwave Technol.* **LT-5**, 478 (1987).
69. G. Nicholson, *Electron. Lett.* **20**, 1006 (1984).
70. G. J. Foschini, L. J. Greenstein, and G. Vannucci, IEEE Trans. Commun. **36**, 306 (1988).
71. G. J. Foshini and G. Vannucci, *IEEE Trans. Commun.* **36**, 1437 (1988).
72. I. Garrett and G. Jacobsen, *IEE Proc.* **136**(pt. J), 159 (1989).
73. W. I. Way, *J. Lightwave Technol.* **LT-7**, 1806 (1989).
74. R. Gross, R. Olshansky, and P. Hill, *IEEE Photon. Technol. Lett.* **1**, 179 (1989).
75. R. Gross and R. Olshansky, *IEEE Photon. Technol. Lett.* **1**, 91 (1989).
76. N. Mohanty, "Signal Processing." Van NostrAND Reinhold, New York, 1987.
77. A. Oppenheim and R. Schafer, "Digital Signal Processing." Prentice Hall, London, 1975.
78. D. Knuth, "The Art of Computer Programming" (vols. 1–3). Addison-Wesley, Reading, MA 1968–1981.
79. K. S. Shanmugan and P. Balaban, *IEEE Trans. Commun.* **COM-28**, 1917 (1980).
80. D. Lu and K. Yao, *IEEE J. Selected Areas Commun.* **JSAC-6**, 67 (1988).
81. P. Balaban, Statistical evaluation of the error rate of the fiberguide repeater using importance sampling. *Bell Sys. Tech. J.* **55**, 745–766 (1976).
82. F. Libbrecht and P. Lagasse, ECOC 90 Amsterdam, paper TuB1.3 (1990).
83. F. Libbrecht and P. Lagasse, IEE colloquium on polarisation effects in optical switching and routing systems (1990), IEE, London.

Chapter 15

PHOTONIC INTEGRATED CIRCUITS

T. L. Koch and U. Koren

AT&T Bell Laboratories
Holmdel, New Jersey 07733

ISBN 0-12-200420-5

1. INTRODUCTION

The term optoelectronic integrated circuit (OEIC) refers to the integration of optoelectronic devices, such as sources and detectors, with electronic devices such as transistors. Photonic integrated circuits (PICs) are a subset of OEICs focusing on the single-substrate or monolithic integration of *optically interconnected* guided-wave optoelectronic devices. Although the balance of OEIC research has leaned heavily toward the electronics, as measured by the relative abundance of electronic versus optoelectronic devices, PICs may contain no purely electronic devices at all.

Most OEICs fabricated to date have been aimed at applications stemming from conventional point-to-point optical communications. Here, the optical device functions only as a terminal device to convert a processed or conditioned electrical signal into an encoded optical signal for fiber or free-space transmission, or vice versa. In contrast, PICs contain optically interconnected devices that somehow successively reroute, condition, or process the signal while still in its optical form.

The driving force for PICs is the expected complexity of next-generation optical communications links, networking architectures, and even possibly switching systems. The elaborate schemes currently being researched, including multichannel incoherent (optical filtered demultiplexing) and coherent (heterodyne electrical demultiplexing) wavelength division multiplexing (WDM), as well as high-speed time division multiplexing, involve a great variety of optically cascaded sources, modulators, filters, amplifiers, switches, detectors, etc. A large component of the cost of such architectures is due to the single-mode optical connections between the guided-wave components. Such devices often have optimized performance with tightly confined waveguides, resulting in difficult submicrometer alignment tolerances when coupling to single-mode optical fibers. By replacing such individually aligned connections with lithographically produced waveguides, PICs offer the promise of cost reduction, dramatically reduced size, and increased packaging robustness.

In some applications, the small interelement reflections inherent to PIC waveguide connections, combined with the short and stable path lengths between elements, may obviate the need for isolation components. Together with the likelihood of more efficient interelement coupling and the resulting higher powers, this may also provide a performance incentive for PICs. Similar arguments have fueled the great progress in OEICs chronicled in the rest of this book. Although the performance advantages have remained elusive for most OEICs, in large volume the increased packaging freedom and cost savings associated with the elimination of discrete electrical connections should prove invaluable. However, if the advanced optical architectures currently under investigation are to become ubiquitous, the PIC replacement of fiber connections with lithographic waveguides may be even more compelling.

A great variety of guided-wave devices have been investigated during the past two decades under the title "integrated optics." This work has mostly been limited to the area of electro-optics and passive waveguide devices, due at least in part to limitations in the relatively immature fabrication technology associated with optical sources and detectors. Only within the past several years have semiconductor crystal growth and processing techniques advanced to the point where the wealth of established integrated optic designs and concepts may be realized in a form truly integrated with semiconductor sources and detectors.

Chief among these advances are the large-area uniformity and reproducibility of computer-automated growth offered by the current generation of vapor and beam growth techniques such as metalorganic vapor-phase epitaxy (MOVPE) and chemical beam epitaxy (CBE). Especially in the InGaAsP/InP system of interest to most telecommunications applications, highly complex vertical layer structures, including large numbers of ultra-thin etch-stop layers, have ushered in a dramatic new level of design and fabrication freedom in the engineering of semiconductor guided-wave devices. Other important advances include the routine fabrication of grating-based distributed feedback (DFB) and distributed Bragg reflection (DBR) lasers for high-Q on-chip resonators without cleaved facets. The ease of semi-insulating Fe-doped InP regrowth with MOVPE has also been important both for high-speed applications and in achieving the required level of electrical isolation between the various PIC components.

This chapter surveys some of the design and fabrication issues encountered with the current generation of experimental PICs. It begins with an introduction to some of the design tools and fabrication techniques useful to the PIC engineer, with illustrations in many cases to clarify the discussion. This is followed by illustrative examples of current research prototype PICs to give a broader view of potential application areas. Although there is a wealth of international work in this field, many of the examples are chosen from the authors' own work because of the completeness of information on fabrication sequences and device performance.

2. GUIDED-WAVE DESIGN TOOLS

In this section we briefly review some materials optical properties and simple guided-wave calculational techniques that are useful to the PIC designer. These are illustrated in some cases with a few simple tutorial design examples.

2.1. Optical Constants

The refractive indices of the $Al_xGa_{1-x}As/GaAs$ and $In_xGa_{1-x}As_yP_{1-y}/InP$ systems have been measured by a number of researchers as functions of x and/or y and as a function of wavelength λ. For the InP system, we have had much success using the expression of Henry *et al.* [1]. Here the index is parametrized by the incident wavelength λ and the experimentally measured quantity λ_{PL} (low excitation photoluminescence wavelength) of the InP lattice-matched InGaAsP composition where $y = 2.917x$. The index of this quaternary alloy is given by

$$n_Q(E, E_{PL}) = \left(1 + \frac{A_1}{1 - \left(\dfrac{E}{E_{PL} + E_1}\right)^2} + \frac{A_2}{1 - \left(\dfrac{E}{E_{PL} + E_2}\right)^2}\right)^{1/2}, \tag{1}$$

where $E = 1.2398/\lambda$ and $E_{PL} = 1.2398/\lambda_{PL}$ are respectively the incident photon energy and photoluminescence peak photon energy for λ in μm, and $A_1(E_{PL})$, $A_2(E_{PL})$, E_1, E_2 are fitted parameters given by

$$A_1 = 13.3510 - 5.4554E_{PL} + 1.2332E_{PL}^2,$$

$$A_2 = 0.7140 - 0.3606E_{PL},$$

$$E_1 = 2.5048 \text{ eV} \quad \text{and} \quad E_2 = 0.1638 \text{ eV}.$$

For application to the binary InP, the value of the photoluminescence peak should be taken as $\lambda_{PL} = 0.939$ μm.

The most common epitaxial material on GaAs substrates is $Al_xGa_{1-x}As$, which is nearly lattice-matched for all values of x, with GaAs at $x = 0$ providing the narrowest bandgap. For photon energies below the absorption edge, the index of refraction of this material system is given by

$$n_{AlGaAs}(E, x) = \left[1 + \gamma(E_f^4 - E_\Gamma^4) + 2\gamma(E_f^2 - E_\Gamma^2)E^2 + 2\gamma E^4 \ln\left(\frac{E_f^2 - E^2}{E_\Gamma^2 - E^2}\right)\right]^{1/2}, \tag{2}$$

where $E = 1.2398/\lambda$ is the incident photon energy,

$$\gamma = \frac{E_d}{4E_0^3(E_0^2 - E_\Gamma^2)} \quad \text{and} \quad E_f = (2E_0^2 - E_\Gamma^2)^{1/2}$$

where

$$E_0(x) = 3.65 + 0.871x + 0.179x^2,$$

$$E_d(x) = 36.1 - 2.45x,$$

$$E_\Gamma(x) = 1.424 + 1.266x + 0.26x^2.$$

Other material characteristics of interest include the gain versus carrier density for gain media as used in lasers and amplifiers and radiative and nonradiative carrier lifetimes. These characteristics, used heavily for both static and dynamic modeling, are the subject of an enormous literature and the reader is referred to a text [3] that contains much of the requisite material.

Also of importance is the index change per unit carrier density or per unit applied field, together with any associated losses induced by the change. For gain media, this is characterized by the linewidth enhancement factor [4] α, which we generalize to be α_X, the ratio of real index change to imaginary index change due to a variation in the parameter X. For $n \equiv n' + in''$, we have

$$\alpha_X = \frac{\partial n'/\partial X}{\partial n''/\partial X}. \tag{3}$$

When X is N, the carrier density, a free carrier plasma contribution is present both in the transparent region $\lambda > \lambda_{PL}$ and in the gain/loss region $\lambda < \lambda_{PL}$, while an anomalous dispersion contribution from the gain/loss changes may dominate when $\lambda < \lambda_{PL}$. In contrast to the AlGaAs/GaAs system, in which the plasma index contribution is much smaller, at longer wavelength (due to a λ^2 dependence) the plasma contribution is comparable to the anomalous dispersion contribution even in a gain medium [5]. This leads to larger α factors, and typical values in long-wavelength InGaAsP bulk material are $\alpha \sim 6$. With a typical material gain coefficient $\partial g/\partial N$ of $\sim$300 cm^{-1} per 10^{18} cm^{-3}, and $g \equiv 4\pi n''/\lambda$, we have at 1.3–1.5 μm a change of $\Delta n' \sim -0.02$ to -0.025 for each 10^{18} cm^{-3} change in carrier density in the gain region, consistent with experimentally measured values [6]. Well below bandgap, this reduces to values of $\Delta n' \sim -0.01$ to -0.015 for each 10^{18} cm^{-3} change in carrier density. However, below bandgap there is much less change in loss, and it is of an opposite sign because increasing carrier density leads to increasing loss via free-carrier and inter-valence-band absorption. From data on transmissive phase-shifting devices, we estimate [7] that $\alpha \sim 40$ for 1.5-μm band light propagating in 1.3-μm λ_{PL} InGaAsP. This suggests that free-carrier injection into heterostructures is a good technique where large index changes are required at moderate modulation speeds ($f_{3dB} \approx 1/2\pi\tau \sim 100$ MHz for good material with typical values of recombination lifetimes τ). An excellent example of this is the InP-based switching array work of Inoue *et al.* [8].

Electro-optic index modulation [9] has still lower, negligible, induced losses unless one is operating near the band edge where Franz–Keldysh or quantum-confined Stark effect (QCSE) losses may be generated [10]. Enhanced field-induced index shifts can be obtained in a variety of quantum-well structures, including the QCSE and band-filling absorption-quenching effects [11, 12]. Although further results are needed to assess the ultimate effective α_X values achievable in refined low-loss device geometries, these devices are likely to be appealing where high-speed operation is desired as in modulators or switching arrays [13, 14].

2.2. Multilayer Leaky/Lossy Slab Structure Mode Evaluation

The utility of a general-purpose multilayer slab waveguide optical mode evaluation routine cannot be overemphasized, and with typical workstations such programs run in seconds. Combined with the index of refraction formula given before, this allows the determination of essential numbers such as the effective index or propagation constant of waveguides required for proper choice of Bragg grating pitches in DFB and DBR lasers or other filter devices. It also provides either confinement factors or detailed mode shapes as required for the evaluation of gain/loss or index changes. Mode overlap integrals are easily executed for evaluation of butt-coupling efficiencies, and field profiles in corrugated grating regions allow the determination of grating coupling constants. These applications are illustrated in the following.

Here we outline the implementation of a particular conceptual approach shown in Fig. 1 that is both easy to understand and easy to implement numerically. Since Maxwell's equations are separable, we need only consider a two-dimensional problem in the y direction perpendicular to the layers and a propagation direction z. The concept of a mode in such a structure is quantified in physical terms as a solution to Maxwell's equations whose sole dependence on the coordinate in the propagation direction z is given by $e^{i\beta z}$. This translates to a requirement that the *shape* of the field distribution in the y direction, perpendicular to layers, remain unchanged with propagation. If we generalize to leaky structures or materials exhibiting loss or gain, β

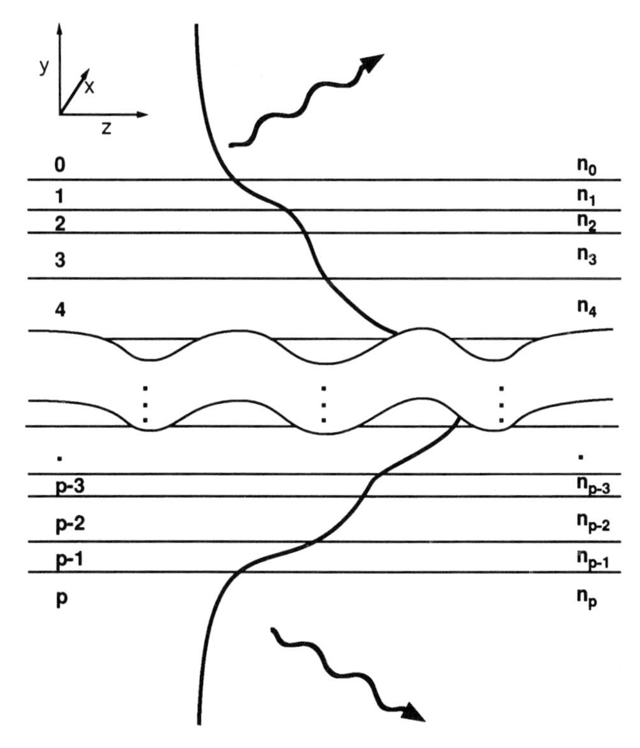

Fig. 1. Layer stack for general slab waveguide calculations. The index of each layer is complex.

may be complex, allowing a scaling of the mode amplitude with propagation, but the relative mode profile in the perpendicular y direction still remains constant. These solutions are not normalizable or "proper" in the sense of mathematical completeness but are very useful in understanding propagation behavior in such structures.

Since the field in each homogeneous layer m is well known to be $e^{+i\mathbf{k}_m \cdot \mathbf{r}}$, with $|\mathbf{k}_m| = 2\pi|n_m|/\lambda$ for the (generally complex) index of refraction n_m, the general solution to the field amplitude in each layer m is

$$\varepsilon_m = [a_m e^{iq_m y} + b_m e^{-iq_m y}]e^{i\beta z}, \tag{4}$$

where $q_m \equiv [(2\pi n_m)^2/\lambda^2 - \beta^2]^{1/2}$. Inspection of the vector Maxwell's equations reveal that the general vector solution in the multilayer slab can be broken down into the superposition of a TE (transverse electric) and a TM (transverse magnetic) solution. The TE (TM) solution is characterized by having only one component of the electric (magnetic) field that points in the x direction, parallel to the layers and perpendicular to the propagation direction z. The field amplitude in Eq. (4) refers to the E_X or the H_X field for the TE and TM case, respectively.

In a very simple exercise, for each of these cases one can successively match boundary conditions for continuous tangential **E** and **H** across the interfaces to provide the coefficients a_{m+1} and b_{m+1} in each layer $m + 1$ based on the value of the coefficients in the preceding layer m,

$$\begin{bmatrix} a_{m+1} \\ b_{m+1} \end{bmatrix} = \begin{bmatrix} \frac{1}{2}\left(1 + \frac{q_m \gamma_m}{q_{m+1}\gamma_{m+1}}\right) e^{-i(q_{m+1}-q_m)y_m} & \frac{1}{2}\left(1 - \frac{q_m \gamma_m}{q_{m+1}\gamma_{m+1}}\right) e^{-i(q_{m+1}+q_m)y_m} \\ \frac{1}{2}\left(1 - \frac{q_m \gamma_m}{q_{m+1}\gamma_{m+1}}\right) e^{i(q_{m+1}+q_m)y_m} & \frac{1}{2}\left(1 + \frac{q_m \gamma_m}{q_{m+1}\gamma_{m+1}}\right) e^{i(q_{m+1}-q_m)y_m} \end{bmatrix} \begin{bmatrix} a_m \\ b_m \end{bmatrix}, \tag{5}$$

where y_m are the coordinates of the interfaces between layers m and $m + 1$, and $\gamma_m \equiv 1$ for TE modes and $\gamma_m \equiv n_m^{-2}$ for TM modes. The wave is assumed evanescently decaying or outward leaking on one initial side of the arbitrary stack of complex-index layers; i.e., $b_0 = 0$ on the uppermost layer. When the lowermost "cladding" layer $m = p$ is reached, one again demands that only the coefficient b_p of the evanescently decaying, or possibly the outward leaking, component be nonzero, which recursively provides an eigenvalue equation $a_p(\beta) = 0$. Arbitrarily letting $a_0 = 1$, this can be written explicitly as

$$a_p(\beta) = [1 \quad 0]\left[\prod_{m=p}^{m=0} \mathbf{M}_m(\beta)\right]\begin{bmatrix} 1 \\ 0 \end{bmatrix} = 0, \tag{6}$$

where $\mathbf{M}_m(\beta)$ is the matrix appearing in Eq. (5). In practice, this is solved numerically in the form of two equations (for the real and imaginary parts of a_p) in two unknowns (the real and imaginary parts of β). Once the complex solutions β_j are obtained using standard root-finding routines, the spatial profiles are easily calculated for each mode j by actually evaluating the coefficients for the solutions using the foregoing relations with $a_0 = 1$, for example.

The utility of such a program is so great that it is well worth the PIC designer's time to write it in an accessible, easily implemented form. An exemplary output from such a routine is shown for an InGaAsP/InP antiresonant-reflecting optical waveguide (ARROW) [15] in Fig. 2. These guides confine in the *lower-index* core using resonant reflections from a multilayer

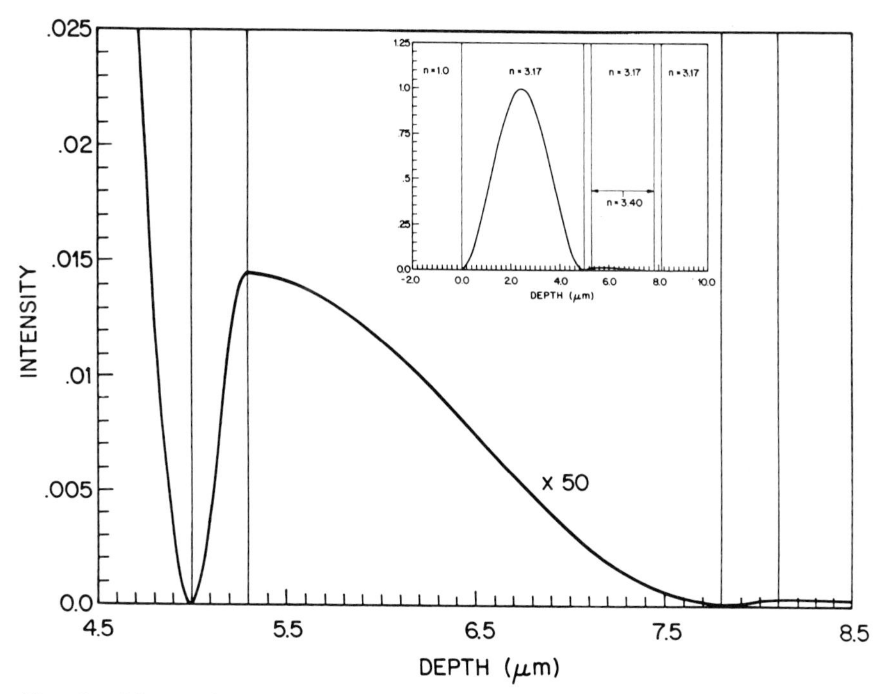

Fig. 2. Rigorously evaluated mode profile for a leaky ARROW guide in the InGaAsP/InP system. A leaky mode numerical routine is required to evaluate the characteristics of these guides, even though losses may be very small (<1 dB/cm).

cladding below the core, and the complex propagation constant available from the routine described before provides a useful description of the propagating leaky modes of this structure. Although the complex nature is also essential for the rigorous evaluation of modes with gain or loss in the layers, these can usually be well approximated using confinement factor techniques, which are essentially perturbative in nature on a loss-free structure. Considerable simplification is possible in the numerical routine described for loss-free and nonleaky structures because the recursive relations can be cast in terms of purely real quantities and the search is only for real solutions of β.

2.3. Lateral Guiding

The preceding discussion was for slab waveguides. For laterally guided or stripe waveguides, a powerful approximation technique known as the *effective index* method is commonly used [16]. This method logically breaks the three-dimensional guide into two sequential slab waveguide problems and becomes more accurate for higher-aspect-ratio guide shapes. (For purposes of discussion, we assume the guide is wider than it is high.) For the TE modes, for example, at each region along the width of the guide the slab effective index $n_{\text{eff}} \equiv \beta\lambda/2\pi$ for the mode profile in the vertical direction is first determined using a routine such as that described earlier using TE boundary conditions. For vertically homogeneous regions the actual material index is used. The sideways problem is then carried out as a slab using these effective indices, this time with TM boundary conditions. The resulting modal index is the effective index approximation to the three-dimensional guide.

As an example, we apply the effective index method to calculate the properties of a directional coupler based on buried-rib waveguides in InGaAsP/InP as shown in Fig. 3. Here 1.5-μm operation is assumed for 1.3-μm λ_{PL} InGaAsP 3000-Å-thick guides, with rib loading provided by 300-Å-thick "floating" ribs of the same material, separated by a 300-Å-thick InP etch-stop layer. The guides are buried on all sides by InP with a 2-μm gap between the two rib-loading layers of 3-μm width each. From Eq. (1) we have $n_{\text{InP}} = 3.1700$ and $n_{1.3\text{Q}} = 3.4130$. Using the mode evaluation routine with TE boundary conditions, the vertical slab is single-moded in all regions, with $n_{\text{eff}} = 3.25833$ in regions 1, 3, and 5 and $n_{\text{eff}} = 3.26652$ in regions 2 and 4.

One then considers a five-layer slab waveguide using these effective indices and TM boundary conditions. One finds laterally an even and an odd "supermode," as shown in Fig. 3, and these are the effective index ap-

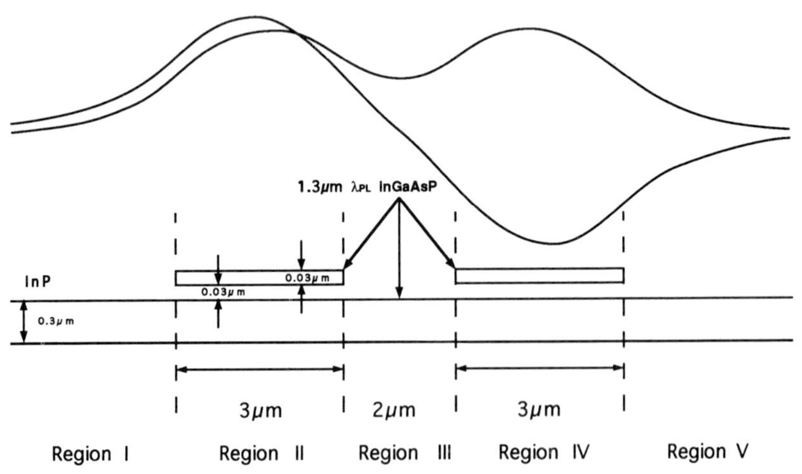

Fig. 3. Schematic structure of coupling region in passive coupler using rib-loaded waveguides. Even and odd lateral "supermodes" of the structure are shown, allowing easy evaluation of coupling coefficients for the coupler.

proximation to the modes of the full problem with $n_{\text{even}} = 3.262937$ and $n_{\text{odd}} = 3.263737$. The spatial qualities of the mode are approximated by multiplying the lateral mode profiles with the vertical mode profile in each region.

This approach can be used to determine the coupling constant of the coupler. If light is incident on one side of the coupler, this initial condition is generated by a superposition of the even and odd modes. Light is transferred to the other guide when the modes propagate enough to generate a π phase shift, or

$$L_{\text{couple}} = \frac{\lambda}{2(n_{\text{even}} - n_{\text{odd}})}. \tag{7}$$

In the usual language of coupled-mode theory based on a view of two isolated guides perturbing each other, complete transfer would occur at $\kappa L = \pi/2$, from which we can ascertain that

$$\kappa = \frac{\pi(n_{\text{even}} - n_{\text{odd}})}{\lambda}. \tag{8}$$

In the case shown, we obtain $\kappa = 16.7\ \text{cm}^{-1}$, and we can easily generate plots such as κ versus guide separation as shown in Fig. 4.

2.4. Butt Coupling

As already mentioned, this simple mode evaluation procedure is also extremely valuable in computing butt-coupling efficiencies between different guide geometries. Once the vertical and horizontal mode profiles $\varepsilon_v(y)$ and $\varepsilon_h(y)$ are determined in axial regions 1 and 2, the effective index approximation for the coupling efficiency is [17]

$$\eta = \frac{|\int \varepsilon^*_{h_1}(x)\varepsilon_{h_2}(x)\,dx|^2 |\int \varepsilon^*_{v_1}(y)\varepsilon_{v_2}(y)\,dy|^2}{|\int \varepsilon_{h_1}(x)|^2\,dx \int |\varepsilon_{h_2}(x)|^2\,dx \int |\int \varepsilon_{v_1}(y)|^2\,dy \int |\varepsilon_{v_2}(y)|^2\,dy} \tag{9}$$

using discrete summations for the integrals.

2.5. Grating Coupling Constants

As a final application we show how these simple numerical techniques can be applied to grating-based devices. For simplicity, we consider a two-dimensional (slab) structure which supports (at least) two modes $\varepsilon_a(y)$ and $\varepsilon_b(y)$, which could just be the forward- and reverse-propagating versions of the same spatial mode. These two modes each have some overlap with a region containing a perturbation in the index given by $\Delta n(y, z)$ which is

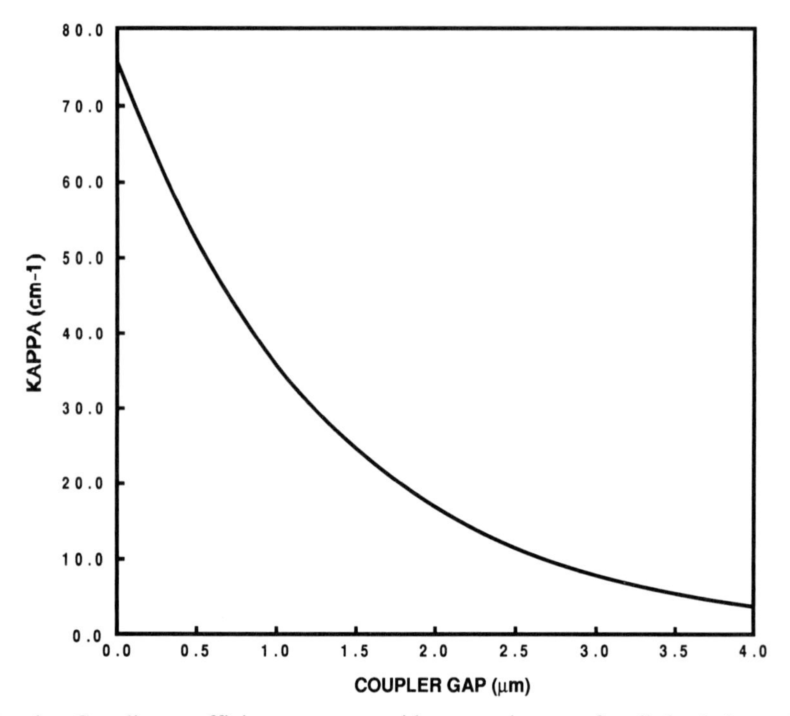

Fig. 4. Coupling coefficient vs. waveguide separation gap for rib-loaded coupler shown in Fig. 3.

periodic in the propagation direction z. When calculating mode profiles, this region, assumed bounded within a slab geometry between y_1 and y_2, is treated as a layer and assigned an index given by the averaged *dielectric constant*,

$$\bar{n} = \left[\frac{1}{\Lambda_g}\frac{1}{(y_2 - y_1)}\right]\left[\int_0^{\Lambda_g} dz \int_{y_1}^{y_2} dy\, [n + \Delta n\,(y, z)]^2\right]^{1/2}. \tag{10}$$

According to coupled-mode theory [18], the coupling constant for TE modes is given by

$$\kappa_m^{ab} = \frac{2\pi}{\lambda}\frac{1}{\Lambda_g}\int_0^{\Lambda_g} dz \int_{-\infty}^{\infty} dy\, e^{-ik_m z}\, \Delta n(y, z)\varepsilon_a^*(y)\varepsilon_b(y), \tag{11}$$

where Δ_g is the grating period, $k_m \equiv 2\pi m/\Lambda_g$, and $m = 1, 2, 3 \ldots$ is the order of the coupling used to achieve phase matching. Here the fields have been assumed normalized such that $\int_{-\infty}^{\infty} \varepsilon^2\, dy = 1$. For TM modes, the equivalent of Eq. (1) is only slightly more complex [19]. The z integral is carried out first to yield a quantity

$$\delta n(y) \equiv \frac{2}{\Lambda_g} \int_0^{\Lambda_g} e^{-ik_m z} \, \Delta n(y, z) \, dz \tag{12}$$

which yields the relevant harmonic content of the perturbation, and we then have

$$\kappa_m^{ab} = \frac{\pi n_m^{ab}}{\lambda} \tag{13}$$

where

$$n_m^{ab} = \int_{y_1}^{y_2} \delta n_m(y) \varepsilon_a^*(y) \varepsilon_b(y) \, dy. \tag{14}$$

In many applications $\Delta n(y, z)$ is confined to a layer bounding the extrema of a corrugated interface between two media of different dielectric constants. If the interface of the corrugation has reflection symmetry about its "teeth" along the direction of propagation z, we obtain

$$\delta n_m(y) = \frac{2\Delta n}{\pi m} \sin\left(\frac{2\pi m Z(y)}{\Lambda_g}\right) \tag{15}$$

where $Z(y)$ is the parametrized shape of one side of the symmetric "tooth" of the corrugation interface between its extrema y_1 and y_2, and Δn is the *total* index difference between the two media. These expressions are easily evaluated in the simple case of a rectangular corrugation where $Z(y)$ = const. $= {}^{f\Lambda_g}/_2$, where f is the "fill factor" of the rectangular wave ($f = {}^1/_2$ for a square wave) to yield

$$\kappa = \frac{2\Delta n}{\lambda} \Gamma_{ab} \sin(f\pi m) \tag{16}$$

where Γ_{ab} is given by

$$\Gamma_{ab} \equiv \int_{y_1}^{y_2} \varepsilon_a^*(y) \varepsilon_b(y) \, dy. \tag{17}$$

For DFB and DBR lasers or Bragg reflection filters, $\varepsilon_a(y) = \varepsilon_b(y)$ and thus $\Gamma_{ab} = \Gamma$, the usual confinement factor of the mode in the layer effectively bounding the rectangular corrugation. Other corrugation shapes can be evaluated by inserting the proper tooth shape in $Z(y)$. In cases where

$\varepsilon_a(y)$ and $\varepsilon_b(y)$ vary little within the corrugation-bounding layer, the fields can be removed from the integrals and approximated with $\Gamma_{ab} \approx \sqrt{\Gamma_a \Gamma_b}$ where Γ_a, Γ_b are the usual confinement factor of each mode separately in the corrugation-bounding layer.

As a specific example, consider a DFB laser in which the guide consists of a 1500-Å-thick, 1.55-μm λ_{PL} InGaAsP active layer with a 1500-Å-thick layer of 1.3-μm λ_{PL} InGaAsP above it, with a 500-Å-deep sinusoidal corrugation etched into the upper surface of the 1.3-μm λ_{PL} layer. Assume the structure is a 2-μm-wide buried heterostructure with InP on all sides. In this case

$$Z(y) = \frac{\Lambda_g}{2\pi} \cos^{-1}\left(2\frac{(y - [y_1 + y_2]/2)}{t}\right) \tag{18}$$

for $y_1 \leq y \leq y_2$ and $t \equiv y_2 - y_1 = 500$ Å. We then find that

$$\kappa = \frac{2\Delta n}{\lambda} \int_{y_1}^{y_2} \left(1 - 4\frac{(y - [y_1 + y_2]/2)^2}{t^2}\right)^{1/2} |\varepsilon(y)|^2 \, dy \tag{19}$$

where $\Delta n = n_{1.3Q} - n_{InP}$. Using Eq.(1), $n_{InP} = 3.1643$, $n_{1.3Q} = 3.3989$, $n_{1.55Q} = 3.5511$, where the last has been reduced by 0.02 to account for some injected carriers. For calculating the mode, the 500-Å-thick grating bounding layer has the index 3.2837. The modal effective index, using the effective index method for the lateral problem, is $n_{eff} = 3.2717$, which yields the required grating pitch for 1.55-μm operation to be $\Lambda_g = 0.2369$ μm.

In the vertical direction, the active layer and grating bounding layer, respectively, have confinements of 34.3% and 6.5%, but these are both reduced by the lateral confinement of 88%. If the grating was a square wave, the simple confinement factor expression Eq. (22) would yield $\kappa = 173$ cm^{-1}. However, with the sinusoidal corrugation, numerical evaluation of Eq. (25) yields $\kappa = 137$ cm^{-1}. This is expected because the square wave has higher fundamental Fourier content.

2.6. Beam Propagation Method

The slab waveguide, effective index, and grating techniques are powerful and easy to use analysis tools for the PIC designer. However, they are unable to answer important, basic questions that arise. At a Y-junction, what is the lateral radiation loss and how does it depend on device geometry? At waveguide intersections, how much crosstalk will arise? How much coupling takes place in the feed-ins to a directional coupler for different coupler geometries? Analytical techniques have been devised for these and other situations, but a commonly used numerical technique is the "beam propagation method" (BPM) [20]. The BPM generates solutions to the scalar Helmholtz equation using numerical treatments of a formal operator equation that generates field solutions at $z + \Delta z$ based on the field at z.

$$\varepsilon(x, y, z + \Delta z) = e^{\Phi \Delta z/2}\, e^{ik_0(n(x,y,z)-n_f)\Delta z}\, e^{\Phi \Delta z/2}\, \varepsilon(x, y, z) + \vartheta(\Delta z)^3, \quad (20)$$

where Φ is the operator

$$\Phi = i \frac{\nabla_t^2}{(\nabla_t^2 + k_0^2 n_f^2)^{1/2} + k_0 n_f} \quad (21)$$

and k_0 is the free-space wavenumber, n_f is the value of the field or background index, and ∇_t^2 is the transverse Laplacian. It can be shown, and is intuitive from the preceding expression, that this is equivalent to free, diffracting propagation for half of a step size, followed by a thin-lens phase correction for the index variation comprising the structure, followed by free diffraction for the remaining half-step. The approximations involve assumptions that the index variations are small, and ignoring the noncommutation of $n(x, y, z)$ and ∇_t^2.

This operator equation is amenable to solutions using fast Fourier transform (FFT) techniques, and this contributes to the popularity of the approach. In the case of semiconductors it is generally also used in conjunction with the effective index method to remove the vertical problem and deal with the in-plane two-dimensional propagation in lateral geometries such as those mentioned earlier, in which case $\nabla_t \rightarrow \partial_x$. The FFT allows the replacement of ∂_x with k_x in the application of Φ in the transform space for easy evaluation of Eq. (26). Figure 5 shows a typical example of the graphical output obtained using the beam propagation method, in this case applied to a y-branch formed in buried-rib waveguides similar to those shown in Fig. 3. This figure illustrates the inherent 3-dB radiation loss that occurs during reverse propagation throught the y-branch as expected from reciprocity considerations. In addition to the FFT approach just outlined, a number of other methods can be used such as finite-difference techniques. The reader is referred to texts and recent conferences for detailed reviews of the subject [21, 22], and other examples of both the effective index method and BPM may be found in Chapter S.4.

3. ACTIVE/PASSIVE WAVEGUIDE COUPLING AND DESIGN ISSUES

One of the most fundamental problems to overcome in the integration of guided-wave photonic devices is the proper engineering and fabrication of

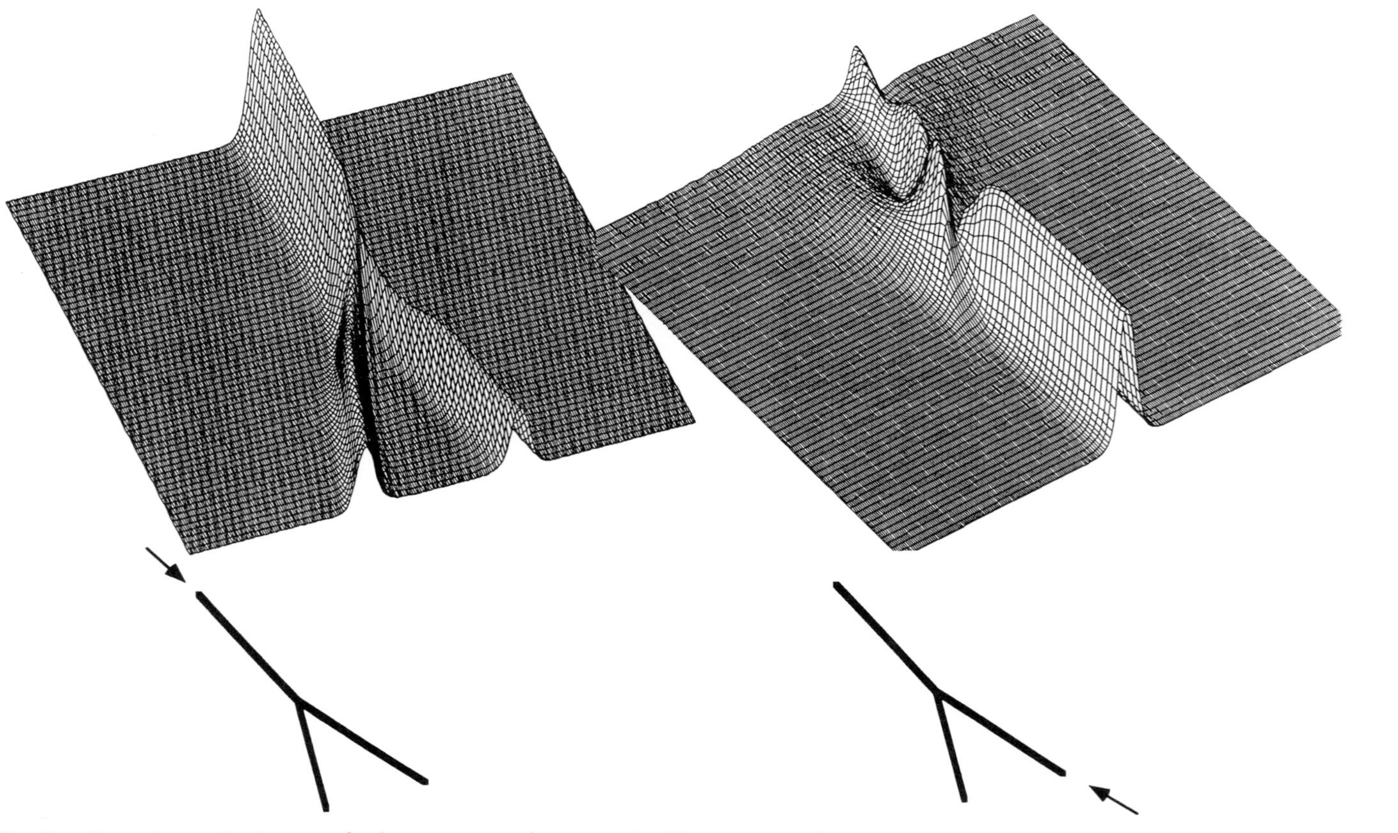

Fig. 5. Typical graphical output for beam propagation analysis. The structure shown, in compressed aspect ratio, is a Y-branch formed using buried-rib waveguides similar to those in Fig. 3. This illustrates the reciprocity of the Y-branch showing the 3-dB radiation loss that occurs in reverse propagation through the Y-branch.

the coupling between active and passive waveguides. In this context, active refers to a guide containing material whose bandgap energy is less than or near to the propagating photon energy. Examples would be a gain medium as in an injection laser or amplifier, the absorbing layer of a waveguide photodetector, or perhaps a medium used as an electroabsorption modulator with a suitable applied field. Passive guides usually have a bandgap energy substantially greater than the propagating photon energy and will exhibit low losses apart from scattering and doping-induced losses such as free-carrier or inter-valence-band absorption. In this section we briefly outline the approaches commonly used, employing some of the simple design tools mentioned in the preceding section. Two general approaches are butt coupling and directional coupling, with several variants possible in each category.

Most PICs demonstrated to date have employed some form of butt coupling, where an active waveguide of one vertical and/or lateral structure mates end-on with a passive waveguide of a different vertical and/or lateral structure. Butt coupling offers design simplicity, flexibility, and favorable fabrication tolerances. The approximate calculation of butt-coupling efficiencies is easily carried out by numerically evaluating Eq. (15).

3.1. Regrown Butt Joint

The most straightforward approach for butt coupling involves selective removal of the entire active waveguide core stack using selective wet chemical etching, followed by regrowth of a mated, aligned passive waveguide structure [23–29]. The principal advantage of such an approach is the independent selection of compositional and dimensional design parameters for the two guides. This can, in principle, be used to attain nearly 100% coupling efficiency between the active and passive guides. Indeed, such coupling joints have shown very good performance in narrow-linewidth DBR lasers [26], which are quite sensitive to intracavity active/passive coupling losses. In other applications, the active laser guide may join onto an electroabsorption modulator where compositional control of the guide is of paramount importance. Here a separate growth of the modulator guide may be preferred because such structures could be lossy even in the transmitting state and could affect laser performance if used as a guide layer inside the laser cavity.

An excellent example of this is the integrated DFB laser/electroabsorption modulator [23, 25]. This device also provides a natural example of an important application of PICs to high-bit-rate dispersive optical fiber transmission. The particular implementation used by Soda *et al.* [25] is shown schematically in Fig. 6. A 1.55-μm DFB laser structure is mated to an elec-

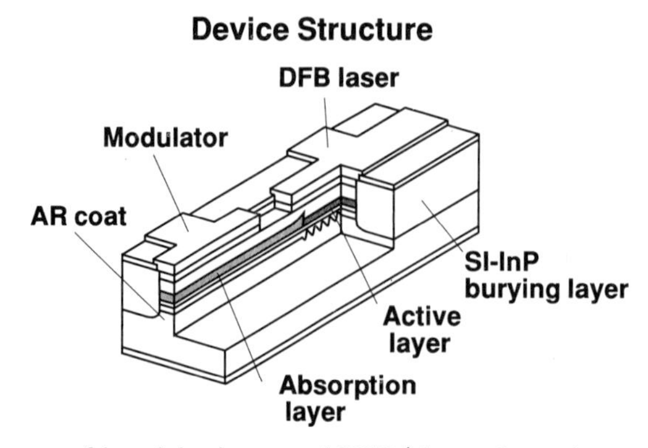

Fig. 6. Structure of butt-joint integrated DFB/electroabsorption modulator .

troabsorption modulator with an InGaAsP core layer having a photoluminescence wavelength of $\lambda_{PL} \sim 1.40$-μm. The entire structure uses a buried heterostructure waveguide with semi-insulated (SI) InP lateral cladding to provide good current blocking with low capacitance for high modulator bandwidth. A scanning electron microscope (SEM) image of an actual cross section of the MOVPE-grown butt joint is shown in Fig. 7, where a smooth transition is clearly seen. Optimization of the modulator core λ_{PL} is very important in this device. With proper design, devices have yielded a good compromise between high output power and high modulator extinction ratios with low voltage drive. Figure 8 shows an example of a class of device designs exhibiting 1.5 mW of fiber-coupled output power with a −10 dB extinction ratio at drive levels as low as 3–5 V. These structures also have low capacitance, with resulting −3 dB bandwidths in excess of 10 GHz, and clean eye diagrams at non-return-to-zero (NRZ) bit rates as high as 10 Gb/s. Finally, one of the principal motivations for this PIC is the low chirp

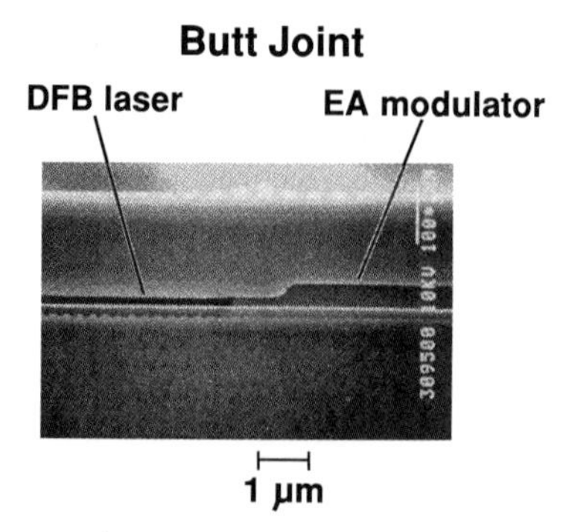

Fig. 7. Scanning electron micrograph of stained cross section of MOVPE-regrown butt joint in integrated DFB/electroabsorption modulator.

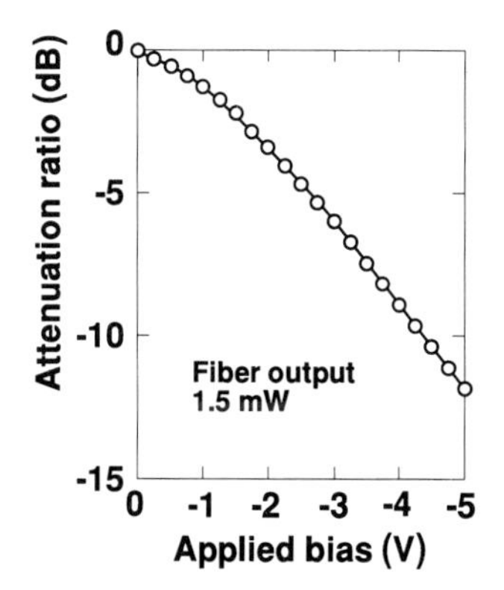

Fig. 8. Extinction vs. voltage applied to electroabsorption modulator, relative to fiber-coupled power of 1.5 mW.

which ensues from the externally modulated continuous-wave (CW) DFB oscillator. For an electroabsorption modulator the α factor is defined through Eq. (3) as α_E where E is the applied electric field. Equivalent α factors of $\alpha < 1$ have been measured in integrated laser/electroabsorption modulator PICs. This has permitted application in long-haul 65-km links with error-free operation at 10 Gb/s [30] with a fivefold reduction in dispersion penalties for a given deviation from the fiber dispersion zero when compared to directly modulated lasers.

Potential disadvantages of the regrown butt-joint waveguides as described are the extra growth step required and the difficulty of getting a highly reproducible joint geometry in each processing run. Butt-joint processing has been studied in this regard by Takeuchi *et al.* [29], who experimentally achieved a maximum coupling efficiency of 90%, with an average coupling efficiency of 60% and a standard deviation of 17% using MOVPE regrowth. The detailed values will of course vary with the exact processing sequence used, and the butt-joint method may become the method of choice for some PIC applications.

3.2. Active-Layer Removal

Another approach to butt coupling employs a largely continuous passive waveguide structure with a thin active layer residing on top, which is selectively removed on the portions of the structure which are to be passive. In the example shown in Fig. 9, a thin multiple-quantum-well (MQW) stack with four quantum wells is separated from an ~0.3-μm-thick passive guide core of 1.3-μm λ_{PL} InGaAsP by an ~200-Å-thick InP etch-stop layer. Using material-selective wet chemical etches, the MQW stack can be removed with

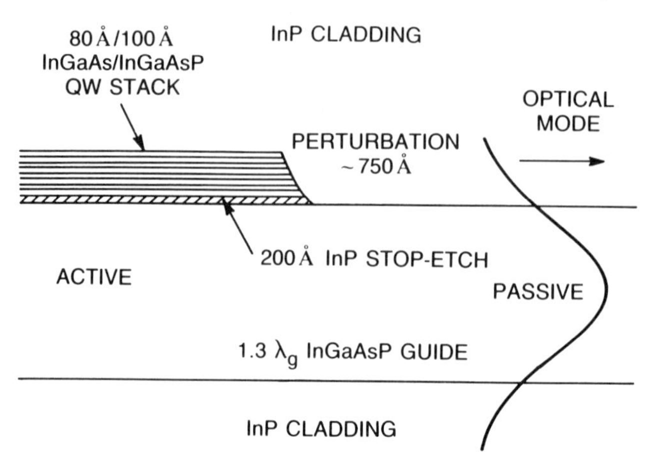

Fig. 9. Schematic of active/passive transition by etch-stop-controlled selective removal of MQW active layer from passive guide core.

very high reproducibility and precision, and the dimensional control is thus placed in the original computer-automated MOVPE growth of the base wafer. Figure 10 shows an SEM image of such a guide core, which contains additional etch-stop layers for subsequent processing of the passive guide itself. The high uniformity of layers and precise etch-stop locations allow tight control of the resulting final active/passive coupling geometry. This design also makes use of the high gain achieved in the small net vertical thickness of an MQW gain layer; the removal of the ~700-Å-thick stack constitutes only a small perturbation of the bulk of the guide core constituted by the lower, thicker 1.3-μm λ_{PL} layer. Numerical evaluation of Eq. (15) suggests that combined lateral and vertical mode mismatches between the active and passive sections in such a joint in a buried-heterostructure geometry can easily yield coupling efficiencies of ~92% with proper design.

This type of butt coupling has been employed in many of the PICs to be illustrated later in this chapter. The simplest example is the continuously tunable multisection MQW-DBR laser [31]. Although the structure comprises only a laser, devices of this type can justifiably be called PICs because the resonators actually consist of a serial coupling of three distinct guided-wave devices; a separate electronically controlled gain medium, variable phase shifter, and tunable Bragg reflection filter. These devices, shown schematically in Fig. 11, have outputs of 20–30 mW, minimum linewidths of 1–2 MHz, and offer continuous, rapid electronically-controlled access to a 1000-GHz (80-Å) tuning range at 1.53 μm while maintaining a linewidth below ~16 MHz. Such devices have served as the basis for computer-controlled

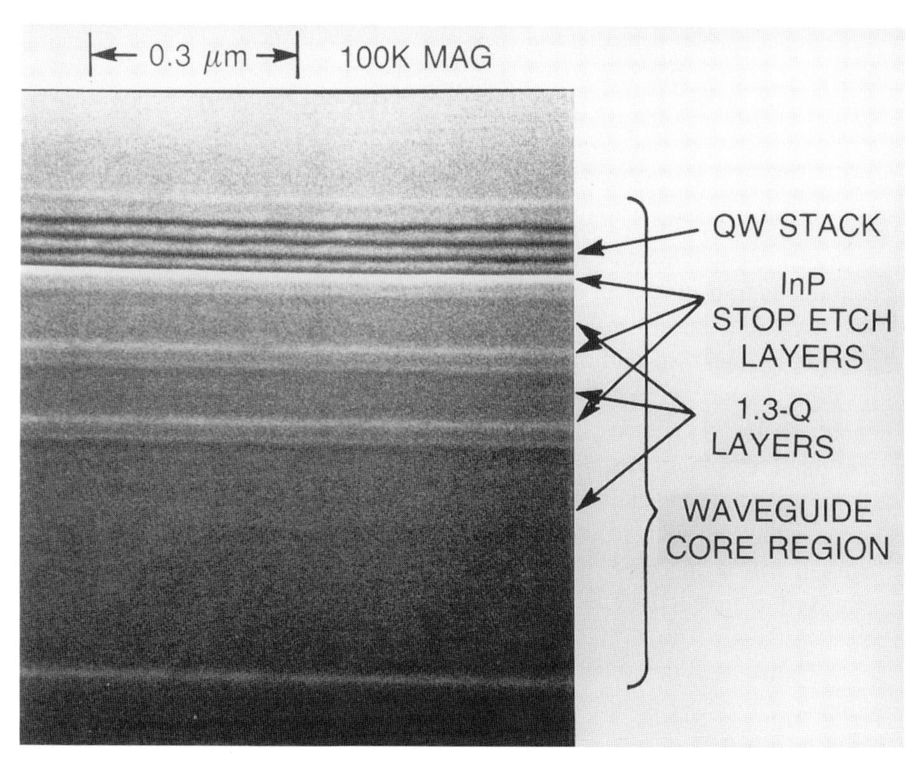

Fig. 10. Scanning electron micrograph of stained cross section of guide structure with a four-QW active stack on top of a passive 1.3 λ_{PL} core containing additional InP etch-stop layers for further passive guide processing.

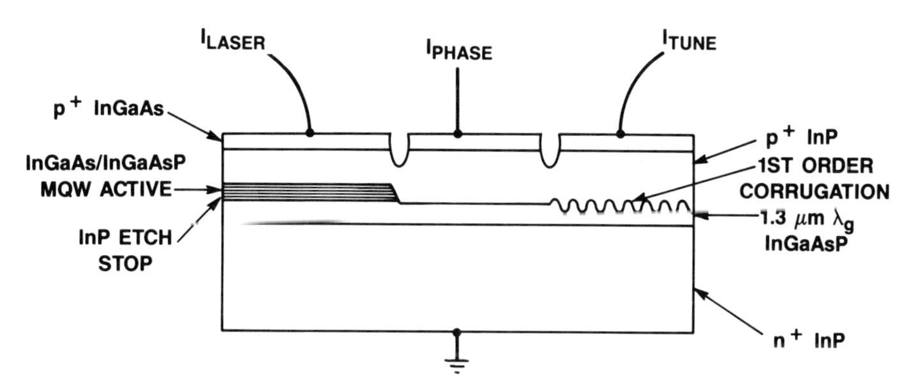

Fig. 11. Schematic of continuously tunable three-section MQW-DBR laser using active/passive transition as shown.

random-access multichannel coherent networks with near-quantum-limited detection sensitivities [32].

3.3. Selective-Area Growth

The foregoing techniques rely on etching and regrowth. Another method has been demonstrated, termed *selective-area growth* (*SAG*), in which smooth, continuous transitions can be made in the waveguide material composition. The SAG method utilizes a variation in growth rates observed in MOVPE induced by masking portions of a surface against epitaxial growth and the resulting impact on quantum well thicknesses and band-to-band transition energies [33–35]. Figure 12 shows an integrated DFB laser/electroabsorption modulator fabricated with the SAG method [34]. In longitudinal regions of the wafer where gain or absorption media are required, lateral regions immediately adjacent to the ultimate location of the waveguide stripe are masked with a dielectric material that inhibits the deposition and nucleation of growth compounds. This produces a vapor above the surface with a higher concentration of growth constituents than in unmasked regions where deposition has been possible. The higher concentration diffuses toward nearby unmasked regions, resulting in effective enhancement of growth in regions adjacent to masked areas relative to a uniform, unmasked surface. Additional mechanisms for enhacement include surface migration of growth constituents that deposit on masked areas that migrate until they form stable

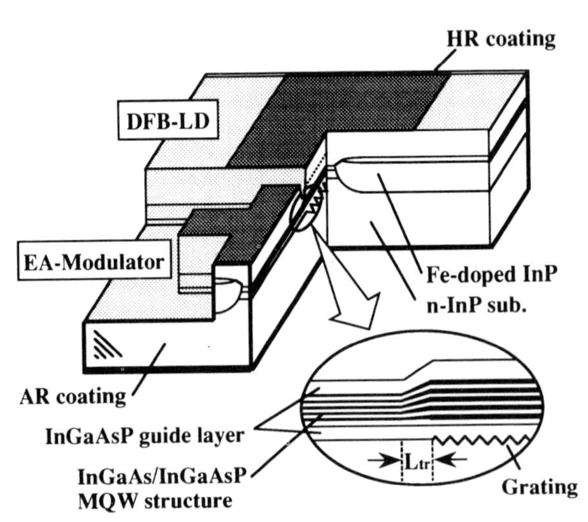

Fig. 12. Structure of an MQW-electroabsorption modulator integrated with a DFB laser using the selective-area growth (SAG) method.

bonds on nearby unmasked surfaces. Both mechanisms have the effect that longitudinal regions with less lateral masking adjacent to the waveguide have slower growth rates.

The longitudinal variation in growth rates induced by longitudinal variations in lateral masking can be used to produce variations in the thicknesses of mutliple-quantum-well stacks along the waveguide. The thicker wells that result from closer or more lateral masking will have a narrower bandgap and thus serve as gain or absorption media. Thinner quantum well regions in unmasked or less masked areas will have higher bandgaps for electroabsorption modulators or even transparent waveguides if the wells are thin enough.

Very high performance integrated laser/modulators have been demonstrated using the SAG technique [34, 35]. External efficiencies as high as 0.19 mW/mA have been obtained out the modulator facet, with extinction ratios of 13 dB with a modulator bias change from 0 to −2 V [34]. Bandwidths as high as 14 GHz have been obtained, allowing good performance at 10 Gb/s. In addition to the laser/modulators, the SAG technique has been applied successfully to tunable multisection MQW-DBR lasers [36].

3.4. Vertical Directional Coupling

Another approach to coupling between two different waveguides employs directional coupling in the vertical plane between epitaxial layers serving as the cores of the two distinct waveguides. This type of vertical coupling can either be accomplished by using the principle of intersecting dispersion curves or by using a grating to achieve phase matching.

In the first case, the two waveguides are chosen to have both different composition and different thickness cores in such a way that the dispersion curves cross at some particular wavelength, i.e., $\beta_1(\lambda_0) = \beta_2(\lambda_0)$ at some particular $\lambda = \lambda_0$. Equivalently, we have $n\text{eff}_1(\lambda_0) = n_{\text{eff}_2}(\lambda_0)$. This has been demonstrated in the InP system by Broberg et al. [37] using $In_{1-x}Ga_xAs_yP_{1-y}$ cores with $y = 0.127$ and $y = 0.078$ and respective thicknesses of 0.42 and 0.91 μm. The InP separation layer had a thickness of 2.19 μm.

The transfer of power in a codirectional coupler is given by [18]

$$\frac{P_2(L)}{P_1(L)} = \frac{\kappa^2 \sin^2(SL)}{\kappa^2 + (\Delta\beta/2)^2} \tag{22}$$

where $S \equiv (\kappa^2 + (\Delta\beta/2)^2)^{1/2}$ and $\Delta\beta \equiv \beta_1 - \beta_2 - K_g$ where $K_g = 2\pi/\Lambda_g$ is the grating k-vector of any grating used for phase matching in the coupler.

The 3-dB bandwidth of a one-coupling-length device, $L = \pi/(2\kappa)$, solved directly from Eq. (28), is

$$\Delta\lambda = 0.3993 \frac{\lambda^2}{L(n_{g_1} - n_{g_2})}, \tag{23}$$

which includes compression effects from the difference in *group* indices rather than the phase indices used for period calculations[38].

In the particular case mentioned, the bandwidth of the coupling was demonstrated to be $\Delta\lambda \sim 220$ Å. The relatively wide coupling bandwidth is due to the small difference in group index that is hard to avoid because the phase indices must match. However, this may be used to advantage in applications in which broadband coupling is desired. The same small slope in the dispersion intersection can lead to high sensitivity of the center wavelength to fabrication parameters such as guide composition or thickness. For example, a 10% change in core thickness can lead to a change in the coupling center wavelength of ~1000 Å.

Another approach to vertical directional coupling employs a grating. This has some potential advantages but carries the price of additional fabrication complexity. One advantage is the steeper dispersion intersection that is possible, leading to narrower bandwidth. The phase-matching equation in this case becomes

$$\beta_1(\lambda_0) = \beta_2(\lambda_0) + \frac{2\pi}{\Lambda_g} \tag{24}$$

where Λ_g is the grating period. Here we can intentionally work with two very mismatched waveguides, where for forward coupling with both β_1 and β_2 positive we require that $\Lambda_g = \lambda_0/(n_{\mathrm{eff}_1} - n_{\mathrm{eff}_2})$. Typical values for Λ_g might be 15 μm for forward coupling [39, 40]; values of 2300 Å would be typical for reverse coupling [41] where β_2 would be negative. A physical realization of the forward coupling is depicted shematically in Fig. 13. The design value for the lower guide is 0.3-μm-thick, 1.3-μm λ_{PL} InGaAsP, with a 0.5-μm-thick, 1.1-μm λ_{PL} InGaAsP layer to support a 600-Å-deep corrugation, with an upper guide of 1.1-μm λ_{PL} InGaAsP with a 0.2-μm thick core separated by 1.5 μm of InP. Using a supermode analysis as described earlier for the directionally coupler example and the grating coupling formalism described earlier, one can easily compute the characteristics of this device. The supermodes of the system are shown in Fig. 14. With a square-wave grating and evaluating Γ_a as given in Eq. (23), we obtained $\kappa = 8.3\ \mathrm{cm}^{-1}$, or a

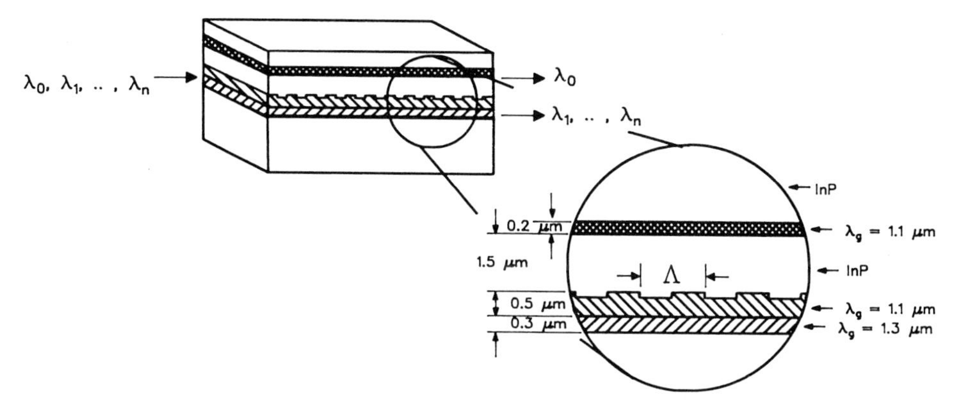

Fig. 13. Schematic of codirectional vertical grating coupler in InGaAsP/InP.

coupling length of $L_{\text{coup}} = \pi/(2\kappa) = 1.9$ mm. Using the theoretical effective indices of the supermodes shown, we obtain a a grating period of 14.5 μm for phase matching at 1.5 μm. The experimental value was 14.4 μm, in excellent agreement with theory [40].

Such devices may be useful for wide-tuning applications, since a change of δn_{eff} in n_{eff_1}, for example, will lead to a tuning of $\Delta\lambda/\lambda = \delta n_{\text{eff}_1}/(n_{\text{group-eff}_1} - n_{\text{group-eff}_2})$, which can be a much larger tuning than would

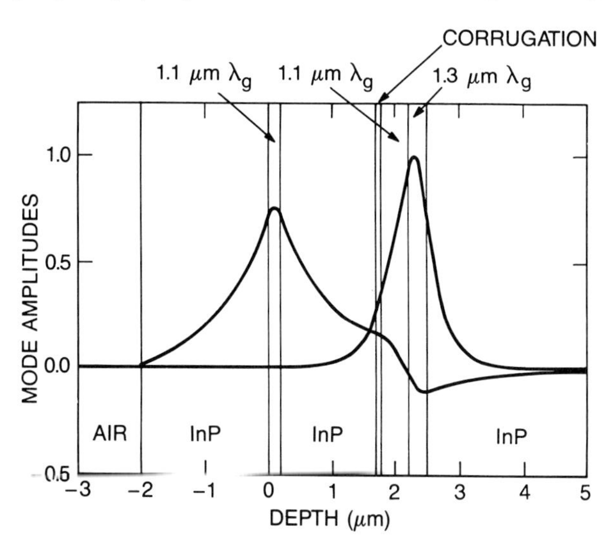

Fig. 14. Mode structure of complete structure as shown in Fig. 12. The grating is taken as a perturbation that couples the two modes of the whole system that are shown.

be obtained in a reverse coupling Bragg filter, for example, where the equivalent equation would be $\Delta\lambda/\lambda = \delta n_{\mathrm{eff}}/n_{\mathrm{group-eff}}$.

3.5. Vertical Impedance Matching into Detectors

Another technique that can be used—not to couple from one waveguide into another but to couple from a waveguide into a bulk medium or absorbing medium as might be used in a photodetector—is vertical "impedance matching" [42, 43]. This technique employs an intermediate layer between the guide and the absorbing medium, which may be above the guide, for example. The characteristics of this layer are carefully chosen to provide an antireflection coating on one side of the waveguide cladding, of course taking into account the angle at which rays are propagating in their zigzag path down the waveguide. Deri *et al.* [43] have effectively used this technique to increase the coupling between large-mode waveguides and detectors that are grown above the waveguide. This has permitted efficiencies of 90% in distances as short as 190 μm for large-mode waveguides that can easily be coupled into from optical fibers and otherwise would have coupling efficiencies into the detectors of only ~30%. This increased coupling in a short distance is important in this instance to allow small-area detectors that are required for low-capacitance, high-speed operation.

3.6. Passive Optics Issues

More complex PICs often require an even larger selection of active and passive guide types and result in additional growth and fabrication demands. One example that imposes new constraints arises in long-path-length truly passive interconnections between different optical devices. Such long interconnections may be required when laterally separated devices are required to allow coupling to different fibers, or where directional couplers or Y-branches are used. Narrow branching angles with long feeds and large-radius curves for bends are both needed to minimize unwanted radiation losses in such PICs. One of the major loss mechanisms in InP-based devices is intervalence-band absorption in *p*-type material. Waveguide losses of ~15–20 cm^{-1} per 10^{18} cm^{-3} *p*-type doping may be encountered [44], corresponding to losses of 30 dB/cm for typical guides and typical confinements in the upper cladding layers used in active structures. For PIC dimensions of several millimeters this is clearly unacceptable, and thus undoped or SI-InP upper cladding layers are required.

Another problem mentioned in the introduction is the desire for precisely controlled phase characteristics and/or strictly fundamental-mode guides in passive structures such as branches, couplers, switches, and filters. Coupling between two buried-heterostructure guides with designs typical for most laser structures requires micrometer-level core proximity, and small submicrometer level lateral dimensional variations from processing will significantly affect the coupling strength. Using the etch-stop methods described earlier, buried rib guides can be fabricated where the rib-loading layer is formed in the original growth step. The cross section of a typical passive buried rib guide is shown in Fig. 15. At 1.5 μm, a 300-Å-thick "floating" rib loading of 1.3-μm λ_{PL} InGaAsP with a 3000-Å-thick core of the same material, embedded in InP, remains strictly single-mode up to a lateral dimension of ~3.2 μm. Also, the loose lateral confinement leads to significantly larger lateral dimensions for directional coupler designs. Not only does this reduce the lithography requirements, but also the actual etching of the guide can be accomplished with high precision because the 300-Å etch-stop-controlled rib depth leads to negligible undercutting. Etching steps such as this are typically carried out with an SiO_2 mask patterned by conventional photolithography and wet or dry etching, followed by a material-selective wet chemical etch of the semiconductor. The SI-InP upper cladding layer growth is then carried out after the oxide mask removal in buried rib portions of the PIC.

3.7 Large-Mode Waveguides

Apart from the performance of the devices themselves on the PIC, one must bear in mind that at some point, light will be coupled into or from the PIC.

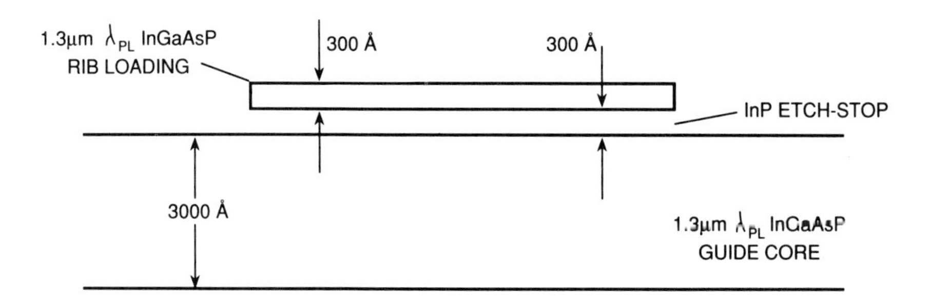

Fig. 15. Cross section of design of passive buried-rib waveguides employing a 300-Å-thick etch-stop controlled "floating rib" loading with SI-InP upper cladding for low loss.

This is typically light from a single-mode optical fiber, where the typical 8-μm mode size is dramatically larger than the ~1-μm mode sizes of the typical semiconductor waveguide of an active device.

One approach to dealing with this is to make large-mode semiconductor waveguides. This can be done through leaky waveguides as in the ARROW [15] example of Section 2, or through conventional bound modes using weak index steps from the core to the cladding and large core dimensions. The latter approach is easy to implement, not by growing an InGaAsP layer with very little Ga and As but by placing dilute quantum wells of InGaAsP or InGaAs in InP to simulate a medium whose index differs only slightly from that of InP. This approach has been used successfully to achieve low-loss, large-mode guides [45–47], with the additional benefit that the absorption edge and the index of refraction are independently engineered paramters. Excellent loss values as low as 0.24 dB/cm have been achieved using this technique [46]. Furthermore, a total 0.8-dB fiber-to-fiber insertion loss has been demonstrated for an 11-mm-long MQW core single-mode guide on InP [47]. In addition, these guides share the features described for the buried rib waveguides in that lateral dimensional control is relieved in making passive devices such as directional couplers, which have also been demonstrated with low loss [48].

This technique has the disadvantage that although fiber coupling has been improved, the resulting guides are now in general not compatible with the preferred geometries of most active guided-wave devices. However, techniques have been devised to circumvent this problem for photodetectors, for example. One technique mentioned earlier is vertical impedance matching. Another remarkable technique is shown in Fig. 16. In this PIC [49], photodetectors have been integrated with a directional coupler made from the large-mode waveguides in a balanced receiver configuration. The photodetectors in this example have been processed and grown on the opposite side of the InP substrate from the waveguide devices, with the light deflected toward the detectors using an etched crystal facet. A combined fiber-coupled detection efficiency as high as 0.7 A/W was obtained in the best devices with common-mode rejection as high as 33 dB.

3.8 Stepped-Taper Adiabatic Mode Expansion

Another principle for coupling from the small guides that optimize the semiconductor waveguides has been termed adiabatic mode expansion (AME). A particular approximation that has been demonstrated is a stepped reduction in the thickness of a passive waveguide core to adiabatically increase the

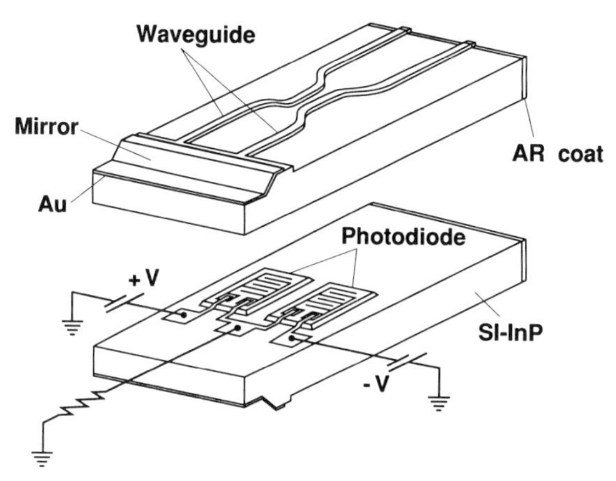

Fig. 16. Shematic of a PIC integrating a passive, large-mode waveguide directional coupler with MSM photodetectors processed on the back surface of the wafer.

mode size and produce a narrow-divergence output beam [50]. This significantly relaxes the submicrometer alignment tolerances usually required for efficient coupling of optical fibers or bulk optical components into the small waveguide modes necessary to optimize semiconductor lasers.

A conventional three-layer waveguide with a high-index core thin enough to have a several-micrometer mode size would have a nonideal double-sided exponential shape with a sharply peaked center and slow decay in the wings. Therefore, in addition to the stepped reduction in core thickness, we employ "outrigger" guide layers which become functional and shape the mode only in the weakly guided, thin-core regions. This is illustrated with a particular example in Fig. 17. The laser structure used here is similar to the standard MQW-DBR laser used in many of the PICs described above, with an MQW active layer on the upper edge of the active section optical mode. The active layer, consisting of four 70-Å InGaAs wells with 150-Å, 1.3-μm InGaAsP barriers, is separated vertically from the rest of the 1.3-μm waveguide core by a 180-Å InP etch-stop layer.

Prior to stripe mesa formation, the active layer is removed in the passive taper region using the etch-stop layer and selective chemical etchants. The 1.3-μm InGaAsP passive waveguide core itself has additional InP etch-stop layers embedded within it to allow the controlled removal of increasingly thinner layers from the core, with sequentially displaced masking, ultimately forming a staircase profile down to a 450-Å-thick waveguide core. In the simple example shown, the first 1.3-μm core step is 280 Å, while the second step is only 160 Å. The distance between the taper steps is typically ~50 μm. The additional 300-Å-thick guide layers shown in Fig. 17 of 1.3- or

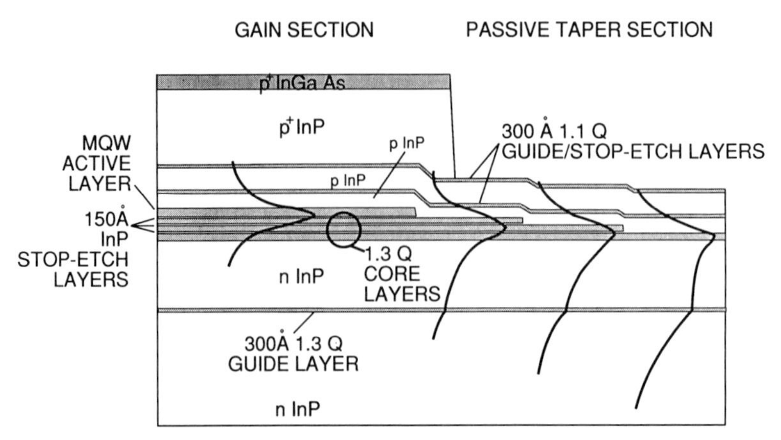

Fig. 17. Longitudinal structure through the buried heterostructure core for a simple two-step passive adiabatic mode expansion taper. Mode profiles show qualitatively the mode evolution in this vertically asymmetric example.

1.1-μm InGaAsP are present below or above the guide core continuously along the length of the structure. The formation of the taper structure does not add additional growth steps because the active stack is already accessible during the "longitudinal" PIC processing, or even as an intermediate growth step of the semi-insulating blocked buried heterostructure laser design employed here. This point will be discussed in detail in the next section.

The resulting far field is shown in Fig. 18 and displays a three- to fourfold reduction in divergence compared to typical 1.5-μm devices with standard active guides as used here. This design permits a dramatic improvement in coupling to fibers. Fig. 19 shows the alignment tolerance to a "cleaved" (flat end) standard single-mode communications fiber, with a loss of −4.2 dB. However, the −1 dB geometric mean alignment tolerance is ±2.8 μm which is an ~fivefold improvement over the case of typical lasers and the lensed-tip fibers needed for comparable coupling loss. This use of the multilayer etch-stop techniques for precise but highly reproducible layer engineering may result in significant cost savings in fiber packaging of PICs or even simple lasers.

4. CRYSTAL GROWTH AND PIC PROCESSING

In this section we briefly review some of the crystal growth and fabrication techniques that are employed in making InP-based PICs.

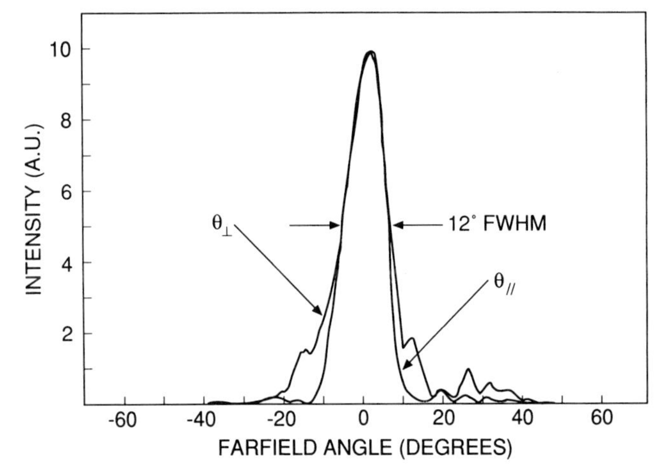

Fig. 18. Experimental far fields parallel and perpendicular to the junction plane. The vertical far field is actually displaced by several degrees at an angle toward the substrate, presumably due to an admixture of the noncoupled light from the taper steps.

4.1 Crystal Growth

Crystal growth technology has evoloved rapidly during the last decade from nearly exclusive use of manually controlled liquid-phase epitaxial (LPE) growth to a variety of computer-automated vapor and beam growth techniuqes. These include atmospheric-pressure and low-pressure metalorganic vapor-phase epitaxy (MOVPE), hydride and chloride vapor-phase epitaxy (VPE), chem-

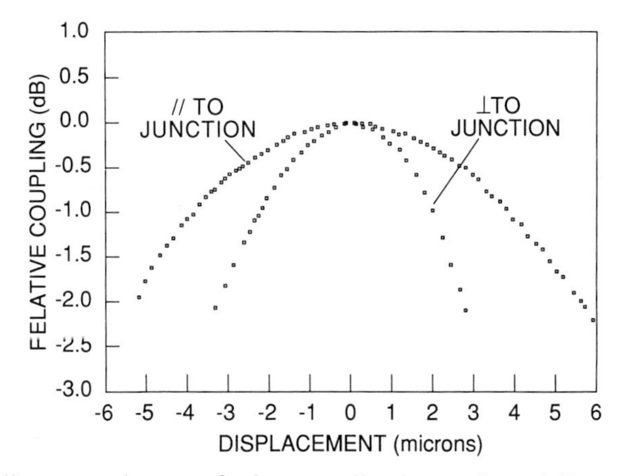

Fig. 19. Alignment tolerances for butt coupling into a cleaved fiber relative to the maximum coupling efficiency of −4.2 dB. The −1 dB tolerance is ±3.8 μm parallel and ±2.1 μm perpendicular to the laser junction plane.

ical beam epitaxy (CBE), and metalorganic molecular beam epitaxy (MOMBE). In the case of the latter two ballistic techniques, CBE and MOMBE differ in that CBE uses metalorganics for both the group III and group V elements, while MOMBE typically uses metallic sources for group III. The interested reader is referred to conference proceedings for the most current information on growth technology [51]. With regard to PICs, these advanced crystal growth techniques offer substantially more versatility and uniformity than LPE. For example, LPE is limited in the number of melts possible and typically requires special layers to avoid "melt-back" problems. LPE also poses difficulties in obtaining reproducible quantum-well structures when the well thicknesses become only a few tens of angstroms.

For the quaternary system, uniformity is required both in the composition (x, y) of the $In_xGa_{1-x}As_yP_{1-y}$ as well as in the layer thicknesses. To get a feel for the uniformity required, one can look at the deviation of the lasing wavelength in a typical DFB structure with thickness and compositional changes. Using the design tools presented in the last section, it is a simple exercise to determine that layer thickness changes of several percent can lead to nanometer-scale wavelength changes. Similarly, a 10-nm shift in the photoluminescence peak wavelength of the guide layers, which is not at all uncommon, can also result in nanometer-scale DFB wavelength shifts in addition to potential undesirable gain-peak mismatches that may result from the λ_{PL} shift itself.

Proper reactor geometry, sometimes with substrate rotation, can also lead to percent-level uniformity in MOVPE [52]. The ballistic geometry of the beam techniques [53], together with substrate rotation, may tend to produce the most highly uniform layers [54]. The difficulty associated with CBE or MOMBE lies with the ballistic "line-of-sight" growth which prevents proper regrowth over reentrant mesa geometries or overhanging mask surfaces often encountered in PIC and laser fabrication, while MOVPE and especially VPE offer outstanding coverage over a wide range of morphologies.

Other criteria to be considered are the doping capabilities. Due to the potential for lower growth temperatures, CBE and MOMBE have demonstrated the ability to produce very highly doped thin layers which are desirable for high-speed HBT-based OEICs, for example [55]. Such structures are difficult with the diffusive Zn and higher growth temperatures commonly used for MOVPE. Both the vapor and beam techniques have successfully grown semi-insulating Fe-doped InP, a material that is playing an increasingly pivotal role in photonic devices.

The typical PIC processing involves the growth of a base structure that is followed by processing and reqrowths. During both the base wafer and regrowths, selective-area growth is often employed where a patterned dielectric film is used to prevent crystal growth over protected areas. This film is typically SiO_2 or Si_3N_4 deposited by chemical vapor deposition (CVD) or plasma-assisted CVD. This technique is readily used with MOVPE, but care must be taken to keep a substantial portion of the field open for growth to aviod the formation of polycrystalline deposits on the dielectric mask.

One area in which great caution must be exercised is during regrowths over mesas or other nonplanar geometries, especially in the vicinity of masked surfaces. Crystal growth rates are very orientation dependent, and this orientation dependence may be a function of growth temperature. For this reason, very different results will be obtained for growth around a mesa stripe running in the [011] direction compared with a stripe in the $[0\bar{1}\bar{1}]$ direction. It is often critical to produce planar surfaces, and growth around masked regions can easily lead to gross deviations from planarity due to overshoots of crytal growth occurring on the different crytal orientation exposed on the mesa walls. Detailed discussion of these phenomena is beyond the scope of this chapter.

All things considered, MOVPE appears to be the most versatile in its ability to produce high-definition multilayered base structures and to provide regrowth capabilities over a range of morphologies. However, PIC processing is invariably broken into several growth stages, beginning with the base wafer, which usually contains the critical quantum well active layers and various guide layers. It is often the case, and is entirely reasonable, that the base wafer and regrowths are done in different reactors using different growth techniques. CBE, for example, could be used for highly uniform base wafers, while MOVPE, VPE, or even LPE could be used for the regrowth stages.

Materials characterization for PICs entails the usual range of epitaxial layer measurements, including X-ray diffraction to obtain the lattice matching and photoluminescence to obtain the bandgap as well as the brightness, which gives a measure of the nonradiative recombination that occurs in the layers. Doping profiles are usually determined by capacitance–voltage measurements during a wet-etching processing through the grown layers. This must be carried out on a test piece, however, because it is a destructive test. A more detailed examination of these and other techniques for materials characterization is beyond the scope of this chapter, however.

4.2. Grating Fabrication Techniques

Many of the PICs employ grating-based resonators or filters, and the most common technique for fabricating these gratings involves a “holographic” or interferometric exposure using a short-wavelength laser source. Here a thin (typically 500–1000 Å thick) layer of photoresist is spun on a wafer surface and exposed with two collimated, expanded beams from a blue or ultraviolet (UV) laser at an appropriate angle to form high-contrast fringes at the desired pitch. Since the illuminating wavelength is precisely known and angles are easily measured in the mrad range, the typical corrugation in the 2000-Å-period range can be fabricated to angstrom-level precision in period. The resist is developed and then functions as an etch mask for the underlying layers. This etching can be either a wet etch (commonly using HBr-based etchants) or a dry reactive ion etch. Commonly used lasers are HeCd at 325 nm or one of the UV lines of an argon ion laser at 364 nm. Figure 20 shows a typical corrugated grating before and after regrowth. This illustrates the most common scenario in which a "bulk" layer of InGaAsP has a corrugation place in it, followed by an overgrowth of InP. Another technique involves a thin layer of InGaAsP, with InP below it, grown just to support the grating. After the corrugation is etched and an InP regrowth is performed, the resulting grating is composed of rectangular "islands" of InGaAsP surrounded by InP. The potential advantage of this technique is that it may yield a more reproducible grating coupling constant because it is less sensitive to corrugation etch depth, and it also provides a means of reproducibly fabricating very weak coupling constants because the InGaAsP layer can be made very thin.

4.3. Reactive Ion Etching Techniques

Another fabrication technology that is useful in PIC work is reactive ion etching (RIE) or other variants such as chemically assisted reactive ion beam etching (CAIBE). This has often been carried out using Cl_2-based mixtures with O_2 and Ar [56], while in other cases the reactive cholorine is derived from compounds such as CCl_2F_2. Excellent results have also been obtained with methane–hydrogen or ethane–hydrogen mixtures [57]. In these latter cases Ar is also often used as a sputtering gas to remove interfering redeposited compounds. Reactive ion etching has been used both to form mesa and facet structures and to transfer grating patterns into semiconductors through an etch mask.

The appeal of reactive ion etching is the lack of mask undercutting that can usually be achieved, allowing very high lateral precision with the promise of reproducible submicrometer mesa features. In addition, the ability to create vertical-wall etched facets through a variety of different compositions of epitaxial layers suggests the possibility of integrated resonator or reflect-

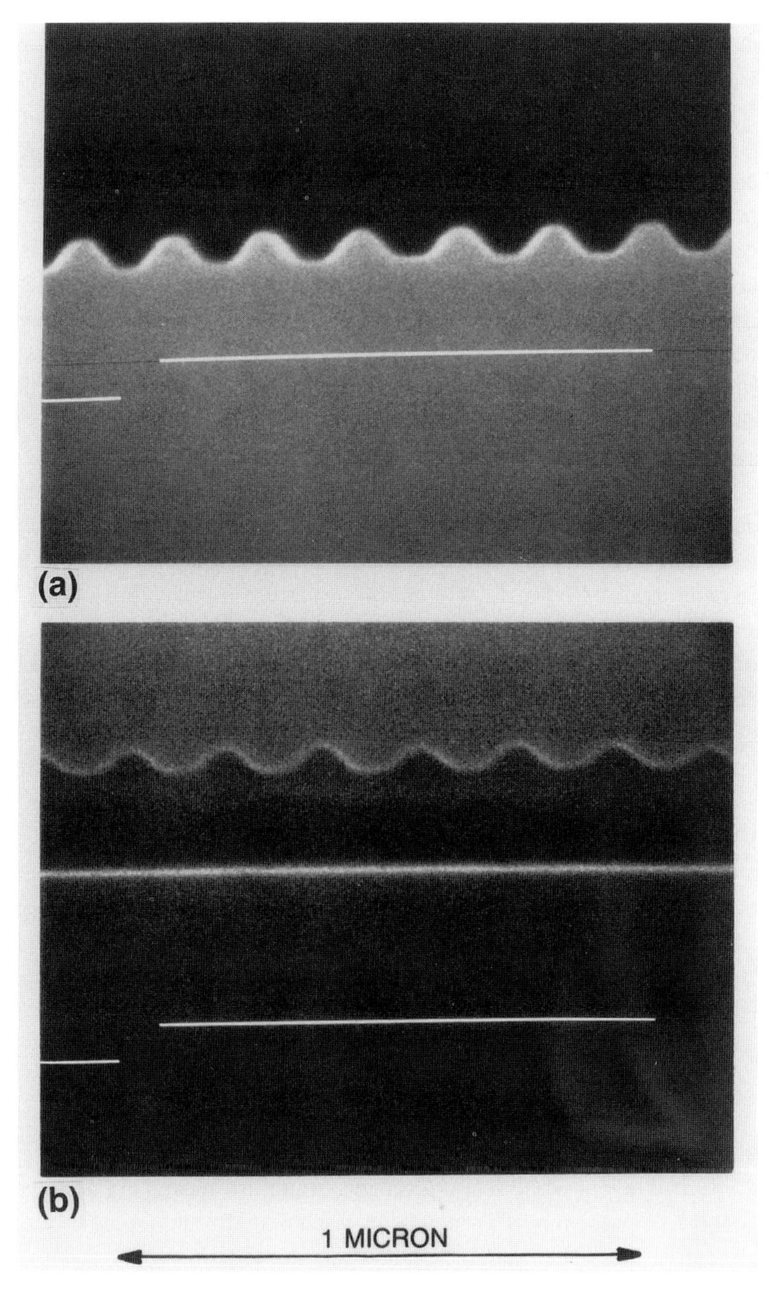

Fig. 20. SEM of a typical holographically fabricated grating (a) before and (b) after regrowth. This particular grating is for a 1.3μm DFB laser, and the corrugation is formed in a 1.1μm λ_{PL} layer on top of the active layer.

ing and coupling structures without the use of gratings. Indeed, this approach has recently been used to form corner reflectors [58], square-geometry ring resonators [59], and a variety of complex waveguide patterns using beamsplitters etc. [60]. Other applications that have highlighted the unique capabilities of RIE include the very high aspect ratio cylindrical mesas demonstrated in some vertical-cavity semiconductor laser geometries [61].

One of the most remarkable applications has been the use of etched-facet technology to create gratings, not as a corrugation *along* the waveguide but as a grating *in the plane* of a waveguide for two-dimensional "free-space" diffraction [62–64]. In this application, a facet is etched all the way through the guide layer and a coating can be applied aimed at 100% reflection. As stated, the grating teeth here are *lateral* in the plane of the slab guide and are usually coarse teeth for higher-order reflection. One demonstration by Soole *et al.* [63] is shown in Fig. 21 and consists of a curved, focusing grating operating in 18th order (5.1-μm teeth spacing). The grating in this case was fabricated by CAIBE using Xe^+ ions with a reactive flux of Cl_2 etched to a depth of 2.3 μm, with a $<4°$ departure from vertical walls. The grating reflects and images the expanding beam from a single input waveguide into one of 78 different ouput waveguides depending on the input wavelength. This $\sim 12 \times \sim 2$ mm device forms a "spectrometer on a chip" and may be useful as a multiplexer–demultiplexer in dense incoherent WDM applications. Thirty-five of the channels had losses within a 3-dB range as shown in Fig. 22. This particular implementation has channel spacings of 10 Å. This chip has operated with on-chip losses of ~16 dB, apart from the coupling losses getting onto the chip. A sizable amount of this is simply Fresnel loss from the uncoated grating facet (5.6 dB) and probable losses from the sloped wall, suggesting that improvements should be realizable. The uniformity of the MOVPE growth in this instance was quite good, with $<1\%$ thickness variation in the waveguide layers and <2 nm guide core λ_{PL}.

4.4. Wet Etching and Etch-Stop Fabrication Techniques

To define the precise geometries required in many PICs one must often remove specific layers while leaving others, or control mesa heights to a very high degree of precision. This has been discussed, for example, in the section on active–passive waveguide coupling. In some instances this may be done with carefully controlled reactive ion etching rates or slow wet chemical etching, but often the resulting precision is inadequate. Other techniques include means of end-point detection in which plasma species can be actively monitored to determine the instant a new layer composition is reached.

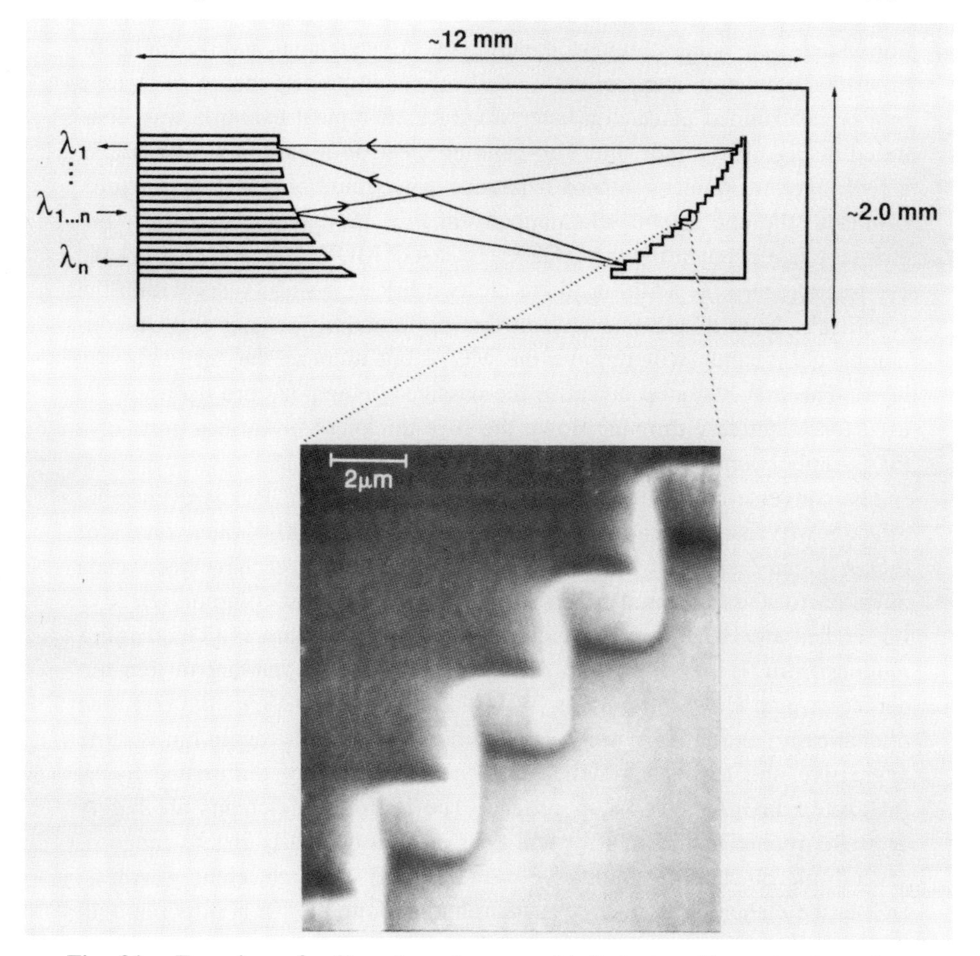

Fig. 21. Top view of a "spectrometer on a chip" device, illustrating how dry etching techniques for facet formation can be used for much more than laser facets.

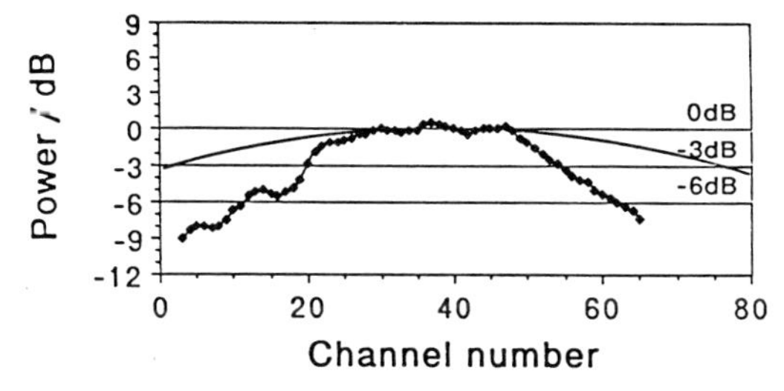

Fig. 22. Loss vs. channel number for WDM mux/demux chip shown in Fig. 20.

Fortunately, the combination of material-selective etchants and the inclusion of special etch-stop layers offers a convenient and precise means of etching to a predetermined place in a layer structure. A typical example was illustrated in Fig. 10, which shows a scanning electron micrograph of a 4-QW active layer residing on top of a passive waveguide core. This structure, similar to the one shown schematically in Fig. 10, has a 200-Å-thick InP etch-stop layer separating the InGaAs/InGaAsP MQW active layer from the 1.3-μm λ_{PL} core. In addition, there are two InP etch-stop layers within the core itself. Here a selective etchant that preferentially attacks the Ga-containing compounds will remove the MQW but halts at the InP etch-stop layer. The InP etch-stop layers in the passive waveguide core itself can be used for selectively thinning down the core thickness in various parts of a PIC. Applications of this will be discussed later.

The converse of this sequence is also possible and in fact is even more effective. Even after many micrometers of etching, 100-Å-thick layers of InGaAsP can easily halt InP etches. Extensive compilations have been made of etches for the InP-based compounds [65], and the most common selective InP etches are HCl based. Typical mixtures are HCl and H_3PO_4 in ratios ranging from 3:1 to 1:3, with the lower HCl content leading to less undercutting and slower etch rates [66]. Addition of H_2O slows the etch rate dramatically, and HCl by itself also functions well in this regard but is more aggressive and leaves a rougher surface in regions not terminated by an InGaAs or InGaAsP etch-stop surface. The HCl-based etchants are highly crystallographic in nature [67] and etch at rates typically in the range of several micrometers per minute. For a (100) InP substrate, etching with an HCl-based etchant with a stripe mask running along the [011] direction will yield a mesa with nearly vertical walls, only slightly reentrant by a few degrees. This is true provided the etching mask does not undercut. The best such a mask might be an epitaxial layer of InGaAsP or InGaAs. Other approximations to this are high-quality Si_3N_4 and SiO_2 masks; in the authors' experience the former has less undercutting. Photoresist masks undercut too much to be useful in producing a vertical profile, and the resulting profile will be outward sloping in a generally nonreproducible geometry. With a high-quality mask, the walls will be nearly vertical all the way down provided an etch-stop plane terminates the etch and sufficient etching time is allowed. Otherwise, an outward-sloping "boot" will remain along the bottom of the wall representing a (111B) plane. This technique, for example, is used for carefully controlled mesa heights and has also been used for very deep, rectangular wave gratings in InP/InGaAsP [68].

For a stripe mask along the $[\bar{0}\bar{1}\bar{1}]$ direction, a relatively shallow outward-sloping wall is produced at an angle of approximately 38° from the surface. This is in contrast to the 54.7° (111A) face that is exposed by many nonselective etchants such as bromine–methanol mixtures. This shallow sloping wall is in general more forgiving in terms of undercutting that the vertical-walled mesa.

Photoresist can be used as an etching mask where either a very short, shallow etch is performed or some undercutting is permissible. This is true, for example, when removing layers in a broad-area fashion or in the field during PIC processing. The performance of photoresist in this application will depend in detail on the spin-on and especially the baking procedures; a longer, higher-temperature bake leads to better mask performance. However, this comes at the expense of more complex resist removal procedures. As mentioned, the best InP mask is an epitaxial layer of InGaAs or InGaAsP, with a good plasma-assisted CVD Si_3N_4 or SiO_2 being the next best thing at thicknesses typically in the 1000–3000 Å range. High-temperature CVD processes (600°C) are to be avoided because they degrade the InP surfaces.

A common selective etch for removing InGaAsP or InGaAs while only weakly attacking InP is a mixture of H_2SO_4, H_2O_2, and H_2O, in a ratio of $X:1:1$ with X typically ranging from 3 to 30. Other mixing ratios offer different performance in terms of selectivity and etch rates; values are deliberately not quoted because we feel that it is imperative to calibrate using the material to be etched and the etchant preparation sequence, age of etchant, and degree and method of agitation. Selectivities in the range of 10:1 and typically much higher are readily achieved, however. Another popular etchant for selectively etching the quaternary compounds is KOH–$K_3Fe(CN)_6$–H_2O in proportions of 4:6:35 by weight, but again other ratios are often used to achieve slower etch rates or different selectivities [69]. Other selective etchants for InGaAsP are based on HNO_3.

In some instances it is required that the etchants be nonselective, uniformly removing layers regardless of composition. This is usually the case when etching a mesa through a multilayer active region to form a buried heterostructure laser. Br-based etchants, such as bromine–methanol (Br_2 and CH_3OH), tend to be very good in this regard. An order of magnitude etch rate is 10 μm/min per percent bromine, but this is only a guide. This etchant ages quite rapidly and should be mixed fresh if depth control is required. This etchant, along with many of the nonselective etchants, will form a reentrant 54.7° (111A) face mesa for stripes along the [011] direction (with a good mask) and an outward-sloping 54.7° walled mesa for stripes along the $[0\bar{1}\bar{1}]$ direction.

Other etchants, with varying degrees of nonselectivity and crystallographic behavior, include mixtures of HBr, CH_3COOH, or HCl, CH_3COOH, and H_2O_2 [70]. The latter, particularly in a ratio of 1:2:1, is referred to as KKI from the authors' names, Kambayashi, Kitahara, and Iga [71], and is a useful multipurpose nonselective etchant that is useful for mesa etching through multilayer stacks.

4.5. Photonic Process II: Application to an MQW Balanced Heterodyne Receiver

Here we will discuss in detail some of the fabrication approaches using the etch-stop fabrication techniques just mentioned. These approaches borrow from many refined processing techniques already developed [72], but in proper sequencing they have offered remarkable freedom in combining the various guide types and device requirements without undue growth complexity.

We will focus on a variant of the fabrication process PPro-2 [73] as applied to an MQW balanced heterodyne receiver PIC [74, 75] shown schematically in Fig. 23. Figure 24 shows a photograph of a cleaved crystal bar containing dozens of these 3-mm-long circuits. This PIC serves as a good vehicle for discussion because it combines five different types and a total of seven, guided-wave optical devices: two tunable Bragg reflection filters, an MQW optical gain section, an electrically adjustable phase shifter, a zero-gap directional coupler switch, and two MQW waveguide photodetectors. In addition, it utilizes a low-loss passive buried rib parallel input port and S-bending feeds and demonstrates self-aligned connections between the buried heterostructure guides, which offer current access and SI-InP lateral current blocking, and the low-loss rib guides with low-loss SI-InP upper cladding. Finally, this PIC is fabricated with no additional growth steps beyond those required for the conventional SIPBH laser process [76, 77]. We note here that impressive results have also been obtained in heterodyne receiver architectures using the butt-joint waveguide technique [28].

Figure 25a and b show cross sections through two very different guide geometries used in the MQW balanced heterodyne receiver PIC. Figure 25a shows a cross section through the laser gain section and waveguide photodetector, and Fig. 25b shows a cross section through the center of the zero-gap directional coupler switch. In both cases, some detail has been omitted for clarity. The local oscillator (LO) laser has the SIPBH laser structure and is almost identical in design to a previously published MQW-DBR device [31]. Since no LO light is used off-chip, a high-reflector tunable Bragg mirror is used at the facet side, with a partially reflecting tunable Bragg mirror

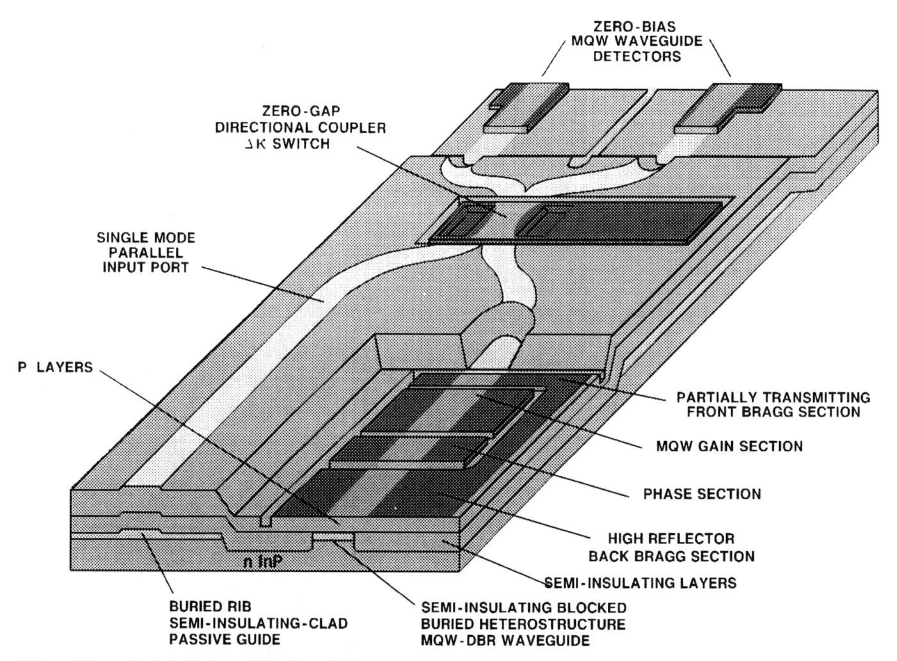

Fig. 23. Schematic of MQW balanced heterodyne receiver photonic integrated circuit, containing a continuously tunable LO, a low-loss buried rib parallel input port, an adjustable 3-dB coupler, and two zero-bias MQW waveguide detectors.

used as the internal output coupler. Also, the underlying passive guide core contains an additional layer which serves no function in the laser but which will act as the rib-loading layer, as discussed earlier, in the passive guides and the directional coupler switch.

The design of the switch, which in this application is used only as an adjustable directional coupler set to 3 dB, has also been discussed previously [78]. It has a double-moded buried rib region with current injection laterally outside the core of the switch. This affects a differential phase shift between the even and odd modes, due to their different evanescent tail overlaps with the free carriers outside the rib-loaded region. This design, although not the most efficient in its use of injected current, is quite compact and offers robustness in performance against fabricational variations.

The processing and growth used in this MQW balanced heterodyne receiver PIC are conceptually broken into two stages. The first consists of growing the base wafer up to and including all the passive core and active layers of the waveguide. In a broad-area processing sequence, the wafer then undergoes "longitudinal processing" as shown in Fig. 26. This consists of

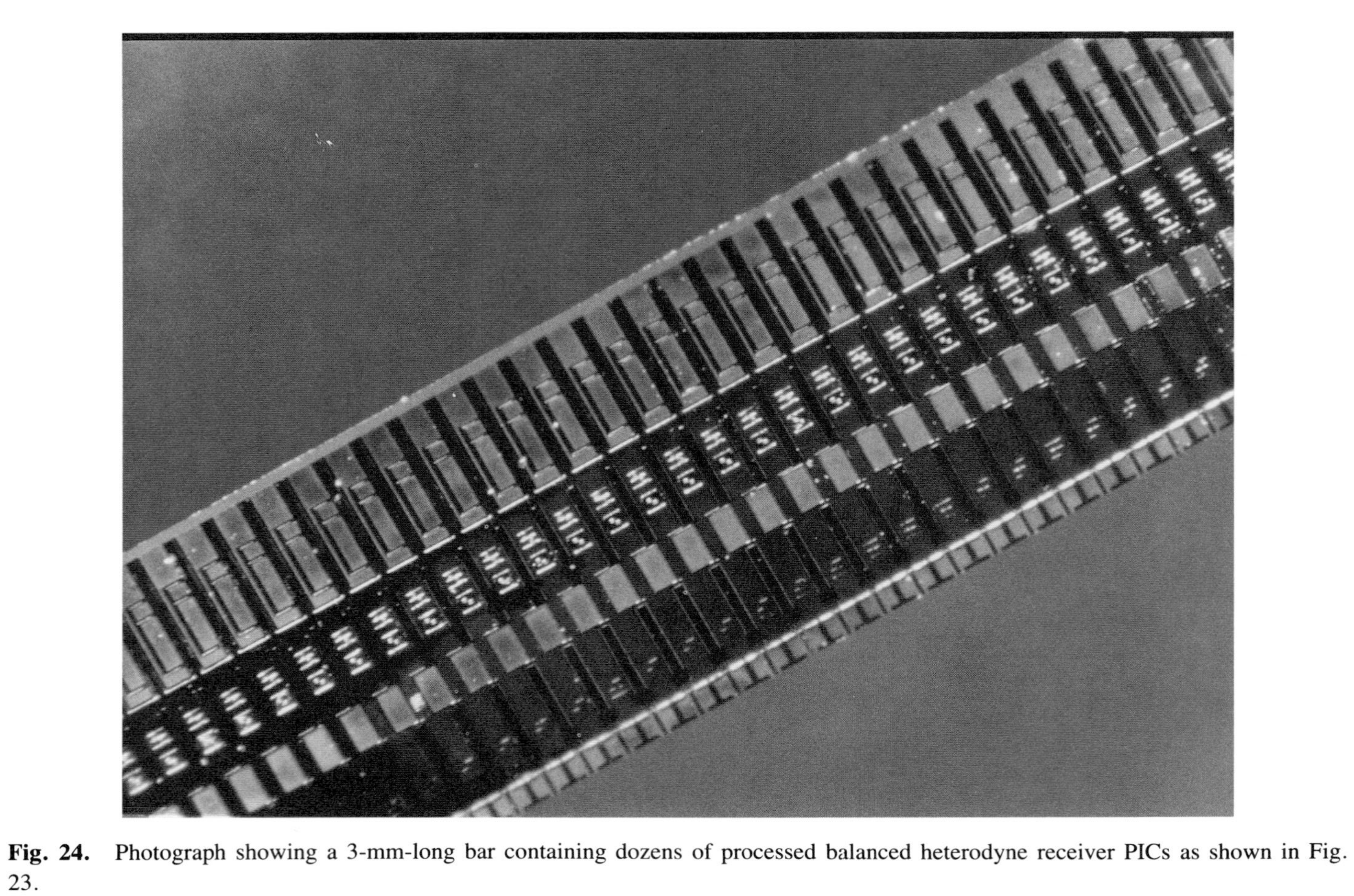

Fig. 24. Photograph showing a 3-mm-long bar containing dozens of processed balanced heterodyne receiver PICs as shown in Fig. 23.

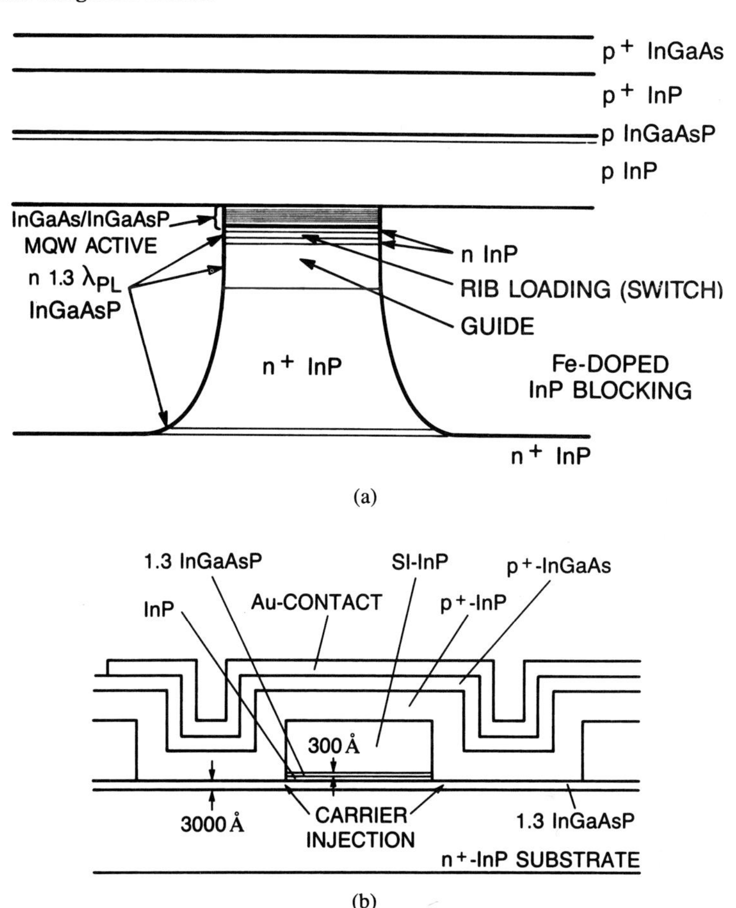

Fig. 25. (a) Cross section of the buried heterostructure laser waveguide in the gain section. (b) Cross section of the buried rib waveguide in the switching region. For clarity, some of the structure in the upper and lower cladding is not shown.

selective removal of the MQW active stack except in the gain and detector portions of the PIC, using an etch-stop-controlled material-selective wet chemical etch; removal of the buried rib-loading layer where it is not needed, also using selective etching; and finally selective placement of first-order gratings where desired using conventional holographic exposure and wet etching. This longitudinal processing thus serves to create a longitudinally continuous slab waveguide with a vertical structure suited to each device type, the only mismatch being due to the perturbation induced by the removal of small layers as discussed earlier. This longitudinal processing is broad-area in the sense that there is no stripe or waveguide definition at this point. However, as seen in the schematic of Fig. 23, there is a parallel input

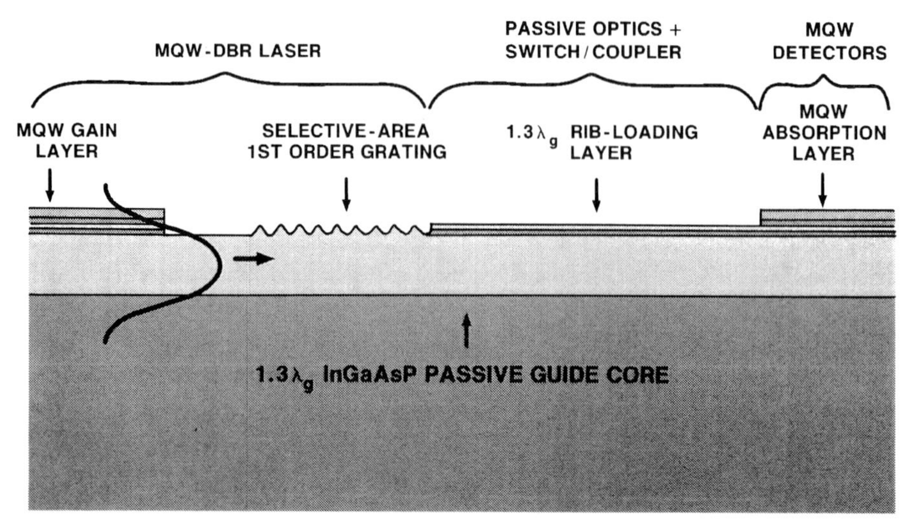

Fig. 26. Longitudinal processing sequence for waveguide stack.

port running next to the laser, for example, so some crude lateral differentiation is required in MQW removal and grating placement. This is accomplished through appropriate masking or resist removal in these steps.

This philosophy is thus quite different from that of the regrown butt joint used for the DFB laser/electroabsorption modulator PIC discussed earlier. All the various guide layers that are needed for the devices incorporated in the PIC are put into the first growth, producing a largely continuous structure from which layers are *removed* where they are not needed. Previous devices have demonstrated that the passive core layer may also contain high-bandgap quantum wells for use in high-speed field-induced electrorefractive modulation [79].

Following the longitudinal processing, the stripe definition takes place in a "lateral processing" sequence which allows the same growths to perform different functions in different portions of the PIC. First, the oxide stripe waveguide pattern etch mask is patterned. Two paths are then followed, one resulting in buried heterostructure waveguide devices (lasers, detectors, modulators, etc), the other resulting in buried rib devices (passive guides, directional couplers, etc.).

This is shown in Fig. 27, where for simplicity the entire complex guiding stack is shown as one layer. After a shallow etch-stop-controlled rib definition everywhere, the oxide is removed in the rib-guide region, and replaced with resist. A deep mesa etch is then performed, controlled in depth

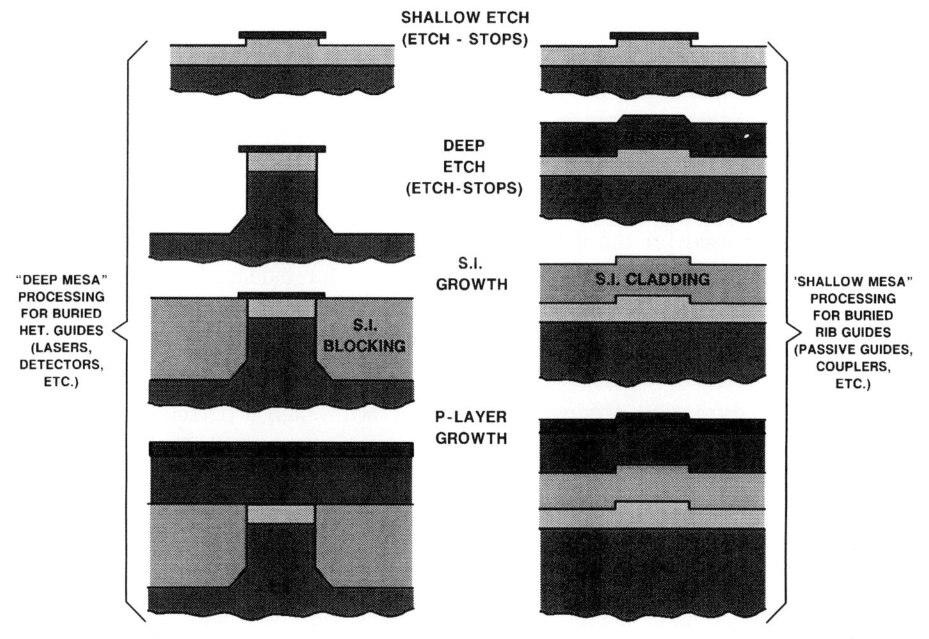

Fig. 27. Simultaneous self-aligned lateral processing of active buried heterostructure and passive low-loss buried rib regions.

by etch-stop layers, followed by a regrowth of an SI-InP blocking layer sequence. The same growth provides low-loss upper cladding in passive regions but forms lateral blocking in the deep mesa active devices. If current or field access is desired in any SI-InP overgrown rib guides, the SI-InP can be locally removed. For example, this option is used on each side of the directional coupler switch shown in Fig. 25b to allow current injection. After the final oxide removal on the deep mesa devices, the final *p*-type upper cladding layers are grown.

An additional advantage of the SI-InP upper-clad buried rib guides is the electrical isolation which becomes possible between different portions of the PIC. Since the upper *p*-layers can be selectively removed over the passive buried rib waveguide regions, essentially complete (megohm-level) isolation can be readily achieved between devices.

This combination of lateral processing and longitudinal processing is thus quite powerful and can result in vertically and laterally self-aligned guides of various types, each suited to a particular device, with only three growth steps.

5. ILLUSTRATIVE EXAMPLES OF PIC APPLICATIONS

5.1. MQW Balanced Heterodyne Receiver

The performance of the MQW balanced heterodyne receiver PIC has been fully characterized [74, 75]. The lasers have thresholds of 20–35 mA, and offer continuous access to an ~60-Å (750 GHz) tuning range at 1.53 μm. The device used for the data described here had a coupling ratio of 1.2:1 with no current, which was deemed adequate for balanced operation. As discussed for the processing steps, the 180-μm-long detectors employ for absorption the same MQW layers used for gain in the laser, and the performance of such detectors has been reported previously [80]. We find that that good absorption efficiency is possible with such a detecting medium and that out to at least ~5 GHz there are no significant speed degradations from hole trapping in the GaInAs/GaInAsP MQW system. Since these detectors use the same current blocking layers and relatively heavy, ~$5 \times 10^{17} cm^{-3}$ doping near the junction, reverse breakdowns are typically only a few volts and leakage currents are high, reaching tens of microamps with only a few tenths of a volt reverse bias. However, we find that good performance can be obtained with zero bias where the dark current is zero, and furthermore dark current is of no concern in heterodyne receivers where milliampere-level local oscillator currents are present. Finally, the capacitance in these structures is dominated by the contact pads rather than the depletion capacitance, and with typical values of several pF, bandwidths of 1–2 GHz into 50-Ω loads are observed. However, if a higher-impedance front end was used for inproved sensitivity, lower-capacitance detectors would be required. While the detectors were used in the system evaluation described below with no bias, a slight improvement of ~1 dB was observed with a reverse bias of ~0.5 V when milliampere-level photocurrents are generated by the LO laser.

The PIC performance was evaluated in a narrow-deviation FSK architecture using only 50-Ω commercial electronics connected directly to the PIC with a test FSK-modulated input optical signal from a three-section MQW-DBR transmitter. Commercial 50-Ω bias tees allow a dc return for the photocurrent from the unbiased detectors, with the signal coupled in each arm to separate "front-end" 2.9-dB noise figure, 50-Ω amplifiers with 1-GHz bandwidths and 25-dB gains. The signals are then subtracted in a 180°-shifted splitter–combiner and fed into a 50-Ω amplifier–variable attenuator chain. This then splits through a directional coupler at the discriminator, with a variable path length in one arm, feeding an RF mixer for discrimination. A bias tee at the mixer output capacitively couples the signal to further amplification, and the signal is finally filtered with a 130-MHz filter for the 200 Mb/s case and an 80-MHz filter for the 108 Mb/s case. The low pass of the bias tee is used in a feedback circuit to tune the LO for automatic locking of the LO to the incoming signal at an IF specified by the variable delay in the discriminator [81].

It should be noted that the architecture used here has separate amplification for each arm of the balanced receiver. Due to the smaller signal at each amplifier input, this effectively *doubles* the thermal noise of the front-end electronics compared with the usual back-to-back diode configuration [82], which provides signal subtraction *before* amplification. The present approach is mandated by the common *n*-type substrate for the two detectors. The impact of this additional thermal noise will be very small with a higher impedance, high-sensitivity front end, but with the 50-Ω electronics used here it resulted in additional thermal noise penalty.

For a measured combined LO photocurrent at the detectors of 1.40 mA, the calculated shot noise current is $i_{\text{shot}} = 21.2\ \text{pA/Hz}^{1/2}$. The combined thermal noise current from both 50-Ω amplifiers with 2.9-dB noise figures is $i_{\text{therm}} = 35.5\ \text{pA/Hz}^{1/2}$. This leads to an excess thermal noise penalty over shot noise of 5.8 dB, in excellent agreement with the measured 6-dB value seen by observing a 1.3-dB drop in the noise floor when the LO laser is turned off. This also indicates that the balancing has effectively eliminated any noise contribution from the excess intensity noise of the LO laser.

With an unmodulated input signal the IF beat electrical power spectrum has a combined −3 dB linewidth of ~13 MHz, and with free-space input beams of several hundred microwatts it rises ~60 dB from the noise floor. It is estimated that the LO laser has a linewidth of ~6 MHz, with an equal contribution from the transmitter laser. This combined linewidth value results from the "linewidth floor" of the transmitter and LO lasers, and the high-frequency phase noise in the band of interest may be less than this value would indicate. The measured quantum efficiency of the signal detection in this PIC from a free-space beam outside the chip is 9.3%. The majority of this loss (~6 dB) is coupling loss into the uncoated facet of the parallel input port buried-rib guide using a simple 40× long-working-distance microscope objective.

Figure 28 shows the sensitivity data and eye diagrams at 108 and 200 Mb/s, with respective free-space beam sensitivities of −42.3 and −39.7 dBm for a $2^{15} - 1$ pseudorandom NRZ code. Very little intersymbol interference is evident in the 108 Mb/s case, while a small penalty is evident in the 200 Mb/s eye. The data shown in Fig. 28 are at $2^{15} - 1$ pseudorandom word length. The IF for these data was set at 600 MHz, and the mod-

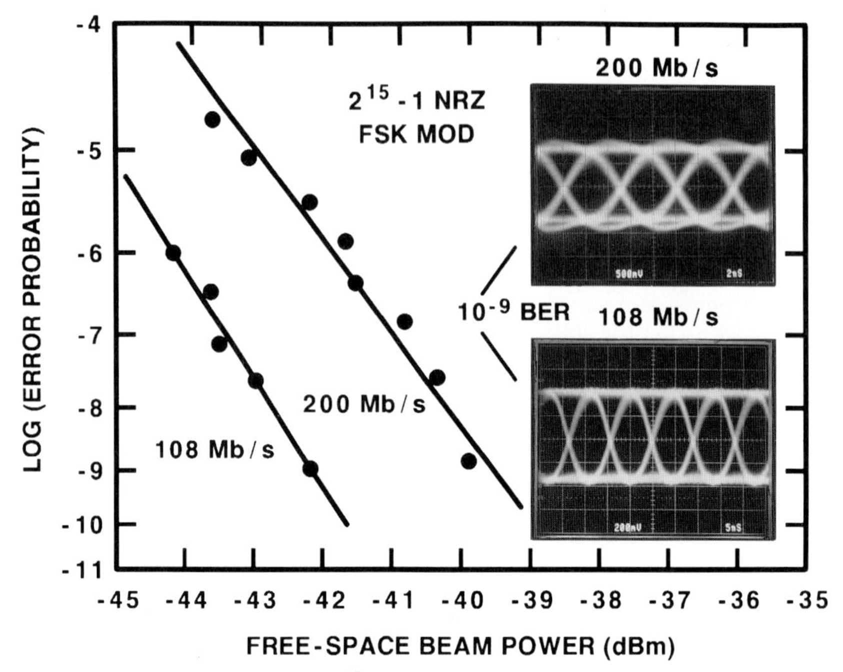

Fig. 28. Sensitivity curves for $2^{15} - 1$ pseudorandom code FSK input data at 108 and 200 Mb/s. Eye diagrams are at 10^{-9} BER in both cases. Power is the free-space input beam average power outside the non-AR-coated facet using only microscope objective coupling. Sensitivities are thus conservatively -42.3 and -39.7 dBm for 108 and 200 Mb/s, respectively.

ulation index for the 200 Mb/s case is $M = 2$, while for the 108 Mb/s case we used $M \sim 3.7$; i.e., the total tone separation in the IF is 400 MHz in both cases.

Table 1 shows an analysis of the various degradations from the shot-noise limit of detection, which is 50 photons/bit for an ideal delay-line discriminator at $M = 2$ [83]. We observe no major degradations beyond the measurable thermal noise from the 50-Ω electronics and obvious coupling losses. This lack of any degradation from LO reflections strongly supports the contention that isolators are not required, as suggested in the introduction.

Also shown in Table 1 are the projected improvements which might be obtained with practical, easily realizable modifications. Chief among these are the use of a higher-impedance, lower-noise front end, the application of an AR coating, and use of a lensed-tip fiber for input. The higher-impedance front end would also have to be accompanied by an insulator layer beneath the detector contact pads for further capacitance reduction. These changes

TABLE 1.

Parameter Values in MQW Balanced Heterodyne Receiver PIC Sensitivity Measurements

<table>
<tr><th colspan="6">MQW BALANCED HETERODYNE RECEIVER PIC
200 Mb/s FSK SENSITIVITY BUDGET</th></tr>
<tr><th>ITEM</th><th>VALUE</th><th>CONTRIBUTIONS</th><th colspan="2">DETERMINED BY</th><th>REALIZABLE VALUE</th></tr>
<tr><td>SHOT NOISE LIMIT</td><td>-58.9 dBm</td><td></td><td colspan="2">THEORY (M=2)</td><td>-58.9 dBm</td></tr>
<tr><td>FREE-SPACE TO CHIP COUPLING</td><td>+6.0 dB</td><td>1.5 dB (no AR coat)
4.5 lens focus mismatch</td><td>calc. & est. from prev. results</td><td rowspan="2">MEASURED
total free-space optical power to photocurrent efficiency:
10.3 dB</td><td>+3 dB
(AR coat + lens fiber)</td></tr>
<tr><td>CHIP EFFICIENCY</td><td>+4.3 dB</td><td>1 dB switch rad. loss
1 dB det. BR/BH coupling
1 dB prop. + bend loss
1.3 dB unassigned:
det. η_d, excess loss</td><td>calc. & est. from prev. results</td><td>+3.3 dB ?
(improved BR/BH coupling)</td></tr>
<tr><td>THERMAL NOISE</td><td>+6.0 dB</td><td>noise from both 50Ω amps w/2.9 dB NF</td><td colspan="2">6.0 dB MEASURED
5.8 dB CALCULATED</td><td>< 1 dB
(high-Z front end)</td></tr>
<tr><td>UNASSIGNED</td><td>+2.9 dB</td><td>RF system flatness, transmitter response, imperfect balance, finite LO and trans. $\Delta\nu$</td><td colspan="2">estimated</td><td>+1.9 dB ?
(improve RF and balance)</td></tr>
<tr><td>TOTAL</td><td>-39.7 dBm</td><td></td><td colspan="2"></td><td>~ -50 dBm
fiber power sensitivity</td></tr>
<tr><td>FREE-SPACE SENSITIVITY</td><td>-39.7 dBm</td><td></td><td colspan="2">MEASURED</td><td></td></tr>
</table>

should be able to produce a 200 Mb/s fiber sensitivity of ~ -50 dBm, or <400 photons/bit. Even in this preliminary experimental prototype, however, the total on-chip losses including propgation losses, bending losses, radiation losses at the coupler and at the active–passive detector transitions, and any departures from 100% quantum efficiency in the detectors, are only 4.3 dB, providing encouragement for even more complex integrations. Hybrid receivers have achieved superior performance, within 1–2 dB of the quantum detection limit. However, the small size and potential for lower cost make the PIC receiver a potential enabling technology for widespread application of coherent lightwave systems.

5.2. WDM Source PIC

Another recently demonstrated PIC, fabricated with the same basic processing sequence, is shown schematically in Figure 29. This is a wavelength

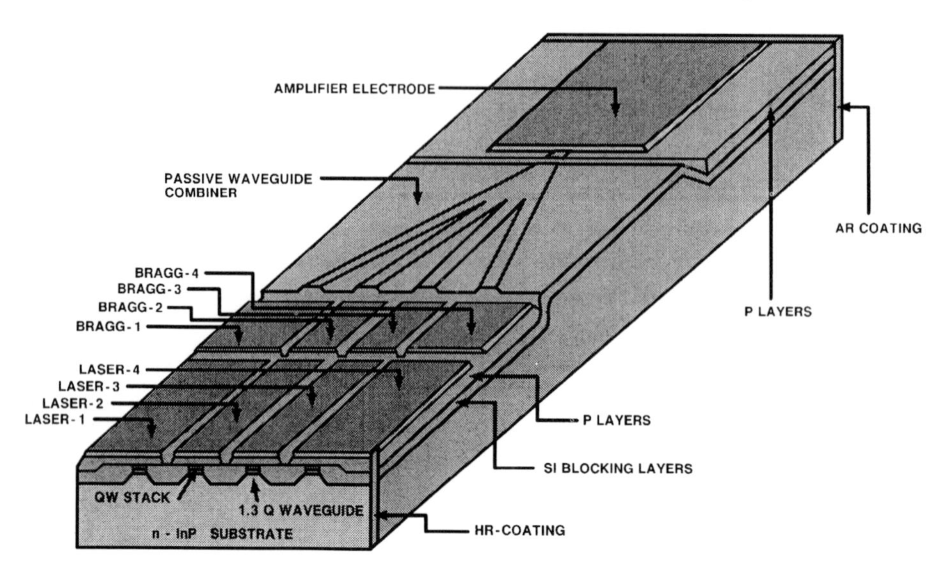

Fig. 29. Schematic diagram of wavelength division multiplexing PIC containing four tunable MQW-DBR lasers combining through an MQW amplifier to a single-waveguide fiber-coupling output port.

division multiplexed (WDM) source PIC, following the concept introduced by Aiki *et al.* [84], where the outputs of a number of single-frequency lasers are combined into a single waveguide output port. The PIC shown here combines the outputs of four independently modulatable and independently tunable MQW-DBR lasers through passive power-combining optics and also includes an on-chip MQW optical output amplifier to recover partially the inherent losses of the power-combining operation [85, 86]. A PIC of similar appearance has been reported in which the center frequency of each successive DBR laser is shifted by changing the guide core thickness, and thus effective index, prior to placing the DBR grating. The cross section of such a device is similar to the SEM shown earlier in Fig. 10. This device exhibited a very large tuning range for a monolithic device, with over 200 Å discrete tuning from a single waveguide port [87].

The WDM source PIC shown in Fig. 29 has been evaluated as a high-speed WDM transmission source [86]. In order to evaluate high-frequency characteristics, a mount was designed with SMA connectors to a 50-Ω microstrip leading to ceramic standoffs with bond wires to the PIC contacts and a thermoelectric cooler as shown in Fig. 30.

One serious area of concern in PICs employing optical almplifiers is the need for good anti-reflection coatings because the source sees the facet re-

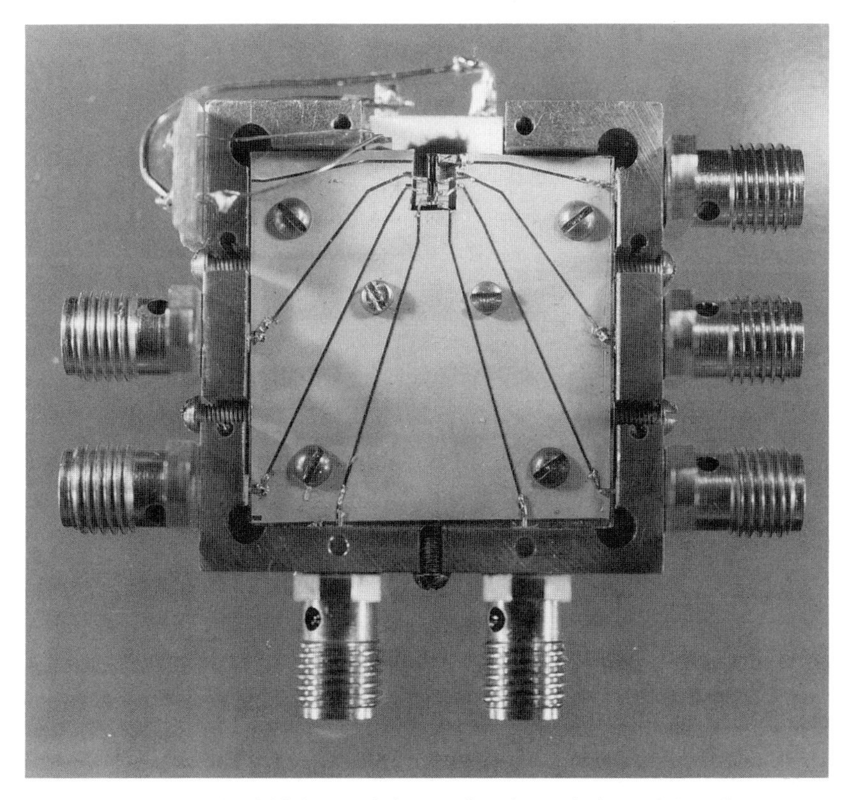

Fig. 30. Photograph of high-speed thermoelectric-cooled mounting fixture, with 3-mm-long WDM PIC between ceramic standoffs where striplines converge.

flection after a double-pass through the amplifier. In the present case, this problem is not severe because the amplifier is a low-gain MQW amplifier (~7 dB), and the beam combiner serves to pad the facet with its own double-pass losses. Another area of concern is the crosstalk between channels, either through electrical leakage on chip or through amplifier saturation effects because all sources share the same amplifier. This latter effect is also minimized in MQW amplifiers, which have been shown to have very large saturation powers [88, 89]. Figure 31a and b show analog small-signal-intensity modulation crosstalk results between two lasers on a WDM source PIC which displayed different effective quantum efficiencies. This difference, whether due to the lasers themselves or the efficiency with which they are coupled into the amplifier, can been seen to lead to a distinct assymetry in the crosstalk behavior. Any electrical leakage from the higher drive required by a low-efficiency laser has a large effect on a high-efficiency laser; the

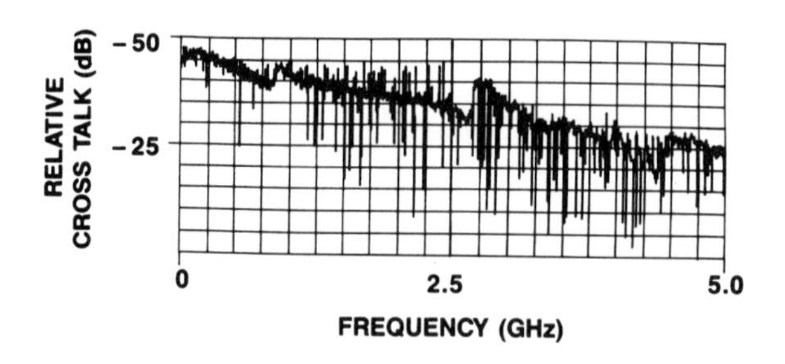

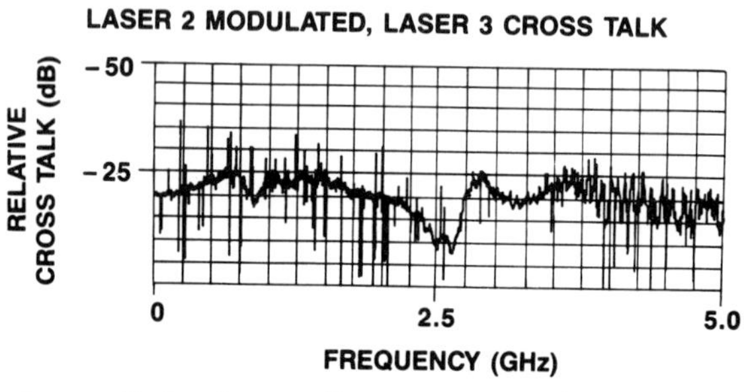

Fig. 31. Analog small-signal crosstalk measurements, i.e., externally detected light on unmodulated channel relative to modulated channel using an external grating to demultiplex signals: (top) crosstalk on lower-efficiency laser when higher-efficiency laser is modulated; (bottom) crosstalk on higher-efficiency laser when lower-efficiency laser is modulated.

converse results in little impact of leakage from the high-efficiency laser to the low-efficiency laser.

To see the impact of this level of crosstalk in a large-signal digital environment, the lasers were modulated with a $2^{15} - 1$ psuedorandom NRZ pattern at 2 Gb/s each, and the combined signal at an aggregate bit-rate of 8 Gb/s was sent over a 36-km transmission path of conventional 1.3-μm dispersion-zero fiber. A fiber Fabry–Perot interferometer was used for channel selection. The PIC used in this experiment had two MQW-DBR lasers at each of two distinct zero-current Bragg frequencies shifted by ~50 Å with etch-stop-controlled core thickness changes as mentioned earlier. One of each pair was then tuned by ~25 Å to yield four channels with a spacing of ~25 Å.

The transmission results are shown in Fig. 32. Laser 4 was the worst-case performer, and even in this case there is virtually no crosstalk penalty without fiber. This indicates that the levels of small-signal crosstalk observed above are not significant in a direct sense, i.e., in their impact on

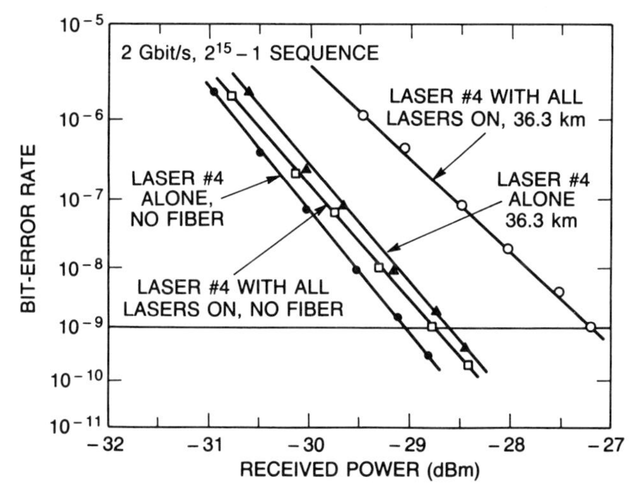

Fig. 32. Sensitivity curves of WDM PIC with 2 Gb/s on each channel for an aggregate rate of 8 Gb/s.

the intensity modulation waveforms at 2 Gb/s. When the 36-km link of dispersive fiber is inserted, a penalty of ~1.4 dB is observed. This is presumably dispersion related, possibly resulting from mode partitioning induced by the cross-modulation from the other laser drive currents on the PIC. Nevertheless, this result indicates that crosstalk is not a severe problem and with some simple design improvements could be reduced to inconsequential levels for digital applications.

5.3. PICs for WDM Transceiver Applications

A great variety of architectures are currently being investigated for application to widely deployed distribution networks, often referred to as "fiber to the home" (FTTH) or "fiber in the loop" (FITL). Most of these architectures are bidirectional, involving some combination of time division multiplexing (TDM) and wavelength division multiplexing (WDM). Some network designs envision 1.3-μm half-duplex bidirectional TDM for low-data-rate services such as telephony and expect to use simultaneously the 1.5-μm band for high-data-rate services such as video, perhaps just in a broadcast mode.

PIC technology may be advantageous in these cost-sensitive links due to the splitters and demultiplexers required in such applications. One approach to such a PIC, shown in Fig. 33, is the work of Williams *et al.* [90, 91]. This PIC integrates a 1.3-μm DFB laser, a 1.3-μm, 3-dB directional coupler, an integrated front-tap monitor photodiode, and an unbalanced Mach–

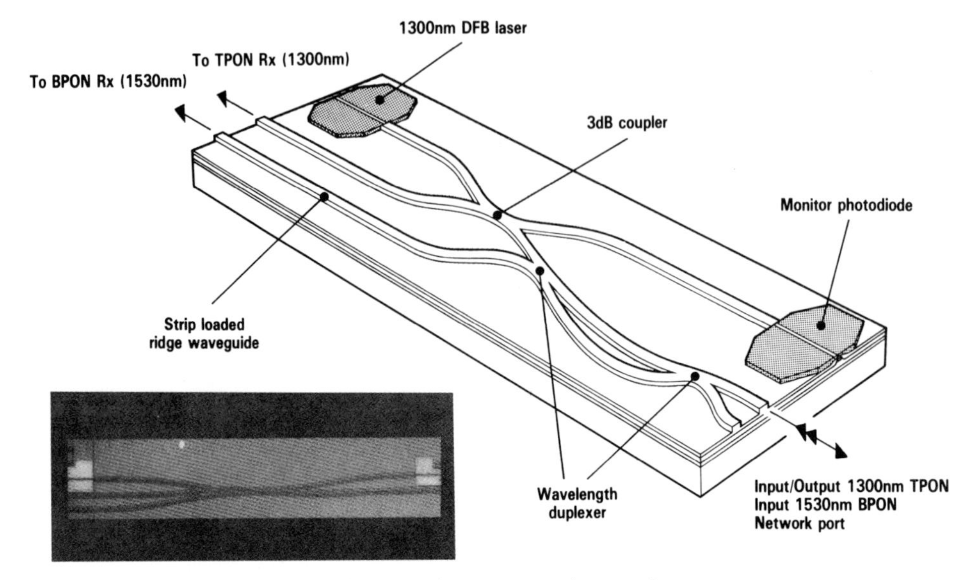

Fig. 33. Configuration of TPON/BPON mux/demux/transmitter PIC for loop transmission.

Zehnder interferometer for 1.3 μm–1.5 μm multiplexing–demultiplexing. This PIC has also incorporated a combination of zero-gap and finite-gap regions in the duplexers and couplers to demonstrate the high degree of polarization insensitivity that would be required in practical applications for PICs of this type.

The active–passive transitions in this PIC are executed with a MOVPE-regrown butt joint similar to that in the integrated DFB–electroabsorption modulator discussed earlier, but uses buried ridge guides for the laser and strip-loaded passive guides for the interconnects and couplers. A 50% coupling has been achieved experimentally in separate straight-guide devices, with laser output powers in excess of 7 mW measured after 1-mm-long passive sections. Crosstalk figures better than −10 dB have been observed in separate wavelength duplexers using the same design as the PIC devices [90].

In the actual PIC devices, the DFB lasers had thresholds in the 10–20 mA range and provided network facet output powers in excess of 0.8 mW. The PICs were packaged into fiber-coupled modules, with network peak pulse power of 164 μW for 60-mA, zero-bias operation. This was sufficient to allow full bidirectional system operation in a four-way split passive optical network at 20.48 Mb/s [91]. This PIC demonstrates a highly desirable architecture from a functional point of view and also poses important research challenges in reducing the fiber-coupled insertion loss for the higher-capacity 1.5-μm channel.

Other network concepts employ 1.3 μm for one direction of digital service and 1.5 μm for the other direction. Here the natural selectivity of WDM is used to help achieve very low crosstalk as required for full duplex operation. Figure 34 shows a schematic of possible in-line bidirectional WDM PIC architectures [92, 7], containing both transmit (Tx) and receive (Rx) functions to form a single-chip transceiver. Both an exemplary in-line 1.5-μm Tx, 1.3-μm Rx PIC and an exemplary in-line 1.3-μm Tx/1.5-μm Rx PIC are shown. Rather than relying on interferometric or grating-based wavelength filtering/routing for separate paths to the source and detector, these PICs make simple use of the absorption/gain characteristics of InGaAsP guide layers of different composition.

PIC architectures such as those in Fig. 34 employ several of the PIC processing techniques discussed earlier but also retain much of the simplicity and low real estate of single-element laser processing. The 1.5-μm Tx/1.3-μm Rx PIC (upper figure) has a design similar to a simple conventional MQW-DBR laser [31], where a four-well MQW stack for gain at 1.5 μm is selectively etched from the upper surface of a 1.3 μm λ_{PL} InGaAsP-core guide to make the active–passive transition. This passive guide, which also contains the DBR grating, is low loss at 1.5 μm but strongly absorbing at 1.3 μm. Thus, with a *p–n* junction at the 1.3-μm input/1.5-μm output end,

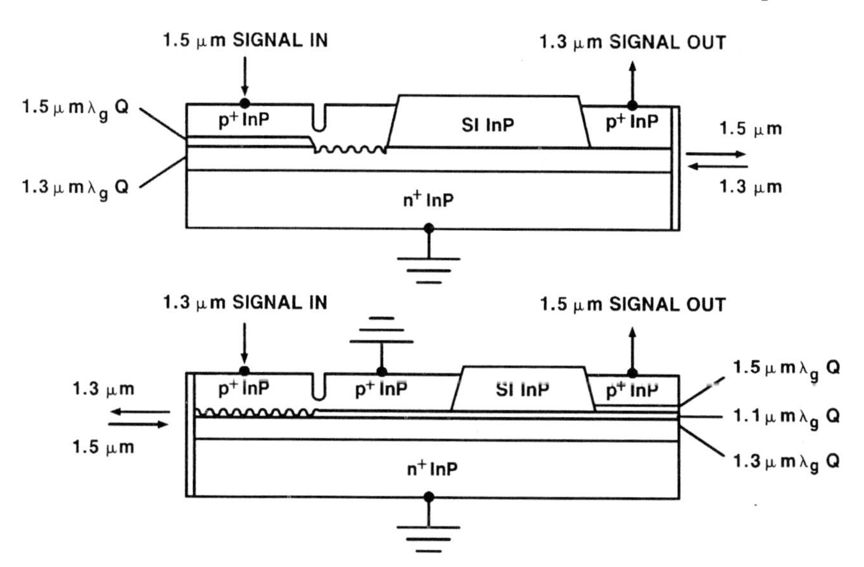

Fig. 34. Configuration of proposed in-line bidirectional 1.5-μm Tx/1.3-μm Rx and 1.3-μm Tx/1.5-μm Rx transceiver PICs.

the incoming 1.3-μm signal is efficiently converted to photocurrent while the local high-power 1.5-μm transmitted signal passes *directly through* the 1.3-μm detector. As discussed in detail with regard to the MQW heterodyne receiver PIC and the WDM source PIC, the upper SI-InP electrical isolation shown in Fig. 34 can be achieved without additional growth steps using established processing techniques such as PPro-2 [73].

The version shown of the complementary chip uses a DFB laser with a 1.3-μm λ_{PL} active layer and a 1.1-μm λ_{PL} grating layer above it. Directly to the right of this is an unpumped section of identical layer structure (no grating needed) which is strongly absorbing at 1.3 μm and electrically terminated to act as a strong sink for 1.3-μm light emitted to the right. Farther to the right is an additional 1.5-μm λ_{PL} upper layer, which has been selectively removed in the first DFB/absorber sections. A 1.5-μm signal incident from the left will then pass *directly through* the transmitting 1.3-μm DFB laser and be absorbed at the far right end due to its mode overlap with the 1.5-μm (possibly MQW) absorbing layer.

For initial experimental evaluation of these concepts, a 1.5-μm Tx/1.3-μm Rx PIC was fabricated. No semi-insulating InP was used for isolation, and the transmitter on the PIC was a 1.5-μm DFB rather than the DBR version shown in Fig. 34. The passive guide was itself composed of two layers, an upper 1.3-μm λ_{PL} and a lower 1.1-μm λ_{PL} InGaAsP layer, used in this case as an asymmetric graded-index structure to improve laser performance. Also, the input/output side had an adiabatic mode expansion (AME) step taper [50] for easy-tolerance fiber alignment. Finally, a 1.5-μm amplifier, identical in structure to the DFB section but without a grating, was also fabricated between the 1.5-μm DFB transmitter and the 1.3-μm detector section.

To evaluate duplex transmission with this 1.5-μm Tx/1.3-μm Rx PIC, it was mated with a hybrid complementary 1.3-μm Tx/1.5-μm Rx station consisting of a simple fiber directional coupler with a conventional pin diode on one arm and a conventional 1.3-μm DFB transmitter on the other arm. The 1.5-μm Tx/1.3-μm Rx PIC had a detection efficiency of 0.3 A/W of fiber power using a nonoptimized lens-tip fiber. The detectors on this PIC are similar to those described for the balanced receiver and were typically run with no bias. With a 6-dB NF 50-Ω amplifier connected directly to the unbiased detector, this PIC had a 10^{-9} BER at 155 Mb/s of -20 dBm, within ~1 dB of the thermal noise limit. In duplex operation, the 1.5-μm transmitter on the PIC was operated at 200 Mb/s to ensure no synchrounous pattern effects, and full duplex operation at 155–200 Mb/s was obtained with 10^{-9} BER over 28 km of fiber using this PIC. Even without SI-InP isolation, dc leakage of the laser drive current to the detector was down by a factor of 30,000. The reduction in the 1.3-μm PIC receiver sensitivity at 10^{-9} BER due to the local 36-mA *p–p* transmitter current drive was only 3 dB. This degradation is believed to result from mount and PIC electrical leakage and can easily be improved both with SI-InP isolation and several options of electrical local transmitter cancellation/isolation.

5.4. DBR Laser/Integrated Backface Monitor

The PICs described are aimed at either high-performance, high-capacity applications or widely deployed low-cost systems expected in the future. It is likely, however, that the first introduction of PIC technology will be in much simpler structures, such as tunable lasers, that form a more incremental evolutionary step from currently manufactured optical devices.

As shown in Fig. 35, PICs similar to the mutlisection tunable DBR laser have been fabricated with another active section, identical to the laser gain section, following the Bragg reflector. In these PICs, the second active section is run with zero or reverse bias and is absorptive at the lasing wavelength, thus functioning as a photodetector for a fraction of light intentionally allowed to leak through a shorter Bragg reflector [93]. This serves as an integrated monitor of the optical power in the laser cavity and may be able to replace the optics and packaging associated with providing a discrete, separate backface monitor photodetector inside today's commercial laser transmitters.

Such devices have shown very high detection efficiency, with the combined efficiency of the laser itself, the passive waveguide coupling, and the photodetector still in the 30–40% range [94]. Other interesting applications of these integrated detectors make use of the variation in the light transmitted through the Bragg mirror as the mirror is current-tuned away from the lasing

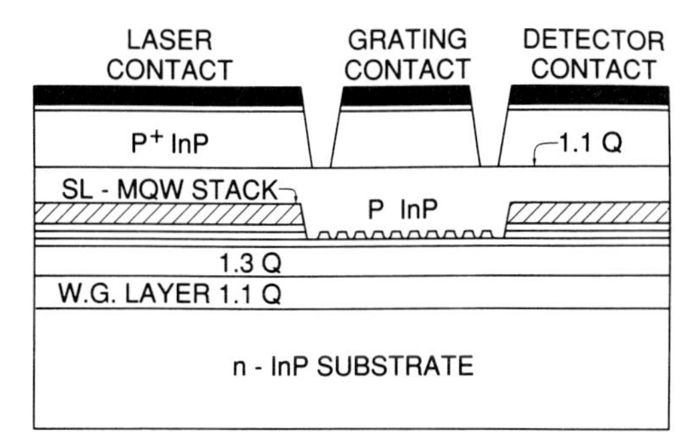

Fig. 35. Laser/backface monitor detector PIC.

wavelength. With a feedback loop, this has allowed a simple external circuit to lock the Bragg mirror right on the center of the lasing longitudinal cavity mode to guarantee single-mode operation in all conditions [95].

5.5. DBR Laser/Integrated Power Amplifier

When high power is required in radio frequency applications, rather than trying to make a very high power single-stage oscillator, a low-power oscillator is followed by a power amplifier. The PIC shown schematically in Fig. 36 carries this philosophy into the optical domain [96]; such devices are called master oscillator-power amplifiers (MOPAs). The structure is very similar to that just described for the integrated monitor detector, but here the gain medium in the specially designed amplifier is run in forward bias to provide optical gain. The oscillator is an MQW-DBR laser, and this feeds through a passive lateral mode expansion section into a wider single-pass optical amplifier. The laser waveguide is maintained at a narrow width to provide stable lateral mode operation; if the entire structure were as wide as the amplifier (~4–5 μm), the beam stability would be poor at even modest output powers.

At the output end, an antireflection coating is applied to prevent reflected light from disturbing the laser oscillator. Also present is an AME structure as discussed earlier to expand the optical beam in the vertical plane to permit less stringent alignment tolerances for coupling to optical fibers. This PIC has demonstrated record-high continuous output powers at 1.48-μm wavelength of 370 mW, and the beam stability is sufficient to permit 117 mW of power to be coupled into a single-mode optical fiber. These properties

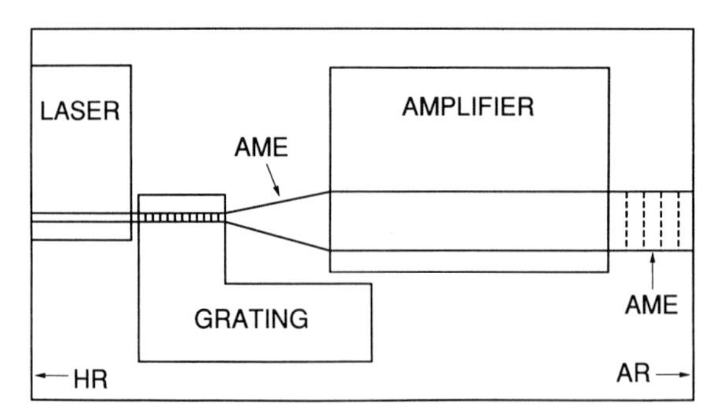

Fig. 36. Schematic of laser/amplifier PIC for high-power 1.48-μm erbium amplifier pump applications. Amplifier has tapers in both dimensions for large-beam, easy fiber alignment at the output. This device has resulted in 120 mW of fiber-coupled power.

suggest its application to optical pumping of erbium fiber amplifiers, which are currently being investigated as digital repeater replacements in long fiber transmission spans.

Devices using an oscillator/amplifier integration have also been successfully implemented in the AlGaAs/GaAs and InGaAs/GaAs systems. Figure 37 shows a 980-nm DBR MOPA in which the amplifier extends in a flared shape to a large emitting aperture for very high output powers [97]. The amplifier efficiency is maintained at a high level through the flare geometry by keeping a relatively constant level of saturation in the amplifier. Diffraction-limited output power as high as 2 W has been obtained in this MOPA architecture, representing a potentially lower-cost alternative to applications served by conventional optically pumped solid-state lasers such as Nd:YAG.

Other MOPA architectures employ near-second-order gratings to achieve vertical emission [98–100]. Figure 38 illustrates how this configuration of output couplers permits a repeating chain of amplifier/output coupler stages; in this case only two stages are shown, but as many as nine have been successfully operated [99]. In a manner similar to the tapered oscillator/amplifier PIC mentioned earlier, the spectral characteristics of such a combination are determined by the master oscillator and the amplifier chain is carefully designed to eliminate undesirable feedback to the oscillator. For example, whereas a true second-order grating is used for the oscillator DBR laser, a near-second-order grating is used for the output couplers to eliminate the phase-matched retroreflection that occurs at second order. This results in an emission angle of $\sim 11°$ from the surface normal, but reflections from the grating are kept below 10^{-5}. By adjusting the amplifier currents to properly phase each output coupler, diffraction-limited output powers of 285 and

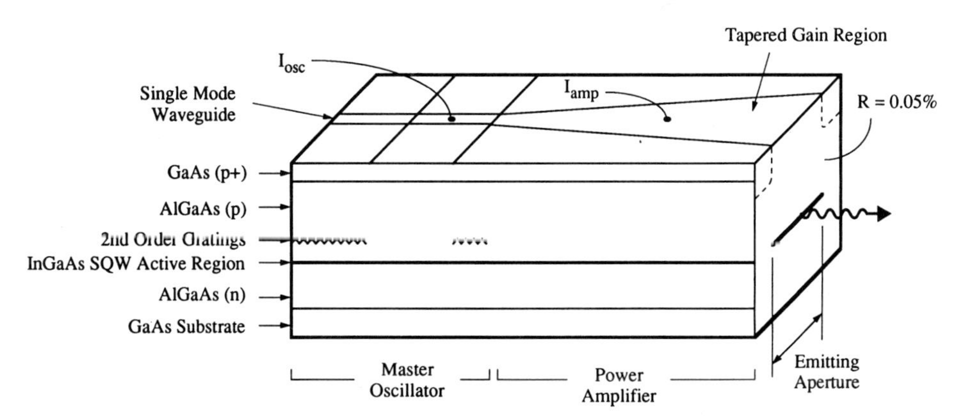

Fig. 37. Schematic of 0.98-μm monolithic master oscillator–power amplifier (M-MOPA) PIC using flared amplifier for high-efficiency saturated amplification.

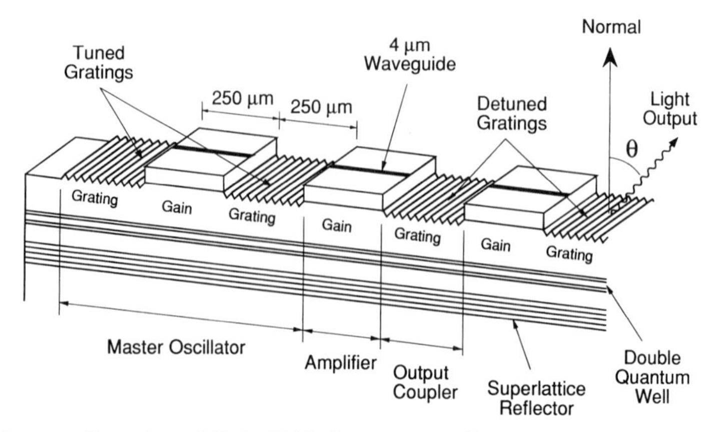

Fig. 38. Configuration of GaAs/AlGaAs master oscillator–power amplifier (MOPA) PIC. Output is vertically coupled by near-second-order gratings following each cascaded amplifier section, with slight detuning to prevent feedback along the waveguide. Reflector layers are also employed beneath the waveguide stack to provide additional top-surface power.

320 mW for seven- and nine-stage amplifiers have been obtained, with the seven-stage MOPA emitting 485 mW of near-diffraction-limited output [99].

To further increase output power, extensions of this concept include branching the master oscillator out into a two-dimensional array as shown in Fig. 39. The master oscillator is split into nine separate guide paths, which each feed a set of cascaded amplifier/output coupler stages [100]. Not shown in Fig. 39 is the full set of amplifier stages, which numbered seven deep, for a total of 63 emitters. This configuration provides the multistripe emission capability of phase-locked laser arrays without the requirement of stable spatial mode operation by lateral mode discrimination. Rather, each amplifier current can be adjusted to control the phase of the next emitting aperture. The particular device shown in Fig. 39 emitted 4.5 W under pulsed conditions, for an average of 70 mW per emitting section.

6. CONCLUSIONS

We have discussed the motivation for semiconductor PIC technology and provided several examples of current experimental PICs from a number of different application areas. These examples were chosen to illustrate current design concepts and current approaches to PIC growth and fabrication.

One important problem that was not discussed is the difficulty associated

2-D M-MOPA

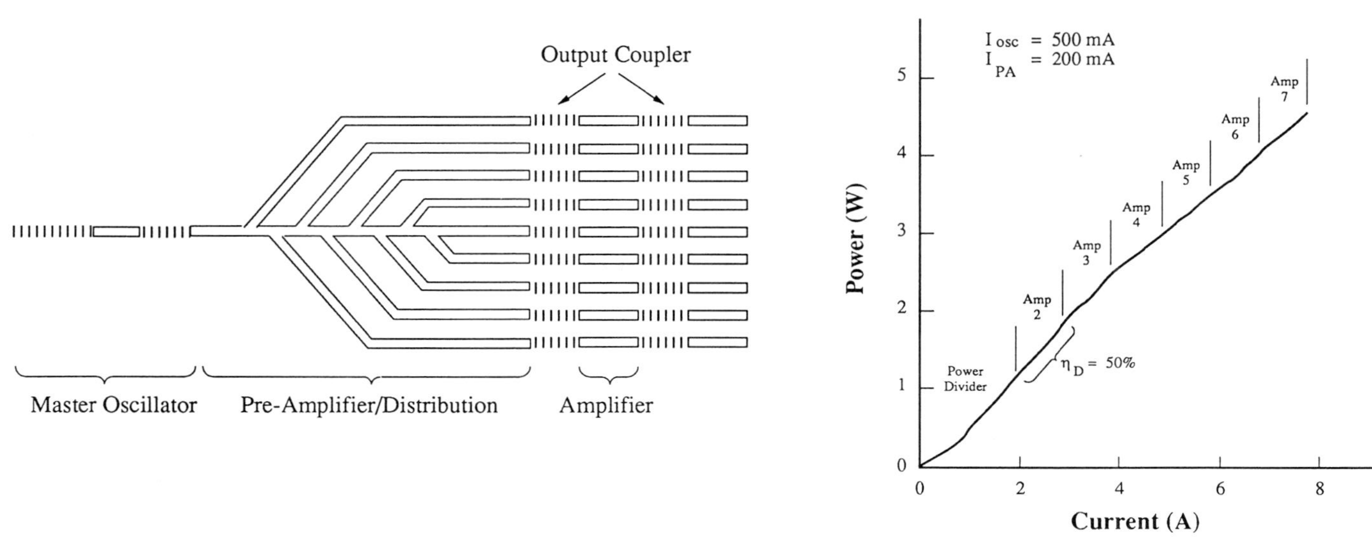

Fig. 39. Variant on PIC shown in Fig. 36 that uses a lateral branching structure in addition to the cascaded amplfiers to provide even more emitting area. This device has yielded 4.5 W under pulsed operation.

with PIC characterization. In some applications or in some PIC elements, it may be geometrically awkward or undesirable from a performance standpoint to couple light circulating on the chip into external optics. For example, the LO laser on the MQW balanced heterodyne receiver PIC is consumed on-chip, yet detailed characterization of this PIC requires precise knowledge of the LO tuning characteristics and linewidth behavior. One solution to these characterization difficulties is based on the observation that the MQW waveguide detectors, when forward biased, become transparent or even amplify. Thus both the LO and the received signal can be passed on through the detector in a transmissive mode in a time-gated fashion for monitoring purposes in a system application. In other PICs, judicious use of amplifiers as detectors, the converse of the preceding approach, can serve as a tremendously helpful diagnostic. However, the general area of PIC diagnostics and characterization, short of a full systems inspection, is still an area in need of research.

Modeling of PICs, as outlined in the description of design tools, provides a fairly accurate description of most optical propagation and waveguide device design issues. However, full *optoelectronic* modeling of three-dimensional current flow in junction devices such as lasers and the transport and parasitic issues associated with high-speed performance are still areas of current research. Reliable modeling of these will be required before full computer-aided design (CAD) support is available for PIC development.

The advances seen in this field in recent years have stemmed largely from basic improvements in crystal growth technology. The freedom in device design and fabrication engineering facilitated by quantum-well growth capabilities, large area uniformity, and complex vertical layer structures has only begun to be explored. Moreover, the pace of improvements in growth technology shows no evidence of slowing down. The coming years should bring even more exciting advances in PIC technology, and the time appears ripe for realizing many of the basic integrated optics concepts [101] envisioned throughout the last two decades.

ACKNOWLEDGMENT

The authors would like to thank their many colleagues and collaborators at AT&T Bell Laboratories, without whose contributions many of the illustrative examples contained here would not have been possible.

REFERENCES

1. C. H. Henry, L. F. Johnson, R. A. Logan, and D. P. Clarke, Determination of the Refractive Index of InGaAsP Epitaxial Layers by Mode Line Luminescence Spectroscopy, *IEEE J. Quantum Electron.* **QE-21**, 1887–1892 (1985).
2. M. A. Afromowitz, Refractive Index of $Ga_{1-x}Al_xAs$, *Solid State Commun.* **15**, 59–63 (1974).
3. G. P. Agrawal and N. K. Dutta, "Long-Wavelength Semiconductor Lasers." Van Nostrand Reinhold, New York, 1986.
4. C. H. Henry, Theory of the Linewidth of Semiconductor Lasers, *IEEE J. Quantum Electron.* **QE-18**, 259–264 (1982).
5. M. Asada, Theoretical Linewidth Enhancement Factor a of GaInAs/InP Lasers, *Trans. IECE Jpn* **E-68**, 518–520 (1985).
6. J. Manning, R. Olshansky, and C. B. Su, The Carrier-Induced Index Change in AlGaAs and 1.3μm InGaAsP Diode Lasers, *IEEE J. Quantum Electron.* **QE-19**, 1525–1539 (1983).
7. T. L. Koch and U. Koren, Semiconductor Photonic Integrated Circuits, *IEEE J. Quantum Electron.* **QE-27**, 641–653 (1991).
8. H. Inoue, H. Nakamura, K. Morosawa, Y. Sasaki, T. Katsuyama, and N. Chinone, An 8 mm Length Non-blocking 4×4 Optical Switch Array, *IEEE J. Selected Areas Commun.* **SAC-6**, 1262–1266 (1988).
9. I. P. Kaminow, Linear Electrooptic Materials, in "CRC Handbook of Laser Science and Technology," Vol. IV, "Optical Materials," Part 2 (M. J. Weber, ed.). CRC Press, Boca Raton, FL, 253–278 (1986).
10. D. A. B. Miller, J. S. Weiner, and D. S. Chemla, Electric-Field Dependence of Linear Optical Properties Quantum Well Structures: Waveguide Electroabsorption and Sum Rules, *IEEE J. Quantum Electron.* **QE-22**, 1816–1830 (1986).
11. J. E. Zucker, I. Bar-Joseph, G. Sucha, U. Koren, B. I. Miller, and D. S. Chemla, Electrorefraction in GaInAs/InP Multiple Quantum Well Heterostructures, *Electron. Lett.* **24**, 458–460 (1988).
12. J. E. Zucker, T. Y. Chang, M. Wegener, N. J. Sauer, K. L. Jones, and D. S. Chemla, Large Refractive Index Changes in Tunable-Electron-Density InGaAs/InAlAs Quantum Wells, *IEEE Photon. Technol. Lett.* **2**, 29–31 (1990).
13. K. Wakita, I. Kotaka, O. Mitomi, H. Asai, and U. Kawamura, Observation of Low-Chirp Modulation in InGaAs-InAlAs Multiple-Quantum-Well Optical Modulators under 30 GHz, *IEEE Photon. Technol. Lett.* **3**, 138–140 (1991).
14. J. E. Zucker, Quantum Well Electrorefraction and Optical Waveguide Device Applications, *SPIE Proc.* **1283**, 94–102 (1990).
15. T. L. Koch, W. T. Tsang, and P. J. Corvini, Spectral Dependence of Propagation Loss in InP/InGaAsP Antiresonant Reflecting Optical Waveguides Grown by Chemical Beam Epitaxy, *Appl. Phys. Lett.* **50**, 307–309 (1987).

16. G. B. Hocker and W. K. Burns, Mode Dispersion in Diffused Channel Waveguides by the Effective Index Method, *Appl. Opt.* **16**, 113–118 (1977).
17. R. G. Hunsperger, "Integrated Optics: Theory and Technology." Springer-Verlag, Berlin, 1985, p. 89.
18. A. Yariv and P. Yeh, "Optical Waves in Crystals." Wiley, New York, pp. 177–201, 425–459, 1984.
19. W. Streifer, D. R. Scifres, R. D. Brunham, TM-Mode Coupling Coefficients in Guided-Wave Distributed Feedback Lasers, *IEEE J. Quantum Electron.* **QE-12**, 74–78 (1976).
20. M. D. Feit and J. A. Fleck, Jr., Light Propagation in Graded-Index Optical Fibers, *Appl. Opt.* **17**, 3990–3998 (1978).
21. D. Marcuse, "Theory of Dielectric Optical Waveguide," 2nd ed., Academic Press, San Diego, 1991, pp. 307–318.
22. See, for example, Sessions TuB & ThB, Electromagnetic Field Propagation I & II, in *Tech. Dig. of Integrated Photonics Research*, 1991, Monterey, pp. 4–7 and 89–94.
23. M. Suzuki, Y. Noda, H. Tanaka, S. Akiba, Y. Kushiro, and H. Isshiki, Monolithic Integration of InGaAsP/InP Distributed Feedback Laser and Electroabsorption Modulator by Vapor Phase Epitaxy, *IEEE J. Lightwave Technol.* **LT-5**, 1277–1285 (1987).
24. Y. Tohmori, F. Kano, M. Oishi, Y. Kondo, M. Nakao, and K. Oe, Narrow Linewidth and Low Chirping Characteristics in High Power Operating Butt-Jointed DBR Lasers Grown by MOVPE, *Electron. Lett.* **24**, 1481–1483 (1988).
25. H. Soda, M. Furutsu, K. Sato, N. Okazaki, Y. Yamazaki, H. Nishimoto, and H. Ishikawa, High-Power and High-Speed Semi-Insulating BH Structure Monolithic Electroabsorption Modulator/DFB Laser Light Source, *Electron Lett.* **26**, 9–10 (1990).
26. Y. Matsui, T. Kunii, H. Horikawa, and T. Kamijoh, Narrow-Linewidth (<200 kHz) Operation of 1.5μm Butt-Jointed Multiple-Quantum-Well Distributed Bragg Reflector Laser, *IEEE Photon. Technol. Lett.* **3**, 424–426 (1991).
27. P. J. Williams, P. M. Charles, I. Griffith, L. Considine, and A. C. Carter, High Performance Buried Ridge DFB Lasers Monolithically Integrated with Butt Coupled Strip Loaded Passive Waveguides for OEIC, *Electron. Lett.* **26**, 142–143 (1990).
28. H. Takeuchi, K. Kasaya, Y. Kondo, H. Yasaka, K. Oe, and Y. Imamura, Monolithic Integrated Coherent Receiver on InP Substrate, *IEEE Photon. Technol. Lett.* **1**, 398–400 (1989).
29. H. Takeuchi, K. Kasaya, and K. Oe, Experimental Evaluation of the Coupling Efficiency between Monolithically Integrated DFB Lasers and Waveguides, paper 20B1–8 in *Tech. Dig. 7th Int. Conf. on Integrated Optics and Optical Fiber Communications*, Kobe (July 1989).
30. T. Okiyama, I. Yokota, H. Nishimoto, K. Hironishi, T. Horimatsu, T. Touge, and H. Soda, A 10 Gb/s, 65-km Optical Fiber Transmission Exper-

iment Using a Monolithic Electro-Absorption Modulator/DFB Laser Light Source, paper MoA1–3 in *Tech. Dig. 15th European Conf. on Optical Communication*, Gothenburg (September 1989).

31. T. L. Koch, U. Koren, R. P. Gnall, C. A. Burrus, and B. I. Miller, Continuously Tunable 1.5 μm Multiple-Quantum-Well InGaAs/InGaAsP Distributed-Bragg-Reflector Lasers, *Electron. Lett.* **24**, 1431–1432 (1988).
32. B. Glance, T. L. Koch, O. Scaramucci, K. C. Reichmann, L. D. Tzeng, U. Koren, and C. A. Burrus, Densely Spaced FDM Optical Coherent System with Near Quantum-Limited Sensitivity and Computer-Controlled Random Access Channel Selection, *Electron. Lett.* **25**, 883–885 (1989).
33. Y. D. Galeuchet and P. Roentgen, Selective Area MOVPE of GaInAs/InP Heterostructures on Masked and Nonplanar (100) and (111) Substrates, *J. Cryst. Growth* **107**, 147–150 (1991).
34. M. Aoki *et al.*, High-Speed (10 Gbit/sec) and Low-Drive-Voltage (1V Peak-to-Peak) InGaAs/InGaAsP MQW Electroabsorption Modulator Integrated DFB Laser with Semi-Insulating Buried Heterostructure, *Electron. Lett.* **28**, 1157–1158 (1992).
35. T. Kato *et al.*, paper B7.1 in *Proc. 1991 European Conf. on Optical Fiber Communications/ International Conference on Integrated Optics and Optical Fiber Communications*, ECOC/IOOC '91, Paris (1991).
36. T. Sasaki, Y. Sakata, N. Kida, M. Kitamura, and I. Mito, Novel Tunable DBR-LDs Grown by Selective MOVPE Using a Waveguide-Direction Bang-Gap Energy Control Technique, paper FB10 in *Tech. Dig. of 1992 Conf. on Optical Fiber Communications, OFC'92*, San Jose, pp. 281–282 (1992).
37. B. Broberg, B. S. Lindgren, M. G. Oberg, and H. Jiang, A Novel Integrated Optics Wavelength Filter in InGaAsP-InP, *IEEE J. Lightwave Technol.* **LT-4**, 196–203 (1986).
38. F. Heismann, L. L. Buhl, and R. C. Alferness, Electro-Optically Tunable Narrowband Ti:$LiNbO_3$ Wavelength Filters, *Electron. Lett.* **23**, 572–573 (1987).
39. T. L. Koch, E. G. Burkhardt, F. G. Storz, T. J. Bridges, and T. Sizer, II, Vertically Grating-Coupled ARROW Structures for III–V Integrated Optics, *IEEE J. Quantum Electron.* **QE-23**, 889–897 (1987).
40. R. C. Alferness, T. L. Koch, L. L. Buhl, F. Storz, F. Heismann, and M. J. R. Martyak, Grating Assisted InGaAsP/InP Vertical Co-directional Coupler Filter, *Appl. Phys. Lett.* **55**, 2011–2013 (1989).
41. T. L. Koch, P. J. Corvini, W. T. Tsang, and U. Koren, Wavelength-Selective Interlayer Directionally Grating-Coupled InP/InGaAsP Waveguide Photodetection, *Appl. Phys. Lett.* **51**, 1060–1062 (1987).
42. T. Baba, Y. Kokubun, and H. Watanabe, Monolithic Integration of an ARROW-Type Demultiplexer and Photodetector in the Shorter Wavelength Region, *IEEE J. Lightwave Technol.* **8**, 99–104 (1990).
43. R. J. Deri, N. Yasuoka, M. Makiuchi, O. Wada, A. Kuramata, H. Hamaguchi, and R. J. Hawkins, Integrated Waveguide/Photdiodes Using Vertical Impedance Matching, *Appl. Phys. Lett.* **56**, 1737–1739 (1990).

44. C. H. Henry, R. A. Logan, F. R. Merritt, and J. P. Luongo, The Effect of Intervalence Band Absorption on the Thermal Behavior of InGaAsP Lasers, *IEEE J. Quantum Electron.* **QE-19**, 947–952 (1983).
45. U. Koren, B. I. Miller, T. L. Koch, G. D. Boyd, R. J. Capik, and C. E. Soccolich, Low Loss InGaAs/InP Multiple Quantum Well Waveguides, *Appl. Phys. Lett.* **49**, 1602–1604 (1986).
46. R. J. Deri, E. Kapon, R. Bhat, and M. Seto, Low-Loss GaInAsP/InP Multiple Quantum Well Optical Waveguides, *Appl. Phys. Lett.* **54**, 1737–1739 (1989).
47. R. J. Deri, N. Yasuoka, M. Makiuchi, A, Kuramata, and O. Wada, Efficient Fiber Coupling to Low-Loss Diluted Multiple Quantum Well Optical Waveguides, *Appl. Phys. Lett.* **55**, 1495–1497 (1989).
48. R. J. Deri, N. Yasuoka, M. Makiuchi, A, Kuramata, O. Wada, and S. Yamakoshi, Low-Loss Optical Directional Coupler on InP, *Electron. Lett. Vol.* **25**, 1355–1356 (1989).
49. R. J. Deri, T. Sanada, N. Yasuoka, M. Makiuchi, A. Kuramata, H. Hamaguchi, O. Wada, and S. Yamakoshi, Low-Loss Monolithic Integration of Balanced Twin-Photodetectors with a 3-dB Waveguide Coupler for Coherent Lightwave Receivers, *IEEE Photon. Technol. Lett.* **2**, 581–584 (1990).
50. T. L. Koch, U. Koren, G. Eisenstein, M. G. Young, M. Oron, C. R. Giles, and B. I. Miller, Tapered Waveguide InGaAs/InGaAsP Multiple Quantum Well Lasers, *IEEE Photon. Technol. Lett.* **2**, 88–90 (1990).
51. See, for example, Conference Proceedings of 3rd International Conference on Indium Phosphide and Related Materials, Cardiff (1991), IEEE Cat. #91CH2950–4.
52. I. Moerman, C. Coudenys, P. D. Demeester, and J. Crawley, Characterization of InP/InGaAs/InGaAsP Using Atmospheric and Low Pressure MOVPE, paper WP.44 in *Conf. Proc. 3rd Int. Conf. on Indium Phosphide and Related Materials*, Cardiff, pp. 472–475 (1991).
53. W. T. Tsang, M. C. Wu, T. Tanbun-Ek, R. A. Logan, S. N. G. Chu, and A. M. Aergent, *Appl. Phys. Lett.* **57**, 2065 (1990).
54. H. Heinecke, B. Baur, N. Emeis, and M. Shier, *J. Cryst. Growth*, (1992).
55. T. Uchida, *et al.*, *J. Cryst. Growth* **120**, 357–361 (1992).
56. L. A. Coldren and J. A. Rentschler, Directional Reactive-Ion-Etching of InP with Cl_2 Containing Gases, *J. Vac. Sci. Technol.* **19**, 225–230 (1981).
57. J. W. McNabb, H. G. Craighead, and H. Temkin, Anisotropic Reactive Ion Etching of InP in Methane/Hydrogen Based Plasma, paper TuD15 in *Tech. Dig. of Integrated Photonics Research 1991*, Monterey, pp. 26–27.
58. P. Buchmann and H. Kaufmann, GaAs Single-Mode Rib Waveguides with Reactive Ion-Etched Totally Reflecting Corner Mirrors, *IEEE J. Lightwave Technol.* **LT-3**, 785–788 (1985).
59. S. Oku, M. Okayasu, and M. Ikeda, Low-Threshold CW Operation of Square-Shaped Semiconductor Ringe Lasers (Orbiter Lasers), *IEEE Photon. Technol. Lett.* **3**, 588–590 (1991).

60. W. J. Grande, J. E. Johnson, and C. L. Tang, AlGaAs Photonic Integrated Circuits Fabricated Using Chemically Assisted Ion Beam Etching, paper OE10.4/ThUU4 in *Conf. Digest of IEEE LEOS Annual Meeting*, Boston (1990), p. 169.
61. J. L. Jewell, J. P. Harbison, A. Scherer, Y. H. Lee, and L. T. Florez, Vertical-Cavity Surface-Emitting Lasers: Design, Growth, Fabrication, Characterization, *IEEE J. Quantum Electron.* **QE-27**, 1332–1346 (1991).
62. S. J. Clements, M. A. Gibbon, G. H. B. Thompson, D. J. Moule, C. B. Rogers, and C. G. Cureton, Fabrication of a Semiconductor Integrated Demultiplexer and Its Polarization Discrimination, paper WM20, *Tech. Dig. of 1990 Optical Fiber Communications Conf.*, Optical Society of America, p. 119.
63. J. B. D. Soole, A. Scherer, H. P. LeBlanc, N. C. Andreadakis, R. Bhat, and M. A. Koza, Monolithic InP-Based Grating Spectrometer for Wavelength-Division Multiplexed Systems at 1.5μm, *Electron. Lett.* **27**, 132–134 (1991).
64. C. Cremer, G. Ebbinghaus, G. Heise, R. Muller-Nawrath, M. Shienle, and L. Stoll, Grating Spectrograph in InGaAsP/InP for Dense Wavelength Division Multiplexing, *Appl. Phys. Lett.* **59**, 627–629 (1991).
65. A. R. Clawson, Reference Guide to Chemical Etching of InGaAsP and $In_{0.53}Ga_{0.47}As$ Semiconductors, NOSC Tech. Note 1206, San Diego (1982).
66. P. Buchmann and A. J. N. Houghton, Optical Y-Junctions and S-Bends Formed by Preferentially Etched Singel-Mode Rib Waveguides in InP, *Electron. Lett.* **xx**, yyyy (1982).
67. L. A. Coldren, K. Furuya, B. I. Miller, and J. A. Rentschler, Etched Mirror and Groove-Coupled GaInAsP/InP Laser Devices for Integrated Optics, *IEEE J. Quantum Electron.* **QE-18**, 1679–1688 (1982).
68. T. L. Koch, P. J. Corvini, and W. T. Tsang, Anisotropically Etched Deep Gratings for InP/InGaAsP Optical Devices, *J. Appl. Phys.* **62**, 3461–3463 (1987).
69. Z. L. Liau and J. N. Walpole, A Novel Technique for GaInAsP/InP Buried Heterostructure Laser Fabrication, *Appl. Phys. Lett.* **40**, 568–570 (1982).
70. S. Adachi and H. Kawaguchi, Chemical Etching Characteristics of (001)InP, *J. Electrochem. Soc. Solid State Sci. Tech.* **128**, 1342–1349 (1981).
71. T. Kambayashi, C. Kitahara, and K. Iga, Chemical Etching of InP and GaInAsP for Fabricating Laser Diodes and Integrated Optical Circuits, *Jpn. J. Appl. Phys.* **19**, 79–85 (1980).
72. A general discussion of typical processing techniques and considerations can be found in: W. C. Dautremont-Smith, R. J. McCoy, R. H. Burton, and A. G. Baca, Fabrication Technologies for III-V Compound Semiconductor Photonic and Electronic Devices, *AT&T Tech. J.* **68**, 64–82 (1989).
73. U. Koren, T. L. Koch, B. I. Miller, and A. Shahar, Processes for Large Scale Photonic Integrated Circuits, paper MDD2, in *Tech. Dig. of the Integrated and Guided Wave Optics Conf.*, Houston (1989).

74. T. L. Koch, U. Koren, R. P. Gnall, F. S. Choa, F. Hernandez-Gil, C. A. Burrus, M. G. Young, M. Oron, and B. I. Miller, GaInAs/GaInAsP Multiple Quantum Well Integrated Heterodyne Receiver, *Electron. Lett.* **25**, 1621–1622 (1989).
75. T. L. Koch, F. S. Choa, U. Koren, R. P. Gnall, F. Hernandez-Gil, C. A. Burrus, M. G. Young, M. Oron, and B. I. Miller, Balanced Operation of an InGaAs/InGaAsP Multiple-Quantum-Well Integrated Heterodyne Receiver, *IEEE Photon. Technol. Lett.* **2**, 577–580 (1990).
76. U. Koren, B. I. Miller, G. Eisenstein, R. S. Tucker, G. Raybon, and R. J. Capik, Semi-insulating Blocked Planar BH GaInAsP/InP Laser with High Power and High Modulation Bandwidth, *Electron. Lett.* **24**, 138–140 (1988).
77. J. L. Zilco, L. J. P. Ketelsen, Y. Twu, D. P. Wilt, S. G. Napholtz, J. P. Blaha, K. E. Strege, V. G. Riggs, D. L. Van Haren, S. Y. Leung, P. M. Nitzsche, J. A. Long, C. B. Roxlo, G. Przyblek, J. Lopata, M. W. Focht, and L. A. Koszi, Growth and Characterization of High Yield, Reliable, High-Power, High-Speed, InP/InGaAsP Capped Mesa Buried Heterostructure Distributed Feedback (CMBH-DFB) Lasers, *IEEE J. Quantum Electron.* **QE-25**, 2091–2095 (1989).
78. F. Hernandez-Gil, T. L. Koch, U. Koren, R. P. Gnall, and C. A. Burrus, A Tunable MQW-DBR Laser with a Monolithically Integrated GaInAsP/InP Directional Coupler Switch, *Electron. Lett.* **25**, 1271–1272 (1989).
79. U. Koren, T. L. Koch, B. I. Miller, and A. Shahar, InGaAs/InGaAsP Distributed Feedback Quantum Well Laser with an Intracavity Phase Modulator, *Appl. Phys. Lett.* **53**, 2132–2134 (1988).
80. F. S. Choa, T. L. Koch, U. Koren, and B. I. Miller, Optoelectronic Properties of InGaAs/InGaAsP Multiple-Quantum-Well Waveguide Photodetectors, *IEEE Photon. Technol. Lett.* **1**, 376–378 (1989).
81. B. Glance and R. W. Wilson, Frequency-Locked Loop Circuit Providing Large Pull-in Range, *Electron. Lett.* **25**, 965–967 (1989).
82. B. L. Kasper, C. A. Burrus, J. R. Talman, and K. L. Hall, Balanced Dual-Detector Receiver for Optical Heterodyne Communication at Gbit/s Rates, *Electron. Lett.* **22**, 413–415, 1986.
83. S. Benedetto , E. Biglieri, and V. Castellani, "Digital Transmission Theory," Prentice-Hall, Englewood Cliffs, NJ, 1987, Section 5.5, p. 231.
84. K. Aiki, M. Nakamura, and J. Umeda, A Frequency-Multiplexing Light Source with Monolithically Integrated Distributed-Feedback Diode Lasers, *IEEE J. Quantum Electron.* **QE-13**, 220–223 (1977).
85. U. Koren, T. L. Koch, B. I. Miller, G. Eisenstein, and R. H. Bosworth, Wavelength Division Multiplexing Light Source with Integrated Quantum Well Tunable Lasers and Optical Amplifiers, *Appl. Phys. Lett.* **54**, 2056–2058 (1989).
86. A. H. Gnauck, U. Koren, T. L. Koch, F. S. Choa, C. A. Burrus, G. Eisenstein, and G. Raybon, paper PD26 in *Tech. Dig. of Optical Fiber Communications Conf.*, San Francisco (1990).

87. U. Koren, T. L. Koch, B. I. Miller, G. Eisenstein, and G. Raybon, An Integrated Tunable Light Source with Extended Tunability Range, paper 19A2–3 at the *7th Int. Conf. on Integrated Optics and Optical Fiber Communications*, Kobe (1989).
88. G. Eisenstein, U. Koren, G. Raybon, T. L. Koch, J. M. Wiesenfeld, M. Wegener, R. S. Tucker, and B. I. Miller, Large and Small Signal Gain Characteristics of 1.5 μm Multiple Quantum Well Optical Amplifiers, *Appl. Phys. Lett.* **56**, 1201–1203 (1990).
89. D. M. Cooper, M. Bagley, L. D. Westbrook, D. J. Elton, H. J. Wickes, M. J. Harlow, M. R. Aylett, and W. J. Devlin, Broadband Operation of InGaAsP-InGaAs GRINSCH MQW Amplifiers with 115 mW Saturated Output Power, paper PD32 in *Tech. Dig. of the Optical Fiber Communication Conf.*, San Francisco (1990).
90. P. J. Williams, R. G. Walker, P. M. Charles, A. K. Wood, N. Carr, R. I. Taylor, and A. C. Carter, Optoelectronic Integrated Circuits for Telephony/Broadband passive Optical (TPON/BPON) Networks: Design and Experimental Results, paper K-3, *Tech. Dig. of the 12th IEEE International Semiconductor Laser Conf.*, Davos (1990).
91. P. J. Williams, P. M. Charles, R. H. Lord, R. Cloutman, D. Gupta, R. Ogden, F. Randle, S. Thomas, G. M. Foster, T. J. Reid, and A. C. Carter, Demonstration of Optoelectronic Integrated Circuits for Bidirectional TPON/BPON Access Links, paper WeA8–2 in *Conf. Digest of IOOC/ECOC '91*, Paris (1991), pp. 485–488.
92. T. L. Koch, U. Koren, A. H. Gnauck, K. C. Reichmann, H. Kogelnik, G. Raybon, R. P. Gnall, M. Oron, M. G. Young, J. L. deMiguel, and B. I. Miller, Simple In-Line Bi-Directional 1.5 μm/1.3 μm Transceivers, paper K-4, *Tech. Dig. of the 12th IEEE International Semiconductor Laser Conf.*, Davos (1990).
93. U. Koren, B. I. Miller, M. G. Young, M. Chien, A. H. Gnauck, P. D. Magill, S. L. Woodward, and C. A. Burrus, Strained-Layer Multiple Quantum Well Distributed Bragg Reflector Lasers with a Fast Monitoring Photodiode, *Appl. Phys. Lett.* **58**, 1239–1240 (1991).
94. U. Koren, B. I. Miller, G. Raybon, M. Oron, M. G. Young, T. L. Koch, J. L. Demiguel, M. Chien, B. Tell, K. Brown-Goebeler, and C. A. Burrus, Integration of 1.3μm Wavelength Lasers and Optical Amplfiers, *Appl. Phys. Lett.* **57**, 1375–1377 (1990).
95. S. L. Woodward, T. L. Koch, and U. Koren, A Control Loop Which Insures High Side-Mode-Suppression Ratio in a Tunable DBR Laser, paper SDL15.3, *Tech. Dig. of IEEE LEOS Annual Meeting*, San Jose, p. 82 (1991).
96. U. Koren, R. M. Jopson, B. I. Miller, M. Chien, M. G. Young, C. A. Burrus, C. R. Giles, G. Raybon, J. D. Evankow, B. Tell, and K. Brown-Goebeler, High Power Laser-Amplifier Photonic Integrated Circuit for 1.48μm Wavelength Operation, post-deadline paper PD10, *Tech. Dig. of Optical Fiber Communications Conf.*, San Diego (1991).

97. R. Parke, D. F. Welch, A. Hardy, R. Lang, D. Muhuys, S. O'Brien, K. Dzurko, and D. Scifres, 2.0 W CW, Diffraction-Limited Operation of a Monolithically Integrated Master Oscillator Power Amplfier, *IEEE Photon. Technol. Lett.* **5**, 297–300 (1993).
98. N. Carlson, J. Abeles, D. Bour, S. Liew, W. Reichart, P. Lin, and A. Godzdz, Demonstration of a Monolithic Grating-Surface-Emitting Laser Master Oscillator-Cascaded Power Amplifier Array, *IEEE Photon. Technol. Lett.* **2**, 708–710 (1990).
99. D. Mehuys, R. Parke, R. G. Waarts, D. F. Welch, A. Hardy, W. Streifer, and D. R. Scifres, Characteristics of Multistage Monolithically Integrated Master Oscillator Power Amplifiers, *IEEE J. Quantum Electron.* **QE-27**, 1574–1581 (1991).
100. R. Parke, D. F. Welch, and D. Mehuys, Coherent Operation of a 2-D Monolithically Integrated Master Oscillator Power Amplifier, *Electron. Lett.* **xx**, yy–zz (1991).
101. S. E. Miller, Integrated Optics: An Introduction, *Bell Syst. Tech. J.* **48**, 2059–2069 (1969).

Chapter 16

PACKAGING INTEGRATED OPTOELECTRONICS

John Crow

IBM, T. J. Watson Research Center
Yorktown Heights, New York 10598

1. INTRODUCTION

Historically, the packaging of discrete optoelectronic (O-E) chips has consisted of simple two- to four-lead submounts (often T0 cans borrowed from the discrete transistor industry). These submounts were assembled, tested, and then placed into larger electronic integrated circuit (IC) modules (dual in-line packages or pin grid array modules) wire or TAB bonded to the IC chip carrier. Examples of this packaging are shown in Figs. 13 and 14 in Chapter 2. Advantages of this approach are that (1) the electrical and optical packaging issues have largely been separated and (2) the required modification of electronic packaging has been minimal, primarily adding fixtures for the alignment of optical components to the O-E device to facilitate coupling of the light from the O-E device to the medium (e.g., lens, optical fiber, or optical disk). The disadvantages of this approach are that (1) the overall module has multiple levels of packaging (added assembly and testing steps which can lead to lower module yield and reliability and higher module cost) and (2) the modules are bulky, often occupying over 20 cm^2 of card space. The cost of optoelectronic modules for data communications has be-

ISBN 0-12-200420-5

come dominated by this high-cost packaging, and the resultant O-E modules have often been uncompetitive with their electrical counterparts.

Integrated optoelectronics, whether in monolithic or hybrid form, offers the promise of greatly reduced package size, with far fewer parts to assemble. However, the chips are more complex, with more optical and electrical input–output (I/O), and have more stringent mechanical and thermal package requirements. Just as in the packaging evolution needed for IC development, OEICs require more advanced packaging, involving multichip modules, more optical and electrical I/O pins, and design for heat dissipation capability, crosstalk suppression, and electrical power distribution. The audio disk integrated read/write head module developed by Sony (illustrated in Fig. 11 of Chapter 12), is an example of one O-E industry making this move to integrated packaging. For Sony, even though the cost of the 10-lead, surface-mounted module was higher than that of the T0 package it replaced, it was less expensive than the cost of the head subassembly it replaced and the size is considerably smaller. It is expected that as this read/write head moves into optical storage and multibeam applications, the leverage of the integrated package will become even higher. It is also expected that the telecommunications and computer industries will follow the consumer industry lead into these integrated O-E packages.

With the added complexity of the integrated O-E package will come added importance of package design simulation tools, and module and chip testability must become part of the entire design process. Some of these topics, as they pertain to the chip itself, were discussed in the previous chapter on integrated chip design. Packaging technology for integrated optoelectronics will require a major R&D effort comparable to that for OEIC chips. This effort is just beginning and is far behind the OEIC chip progress.

Some of the specific considerations in an integrated package are:

1. The high-speed, analog, electrical signal lines are in close proximity, in the case of on-chip arrays, as well as in close proximity to the digital and signal processing circuitry on other chips in the module. This leads to a need for careful package simulation and layout and for lower and better-controlled values for the parasitics of the package. As the OEIC will generally have lower parasitics and on-chip devices customized for I/O, a module performance enhancement will result from keeping package parasitics low.
2. The large number of electrical and optical I/Os lead to the requirement for automated wiring techniques. The presence of optical elements can lead to special problems in the automated techniques usable. The pre-

cision alignment required for the optical components (e.g., a few micrometers for multimode fiber links and a fraction of a micrometer for single-mode links) coupled with the multiple optical connections means that self-alignment techniques are valuable in chip attach.

3. Multiple optical components and connections will lead to a desire for an integrated optical subassembly, initially by combining multiple optical devices (e.g., lenses, lightguides, splitters) into one component to be hybrid attached to the module. This would evolve to the processing of the optical wiring and components onto the chip carrier itself, as is done with electrical components in IC packaging today.
4. The small size of the OEIC and support chips requires that the package dimensions also be kept small to realize the advantage of chip integration.
5. Integration of many high-speed functions into a small module implies that chip heat can become a problem, and eventually chip thermal expansion matching to the substrate might be a concern. Essentially all the applications discussed in Part I require an air-cooled environment for the module and operation up to about 50°–60° C ambient temperature (even higher for military applications). A special concern is the potential sensitivity of the laser device's bias point, performance, and reliability to temperature, so that proximity to heat-generating electronics in a multichip package might be a problem. Package design needs to conduct heat away from the O-E devices, not just provide a good heat flow path among the various elements of the package and the environment. In arrays of lasers, there is added concern because heat gradients can cause different device behavior across the array. For example, if the array is transmitting synchronized bytes of data (as in a classic processor data bus application), the byte error rate may become data pattern dependent and thus more difficult to isolate. This may lead to the need for individual device optical monitoring and control.
6. As in the IC industry, multichip OEIC modules will be expensive to develop. To keep development costs to a minimum, it is important to base optical multichip packages on existing electronic industry packaging technology and to try to share the design tools.

As mentioned, packaging technology for OEICs is far behind chip development and application dependent. Because of this lack of maturity, there is little industry experience in the foregoing topics. Some results have been mentioned in the other sections of Part V. The general topic of microelectronics packaging can be found in *The Microelectronics Packaging Handbook* [1]. In this chapter two general issues from the preceding list will be discussed in relation to large-scale integration: electrical crosstalk in hybrid and monolithic array packages, and the simultaneous self-alignment and electrical connection of an OEIC to its package wiring. These are topics of importance to the batch fabrication of integrated modules.

It cannot be overstressed that chip and package development must go hand in hand for optimum module performance and cost. Often chip design tradeoffs must be made solely for improved packaging (or package testing), resulting in enhanced module-level cost or performance. The R&D phase of integrated optoelectronics has been driven, to date, by device and chip designers. Now, as the OEIC matures, there will likely be much more attention to the packaging and functional requirements of these chips.

2. Electrical Noise and Distortion in Array Optical Packages

The electrical signal levels are analog and small at the front end of optical receivers, so electrical noise can easily occur in the receiver. This section therefore concentrates on the receiver.

The integrated optical receiver was discussed in Chapters 12 and 13 in both hybrid and monolithic forms. Figure 12 in Chapter 12 showed performance data for both monolithic receivers and the best hybrid receivers. As pointed out in Section 3 of Chapter 13, the performance of the hybrid receiver is strongly influenced by the input capacitance to the preamplifier. The packaging can dominate this capacitance, and the magnitude of packaging capacitance and its control in a manufacturing environment can strongly influence module cost and performance. It is interesting to note that commercial hybrid receivers (for example, for the FDDI data link application) are about 10 dB lower in sensitivity than the "best hybrid" curve of Fig. 12 of Chapter 12, in part due to the variability in package parasitics in a hybrid design.

In packaging high-speed hybrid receivers (100–1000+ Mb/s), the design of the amplifier itself must take into account the packaging parasitics. For example, a variation of 50 fF in packaging parasitic capacitance coupling the input to the output of the amplifier for an FDDI receiver can lead to a variation of 10 MHz in bandwidth [2]. Thus, the receiver bandwidth as measured on the wafer using a high-speed probe station may not be the same as the actual receiver bandwidth when packaged with a photodiode in a hybrid package, due to the variability of parasitics. This makes it difficult to automate the screening of the receiver die on the wafer. There are two approaches to overcoming these discrepancies, both adding cost to the receiver module. One approach is to design a number of different receiver sites on the wafer with a cluster of values of feedback transimpedance and select the one matched to the parasitics measured in a particular package. This, of course, results in a large effective yield loss. The other approach is to design margins into the receiver and add scaling factors into the wafer screening tests to allow a prediction of the receiver module bandwidth based on automated wafer-level tests. This adds extra ac testing expense at the wafer level and sacrifices performance. For small-volume production, these "match and select" approaches are practiced today and even offer flexibility for variable chip and package parameters. They also help explain the high cost of hybrid receivers. In a monolithic receiver design, control of the chip capacitance and transistor parameters is resolved in chip process development. Once the process yields good chips, the performance of the chip and that of module are the same. Thus, with equivalent chip yield, the integrated receiver should be comparable to an amplifier at the chip level and lower in cost than the hybrid receiver at the module level.

As integration moves from single-channel to multichannel chips (e.g., arrays of O-E devices), package parasitics can also lead to interchannel crosstalk. For example, AT&T researchers demonstrated that 220 nF of wirebond coupling capacitance between photodiodes of a 12-channel monolithic array resulted in over 5 dB of receiver sensitivity degradation at high ($\geq$ 50 μW) average input optical power and 45 Mb/s [3]. The capacitive crosstalk was subsequently reduced to 10 fF by flip-chip mounting the array and dressing alternate wirebond leads in opposite directions [4], resulting in crosstalk of <-40 dB at 200 Mb/s.

NEC has also used solder bump technology for mounting light-emitting diode (LED) and photodetector (PD) arrays in 150 Mbit/s/line array transmitter and receiver modules [5], reporting a dc crosstalk between adjacent devices of -39 dB (LED) and -29 dB (PD) for monolithic devices on 250-μm centers.

Chip-on-chip solder bump packaging (see Fig. 13 of Chapter 12) has been shown in the VLSI IC world to reduce the size of package parasitics, as well as control the uniformity of parasitic values, as pointed out in Chapter 12. For example, NTT researchers have used this technique to attach an InGaAs pin photodiode to a GaAs MESFET preamplifier chip [6]. Five, 26-μm-diameter solder bumps were used to flip attach the photodiode (illumination was through the back surface of the chip). Although the parasitics of this attachment were not reported in detail, the photoreceiver was operated at 10 Gbit/s. This receiver package also employed coplanar electrical waveguides on polyimide to make the electrical connections to the amplifier chip.

An example of how receiver module cost can be reduced by using an integrated receiver chip is illustrated in Figure 1. This flow chart shows the steps in the assembly and testing a receiver module. The assumed relative cost and yield of each step are shown in the accompanying table. It assumes that the cost of the chips (OEIC in the monolithic case, photodiode and amplifier chips in the hybrid case) are the same (Chapter 12 points out that this can be the case today). It also assumes that dc and ac functional testing must be done on either package and this must be added to the module cost. The yield degradation in testing the hybrid module could be due to the lack of parasitic capacitance control, discussed above, while the yield degradation in the OEIC case could be due to lack of process uniformity across the wafer. Furthermore, it assumes that the cost associated with package assembly and test dominates the chip cost by 3X. What this flow chart points out is that an OEIC that can be fully ac tested at the wafer level and screened before packaging, can result in a 50% reduction in module cost, compared to a hybrid package that can only be ac tested after packaging into the module.

The cost and yield of chip-on-chip soldering still require investigation.

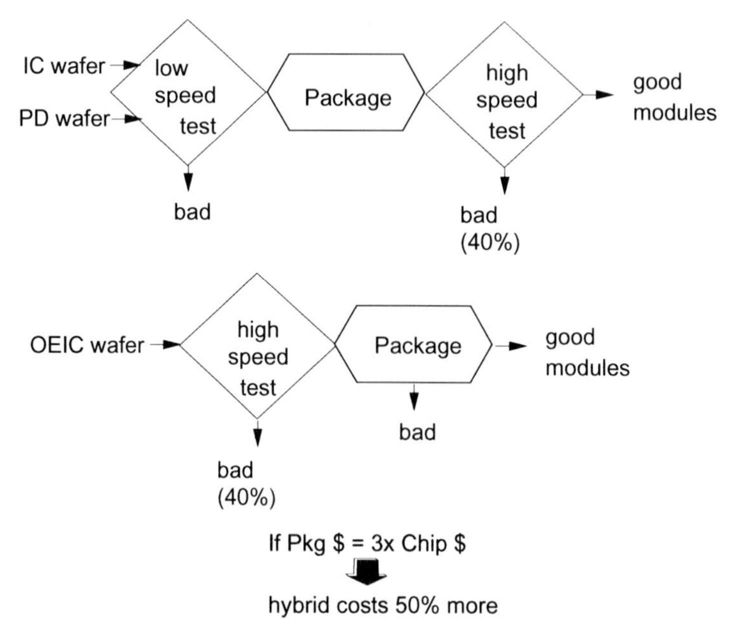

Fig. 1.

Currently no known commercial packages in electronics employ such techniques, although the area has been researched extensively.

3. MANY SIMULTANEOUS OEIC CHIP CONNECTIONS, BOTH ELECTRICAL AND OPTICAL

A package for an optical switch, from Fujitsu, was described in Section 3 of Chapter 12. Wire bonds were used for chip connections off the module, and micro-stripline TAB was used for chip-to-chip high-speed lines (see Fig. 15 in Chapter 12). Optical connections were made by manually aligning and attaching optical fibers in front of, or over, the chips. In another example of multichannel packaging, from IBM [7], fibers were embedded in a precision-grooved holder and then the holder was manipulated over the multichannel chip, shown in Fig. 20 in Chapter 13. This coupling arrangement is shown in Fig. 2. The vertical degree of freedom in optical alignment was removed by using a precision spacer, thus simplifying package assembly. Nevertheless, the assembly of these packages was very complex, with testing required at many steps, as the flowchart in Fig. 3 indicates.

A trend in multichip IC packaging has been to migrate to solder bump technology to attach ICs to chip carriers. This was done to increase the number of I/O connections to the chip, lower the wiring parasitics, and allow simultaneous electrical and mechanical attachment. Recently, flip-chip, solder bump technology has been utilized to self-align O-E devices to the chip carrier. Self-alignment is achieved because the solder bump (or solder ball) tends to pull itself into a shape which minimizes surface energy during the reflow process—often a spherical shape over the bonding pad. Thus, if the bonding pads on the chip are lithographically aligned to the optical device's emitting or detecting surface and the corresponding bonding pad on the carrier is aligned to the optical component (e.g., lightguide or fiber-in-groove) on the carrier, then the solder reflow process can self-align the optical path between chip and carrier. In ref. 8, GEC-Marconi researchers demonstrated that many small (10 μm in diameter) balls could make alignments to submicrometer tolerances, suitable for single-mode modules. GTE Labs has demonstrated [9] that precision standoffs can be used on the chip carrier to achieve submicrometer vertical height control, even when larger and fewer solder bumps are used.

Some practical considerations in using solder bumps for simultaneous optical and electrical attachment are that (1) the solder technology must make reliable, low-electrical-loss bonds across the entire chip and to all chips in

Fig. 2. An SEM photograph of four fibers with beveled ends mounted over a four channel receiver OEIC. The fibers are on 250-μm centers and held in a V-groove etched piece of silicon. The photodiodes are below the beveled ends of the fibers. (Courtesy of IBM.)

the package, (2) the deposited solder must be uniform enough to ensure precise alignment when reflowed, (3) the solder must not creep, (4) excessive stress must not be introduced which might alter or degrade chip performance, and (5) heat flow is sufficient for reliable and stable chip operation. Heat flow can be addressed by solder ball placement or by removing heat from the back of the chip (provided this is not in the optical path).

NTT researchers have investigated some of the manufacturing issues associated with using solder bumps in optical self-alignment [10]. They indicate that average in-plane alignment accuracy improved from 2.6 to 1.0 μm as the solder bump diameter decreased from 260 to 75 μm (4 bumps/chip) and from 1.5 to 0.8 μm as the number of bumps increased from 4 to 16 (130 μm bump diameter). A test package was cycled 300 times over a temperature between −10 and +60°C, and <0.5 μm of relative movement between the Si chip and glass chip carrier was measured (for the four 75-μm bump test vehicle). This indicates that large numbers of small solder bumps would be required for single-mode alignment tolerances, and thermal stability limitations would remain. However, multimode alignment tolerances should be routinely achievable.

IBM researchers have demonstrated what many consider to be the current state of the art in OEIC multichip module packaging: the four-channel transmitter–receiver module shown in the photograph of Fig. 16 in Chapter 12. This module combines many of the concepts necessary for an O-E MCM package, and a detailed description of it can serve to illustrate some of the issues in OEIC packaging.

A schematic of the OEIC package is shown in Fig. 4. Solder bumps are used for simultaneous self-alignment of the transmitter OEIC chip and the receiver OEIC chip to planar lightguides on the chip carrier in one solder reflow operation. Glass mechanical stops, coprocessed with the chip carrier's lightguide wiring, provide the final precision self-alignment of the chips to the lightguides, both vertically and in plane. Electrical coplanar transmission lines and glass optical lightguides are compatibly processed on the same chip carrier. Both silicon and ceramic chip carriers are possible [11]. The lightguides are used to conduct signals to or from the O-E chips to the edge of the module, also providing a fan-in/fan-out spatial transformation so that O-E device spacing on chip does not have to match fiber array spacing. This flexibility in component placement might allow an upgrade for one component to be done without affecting another component. Besides the lightguides, there is a single level of metallic wiring traces to carry the electrical signals to the carrier edge.

To align a fiber-optic array connector to the module, pairs of glass mechanical ridges were fabricated on each side of the lightguide array on the chip carrier. The steel pins of the MT connector were placed between these ridges, thus aligning the pins to the lightguides. The pins were aligned to the fibers in the MT connector by a plastic molding process. As in the case of the chips, < 2 μm misalignments were demonstrated, yielding ~ 1 dB in multimode coupling loss (due primariliy to beam shape mismatch).

The modules were operated at 1 Gbit/s, with more than adequate link power budget to ensure low-error-rate operation over 200-m optical buses. The laser-to-lightguide to fiber insertion loss was 6.6 dB, and a fiber-to-lightguide-to-photodiode insertion loss was 2.8 dB.

The solder bumps used were typical of manufacturable IC solder bumps (PbSn, 75 μm in diameter), and 15–16 bumps per chip were employed for four-channel differential electrical connections. Solder pads were 100–125

(a)

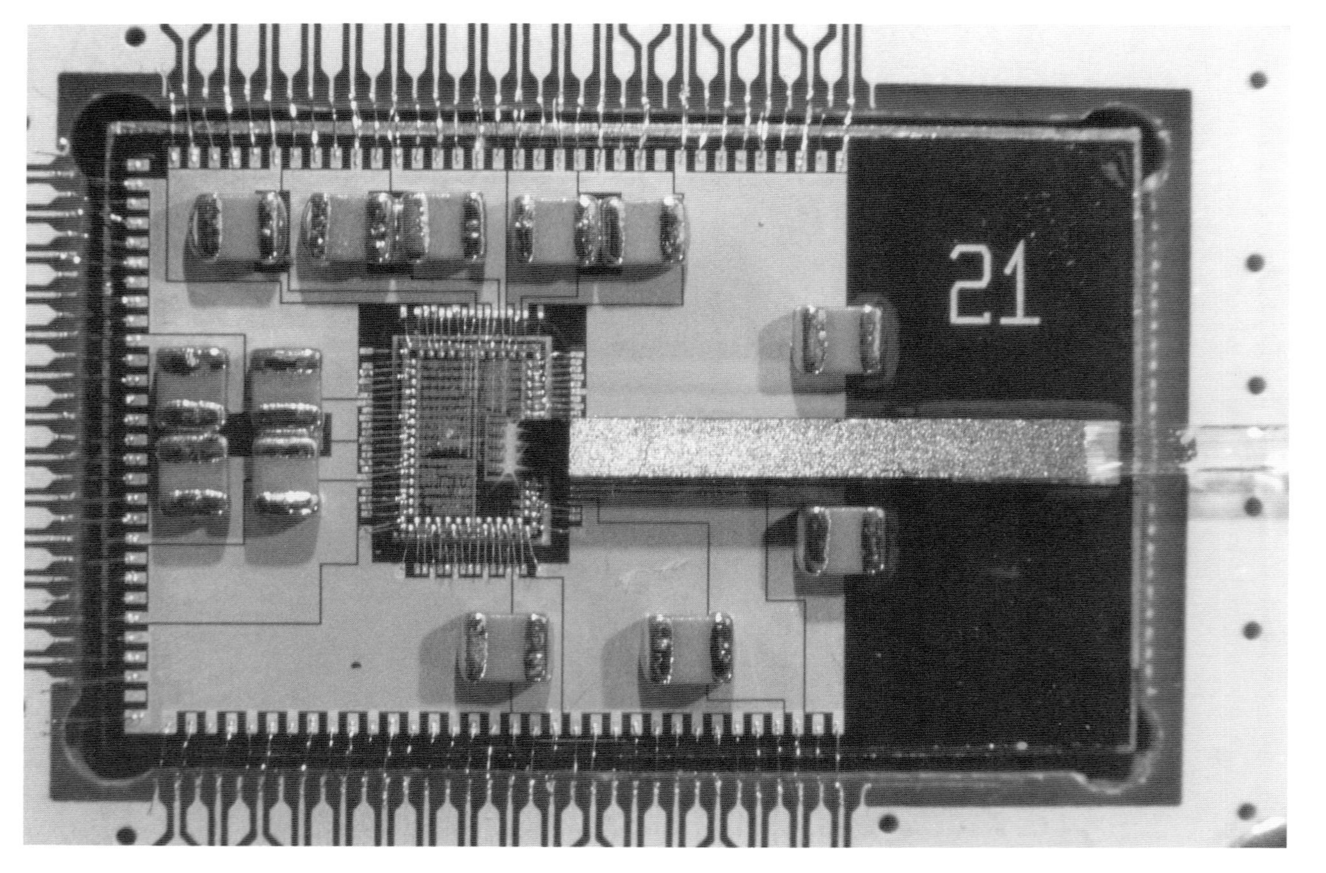

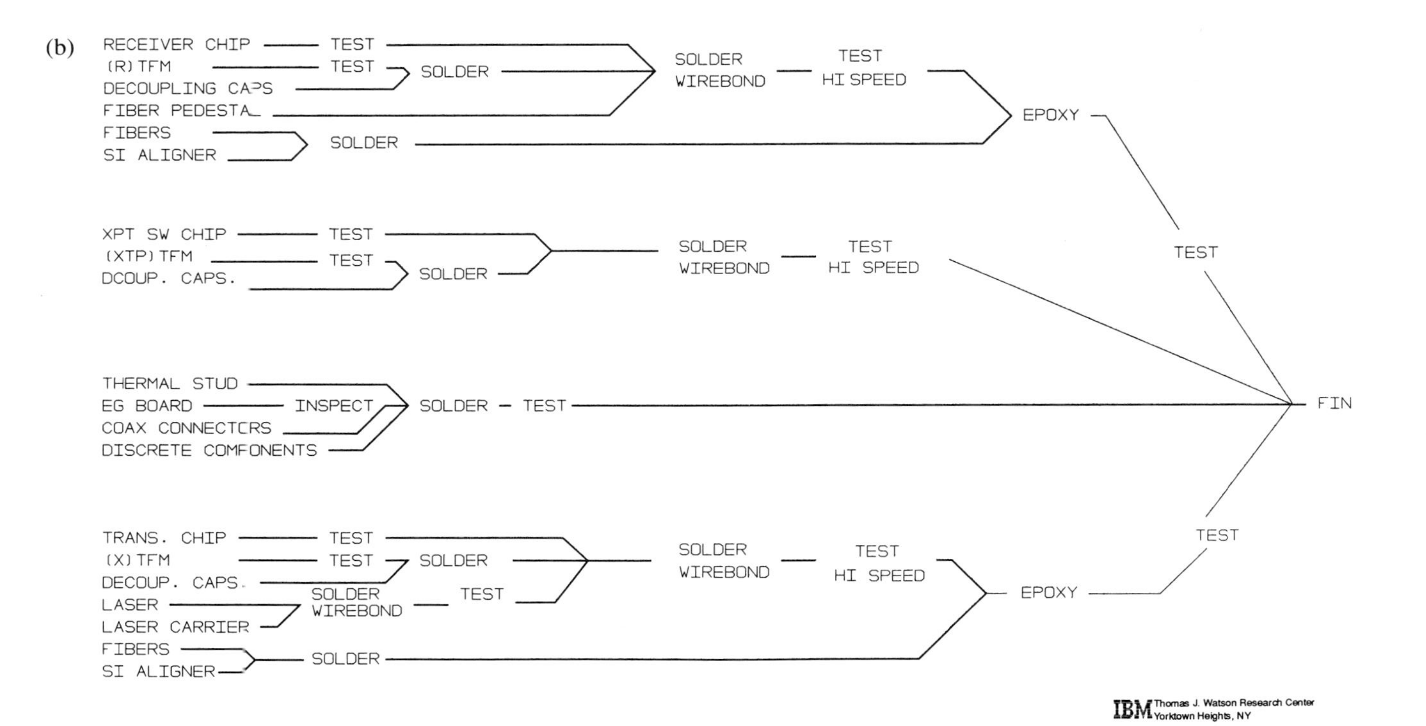

Fig. 3. The receiver module assembled from the receiver chip of Fig. 20 in Chapter 13 and the fiber array of Fig. 1. (a) Top view of the module. (b) Assembly flowchart for the packaging of the module and a companion transmitter module. It describes both assembly and testing steps. (Courtesy of IBM.)

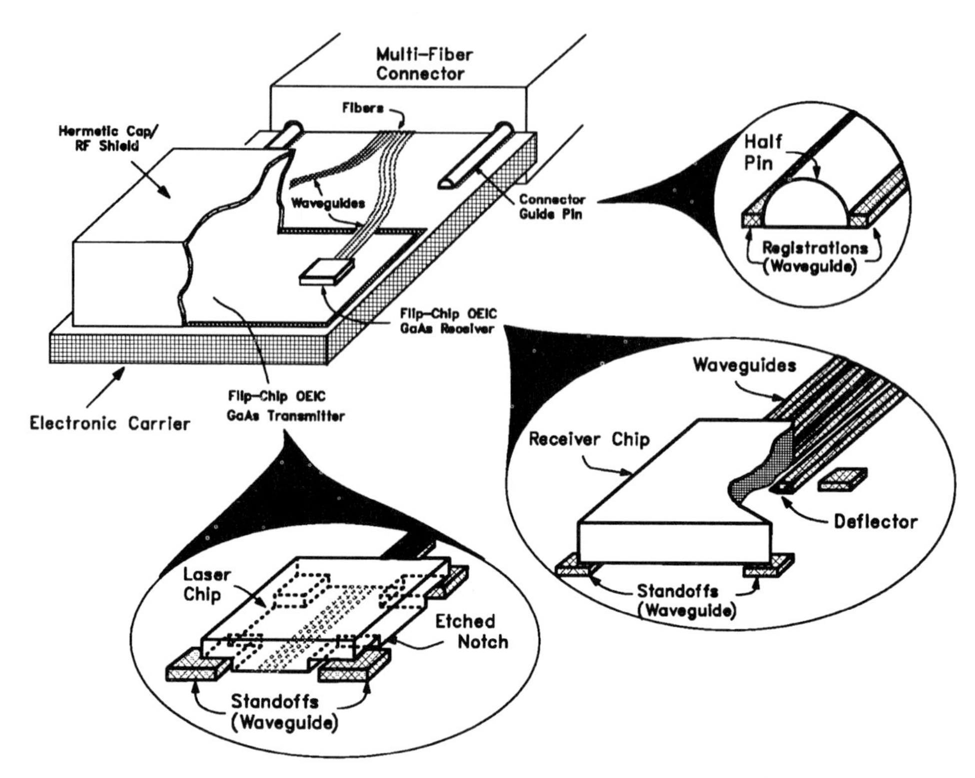

Fig. 4. Schematic of an optoelectronic multichip module (OE-MCM). The blow-ups illustrate the self-alignment approaches in chip and connector assembly. The standoffs are processed in common with the lightguides.

μm in diameter. Repeatable alignment of less than 2 μm in all three planes was achieved, sufficient for multimode lightguide modules. Solder bump reflow volume is often difficult to control to better than about 30% in a manufacturing environment—because of the solder density variations and lack of dimensional control in deposition (as mentioned earlier)—so the solder reflow was employed solely to pull the OEICs against the mechanical stops on the IBM package. Vertical pull was accomplished by the usual collapse of the solder during reflow. In-plane pull was accomplished by intentionally offsetting some of the solder bump pads on the chip from their corresponding pad on the carrier. Then, the surface tension of the solder reflow in attempting to align the pads creates the lateral force necessary to pull the chip against the in-plane stops on the chip carrier, as illustrated schematically in Fig. 5. A scanning electron microscope (SEM) photo of a laser array OEIC attached to the carrier using solder reflow is shown in Fig. 6.

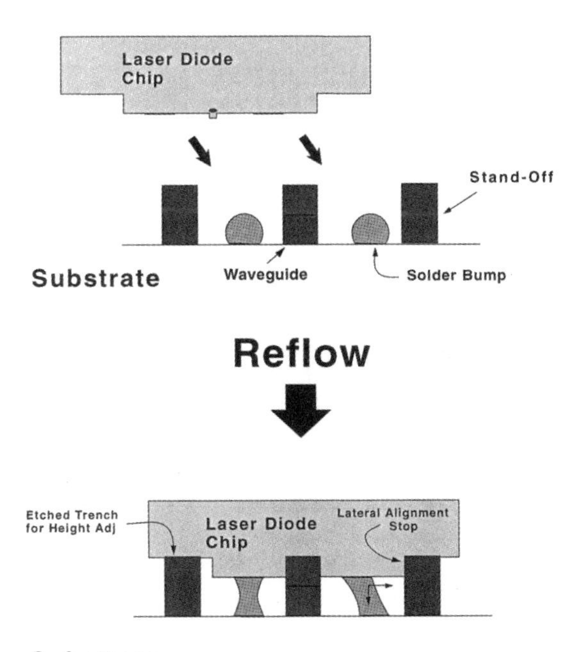

Fig. 5. Diagram of the microstop alignment technique. One of the solder bumps shown is purposely misaligned with the chip bonding pad during the reflow process to ''pull'' the chip against the lateral stop alignment. The arrows indicate the sideways and downward forces on the chip during and after the reflow.

The lightguides were fabricated from silica glass, the material used for low-loss optical fibers, because of the demonstrated precision and stability of this technology [12]. The high consolidation temperature for this "optical wire" (1250°C) might degrade the chip carrier or any metallic wiring (e.g., raise resistivity). For the ceramic carrier containing multilayer embedded wiring. it was found that the glass consolidation raised the resistivity <5% because refractory metal (Mo) was used. However, on the Si carrier, which employed nonrefractory metals (e.g., Cu, Al), the glass had to be processed before fabricating the electrical controlled impedance lines. The nonplanarity of the Si carrier after lightguide processing limited the size of the metal lines that could be processed.

Other lightguide materials have been explored for integrated O-E modules. Polyimide polymers are attractive because of their ease of processing and low processing temperatures, high-temperature stability, low dielectric constant for electrical transmission lines, and potential for low loss. Ho-

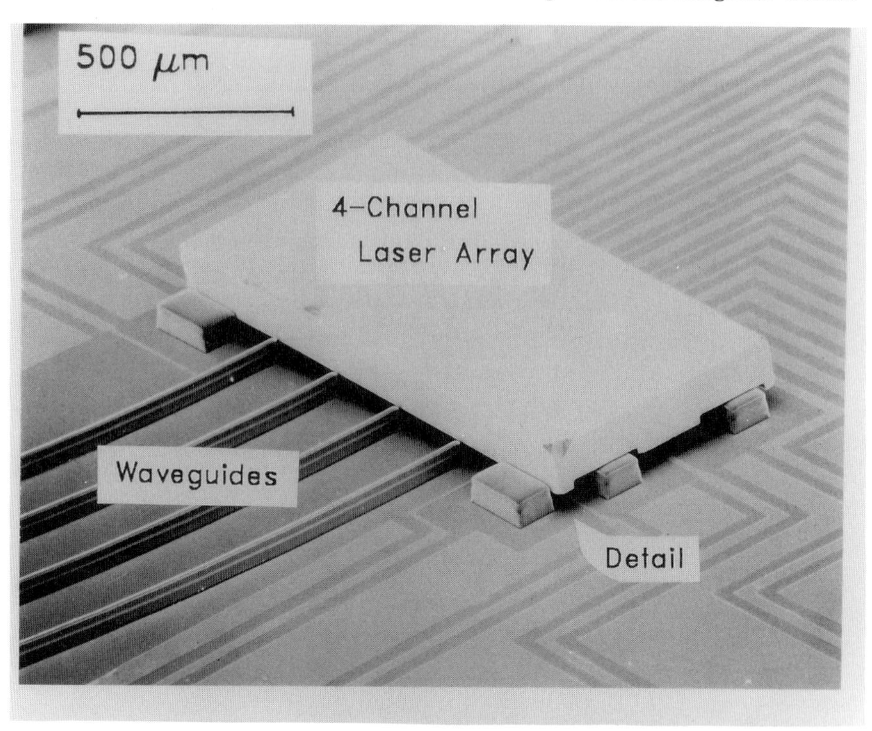

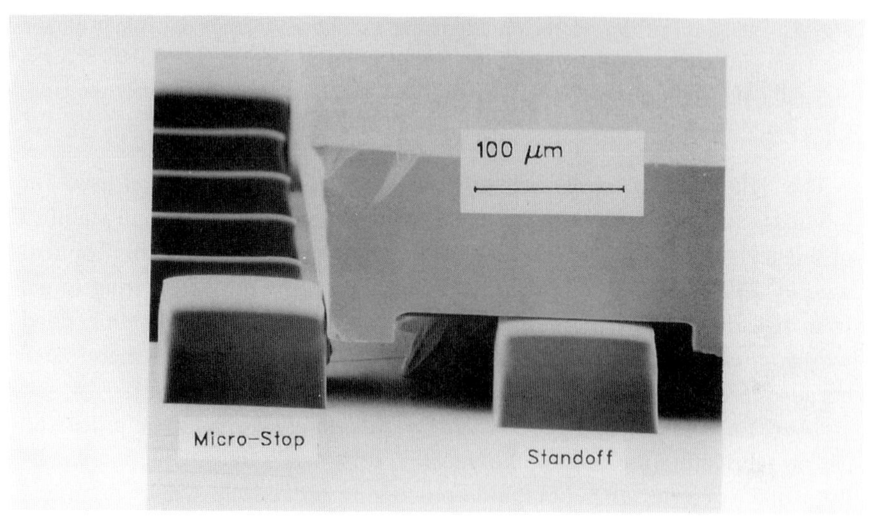

Fig. 6. An SEM photograph of a flip-chip, solder bump attached laser array chip, using the microstop alignment technique. The smaller photo shows more detail of the etched wells in the laser chip and both the vertical and one lateral standoff.

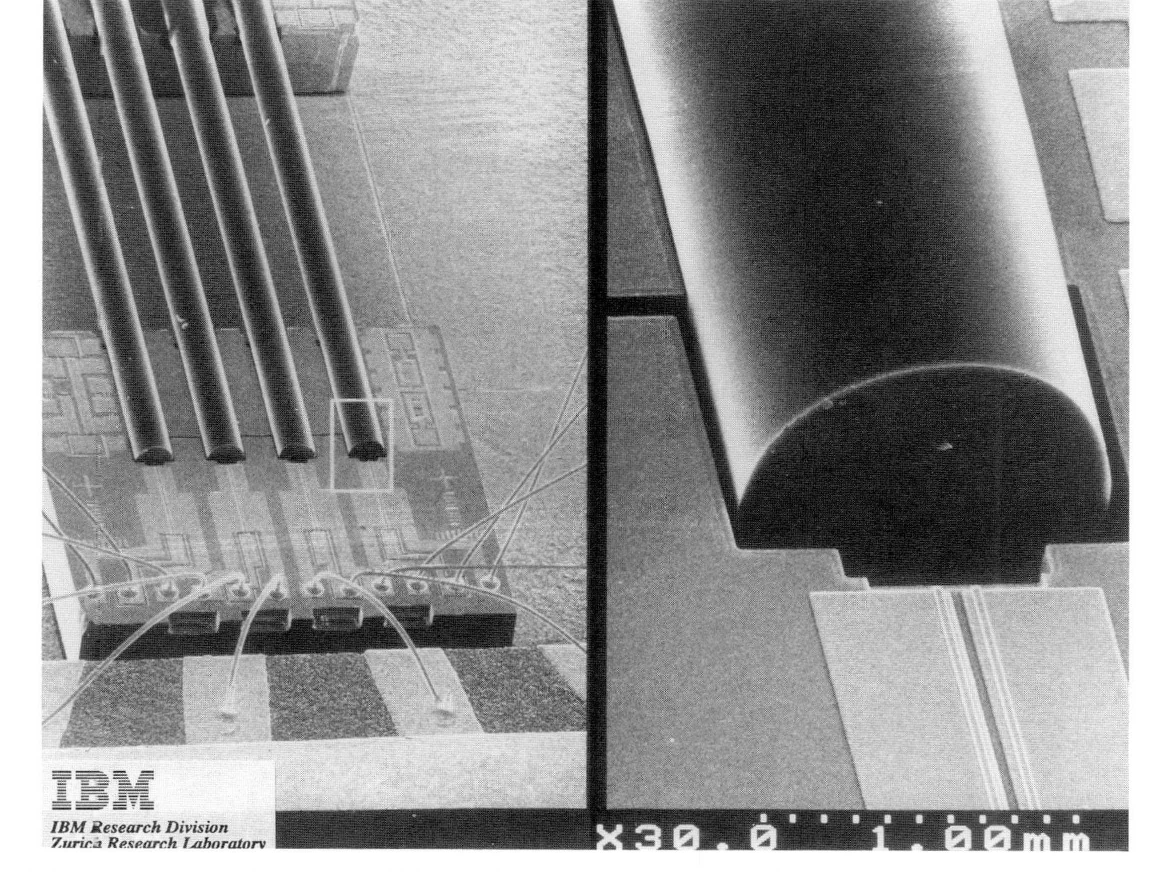

Fig. 7. An SEM photograph of a laser array OEIC with four etched facet lasers on 250-μm centers. The inset shows in more detail the deep U-trough etch in the laser chip which aligns the multimode fiber to the laser facet. (From IBM, Zurich Laboratory.)

neywell researchers have demonstrated polyimide lightguide loss of less than 0.5 dB/cm at 850 nm for temperatures up to 250°C and fabricated coupling and distribution networks using these lightguides [13]. Dupont researchers have developed TAB-like stencils for off-module lightguides [14]. They report 0.15 dB/cm loss at $\lambda = 830$ nm. IBM has demonstrated photoprocessed lightguides of epoxy also with low loss [15]. The reader is referred to the literature for further information.

It was pointed out earlier that OEIC chip design will likely have to be modified in order to facilitate simplified packaging. The IBM optoelectronic MCM discussed earlier demonstrates this in the following ways: (1) a pad metallurgy compatible with solder bump technology is used; (2) on the transmitter OEIC, integrated photodiodes are incorporated to monitor each laser individually and etched facet lasers are used to achieve laser uniformity in optical properties; (3) etched recesses are fabricated into the transmitter chip to provide lithographically precise stops for the glass bumps on the chip carrier; and (4) large-area MSM photodiodes are used on the receiver OEIC to relax the tolerances between the lightguide and the photodiode.

To avoid optical wiring on the chip carrier, the etched features on the OEIC might be used to align directly to the off-module fiber wiring. For example, deep groove etching in the front of the lasers themselves has been demonstrated in another project at IBM Zurich Laboratory and used to align multimode fibers to the laser facet, as shown in Fig. 7 [16]. This would eliminate the need for optical wiring on the chip carrier, at the expense of OEIC chip real estate for optical registration. The yield, reliability, and reworkability of directly attaching optics to OEIC chips are subjects for further research.

An alternative to planar processed optical wiring or alignment stops might be to use a separate molded or formed component made of glass or plastic containing all the optical components. Then only one precison alignment per package would be required. For example, Corning Glass Works [17] has demonstrated an ion-exchange technology for the simultaneous fabrication of lightguides, splitters, and molded alignment grooves. Alternately, Dupont has developed arrays of lightguides and optical components in polyimide sheets, leading to the possibility of optical TAB connections between multichannel optoelectronic chips and electronic cards.

This chapter has outlined some of the important developments needed for OEIC packaging, especially for large-scale OEICs. As this level of packaging is very early in development and will become application specific as it develops further, we have confined the discussion to listing some issues to consider for the chip and package developers. The research examples discussed illustrate how existing integrated electronics packaging technology can be extended to encompass optoelectronics. But they also illustrate the challenges remaining in developing a manufacturable, low-cost OEIC or PIC package.

REFERENCES

1. R. R. Tummala and E.J. Rymaszewski, eds., "Microelectronics Packaging Handbook," Van Nostrand Reinhold, New York, 1st ed. 1989, 2nd ed. to be published 1994.
2. D.L. Rogers, IBM, private communication.
3. D. R.Kaplan, et al., "Electrical Crosstalk in PIN Arrays," paper WCC4, *Proceedings of IEEE/LEOS OFC'86,* pg. 100, (Feb. 1986).
4. Y.Ota, et al., "Twelve-Channel Individually Addressable InGaAs/InP pin photodiode and InGaAs/InP LED Arrays in a Compact Package, *"IEEE Jnl. of Lightwave Tech., Vol. LT-5,* p. 1118, (Aug, 87).
5. M. Itoh, et al., "Compact Package Multi-channel LED/PD Array Modules for Hundred Mbps/ch Parallel Optical Interfaces," NEC Research and Development, Vol. 33, pp.153–160 (4/92).
6. H. Tsunetsugu, et al., "A new Packaging Technology for High Speed Photoreceivers using Micro-Solder Bumps," *Proceedings of the IEEE ECTC'91,* pg. 479
7. See Ref. 50 in Chapter 13
8. M. J. Wale and C. Edge, Self-Aligned Flip-Chip Assembly of Photonic Devices with Electrical and Optical Connections, *IEEE Trans. on Components, Hybrids Manuf. Tech.* **13,** 780–786 (1990).
9. C. A. Armiento, et al., Four-Channel, Long Wavelength Transmitter Arrays Incorporating Passive Laser/Singlemode Fiber Alignment on Silicon Waferboard, *Proc. of Electronics Components and Tech. Conf.,* pp. 108–114 (1992).
10. T. Hayashi, An Innovative Bonding Technique for Optical Chips Using Solder Bumps That Eliminate Chip Positioning Adjustments, *IEEE Trans. Components Hybrids Manuf. Tech.* **15,** 225 (92).
11. T. Miyashita, Planar Waveguide Devices on Silicon, *Proc. Electronic Components and Tech. Conf.,* pp. 55 (1990). C. J. Sun, et al., High Silica Waveguides on Alumina Substrates for Hybrid Optoelectronic Applications, *IEEE Photon. Technol. Lett.* **4** (6), 630–632 (1992).
12. T. Miyashita, et al., Integrated Optical Devices Based on Silica Waveguide Technologies, *Proc. SPIE* **993,** 288 (1988).
13. J. P. G. Bristow, et al., Polymer Waveguide-Based Optical Backplane for Fine-Grained Computing, *Proc. SPIE* **1178,** 103–114 (1989).
14. B. L. Booth, et al., Planar Polymer Waveguides for Optical Interconnects, *Proc. American Chemical Society 4th North American Chemical Congress,* New

Chapter 17

FUTURE OEICs: THE BASIS FOR PHOTOELECTRONIC INTEGRATED SYSTEMS*

Izuo Hayashi

Optoelectronics Technology Research Corporation,
Ibaraki, Japan

1. INTRODUCTION

The optoelectronic integrated circuit (OEIC) will evolve as basic hardware which will be used for information processing systems in the next century. Electronic integrated circuits, LSIs, have been the basic hardware for today's information systems. The progress of electronics toward higher speeds and density of integration, however, will hit fundamental barriers during the coming decades, which cannot be overcome by systems using only elec-

*The major part of this chapter was written in December, 1990. Significant advances in technology have been made and some revisions to the chapter have been added to reflect this. However, the basic concept of the chapter which is described in the introduction, stays the same. That is, the new kind of IEICs, in which micrometer-size optoelectronic devices are integrated into electronic ICs, will become the basic hardware for future information processing systems.

ISBN 0-12-200420-5

trons. The collaboration of photons with electrons will be essential for the barrier breakthrough. New kinds of optoelectronic integrated circuits, which contain both electronic and photonic micrometer-size devices and are closely integrated within a single substrate, will become basic hardware for future "photoelectronic integrated systems," in which photons and electrons can be used complementarily to obtain the best system performance.

Electrons have been very useful for logic operations because of their large charge/mass ratio (e/m). Semiconductors have provided excellent working environments for electrons, particularly in microelectronics, which constitutes the basis of today's computers. However, along with the progress of computers, a problem arises involving the interconnection between distant points in electronic circuits. The very characteristics of electrons, charge-carrying particles, are now becoming the origin of interconnection problems, signal delay, mutual coupling and excess power dissipation.

"Optical interconnections"—using light instead of electrons for signal interconnections—have now appeared as a technology which may be able to solve the interconnection problems [1]. Photons can carry signals at the highest possible speed without using any wires. Because they have no charge, their interaction with materials is weak and they have no interaction with each other. They are ideal carriers of signals, just opposite to the situation for electrons. Optical interconnections in electronic circuits are therefore an appropriate solution for problems involving signal interconnections in electronic circuits.

The idea of using electrons with photons in a single system has existed for some time [2]. However, the technologies were too immature to realize the full collaboration of electrons and photons in the past.

Optical interconnections are now being introduced inside computer systems [3] and will proceed much deeper inside, eventually entering microelectronic circuits LSIs. In photoelectronic integrated systems the conversion of electrons into photons, and vice versa, can take place at any time and at any place, in order to take full advantage of a photon–electron collaborating system. As can be seen from this chapter the best photoelectronic integrated system will be realized by using new, highly integrated OEICs. Many micrometer-size optoelectronic devices will be integrated into LSIs, which may be called U-OEICs, for ultra-large-scale OEICs. In the next century U-OEICs will become the hardware basis for many kinds of computers, optical information processing systems, and many other systems such as neural systems.

In this chapter, the feasibility of introducing optical interconnections in microprocessors (μ-P) is described. Numerical analysis of electrical interconnection problems in (μ-P)s and how they will be solved by using optical interconnections are presented in Section 2. [6] The feasibility of micrometer-size lasers (μ-LD) and photo-detectors (μ-PD) is discussed. Device and material technologies necessary for these microoptoelectronic devices are discussed in Section 2.3. Possible structures for lightguiding circuits are shown in Section 2.2. Three-dimensional integrated circuits using vertical optical interconnections are presented in Section 3. Future developments concerning photoelectronic integrated systems are reviewed, based on various technologies used for U-OEICs as well as recent studies which indicate new developments concerning future photoelectronic integrated systems (Section 4).

The feasibilities of ULSI optical interconnections discussed in this chapter are based on study results obtained by the U-OEIC Research Group since 1991 [4]:

> Historically, extensive research efforts concerning OEICs started in 1979 by the so-called large scale project for optoelectronics sponsored by MITI. The idea was that OEICs will become useful for communication systems, such as one-chip switching circuits in optical local area networks. Device and material research concerning OEICs were continued until 1986 in this project. Even though many of the benefits of OEICs were expected to find application in communication's field, technical barriers to make OEICs have continued to be too high to compete with conventional methods; e.g., hybrid sets of discrete optoelectronic devices with electronic ICs. No distinct benefits in performance or cost reduction were realized in OEIC schemes. Small markets in the area of communication were another factor. However in many laboratories throughout the world, particularly in Japan, expectations for OEICs have continued up to today, due to expectations for unspecified future applications using OEICs.[2] It is also true that the OEIC is a very appropriate "vehicle" for developing semiconductor technologies.
>
> Meanwhile, the need for optical interconnections in computers has become clear. And, it has also gradually been recognized that the problem is due to the intrinsic defective characteristics of high-density and high-speed of all-electronic systems. The interconnection problems are becoming more severe as electronic technology makes more progress. It will soon reach into LSI circuits, where the integration of optoelectronics with electronics is greatly needed. It is rather different compared with the case of the OEICs for communication. As explained, the requirement for light interconnections is intrinsic, being based on the fundamental physics of electrons and photons.
>
> The efforts taking place in a new OEIC project (Optoelectronic Technology

[2]This was also the conclusion in the first OEIC workshop in the United States at Hilton Head in 1989. See M. Dagenais, R. F. Leheny, H. Temkin, and P. Bhattacharya, Applications and Challenges of OEIC Technology: A Report on the 1989 Hilton Head Workshop, *J. Lightwave Technol.* **8**(6), 846–862 (1990).

research, (OTL)), are different from its predecessor laboratory, Optoelectronics Joint Research. The OTL has been focussing on the creation of new semiconductor material technologies for future optoelectronic devices [7]. Process technologies for atomic-scale structures, such as Quantum devices, are thought to be important for future optoelectronics technology. The phrase Quantum Optoelectronic Integrated Circuit, (Q-OEIC) is its symbol, indicating the direction of research at OTL.

Although the applications of Q-OEIC technology have generally been believed to be somewhat far in the future, it is now becoming clear that OTL's efforts will also essential for providing breakthroughs for micro-optoelectronic devices as well as their integration into large-scale optically interconnected LSIs, (U-OEICs) in the near future.

2. DESIGN TRIAL OF THE OPTICAL INTERCONNECTION IN ULSI MICROPROCESSORS

Theoretically, it is evident that optical interconnections could solve problems involving signal interconnections, "communication crisis" in computer systems [1]. Signal delays due to the capacitance charging of interconnected metal wires have become a significant part of delays in computer systems. Also, the mutual coupling of current and voltage between metal wires is becoming a severe problem. These problems become more significant when the speed of computers increases, and eventually they will limit the performance of the system.

How to utilize optical interconnections in computers is theoretically clear and efforts to install optical interconnections are under way [3]. However, there have been no trials to solve interconnection problems in ULSI microprocessors by optical means, where the effect of optical interconnections will be more significant than those in large computer systems.

For the "long-distance" optical interconnections in computers—board-to-board interconnections, for example—length ranges from meters to tens of centimeters, and conventional techniques for optical communication could be employed. For this application, arrays instead of discrete optical devices, ribbons instead of discrete fibers, will be used. However, when the distance of the optical interconnection decreases and the number increases, one may need optical devices installed on the "bonding pads" of LSIs, where interconnection signals are generated or received. This will be the first version of OEICs in computer systems. A technique called flip-chip bonding, in which chips of optical devices made of III–V materials are solder bonded on bonding pads of the LSIs, will be appropriate at first. However, for optical interconnection inside LSIs, such more or less conventional techniques cannot be effective and real "integration" of optoelectronic devices with electronic integrated circuits (ICs) is needed.

In addition, the size and power of optoelectronic devices must be minimized in order to match those of electronic devices in LSIs, as described later in this chapter. Therefore, OEIC techniques required for optical interconnection in LSIs are far advanced compared with those presently available. In this sense, such OEICs will represent the first step of the "future OEICs" in the title of this chapter.

In ULSI microprocessors, interconnection problems are more concentrated and more difficult to solve than those in large computer boxes. With decreasing device size and increasing device density in a chip, charging delays due to interconnecting metal wires increase. The capacitance of metal wires does not decrease with decreasing dimension below 1 μm due to wire-to-wire sidewall capacitance (Fig. 1). Their increased resistance results in an increased *RC* time constant of metal wires (Fig. 2) [6]. Therefore, long signal interconnections between circuit blocks in a microprocessor are due to the large *RC* delay (Fig. 2) [6]. Significant improvement is seen with optical interconnections above 10 mm, and the difference is small below 2 mm. At 20 mm, optical interconnection delay is less than one-tenth of the electrical delay, even with large wire (1 μm). Operation at a few gigahertz is feasible if one uses optical interconnections for global lines, whereas it is limited at a few hundred megahertz with electrical wiring, even using thick wires. Therefore, use of optical interconnections for global lines will improve the ULSI performance a great deal (Fig. 3). Inside circuit blocks,

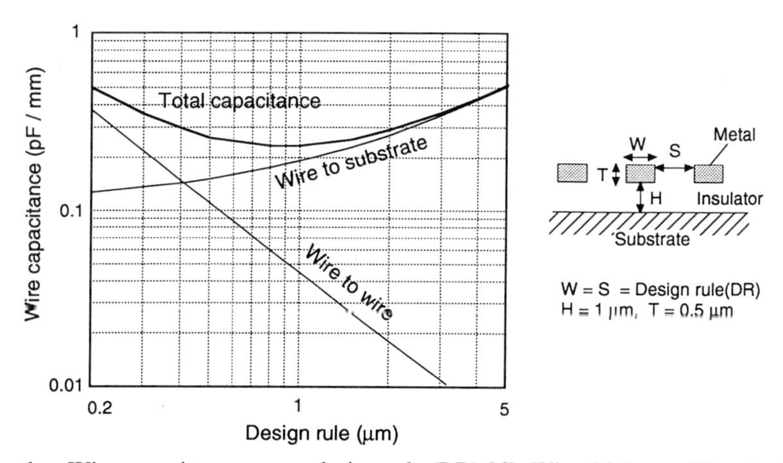

Fig. 1. Wire capacitance versus design rule (DR) [6]. Wire thickness (T) and wire height from the ground (H) are fixed. Capacitance increases with decreasing dimension (DR) below 1 μm.

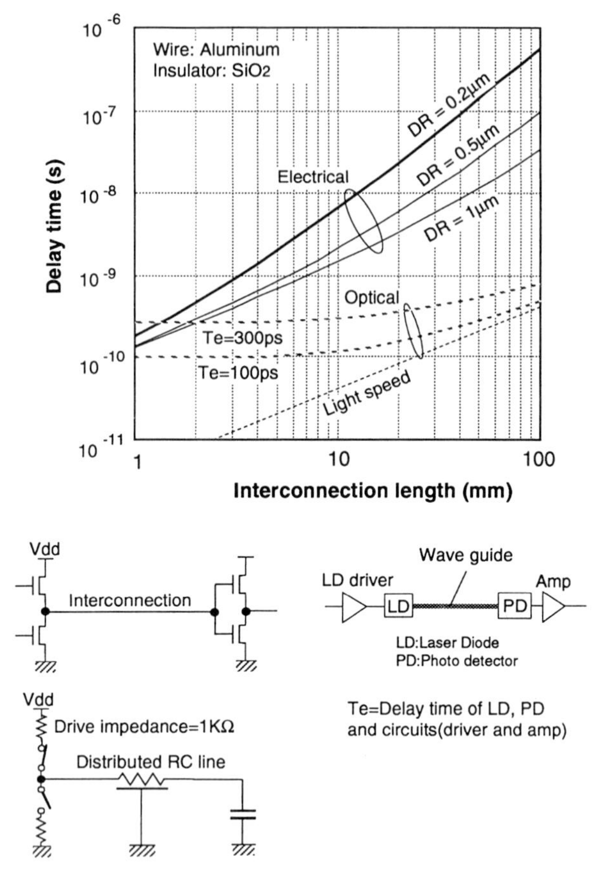

Fig. 2. Delay time of electrical and optical interconnections versus interconnection length [6]. Electrical interconnections for three different wire widths (DR). Optical interconnection delay includes extra delay time, Te, due to (driver and LD) plus (PD and amp). Te = 300 ps is a feasible value using CMOS transistors today. Te = 100 ps will be possible in ther future.

with decreasing design rules, device dimension decreases and the interconnection length between neighboring transistors decreases, thus increasing the operating speed of circuits in individual circuit blocks.

Superconducting wires cannot solve this problem, because the capacitance of the wire remains the same. Thus, charging current stays at the same level and driving power does not decrease. With scaling, the decreased output current of small transistors limits the charging time; in addition, the mutual coupling between wires remains the same. It is clear that optical interconnection is the correct solution.

It should be noted that optical interconnection results in not only a short

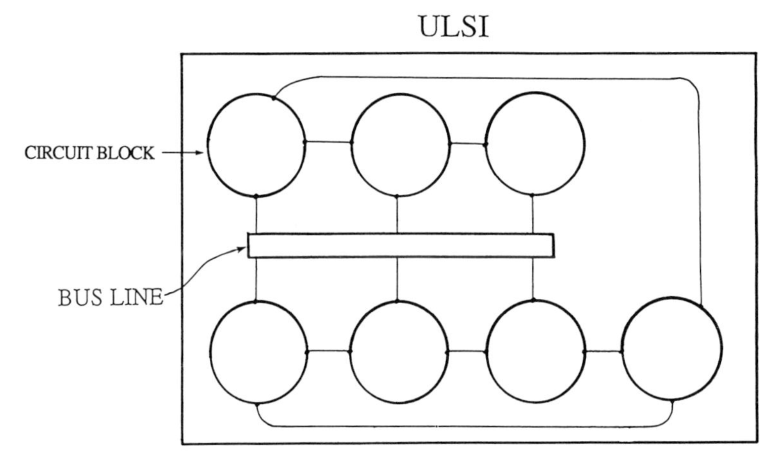

Fig. 3. Schematic drawing of a microprocessor circuit. It has many circuit blocks (circles). Interconnection is short (less than a millimeter) in the blocks. There are long (global) interconnection path between circuit blocks. A large fraction of global lines are in bus lines and clock distribution lines.

delay time but also low power for interconnections. Figure 4 shows diagrams of electrical versus optical interconnection. For a global electrical interconnection ($l \sim 1$ cm), the line capacitance (C_L) is much larger than the transistor input capacitance (C_{in}). Therefore the main part of the driving power is lost by charging wires (Fig. 4a). It is 0.2 mW for $C_L = 2$ pf at 100 MHz and 2.0 mW at 1 GHz.

In the optical interconnection, photons generated at a laser diode (LD) mostly reach a photodiode (PD) (Fig. 4c). Assuming 10,000 photons per pulse, optical power is about 1 μW at 1 GHz (0.1 μW at 100 MHz). The 10,000 photons per pulse is a sufficiently large number to give a low enough error rate (less than 10^{-15}) [8].[3] The power efficiency of a laser diode can be made several tens of a percent. For an electrical transmission line (Fig. 4b), the delay time is the same as for an optical interconnection; however, the power needed for interconnection is much larger. It is 10 mW for a 1-V signal in a 100-ohm line.

In order to realize optical interconnections in ULSI, extremely small optical devices are needed to match transistors, which already have decreased dimensions. The minimum possible size of lasers is about 1 μm, which is their wavelength. The threshold of these microlasers (μ-LD) is hoped to be

[3]However, if one estimates amplification of 1 μW PD input, the power needed for the amplifier exceeds that of the LD power. For 10^5 photons for pulse, 10 μW at 1 GHz is assumed in this chapter [6(b)].

a) Electrical Interconnection

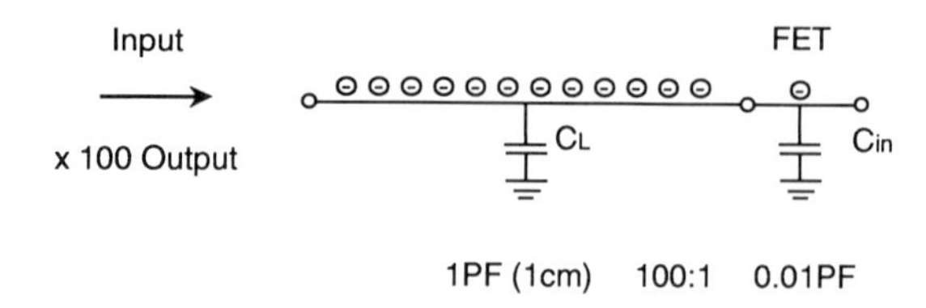

b) Transmission Line

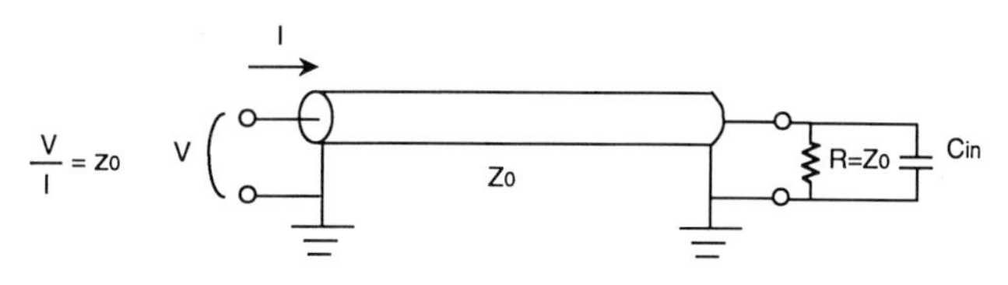

C) Optical Interconnection

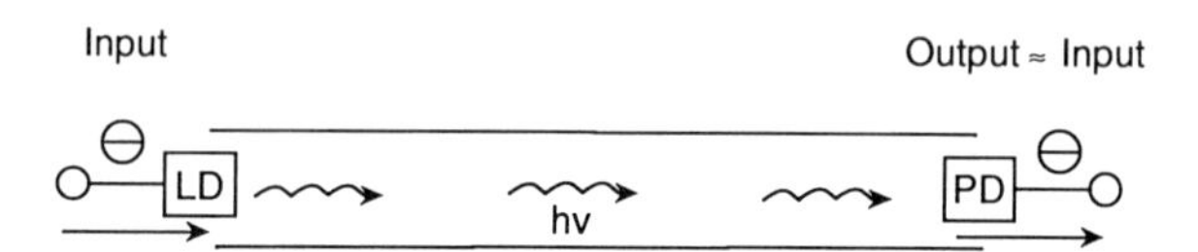

Fig. 4. Comparison of electrical and optical interconnections. (length = 10 mm). (a) Metal wire interconnection in an ULSI Line capacitance C_L = 2 pF (DR = 1 μm). Transistor input capacitance C_{in} = 0.02 pF. Driving power $P = fCV^2$ = 100 MHz · 2 pF · 1 V^2 = 0.2 mW (2 mW for 1 GHz). Efficiency = $C_{in}/C_L \sim 10^{-2}$. (b) Transmission line (coaxial, coplaner, etc.). Characteristic impedance ≈ 100 ohms. Driving power = (V^2/Z_0) = 10 mW (independent of frequency and length). (c) Optical LD, PD, waveguide (WG) interconnection. No intrinsic power loss in the WG. OE conversion losses in LD, PD are small. Optical input power at PD ~ 10 μW at 1 GHz. Driving power of LD ~ 100 μW (possible minimum power).

of the order of microamperes. Photodetectors need to have similar dimensions. These microoptoelectronic devices will be discussed in detail later.

2.1. Design Examples of Optical Interconnection in a ULSI

In this section, a design study of the optical interconnection in a ULSI microprocessor chip is presented. Some details of μ-LDs and optical waveguides (WGs) are also discussed. The discussion here is based on studies done by the U-OEIC Research Group [4, 5].

Assuming a ULSI of about 1 cm square, long-distance interconnects are about 1 cm (Fig. 5). It takes about 50 ps for a light signal to cross between two points (A and B) in the wafer. Here, an optical WG with a refractive index of 1.5 is assumed. In order to convert an electrical signal from the ULSI circuit into an optical signal, a driver circuit and a small semiconductor laser (μ-LD) must be installed at the transmitting end of WG A (Fig. 5). At the receiving end, B, a photodetector and an amplifier are placed to convert the signal back into an electrical circuit. The laser diode delay, t_{LD}, depends on the response time of the LD (cutoff frequency) and the speed of the electrical signal of the driver circuit. A response time of 10 ps will be obtained when an LD of micrometer-size dimension (μ-LD) is operated at a current level much higher than its low threshold. The response speed of the detector is assumed to be 10 ps, which can be achieved for a detector

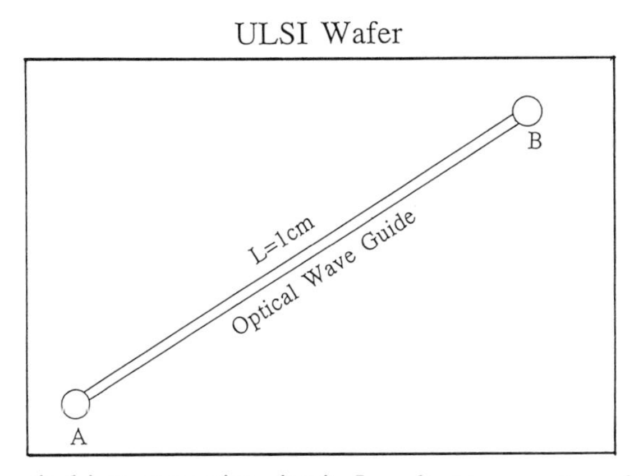

Fig. 5. An optical interconnection circuit. In order to convert an electrical signal into an optical one, a laser and its driver circuit are installed at the transmitting end A. At the receiving end B, a photodetector and an amplifier are placed to convert the signal back to an electrical signal. A total delay of a few hundred ps will be obtained using CMOS circuits, when the distance between A and B is 1 cm (see text).

of small size (micrometers). However the speed of the LD driver (t_D,) and the amplifier (t_{AMP}) for the PD will be slow. An overall delay of (300) ps is assumed in Fig. 2 using C-MOS transistor circuits [6]. An overall delay of 100 ps will be possible in the future.

Although the minimum input power to the PD is calculated to be 1 μW at 1 GHz,[4] here the desirable input power to the PD is assumed to be 10 μW at 1 GHz as described in Section 2.2. The threshold power is proportional to the signal frequency, so even lower power can be used for slower signals. The new concept of a microdetector (μ-PD) with microwatt receiving power is another challenge to match the microlaser. The photodetector efficiency, η_{PD}, depends on both the material and structure of the photodiode and can be made 50% or higher.

A distinct feature of optical interconnections is their potential for low power, contrary to common belief. The reason for this misunderstanding is that today's lasers used for optical communications or memory disks have high thresholds of tens of milliamperes or more. For these applications, power consumption is not a major concern; their modes, spectrum, or response speeds are of major interest. Because they are still made using the old traditional techniques, such as mirrors made by cleavage of planar multilayer epitaxy, their size is unnecessarily large, which is limited by the fabrication technique. As discussed later, it is feasible to make low-power lasers if one considers the basic operating principles of lasers and advanced fabrication techniques.

For optical interconnections, it is possible that most of the photons emitted from the laser will reach the detector and be efficiently converted into electrons, which are then supplied to the receiving transistor. Counting the conversion efficiencies of the laser, $\eta_{LD} = 0.3$; detectors, $\eta_{PD} = 0.5$; and their coupling in and out of the optical WG, $\eta_C = 0.7$, the total transmission, electrical input to electrical output, can be made 10%.

For a fan-out of N, the optical power must be divided into N branches of the WG; therefore, the laser must deliver at least N times more power, which is required at each receiving end. More laser power is needed if the power division is not equal and extra absorption is provided in order to prevent multiple reflection in the WG systems. A nice feature of optical interconnection is that the overall speed is the same as that of a one-to-one (N =

[4]For conventional optical communication, PD input power of 0.1 mW is a safe threshold value for 10 Gbit with an error rate of 10^{-12}. However, if one counts difference in device structure for the optical interconnection in LSIs, much smaller size (~μm), and possible integrated PD-FET structures, it will be possible to lower the threshold by at least an order of magnitude.

1) interconnection, since photons from the laser are divided and travel independently before being absorbed individually at photodetectors (Fig. 6). On the contrary, with electrical interconnection, fan-out changes load impedance. If fan-out N is in parallel, load impedance decreased to Z/N (Fig. 6). It distorts the output waveform, slows down its rise time, and increases output delay. In clock circuits or bus lines, many fan-outs (N = 10–100) are needed, thus creating trouble.

Figure 7 shows light guiding for optical interconnections on a ULSI chip. The use of optical WGs is preferable for many reasons. Small light loss is expected in WGs, rather than in free space, and coupling between different interconnecting WGs can be made small. A WG is made of a transparent medium of a variety of dielectric materials. Major concerns are the fine

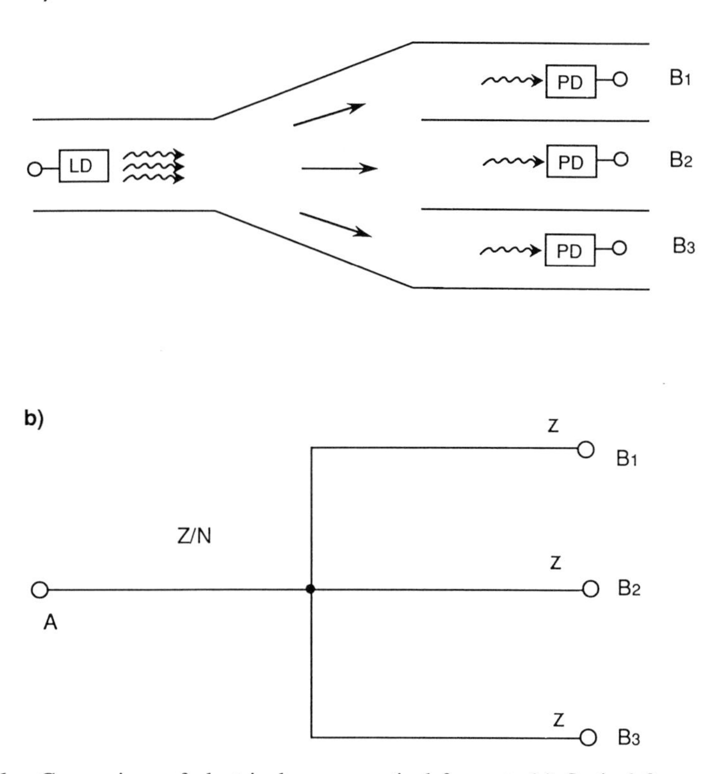

Fig. 6. Comparison of electrical versus optical fan-out. (a) Optical fan-out. Photons from a laser diode (LD) travel divergently into several photodfetectors (PD). There is little interaction between these photons. No reaction from PDs back to the LD. (b) Electrical fan-out. The output impedance changes due to the parallel loading (Z_{out} = Z/N). Output waveform slows down.

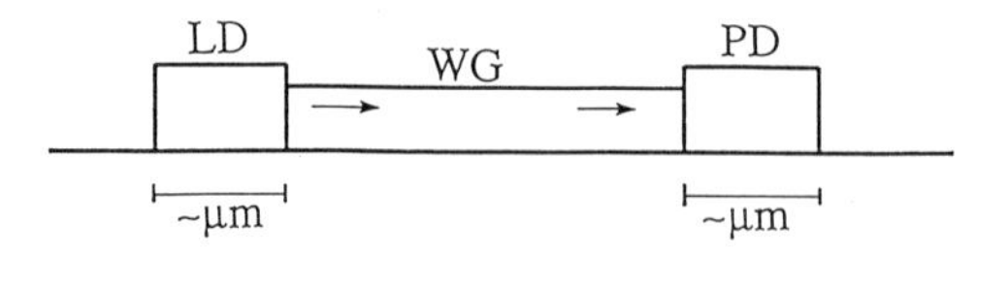

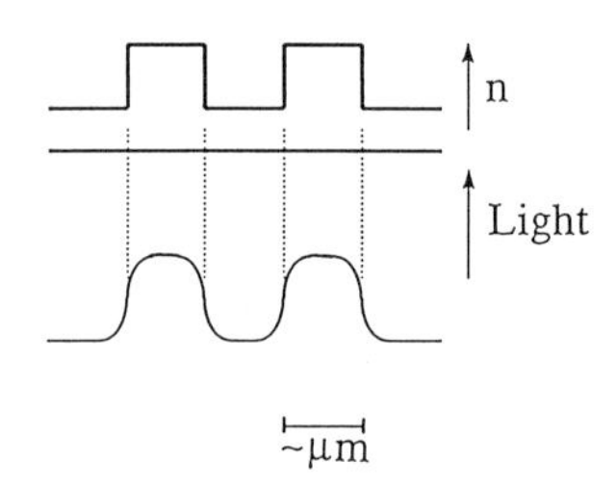

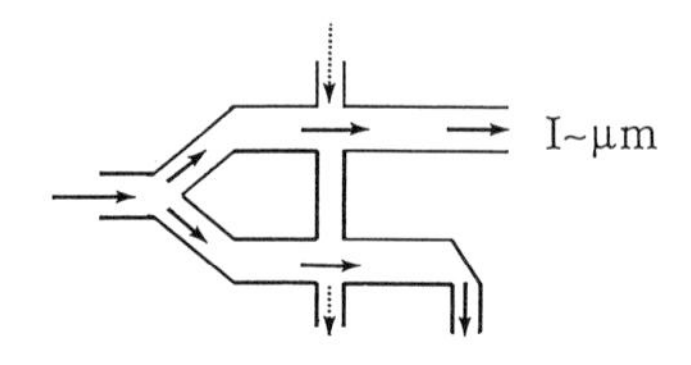

Fig. 7. (Top) Optical waveguide (WG) for interconnection. The WG has high coupling efficiency (~1.0) between LD and PD. (Middle) In plane crossing and sharp bending are possible. (Bottom) With a high index step ($\Delta\ n/n$), mutual coupling through leakage can be suppressed. Metal film can be used for decoupling if necessary.

microstructure of the WGs (submicrometers) and ease of fabrication, as these influence the loss and crosstalk between optical wires.

The width of a WG is about the same as the wavelength of the light used, which is in the order of micrometers. It varies with the choice of waveguide materials and modes. Either single-mode or multimode lightguides are possible due to the short transmission distance. In order to avoid any coupling between adjacent WGs, leakage of the optical field outside the WG must be kept small, which requires a large refractive index step (Fig. 7). There is no intrinsic coupling between WGs like that between two parallel wires in electrical interconnections. A nice feature of the optical WG coupling is that crossing of two WGs in a single plane is possible. Sharp bending of a WG is possible, just like the mirror reflection of a light beam. Precision of the WG dimensions is not so stringent, except at places related to couplings between WGs and detectors or LDs. Furthermore, a small amount of waviness would be allowed. Simple direct coupling of a WG with a laser and a detector will be used. This is not difficult to do with micro-fabrication techniques, which must be used to form the whole interconnection system. The WG system will be described further later in this chapter (see Fig. 10).

2.2. Optically Connected Clock Distribution and a Bus Line in a ULSI

In this section, examples of long-distance global optical interconnections on a ULSI chip are described. [5, 6] A clock distributing circuit has one laser diode for each system (Fig. 8). The output of the laser diode is fed into a WG, which is divided into many branches. At the end of each branch, a photodetector is connected. The output of each amplifier for the photodiode is fed into several electrical circuits, where the clock signal is required.

Several conditions must be satisfied in this circuit in order to obtain accurately timed clock signals at all output points of the circuit. First, the length of the WG, including branched parts, must be equal. If the total length is 1 cm, the delay corresponding to this length is about 50 ps, so a 1-mm difference makes a 5-ps difference, which results the clock skew. Second,

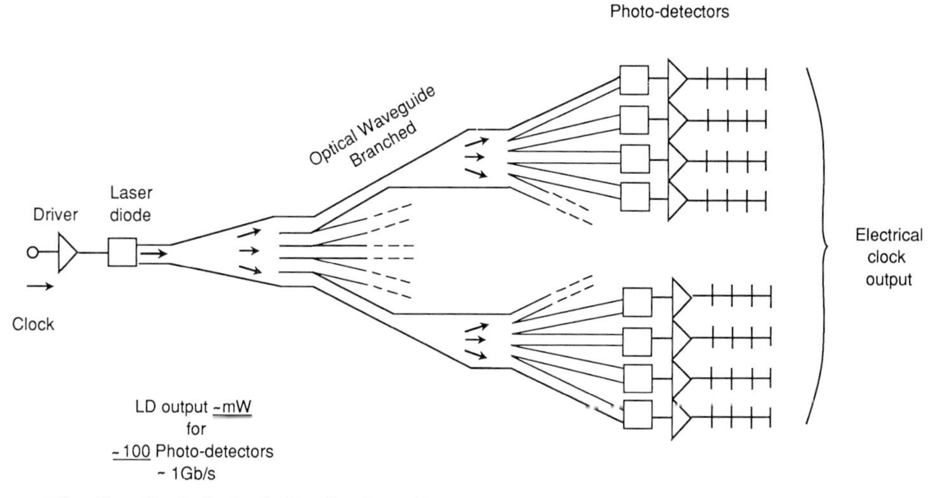

Fig. 8. Optical clock distribution. One example of N (=100) multiple fan-out for an optical clock distribution. Other branching designs would be possible. Estimated laser power is 3 mW for 300 MHz, which is much smaller than for the equivalent electrical system (see text). N = 100, detector input ~3 μW, clock skew <0.01 ns.

optical power should be divided into equal amounts at each branching. Different optical methods will be available to accomplish this. Third, reflections at various locations of the WG (at branching or at the detector ends) should be minimized. Reflections disturb the optical signal and produce additional delays and unbalanced outputs. It is easier to install terminating absorbers to avoid reflection; however, it requires an extra amount of laser power. In this clock circuit, if the clock frequency is 300 MHz, the minimum input power of each photodiode is 0.3 μW. However, if one counts power needed for amplification of the PD output, a larger PD input power of 3 μW instead of 0.3 μW is desirable, in order to minimize total clock system power. Including optical coupling loss and other margins, a laser output of 3 mW will be needed for a fan-out of $N = 100$.

In comparison, the equivalent electrical clock circuit is much more elaborate. The delay time of electrical interconnection (Fig. 2) is a few nanoseconds at 20 mm, even using thick connecting wires. This means large attenuation and phase shift in the clock circuits at the 300-MHz signal. Actually, a few additional stages of extra amplifiers are needed between the clock driver and the final amplifier for local distribution of the clock signal (Fig. 8 in [6]). Therefore the power required for the optical clock system is much smaller even with a conservative estimate, and there is no source of extra phase shift compared with the equivalent electrical system. A numerical estimate for a 0.2-μm CMOS microprocessor, 250-MHz clock frequency, showed that the total clock system power including all electronic circuits decreased less than half, from 5 to 2 W [6]. It would be difficult to obtain a clock frequency over a few hundred megahertz using electrical circuits, but at least a few gigahertz is feasible with optical techniques.

A schematic drawing of an optical bus line is shown in Fig. 9. This bus line connects registers and an arithmetic logic unit (ALU). The signal is transmitted in order to exchange information between the ALU and the registers. The bus line also carries signals between the registers. These signal exchanges are able to share the same bus line because timing, which is given by a central control unit. As shown in Fig. 9, the bus line waveguide has many optical inputs and an equal number of outputs. Laser diodes (LDs) are connected with the WG through couplings. The coupling should be designed in such a way that the power from an LD is fed into the WG efficiently, while not coupling any optical waves traveling in the WG back to the LD. This should be possible using loose coupling, since the laser cavity has a large Q value and that of a WG is low.

The detectors are coupled loosely with the WG to pick up a small amount of light in the WG. If there are 20 registers, which is typical in micropro-

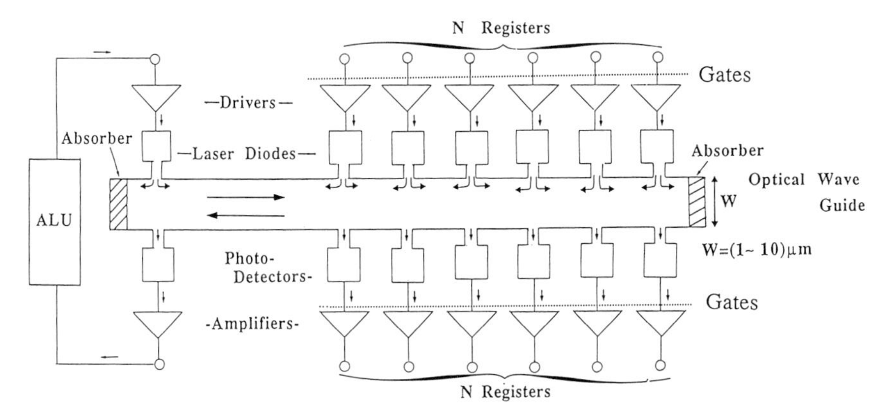

Fig. 9. An optical bus line, example of an optical bus line connecting ALU and registers. Each has an LD and a PD. A waveguide has many couplings with LDs and PDs. These couplings would require special care in order to make efficient and balanced connections. Bus line length ~1 cm (0.05 ns); total Delay ~0.3 ns; N ~30; laser ~1 mW.

cessors, the detector coupling will pick up about an order of a percent of the optical signal propagating in the WG. The signal will decay along the way due to couplings. The absorbers at the end of the WG are necessary in order to suppress reflection. A laser output of 1 mW would be sufficient for a 300-MHz signal under the same design principle as before. The total delay between any two points in the bus line can be made to be a few hundred picoseconds or less. Although the bus line requires many LDs, since only one of them operates at one time, the total power is small. The performance of this optical bus line is much superior to that of conventional electrical bus lines. An operating speed of a few gigahertz will be possible. A bus line with an advanced architecture has been proposed, in which an equal amount of optical power from any light source is distributed into every receiveing channel. [9]

In an electrical bus line, all outputs and inputs of amplifiers are connected to a single metal wire (the bus line) in parallel (Fig. 8 in [6]). Improvements in speed and power by replacing electrical bus lines with optical ones are shown for CMOS microprocessors in ref. 6. Operating speeds in the gigahertz range can be obtained with optical bus lines, which will make it possible to realize new system architectures of parallel processor operation.

In order to make such a bus line, a variety of studies are needed; in particular, the design and fabrication of the WG with optical couplings will be a great challenge. For a 64-bit μ-processor, 64 such bus lines are necessary and the total number of LDs for several sets of bus lines would become

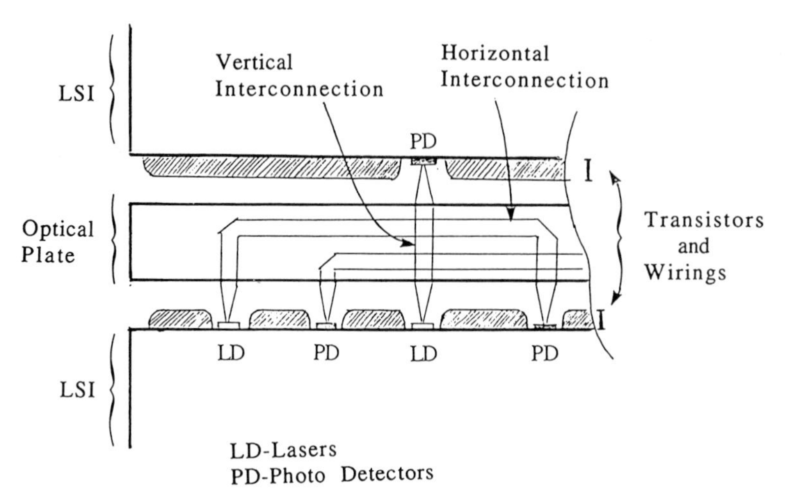

Fig. 10. The figure illustrates the principle of the optical plate, which accommodates all optical waveguides and associated passive optical components. Optical circuits will enable interconnection between lasers and photodetectors. Mechanical positioning of the optical plate with respect to the LSI substrate will complete all positioning of optoelectronic devices with inputs and outputs of corresponding optical circuits. Not only horizontal interconnection between LDs and PDs on the same LSI but also vertical interconnection between the two facing chips will be achieved with this scheme. It is also possible to interconnect spots in different LSIs which are placed horizontally on the same plane. Such optical interconnections will be important in multichip modules [6].

several thousand in a microprocessor with a 0.2-μm design rule [6]. It has been shown by circuit design experts that the total number of long-distance interconnections in a μ-processor will be of the order of several thousand. It is, therefore, fortunate that a relatively small number of optical devices, about 10,000, will be sufficient to realize drastic improvements in μ-processor systems. [6]

So far we have not discussed how to install these optical waveguide circuits in a microprocessor chip, which is already sufficiently crowded. The chip must have a large amount of interconnecting wiring, which occupies the main part of the chip area, even using multilayer interconnecting wires.

In the new concept shown in Fig. 10,[5] the whole passive optical circuit, including WGs and associated optical components, is accommodated separately in a substrate which is completely separated from the microprocessor

[5]Mitsuo Takeda, U-OEIC research group, private communication. Also see Proceedings of the 9th Micro-optics Seminar (MOS '92) an Optical Interconnects, published by the Japanese Optical Society, pp. 22 and 33 (5/92) (in Japanese).

chip. [9] The optical substrate, which may be called an "optical plate" (optical plane), is located several micrometers "in the sky" like an express highway in a crowded city. [9] It takes advantage of the free space propagation of light while preserving high efficiency of couplings. The nice feature of this structure is that the optical plate can be fabricated separately from the semiconductor substrate, yet all optical inputs and outputs can be aligned precisely with corresponding optoelectronic devices, using the photolithographic technique. The optical plate must be mechanically aligned accurately with the semiconductor chip. [5, 9]

This optical plate concept is useful not only in the single microprocessor chip but also in systems in which number of chips are located vertically or horizontally, connected by the optical technique and constructing a total system [6]. A large amount of R&D work will be necessary to make this type of optical plate, but there seems no intrinsic difficulties in installing them.

The estimates used for optical interconnection circuits are not accurate, yet they predict a definite advantage of optical interconnection in ULSI systems. Experimental and theoretical studies will be needed to determine the parameters in many parts of the system. Techniques of microwave waveguides will be useful for the design of optical WGs and new microfabrication techniques will be needed. Although no basic problems should exist regarding WGs, making them commercially will be an engineering challenge.

Optical interconnections will remove the interconnection bottleneck by drastically decreasing the signal interconnecting delay and power, utilizing the fundamental features of light. The time required to interconnect any points on a ULSI can be made small, a fraction of a nanosecond, which is almost negligible compared to the general operation time of ULSIs. This should have a drastic impact on circuit architecture. Circuit blocks can be moved to any location on the wafer without delay time restrictions.

2.3. Design and Fabrication Techniques for Optoelectronic Devices for ULSI Interconnections

The important subjects regarding optical interconnections in ULSI are how to make micrometer-size optoelectronic devices with desired characteristics.

The basic design of a small laser diode of micromater size and with microampere thresholds can be achieved comparatively easily. If one considers a laser diode with μm dimensions, a threshold current in the μA range could be expected by "scaling." The threshold current density, I_t = 1000 A/cm^2, is the standard value for double heterostructure (DH) lasers, and a value as low as 100 A/cm^2 has been obtained using a quantum well (QW) active region. In order to obtain a 10-μA threshold, an area of 1 or 10 μm^2 should be used for DH or QW laser diodes, respectively.

The conditions necessary to achieve such thresholds are that (1) the mirror loss of the cavity must be made small to match the volume loss of the active region; (2) the side-wall loss of the cavity must be small, equal to that of the long conventional laser per unit length; and (3) there should be no excess leakage current.

Because of the small dimension between mirrors (1), high reflectivity is required to satisfy condition (1): for $l \sim 1$ μm, $R \sim 0.99$ or larger. Using GaAs and AlAs multilayer mirrors, laser oscillation was actually demonstrated in surface-emitting (vertical) lasers with a very short active region thickness [10]. These lasers had a large mirror thickness (a few μm) due to the small refractive index difference between GaAs and AlAs. The dimensions can be made small by choosing dielectric materials with a large index difference, such as TiO_2 and SiO_2 [11]. For a vertical cavity laser, however, growing the active material on the mirror layer on the substrate side (Fig. 11) is a problem. Techniques such as atomic layer epitaxy (ALE) will be useful for controlling the thickness of the active layer and the multilayer mirrors. Loss control of the side walls is an unsolved problem. Although

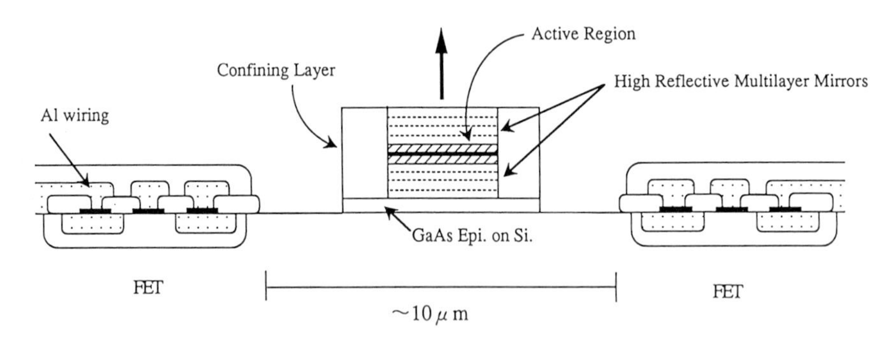

Fig. 11. A microlaser (μ-LD) on a silicon LSI; atomic layer epitaxy will be useful for making multilayer mirrors and an active region with a quantum well. A μ-LD must be made in a small vacant place in an Si LSI after Si processes at low temperatures. High-quality epitaxy of GaAs is needed on the Si substrate. The μ-LD must be fabricated without defects around the active region. A low threshold of 10 μA will be required.

trials involving vertical microsize lasers have been made [12], they were all made by cutting the laser structure layers down into planar stacked layers which include both mirrors and an active layer. This cutting was done with dry etching techniques. In such techniques, the side wall of the cavity is damaged by ion bombardment, which creates a large cavity loss and significantly increases threshold. In addition, if the side wall of the active region is exposed to air, nonradiative recombination center is created at the surface, which kills injected carriers in and around the active region. It is well known that the density of such recombination centers at the air-exposed surface is sufficiently high to decrease the level of injected carriers almost to zero at the surface [13]. These effects should result in exceedingly high thresholds, particularly when the diameter of the laser becomes small, in the range of a few μm.

To solve this side-wall problem, it is first necessary that cutting of the laser periphery be done without damage. Conventional dry etching with ions cannot be used. The new technique of electron beam–defined etching is one candidate [14]. Second, an etched surface should be protected from air exposure until the active region is buried by a high-bandgap low-index layer (Fig. 11). Sequential processes of etching and burying by epitaxy, all in an ultrahigh vacuum (UHV), are appropriate for this purpose. The feasibility of such UHV *in situ* processing has already been demonstrated [15].

Technical breakthroughs for μ-LDs not only are the key to success with optical interconnection on ULSI but also should provide a base for a new generation of laser diodes, which will open up a new world of laser technology.

Instead of laser diodes, light-emitting diodes (LEDs) can be used as light sources for optical interconnections. LEDs have lower efficiency and a slower response in general because they utilize spontaneous rather than stimulated emission. The efficiency of LEDs is generally less than one-tenth that of LDs. They are also slower, in the range of nanoseconds. However, by using a specially designed small cavity, which ideally has only one mode, the efficiency can be made much higher than that of ordinary LEDs, since spontaneous emission falls into the single mode whereas in large LEDs it is divided into many modes. This is called a microcavity LED [16], and because it has no threshold it is almost like a μ-LD with a very low threshold. If the characteristics required for ULSI are not as stringent as for LDs, the microcavity LEDs or even normal LEDs, will find a place in optical interconnections, since fabrication will be easier than that for LDs.

The use of silicon for photodetectors is very attractive. However, owing to the indirect nature of silicon, its light absorption length is large. It is about

30 μm for 0.8-μm light, but it decreases quickly with decreasing wavelength and is about 3 μm at 0.6 μm. A short-wavelength laser made of InGaP on GaAs might be appropriate for this purpose. In addition, this material is less defect sensitive than GaAs [17]. The use of a short wavelength is also attractive in terms of leakage light suppression by high absorption of surrounding silicon materials.[6] Of course, a short wavelength is preferred because the size of all optical components shrinks proportionally.

The growth of high-quality material, mainly GaAs on silicon, is another big challenge [18]. Steady progress is being made, and I believe it should be possible to obtain materials sufficient to meet future requirements [18]. The choice of laser device material may also be helpful for increasing reliability [17].

In order to install optoelectronic devices of III–V compounds on silicon ULSIs, process compatibility must be considered. Silicon devices occupy the majority of a wafer's area and use high processing temperatures; it is, therefore, natural to install optoelectronic devices after finishing processes for ULSI silicon devices. Complex metal wire interconnections of silicon transistors should also be made with silicon transistor processing. Lasers and photodetectors must, therefore, be made afterward in small vacant places scattered over the sea of silicon transistors and wirings. The maximum temperatures for laser fabrication, including epitaxy on the silicon substrate, must be carried out at low temperatures at around 400°C, which does not disturb the silicon part of a U-OEIC, if Al wires are used for interconnection. According to recent studies, it is likely that processes for optical devices can be performed at around such low temperatures. The UHV *in situ* process of maintaining a clean surface between fabrication steps promises to be useful for this purpose.

The feasibility of optical interconnections on ULSI was discussed. It will become possible to install light-guiding systems used to carry signals between points in a ULSI. Interconnection delay can be made very small (a fraction of a nanosecond) between any two points in a ULSI, including E-O conversion.

Once this U-OEIC technology has been established, any two points in a ULSI can be interconnected with a delay comparable to that of neighboring circuits, so that the freedom of circuit arrangement will expand enormously and new system architectures will become possible, taking advantage of ultrafast bus lines and other features of optically interconnected circuits.

[6]Long-wavelength light will pass through a silicon wafer, so it will have some use when signal passage through the wafer is needed. However, low absorption will result in light spilling all over the place, while perhaps creating background noise.

3. THREE-DIMENSIONAL INTEGRATED CIRCUITS USING VERTICAL OPTICAL INTERCONNECTIONS

In the previous section, optical interconnections between points in one integrated circuit, "horizontal" interconnections, were discussed. There is another approach of using light to interconnect "vertically" between points in stacked electronic circuits. Three-dimensional systems can be realized.

A new class of parallel-processing systems will become feasible using vertical optical interconnections between multilayer processor circuits. A prototype of such a system has been under construction by Koyanagi et al. [19] (Fig. 12). In this system four sets of 4K bit memory circuits are connected vertically by (LED–photodiode) pairs (Fig. 12). Thus these four set memories always keep the same data, and this "common memory" connects four CPUs for parallel operation [19]. The speed of data transfer is 16 ns, which corresponds to 32 G/s when 512 bits are operated in parallel [19]. Such three-dimensional interconnecting systems will become practical only after replacing electronic interconnections by optical interconnections: thus, each layered circuit will become free to be constructed individually (Fig. 12).

Figure 13 shows a conceptual drawing of a three-dimensional integrated circuit system (3DIC) utilizing both vertical and horizontal optical interconnections between and within wafers. The delay time of signal interconnections between any two points in the system would be less than a nanosecond using lasers instead of LEDs and using techniques similar to that used in a single ULSI in the horizontal direction. This three-dimensional configuration is superior to wafer-scale integration because of the short distances of interconnections; thus much higher speed is expected. In addition, high yields should be achieved due to the possibility of testing individual wafers before total assembly.

A new "optical system" has been constructed using the photoelectronic scheme by Ishikawa et al. [20]. A large two-dimensional array of 4096 photodetectors has been connected individually to small processor elements (Fig. 14a). A group of processors connected with each other's nearest neighbors is capable of processing incoming pictures. Although today's system is a prototype with large dimensions, it is possible to reduce the size greatly to the LSI level.

According to Ishikawa, by connecting the two-dimensional LED output of such a system to the next-stage system, which also has a two-dimensional photodetector array, sophisticated image-recognition systems could be built based on an extension of this system in the future (Fig. 14c).

a)

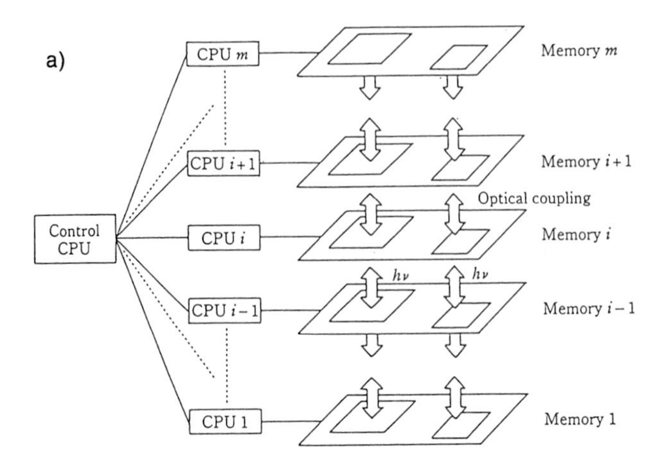

b)

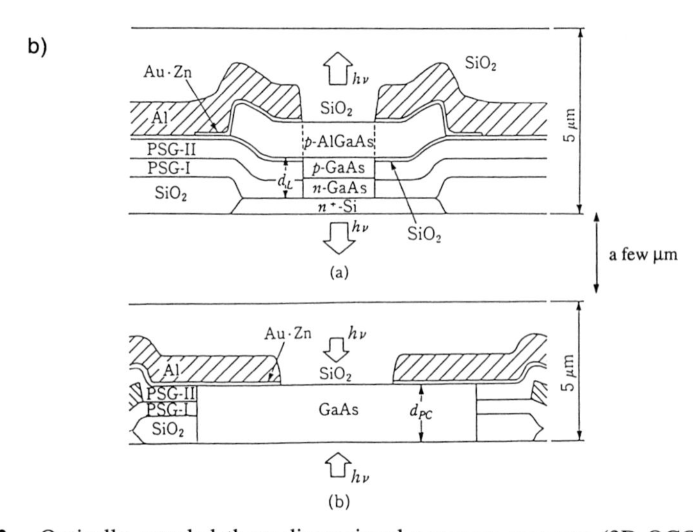

Fig. 12. Optically coupled three-dimensional common memory (3D-OCC) [19]. (a) Parallel processor system with 3D-OCC memory. Each layer in the 3D-OCC memory is connected to the respective processor (CPU) and each memory layer is optically coupled with the upper and lower layers. In this system a large block of data can be simultaneously transferred with very high speed by optical couplings in the vertical direction. Memory cells with identical addresses in all memory layers have the same data; the 3D-OCC memory with 4 Kbit × 4 layers (m = 4) is under construction with 2-μm CMOS technology [19]. (b) Cross-sectional structure of the 3D-OCC memory. LEDs and photoconductors are produced by heteroepitaxial growth of GaAs on silicon. Light from the LED is emitted upward and also downward through the thin silicon substrate into PCs in both layers.

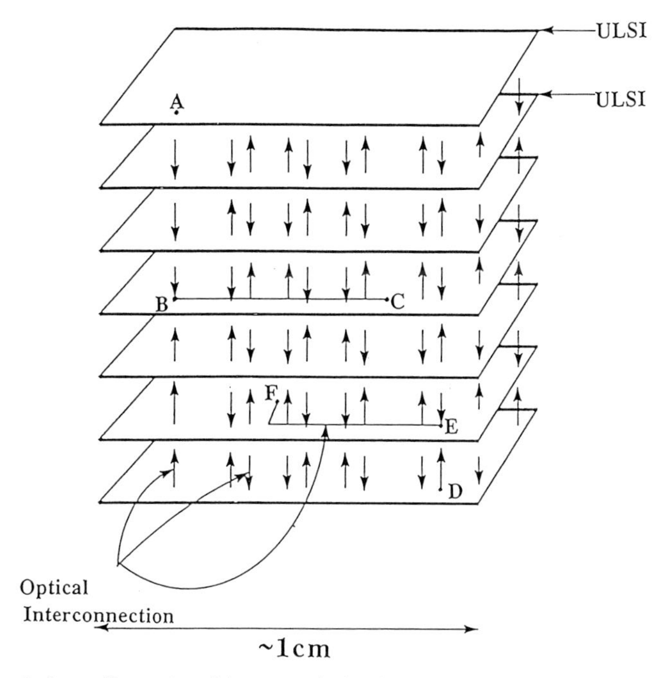

Fig. 13. A three-dimensional integrated circuit (3DIC) using vertical and horizontal optical interconnections. Many ULSI wafers ($10-10^2$) can be stacked in a small volume. Utilizing vertical and horizontal optical interconnection, any two points in the volume can be connected with short delay (a few tenth of nanoseconds). There is no need for bonding. Distance between wafers is a few tens of μm; optics needed for large distance. ULSI wafer thickness is a few μm to 100 μm.

These systems demonstrate another important features of the three-dimensional photoelectronic integrated system. Demountability of wafers at the vertical interconnection is very important for such 3D systems. If they must be assembled all at once by metallic interconnection, no checking or corrections can be made. Block-by-block assembly can be done. Modification or expansion of the system will become practical

4. SUMMARY AND FUTURE PROSPECTS

In this chapter, the feasibility of optical interconnections in ULSI microprocessors (U-OEICs) has been discussed. It was shown that by using micrometer-size lasers and photodetectors, interconnection delays and excess power losses are greatly suppressed. Fabrication technologies for these micro-optoelectronic devices and their integration onto ULSI will be a great

a)

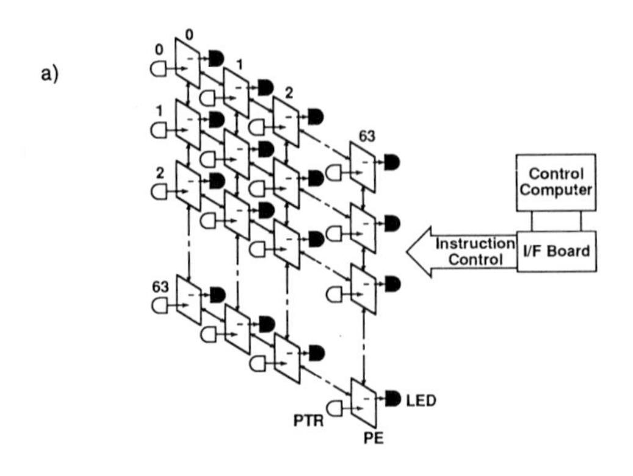

b)

Processings	No. of steps	Processing time maximam speed
Edge detection (4neighbors)	33	3.3μs
Skeletonization (4neighbors)	360	360μ s
Detection of moving objects	7	0.7μs
Trace	4	0.4μs
Poisson's equation	125	2.5ms

c)

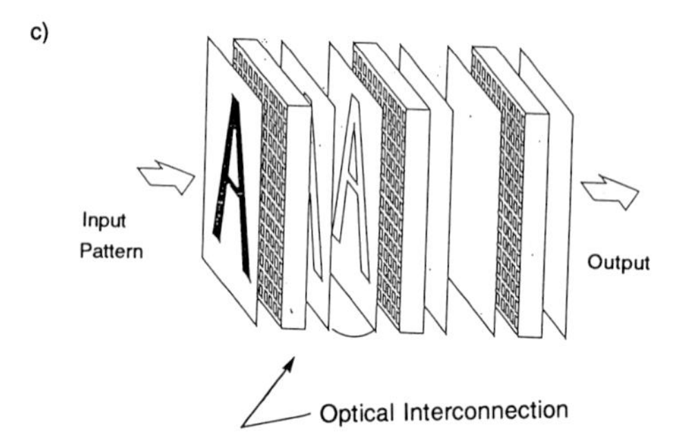

Fig. 14. Massively parallel 2D image processing system [20]. (a) A large two-dimensional array of phototransistors (PTR) ($64 \times 64 = 4096$) are connected individually to small processor elements (PEs), each of which has 300 gates. These processors are connected with each other's nearest neighbors [20]. The output of each processor is shown by light-emitting diodes (LEDs). (b) Processing time of the system for several applications. Processing time is limited by the processor element used (100-ns cycle time). Because the processing is all in parallel, the speed is about 10^5 times faster than in the conventional method based on video scanning. It is 3.2 gigaoperations per second [20]. (c) Future version of the three-dimensional image processing system. Sophisticated image systems could be built using optically interconnected 2D systems similar to the ones shown above.

challenge. However, given a continued high level of research in optoelectronic devices and progress in material technologies, it should be possible to develop the U-OEIC technologies within the coming decade

Feasibility for three-dimensional systems is also described. These systems benefit from using a 3D microelectronic structure, which can be obtained only with optical interconnection. The common memory architecture for the multiprocessor system is under construction [19].

The idea of using common memory for parallel operation of microprocessors in a hypercube system was proposed by Ae [21]. This system is suitable for highly parallel operation of microprocessors and can be constructed by using vertical optical interconnection as described before [19]. A three-layer hypercube structure is shown in Fig. 15. It should be noted that with vertical optical interconnection, processors in each layer are connected with each other only through common memories in the same plain. No metal wiring connecting with other planes is necessary (Fig. 15) [22]. A hypercube system with 25,000 microprocessors can be designed with a 13-layer common memory structure [22]. Interconnection speeds between processors in a system of this type can be very high. In an n-layer system, the longest delay time is n times the interconnection delay between adjacent memory layers [22]. Using lasers, the longest delay can be made less than 10 ns for a 13-layer, 25,000-microprocessor system. Such hypercube systems will also be suitable for neural computing [22].

It was proposed that an optically coupled three-dimensional memory can also serve as an ultrafast memory for a processor system [23]. With this (conventional DRAM) RAMbus scheme the data transfer rate can be orders of magnitude higher than in the conventional DRAM scheme, using the parallel transfer capability of the face-to-face optical interconnection [23].

A two-dimensional image processing system using vertical optical interconnections was described in Section 3 [20]. Many kinds of "optical" image manipulation can be carried out efficiently using three-dimensional photoelectronic integrated systems, based on U-OEIC technology. Considering the rapid progress of ULSI technology, a sufficient number of electronic circuits in an optically interconnected 3D system will be available for parallel processing of 2D images with a low price. Therefore, I believe this kind of photoelectronic integrated system approach will become the main direction of evolution in digital optical image processing systems. An all-optical approach will have benefits mainly in processing utilizing the phase and amplitude of light signals.

So far we have not discussed the use of multiple wavelengths in a system, which is another distinct feature available in optical interconnection. Use of

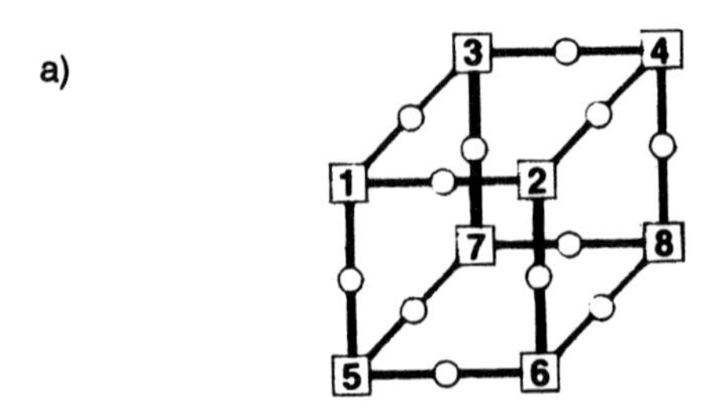

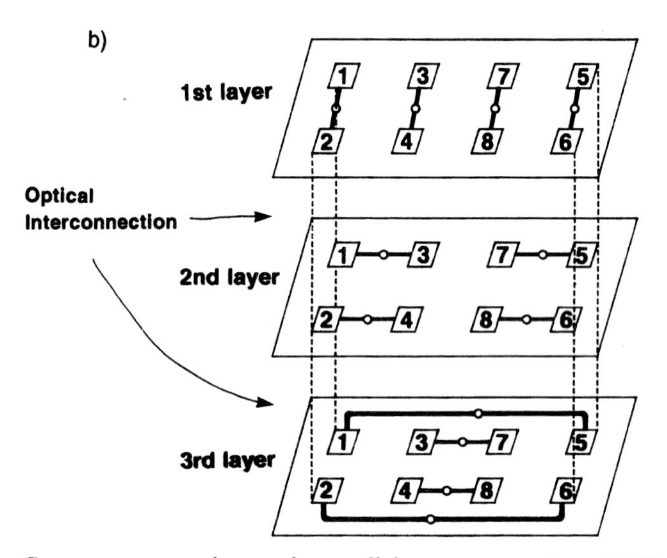

Fig. 15. Common-memory hypercube parallel processor system. (a) Three-port common memory hypercube. (□) 8 memories; (○) 12 processors. (b) Multilayer optically coupled common memory hypercube structure. Memories in one set (1,1,1, 2,2,2, etc.) are coupled to each other by optical interconnections and always have the same memory contents. Each processor is connected only with its two memories within its layer. No signal interconnection is needed except the vertical optical ones, so each layer can be constructed individually and the total system is assembled layer by layer.

multiwavelength light is certainly beneficial in many systems. In bus lines, for example, by using *n* colors, *n* bus lines can be operated on a single waveguide [22]. Many other "optical" systems will be conceived using multicolor interconnection.

Micro-optoelectronic devices are particularly appropriate for multicolor circuits. In microdevices with a low mode cavity, the wavelength is directly controlled by the cavity dimension. Studies of fabrication on the atomic level will make it possible to fabricate series of devices, which have wavelengths precisely adjusted by atomic layer control.

Broadcasting-type interconnection is of course, an intrinsic feature of optical interconnection. Combined with multicolor devices, broadcasting circuits are attractive for applications such as neural computing systems.

All of these developments are based on the U-OEIC technology, in which small optoelectronic devices are integrated with high-density electronic circuits and optical interconnections can be used at any point in the system.

The U-OEIC technology will also have an impact on technologies in many fields. OEICs for communication purposes will become practical when technological barriers are removed by the development of U-OEICs for information processing.

For optical interconnection in multichip modules (MCMs),the U-OEIC technology will provide excellent interconnection technologies, which had been difficult using extended technologies from conventional optoelectronics. Problems of increasing the number of terminals by Rent's rule will also be solved.

Many types of measurement apparatus using electronics as well as optoelectronics will benefit from the developments of the U-OEIC technologies. A variety of one chip measurement instruments for environments, factories, or home use will be developed.

The U-OEIC technology will also have a large impact on discrete optoelectronic devices. Extremely fast lasers will be developed using μ-LD techniques. Stable single-mode oscillation will become available without using such complex structures as DFBs. Development of microcavity lasers may result in a new kind of light emitter. These new-generation light emitters will open up a new era of optoelectronics associated with the development of microphoto detectors.

In some distant future, photoelectronic integrated systems using the U-OEIC technology will continue to be the basis for new systems which will contain new kinds of devices such as three-dimensional quantum devices or tunnel junction devices. It is likely that these new devices will occupy some part of the total system, and microelectronic circuits with optical interconnections will continue to provide the basic hardware of the whole system for many varieties of applications.

ACKNOWLEDGMENT

The idea behind this chapter has developed from discussions with many people over many years. In particular, I acknowledge the following people: all members of the U-OEIC Research Group[7] for developing concepts of U-OEIC technologies; Dr. Junichi Shimada for the first introduction of the OEIC concept and Drs. J.W. Goodman, Roy Lang, Yoshifumi Katayama, Teruo Sakurai, Osamu Wada, Rod C. Alferness, L.A. Coldren, P.W. Smith, and R.F. Leheny for the conceptual development of optoelectronic integration; Dr. Yoshifumi Katayama and members of the Optoelectronic Technology Research Laboratory and the Optoelectronic Joint Research Laboratory for providing my confidence in future developments of material technology for U-OEICs; Drs. Akihiko Morino, Yasushi Ohi, Atsushi Iwata, Masataka Hirose, Mitsumasa Koyanagi, Shojiro Asai, Hideo Sunami, and Takashi Ito for the introduction of ULSI microprocessor technologies and needs for optical interconnections; Drs. Roy Lang, Katsuhiko Nishida, Minoru Shikada, Koichi Dazai, Osamu Wada, Masamichi Yamanishi, Yoshihisa Yamamoto, Hiroyuki Yokoyama, Wataru Suzaki, and Kouichi Iga for discussion of micronmeter-size lasers and detectors; Drs. Mitsuo Takeda and Yasuo Kokubun for optical waveguide techniques; Drs. Masatoshi Ishikawa, Junpei Tsujiuchi, and Yoshiki Ichioka for discussions on the use of optoelectronic techniques for computing systems; and Drs. Tadashi Ae and Nobuhiko Koike for computer system architecture.

REFERENCES

1. J. W. Goodman, F. I. Leonberger, S. Y. Kung, and R. A. Athale, Optical Interconnection for VLSI Systems, *Proc. IEEE* **72**, 850–866 (1984).
2. E. E. Loebner, Opto-Electronic Devices and Networks, *Proc. IRE* 1897–1906 (December 1955).
3. (a) J. W. Parker, Optical Interconnection for Advanced Processor Systems: A

[7]Member list of U-OEIC Research Group: Masayuki Abe, Eisuke Arai, Hideo Ohno, Masaki Ogawa, Masahiro Kashiwagi, Yoshifumi Katayama, Mitsumasa Koyanagi, Hitoshi Satoh, Wataru Suzaki, Mitsuo Takeda, Koichi Dazai, Tadashi Nishimura, Izuo Hayashi, Masataka Hirose, Tohru Nakamura, Osamu Wada, Atsushi: Iwata (at 1990).

Review of the ESPRIT II OLIVES, *J. Lightwave Technol.* **9**(12), pp. 1764–1773 (1992).
(b) K. P. JACKSON, E. B. FLINT, M. F. CINA, D. LACEY, AND J. M. TREWHELLA, Compact Multichannel Transceiver Module Using Planar-Processed Optical Waveguides and Flip-Chip Components, *IEEE 42nd Electronic Components and Technology Conf.*, pp. 93–97 (1992).
(c) M. R. FOLDMAN, "Holographic Optical Interconects for Multichip Modules," *SPIE Proc. 1390,* Int. Conf. on Advances in Interconnection and Packaging, pp. 427–433, (1990).

4. The U-OEIC Research Group is a study group composed of 17 individuals in the Optoelectronics Industry and Technology Development Association. It includes experts in microelectronics, optoelectronics, materials, and systems.
5. (a) I. HAYASHI, Optoelectronic Devices and Material Technologies for Photo-Electronic Integrated Systems, *Jpn. J. Appl. Phys.*, 32(1B) pp. 266–271 (1993).
(b) I. HAYASHI, Photo-Electgronic Integrated Systems-Basic Concept of U-OEIC and their Feasibilities, *Optoelectronics-Devices and Technologies,* 9(1), pp. 1–14 (1994).
6. (a) A. IWATA AND I. HAYASHI, Optical Interconnections as a New LSI Technology, *IEICE Trans. Electron.,* E76-C(1) pp. 90–99, (1993).
(b) A. IWATA, Optical Interconnection for ULSI Technology Innovation, *Optoelectronics Devices and Technologies,* 9(1), pp. 39–54 (1994).
7. (a) Y. KATAYAMA, K. AKITA, Y. OHKI, Y. SUGIMOTO, Y. HIRATANI, AND H. KAWANISHI, Epitaxy and Pattern Formation of III-V Semiconductors as a Through UHV Processing Towards 2- and 3-Dimensional Nanostructures, *Extended Abstracts of the Int. Conf. on Solid State Devices and Materials,* pp. 281–283, Yokohama (1991).
(b) Y. KATAYAMA, Process Technologies for 3-Dimensional Optoelectronic Integrated Circuits, *Trans. Inst. Electron. Inform. Commun. Eng.* **J75-C-I,** 278–284 (1992).
8. H. OHNO, Minimum Light Power for Optical Interconnection in Integrated Circuits, *Optoelectronics-Devices and Technologies,* 9(1), pp. 131–142 (1994).
9. M. TAKEDA, Stack-Type Optical Buslines Based on a Modified Dragone Star Coupler: A Concept, *Optoelectronics-Devices and Technologies,* 9(1), pp. 137–142 (1994).
10. P. L. GONRLEY, T. M. BRENNAN AND B. E. HAMMONS, S. W. CORZINE, R. S. GOELS, R. H. YAN, J. W. SCOTT, AND L. A. COLDREN, High Efficiency TEM_{10} Continuous-Wave (Al, Ga) as Epitaxial Surface-Emitting Lasers and Effect of Half-Wave Periodic Gain, *Appl. Phys. Lett.* **54**(13), 1209–1211 (1989).
11. Reflectivity of over 95% is obtained using four pairs of TiO_2 (n = 2.5) and SiO_2 (n = 1.45) (Kenichi Iga, Tokyo Institute of Technology, private communication).
12. J. W. JEWELL, Y. H. LEE, S. L. MCCALL, A. SCHERER, N. A. OLSSON, J. P. HARBISON, AND L. T. FLOREZ, Two-Dimensional Array Microlasers for Photonic

Switching, Technical Digest of Photonic Switching, pp. 14–19, Kobe, April 1990.

13. D. E. Aspnes, Recombination at Semiconductor Surfaces and Interfaces, *Surf. Sci.* **132**, 406–421 (1983).
14. K. Akita, Y. Sugimoto, M. Taneya, Y. Ohki, H. Kawanishi, and Y. Katayama, Pattern Etching and Selective Growth of GaAs by in-situ Electron-Beam Lithography Using an Oxidized Thin Layer, *Proc. SPIE Advanced Techniques for Integrated Circuit Processing,* pp. 576–587, Santa Clara, CA (October 1990).
15. H. Kawanishi, Y. Sugimoto and T. Ishikawa, In-situ Patterning and Overgrowth for the Formation of Buried GaAs/AlGaAs Single Quantum-Well Structures, *Appl. Phys. Lett.* **60**(3), 365–367 (1992).
16. (a) Eli Yablonovitch, Inhibited Spontaneous Emission in Solid State Physics and Electronics, *Phys. Rev. Lett.* **58**(20), pp. 2059–2062 (1987).
    (b) Y. Yamamoto, S. Machida, K. Igeta, and Y.Horikoshi, Enhanced and Inhibited Spontaneous Emission of Free Excitons in GaAs Quantum Wells in a Micro-cavity, *Coherence and Quantum Optics VI*, (L. Mandel, E. Wolf, and J. H. Eberly, eds.), pp. 1249–1257. Plenum Publishing, New York, 1990.
    (c) H. Yokoyama, *Science* **256,** 66 (1992).
17. S. Kondo, H. Nagai, Y. Itoh, and M. Yamauchi, InGaP orange Light-Emitting Diodes on Si Substrates, *Appl, Phys. Lett.* **55**(19), 1981–1983 (1989).
18. (a) K. Nozawa and Y. Horikoshi, Low Threading Density GaAs on Si(100) with InGaAs/GaAs Strained-Layer Superlattice Grown by Migration-Enhanced Epitaxy, *J. of Electronic Materials*, **21**(6), p. 641, (1992).
    (b) Y. Okada, H. Shimomura, M. Kawabe, Low Dislocation Density GaAs on Si Heteroepitaxy with Atomic Hydrogen Irradiation for Optoelectronic Integration, *J. Appl. Phys.,* **73**(11), p. 7376, (1993).
    (c) T. Kawai, H. Yonezu Y. Ogasawara, D. Saito and K. Pak, Suppression of Threading Dislocation Generation in Highly Lattice Mismatched Heteroepitaxies by Strained Short-Period Superlattices, *Appl. Phys. Lett.*, **63**(15), pp. 2067–2069 (1993). The similar technique has applied to GaAs on Si.
    (d) J. E. Palmer, T. Saitoh, T. Yodo and M. Tamura, Growth and Characterization of GaAs/GaSe/Si Heterostructures, *Jpn. J. Appl. Phys.*, **32**(8B), p. L1126 (1993).
    (e) J. E. Palmer, T. Saitoh, T. Yodo and M. Tamura, Growth and Characterization of GaSe and GaAs/GaSe on As-Passivated Si(111) Substrates, *J. of Appl. Phys.*, **74**(12), (1993).
    (f) M. Tamura, "Recent Progress in GaAs on Si Technology—Concerning the Reduction of Threading Dislocation—*Optoelectronics-Devices and Technologies*, 9(1), pp. 95–118 (1994).
19. (a) M. Koyaragi, H. Takata, H. Mori, and J. Iba, Design of 4k bit × 4 Layer Optically Coupled Three-Dimensional Common Memory for Parallel Processor System, *IEEE J. Solid State Circuit*, 25(1), pp. 109–116 (1990).

(b) M. Koyanagi, K. Miyazaki, H. Krotani, S. Yokoyama, Y. Horiike, and M. Hirose, Fundamental Characteristics of Optically Coupled Three-dimensional Common Memory, to be published in *Optoelectronics-Devices and Technologies*, **9**(1), pp. 119–130 (1994).

20. (a) M. Ishikawa, A. Morita, and N. Takayanagi, High Speed Vision System Using Massively Parallel Processing, Proc. of the 1992 EEE/RS Int. Conf. on Intelligent Robots and Systems, pp. 373–377, Raleigh, N.C., July (1992).
(b) M. Ishikawa, System Architecture for Integrated Optoelectronic Computing, *Optoelectronics-Devices and Technologies*, **9**(1), pp. 29–38 (1994).
21. T. Ae, S. Fujita, T. Yamanaka, and R. Aibara, Hypercube Is Better than De Bruijn for Connectionist, *Proc 5th ISMM Int. Conf. on Parallel and Distributed Computing and Systems,* pp. 370–372, Pittsburgh (October 1992).
22. (a) I. Hayashi, T. Ae and M. Koyanagi, Optical Interconnection *J. Inst. Electronics, Inf. and Com. Engineers,* **75**(9), 951–961 (1992) (In Japanese).
(b) T. AE, Optical Interconnection for Parallel Data Processing, *Optoelectronics-Devices and Technologies,* **9**(1), pp. 15–28 (1994).
23. (a) M. Koyanagi, Optically Coupled Three-Dimensional Memory System, *Proc. Int. Symp. on Advances in Interconnection and Packaging, Microelectronic Interconnects and packages: System and Process Integration, 1390,* pp. 467–473 (1991).
(b) M. Koyangi, K. Miyake, H. Kurotaki, S. Yokoyama, Y. Hirose and M. Hirose, Fundamental Characteristics of Optically Coupled Three-Dimensional Common Memory, *Optoelectronics-Devices and Technologies*, 9(1), pp. 119–130 (1994).

Index

Quantum Electronics—Principles and Applications

Edited by Paul F. Liao, *Bell Communications Research Inc., Red Bank, New Jersey*
Paul L. Kelley, *Electro-Optics Technology Center, Tufts University, Medford, Massachusetts*

N.S. Kapany and J.J. Burke, *Optical Waveguides*
Dietrich Marcuse, *Theory of Dielectric Optical Waveguides*
Benjamin Chu, *Laser Light Scattering*
Bruno Crosignani, Paolo DiPorto and Mano Bertoltti, *Statistical Properties of Scattered Light*
John D. Anderson, Jr., *Gasdynamic Lasers, An Introduction*
W.W. Duly, *CO_2 Lasers: Effects and Applications*
Henry Kressel and J.K. Butler, *Semiconductor Lasers and Heterojunction LEDs*
H.C. Casey and M.B. Panish, *Heterostructure Lasers: Part A. Fundamental Principles, Part B. Materials and Operating Characteristics*
Robert K. Erf, editor, *Speckle Metrology*
Marc D. Levenson, *Introduction to Nonlinear Laser Spectroscopy*
David S. Kliger, editor, *Ultrasensitive Laser Spectroscopy*
Robert A. Fisher, editor, *Optical Phase Conjugation*
John F. Reintjes, *Nonlinear Optical Parametric Processes in Liquids and Gases*
S.H. Lin, Y. Fujimura, H.J. Neusser and E.W. Schlag, *Multiphoton Spectroscopy of Molecules*
Hyatt M. Gibbs, *Optical Bistability: Controlling Light with Light*
D.S. Chemla and J. Zyss, editors, *Nonlinear Optical Properties of Organic Molecules and Crystals, Volume 1, Volume 2*
Marc D. Levenson and Saturo Kano, *Introduction to Nonlinear Laser Spectroscopy, Revised Edition*
Govind P. Agrawal, *Nonlinear Fiber Optics*
F.J. Duarte and Llyod W. Hillman, editors, *Dye Laser Principles: With Applications*
Dietrich Marcuse, *Theory of Dielectric Optical Waveguides, 2nd Edition*
Govind P. Agrawal and Robert W. Boyd, editors, *Contemporary Nonlinear Optics*
Peter S. Zory, Jr., editor, *Quantum Well Lasers*
Gary A. Evans and Jacob M. Hammer, editors, *Surface Emitting Semiconductor Lasers and Arrays*
John E. Midwinter, editor, *Photonics in Switching, Volume I, Background and Components*
John E. Midwinter, editor, *Photonics in Switching, Volume II, Systems*
Joseph Zyss, editor, *Molecular Nonlinear Optics*
William Burns, editor, *Optical Fiber Rotation Sensing*
Mario Dagenais, editor, *Integrated Optoelectronics*

Yoh-Han Pao, *Case Western Reserve University, Cleveland, Ohio*, Founding Editor 1972–1979.